Computed Tomography and Magnetic Resonance Imaging of the Head and Neck

Second Edition

Computed Tomography and Magnetic Resonance Imaging of the Head and Neck

Second Edition

Anthony A. Mancuso, M.D.

Associate Professor of Radiology
University of Florida College of Medicine
Clinical Director MRI Shands Hospital
Gainesville, Florida

William N. Hanafee, M.D.

Professor, Department of Radiology
Center for Health Sciences
University of California
Los Angeles, California

With a special contribution from
John A. Kirchner, M.D.
Professor and Chief of Otolaryngology
Yale University School of Medicine

WILLIAMS & WILKINS
Baltimore • London • Los Angeles • Sydney

Editor: George Stamathis
Associate Editor: Carol Eckhart
Copy Editor: Caral Shields Nolley
Design: Bob Och
Illustration Planning: Reginald R. Stanley
Production: Anne G. Seitz

Williams & Wilkins
428 E. Preston Street
Baltimore, Md. 21202, U.S.A.

Made in the United States of America

First Edition, 1982

Library of Congress Cataloging in Publication Data

Mancuso, Anthony A.
Computed tomography and magnetic resonance imaging of the head and neck.

Rev. ed. of: Computed tomography of the head and neck. c1982.
Includes index.
1. Head—Radiography. 2. Neck—Radiography. 3. Tomography. 4. Nuclear magnetic resonance—Diagnostic use. I. Hanafee, William N., 1926– . II. Mancuso, Anthony A. Computed tomography of the head and neck. III. Title. [DNLM: 1. Head—radiography. 2. Neck—radiography. 3. Nuclear Magnetic Resonance—diagnostic use. 4. Tomography, X-Ray Computed. WE 700 M269c] RC936.M33 1985
617′.5107572 84-11940 ISBN 0-683-05476-7

Composed and printed at the
Waverly Press, Inc. 86 87 88 89 5 4 3 2

To Toni, Nick (my partner) and Gina (my most favorite girl)
and the scientists, engineers and support personnel
who made CT and MR imaging possible.

A.A.M.

Preface to the Second Edition

The promise of CT in head and neck imaging presented in the first edition has been more than fulfilled. Now we have nuclear magnetic resonance (NMR) proton imaging which promises even further refinements in our diagnostic capabilities. NMR spectroscopy remains an exciting, although strictly experimental, prospect at present.

Magnetic resonance imaging (MRI) is a very rapidly evolving field. The material for this portion of the text was gathered from early 1983, until February of 1985. The quality of the images here represents the evolving state of the technology. In 3 or 4 years it will be possible to produce a more definitive statement about the impact of MRI on head and neck imaging. Many of the images seen here were produced on prototype rf receiver coils developed at the University of Florida College of Medicine. Other developments such as paramagnetic contrast agents will alter the efficacy of MRI. We bring to you in this volume a look at the potential of MRI, much as was done with CT of the temporal bone in the first edition. None of us will be disappointed with MRI's eventual contribution to medical imaging. MRI is already proving efficacious for diagnosing acoustic neuromas of all sizes. Suprisingly it is excellent for lesions involving the skull base and showing bone detail quite nicely, if somewhat differently, from CT. It is demonstrating some added specificity in distinguishing post-treatment scarring from tumor. The list goes on (read the text). At UCLA, MRI costs less than a contrast-enhanced CT. At the University of Florida-Shands Hospital it is only slightly more expensive; this reflects our involvement with the development of low field-low cost systems.

We have retained our format of splitting the text and illustration sections (see Preface to the First Edition). The text has been nearly completely rewritten and more tables and line diagrams have been added. The pattern of spread diagrams in Chapters 9–11 should prove especially useful to those unfamiliar with the natural history of tumors in the pharynx and larynx. The CT illustrations represent the state-of-the-art in almost all instances. This is most apparent in the temporal bone chapter. Very little old case material has been kept. A chapter on the orbits has been added and sections on the skull base and posterior fossa (the crossover area between classical neuroradiology and the head and neck subspecialty) have been strengthened. A separate chapter on aspiration cytology and biopsy techniques was added to avoid the redundancy of repeating these comments in each anatomic area where they may be applied.

While this book is a joint venture, we have found it logistically necessary to split the chapter assignments. Dr. Hanafee is primarily responsible for Chapters 1–7 and Dr. Mancuso for Chapters 8–11, respectively. This accounts for slight differences in style and organization which we hope will not disturb the reader. This also proves to be of some benefit because the reader is exposed to two different spheres of experience with CT and MRI equipment.

Lastly, we hope that this text will be used to update the standard of care for patients with head and neck problems. At the time of the first edition, many people were unsure of CT's ability to contribute information that could impact the outcome for the patient. Now we see this every day in our practices and observe an increasing demand in the nonacademic setting for such sophisticated imaging data and interpretation. We intend that the text push everyone a little farther along on the learning curve and that this will be translated into more accurate diagnosis and treatment of patients.

Anthony A. Mancuso, M.D.
William N. Hanafee, M.D.

Preface to the First Edition

Computed tomography vastly expands the role of radiology in the management of head and neck cancer. The radiologist continues to perform a valuable function in diagnosis but now CT places him as an integral partner in the team managing cancer along with the surgeon, radiation oncologist, and chemotherapist. The reader cannot help but recognize how frequently the emphasis of chapters and illustrations will be devoted to determining extent of disease as it relates to surgical approaches or the planning of treament ports.When appropriate, we have tried to emphasize all of the situations in which CT offers new insights into diagnosis.

Our scanner was installed in the University of California at Los Angeles in June, 1976. Much of the illustrative material used in this book has been the subject of exhibits and/or papers presented at radiology and head and neck society meetings. The lively discussions that followed from surgeons, oncologists, radiologists, and other interested parties convinced us that a concise, liberally illustrated text on CT of the head and neck would serve a useful function to alert all concerned physicians to the contributions made by computerized tomography.

The very nature of roentgen diagnosis is to define the anatomy of hidden areas. In the head and neck, details of air-mucosal interfaces and the delicate bones of the facial structures are available by less sophisticated techniques. In many instances, the same information is also within easy reach of the probing finger or mirror of the examining physician. The same delicate bony and mucosal-air interfaces also serve as obstacles for the evaluation of underlying soft tissue fascial planes. CT has led us out of the era of conjecture as to what causes a bulging surface or to what extent lack of mobility is due to infiltrating tumors. Tissue diagnosis is still an elusive goal for this primarily anatomic, diagnostic discipline. Inflammatory or reactive changes remain indistinguishable from neoplastic ones. Nevertheless, CT provides enhanced contrast resolution which gives us access to the anatomy of the fascial planes and provides clues for the accurate staging of spreads of tumor mass. This help provided by CT is most evident in the areas that are at the boundaries of surgical resectability, i.e. infratemporal fossa, parapharyngeal and paranasopharyngeal space, carotid sheath, cricoid cartilage destruction, and extensions into the preepiglottic space. Without a doubt, as technical advances occur in mechanics of CT scanning, staging of primary cancers will become the realm of CT scanning.

With the severe limits on space available to radiology departments in modern hospitals, some equipment will become obsolete, not through providing misinformation but because CT will be able to do the job equally well and provide additional information regarding the soft tissues deep to bone or air interfaces. For example, it will be extremely difficult to justify facilities for studies such as temporal bone tomography when expanded gray scale, edge-enhanced CT scanning can provide accurate diagnosis of middle ear structures as well as some suggestion of characteristics of soft tissue versus fluid infiltrations of middle ear cavities.

The format of this book is slightly different than most similar works. The text contains line diagrams and tables, in lieu of CT images, which help emphasize the anatomy and clinical significance of CT findings; the illustrative CT material is presented in a Plates section at the end of each chapter and follows the order of presentation of the subject matter in the text. These illustration sections also may be read as a separate "book" since the legends are narrative and contain pertinent clinical highlights of the problem under consideration. We hope that this format will allow the book to serve both as a reference work for those surgeons, radiation oncologists, and others interested in an in depth imaging approach to head and neck disease as well as an atlas for the radiologist in everyday practice.

A. A. M.
W. N. H.

Acknowledgments

We extend a very special thanks to David Maxwell, Ph.D., Professor of Anatomy at University of California at Los Angeles Medical School for his guidance and encouragement in planning the original text.

The value of the book to clinicians and radiologists has been markedly increased by the thoughts and materials of John A. Kirchner, M.D., Professor and Chief of Otolaryngology at Yale University. All of the pathologic whole organ sections shown to correlate with the CT sections were loaned to us by him from work that was supported by United States Public Health Service Grant 2-P01-CA-22 101 Award by the National Cancer Institute.

Many diagnostic radiologists have loaned us material and referred us difficult cases since the first edition. These are acknowledged where they appear. A special mention, however, goes out to Gordon Gansu, M.D. and his colleagues at UCSF with whom we have exchanged material many times over the last 5 or 6 years.

Our debt to our clinical colleagues is now larger. Paul Ward, M.D., Chief of Head and Neck Surgery at UCLA let us get started and remains a good friend and supporter. James Parkin, M.D. and Mike Stevens, M.D. at the University of Utah were equally encouraging. Nick Cassisi, M.D., Chief of Head and Neck Surgery and Rod Million, M.D., Chief of Radiation Therapy at the University of Florida have taught us how well and humanely patients with head and neck cancer can be treated when there is a good working relationship between these areas of discipline. All of these people and more have provided us with the most useful perspective of what our work means; namely its relationship to patient care.

At the University of Florida College of Medicine, Clyde M. Williams, M.D., Ph.D., Professor and Chairman of Radiology has accumulated the best NMR physics group with which one could hope to be associated. Kate Scott, Ph.D. heads the group. Professor Raymond Andrew is a constant source of information, experience and inspiration. Jeff Fitzsimmons, Ph.D. has built the rf receiver coils which make our 0.15 T resistive unit realize MRI's full potential at this low field strength. Tom Mareci, Ph.D. rewrites everyone's software to make it work better and is a prime mover in our spectroscopy effort. These people are a blast to work with.

Gwynne Gloege at UCLA and Debra Niel-Mareci (Tom Mareci's wife) at the University of Florida have produced beautiful and intelligent illustrations to help us present our work more clearly. Barbara Adashek, Helen Thompson and Anne Erwin worked tirelessly on this and the first edition manuscripts.

The Siemens Corporation provided a substantial grant for the art work and reproductions in the text which is greatly appreciated. Much of the work in the first edition was supported by a National Research Service Award (IF 32 NS05J 771). In this edition, the MRI work from the University of Florida College of Medicine was supported in part by an NIH resource grant (IP41RR022278-01). We gratefully acknowledge the granting federal agencies.

Lastly, we thank our publisher, Williams & Wilkins, and the Waverly Press. This is the third project we have had with them and all have been a pleasure. Ruby Richardson started all of this in 1977. Since then, we have worked with several editors and production people, and the same high standards and easy communications have always been available. On this project, George Stamathis and Carol Eckhart have been especially helpful with Anne Seitz pinch-hitting while Carol had a baby. Thank you all!

Anthony A. Mancuso, M.D.
William N. Hanafee, M.D.

Contents

Malignant Sinus

INTRODUCTION

As the role of computed tomography (CT) and magnetic resonance imaging (MRI) expands we can anticipate these two imaging modalities assuming all of the roles that were formerly contributed by conventional pluridirectional tomography. Unsuspected inflammatory disease of the paranasal sinuses frequently can be seen by pluridirectional tomography that is not visible on routine films (5). This same statement can now be made about CT and to a large extent MRI. More aggressive inflammatory lesions that tend to spread to the central nervous system such as mucor mycosis may cause life-threatening cerebritis and meningismus (3). The ability to examine the brain as well as paranasal sinuses makes CT and MRI an extremely valuable tool in the management of this potentially lethal disease.

Malignant tumors of the paranasal sinuses constitute slightly less than 1% of all malignancies (2). The average delay from the onset of symptoms to diagnosis is approximately 6 months (9). Despite the ease with which the antrum can be viewed radiographically, approximately 80% of the patients have evidence of bone destruction at the time diagnosis of malignancy is made (4). At least 80% of all paranasal sinus cancers involve the maxillary antrum. The ability of CT to show bone destruction as well as invasion of soft tissues has great potential for the planning of radical surgical procedures or precision radiation therapy (16, 28). Improvements in MRI that allow more tissue specificity also may offer the opportunity of a major breakthrough in the early diagnosis of paranasal sinus cancer.

Occasionally, a tissue-specific diagnosis is possible with CT either because of calcification within the tumor mass such as seen in chondromatous tumors, or the intense contrast enhancement found in juvenile angiofibromas (30). MRI would seem to have great potential for tissue specificity but sufficient clinical experience is not as yet available to accurately appraise its differential diagnostic capabilities.

ANATOMY, VARIANCE AND PITFALLS

Only those factors of anatomy that directly relate to CT and MRI scanning will be reviewed. The majority of the problems encountered are related to anatomical variations and unfamiliarity of the tomographic anatomy in the axial projection. With MRI, selection of proper pulse sequence has a profound effect on the strength of signal received and spatial resolution available within the image.

Maxillary Sinuses

The bony margins of the antra are quite variable in thickness. The anterior surfaces are made up of dense bone to support chewing and resist trauma. The thinnest bone lies along the posteromedial boundary of the antrum. At times, this very thin area is spoken of as the membranous wall of the antrum. CT, with its greater ability to image calcium, will usually demonstrate the membranous wall as a "bony wall."

The infraorbital nerve, artery and vein lie in a groove in the posterior portion of the roof of the antrum. The groove becomes a canal anteriorly and later exits to the cheek as a foramen. Since this roof lies in the plane of an axial section it may be partially visualized and appear as soft tissue density closely simulating tumor.

The maxillary sinuses are surrounded by low density fat along all margins except the infraorbital rim and the medial margin, which lies adjacent to the nasal mucosa. The levator labii superioris muscle is adjacent to the bone surrounding the inferior orbital rim. Elsewhere the low density plane can be used to good advantage to identify infiltrating processes that have escaped from the sinuses. The posterior wall and posterolateral wall of the antrum form the anterior boundary of the pterygopalatine fossa and the infratemporal fossa, respectively (Fig. 1.1). This region is of extreme importance in the management of malignant as well as inflammatory disease and will be discussed in some detail.

Infratemporal Fossa

The infratemporal fossa lies beneath the base of the skull with the medial and lateral boundaries being the walls of the pharynx and the ramus and body of the mandible, respectively. The anterior boundary is the maxillary sinus and, posteriorly, the sheath of the carotid artery and prevertebral fascia limit the space. Along the inferolateral surface of the greater wing of the sphenoid is the infratemporal crest which forms a boundary between this space and the deep temporal fossa.

Within the infratemporal fossa lie the medial and lateral pterygoid muscles, mandibular nerve segment, the posterior superior alveolar nerve, the terminal branches of the internal maxillary artery, and the pterygoid venus plexus.

Pterygopalatine Fossa

At the anteromedial extremity of the infratemporal fossa one can enter the pterygopalatine fossa through the sphenomaxillary fissure. The pterygopalatine fossa

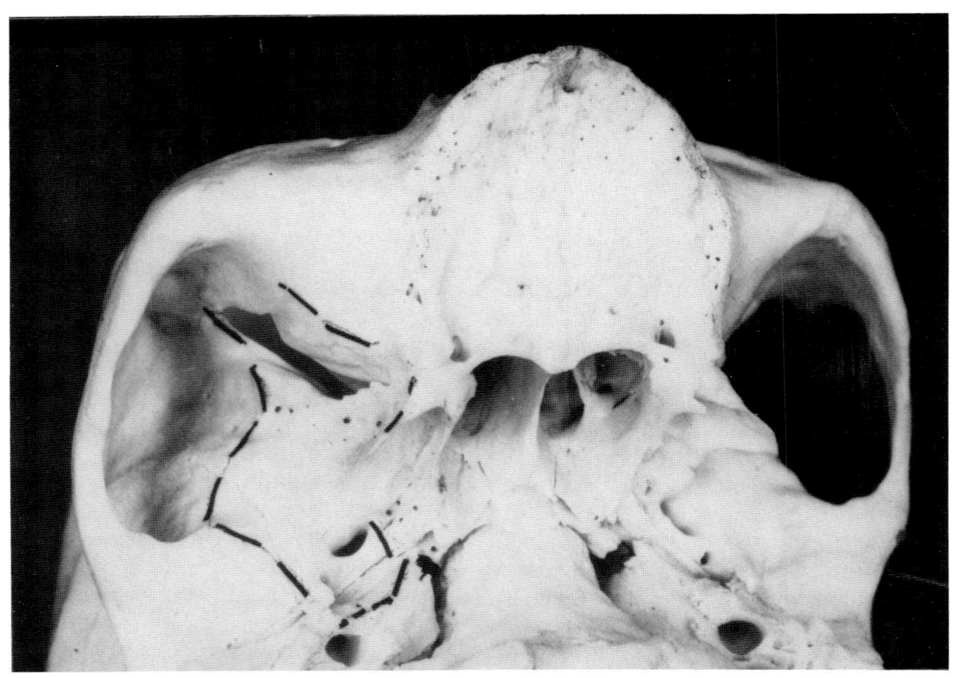

Figure 1.1. Infratemporal fossa. The infratemporal fossa is outlined on the base of the dried skull by a *dashed line.* Anteriorly, it is bounded by the posterior wall of the antrum. Laterally, the infratemporal crest of the greater wing of the sphenoid separates the infratemporal fossa from the more laterally placed temporal fossa. The mandible is the main lateral wall. Medial, the fascia extends to the pharyngeal wall. Posteriorly, the boundary is formed by the carotid sheath and styloid mechanism. Inferiorly, the fossa is open.

lies in close proximity to the skull base between the junction of the medial lateral pterygoid plates and the posterior margins of the maxillary sinus. It contains the sphenopalatine artery, sphenopalatine ganglion and branches of sympathetic and parasympathetic nerves. The posterosuperior alveolar artery and nerve rise in this fossa and course anteroinferiorly along the posterolateral margin of the maxillary sinus. The vessels and nerves of this region can be studied quite nicely by MRI or by CT with intravenous contrast and bolus injections.

Frontal Sinus

The frontal sinuses are quite variable in their configuration as well as the thickness of their bony wall. Regardless of their degree of development, there should be an intact bony wall anteriorly as well as posteriorly. The frontal sinuses drain via the nasofrontal duct beneath the middle turbinate. Extensions of the frontal sinuses, or ethmoid sinuses, into the roof of the orbits may produce some confusion when the bony coverings of these sinuses lie parallel to the plane of the section. In the latter situation, the margin may give a spurious impression of dehiscence and direct coronal scanning or coronal reconstruction may be quite helpful in elucidating the true anatomy. Many times it is impossible to tell if a small sinus is frontal or ethmoid in origin.

Ethmoid Sinuses

The ethmoid air cells form the medial and inferior boundaries of the orbit. The two sides are arranged parallel in the upper portions of the orbit, whereas posteriorly and inferiorly they tend to diverge to form the medial floor of the orbit. The ethmoid cells are approximately 10 in number and they are divided by two vertical plates into anterior, middle and posterior air cells (13). The medial and lateral boundaries of the ethmoid air cells are thin but distinctly visible on CT scanning. Definition of the superior boundary of the ethmoid air cells is vital information yet, sometimes, difficult to obtain unless coronal scans are done. The cribriform plate forms the roof of the nasal cavity but is difficult to visualize on routine roentgen examinations because it is a perforated plate and does not cast a dense shadow. The cribriform plate is not on a level with the roof of the ethmoid sinuses but lies more caudal in location thus creating a depression or fossa to either side of the crista galla. The paired fossa so created contain the olfactory nerves and are called the olfactory fossae.

Sphenoid Sinus

The sphenoid sinuses are truncated in configuration and quite variable in the amount of pneumatization. The sinuses are divided by a vertical septum that rarely stays in the midline. There may be a deviation of this vertical septum into a horizontal component that also divides the sinuses into a superior and inferior portion (23). The lateral walls of the sphenoid sinus diverge laterally as they pass caudidly and are difficult to visualize at axial scanning. Coronal sectioning is extremely valuable to see the lateral walls of the sphenoid sinuses but frequently dental amalgums interfere with visualization. This problem can be overcome by proper gantry angulation based on preliminary scout views. Particular attention should be paid to the inferolateral extensions of the sphenoid sinus that occupy the floor of the middle cranial fossa. The amount of extensions are quite variable and at times the entire greater wing of the sphenoid may be occupied by extensions of the sphenoid sinus. Medullary cavity containing fat and red marrow occupy that portion of the basi sphenoid and clivus that are not pneumatized. AT MRI, a bright signal will come from the marrow spaces. Differentiation of a large fluid-filled sphenoid sinus from bone marrow is difficult.

Variants

Hypoplasias of the sinuses are most troublesome in the maxillary antrum. The asymmetry of the roof of the antrum may cause partial voluming that, on CT section, closely simulates bone destruction. The underdevelopment of the maxilla may lead to an enophthalmous which clinically is confusing and is usually diagnosed as exophthalmous on the side opposite the hypoplasia (18, 22). The clues to maxillary hypoplasia are as follows:

1. Lateral displacement of the lateral wall of the nasal cavity
2. Thickened walls of the antrum
3. Hypertrophy of the inferior turbinate on the side of the hypoplasia
4. Underdevelopment of the zygomatic recess of the antrum
5. Inferior displacement of the orbital floor
6. Enophthalmous of the ipsilateral globe

Hypoplasia of the remaining sinuses is usually not difficult to diagnose, except in the sphenoid sinus one may confuse the increased bony density as being a soft tissue mass when partial voluming has occurred. The sphenoid sinus may also cause some difficulty due to its inferolateral recess that may extend into the greater wings of the sphenoid or into the base of the pterygoid plate. These sinus extensions may be partial volumed and give the appearance of bone destruction. The inferolateral extension of the sphenoid sinus may completely surround the foramen ovale and foramen sphinosum as they extend lateral. Fluid within the sphenoid sinuses may collect in the inferolateral recesses and not be visible on routine films. During axial CT scanning in the brow up position, air fluid levels in inferolateral recesses are easily seen on sections taken through the skull base.

Lying lateral to the sphenoid bone, is the foramen lacerum which is largely made up of cartilagenous walls. The amount of cartilage present is quite variable and irregular margins of the bony portions of the foramen may be mistaken for bone destruction.

CT AND MRI SCANNING TECHNIQUES

Whether the CT scan is being performed for malignant disease or benign inflammatory disease, intravenous contrast enhancement techniques are almost invariably utilized. We use 300 ml of 30% contrast or 150 ml of 60% contrast as a rapid intravenous drip. The intravenous contrast will assist in determining the contents of sinus cavities, the mucosal surface and vascularity of normal tissues versus tumors.

Our technique for scanning is as follows:

1. The patient is placed supine on the table with Reid's baseline approximately perpendicular to the table top.
2. Scanning is begun at the level of the margins of the teeth roots in the maxilla. The scans are extended cephalad to the levels of the frontal sinus.
3. Sections are taken at 5-mm intervals, with 5-mm thick sections. When the clinical situation requires detailed information regarding thin bony septun, the sections are taken at 1.5-mm thickness, at 5-mm intervals and then viewed at bone window settings.
4. When examining for malignant disease within the ethmoid sinuses or sphenoid sinuses, direct coronal scans are performed through the anterior and/or middle cranial fossa to identify intracranial extensions. If the patient is unable to maintain the extended position, reformatted images following scanning with 1.5-mm thick section at 1.5-mm intervals can be used.
5. In the majority of patients the scans are photographed at both soft tissue and bone window settings.
6. Rarely do we examine the cervical region for lymph nodes even in the presence of malignant sinus disease unless the tumor is very advanced or known to be something other than squamous cell carcinoma.

MRI is most useful in demonstrating intracranial extensions but lack of signal from bone limits its use in sinus disease.

Children

Examination of cooperative children is really quite simple if one keeps in mind the development of the paranasal sinus. The frontal sinus is absent at birth while the remaining sinuses are rudimentary. They develop from budding respiratory epithelium and show rapid growth from approximately 6 to 7 yr of age to the postpuberty period. At approximately 8 to 10 yr of age, the maxillary sinuses are usually well aerated and can be studied completely.

The sphenoid sinuses are absent at birth and do not extend beneath the planum sphenoidale until 2 to 3 yr. Pneumatization beneath the sella becomes apparent at 4 yr and is present in the majority of children by 8 to 10 yr of age (8). Even without a tilting gantry, coronal sections can be obtained much more readily in children providing the child is cooperative. Their necks are much more flexible and they are able to lie prone with the chin extended. The foramina of the base of the skull and about the bony orbit are best demonstrated by a combination of axial and coronal scanning.

Pitfalls

CT scanning mimics the limitations of pluridirectional tomography with regard to "the law of tangents" (29). This physical rule states that a bony plate will only be visualized during tomography if the x-ray beam passes tangent to the cortical surface sometime during the tomographic sweep. In CT, bony septa that run parallel to the axis of scan are difficult to visualize and frequently appear as soft tissue densities rather than true bony margins. For this reason, coronal and axial scanning are complimentary rather than competitive views. At times, sagittal projections (20) can be used to good advantage.

With MRI the bony margins of the sinus appear merely as a plane of absent signal. Some signal may be present about the sphenoid sinus because of the bone marrow present. This bright signal from marrow surrounding the sphenoid sinus or about the frontal sinuses can be extremely troublesome and may closely simulate mucosal disease. The mucosa of the turbinates and mucosa of the paranasal sinuses cast extremely bright shadows on spin echo MRI. Differentiating bright signals from edematous mucosa, inflammatory disease and even malignancy is a perplexing problem that is going to require a great deal of study with varied pulsing sequences.

With conventional tomography, the totally opacified sinus presents a diagnostic dilemma. The cause may be fluid, tumor, scarring, or chronic infection, or a combination. Considerable assistance is possible when intravenous contrast is added to the CT examination. Fluid without cellular debris will give low density readings while "masses" associated with hypertrophied mucosa and fluid shows varying density within the sinus cavity. Solid tumors usually stain brightly but not to the extent that one sees in the extremely vascular juvenile angiofibromas which are in a class by themselves. Unfortunately when a sinus is totally filled with inflammatory disease or tumor, differentiation is not possible by CT numbers alone. Surgical exploration or biopsy continues to be the mainstay of diagnosis for definitive treatment.

INDICATIONS

Regardless of whether the process is an aggressive inflammatory disease or a malignant lesion, CT offers the maximum advantage in demonstrating soft tissue as well as bony infiltrations by expansive processes. The most important function is in demonstrating extensions that are beyond the limits of surgical resection. Equally important is the demonstration of involvement of vital structures that require more radical surgical approaches such as orbital exenteration. Some limits of tumor masses that indicate *relative* inoperability and which therefore constitute indications for CT or MRI are as follows:

1. Involvement of the infratemporal fossa
2. Involvement of the high pterygopalatine (pterygomaxillary) fossa
3. Extensions of the tumor masses to the mucosa of the nasopharynx
4. Significant intracranial extension through the cribriform plate or into the middle cranial fossa.
5. Orbital invasion and spread both inside or outside the muscle cone.

CONTRAINDICATIONS

There are very few contraindications to CT scanning of the paranasal sinuses other than patients who are totally uncooperative or have motion disorders that prevent adequate scanning. Some lesions can be studied adequately with routine films and some form of tomography. Since CT gives so much more information regarding soft tissues there is little to recommend the continued use of pluridirectional tomographic examination. In patients who have an extreme fear of x-ray radiation, MRI may prove to be a useful alternative. The patient with acute sinusitis or the headache patient with no other evidence of sinus disease will probably give very low yield of new information by any scanning technique. In general, one might say that CT is now indicated in any situation where conventional tomography was formerly used. MRI shows promise of greater tissue sensitivity related to the status of bound water in the cellular and extracellular spaces. Specificity remains to be seen.

TUMOR

Pathology

Numerous classifications of the tumors of the paranasal sinuses have been proposed based on involvement of the structures cephalad or caudad to the sinuses (6, 14, 19). All of these classifications seem to emphasize radiation therapy as being the primary modality of treatment. CT offers the possibility of accurately delineating the extent of tumors and correlating the areas of tumor involvement for surgery as well as radiation management. Classification based on histologic characteristics has some bearing on the conduction of the CT scan because some tumor types show a high incidence of nerve invasion while others spread by direct extension. Metastases to regional lymph nodes are rare except in the nonepithelial tumors (10).

In general, primary malignant tumors of the paranasal sinuses may originate from either epithelial elements, the supporting stroma or osseous structures. Secondary tumors either invade the sinuses from adjacent sites or metastasize from remote primaries. A listing may be as follows:

- Epithelial malignancies
  - Squamous cell carcinoma
  - Adenocarcinoma
  - Adenocystic carcinoma
  - Melanoma
  - Esthesioneuroblastoma
- Nonepithelial malignancies
  - Rhabdomyosarcoma
  - Chondrosarcoma
  - Osteogenic sarcoma
  - Reticulum sarcoma
  - Malignant fibrous histiocytoma
  - Lymphoma
- Secondary tumors
  - Direct extension oral tumors
  - Direct extension of dental malignancies
  - Metastases from distant sites

Epithelial Tumors

The epithelial tumors show considerable similarity in their gross morphology and clinical spread that permits them to be discussed as a group with regard to their CT and MRI diagnosis. Only those peculiarities that are pertinent to the roentgen findings will be discussed under the individual histologies. For further details the reader is referred to some excellent texts on pathology such as Batsakis (3), Papparella and Shumrick (21) and others (11).

SQUAMOUS CELL CARCINOMA

Over 80% of the malignancies of the paranasal sinuses are epithelial in origin: squamous cell being the most common. The tumors are moderately well differentiated as a general rule and metastases occur quite late in the illness. In the past 8 yr none of our squamous cell carcinomas have had lymph node metastasis at the time of initial presentation. Squamous cell carcinomas are more likely to follow chronic inflammatory disease than are the adenocarcinomas or adenocystic carcinomas (12).

Because of an antecedent history of chronic inflammatory disease the patients usually seek medical assistance late in the course of their tumor. Between 80 to 90% of the patients have evidence of bone destruction and the tumor has usually spread to involve the nasal cavity or multiple sinuses in 75% of the patients. At times the exact sites of origin of squamous cell carcinomas are difficult to determine. Classifications of paranasal sinus cancers are frequently cumbersome. The tumors show an increased severity of stage, based primarily on the surgical resectability or involvement of vital structures rather than a measurement of the tumor size.

If the tumors of the paranasal sinuses have metastasized the situation for clinical cure is usually almost hopeless. The metastases from ethmoids and, to some extent, maxillary sinus cancers goes to retropharyngeal nodes as well as deep cervical nodes. They are only discovered late in the illness. Even when discovered, the metastasis proves very resistent to management.

ADENOCARCINOMAS

Adenocarcinomas constitute less than 10% of all carcinomas of the paranasal sinuses (27). They may be slightly more common in the ethmoid regions (1) but their growth rates and method of spread are very similar to that of squamous cell carcinoma. Nodal metastasis occurs earlier.

ADENOCYSTIC CARCINOMA

Adenocystic carcinomas arise from salivary gland rests and grow with the same propensity for invading nerves as is found with adenocystic carcinoma from other sites. These tumors may grow rapidly but most of the time they tend to be slow, relentless, locally invasive, tumors that also disseminate widely to the lungs and liver. Microscopically, this tumor is composed of basaloid cells with very little cytoplasm that are arranged in cords and nests. Characteristically the tumor forms glandlike spaces and frequently shows a tubal glandular type of configuration which caused them to be called "cylindromas" in the past. These tumors are definitely not a benign lesion and the term adenocystic carcinoma more correctly reflects their true nature.

Adenocystic carcinomas have a persistent recurrence rate that is almost a straight line of declining survival over a period of 17 to 20 years (26).

MELANOMAS

The clinical course of melanomas arising in the nasal cavity and paranasal sinuses are quite variable. At times the tumors disseminate widely by blood vessel invasion with distant metastasis. Approximately 10 to 12% of the tumors seem to have a very prolonged course and remain localized for long periods of time when managed with repeated surgical resections. No local growth patterns can be identified which will predict the ultimate biology of this tumor. Eventually widespread blood-borne metastases supervene even in the localized form of melanoma.

ESTHESIONEUROBLASTOMAS

Esthesioneuroblastomas arise from the primitive neuroectoderm that can be found in the high primitive nasal epithelium. Two distinct age peaks are evident in the incidence of esthesioneuroblastomas. Silva et al. (25) divide the lesion into neuroendocrine carcinomas and neuroblastomas. The majority of the lesions tend to be of the neuroendocrine carcinoma type and reach a peak incidence in the 50- to 60-yr age group. The remaining patients tend to center in the 15- to 25-yr age group and are associated with a more malignant course.

The CT and MRI diagnosis of this tumor is primarily

related to location. Because the cribriform plate is a perforated plate the tumor follows the olfactory nerves intracranially into the anterior cranial cavity. They rapidly spread in a circumscribed fashion on either side of the falx in the interhemispheric fissure. The superior margins of these tumors can be shown to be rounded and relatively sharply delineated from the adjacent brain. They are easily mistaken for subfrontal meningiomas because of their location and vascularity. Involvement of the nasal cavity and ethmoid air cells is the valuable clue that the tumor is primarily an esthesioneuroblastoma rather than a primary intracranial tumor.

Nonepithelial Tumors

The growth patterns of some of the nonepithelial tumors is distinctive enough that a diagnosis can be suggested on the basis of their roentgen appearance. These changes include chondromatous matrix calcification, bulky tumor masses, aggressive bone infiltration, and widespread lymphatic involvement.

RHABDOMYOSARCOMAS

Rhabdomyosarcomas are uncommon in the nasal cavity and paranasal sinuses. They are more likely to involve this region by direct extension from the nasopharynx or orbits. The majority of the patients are under 12 yr of age. The tumor is characterized by its very aggressive growth patterns and its ability to infiltrate widely through bony septa. The local infiltrations are usually so extensive that hematogenous spread and local disease kills before lymphatic metastases occur.

CHONDROSARCOMA

Chondromatous tumors have a predilection for originating from the ethmoid air cells, nasal septum and hard palate regions. The malignant variety may start out initially as malignant or may represent malignant transformations of the cartilagenous cape of osteochondromas (17).

Chondrosarcomas arising from the nasal cavity and paranasal sinus tend to be of low grade malignancy (7). They metastasize late and the ultimate poor prognosis is primarily related to inadequate primary surgical resection. Because of their low grade invasive nature they tend to push and displace local bony septum rather than invade and destroy.

At CT, calcification of matrix can be demonstrated within the lesion. The tumors are usually bulky and very little enhancement is demonstrated except around the periphery of the lesion. Experience with MRI is limited but, in general, the signals are alternate areas of brightness and diminished signal on the spin-echo technique because of the calcified matrix throughout the tumor.

OSTEOGENIC SARCOMA AND RETICULAR CELL SARCOMA

Osteogenic sarcoma and reticulum cell sarcomas are usually not primary lesions of the paranasal sinuses but begin along the alveolar ridge or hard palate and invade into the sinuses. Because of this site of origin, separate from the sinuses, the manifestations of these lesions is usually that of dental problems or oral tumor masses rather than sinus disease. Only in advanced lesions can one demonstrate opacified sinuses and destroyed bony margins of the paranasal sinuses.

MALIGNANT FIBROUS HISTIOCYTOMAS

This entity is a recently described variety of soft tissue sarcoma. It occasionally affects the nose, nasopharynx and paranasal sinuses. The tumor tends to form localized bulky masses that remain masked by inflammatory changes. The tumors strike all age groups and have been described from age 3 to 80 (24).

CT and MRI Findings

At CT the lesion may simulate a mucocele of the affected sinus because of the amount of expansion and opacification that takes place. Metastases occur late so that the majority of the patients present with symptoms suggesting inflammatory disease and an increase in bulk to the affected sinus. As yet no calcifications have been found in the center of these tumors.

LYMPHOMA

Lymphoma of the paranasal sinuses is an unusual occurence and is usually related to an extranodal origin of lymphoma (15). Histiocytic lymphomas are the most common histology and the lesion is usually confined to the antroethmoidal complex. The tumor infiltrates into the bony margins and frequently escapes into the cheek or infratemporal fossa region. Nodal disease is rare when the patient presents.

Spread to the sinuses from the nasopharynx is not at all unusual because lymphoma crosses fascial planes quite freely and dissects into the infratemporal fossa.

CT and MRI Findings

Lymphoma around the nasopharyngeal region and the paranasal sinuses is usually characterized by large bulky lesions. The lymphoma tends to push bony septa and, in particular, the posterior wall of the antrum. In the paranasopharyngeal space and parapharyngeal spaces lymphoma widely disseminates through the fascial planes, much more so than in the epithelial tumors. The response to radiation therapy or chemotherapy can be followed very nicely by scanning. Fascial planes and bony septa that have been demineralized will return to normal as the lesions regress.

Secondary Tumors

Secondary tumors originating in the oral cavity or alveolar ridges means that any tumor of these locations can be mistaken for a primary sinus tumor unless careful history and examination is conducted. Loosening of teeth is a presenting complaint common to both primary and secondary antral tumors.

Isolated metastases to the paranasal sinuses can originate from the kidneys, breast, lungs, or even the GI tract. They have a tendency to center their expansion about the margins of the sinus rather than a mucosal thickening but this finding is far from precise.

SUMMARY

In summary, one must constantly remember that CT and MRI does not give histology but both modalities are superb methods of looking at altered anatomy and physiology. Pain in the sinus after the age of 60 is malignant until proven otherwise. One does not diagnose malignancy of the sinuses unless it is thought of. The role of CT is not only in diagnosis but plays a major function in the planning of precision radiation therapy or radical surgery for malignancies of this region. Only by carefully outlining the full extent of tumors will adequate surgery be carried out on the first attempt. The first chance for cure is always the best chance.

References

1. Batsakis JG: Mucous gland tumors of the nose and paranasal sinuses. *Arch Otolaryngol* 79:557, 1970.
2. Batsakis JG: *Tumors of the Head and Neck: Clinical and Pathologic Considerations*, Baltimore, Williams & Wilkins, 1979, chap 6.
3. Blitzer: Paranasal sinus mucor. *Laryngoscope* 90:635–645, 1980.
4. Conley J: *Concepts in Head and Neck Surgery.* Stuttgart, Georg Thieme Verlag, 1979, p 570.
5. De St Jeor, Konrad H, Hanafee WN: Tomography of paranasal sinus disease undetected by routine views. *Appl Radiol*, 9:73–80, 1980.
6. Dodd GD, Collins LC, Egan RL, Herrera JR: The systemic use of tomography in the diagnosis of carcinoma of the paranasal sinuses. Radiology 72:379, 1959.
7. Fu YS, Perzin KH: Non-epithelial tumors of the nasal cavity, paranasal sinuses, and nasopharynx: a clinicopathological study. III Cartilagenous tumors (chordroma, chondrosarcoma). *Cancer* 34:453, 1974.
8. Fujioka M, Young LW: The sphenoidal sinuses: radiographic patterns of normal development and abnormal findings in infants and children. *Radiology* 129:133–136, 1978.
9. Harrison DFN: The management of malignant tumors of the nasal sinuses. *Otolaryngol Clin North Am* 4:159, 1971.
10. Jackson RT, Fitz-Hugh GS, Constable WC: Malignant neoplasms of the nasal cavities and paranasal sinuses (a retrospective study) *Laryngoscope* 87:726–736, 1977.
11. Karmody CS: *Textbook of Otolaryngology.* Philadelphia, Lea & Feibiger 1983.
12. Larsson LG, Martensson G: Maxillary antral cancer. *JAMA* 219:342, 1978.
13. Last RJ: *Anatomy: Regional and Applied*, ed. 6. New York, Churchill Livingstone, 1978.
14. Lederman M: Cancer of the upper jaw and nasal chambers. *Proc R Soc Med* 62:55, 1969.
15. Lehrer S, Roswit B: Primary malignant lymphoma of the paranasal sinuses. *Ann Otol Rhinol Laryngol* 87:81, 1978.
16. Mancuso AA, Hanafee W, Winter J, et al: Extensions of paranasal sinus tumors and inflammatory disease as evaluated by CT and pluridirectional tomography. *Neuroradiology* 16:449–453, 1978.
17. Mira J: *Bone tumors: Diagnosis and Treatment.* Philadelphia, Lippincott, 1980.
18. Modiac MT, Weinstein MA, Berlin J, Duchesneau P: Maxillary sinus hypoplasia visualized with computed tomography. *Radiology* 135:383–386, 1980.
19. Ohngren LG: Malignant tumors of the maxillo-ethmoidal region: a clinical study with special reference to the treatment with electrosurgery and irradiation. *Acta Otolaryngol (Suppl)* 19:1, 1933.
20. Osborn AG, Anderson RR: Direct sagittal computed tomographic scans of the fact and paranasal sinuses. *Radiology* 129:81–87, 1978.
21. Paparella, Shumrick: Otolaryngology; Philadelphia, Saunders, 1980.
22. Potter GD: Case 11 Question 37039. In LF Rogers (ed): *Disorders of the Head and Neck Series 2, Syllabus.* Chicago, American College of Radiology, 1977.
23. Radberg, C: Appearance of sella turcica following transphenoidal hypophysectomy. *Acta Radiol* (Excerp) 1:140–151, 1963.
24. Rice DH, Batsakis JG, Headington JT, et al: Fibrous histiocytomas of the nose and paranasal sinuses. *Arch Otolaryngol* 100:398–401, 1974.
25. Silva EG, Butler JJ, Mackay B, Goepfert H: Neuroblastomas and neuroendocrine carcinomas of the nasal cavity. A proposed new classification. *Cancer* 50:2388–2405, 1982.
26. Spiro RH, Huvos AG, Strong EW: Adenoid cystic carcinoma of salivary origin. A clinicopathologic study of 242 cases. *Am J Surg* 128:512, 1974.
27. Weaver DF: Cancer of the ethmoid sinuses *Arch Otolaryngol* 74:333, 1961.
28. Weber L, Tadmor R, David R, et al: Malignant tumors of sinuses. *Neuroradiology* 16:443–448, 1978.
29. Ziedes Des Plantes BG: Een Byzondere methode voor bet maken van roentgnfot's Van Schedelen. *Inverrelkelm Med T Geneesk* 75:5219, 1931.
30. Zimmerman RA, Bilaniuk LT: Computed tomography of sphenoid sinus tumors. *Comput Axial Tomogr* 1:23–32, 1977.

Chapter 1 Plates

Malignant Sinus

The abbreviations used on Plates 1.1–1.5 are: *ZR*, zygomatic recesses; *GPC*, greater palatine canal; *PPF*, pterygopalatine fossa; *UP*, uncinate process; *IT*, inferior turbinate; *NLD*, nasolacrimal duct; *IOF*, infraorbital foramen; *MT*, middle turbinate; *BE*, bulla ethmoidalis; *SH*, semilunar hiatis; *S*, sphenoid sinuses; *ILR*, inferolateral recesses; *ST*, superior turbinate; *LF*, lacrimal fossa; *AE*, anterior ethmoid air cells; *ME*, middle ethmoid air cells; *PE*, posterior ethmoid air cells; *V*, volmer; *PPE*, perpendicular plate of the ethmoid; *LP*, lamina papyracea; *SOF*, superior orbital fissure; *CA*, carotid artery; *FS*, frontal sinus; and *CG*, crista gala.

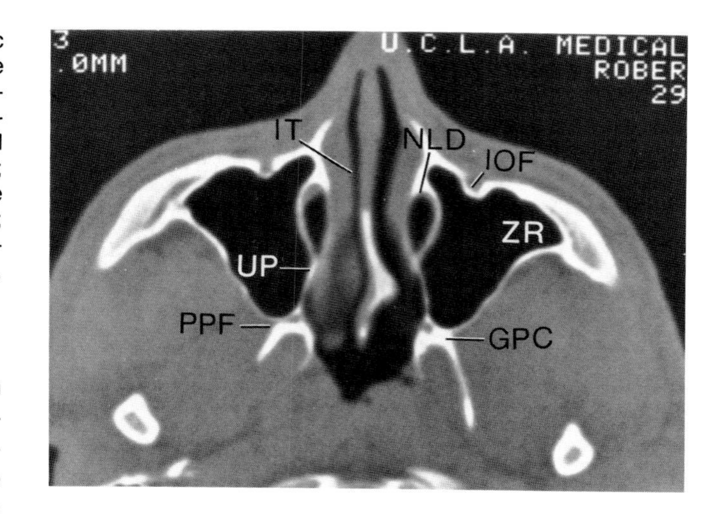

Plate 1.1. Normal—midantrum level. This 1.5-mm thick section through the midantrum viewed with extended scale shows well developed zygomatic recesses bilaterally. On the right the pterygopalatine fossa is becoming slitlike. On the left the fossa has changed into the greater palatine canal.

Within the nasal cavity the inferior turbinate covers the region of the inferior opening of the nasolacrimal duct.

The anterior wall of the maxillary sinus is thick and the opening of the infraorbital foramen is partially volumed.

The uncinate process is arising to form the medial wall of the antrum and anterior border the ostiom.

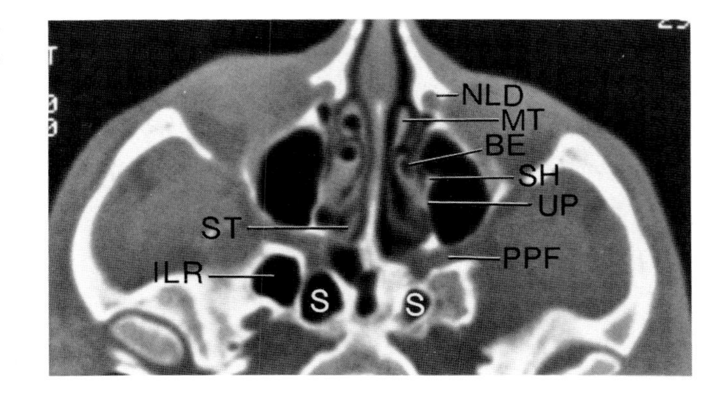

Plate 1.2. Normal—section through the highest portion of the antrum and low portion of the orbit. The nasolarimal ducts are sectioned immediately inferior to the lacrimal sac. The middle turbinate is extending inferiorly to cover the semilunar hiatus which is partially visible in this section. The bulla ethmoidalis bulges prominantly beneath the apron of the middle turbinate. The uncinate process extends superior and posterior, forming a margin for the hiatus of the maxillary sinus which lies immediately posterior to this process.

On the right the origin of the superior turbinate is visible. The base of this superior turbinate denotes the region of the sphenoid recess which receives the openings of the sphenoid sinuses. Lateral to the sphenoid sinuses are inferolateral recesses which extend into the base of the pterygoid plates and into the greater wing of the sphenoid.

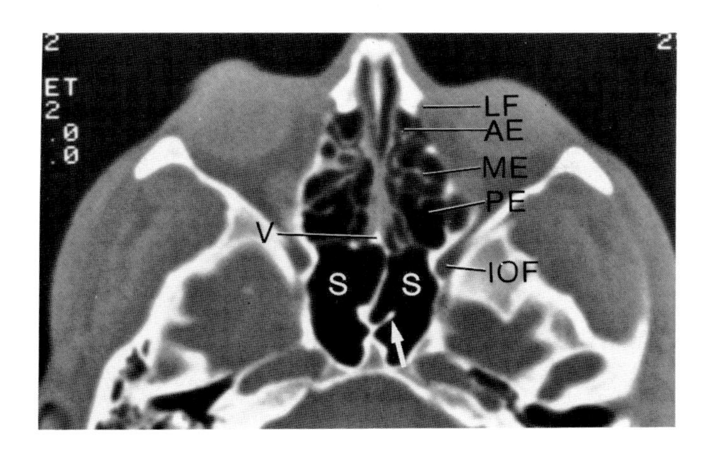

Plate 1.3. Section through the lower ethmoid and lower orbits shows how the anterior, middle and posterior ethmoid air cells expand inferolatrally to form part of the floor of the orbit. The lacrimal fossa is made up by the anterior and posterior crests of the lacrimal bone. The inferior orbital fissure is seen as a dehiscence in the posterior part of the orbit.

Near the midline the volmer with its ala gives support to the nasal septum. Posteriorly the paired sphenoid sinuses are divided by a vertical septum. In the posterior portion of this septum a second septum can be seen arising (*arrow*) which frequently divides the sphenoid sinuses into a superior and inferior compartment. The arrangement of these septa in the sphenoid sinus is quite variable.

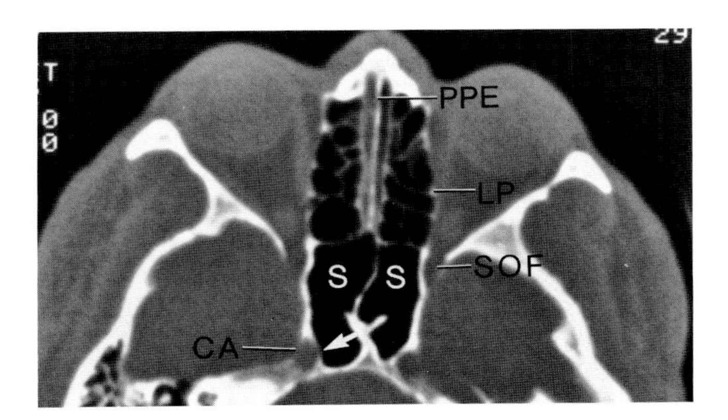

Plate 1.4. Section through the high ethmoid air cells shows that the medial walls of the orbit are now parallel. The wall itself is made up of the extremely thin lamina papyracea. The section is immediately below the level of the cribriform plate which can be confirmed by the fact that the perpendicular plate of the ethmoid extends all the way from the nasal bones to the anterior wall of the sphenoid sinuses.

The carotid artery at this level is in the posterolateral wall of the sphenoid sinus. The medial wall of this carotid artery canal is extremely thin (*arrow*). In some patients it may be totally dehiscent and the carotid artery may lie within the sphenoid sinus.

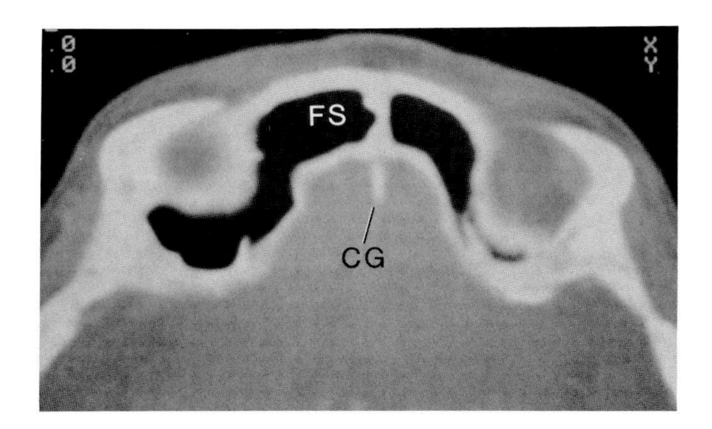

Plate 1.5. Frontal sinuses. This 5-mm thick section in another patient shows how the frontal sinuses may extend posteriorly and laterally to pneumatize the entire orbital roof. In some rare instances the pneumatization of the orbital roof is from ethmoid sinuses. The superior tip of the crista gala is visible in the midline.

In this patient the septum between the two frontal sinuses is well formed. Occasionally it will contain air cells which are prone to disease because of inadequate drainage.

Some important spreads of malignancy that limit the ability to perform curative surgical resections are as follows:

1. Infratemporal spread by the posterolateral wall
2. Invasion of the pterygopalatine fossa
3. Pterygoid fossa extension
4. Orbital extension particularly involving the orbital apex
5. Anterior cranial fossa invasion by the cribriform plate
6. Middle cranial fossa by the sphenoid sinus (either lateral wall or inferolateral recess extensions
7. Nasopharyngeal involvement
8. Skin lymphatic involvement

These regions will be stressed in the figures.

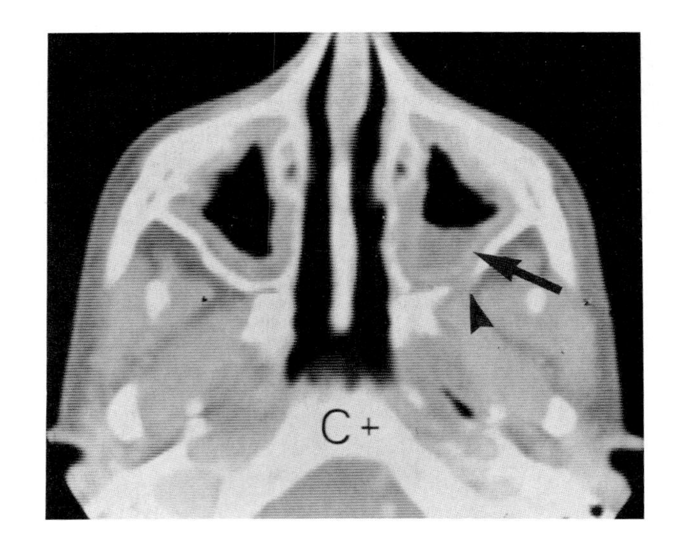

Plate 1.6. Normal dehiscence of posterolateral wall. The posterolateral wall of the left antrum is dehiscent at the site of the groove for posterosuperior alveolar artery and nerve (*arrowhead*). By giving intravenous contrast the capillary plexus adjacent to the mucosa stains intensely (*arrow*). In this patient suspected of having malignant sinus disease, one notices that the mucosa is not interrupted at the level of the bony dehiscence and would lessen the likelihood of a spread of the infratemporal fossa had this lesion been malignant.

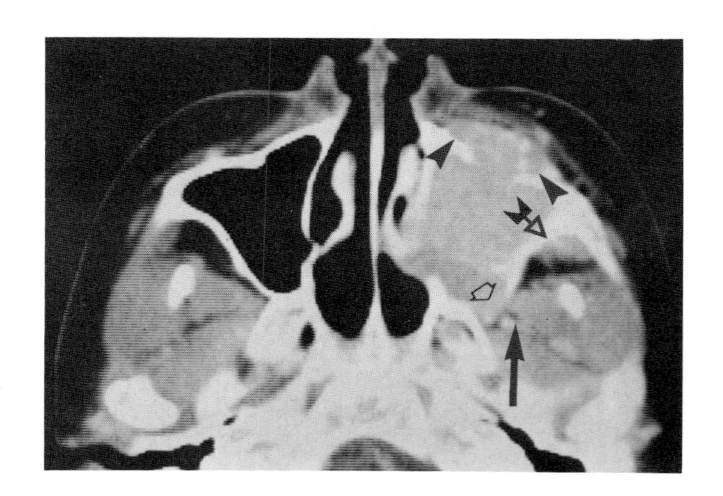

Plate 1.7. Expansion into the infratemporal fossa. The posterolateral wall of the antrum is expanding into the infratemporal fossa but not definitely infiltrated (*open arrowhead*). The posterior superior alveolar artery has been displaced posterior (*arrow*) and there is no disruption of the fat surrounding the vessel.

On the otherhand this tumor has invaded through the anterior wall of the antrum (*arrowheads*) and a small amount of tumor is present in the anterior portion of the deep temporal fossa (*arrow with tail*).

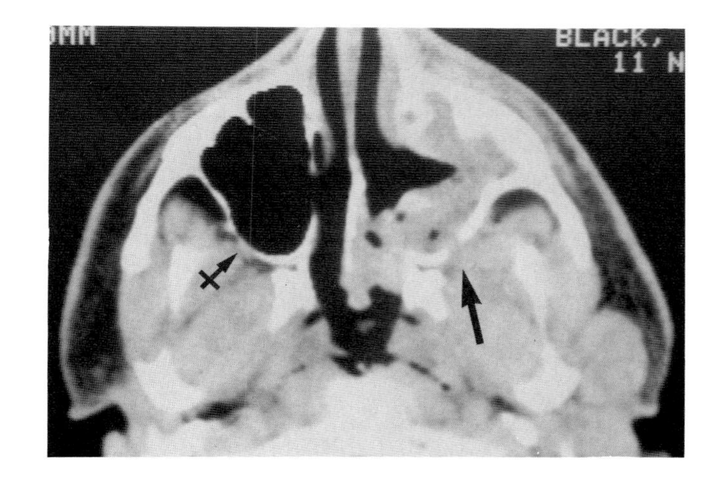

Plate 1.8. Early infiltration of the infratemporal fossa. In this patient with dehiscence of the posterolateral wall there is infiltration into the infratemporal fossa as evidenced by loss of the normal low density fat that covers the region of the posterosuperior alveolar artery and nerve (*arrow*). Notice the artery and the retained low density fat on the patient's normal right side (*crossed arrow*).

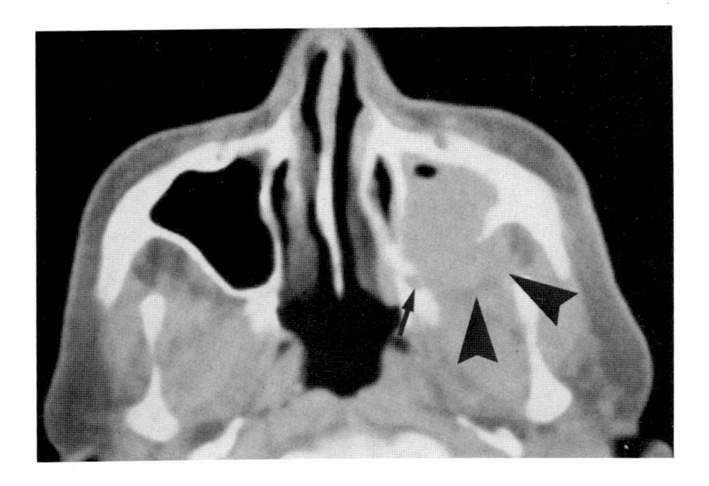

Plate 1.9. Pterygopalatine fossa invasion. This patient with squamous cell carcinoma of the left antrum shows destruction of the posterolateral wall with the tumor mass invading into the infratemporal fossa (*arrowheads*). Note that medially there is destruction of the wall of the antrum and the tumor has involved the pterygopalatine fossa (*arrow*).

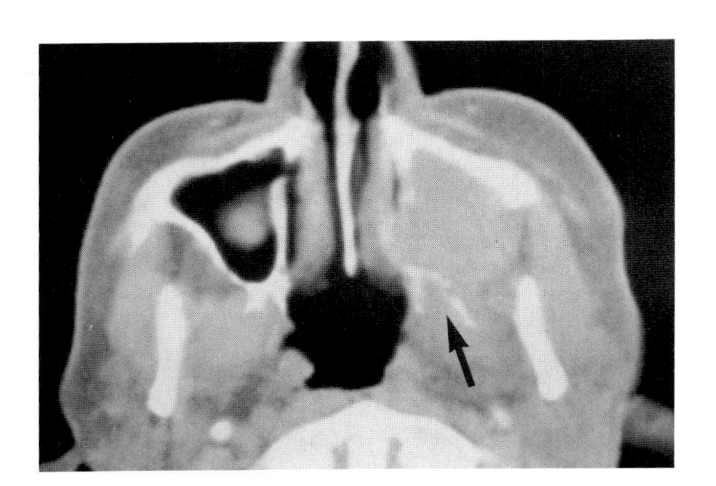

Plate 1.10. Pterygoid fossa invasion. This squamous cell carcinoma has partially destroyed the medial pterygoid plate and the anterior portions of the lateral pterygoid plate. With this degree of pterygoid plate destruction one can be certain that the tumor has extended from the pterygopalatine fossa into the pterygoid fossa (*arrow*).

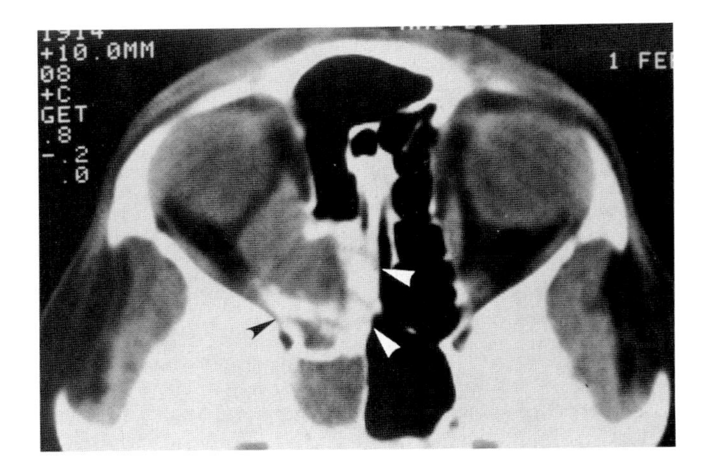

Plate 1.11. Orbital apex extension. This chondrosarcoma arising in the ethmoid shows typical matrix calcification especially around the periphery of the tumor (*arrowheads*). The tumor has invaded into the medial-inferior orbital apex and involves the tendenous ring of Zinn. These tumors can be very slow growing. The patient is clinically free of disease even though the tumor was only partially removed over a year ago.

Plate 1.12. Anterior cranial vault involvement.

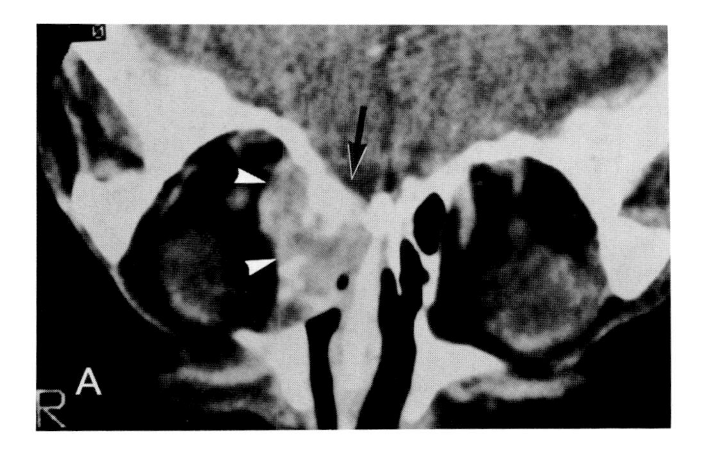

Plate 1.12*A*. Direct coronal scan viewed with brain setting shows a minimum amount of intensely staining tumor extending into the olfactory fossa on the right (*arrow*) due to an adenocystic carcinoma of the ethmoid sinuses. There is also invasion into the medial portions of the orbit (*arrowheads*) but the tumor appears to be respecting the muscle cone region.

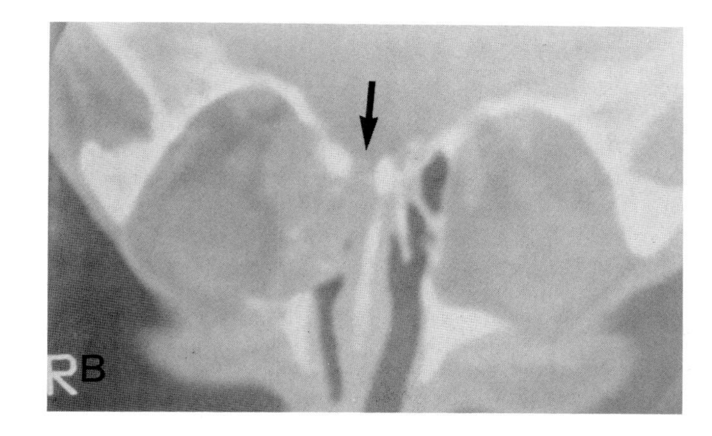

Plate 1.12*B*. Bone window settings on this same patient confirms that there is bone destruction of the roof of the ethmoid air cells on the right as well as the medial wall of the ethmoid air cells (*arrow*). Although disabling, the orbital invasion does not have the ominous significance of the intracranial extension. The adenocystic carcinomas are prone to follow nerves, and in this instance the tumor is accompanying the olfactory nerve.

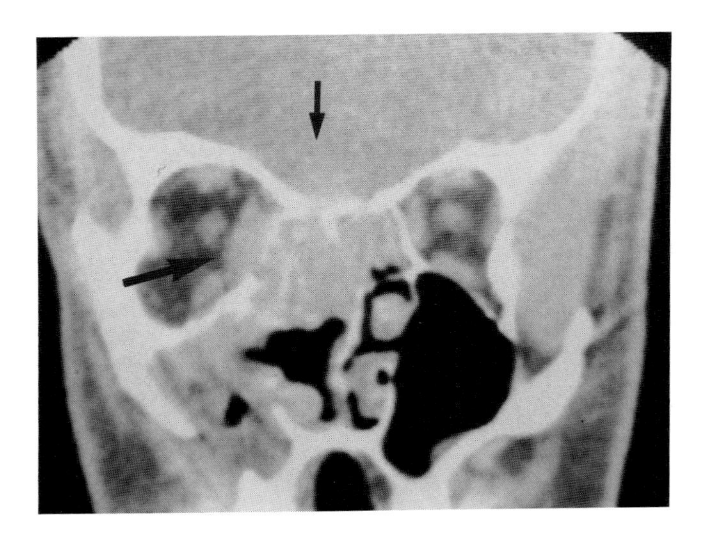

Plate 1.13. Anterior cranial vault invasion. This 22-yr-old male with a squamous cell carcinoma involving the antrum and ethmoid air cells shows destruction of the anterior portion of the planum sphenoidale with a tumor extending intracranially (*small arrow*). There is also destruction of the posteromedial wall of the orbit with tumor extending into the right orbit into the space peripheral to the muscle cone region (*large arrow*).

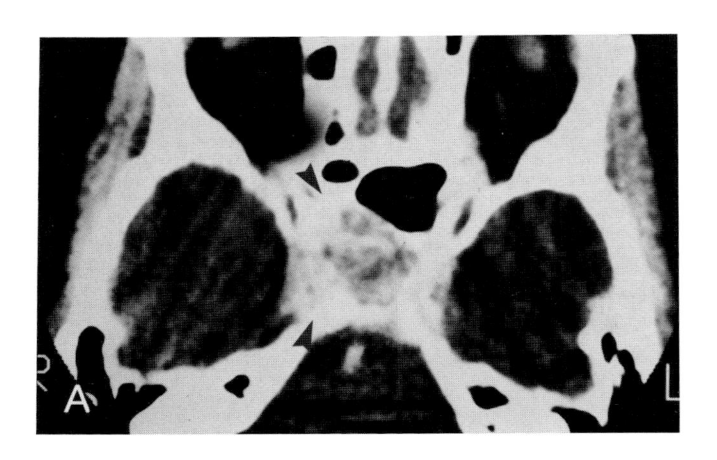

Plate 1.14. Squamous cell carcinoma of the sphenoid sinus.

Plate 1.14*A*. Squamous cell carcinoma of the sphenoid sinus shows a mottled enhancement with involvement of the right side of the sphenoid sinus to a greater extent than on the left (*arrowheads*).

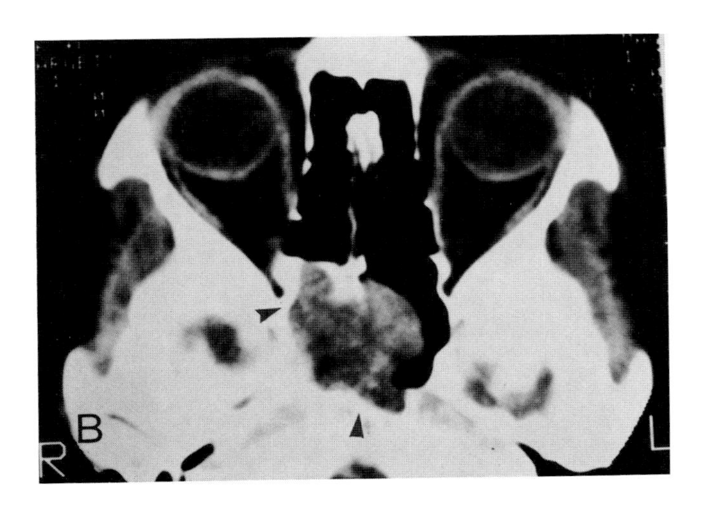

Plate 1.14*B*. At a slightly higher level tumor can be seen to extend through the lateral and anterior wall of the sphenoid sinus (*arrowheads*). No tumor can be identified lateral to the cavernous sinus.

Plate 1.15. Middle cranial fossa extension from antral carcinoma.

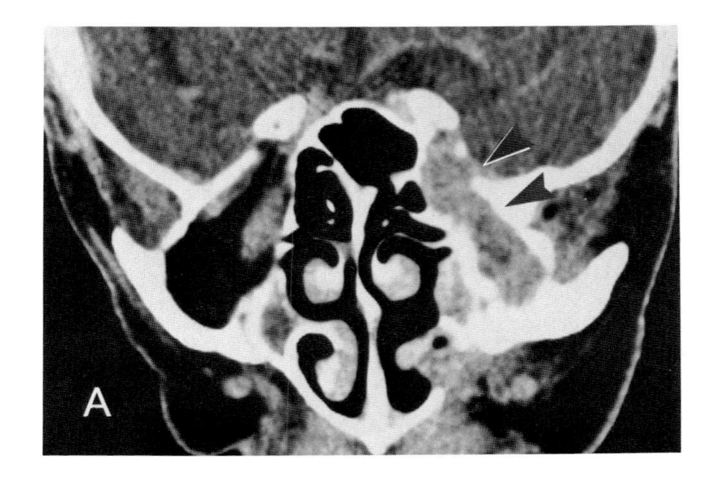

Plate 1.15*A*. Coronal scan shows the tumor lying within the inferior orbital apex (*arrowheads*). This tumor gained access to the middle cranial fossa by extending into the infratemporal fossa. From the infratemporal fossa the tumor spread through the inferior orbital fissure to the orbital apex. It extended posteriorly through the superior orbital fissure from the orbital apex into the middle cranial fossa.

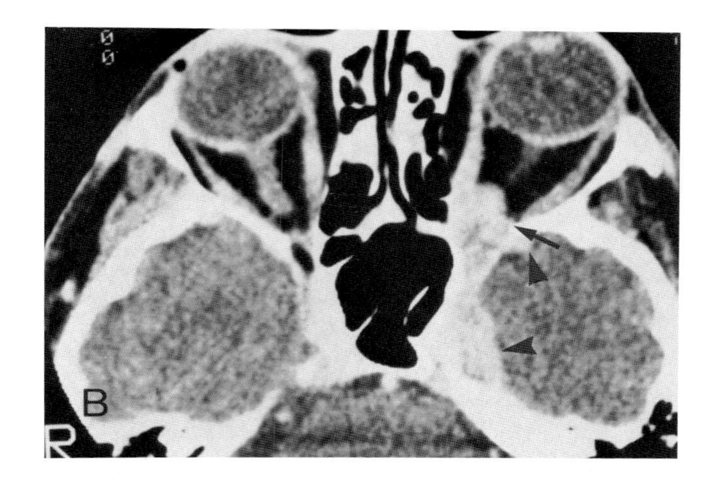

Plate 1.15*B*. In the axial plane tumor can be identified in the orbital apex (*arrow*) and extending through the superior orbital fissure to involve the medial portions of the middle cranial fossa (*arrowheads*).

Facial Deformities

Midface hypoplasias and mandibulofacial dysplasias can be diagnosed by clinical examination but the degree of underdevelopment can be documented by CT. The entire management of these individuals with facial deformities is undergoing radical change but CT would seem to add a significant dimension to understanding arrested growth and the degree of correction that has been possible through midface advancements.

CT and MRI Findings

Following surgery, the healing of bone defects may be difficult to evaluate. The postsurgical changes in paranasal sinuses and midface may stimulate inflammatory disease. This research work is in its infancy and the contribution of CT to day-to-day clinical management remains to be sorted out.

INFLAMMATORY DISEASE: ACUTE

Bacterial Infections

The clinical presentations of acute bacterial infections are usually sufficiently typical that plain films are the only diagnostic modality needed to compliment the clinical examination. If developmental variations of the sinus are present CT may add to this information. For example in patients with symptoms of sphenoid sinus disease air fluid levels may escape detection on the conventional films because the fluid has collected in inferolateral recesses. When CT is performed the patient lies in a supine position. Fluid levels that are normally obscured by the skull base can be seen in the inferolateral recesses.

More frequently CT is used to "map" a region prior to surgical management. Dehiscence of the posterior wall of the frontal sinus prior to drainage procedures or a carotid artery that does not have a bony covering in the sphenoid sinus could lead to disastrous complications (10).

CT and MRI Findings

MRI studies of acute sinus disease will show bright signal coming from edematous mucosa and fluid-filled sinuses on T_2-weighted images. Chronic thickening and scarring of the mucosa give a bright signal but less than that of fluid-filled sinuses.

The CT studies can be extremely helpful in identifying sequestra within inflammatory disease or changes in mineralization in the bony wall associated with osteitis or frank osteomyelitis. Unfortunately the CT findings, like x-ray findings, lag so far behind clinical examination and symptomatology that the CT diagnosis of osteomyelitis is usually made after the patient is on the road to recovery from antibiotic therapy.

Fungal Infections

Aspergillosis, mucormycosis and actinomycosis may affect the paranasal sinuses, especially in diabetics or immune suppressed patients (5, 18, 19, 26). The acute forms of fungus infections show nonspecific findings within the paranasal sinuses but the ability of funguses to invade blood vessels is very important for their spreads. In the ethmoid air cells they tend to follow the anterior and posterior ethmoidal vessels into the orbit. Unlike bacterial infections, the fungal infections tend to "by-pass" the periosteum of the medial orbit wall to involve the muscle cone region. They do not form subperiosteal abscesses along the medial wall of the orbit as occurs in children with acute ethmoid sinus disease (2, 15). In the maxillary sinuses they follow the posterior alveolar artery branches into the infratemporal fossa and inferior orbital fissure or invade the orbit directly. From the orbital apex the spread may extend posteriorly through the superior orbital fissure into the middle cranial fossa and cavernous sinus. If neurological deficits and brain involvement become evident the prognosis is extremely guarded.

A chronic form of actinomycosis and aspergillosis can be found within the paranasal sinuses which may be related to long-standing antibiotic therapy.

CT and MRI Findings

In acute forms of fungal infections one may see opaque sinuses and distortion of fascial planes in adjacent regions (27). If the orbit becomes involved secondary to ethmoid disease there is increase in density of the fat involving the orbital apex and frequently the medial rectus muscle may become enlarged. Destruction of the lamina papyracea about the ethmoid artery branches in frequently seen. Unlike acute bacterial sinus disease, sequestra are rarely found in the fungus infections. CT can trace spread to the orbits from maxillary sinusitis via two routes; either via ethmoid air cells or from the maxillary sinus into the infratemporal fossa and infraorbital fissure.

In more chronic forms of fungus infection, bony proliferation of the margins of the involved sinuses becomes the rule. Thicknesses of bony wall ranging from 3 to 5 mm are not uncommonly seen. The central content of the sinuses show the thickened mucosal lining and epithelial debris and/or fluid in the contracted remaining sinus lumen. No specific clues are present in the chronic disease to definitely indicate fungus infection but greatly thickened sinus wall in a patient who has received long-term antibiotic therapy is highly suspicious of chronic aspergillosis.

INFLAMMATORY DISEASE: CHRONIC

Retention Cyst

Retention cysts have been called nonsecreting cysts, noninflammatory cysts, serous cysts, and mucous gland cysts. They are more common in the inferior recesses of the maxillary sinuses but may be found in the frontal and sphenoid sinuses. They are quite mobile and usually fall into the dependent portions of the maxillary sinus when the patient is examined in the supine position.

CT and MRI Findings

CT examination is not indicated in a diagnostic evaluation of a patient with retention cysts. Only occasionally retention cysts may almost completely fill the maxillary sinus and simulate a mucocele. Under these circumstances CT can verify that a miniscule amount of air is retained within the sinus capping the retention cyst. The cysts contain low density fluid. The extremely thin walls do not enhance on intravenous contrast. They are to be differentiated from polypoid hypoplasia of the lining of the sinuses which is a much more symptomatic lesion.

Allergy

The manifestations of allergy in the paranasal sinuses is quite a spectrum of disease that usually originates in early childhood or young adults. Since it is a systemic disease of respiratory surfaces a pansinusitis is almost invariably present. When associated with polyp formation they originate near the ostium of the maxillary sinus on the sinus side of the nasal wall.

The allergy may persist unabated leading to chronic inflammatory disease superimposed on the antral polyps. The combination of inflammation and allergy may lead to formation of polyps from other structures within the nasal vault including turbinates and nasal septum. Because of the dominance of inflammation in the polyps in some of the latter locations, these lesions are frequently spoken of as inflammatory polyps.

The most severe form of allergy with polyp formation bears the title of chronic hypertrophic polypoid rhinitis (33). These individuals have a fulminating form of polyposis and chronic infection. All of the mucosal surfaces are affected so that polypoid hypoplasias occur both within and outside of the sinus structures. Expansion of the associated sinuses filled with polyps cause hypertelorism, fullness of the malar eminence and brow prominance.

CT and MRI Findings

On CT scanning with intravenous contrast enhancement the sinuses are shown to be filled with combinations of contrast-laden mucosal thickening and some areas of low density fluid due to the obstructed sinuses. The bony distortions may reach grotesque proportions if not aggressively managed. In some patients the nose becomes "piglike" and occupies the major portion of the midface.

When mucoceles form they are more commonly located in the ethmoid sinuses followed by frontal, then sphenoid and, lastly, maxillary sinuses. Multiple sinuses containing mucoceles are so common that a radiologic diagnosis of panmucoceles would seem to be warranted (35).

Choanal Polyps

Choanal polyps are a special form of polyps that are found predominantly in young males. Only occasionally are choanal polyps associated with polypoid disease of a generalized nature. Choanal polyps begin in the antrum and have a component outside the antrum. The mass extends into the nasal cavity and eventually presents posteriorly in the posterior nasal choana. When removed they have three lobes and two distinct constrictions conforming to the ostium of the maxillary sinus and the choana of the nasal cavity. They are usually not associated with allergy (1).

CT and MRI Findings

Choanal polyps may be diagnosed quite nicely with plain films showing the opaque sinus and a tumor mass within the posterior nasal cavity. At times it might not be easy to identify all components of the polyp on clinical examination and only a mass presenting in the anterior nares is available to the physician. CT can be used effectively to rule out a juvenile angiofibroma in a young boy (4, 11).

Characteristically choanal polyps do not stain to any great extent by intravenous contrast except along their surface. They do not invade into the pterygopalatine fossa which is so characteristic of juvenile angiofibromas. Choanal polyps do not extend beyond the mucosal limits of the nasopharynx (see nasopharynx for juvenile angiofibroma). The choanal polyp can be traced by CT in its entirety from the maxillary sinus to the anterior nasopharynx without difficulty.

Mucoceles

Mucoceles of the paranasal sinuses most commonly affect the frontal sinus, followed by ethmoid sinuses, sphenoid sinuses and lastly maxillary sinuses, in that order (9). They form when the natural ostium of a sinus becomes totally obstructed. The occlusion may be related to previous trauma, structural abnormality, chronic infection or, at times, the exact etiology may be obscure. When infected a pyomucocele is formed which is potentially lethal due to intracranial extension or septicemia.

The epithelial lining of mucoceles continues to secrete causing pressure erosion of adjacent bone although months or even years is involved in the process.

Mucoceles of the frontal sinus frequently extend inferolaterally into the orbit producing proptosis or erode the posterior table of the frontal sinus producing intracranial extension. If the anterior table of the frontal sinus is eroded, a mass is formed called the "Pott's puffy tumor." The subcutaneous tissues of the brow become doughy in consistency due to a lack of bony wall. Sphenoid sinus mucoceles are especially ominous because they erode the bony coverings of the carotid artery lying in the lateral wall of the sinus and may erode the cortex of the dorsum thus exposing the dural covering of the clivus. Surgical drainage must be conducted with extreme caution.

Surgical management of mucoceles of the frontal sinuses is vastly different when the ethmoid sinuses are also involved (20, 29). The surgical exposure of the frontal sinus is excellent using coronal scalp incision within the hairline (Fig. 2.1*A*). The flap of scalp tissue can be retracted inferiorly to expose the entire brow region. Unfortunately access to the ethmoid region is quite limited by this approach. If the ethmoid and frontals are both involved an incision through the brow and extending along the lateral border of the nose can be used (Fig. 2.1*B*). This gives good exposure to both regions but unfortunately leaves a scar on the nose. The extent of a lesion can be nicely appraised by CT to include frontal and ethmoid, as well as sphenoid sinus mucoceles.

Maxillary sinuses may be loculated by bony septum that lead to mucocele formation in only a portion of the sinus (31). This latter condition is extremely difficult to diagnose by routine films whereas the usual maxillary sinus mucocele that occupies the entire sinus is within the realm of plain film diagnosis.

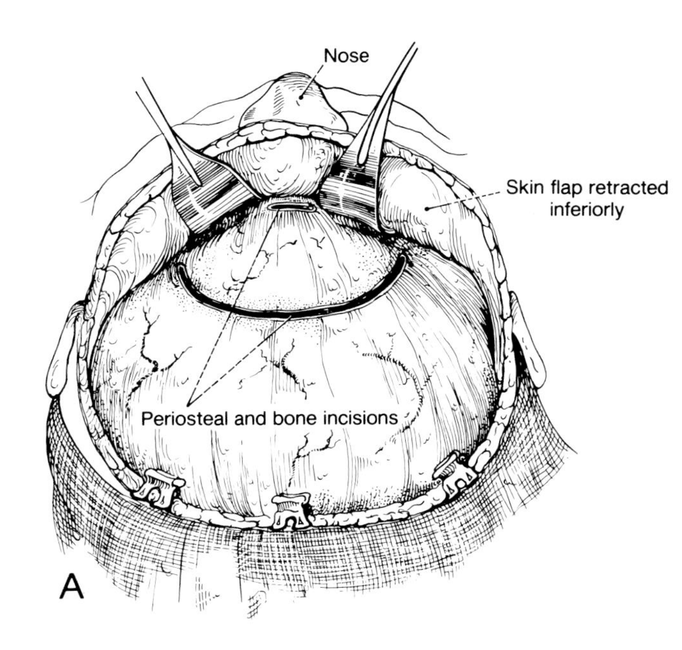

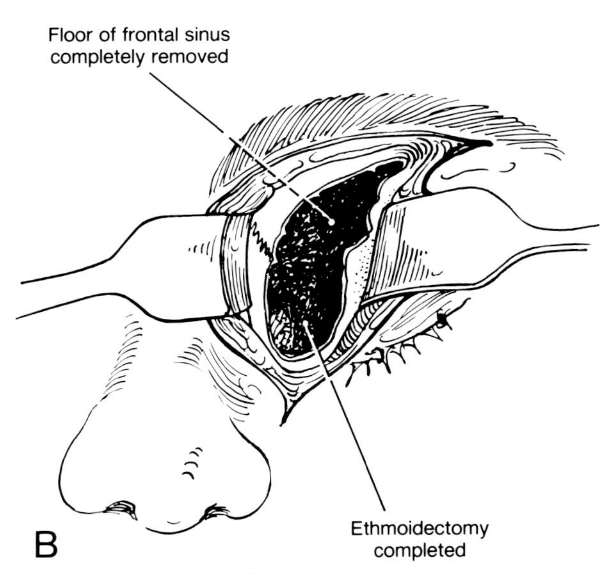

Figure 2.1. Surgical approaches to the frontal and ethmoid sinuses. *A*, coronal incision for osteoplastic flap. The coronal incision through the hairline frees up the entire forehead as a flap which can be pulled down giving exposure to the frontal sinus. The anterior wall of the frontal sinus is removed and all mucosa can be removed. The sinus cavity is obliterated with fat and the bone flap replaced. This osteoplastic flap operation gives excellent exposure to the frontal sinus but does not permit axis to the ethmoid sinuses. There is no deforming scar. *B*, Lynch incision (brow incision) for access to ethmoid frontal sinuses. The incision is made through the brow and extended inferiorly and lateral to the nose giving excellent access to the lower frontal sinus and ethmoid sinuses. It can be extended across the midline to the opposite brow for wider exposure. This incision gives excellent visualization of both sinus regions but does leave a scar on the face. (*A* and *B* from Montgomery WW: *Surgery of the Upper Respiratory System*, ed 2. Philadelphia, Lea & Febiger, 1979, chap 4.)

CT and MRI Findings

The MRI signal from mucoceles of the sinuses is quite bright whether the contents are fluid or a combination of hypertrophied mucosa, polyps and retained secretions. As yet insufficient information is available to differentiate with confidence whether a lesion is due to malignancy or inflammatory disease. For the time being, CT remains the mainstay of mucocele diagnosis (12, 35). CT should show expansion of the bony walls, erosion of adjacent tables, fluid low density material within the sinus, or a combination of fluid and hypertrophied mucosa when polypoid disease is the etiology of the mucocele.

In frontal sinus lesions extreme care should be taken to see if the lesion has extended posteriorly into the ethmoids or into the orbital roof. Careful search should be made for delicate bony septa that indicate frontal as well as ethmoid mucoceles rather than one single cavity.

In ethmoid mucoceles, coronal scanning is extremely helpful to see the status of the cribriform plate and bony covering of the roof of the ethmoids so that the surgeon may be forewarned about exposed dura or cribriform plate.

Sphenoid sinus mucoceles may extend into the inferolateral recesses involving the floor of the middle cranial fossa. Venous plexuses in the pterygoid fossa may communicate intracranially and thus lead to dural defects when expanding sphenoid sinus mucoceles are present. Coronal scanning at times is difficult in the middle cranial fossa because molar teeth with metallic restorations interfere with high resolution. Reformatted images, although not perfect, have proven quite useful in this location.

Structural abnormalities may lead to obstructive disease and sinus expansion. This is particularly important in deviated nasal septum or benign tumors originating near the ostia of paranasal sinuses. Remembering that there is continuity of respiratory epithelium throughout the nasal cavity and paranasal sinuses, one should make a careful search for multiple etiologies when isolated sinus disease appears. In the older age groups malignancy must be suspected and appropriate clinical investigation undertaken.

BENIGN TUMORS

Inverting Papillomas

Inverting papillomas are probably one of the most interesting of the "benign" tumors of the paranasal sinuses. They derive their name from the histologic pattern of the tumor wherein nests of squamous epithelium lie within the substance of the polypoid lesion (1, 9). This is probably related to squamous metaplasia within ductal systems of mucous and/or serous glands of the mucosa. Inverting papillomas characteristically affect individuals in the 20- to 40-yr-old-age group with male population dominating. The lesion most generally starts near the junction of maxillary and ethmoid sinuses along the lateral nasal wall. By the time patients seek medical assistance the tumor is usually bulky and has produced distortions of sinus cavities and may have invaded into some of the adjacent sinuses. Inverting papillomas are known to be locally aggressive and malignancy is reported in 4 to 15% of the tumors. This number has been suspected of being artifically low merely because of insufficient clinical follow-up. Malignancy of the nasal vault or paranasal sinuses is being reported with increasing frequency 10 to 20 yr following conservative surgery for inverting papillomas.

CT and MRI Findings

The bulky infiltrating nature of this tumor growth dominates the CT appearance. Although the tumor does not calcify it frequently distorts the turbinates and destroys their normal architecture so that dense calcification can be identified within the tumor substance. When the tumor originates solely within the nasal cavity, the medial wall of the sinus is usually displaced laterally. Intravenous contrast will frequently show that the tumor has crept along the mucosal surfaces producing a mottled high density mass lesion within maxillary sinuses. Frequently the ethmoid and frontal sinuses are also involved. The bulk of the tumor mass may also obstruct ostia-producing mucoceles of involved sinuses. The mucoceles when of an obstructive nature show low density fluid rather than high density hypertrophy mucous membrane as is seen in polypoid type lesions (7). A histologic diagnosis of inverting papilloma cannot be made by CT but accurate information can be given to the operating surgeon of the extent of lesions and involvement of adjacent paranasal sinuses.

Squamous Papilloma

Squamous papillomas are warty growths characteristically arising on the nasal septum (29). Chronic irritation of trauma has been cited as an etiologic agent although some occur spontaneously as evidenced by their extreme posterior location on the septum. Rarely do the squamous papillomas reach sufficient size to obstruct the sinus ostia.

CT and MRI Findings

At CT the squamous papillomas may grossly resemble more malignant lesions such as melanomas or squamous cell carcinomas. When the papillomas are largely exophytic they resemble inflammatory polyps in their morphology. The underlying nasal cartilages can usually be shown to be intact by CT. Rare instances occur when the papillomas obstruct adjacent sinuses producing opacified sinus.

Granuloma Gravaderum

Closely akin to squamous papillomas are granulomas of the nasal septum which occur or exhibit accelerated growth during pregnancy. Growth is very rapid during the second and third trimester and these lesions may fill the nasal cavity (8). The granulomas do not cross the nasal septum but may be bilateral in origin.

CT and MRI will show soft tissue mass filling the nasal cavity. The soft tissues enhance with intravenous contrast. The bulky nature of the lesions may cause obstructive sinus disease.

Angiomatous Polyps

An important differential diagnostic problem facing the clinician in any bulky polypoid lesion is whether the tumor is juvenile angiofibroma with a tremendous propensity for bleeding or some other type of tumor that can be safely biopsied. The angiomatous polyps closely resemble the gross morphology of juvenile angiofibromas presenting in the anterior nasal cavity and maxillary sinuses. Even on angiography the lesions are extremely vascular. Angiomatous polyps can be found in either sex, however, their peak age of appearance is more in the young adult age group. The major importance of preoperative diagnosis in the angiomatous polyp has to do with the ease of management. These lesions tend to "shell out" readily at surgery and bleeding problems are minimal despite their very vascular nature (30).

CT and MR Findings

The point of origin of angiomatous polyps is near the natural ostium of the maxillary sinus with the mass lying within the sinuses and within the nasal vault. CT demonstrates staining following intravenous contrast but not to the extent that one finds with juvenile angiofibromas. The tumors may bulge into the nasopharynx but do not involve the pterygopalatine fossa. The medial margins of the antrum are usually largely destroyed by the bulk of the tumor growth and collections and secretions that do not enhance can be identified in the lateral portions of the maxillary sinus.

Both T_1- and T_2-weighted images show less signal with polypoid lesions than with retension cysts.

Fibrous Lesions

A rather confusing group of nonepithelial lesions that could be considered a spectrum of disease is identified by the fibrous-histiocytic group. The fibromyxoma is a benign lesion which has a propensity to recur unless aggressively excised. A gradation toward malignancy can be found as one proceeds from a fibrous histiocytoma to a frank fibrosarcoma (3).

CT and MRI Findings

The fibromyxoma has a tendency for bony proliferation around the margins of the sinus with low density myxomatous centers. Expansion of the sinus walls occurs due to the cystic nature of the lesion. The fibromyxomas are usually sharply outlined while the more aggressive lesions have fingers projecting into surrounding fat and muscle. The term "protrusion" may be added to their title to explain their tenacious nature of recurrence following surgical extirpation.

Chondromas and Osteomas

Chondromas and osteomas occur most commonly about the ethmoid air cells and the inferior portions of the frontal sinuses. It is surprising how frequently these lesions may reach considerable size without producing mucoceles. Interference with sinus drainage does occur and localized sinus inflammatory disease is usually the presenting symptom of these patients. When multiple osteomas are found within the frontal sinuses and other sinuses a search should be made for colon polyps because of the familial tendency of this syndrome (Gardner's syndrome).

CT and MRI Findings

The osteomas are not visible at MRI because of their bony nature. They appear as a region of low signal which cannot be differentiated from air-containing sinuses in an MRI experiment.

CT shows the extreme dense nature of osteomas and the matrix calcification of chondromas to excellent advantage. The full extent of circumscribed rounded masses associated with these two lesions can be readily appreciated and the histology of the tumors suggested. Occasionally chondromatous tumors of low malignancy potential will be contained in a circumscribed tumor mass. CT can only suggest the diagnosis of malignancy when there is aggressive infiltration of adjacent regions surrounding the main tumor mass. Intravenous contrast can be helpful in separating malignant from benign chondromatous tumors by showing a large soft tissue component with invasive characteristics. It also makes evaluation of mucoceles in adjacent sinuses readily diagnosable. Many times the confirmation of a truly normal sinus despite an adjacent osteoma is as important as demonstrating pathology.

Fibrous dysplasia must be differentiated from osteomas. Fibrous dysplasia has a tendency to creep along the orbital boundaries and involve the optic canal. Osteomas, on the other hand, have a more rounded growth pattern and do not tend to grow by extension along bony plates but rather by expanding as a rounded or oblong mass.

Epidermoid Tumors

Epidermoid tumors may arise as cellular rests within the lateral margins of the frontal sinuses or in the lateral portions of the frontal bone. These tumors will expand into the frontal sinus thus simulating a localized mucocele. The tumors have sharp margins in the frontal bone and do not excite significant adjacent bony sclerosis.

They are important to recognize radiologically because management of an epidermoid tumor of the frontal bone is vastly different than that of a frontal sinus mucocele. The management of a frontal sinus mucocele is an osteoplastic flap which carries with it a significant percentage of recurrent mucoceles. The management of an epidermoid tumor is to remove the epidermoid but leave the natural drainage of the frontal sinus intact (23).

CT and MRI Findings

CT of a frontal epidermoid tumor will show a sharply circumscribed soft tissue mass filling a bony defect in the lateral portions of the frontal sinus. Usually the frontal sinus cavity itself is clear with normal mucosal linings. The lesions must be differentiated from retention cysts which characteristically are located in the inferior portions of the frontal sinus and are sharply circumscribed mucosal lesions within the cavity of the frontal sinus as opposed to the epidermoid tumors which are truly outside the main lumen of the frontal sinus.

Dental Origin Tumors

The entire subject of dental-origin tumors is beyond the scope of this discussion of sinus disease, however, some mention must be made of their presence.

It is not surprising that the thin bone separating sinuses from tooth roots can be disrupted or displaced by dental processes. Extractions may lead to fracture of this bony plate and an oral antral fistula. Inflammations may extend from the periapical areas into the antrum producing an antral sinusitis. Since the response of the maxillary sinus to intrinsic inflammatory disease is mucoperiosteal thickening, it is extremely difficult to determine the origin of a process involving the floor of the antrum without careful dental evaluation.

In general, inflammatory disease of maxillary sinus does not become significantly symptomatic until partial obstruction to drainage has occurred. Since the natural ostium is high on the posteromedial wall, dental tumors do not produce obstruction until late in their course. By this time there are usually other manifestations of involvement of the alveolar ridge.

CYSTS OF THE MAXILLA

Cysts of the maxilla may be odontogenic or nonodontogenic in origin. Dentigerous and radicular cysts represent 85% of the expanding cysts that occur in the maxilla and involve the antrum (14). Primordial cysts, odontogenic keratocysts, and globulomaxillary cysts are occasionally found but with a much less frequency.

DENTIGEROUS CYSTS

Dentigerous cysts may form over the crown of an uninterrupted tooth or, rarely, over the root. As they expand in the maxilla they contain the uninterrupted tooth and a thin cortical margin. They are most commonly associated with the third molar.

RADICULAR CYSTS

These cysts begin about the roots of the tooth and are almost invariably associated with previous inflammatory disease. As they expand, over half of the cysts produce tooth erosions.

Other lesions may begin in the walls of dentigerous, radicular or primordial cysts. These complex lesions then take on the characteristic of the most active cell type such as ameloblastoma or mural cholesteatoma.

ADENOAMELOBLASTOMA

Adenomatoid odontogenic tumor (adenoameloblastoma) bears special consideration since it is most commonly located in the maxilla in the central regions. Early, the tumor is cystic in appearance but later it may have focal areas of calcification. The tumor affects young females in the second decade over males by a ratio of 2 to 1 (34).

INCLUSIVE CANAL CYSTS

Of the nonodontogenic tumors, the inclusive canal cyst is the most common. They are usually quite small and originate about the ends of the primitive nasal palatine duct. Their position is usually about the lateral aspect of the nasal septum and rarely do they reach a size sufficient enough to encroach upon the paranasal sinuses.

AMELOBLASTOMA

Ameloblastoma is an odontogenic tumor which tends to be more aggressive than the adenomatoid tumors. They may be multiloculated or uniloculated. Their aggressiveness is evident by the fact they break through cortical margins early in the course of the disease.

CT and MRI Findings

The major contribution of CT to dental origin tumors is to demonstrate whether lesions are fluid containing and where solid components of dental tumors are located. Uninterrupted teeth and erosions of tooth roots are better demonstrated by routine radiography. Occasionally the thin cortical margins of a dental cyst expanding into the sinus may be difficult to image when there is a superimposed sinus infection. Intravenous contrast will enhance the differential between fluid components and solid elements within dental tumors.

Magnetic resonance at the present time plays no significant role in the management of these mixed lesions.

SYSTEMIC DISEASE

Sarcoid, Midline Granuloma and Wegener's Granuloma

Within the nasal cavity and paranasal sinuses the CT appearance of these three conditions are so similar that they will be discussed together.

Sarcoid produces masses of granulomas that fill the sinus cavities and expand the bony margins. They tend to be bulky lesions that are relatively asymptomatic for long periods of time. The changes may closely resemble lymphoma or other malignant diseases. Only the total clinical picture of the patient will provide the diagnostic clues for sarcoid.

Midline granuloma is a destructive inflammatory process that affects the nasal cavity (28). It closely resembles lymphoma and may even be a localized form of lymphoma. Patients with midline granuloma are much more symptomatic than sarcoid, showing manifestations of acute and chronic infection of the upper respiratory passages. Bone destruction, when present, is not that of expansion but more of infiltration and local erosion.

Wegener's granuloma is characterized by a vasculiatis which may be masked by superimposed iflammatory disease (16, 22). The triad of glomerular nephritis, lung parenchymal infiltrations, and upper respiratory inflammatory disease associated with vasculitis distinguishes Wegener's granuloma as a distinct entity. Patients with Wegener's granuloma tend to be more symptomatic than sarcoid patients and bone destruction does not occur until late in the disease. The vasculitis of Wegener's granuloma frequently produces a dry gangrene of the turbinates which is quite distinct from the manifestations of sarcoid but can be found in midline granuloma.

CT and MRI Findings

All three entities frequently present with opacified sinuses and local destruction centering about the ethmoid air cells and the nasal septum. When the expansion of ethmoid air cells extends intracranially, and the patient is relatively asymptomatic, this is more likely to be found in the patients with sarcoid. In the late manifestations of the three diseases with grotesque facial defects the diagnosis is more likely midline granuloma or Wegener's and the two can be differentiated by clinical means rather than CT or MRI.

Immune Deficiency Syndrome

A group of conditions may lead to immune deficiency which predisposes to sinus inflammatory disease. CT examinations may be performed to identify early infections since sinusitis is frequently the initiator of a fatal central nervous system infection. This is especially true with bone marrow transplant patients and other individuals on high dose chemotherapeutic agents. AIDS patients usually exhibit pulmonary infection or toxoplasmosis of the brain so that sinus disease is just one manifestation of their systemic infections.

CT and MRI Findings

The CT findings are merely those of chronic inflammatory disease particularly in the ethmoid sinuses. Bone destruction is not a common feature but when the infections extend from ethmoids into the orbits subperiosteal collections of fluids form unless the etiology is primarily that of a fungus infection. The CT findings of central nervous system are that of a meningitis or of septic emboli in the case of AIDS.

Fibrous Dysplasia

Fibrous dysplasia is a proliferative disorder of bone that may affect either a single bone or multiple areas (3). The condition affects the paranasal sinuses of young teenagers giving rise to gross facial deformities spoken of as leontiasis ossea or "lion face." The maxillary sinus cavities may become totally obliterated by the bony proliferation. The changes may extend to temporal bones and orbits. When the ethmoid sinuses become involved visual fields must be watched. Posterior spread of the fibrous dysplasia involves the optic canals. It is surprising how much bony involvement may be present with little in the way of nerve compression symptoms.

CT and MRI Findings

CT will show the margins of bony thickening quite accurately and assist in planning surgical management. In general the surgical procedures are not curative but merely cosmetic and paliative because the disease process is unrelenting. Despite the degree of bony overgrowth mucocele formation and obstruction of the paranasal sinuses is not a major problem.

SUMMARY

In our attempt to achieve brevity, some of the less common lesions have not been mentioned. Undoubtedly CT and MRI will be used to resolve many perplexing clinical situations that are difficult to describe or anticipate. It is probably more important for the reader to be armed with good anatomy and an understanding of the limitations of the imaging technique than to try to provide an atlas of all of the known clinical entities.

References

1. Batsakis JG: *Tumors of the Head and Neck*, ed. 2. Baltimore, Williams & Wilkins, 1979, chap 5.
2. Bilaniuk LT, Zimmerman RA: Computer assisted tomography: sinus lesions with orbital involvement. *Head Neck Surg* 2:293–301, 1980.
3. Biller HF: Juvenile nasopharyngeal angiofibroma. *Ann Otolaryngol* 87:630–632, 1978.
4. Bohman LL, Mancuso AA, Thompson J, Hanafee W: CT approach to benign nasopharyngeal masses. *AJNR* 1:513–520, 1980.
5. Centeno RS, Bentson JR, Mancuso AA: CT scanning in rhinocerebral mucormycosis and aspergillosis. *Radiology* 140:383–389, 1981.
6. Duke-Elder S: *Textbook of Ophthalmology*. St Louis, Mosby, 1952, vol 5, p 5427–5443.
7. Dunn W: Comparison of polyposis and inverting papillomas by CT. (Presented at the American Society of Head and Neck Radiology meeting, May 30, 1981, Los Angeles, Calif.)
8. Fechner RE, Fitz-Hugh S, Pope TL: Extraordinary growth of giant cell reparative granuloma during pregnancy. *Arch Otolaryngol* 110:116–120, 1984.
9. Fu YS, Perzin KH: The nasal cavity, paranasal sinuses, and nasopharynx. Silverberg S (ed): *Principles and Practice of Surgical Pathology*. New York, Wiley, 1983, ch 17.
10. Fujii K, Chambers S, Rhoton A: Neurovascular relationships of the sphenoid sinus. A microsurgical study. *J Neurosci* 50:31–39, 1979.
11. Gonsalves CG, Briant TDR: Radiologic findings in nasopharyngeal angiofibromas. *J Can Assoc Radiol* 29:209–488, 1978.
12. Hesselink JR, Weber AL, New PFJ, et al: Evaluation of mucoceles of the paranasal sinuses with computed tomography. *Radiology* 133:397–400, 1979.
13. Jing BS, Goepfert H, Close LG: Computerized tomography of paranasal sinus neoplasms. *Laryngoscope* 88:1485–1502, 1978.
14. Kiley CH, Kay LW: Benign mucosal cysts. In Kay, LW (ed): *Oral Surgical Transactions of the Fourth International Conference on Oral Surgery*, Copenhagen, Munksgaard, 1973, pp 169–174.
15. Krohel GB, Krauss HR, Winnick J: Orbital abscess—presentation diagnosis, therapy and sequelae. *Ophthalmology* 89:492–498, 1982.

16. Lawson VG, Reid AJ, Cardella CJ, et al: Wegener's granulomatosis and the respiratory system. *J Otolaryngol* 11:60–64, 1982.
17. Mancuso A, Hanafee WN, Winter J, Ward P: Extensions of paranasal sinus tumors and inflammatory disease as evaluated by CT and pluridirectional tomography *Neuroradiology* 16:449–453, 1978.
18. McGill TJ, Simpson G, Healy GB: Fulminant aspergillosis of the nose and paranasal sinuses: a new clinical entity. *Laryngoscope* 90:748–754, 1980.
19. McGuirt WF, Harrill JA: Paranasal sinus aspergillosis. *Laryngoscope* 89:1563–1568, 1979.
20. Meyers D, Myers E: The medical and surgical treatment of nasal polyps. *Larngoscope* 84:833–847, 1974.
21. Modic MT, Weinstein MA, Berlin AJ, Duchesneau PM: Maxillary sinus hypoplasia visualized with computed tomography. *Radiology* 135:383, 1980.
22. Paling MR, Roberts RL, Fauci AS: Paranasal sinus obliteration in Wegener's granulomatosis *Radiology* 144:539–543, 1982.
23. Paparella MM, Shumrick DA: *Otology, vol 3, Head and Neck*, Philadelphia, Saunders, 1973.
24. Parsons C, Hodson N: Computed tomography of paranasal sinus tumors. *Radiology* 132:641–645, 1979.
25. Potter GD: Questions 37 through 39. In Rogers L (ed): *Disorders of the Head and Neck (2nd Series) Syllabus.* Chicago, American College of Radiology, 1977, pp 126–136.
26. Price JC, Stevens DL: Hyperbaric oxygen in the treatment of rhinocerebral mucormycosis. *Laryngoscope* 90:737–746, 1980.
27. Robson TW, Bloom VR, Swann GR, et al: Fungal infections of the base of the skull: two case reports. *J Laryngol Otol* 95:109–114, 1981.
28. Scully RE, Gladabini JJ, McNeeley BU: Case records of the Massachusetts General Hospital. Weekly clinicopathological exercises. *N Engl J Med* 304:1217–1224, 1981.
29. Snyder RN, Perzin KH: Papillomatosis of nasal cavity and paranasal sinuses (inverted papilloma, squamous papilloma). A clinicopathologic study. *Cancer* 30:668–690, 1972.
30. Som PM, Cohen BA, Sacher M, Choi IS, Bryan NR: The angiomatous polyp and the angiofibroma: two different lesions. *Radiology* 144:329–344, 1982.
31. Som PM, Shugar JMA: Antral mucoceles: a new look. *J Comput Assist Tomogr* 4:484–488, 1980.
32. Weber AL, Tadmor R, Davis et al: Malignant tumors of the sinuses. *Neuroradiology* 16:443–448, 1978.
33. Wilson M: Chronic hypertrophic polypoid rhinosinusitis. *Radiology* 120:609–613, 1978.
34. Wood NK, Goaz PW: Differential diagnosis of oral lesions. St. Louis, Mosby, 1980.
35. Zizmor J, Noyek A, Chapnik J: Mucocele of the paranasal sinuses. *Can J Otolaryngol* (Suppl 1) 1974.

Chapter 2 Plates

Benign Sinus

Plate 2.1. Congenital hypoplasia. The bony margins of the sinus are thicker than normal (*crossed arrow*) in congenital hypoplasia. On conventional films this increase in bony thickness may give an overall haziness to the sinus-stimulation fluid. This scan shows that the hypoplastic sinus is clear. The other features denoting hypoplasia are the laterally placed nasal wall, hypertrophy of the inferior turbinate, and the slight asymmetry of the cheek. Higher scans showed the lower portion of the orbital floor.

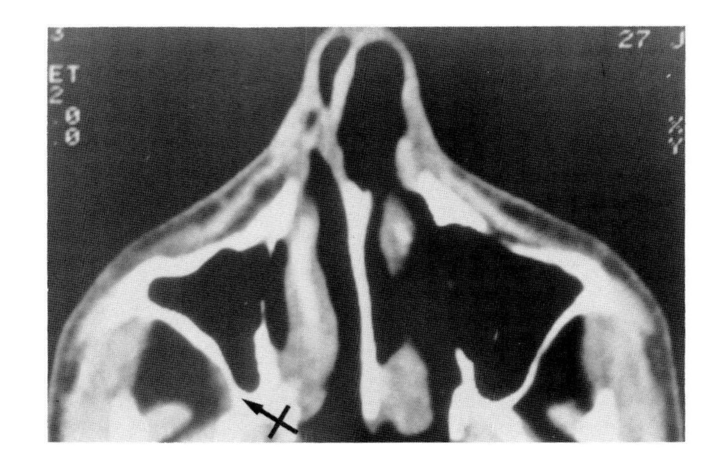

Plate 2.2. Choanal atresia. In this young adult patient the atresia is bony (*arrow*). Retained secretions can be seen anterior to the bony plate because the patient is lying supine. When surgery is performed CT is excellent for following the patient and noting the completeness of surgical correction. Restenosis is common if a significant bony ridge is left behind.

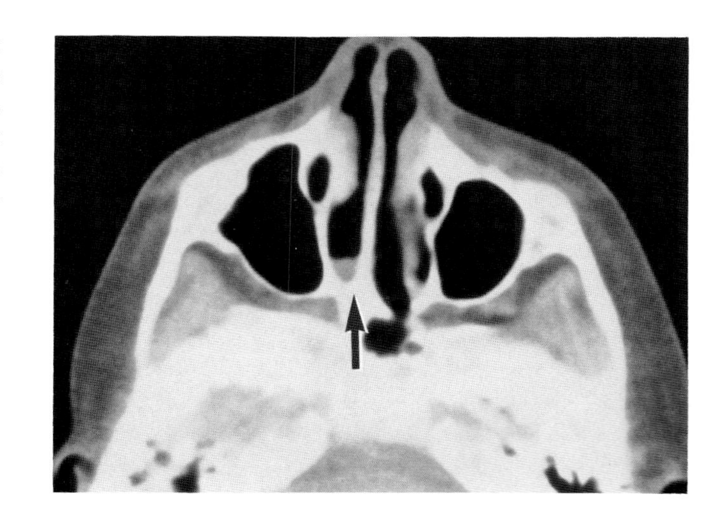

Plate 2.3. Acute sinusitis.

Plate 2.3*A*. Scan through the maxillary sinuses shows an air fluid level present on the right (*arrow*) and total opacification of the left antrum. The bony margins are intact. The irregularity of the left anterior antral wall is merely due to partial volumning of the roof of the antrum and orbital floor (*arrowheads*).

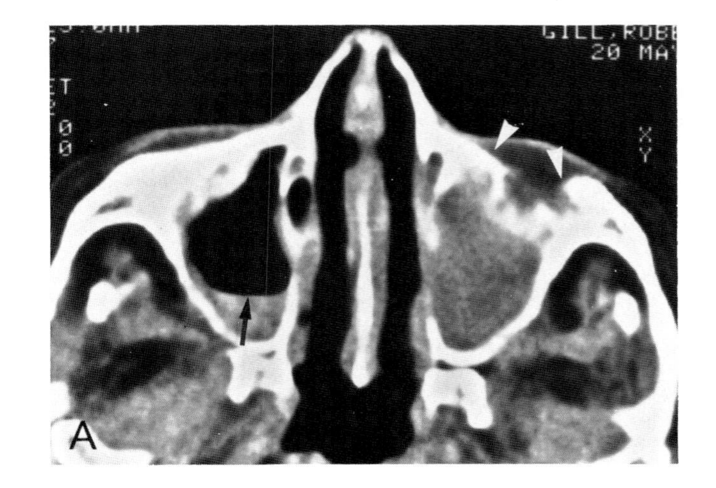

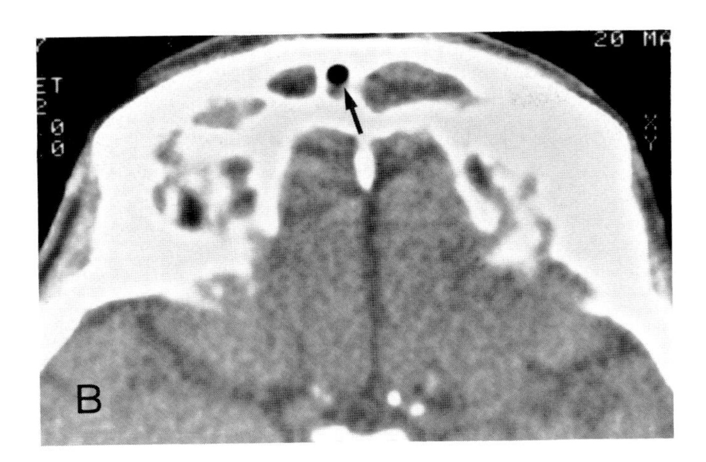

Plate 2.3*B*. The right and left frontal sinuses are totally opacified. A sinus cavity is present in the septum between the two sinuses (intersinus septum). This small sinus contains an air fluid level (*arrrow*).

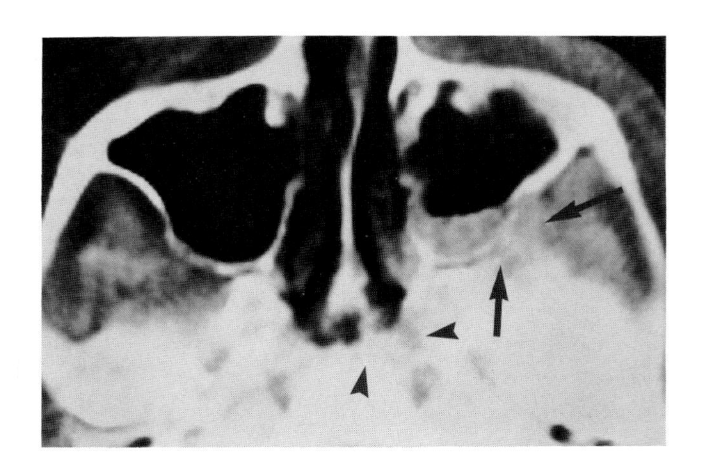

Plate 2.4. Aspergillosis—acute. In this immune-suppressed patient with maxillary sinusitis on the left, the inflammatory process has broken out through the posterolateral wall into the infratemporal fossa (*arrows*). Infection is also present in the vault of the nasopharynx (*arrowheads*). This process spread from the infratemporal fossa into the orbital apex and later into the middle cranial fossa.

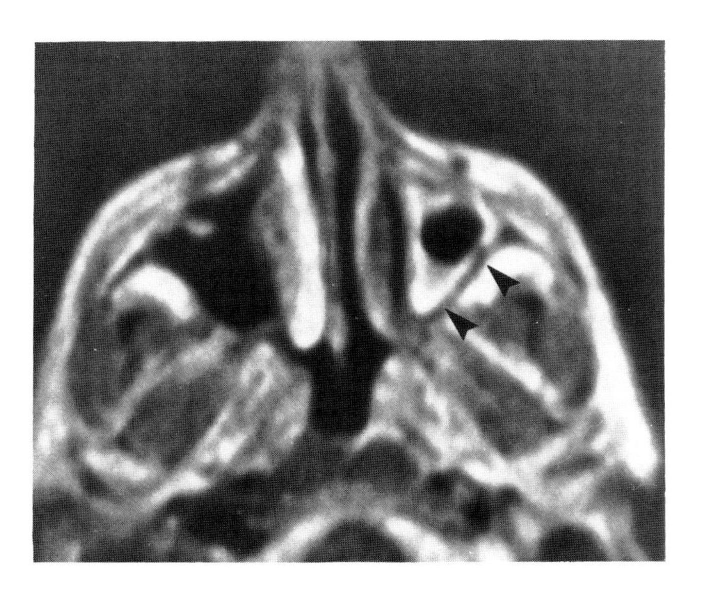

Plate 2.5. Chronic sinusitis—MRI spin echo TR 1.5 sec TE 110 msec. The left antrum is much smaller than the right. The hypertrophied mucosa produces a bright signal. Fat and areolar tissue in the infratemporal fossa provides contrast so that the thickness of the bony wall can be clearly delineated as absence of signal (*arrowheads*).

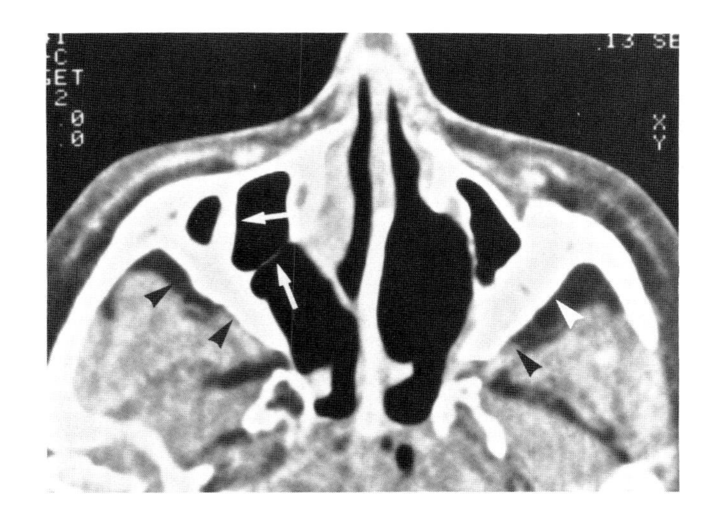

Plate 2.6. Chronic aspergillosis. This patient with long-standing chronic sinus disease and drainage was treated by bilateral Caldwell-Luc procedures. Drainage persisted and aspergillosis has been repeatedly cultured from the sinuses.

The CT shows massive amount of bony thickening of the antral walls (*arrowheads*). Multiple septa within the right antrum (*arrows*) are present which may interfere with drainage and contribute to the persistent inflammatory disease.

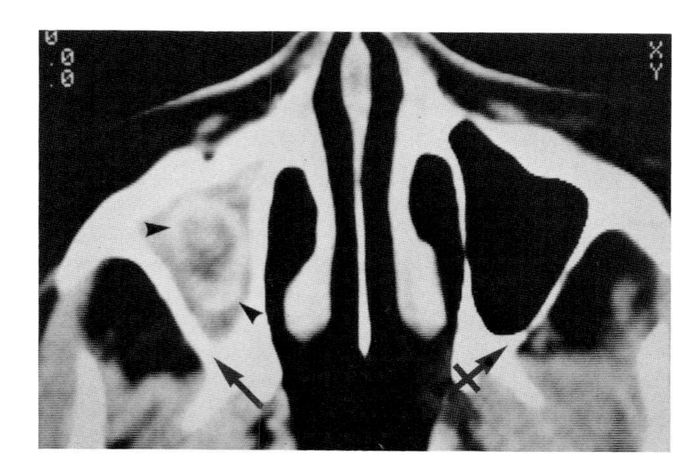

Plate 2.7. Chronic streptothrix (*Actinomyces israelii*). This sinus infection present on the right spread from the lacrimal sac. The sinus wall is markedly thickened (*arrow*) which is characteristic of long-standing disease as compared to the normal right sinus (*crossed arrow*). The mucoperiosteium of the sinus has also thickened but some fluid is present centrally. The level of the mucosal surface can be identified because of intravenous contrast (*arrowheads*). The marked degree of bony thickening of a sinus is suggestive but not absolutely diagnostic, of chronic fungus infection.

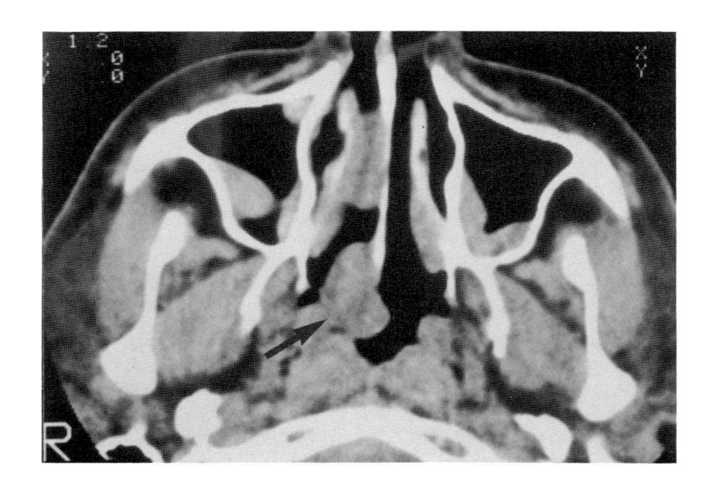

Plate 2.8. Inflammatory polyps. With persistent allergy and infection polypoid changes occur throughout the mucosa of the sinuses and nasal cavity. The polyps may originate in this later stage from the nasal septum as well (*arrow*). Because of the intense inflammatory reaction seen at microscopy, these lesions are frequently spoken of as inflammatory polyps.

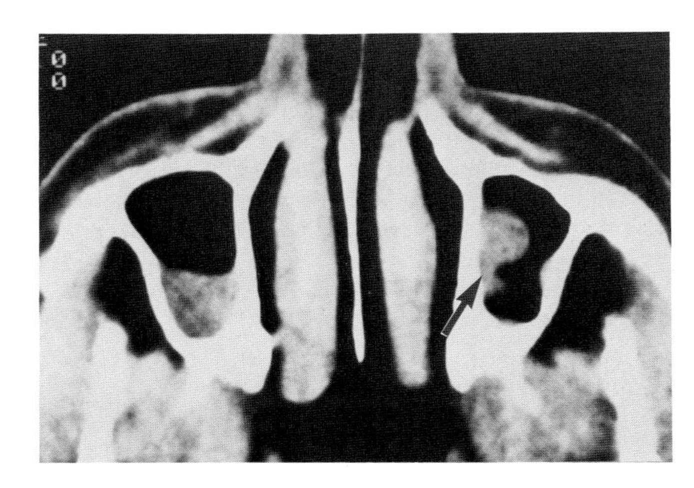

Plate 2.9. Allergic polyposis. Allergic polyps characteristically originate within the maxillary sinus adjacent to the natural ostium (*arrow*). With increasing disease the polyps become extruded into the nasal cavity and nasopharynx. Later polys may develop from turbinates or septum.

Plate 2.10. Rhinopolyposis.

Plate 2.10*A*. In this patient with marked polyposis all of the sinuses are almost totally opaque. The frontal sinus margins are all intact (*arrowheads*) but the sinuses were filled with polypoid material and retained secretions.

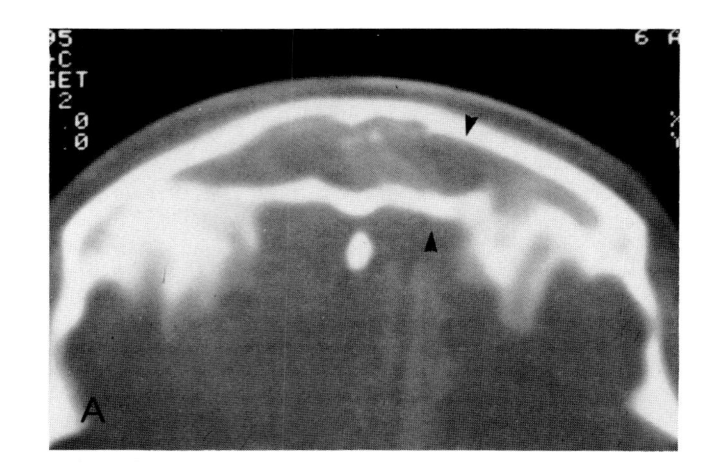

Plate 2.10*B*. The ethmoid air cells are totally filled with polypoid material and stain intensely with intravenous contrast. Expansion into the orbit is present from the anterior ethmoid air cells (*arrows*). Note that the sphenoid sinus content is of low density consistent with an obstructive sinus filled with fluid (*arrowheads*).

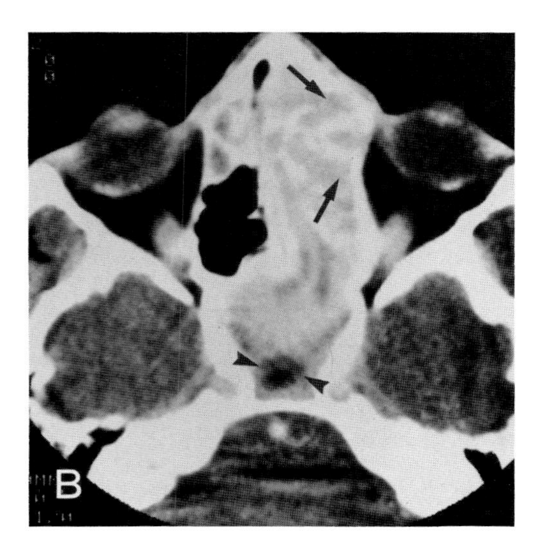

Plate 2.10*C*. In this same patient the right antrum is filled with a mixture of polyps as evidenced by staining and retained secretions which are the low density areas. The natural ostium of the antrum is greatly widened (*arrowheads*) and polyps fill the nasal cavity. The nasal bones are somewhat displaced laterally by the polyps in the anterior nasal vault.

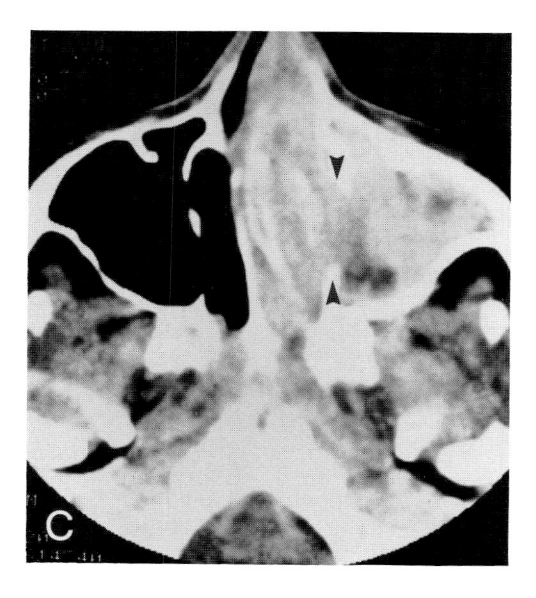

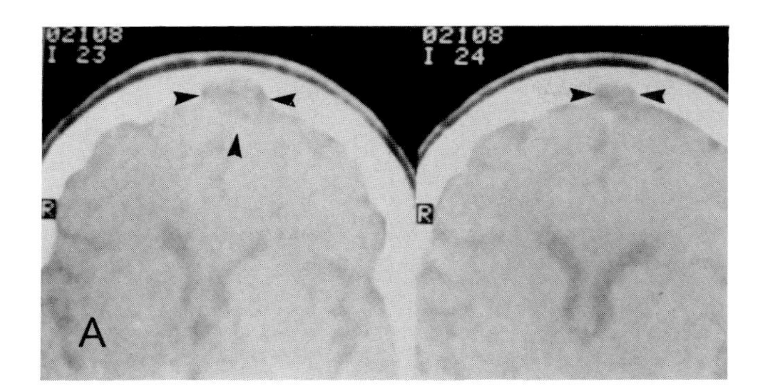

Plate 2.11. Frontal and ethmoid mucocele.

Plate 2.11*A*. Scan through the frontal sinus shows dehiscence of the posterior wall of the frontal sinus (*arrowheads*).

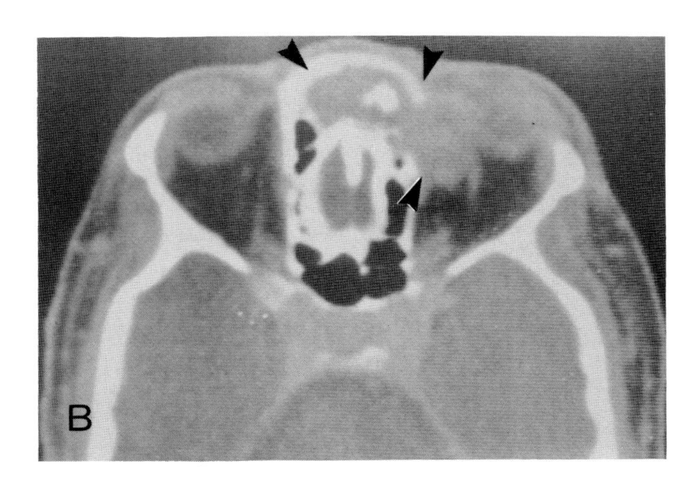

Plate 2.11*B*. The mucocele involves the inferior portions of the frontal sinus and extends laterally into the orbit on the left (*arrowheads*). The left globe is displaced inferiorly and laterally.

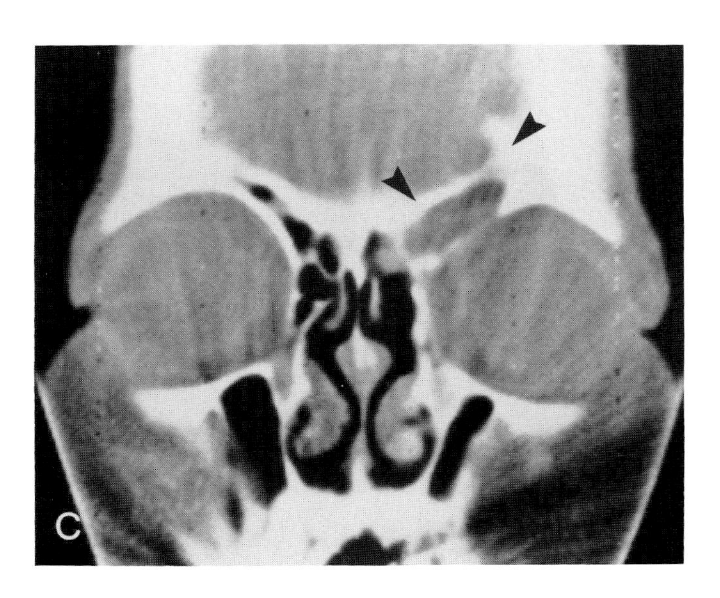

Plate 2.11*C*. Coronal scans shows that the mucocele involves a major portion of the left orbital roof (*arrowheads*). A small amount of the mucocele is involving the anterior ethmoid sinus.

A coronal incision would not be sufficient to get surgical exposure of all of the posteriorly placed disease.

Plate 2.12. Ethmoid mucocele.

Plate 2.12*A*. Ethmoid mucocele destroys the lamina papyracea and extends into the right orbit (*arrowheads*).

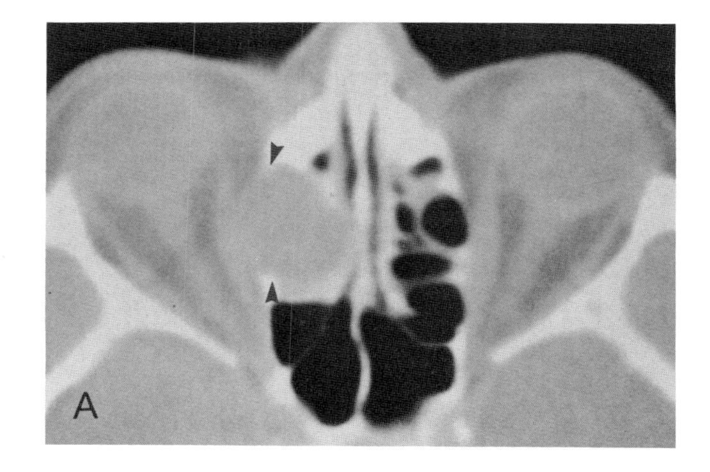

Plate 2.12*B*. Reformatted image in this same patient shows that the cribriform plate region and lateral wall of the olfactory fossa on the right has been destroyed (*arrowhead*). The surgeon is forewarned that a cerebrospinal fluid leak is imminent in this type of case. The exact etiology of this ethmoid mucocele is obscured since there is only minimal evidence of generalized sinus disease (*arrow*).

Careful search must be made for some type of tumor or structural abnormality.

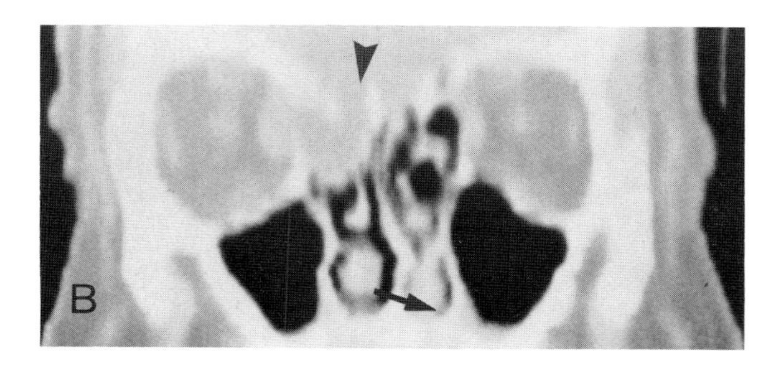

Plate 2.13. Recurrent disease. This patient had an osteoplastic flap procedure (*arrowheads*) with obliteration of the frontal sinus 8 months prior to the CT scan. The obliteration procedure was performed because the frontal sinus had been entered during a neurosurgical procedure. The patient complained of pain and it was difficult to determine whether this was due to intracranial problems or the frontal sinus. The extremely small amount of air present in the frontal sinus (*arrow*) could not be visualized on routine films. The fact that some air was present indicated that a sinus cavity had reformed and his pain was probably due to residual frontal sinus disease rather than recurrent intracranial tumor.

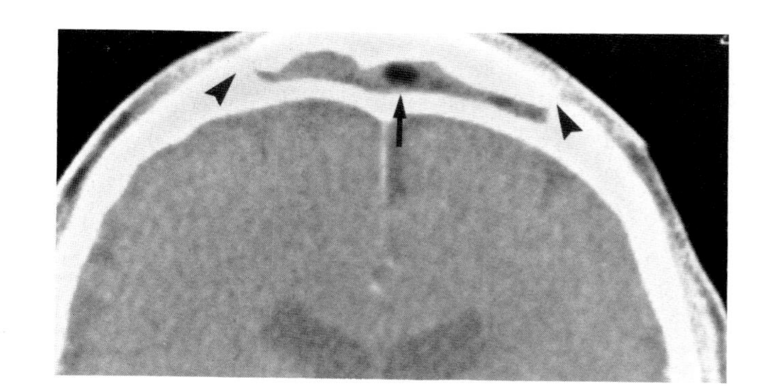

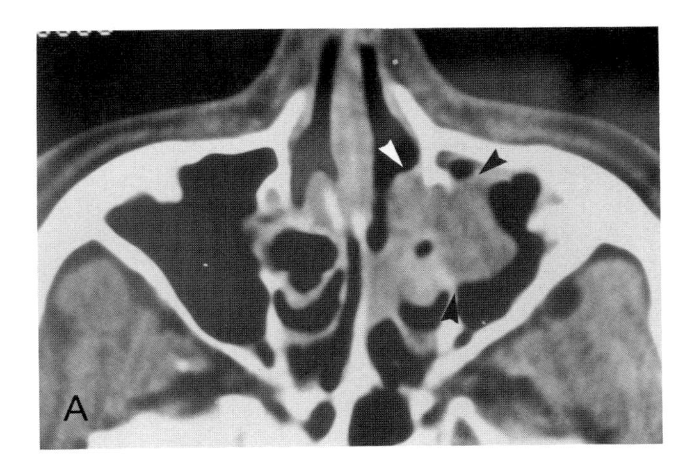

Plate 2.14. Inverting papilloma.

Plate 2.14*A*. On the left a mass (*arrowheads*) is seen arising from the lateral wall of the nasal vault beneath the middle turbinate. The tumor also extends into the antrum through the ostium.

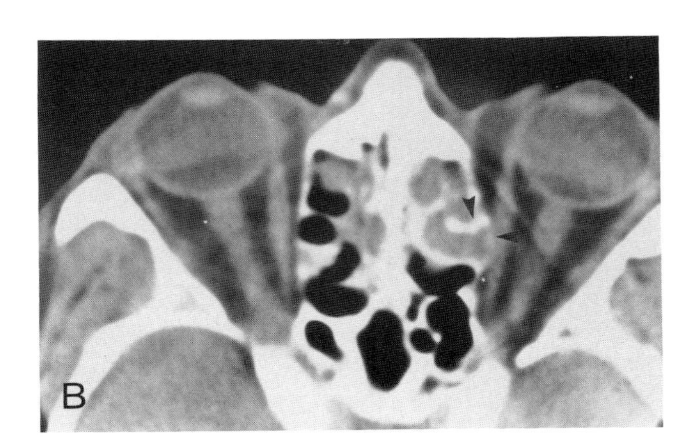

Plate 2.14*B*. Same patient. The inverting papilloma has invaded into the ethmoid sinuses because the ethmoid sinuses drain beneath the middle turbinate. The bulky mass expands the lamina papyracea into the left orbit (*arrowheads*) simulating a mucocele.

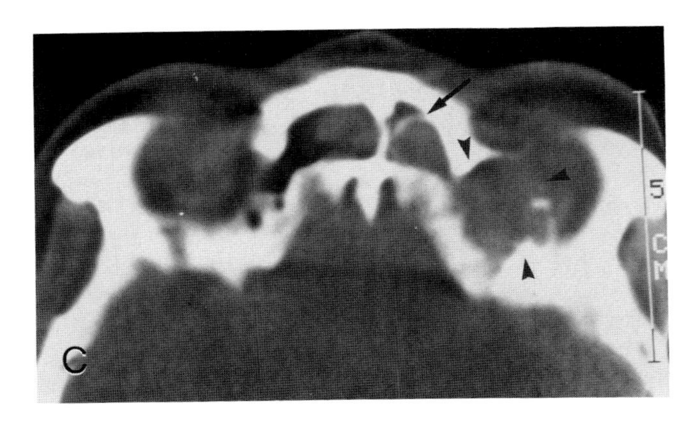

Plate 2.14*C*. Same patient. The lesion has expanded ethmoid air cells into the lower frontal sinus (*arrow*). The expanding anterior ethmoids extend well into the roof of the left orbit (*arrowheads*).

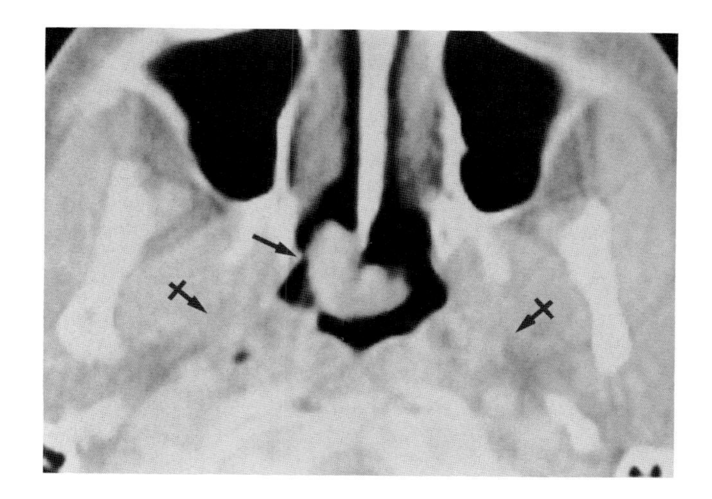

Plate 2.15. Squamous papilloma. These tumors are more commonly located along the anterior nasal septum but this polypoid lesion from the posterior nasal septum (*arrow*) proved to be squamous papilloma. The normal appearance of the paranasopharyngeal tissue planes (*crossed arrows*) is comforting in confirming the benign nature of this lesion.

The clinical appearance of the tumor is more like a "wart" as opposed to the "fleshy" appearance of polyps.

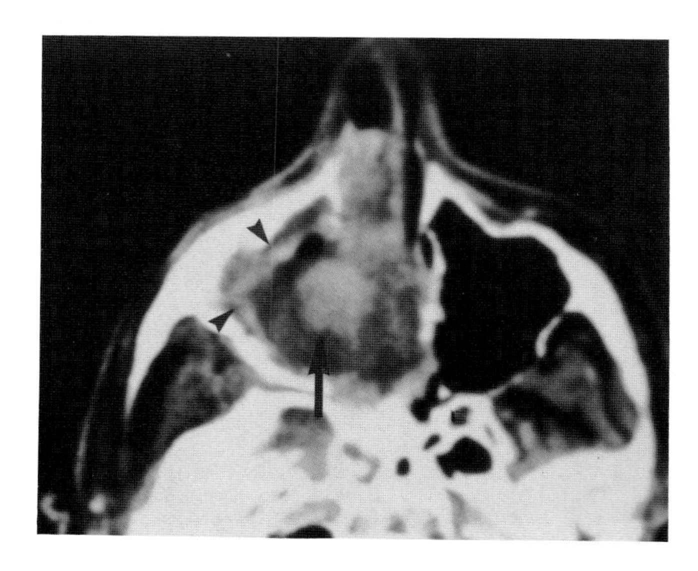

Plate 2.16. Angiomatous polyps. The bulky tumor of the right nasal cavity and antrum stains in an irregular fashion (*arrow*) with some thickened mucosa being retained laterally (*arrowheads*). The low density material between the mucosa and the main tumor mass is probably retained secretions. This pattern of staining indicates a rather vascular lesion but is not the massive amount of staining that one would see with a juvenile angiofibroma.

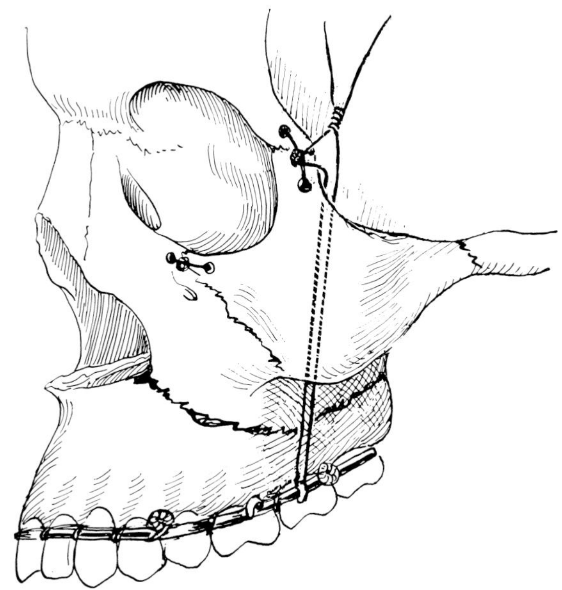

Figure 3.1. Principles of wiring for facial trauma. The principle of treatment of facial fractures involves immobilization of all fracture fragments to something that is stable. Sutures should be placed across all nonstable fracture lines. In this illustration of a complex midface fracture the maxilla is stabilized to the cavarium through arch bars and wiring to the orbital process of the frontal. Fracture lines that are through the fronto-zygomatic suture and through the floor of the orbit are also wired to each other. Packing and external splints can be used to augment stability. (From Kazanjian VH, Converse JM: *The Surgical Treatment of Facial Injuries*, ed 2. Baltimore, Williams & Wilkins, 1959, p 210.)

The medial pillar extends from the canine tooth through the anterior medial wall of the maxillary sinus to the medial wall of the orbit. This support is provided by a comination of the maxilla, the frontal process of the maxilla, the lacrimal bone, and the nasal process of the frontal bone (Fig. 3.2).

Posterior to the surface bones of the face lie delicate bony septa which do not support muscle attachment. The septa give way easily after fracture of the heavy bones of the anterior face. Major deformities result when bilateral fractures occur through both medial and lateral pillars of support. Fractures of thin septa or lesser structures may be cosmetically disfiguring but the functional disability is not life threatening.

TYPES OF INJURIES

Nasal Fracture

By far the most common fracture of the facial skeleton is the nasal fracture. The vast majority of nasal fractures do not require x-ray examination of any type. The two major questions to be answered are: (*a*) is

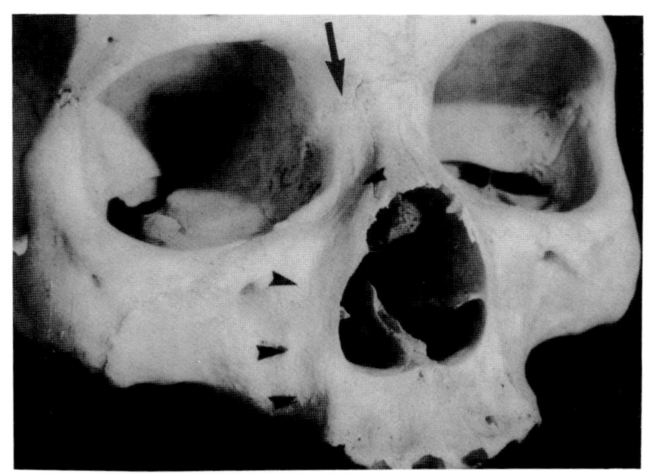

Figure 3.2. Facial pillars. The medial pillar of support is outlined by the *arrowheads* on this dried skull. It extends from the incisor tooth to the nasal process of the frontal (*arrow*). It is composed of the medial wall of the antrum, lacrimal bone and frontal process of the maxilla which articulates with the nasal process of the fontal bone. Laterally, the lateral wall of the orbit, zygoma, and lateral wall of the maxillary sinus forms the lateral bony strut.

there facial deformity that is unacceptable by the patient? and (*b*) does the resultant deformity interfere with respiration or sinus drainage? The above questions can be answered quite nicely by having the patient return 4 to 5 days following the injury with a picture of himself. If the cosmetic aspects can be settled then clinical examination will determine the status of the airway. A decision can be made at that time whether to realign the nasal structures. Unfortunately medicolegal considerations frequently take precedent over logic in many situations.

At times the deformity may be so gross that immediate reduction is indicated. Depending on the force of the trauma further x-ray studies may or may not be indicated. The nasal bones are supported by the nasal processes of the maxilla and the medial pillars. Fracture of the medial pillar can be easily overlooked on routine films. If any question exists regarding the status of the medial wall of the maxillary sinus, frontal process of the maxilla and medial wall of the orbit, CT is the examination of choice to define displacements of these structures.

Medial Pillar Fractures

Medial pillar fractures are usually part of more complex midface fractures and rarely occur as an isolated event (11). A blow by a narrow object such as a policeman's club to the side of the base of the nose is a typical type of injury. They usually accompany nasal fractures under this type of circumstances and the overlying nasal fracture may obscure the disruption of the medial wall of the antrum or the medial wall of the orbit. The fracture fragments become displaced posteriorly and are usually rotated. The lacrimal apparatus may be disrupted through the nasolacrimal duct or at the junction of the canaliculi with the lacrimal sac. The entire medial support of the globe may be lost.

CT Findings

Diagnosis of medial pillar fractures is extremely easy by axial scanning but can be very difficult with routine plain films. The fracture fragments are displaced posteriorly so that the entire nasal support is lacking. Asymmetry between the two sides and disruption of the medial wall of the maxillary sinus is quite apparent. The injury usually causes major deformities of the nasal septum with displacements or even total disruptions of the nasal septum.

If there is an associated fracture of the orbital rim recognition of a medial pillar fracture is imperative. Modern surgical reductions require anchoring of orbital rim fractures to the medial pillar and to the orbital process of the frontal bone. If instability occurs at one of the anchoring sites total loss of support of the orbit can result. For this reason CT scans for midface injuries should always extend well up into the orbit roof and frontal bone.

Trimalar Fractures

The three components of a trimalar fracture are as follows:

1. Fracture or disruption of the lateral wall of the orbit usually through the frontozygomatic suture.
2. Fracture of the floor of the orbit involving the orbital rim and extension of the fracture line into the maxillary sinus. This fracture line will involve the lateral wall of the maxillary sinus.
3. Fracture of the zygomatic arch which may be complete or incomplete.

The trimalar fracture is caused by a blow to the malar eminence which may cause the fracture fragments to be displaced posteriorly and rotated. This injury disrupts the lateral pillar of support and results in a variety of clinical symptoms.

The patient may present with facial numbness due to damage to the infraorbital nerve (9) (Fig. 3.3). Thirty percent of these nerve injuries remain permanent. If the zygomatic arche is depressed it may impinge upon the coronoid process of the mandible. The patient may be unable to close the mouth or chew.

Lateral orbital wall disruption releases the normal support of the globe. An unequal height of the pupils can lead to diplopia. The fracture through the floor of the orbit may also entrap the inferior rectus or inferior oblique muscles reducing globe mobility.

Bleeding into the maxillary sinus almost invariably occurs and results in a bloody nose.

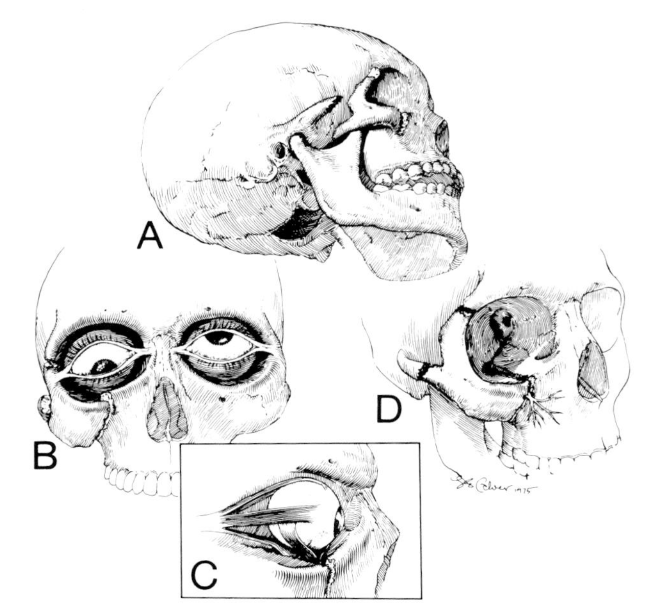

Figure 3.3 Clinical findings associated with trimlar fracture. *A*, depression of the zygomatic arch may encroach upon the coronoid process to cause the mouth to be fixed in a semiopen position, *B*, the fracture of the lateral wall of the orbit may release the lateral support for the globe and cause inferior displacement of the orbital contents, *C*, the inferior rectus muscle may become entrapped by the fracture through the floor of the orbit, and *D*, the inferior orbital nerve may be compressed by the fracture of the inferior orbital roof producing anesthesia of the cheek. (From) Gerlock AJ, Sinn DP: Zygomatic and complex fractures. *American Journal of Roentgenology*, 128:236, 1977).

CT Findings

Cephalocaudad displacements with a trimalar fracture can be appraised very nicely by a digital anteroposterior scout view. Of all of the projections available by any roentgen technique the axial display of CT scanning is the best for demonstrating anteroposterior and rotational displacements. The malar eminence may displace posteriorly or inferiorly. Comparison of the fracture fragments to the normal side is ideally suited to evaluating facial asymmetries. CT shows the lateral wall fractures and a degree of displacement. It is usually quite a surprise to see how far posteriorly the fracture lines extend into the lateral wall of the orbit.

Since the axial scan is usually taken very close to the plane of the zygomatic arch, depression of the arch can be accurately assessed.

One should carefully evaluate the contents of the maxillary sinus for any free fragments that have been displaced from the orbital floor or lateral wall of the orbit. Free fragments from the lateral wall of the orbit are an ominous sign and suggest a very unstable fracture. Free fragments in the floor of the orbit usually indicate that globe support has been seriously impaired. The clinician should be warned that a major floor deficit exists and must be restored either by packing of the antrum or prosthetic material insertions into the floor of the orbit.

Isolated Zygomatic Arch Fracture

Direct blows to the lateral side of the face will result in a depression of the zygomatic arch. If sufficient displacement is present the fracture fragments interfere with the movements of the condyloid process. The patient presents in an open mouth fixed position.

CT Findings

A single section through the zygomatic arch will demonstrate at least two fracture sites when the zygomatic arch fracture is an isolated injury. The blow may result in three fracture lines. A central **V** type of injury results at the site of impact and two fractures occur on either side of the point of impact. The axial display of CT is very helpful in showing an intact lateral wall of the orbit confirming that one is dealing with an isolated zygomatic arch fracture.

Orbital Blow-Out Fracture

When the force of a blow is distributed to the globe rather than to the rim of the orbit, pressure is raised within an enclosed space. The force is transmitted by soft tissues to all walls equally and those areas of physiologic weakness will give way to the applied pressure. A second type of fracture is the "blow-in" fracture resulting when a more severe force strikes the rim or anterior wall of the maxillary sinus. The rim is displaced posteriorly crushing the orbital walls (4).

ORBITAL FLOOR FRACTURES

The wide expanse of the orbital floor makes soft tissue injury imminent. The segment of bone is extremely thin and when depression occurs extraoccular muscles are easily entrapped. Many times it is not the muscle itself that is caught by fracture fragment, but the elastic tissue and fascial planes that envelope extraoccular muscles. This type of "entrapment" will improve without surgical intervention (5, 17). The fact that a large area of support of the orbit gives way often dictates surgical intervention (15).

CT Findings

If orbital floor fractures are the only injuries plain roentgenograms are superior to CT in most instances. On the other hand when the CT study is being performed for other reasons or the injuries are through to be more extensive, accurate appraisal of the orbital floor is possible. The CT examination should begin by a digital anteroposterior scout view taken at an angle that is halfway between a Caldwell and a Water's projection. The inferior rim of the orbit is made to project approximately 5 to 10 mm cephalad to the external auditory canal. This will project the floor of the orbit slightly above the petrous pyramid. Cephalocaudad displacements of a minor nature are easily detected by this projection. Axial scanning is performed either by 1.5-mm-thick sections at 1.5-mm intervals for fine detail or 5-mm thick sections at 5-mm intervals for a more survey type of exam. Regardless of the technique used for axial sectioning displacements of the orbital rims will be readily apparent and the size of a defect in the floor of the orbit can be appreciated from these scans. In the more cooperative patient or individuals who are not in the acute postinjury phase, coronal sectioning can be performed. The coronal scans will definitely identify the outline of the individual extraoccular muscles and demonstrate any displacements or entrapments.

CT should be carefully evaluated to see the diameter of a soft tissue mass that has protruded into the floor of the orbit. By comparison of the size of any free bony fragments to that of the volume of soft tissue components, one can begin to approximate the altered physiology or orbital content. Bear in mind that a sizable hematoma and some edema accompanies the fracture.

MEDIAL WALL BLOW-OUTS

The lamina papyracea is one of the thinnest bones of the body and is frequently dehiscent in the normal state. It is said to give way in 20% of orbital floor blow-out fractures (15). In reality fractures of the medial wall probably occur with equal frequency to floor fractures and possibly even a greater frequency. Because muscles rarely become entrapped in the small ethmoid sinuses these lesions usually escape diagnosis. Attention is directed to a medial wall blow-out usually on the basis of orbital emphysema. By far the most common cause of severe orbital emphysema is the medial wall blow-out fracture. The patient may recognize the relationship of nose blowing to progressive emphysema.

CT Findings

The CT findings of medial wall blow-out are readily recognized because of the axial display. Orbital emphysema extends into the central surgical space because the periosteum of the orbit is usually disrupted. Displaced free bony fragments rarely occur. Only rarely is the medial rectus muscle entrapped into fracture defects caused by medial wall blow-outs.

Orbital Roof and Optic Canal Fractures

Blow-out fractures of the orbital roof are very rare (20). They are almost always associated with extensive pneumatization of the orbital roof. When blow-out fractures of the orbital roof occurs in the absence of an orbital rim fracture, management is greatly altered.

Entrapment of the levator muscle has not as yet been reported.

Fractures involving the superior orbital rim are related to more violent trauma. They are usually associated with frontal sinus fractures or of the lateral wall of the orbit. When the rim is fractured, extensions can be found into the anterior and/or middle cranial fossa. A common extension of an orbital roof and rim fracture is to the region of the optic canal or superior orbital fissure. Any time a superior orbital rim fracture is encountered great care should be taken to demonstrate the optic and soft tissues of the orbital apex.

CT Findings

CT demonstration of blow-out fractures of the orbital roof is really quite simple. A separate fracture fragment is usually visible within the sinus. The sinus may or may not be opaque secondary to accumulation of blood from the injury. A careful search should be made for associated fractures.

The orbital rim injuries require extremely thin sections at frequent intervals to carefully visualize the optic canal, superior orbital fissure and status of the optic nerve and globe. All scans shoulds be reviewed at bone window settings and soft tissue window settings. If the rim of the orbit is displaced posteriorly there is frequently associated free fragments of bone that have entered the intracranial cavity or the muscle cone region of the orbit. The location of the free fragment prior to surgical reduction is imperative since the fragment may be mobilized at the time of reduction and cause further injury to the optic nerve.

Extensions of orbital roof fractures may pass into the medial wall of the sphenoid sinus. Here they involve the carotid artery and/or cavernous sinus with its small arterial branches. Aneurysms and arteriovenous fistula are a complication of this injury.

Midface Fractures

LeFORT FRACTURE I

A more massive type of trauma to the midface results in disruption of the anterior pillars of support plus the posteriorly located delicate septa. These are the familiar LeFort fractures and are staged as I, II and III in increasing severity. A LeFort I fracture is a transverse fracture of the midface through the inferior portions of the maxillary sinuses (Fig. 3.4). Clinical examination of a patient with a LeFort I fracture will demonstrate the hard palate and superior alveolar ridge to be a separate fragment when the examiner grasps the base of the upper central incisor teeth and shakes. The hard palate and upper alveolar ridge will move while the cheeks, nasion and orbits will remain stable. This type of a vigorous clinical examination is only possible in the unconscious or anesthetized patient.

LeFORT FRACTURE II

In the LeFort II fracture the hard palate and nasal structures become a separate fragment from the re-

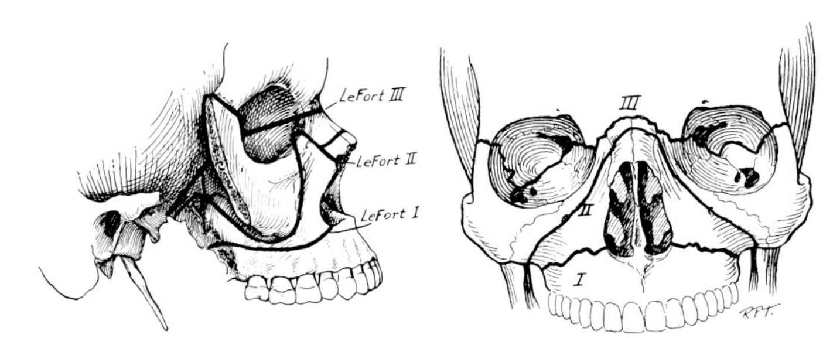

Figure 3.4. LeFort fractures. Line diagrams show the sites of the three types of midface fractures as described by LeFort. *LeFort I fracture.* A transverse fracture separates the midface just above the hard palate at the level of the floors of the maxillary sinuses. The fracture line extends through the lower nasal septum. In this injury the hard palate and maxillary ridges are a separate fragment. *LeFort II fracture* showing the oblique fracture lines through the maxillary sinus and extending into the nasion. In this type of fracture the hard palate and the nose constitute one single fracture fragment that is separate from the face. *LeFort II fracture.* Transverse fracture through the orbits and base of the nose. In this injury the hard palate, nose and malar eminence all comprise a single fracture fragment. LeFort type of fractures may be mixed, i.e., LeFort II on one side and III on the other depending on the type of trauma. (Modified from Kazanjian VH, Converse JM: *The Surgical Treatment of Facial Injuries,* ed 2. Baltimore, Williams & Wilkins. 1959, p 193.)

mainder of the face. The fracture lines extend obliquely through the floor and medial walls of both orbits. A transverse fracture line occurs through the superior portions of the nasal vault in the region of the cribriform plate. The lateral walls of the orbits remain intact. Fractures may be of a mixed variety with a LeFort I on one side and a LeFort II on the opposite side. If a LeFort I and a LeFort II occur on the same side, the fracture is usually described according to the more severe injury but all fracture lines themselves must be reported.

On clinical examination of the classic LeFort II type of fracture, the examining physician may grasp the premaxilla (maxillary ridge adjacent to the incisor teeth) and test for stability. If the nose and hard palate move independently of the orbits and malar eminences this is the diagnostic feature of a LeFort II fracture.

Le FORT III

LeFort III fractures are characterized by a transverse fracture through both orbits involving both medial and lateral walls. The fracture line extends through the region of the cribriform plate. On clinical examination the hard palate, nose and malar eminences are all one large free fragment that can be shaken by grasping the maxillary alveolar ridge. Rarely is the LeFort III fracture an isolated injury and multiple fracture lines are found throughout the facial structures. The massive amounts of trauma required to produce this fracture are usually accompanied by brain damage, subdural hematomas and other systemic injuries. If the patient will maintain a position for the CT study, surveys of intracranial structures and facial structures can be conducted very nicely at a single setting.

CT Findings

Anterior and lateral digital radiographs or plain films are extremely helpful in evaluating midface injuries. The LeFort I fracture can only be diagnosed if displacement is present or free bony fragments lie adjacent to the fracture fragments. Fluid is invariably present in the maxillary sinus. If a palatal fracture is present axial CT may be the only modality demonstrating the injury. The fracture line runs in an anteroposterior direction and frequently involves the incisive canal (8).

In LeFort II injury the midface fractures are displaced posteriorly. Varying degrees of disruption of the medial pillars are frequently present. If they are displaced as well as rotated this information should be conveyed to allow for better reduction by the clinician. Free bony fragments may be found within the maxillary sinuses. They may originate from any of the walls of the maxillary sinus. They indicate a total disruption of support in the region of origin which must be considered in stabilization procedures.

The extent of injury of the cribriform plate region should be carefully documented and any free fragments noted. This information will be extremely valuable should a cerebrospinal fluid leak develop and persist.

Occasionally orbital emphysema may be encountered and the site of origin not be readily apparent. This may provide the clue that a fracture line has extended into the ethmoid air cells and cribriform plate. CT scanning techniques can be altered to provide high quality reformatted images or, if possible, direct coronal scanning can be done to better demonstrate the entire extent of injury.

Magnetic resonance will probably become the future modality of choice for studying intracranial trauma but it will not replace CT in defining bony fragments in midface fractures. For this reason we will continue to see CT play a major role in demonstrating the displacements of LeFort III type fracture. The digital anteroposterior and lateral projections are invaluable to identify the complexity of these fractures which usually will include fractures through the orbital floor and medial wall of the antrum as an accompaniment of the severe midface trauma. Rotational injuries of the malar eminence, free bony fragments lying within the maxillary and pillar injuries are all present with this type of injury. It is very important to determine whether the inferior orbital rim is intact or not. One of the points of fixation of this injury will be an open reduction of the lateral wall of the orbit. If the inferior orbital rim is fractured, fixing the lateral orbital rim will not provide stability to the medial orbital structures and base of the nose (11).

If the medial struts are fractured and the nose is comminuted surgical reduction and fixation requires careful consideration. Usually the nasal passages are packed to exert lateral pressure against the medial pillars and nasal bone. The nasal bones then must be kept from being displaced laterally by the packing by splints and through wires to hold the lateral nasal walls in place (12, 13).

Frontal Sinus Fractures

The frontal sinus involvement by fracture creates a potential path for infections from the face to the intracranial cavity. The fractures through the anterior wall of the frontal sinus lead to the "dish" deformity of the forehead. Fractures of the posterior table open the way for CSF leaks and possible herniation of brain tissue into the frontal sinus. Pneumoceles are another complication of this type of injury.

CT and MRI Findings

For the most part axial scanning is quite adequate for demonstrating displacement of fracture fragments about the frontal sinuses. This projection will also show any bony fragments that have been displaced into brain parenchyma. When the posterior wall is dehiscent herniations of brain into the frontal sinuses can be seen as soft tissue masses. Either direct or reformatted coronal scans may be required to delineate the cribriform plate regions and floor of the anterior cranial fossa. Fractures in this region are extremely difficult to identify on purely axial scanning.

Mandibular Fractures

The most common sites of fractures of the mandible are near the angle, condylar neck and parasymphyseal regions. Fractures of the coronoid process are uncommon.

The configuration of the mandible is such that it approaches the ring when coupled with the base of the skull. Fractures of the parasymphyseal region or angle on one side are frequently accompanied by a subcondylar fracture on the opposite side.

The management of mandibular fractures is predicated on information that might not be visible by CT studies. These factors include:

1. Fractured tooth roots within the line of fracture
2. Compound versus simple closed fracture
3. Status of oral hygiene and remaining teeth
4. Favorable versus nonfavorable angulation of fracture

By favorable we mean that the fracture lines are at such an angle that normal muscle pulls would tend to pull the fracture fragments together. If the angle is unfavorable, the normal muscle pulls would tend to separate the fracture fragments.

CT and MRI Findings

CT is much better suited to tumors of the mandible rather than traumatic lesions (16). At least some survey is possible with CT.

The fractures through the symphysis and ramus of the mandibles can be demonstrated quite nicely on axial CT but evaluations of neck and condylar fractures are extremely difficult. If the patient is cooperative and there are no spinal injuries, extension of the chin may place the neck and body of the mandible within the plane of scan for identification of the more posteriorly located fractures. In general patients with mandibular fractures are best examined by routine radiographs or some type of panoramic tomography. Dental films taken transorally are extremely helpful in some of the more difficult cases.

Cerebrospinal Fluid Leaks

Fracture with resultant cerebrospinal fluid leak may occur almost anywhere along the floor of the cranial vault. Leaks are most common through the cribriform plate and ethmoid air cell regions. Fractures with resultant cerebrospinal fluid leaks occur through the frontal sinuses, sphenoid sinuses, middle cranial fossa, temporal bones, and even through the clivus. Leaks have been associated with tumor masses of the nasal vault, ethmoid or sphenoid sinuses. Spontaneous cerebrospinal fluid fistula are surprisingly common. Approximately half of the patients presenting at UCLA have a spontaneous onset.

Careful history gives some clues as to the site of leak. Temporal fractures that extend into the middle cranial fossa allow a communication to take place between inferolateral recesses of the sphenoid sinus and the subarachnoid space. If a fracture line extends into the temporal bone the communication may extend from the middle ear cavity and subarachnoid space with drainage via the eustachian tube to the nasopharynx.

The use of dye injection such as fluorescein into the lumbar cistern has met with varying degrees of success. Meticulous technique with shrinkage of nasal mucosa and good light for visualization of the placement of the pledgets seems to be crucial to making this very simple method reliable (3).

CT and MRI Findings

Air fluid levels within the sphenoid sinus do not mean that a CSF leak is taking place directly to the sphenoid sinus. Any patient that lies on his back with a drip of CSF occurring into the vault of the nasal cavity will have a fluid level within the sphenoid sinus. This is because the natural opening of the sphenoid sinus is so arranged that its inferior margin is more anterior than its superior margin and fluid will naturally flow into the ostium of the sphenoid sinus (22). Blood and secretions will follow the same pathway. With this in mind one must look for the direct sign of bone interruption for acute injuries and indirect signs of trauma in the delayed cerebrospinal fluid leak patients. Without the use of contrast agents cerebrospinal fluid tends to be somewhat irritating to mucosa, possibly related to its sugar content. Opacification of ethmoid sinuses or sphenoid sinuses associated with the site of leak may be one of the early clues that can be identified on routine CT scanning.

When other methods have failed CT combined with metrizamide offer the greatest chance of demonstrating the site of a leak. Metrizamide is injected either translumbar or transcervical into the subarachnoid space as a contrast agent. With the patient prone, the contrast is run into the cranial cavity and made to layer over the suspect area such as the cribriform plate region. If the patient is able to produce a flow of CSF fluid through straining or coughing, CT scanning is performed during such a maneuver and the contrast agent can be detected in the defect (14). Otherwise, the patient must be left in position for periods of 8 to 10 min to cause contrast to flow into the appropriate sinus. The latter technique can be quite helpful if the cerebrospinal fluid leak happens to occur into the mastoid region with flow occurring from the middle ear cavity into the nasopharynx.

SUMMARY

Modern CT scanning is able to show fracture lines and displacements that are frequently hidden by overlying soft tissues and associated injuries. Even if one discounts the discovery of additional injuries, CT in trauma is still a highly desirable examination because of its ease of performance in the severely traumatized patients. The improved contrast resolution and improved patient comfort during a CT scan will probably make this modality the examination of choice for facial trauma.

References

1. Adams WM: Internal wiring fixation of facial fractures. *Surgery* 12:523, 1942.
2. Brandt-Zawadzki M, Minagi H, Federle M, Rowe LD: High resolution CT with image reformation in maxillofacial pathology. *AJNR* 3:31–37, 1982.
3. Calcaterra TC, Mosely J, Rand RS: Cerebrospinal rhinorrhea: extracranial surgical repair. *West J Med* 127:279–283, 1977.
4. Converse JM, Smith B, Obear MF, Wood-Smith D: Orbital blow-out fractures. A ten-year survey. *Plast Reconstr Surg* 39:20, 1967.
5. Crikelair CF, et al: A critical look at the "blow-out" fracture. *Plast Reconstr Surg* 49:374–379, 1979.
6. Cromwell LD, Mack LA, Loop JW: CT scout view for skull fracture: substitute for skull films: *AJNR* 3:421–423, 1982.
7. Gentry LR, Manor WF, Turski PA, Strother CM: High resolution computed tomographic analysis of facial struts in trauma. Part I. normal anatomy. (Accepted for publication AJR)

8. Gentry LR, Manor WF, Turski PA, Strother CM: High resolution computed tomographic analysis of facial struts in trauma. Part II. osseous and soft tissue complications. AJR
9. Gerlock AJ, Sinn DP: Anatomic, clinical, surgical, and radiographic correlation of zygomatic complex fracture. *Am J Roentgen* 128:235–238, 1977.
10. Grove AS, Jr, Tadmor R, New PJ, Momose J: Orbital fracture evaluation by coronal computed tomography. *Am J Ophthalmal* 85:679–685, 1985.
11. Johnson D, Coleman M, Larsson S, Garner OP, Hanafee WN: CT in medial maxilla-orbital fractures. (Accepted for publication in *J Comput Assist Tomogr*)
12. Kazanjian VH, Converse JM: The Surgical Treatment of Facial Injuries, ed 2. Baltimore, Williams & Wilkins, 1959, pp 191–220.
13. Mustarde JC: *Repair and Reconstruction in the Orbital Region*. Baltimore, Williams & Wilkins, 1971, pp 292–293.
14. Naiditch TP: Personal communication.
15. Noyek AM, Kassel EE, Wortzman G, Jazrawy H, Greyson D, Zizmor J: Contemporary radiologic evaluation in maxillofacial trauma. *Otolaryngol Clin North Am* 16:473–508, 1983.
16. Osborn A, Hanafee W, Mancuso A: Normal and pathologic CT anatomy of the mandible. *AJR* 139:555, 1982.
17. Putterman AJ, Stevens T, Urist MJ: Nonsurgical management of blow-out fractures of the orbital floor. *Am J Ophthalmol* 77:232–239, 1974.
18. Rabuzzi D: Revision of surgery of malaligned mid-facial fractures. *Otolaryngol Clin North Am* 7:107–117, 1974.
19. Rowe LD, Miller E, Brandt-Zawadzki M: Computed tomography in maxillofacial trauma. *Laryngoscope* 91:745–757, 1981.
20. Sato O, Kamitani H, Kokunai T: Blow-in fracture of both orbital roofs caused by shear strain to the skull. Case report. *J Neurosurg* 49:734–738, 1978.
21. Shapiro HH: *Maxillofacial Anatomy*. New York, Lippincott 1954.
22. Tamakawa Y, Hanafee WN: Cerebrospinal fluid rhinorrhea: the significance of an air fluid level in the sphenoid sinus. *Radiology* 135:101, 1980.
23. Zilkha A: Computed tomography in facial trauma. *Radiology* 144:545–548, 1982.

Chapter 3 Plates

Facial Trauma

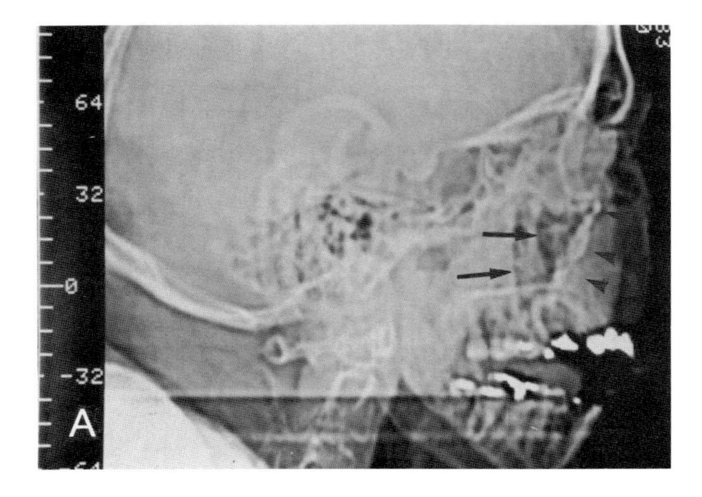

Plate 3.1*A*. LeFort I and II and trimalar fracture. Digital lateral projection shows posterior displacement of the midface (*arrowheads*). Air fluid levels are present in the maxillary sinus (*arrows*) but fine detail is difficult to determine from this single digital view.

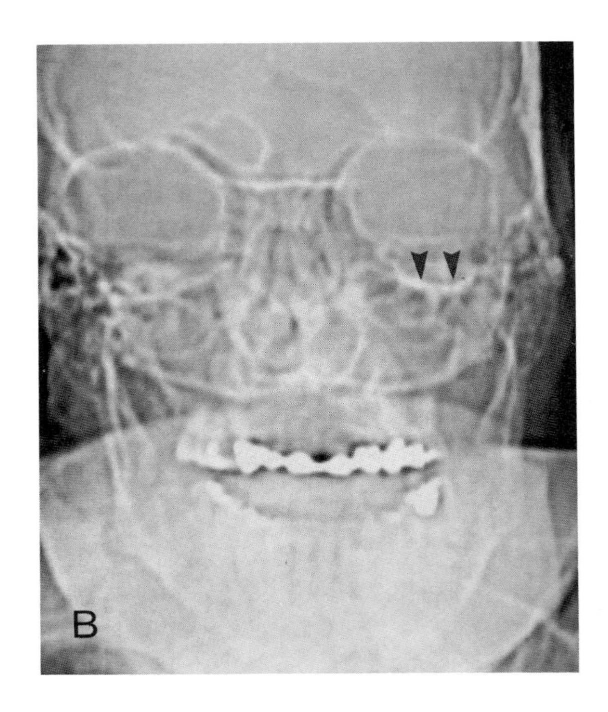

Plate 3.1*B*. LeFort I and II and trimalar fracture (same patient). Digital anteroposterior projection demonstrates inferior displacement of the left orbital floor (*arrowheads*). Maxillary sinus fractures are partially hidden by the overlying occiput.

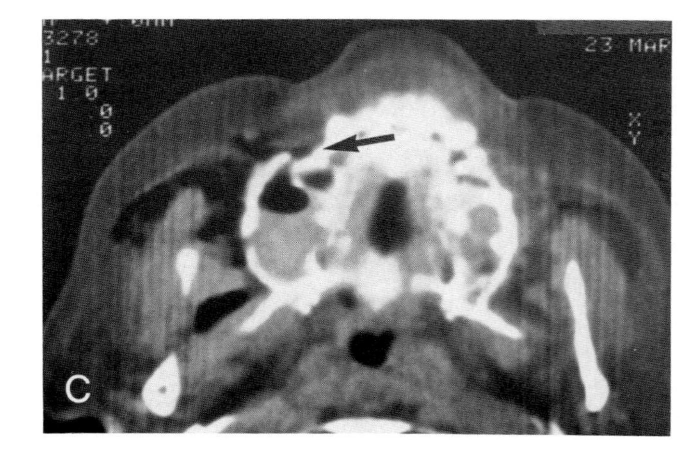

Plate 3.1*C*. LeFort I and II and trimalar fracture (same patient). Scan through the inferior portion of the maxillary sinuses demonstrates a definite fracture almost involving the anterior alveolar ridge on the right (*arrow*). The entire palate and alveolar ridge are slightly rotated to the left.

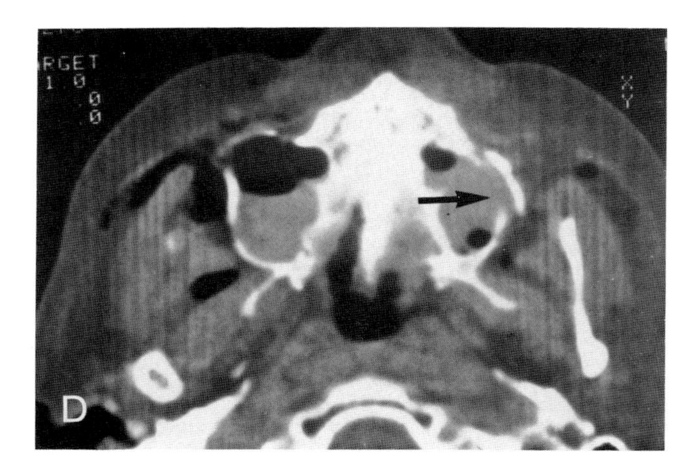

Plate 3.1*D*. LeFort I and trimalar fracture (same patient). Scan 5 mm cephalad to the previous scan shows the fracture through the left lower maxilla with a free fragment being displaced laterally (*arrow*).

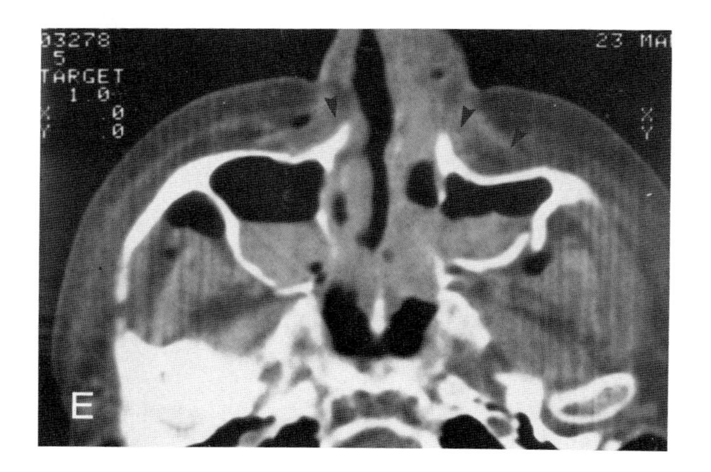

Plate 3.1*E*. LeFort I and II and trimalar fracture (same patient). Section throught the midantrum shows fractures through the anterior and posterior walls of both antra with posterior displacement of the medial pillars of support (*arrowheads*).

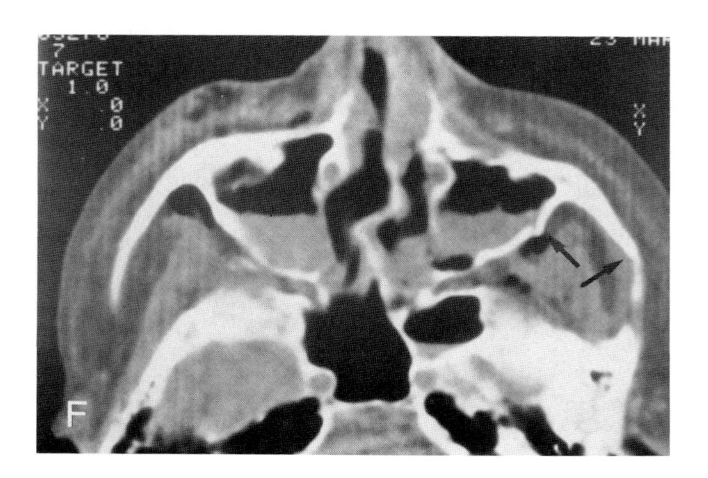

Plate 3.1*F*. LeFort I and II and trimalar fracture (same patient). Section at the level of the malar eminence shows the fracture lines have extended into the posterior wall of the antrum and the zygomatic arch on the left (*arrows*). The nasal septum is disrupted and nasal bone fractures are also present.

Plate 3.1*G*. LeFort I and II and trimalar fracture (same patient). Scan through the floor of the orbit shows a free fragment lying in the superior portion of the antrum (*arrow*). The fractured nasal bones are visible anteriorly. Air is present in the temporal region on the right and both orbits secondary to medial wall blow-out fracture.

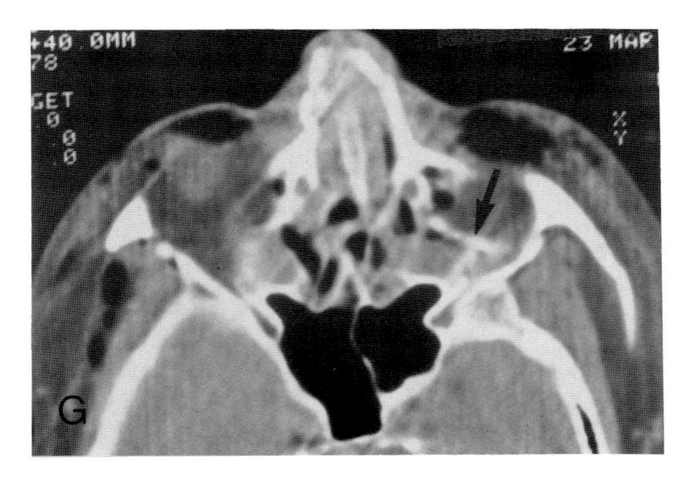

Plate 3.1*H*. LeFort I and II and trimalar fracture (same patient). Scan through the oribt shows the fracture of the lateral wall of the orbit (*arrow*) and posterior displacement of this entire orbital rim. The medial pillar of support on the left is intact (*crossed arrow*) while the right side shows some disruption. The right lateral wall of the orbit is intact and the normal suture should not be mistaken for fracture.

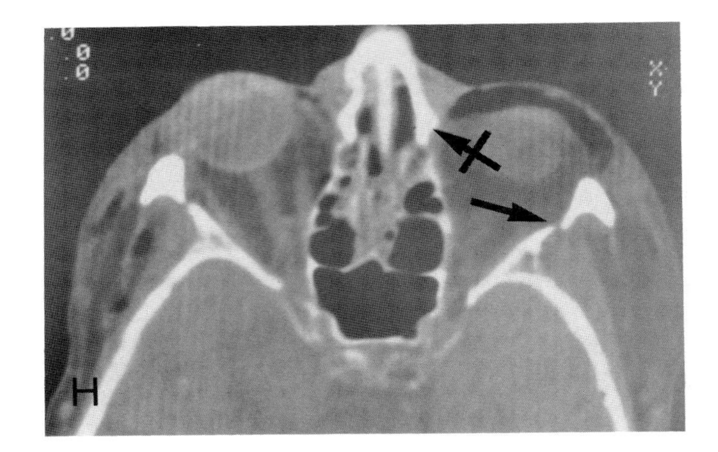

The combination of digital anteroposterior and lateral projections to give a gross appreciation of displacements and the detailed CT scan provide an excellent appraisal of this severe midface fracture. In summary, with both lower maxillary sinuses fractured, there must be a LeFort I fracture. On the right the fracture becomes a LeFort II because the fracture extends to the cribriform plate region. On the left is a trimalar fracture. The intact medial pillar prevents this from being a LeFort III.

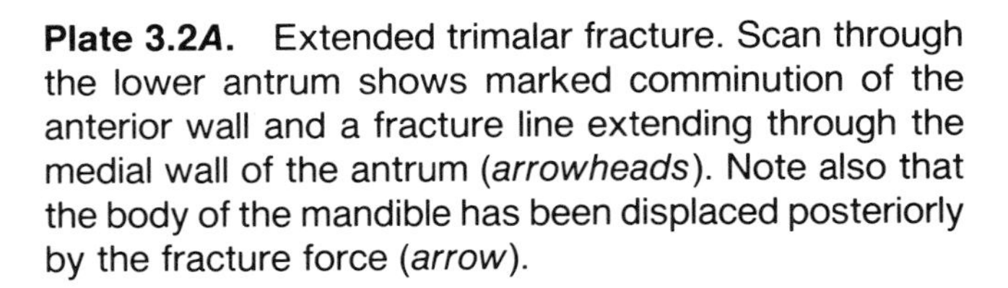
Plate 3.2*A*. Extended trimalar fracture. Scan through the lower antrum shows marked comminution of the anterior wall and a fracture line extending through the medial wall of the antrum (*arrowheads*). Note also that the body of the mandible has been displaced posteriorly by the fracture force (*arrow*).

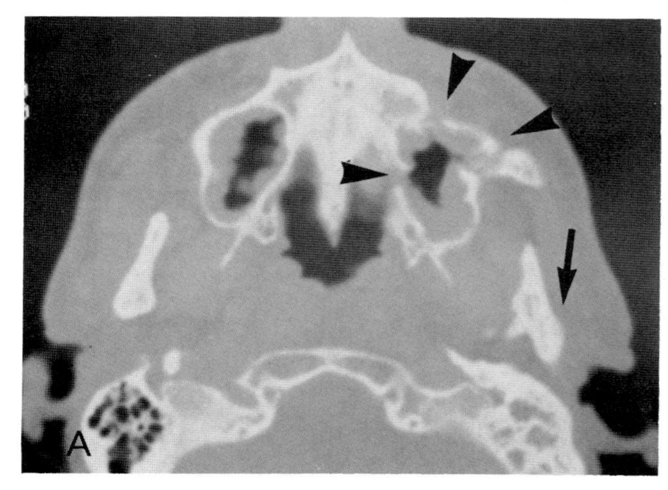

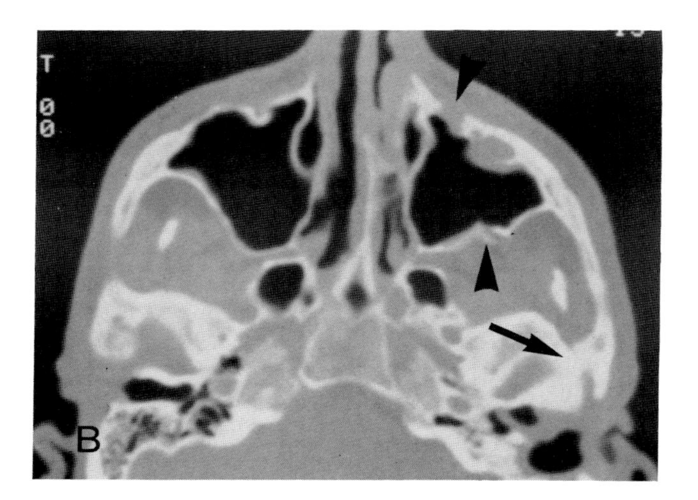

Plate 3.2*B*. Extended trimalar fracture (same patient). Section through the midantrum shows the fracture line through the anterior and posterior wall of the left antrum (*arrowheads*). The entire zygoma has been displaced posterior. In this patient the zygomatic arch remained intact and a fracture occurred through the temporomandibular joint (*arrow*).

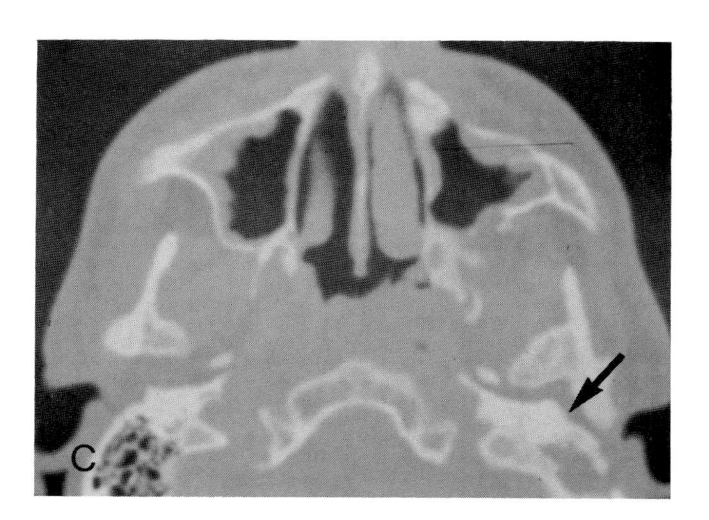

Plate 3.2*C*. Extended trimalar fracture (same patient). Scans through the upper antrum and temporomandibular joint shows disruption of the temporomandibular joint, external auditory bony canal (*arrow*), and deformity of the normal relationships of the neck and condyle of the mandible.

The displacements of the zygoma are clearly identified by the CT scan. This extension of the fracture in the condyle region was extremely difficult to appreciate on routine films.

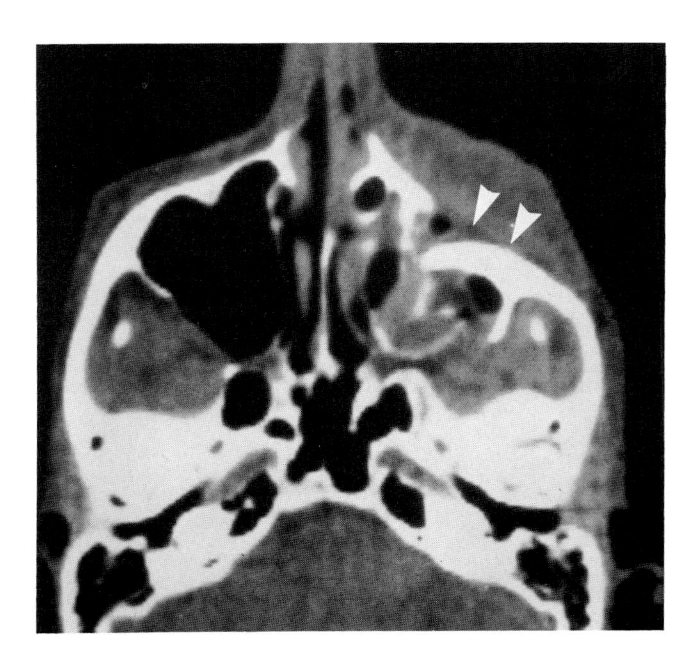

Plate 3.3. Rotated trimalar fracture. This single scan provides a tremendous amount of information. The malar eminence is displaced posteriorly and rotated medially into the antrum indicating a trimalar fracture (*arrowheads*). A free fragment of bone is visible in the antrum indicating that the floor of the orbit has been severely disrupted. The zygomatic arch is sprung but not greatly displaced. The medial pillars are intact.

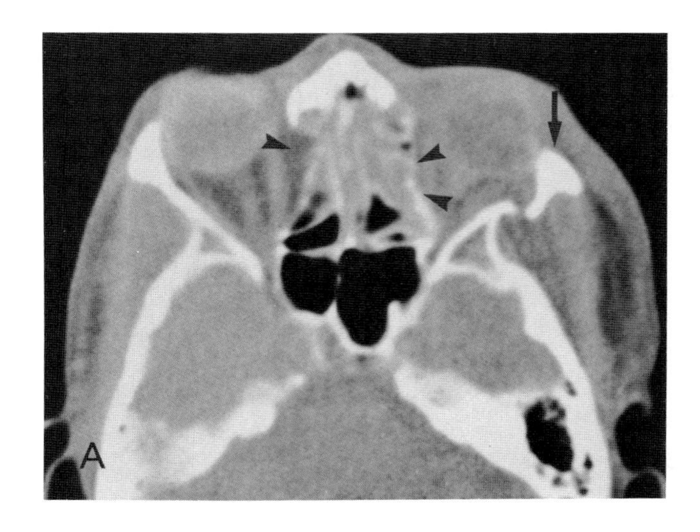

Plate 3.4*A*. Left LeFort III and right LeFort II. Scan through the midorbit shows the ethmoid air cell complex is disrupted bilaterally (*arrowheads*) and the crista galla alignment is oblique. On the right the malar eminence and lateral wall of the orbit appear intact whereas on the left the malar eminence is displaced posteriorly (*arrow*).

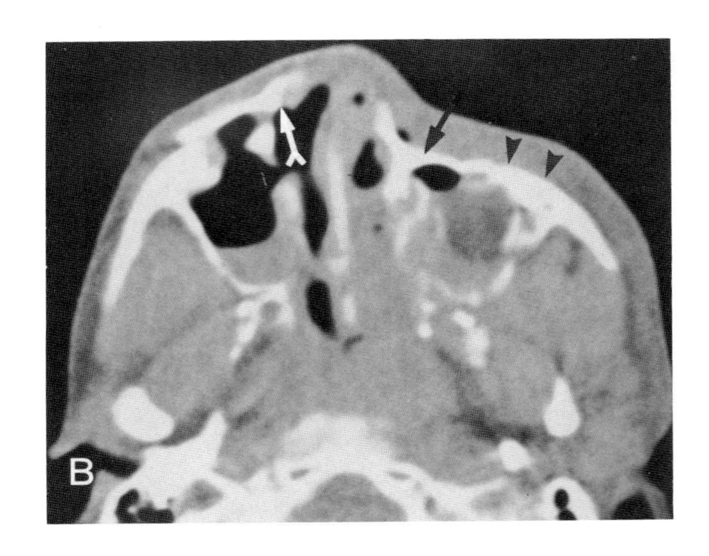

Plate 3.4*B*. LeFort II on the right and LeFort III on the left (same patient). At the level of the midantrum comminuted fractures are noted bilaterally. The medial pillar of support is displaced posteromedially on the left (*arrow*) and a similar structure is displaced anteriorly on the right (*arrow with tail*). The entire malar eminence, which constitutes a major portion of the lateral pillar, is displaced posteriorly on the left (*arrowheads*) but is in normal position on the right.

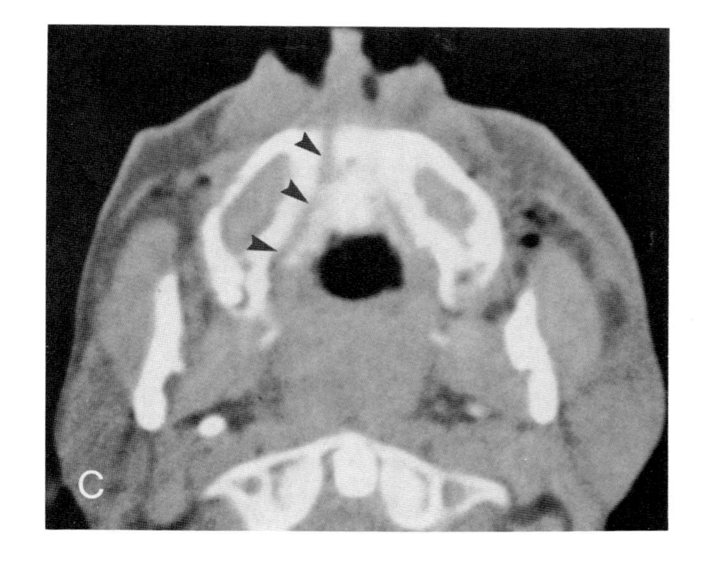

Plate 3.4*C*. LeFort III on the left and LeFort II on the right (same patient). At the level of the alveolar ridge a linear fracture can be seen extending through the right side of the hard palate (*arrowheads*). In this type of midface injury displacements of the medial pillar are extremely difficult to appreciate and the hard palate fracture is easily overlooked unless suspected from clinical examination.

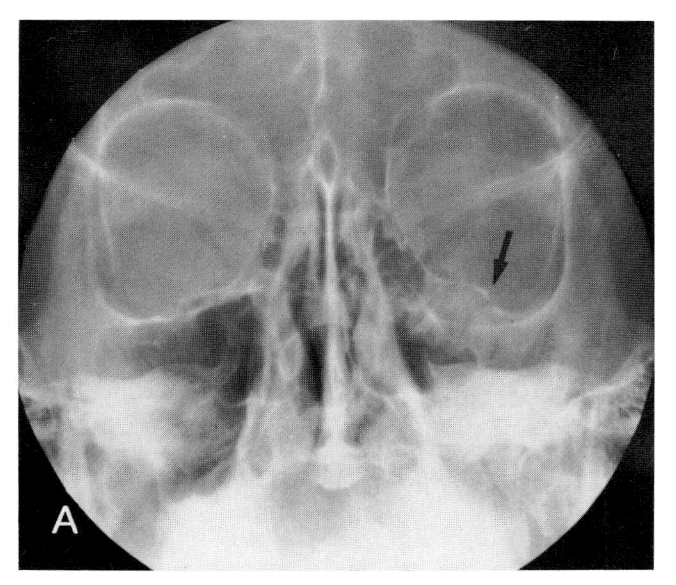

Plate 3.5*A*. Blow-out fracture of the orbital floor and displaced fracture of the medial pillar. The shallow Water's projection shows a linear fracture extending through the floor of the orbit (*arrow*). The posterior portion of the orbital floor adjacent to the arrow tip is not displaced whereas the anterior portion of the floor is depressed. In this view and other routine films, the medial wall of the antrum and medial pillar appear to be intact.

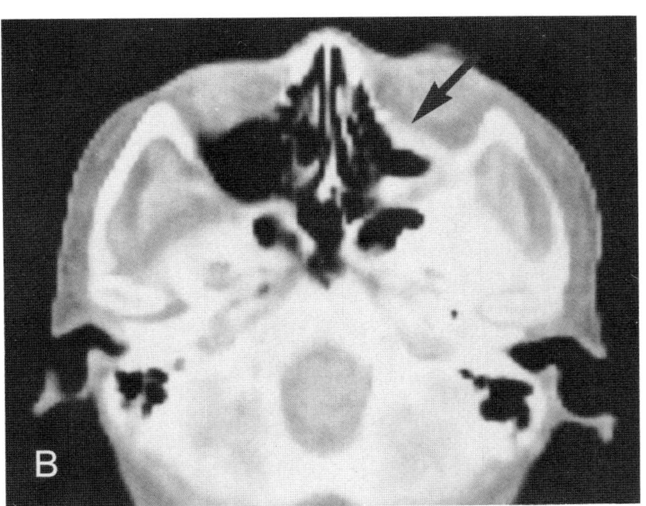

Plate 3.5*B*. Blow-out orbital floor and medial pillar fracture (same patient). Scan through the floor of the orbit shows the floor to be depressed on the left (*arrow*) by comparison to the normal appearance on the right. This particular section which is exactly through the floor combines the antrum in its posterior portion with the floor of the orbit and orbital contents in the anterior portion. Note that the zygomatic arches are both at equal levels so that the unequal height of the floor of the orbit is a real finding.

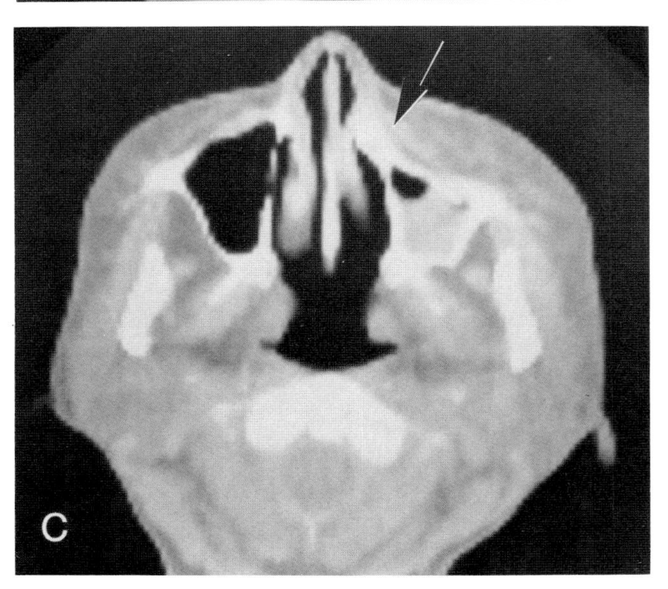

Plate 3.5*C*. Blow-out of the orbital floor and medial pillar fracture (same patient). At the level of the midantrum the medial pillar is displaced posteriorly and medially (*arrow*). The antrum is opaque but no free fragments can be demonstrated within the lumen of the antrum. The lack of free fragments or the presence of orbital floor in the midantrum would indicate that it is a minimal blow-out, but the principle finding to be corrected is the displaced medial pillar. This injury was not appreciated on the routine films.

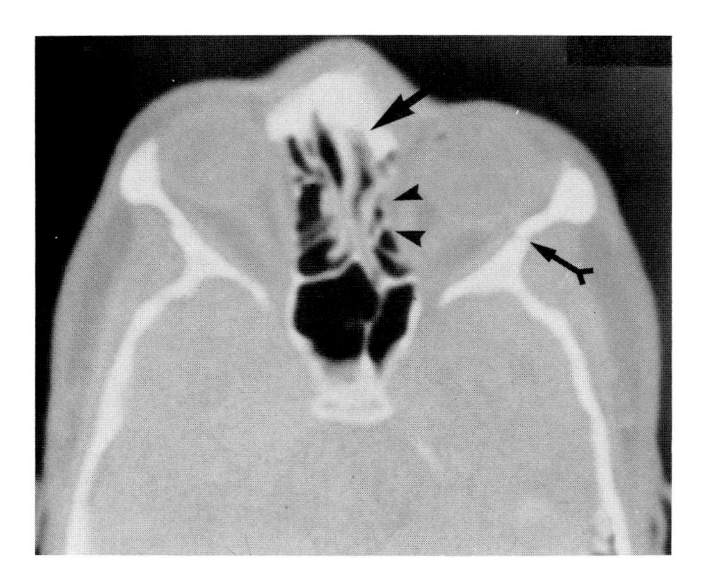

Plate 3.6. Medial wall blow-out and frontal process of the maxilla as a result of a direct blow. The fontal process of the maxilla is displaced posteriorly (*arrow*). In addition there is a blow-out of the medial wall of the orbit (*arrowheads*). The lateral wall of the orbit is also fractured (*arrow with tail*) indicating that this patient has an associated trimalar fracture.

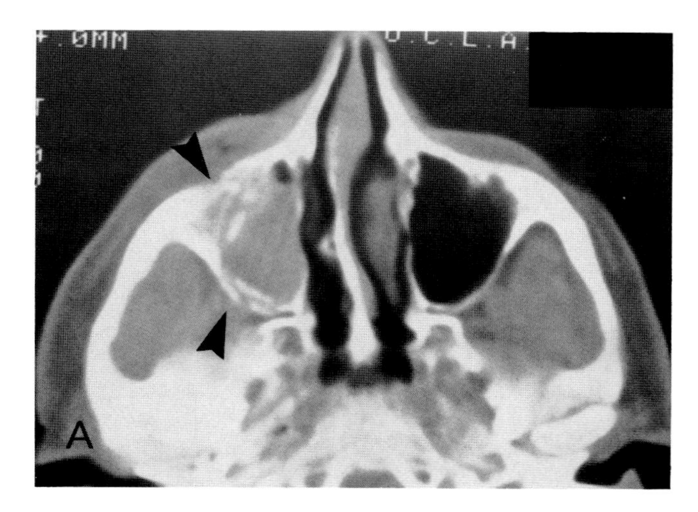

Plate 3.7*A*. Rotated trimalar fracture. Scan through the midantrum shows fractures through the anterior and posterolateral wall of the antrum on the right (*arrowheads*) with some posterior displacement of the malar eminence. The zygomatic arch is distorted but a fracture line cannot be identifed. Several linear fracture fragments can be seen within the lumen of the antrum.

erode the bony margins of the orbit. Those that begin within the orbit produce erosions of the bony margins from within and have sclerotic margins. Calcifications within the tumor mass may be present and visual loss is a late occurrence.

CT and MRI Findings

MRI may show the signals coming from the lesion to be bright or isodense with muscle tissue depending on fat content. Dermoids rarely enhance on CT and their diagnosis is best typified by sharp margins and low density due to fat content. When mixed with lymphoid tissue the tumors may be isodense with muscle or lacrimal gland tissue.

RHABDOMYOSARCOMA

Rhabdomyosarcoma is a disease of childhood with a peak incidence in the 8- to 10-yr age group. The lesions are characterized by their ability to invade into bone as well as soft tissue. They may originate outside the muscle cone or within the muscle cone itself. Rhabdomyosarcomas are rapidly growing, producing pain and profound exophthalmus. They invade the central nervous system and the adjacent temporal fossae.

CT and MRI Findings

Serial CT and MRI may show the rapid growth of the tumor. Tumor enhancement by intravenous contrast can be anticipated. On MRI the late spin echo signals are bright and cross bony margins. Rhabdomyosarcomas invade intracranially as well as inferiorly into the infratemporal fossa. Part of the bright signal coming from rhabdomyosarcomas is from edema in adjacent brain or fascial planes. Metastases to lymph nodes is a late occurrence and usually the tumor spreads by direct extension.

PSEUDOTUMOR

Any reactive process that produces expansion of the retroorbital tissues could be considered pseudotumor. Characteristically the diagnosis must be one of exclusion of known disease entities. Unilateral, rapidly progressive, painful proptosis with some decreased ocular mobility are more or less specific for pseudotumor and give the lesion a more distinctive diagnostic place. Unfortunately, specific inflammations and some lymphomas may mimic this presentation. Some authors believe a subgroup of pseudotumor exists which is a forerunner of lymphoma (16). Pseudotumor is thought to be related to autoimmune disease. Histologically one sees plasma cells, eosinophils and an abundance of lymphocytes that have a tendency to form follicles. Isolated granulomas may be present.

Grossly the pseudotumors may be circumscribed but are usually diffuse. They may lie anywhere within the orbit, however, a greater percentage of the orbital pseudotumors seem to be more anteriorly located. They tend to involve the tendinous insertions of the extraocular muscles and the reflections of Tenon's capsule along the posterior margins of the globe. The lacrimal gland region is another common site for pseudotumor to occur.

The clinical course of a patient with orbital pseudotumor is usually initiated by a painful proptosis that is progressive over a period of several weeks. Young females are affected more commonly than other age groups. The process is usually unilateral but may become bilateral, especially if treatment is terminated prematurely. A response to steroids is dramatic and has even been used as a diagnostic therapeutic test (14).

The more localized forms of pseudotumor cause a diagnostic dilemma. If response to steroids is minimal or there is persistent proptosis the differential diagnosis between pseudotumor and a true tumor becomes extremely difficult. Under these circumstances it is necessary to rule out such entities as metastatic tumor, inflammatory disease, histiocytosis, lymphoma, Wegener's granuloma, periarteritis, sarcoid, and other lymphoepithelial proliferative disorders. In the proper clinical setting, orbital exploration may be indicated to obtain histological material.

CT and MRI Findings

MRI will show a bright image which is nonspecific, however, the extent of involvement is well demonstrated. The orbital fat may cause MRI to overestimate the size of the lesion. If CT is used, contrast enhancement is a prominent feature of pseudotumor. Involvement of the posterior reflections of Tenon's capsules about the posterior margin of the globe and about tendinous insertions is almost specific. When the changes are diffuse throughout the muscle cone region, CT diagnosis is impossible.

The localized forms of pseudotumor may be within the muscle cone region or in the lacrimal fossa. The margins are usually not as sharp as one encounters with cavernous hemangioma. Given a lesion that is poorly marginated but fairly well localized one must consider metastatic tumor, venous malformation, lymphoma, infection and other tumors of blood vessel origin. Occasionally the localized form of pseudotumor is confined to a single extraocular muscle. Some consider this is a myositis while others believe that this is merely a variant of orbital pseudotumor.

Regardless of the theories concerning the CT appearance of extraocular muscle involvement, differential diagnosis between the pseudotumor family of lesions and primary or metastatic tumors involving the extraocular muscles remain a difficult differential diagnostic problem that may ultimately require biopsy for definitive diagnosis. The differentiation of pseudotumor with multiple extraocular muscle involvement as compared to endocrine exophthalmus is not 100% accurate but some important clues can be derived from the CT. Orbital pseudotumor centers about the anterior tendinous portions of the extraocular muscles while the myositis of endocrine exophthalmus involves the more posterior muscle bellies of the muscles. Endocrine exophthalmus involves the medial and inferior rectus most commonly and is frequently bilateral whereas pseudotumor is unilateral and does not have this predilection for specific muscles. Pseudotumor also causes a change in density of the fat of the muscle cone region whereas in endocrine exophthalmus fat changes are not present until very late in disease.

In the differential diagnosis from metastatic tumor, bone involvement can occur with either pseudotumor or metastatic tumor. Most malignant tumors produce a destructive change in bone with exception of some breast and prostate tumors whereas pseudotumor generally produces a productive change. Exophthalmus is usually proportional to the amount of orbit involvement with pseudotumor whereas some of the more sclerotic metastatic tumors are not accompanied by a significant exophthalmus. In general the clinical course of the patient is more helpful than the CT appearance. When changes are present in the orbital apex as well as the cavernous sinus, the condition is known as Tolosa-Hunt syndrome. These filling defects within the cavernous sinus are the result of granulation tissue that responds promptly to steroids. The clinical course and CT findings resemble pseudotumor and may even be part of a spectrum of this disease.

LYMPHOMA

Lymphoma of the orbit may be localized or diffuse. The localized form is most commonly located in the region of the lacrimal fossa. Infiltrations of the lacrimal gland produce an orange-colored palpable mass beneath the upper lid (see Lacrimal Gland Tumors). More diffuse forms are occasionally seen which involve the entire muscle cone region including Tenon's capsule. The anterior infiltration causes an intense conjunctivitis which is frequently bilateral.

CT and MRI Findings

CT may show enhancing cellular infiltrations that spread in an irregular fashion from the upper outer quadrant of the orbit along the lateral bony walls. When centrally, the infiltrations permeate the fat surrounding the optic nerve and concentrate between the reflections of Tenon's capsule and the posterior surface of the globe. Diffuse infiltrations throughout the posterior orbit may be indistinguishable from other diffuse processes such as metastases, pseudotumor and some aggressive inflammatory reactions.

Endocrine Exophthalmus

Endocrine exophthalmus is one of the most common causes of proptosis which may be unilateral or bilateral. Endocrine exophthalmus may occur in a completely euthyroid state which makes CT an extremely important modality in the diagnosis. The disease is primarily a myositis. The swelling and deposits of glycoproteinous material centers in the muscle bellies and retroorbital fat rather than in the tendons' insertion of the extraocular muscles. The myositis of endocrine exophthalmus most frequently affects the inferior and medial rectus muscles. The disease is bilateral in over 50% of the cases but not necessarily symmetrical. It is rare for endocrine exophthalmus to occur as isolated disease in the lateral rectus muscle.

The early disease is manifest by proptosis and lid lag. As swelling causes increased pressure within the orbital apex, venous obstruction adds to the patient's problem. The conjunctiva and sclera become edematous and infected. With continued swelling of the posterior portions of the extraocular muscle bundles, pressure takes place on the optic nerve immediately prior to its entrance into the optic canal. This may lead to diminished visual acuity and eventually blindness (17).

CT and MRI Findings

CT in the axial plane will demonstrate the thickness of the muscle but coronal scans or reformatted images are necessary to accurately compare diameters of all of the muscle bundles. Specific numbers and measurements of extraocular muscles are not as important in diagnosis as a comparison of the two sides and muscle bundles with each other. Coronal and sagittal MRI scanning is ideal for this task (13). The lesions enhance with intravenous contrast. The fat within the muscle cone may be affected by the endocrine exophthalmus but only in advanced diseases is it possible to demonstrate density changes by CT. CT can show the thickened muscle bundles and precisely how optic nerve compression is produced in the orbital apex. CT is also excellent for demonstrating whether adequate decompression of the optic nerve has been accomplished by surgical intervention.

Venous Lesions

A spectrum of lesions exists that vary in etiology from traumatic lesions to true tumors. The common denominator is the presence of venous structures and fluctuating exophthalmus which is position related. Characteristically these patients can exaggerate their exophthalmus by bending forward or straining. The list includes capillary hemangioma, cavernous hemangiomas, orbit verix, blood cysts, and true arteriovenous fistulae. True arterial lesions are quite rare but could be considered within this same grouping. The type of venous lesions may be diagnosed by age of onset, history of trauma, or possibly only after extensive CT and angiographic work-up.

CAPILLARY HEMANGIOMA

Capillary hemangiomas are usually present at birth or shortly thereafter and are associated with skin changes of blood vessel origin. The lesions within the orbit are diffuse and consist primarily of dilated capillaries with one or two large draining veins. Lymphangiomatous elements may be present. They are usually in apposition with the overlying skin changes. The lesions do not pulsate but are exaggerated with forward bending. They may be confined to the structures anterior to the orbital septum, but occasionally extend to the retroorbital tissues.

CT and MRI Findings

At CT the capillary hemangiomas enhance to some extent following intravenous contrast. The dilated veins do not fill completely with opacified blood but remain hazy in outline due to a mixture of nonopacified blood. This slow flow leads to a bright MRI signal rather than absent signal that one expects in more rapidly moving blood. Their margins are usually somewhat diffuse. The main bulk of the mass is not of sufficient consistency to cause major bone erosions. There may be some increased volume to the orbit at the expense of the ethmoid air cells but the more firm bony margins do not become modified. The diagnosis of the lesion is not usually a problem because there is an overlying skin change. Only the extent of the mass becomes important when a surgical intervention is being considered.

ORBITAL VARICES

Orbital varices are localized dilatations of venous channels much like any other verix within the body. The blood remains confined to a venous channel with well formed walls. Proptosis is usually the patient's only complaint although some diminution of visual acuity may occur in far advanced lesions. The varix may produce erosions of bony margins of the orbits due to their long-standing pressure. The erosions are characteristically located in the anterior part of the roof of the orbit where the superior ophthalmic vein exits from the muscle cone.

CT and MRI Findings

Orbital venography is probably a better examination than any computerized exam for showing all of the ramifications or orbital varix. CT will demonstrate only the portions of the lesion containing contrast material. Placing the patient in the prone position and having the individual strain during CT scanning is a maneuver that will improve venous filling. By this maneuver contrast will pool in greater concentrations throughout the lesions (28).

CAVERNOUS HEMANGIOMAS

Following orbital pseudotumor, cavernous hemangiomas are one of the most common tumors of the orbit in adults. They are true tumors composed of venous channels that are encapsulated in a fibrous capsule. The lesions are most frequently located in the muscle cone region. They may indent the posterior surface of the globe producing retinal stria but do not deform the bony walls of the orbit to any significant extent. They produce a central type of proptosis. Because of their soft nature visual loss is minimal.

CT and MRI Findings

At CT the findings of a sharply circumscribed mass that displaces the optic nerve but is not part of the optic nerve shadow is practically pathognomonic for this tumor. The tumor stains homogeneously on intravenous contrast. The intensity of the stain varies considerably but not its homogeneousness. This is in contradistinction to neuromas which may mimic cavernous hemangiomas by their circumscribed outline but generally have a mottled appearance and are usually located either in the roof of the orbit or in the extreme orbital apex.

MRI is not as useful in cavernous hemangiomas as in other lesions. The slow flowing blood gives a bright signal which may be indistinguishable from the fat in the muscle cone region. Nevertheless, by using surface coils, the fibrous capsule of the cavernous hemangiomas can be identified as a lower intensity signal region.

BLOOD CYSTS

Blood cysts are merely endothelial-lined lakes of blood that communicate with an orbital vein. They commonly communicate with the superior or inferior ophthalmic vein in the extreme orbital apex. Blood cysts may be of traumatic origin or secondary to inflammatory disease with rupture of one of the orbital veins. The blood cysts tend to permeate through muscle bundles and are extremely resistive to management.

CT and MRI Findings

The blood cysts may be made to fill with contrast provided long injection times are given and scanning is performed late during the injection. With the patient supine or prone, only the dependent portions of the lesion will be outlined.

ARTERIOVENOUS FISTULA

Arteriovenous fistula have their origin within the intracranial cavity. The communication may be from the internal carotid artery to the cavernous sinus as a major shunt. A minor shunt can occur from the external carotid system via the meningeal vessels and dural veins.

The internal carotid artery-cavernous sinus fistula are almost invariably the result of trauma or rupture of an infraclinoid artery aneurysm. The external carotid system fistula may follow trauma or may occur spontaneously. Either type of fistula will cause dilatation of the normal venous channels within the orbit and, in particular, the superior ophthalmic vein. The diagnosis of A-V fistula is readily made on a clinical basis because of pulsation of conjunctival vessels and the history of a sudden onset of objective bruit. Differential diagnosis between the internal carotid-cavernous sinus fistula and dural communications may be suggested by the loudness of the objective bruit and the magnitude of the shunt but usually angiography is required for confirmation.

CT and MRI Findings

With AV fistula the superior ophthalmic vein enlarges and is readily demonstrable.The inferior ophthalmic vein can be prominent in scans through the lower orbit but usually the superior ophthalmic vein deformity dominates the picture. The extraocular muscles also enlarge due to venous congestion of the orbit. Conjunctivitis may be present and there may be contrast enhancement of the sclera of the globe and Tenon's capsule.

At MRI the high flow through the superior ophthalmic vein may make it readily visible as a low signal region. The extraocular muscles may also show varying degrees of high signal due to edema and low signal due to the increased blood flow to the regions.

ARTERIAL MALFORMATIONS

True arterial malformations are quite rare and vigorous pulsating masses are present which distinguish the lesions from arteriovenous malformations. They differ from fistula in that there is substance to the lesion consisting of a capillary bed or multiple arterioles and veins that form part of the mass of the lesion. This transition zone is extremely important in intervention therapy and determines whether a balloon can be used to plug a simple fistulous tract or whether some type of material must be injected to obliterate a capillary bed.

The arterial malformations may be congenital in origin (Wyburn-Mason syndrome) and follow the optic pathways intracranially into the occipital area. The arterial malformations are more likely to produce bony erosive changes because of the pulsating blood flow.

CT and MRI Findings

At CT or MRI the extensive arterial changes can be identified and the evidence of a tumor mass indicating the expansion which frequently accompanies these lesions. Contrast enhancement is indicated to outline the full extent of the tumor mass. The Wyburn-Mason malformations can frequently be traced from the retina to the occipital lobe.

HEMANGIOPERICYTOMAS AND GLOMUS TUMORS

Hemangiopericytomas and paragangliomas originate from the cells of blood vessel walls. The hemangiopericytoma is a rare tumor of endothelial origin and is characterized by round or spindle cells in a rich capillary network. They are an aggressive infiltrating tumor that rarely metastasizes. The glomus tumors or nonchromaffin paraganglioma is of chemoreceptor origin and also shows a rich capillary bed plexus similar to hemangiopericytoma. The lesions may occur anywhere within the orbit, either retroorbit, muscle cone, or even within the extraocular muscles themselves. They affect all age groups.

Glomus tumors have both a familial tendency and the propensity to affect multiple sites but rarely metastasize. Specific diagnosis is possible if there is sufficient endocrine function.

CT and MRI Findings

At CT both hemangiopericytoma and paragangliomas enhance richly with intravenous contrast. Their margins may be sharp and distinct or they may show some haziness in outline. Bone invasion and destruction is common for both tumors. There may be enlargement of natural foramena when the lesions originate in the orbital apex. A histological diagnosis is not possible by CT or MRI.

METASTATIC TUMOR

Metastatic tumor to the globe or to the orbit occurs with a surprisingly high frequency (15) but other manifestations of the primary tumor usually dominate the clinical picture. This is not quite the case in neuroblastoma of children where the metastases to the orbit may be the dominant feature. Carcinomas of the kidney, breast, lung, and colon may go to individual extraocular muscles early in the course of the disease so that differential diagnosis can be a problem (23). The presence of an isolated muscle bundle enlargement is extremely rare in endocrine exophthalmus but may be seen in myositis or pseudotumor (21). Differential diagnosis may only come about after a period of follow-up or possibly after orbital biopsy. One of the distinctive features of metastatic scirrhous carcinoma of the breast is the production of diffuse orbital changes without the presence of exophthalmus. This is because the fibrous contractive scarring of the tumor prevents proptosis.

CT and MRI Findings

The CT and MRI findings of metastatic tumor are not specific. Diffuse enhancing masses that cross fascial planes are commonly encountered. The orbital apex is a favorite location. CT usually plays a dominant role because of the ability to see the bony margins of the orbit as well as the soft tissue contents. Metastases such as carcinoma of the prostate can be totally missed if one is only performing MRI and not CT or plain films to complement the information obtained from proton signals.

RETINOBLASTOMA

Retinoblastoma is a tumor of infancy and childhood characterized by a mass arising from the retinal surface of the globe and projected into the vitreous. The tumors

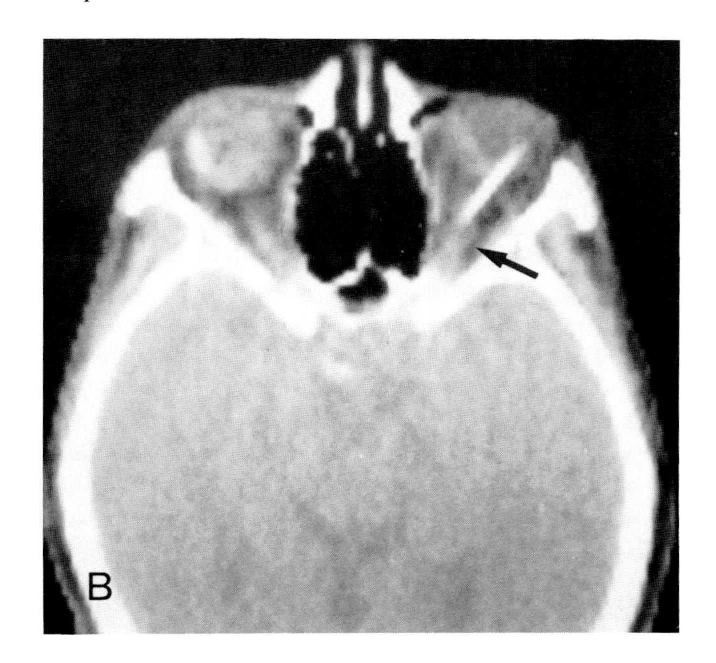

Plate 4.3*B*. Same patient. Because the patient was already blind it was believed that percutaneous cell aspiration would not in any way damage the patient's vision. We wished to rule out meningioma. Scan shows a needle has been placed in an optic nerve (*arrow*) for aspiration cytology. Only fibrotic tissue was obtained. Open biopsy of a 15-mm section of nerve confirmed sclerosis and granuloma formation. Although the patient gave a history of syphilis, no active lesions could be demonstrated in the histological material.

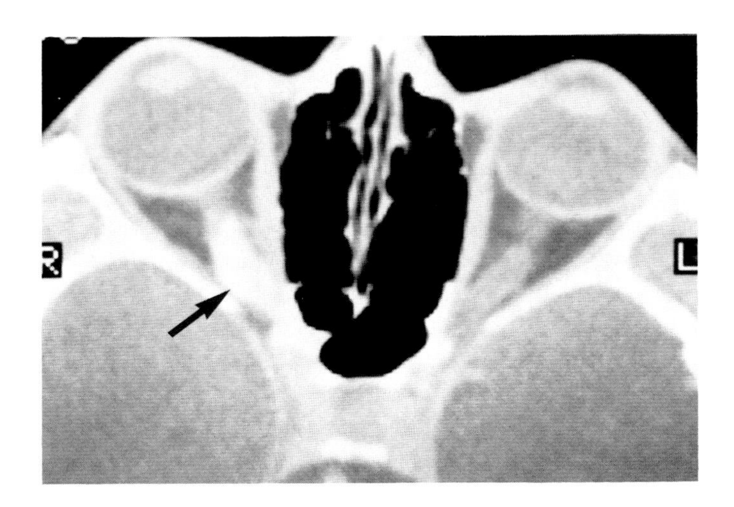

Plate 4.4. Meningioma of optic nerve sheath. Extensive calcifications are present in this meningioma of the optic nerve sheath (*arrow*). This particular lesion is quite small although the patient was almost totally blind. The configuration of the optic nerve meningiomas may be identical to that of gliomas. When calcifications are present in gliomas they are very minimal and tend to be punctate. Dense calcifications of this nature are almost pathognomonic of a meningioma.

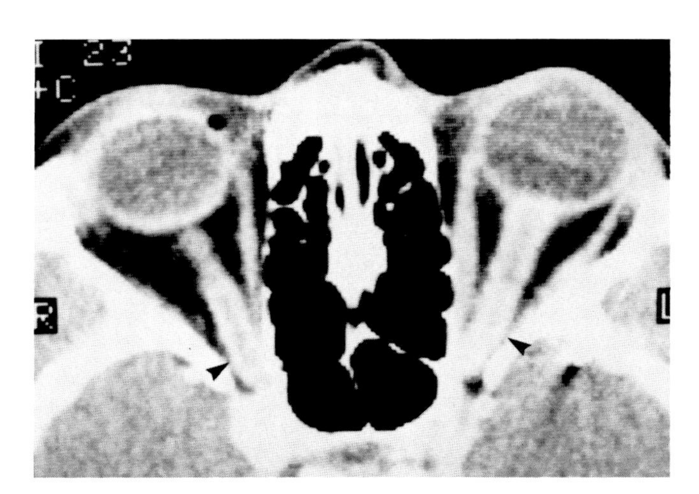

Plate 4.5. Meningeal carcinomatosis. This patient had diffuse meningeal carcinomatosis from an amelanotic melanoma. Scan of the orbits with intravenous contrast shows bilateral thickening of the optic nerves with intense staining of the meningeal sheaths of the optic nerve (*arrowheads*). Note that on the left the staining extends all the way to the optic nerve head. Presumably tumor cells followed the optic nerve in the subarachnoid space. The tumor cells and the increased intracranial pressure combined to form an enlargement of the optic nerve shadow.

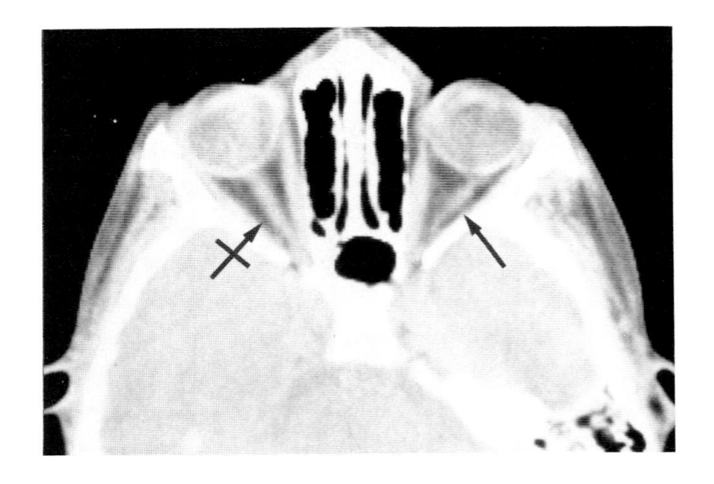

Plate 4.6. Septo-optic dysplasia. The left optic nerve is congenitally hypoplastic (*arrow*) compared to the normal right (*crossed arrow*). These patients have diminished visual acuity and the intracranial scan showed absence of the septum pellucidum. Occasionally more severe midline dysrhaphism is present.

VASCULAR LESIONS

Plate 4.7. Cavernous hemangioma.

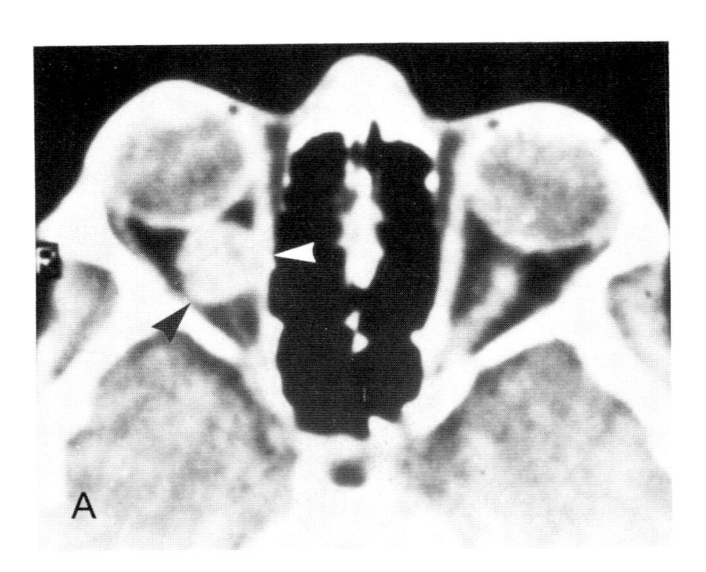

Plate 4.7*A*. Scan through the orbits without contrast shows a well circumscribed mass (*arrowheads*) within the muscle cone region displacing the optic nerve but not indenting the globe or the bony margins of the orbit.

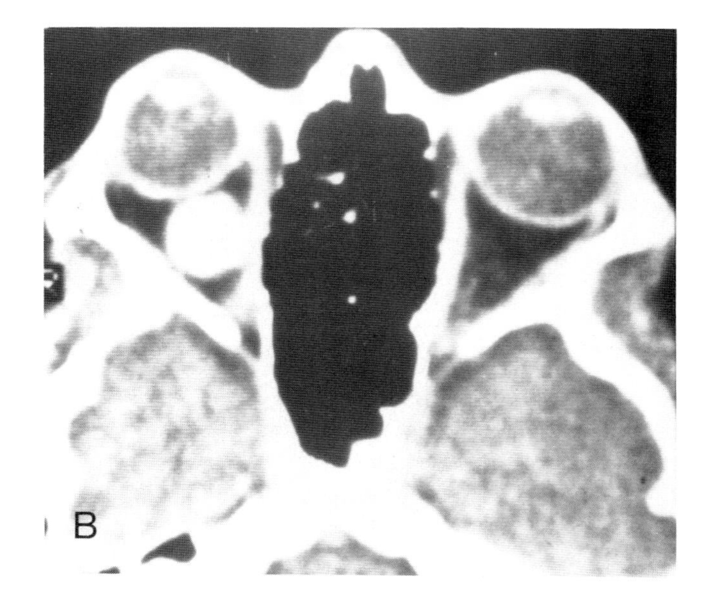

Plate 4.7*B*. Following intravenous contrast, the tumor is enhanced and again sharply marginated. The homogeneous nature of the staining plus the absence of any expansion of bony walls is almost pathognomonic for a cavernous hemangioma. Occasionally even these rather soft tumors will indent the ethmoid air cells.

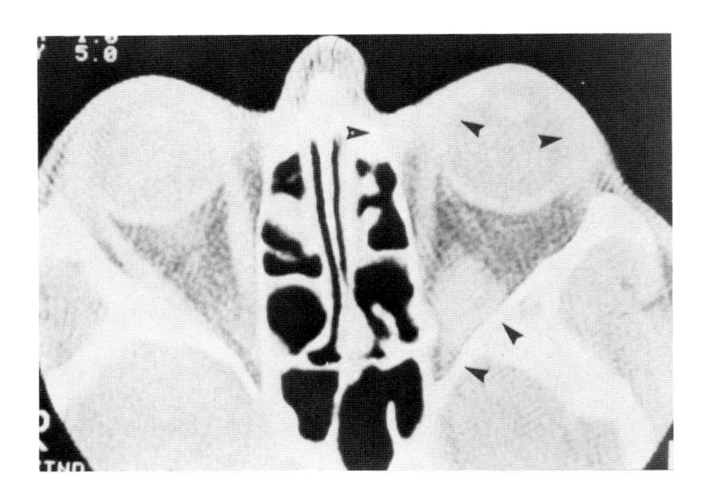

Plate 4.8. Venous malformation. This venous malformation appears quite localized (*arrowhead*) in the orbital apex. Unfortunately the lesion is more of an epithelialized blood "lake" that is sharply outlined because of its being confined by the muscle cone. One of the orbital veins has ruptured and over time (months) the "venous lake" produces some obstruction to flow with dilatation of conjunctival veins (*anterior arrowheads*).

LESIONS OF THE MUSCLE CONE REGION

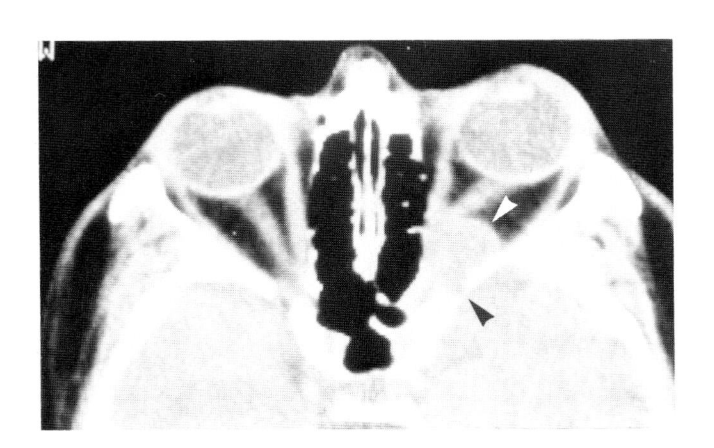

Plate 4.9. Schwannoma. Scan following intravenous contrast shows a sharply circumscribed tumor mask confined to the orbital apex but within the muscle cone (*arrowheads*). The tumor stains approximately the same as muscle and has a mottled appearance. No calcifications are present and there is no significant deformity of the bony wall. The fact that the tumor is located in the orbital apex suggests a lesion of neurogenic or blood vessel origin. Meningioma or hemangiopericytoma should stain more intensely. Histologic diagnosis becomes a little more than guesswork.

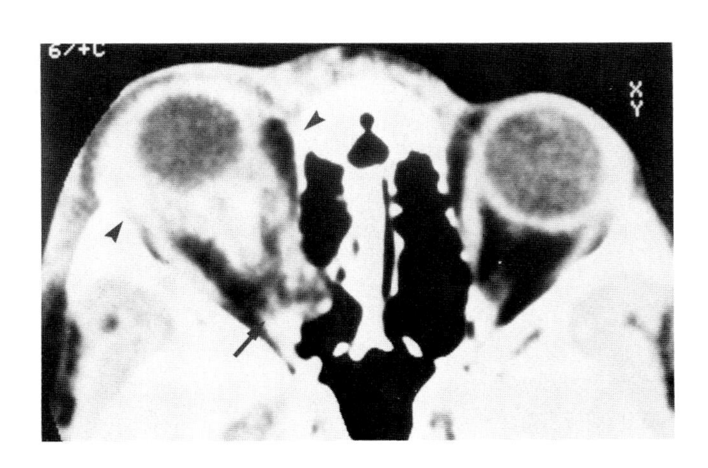

Plate 4.10. Orbital pseudotumor. Following intravenous contrast there is intense staining of an ill defined process involving the sclera, extraocular muscles and optic nerve. The involvement of the posterior surface of the globe and distal optic nerve corresponds to the reflections of Tenon's capsule. The major portion of this inflammatory process involves the more anterior portions of the orbit (*arrowheads*). Some inflammatory reaction is also present in the extreme orbital apex (*arrow*), however, this posterior involvement is minimal which corresponds to the usual pattern of pseudotumor.

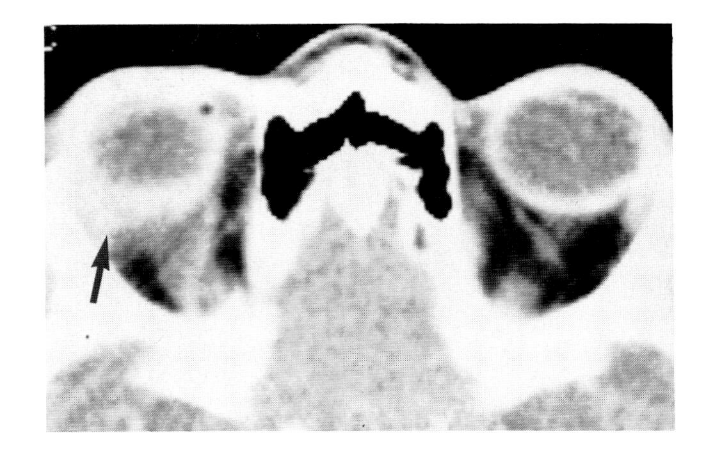

Plate 4.11. Orbital pseudotumor. This enhanced CT scan shows inflammatory reaction in the lacrimal fossa region that involved the lacrimal gland (*arrow*) and extended into Tenon's capsule region along the posterior surface of the globe. Conjunctival thickening is also present which contributes to the prominence of the superior lid on the right. The left orbit remains normal in this patient.

The lacrimal fossa region is a common place for pseudotumor as well as other lymphoepithelial disorders. This CT configuration is not pathognomonic for pseudotumor.

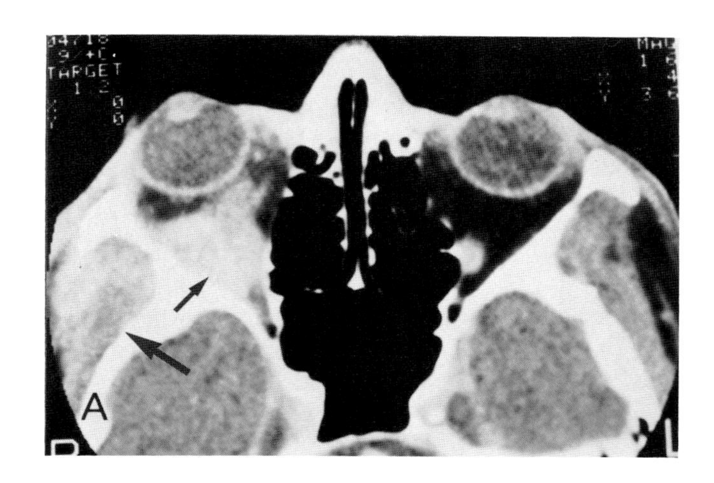

Plate 4.12. Lymphoma.

Plate 4.12*A*. Diffuse infiltration which shows that some contrast enhancement is present throughout the muscle cone region (*small arrow*). There does not appear to be any significant involvement of the tendonous insertions of the extraocular muscles. The infiltrative process extends outside of the orbit into the infratemporal fossa and the deep temporal fossa (*large arrow*).

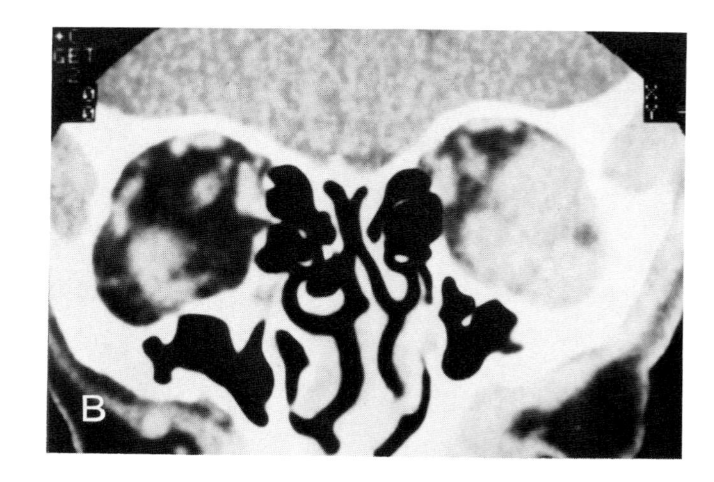

Plate 4.12*B*. Coronal scan shows that the involvement tends to spare the upper medial quadrant of the muscle cone. The muscles themselves are normal in this quadrant but are infiltrated in the remaining three quadrants of the orbit. The lacrimal fossa region is relatively spared. No bone involvement could be identified on extended window settings. This pattern of infiltration is not really specific for any disease process. Pseudotumor remains confined to the orbit so that one must think of either inflammatory disease or small cell abnormality such as lymphoma.

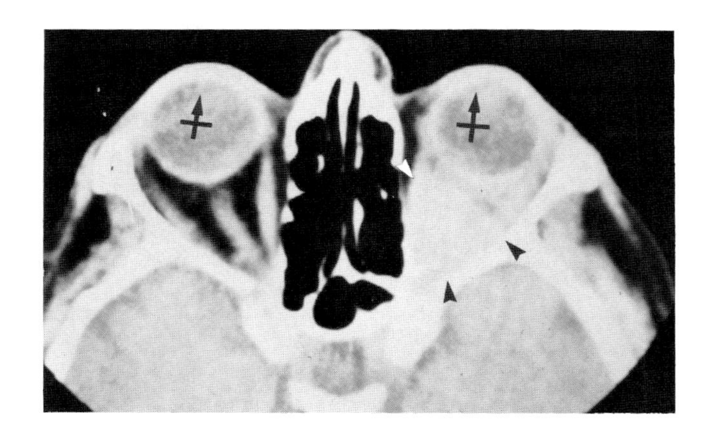

Plate 4.13. Metastatic carcinoma. Several interesting features of this diffuse infiltrative process of retroorbital process permit considerable specificity in histologic diagnosis. The enhancing infiltrative process is diffuse throughout the muscle cone region (*arrowheads*). The paranasal sinuses are clear and the process is confined by the muscle cone. Despite this extensive disease there is no discrepancy in the position of the two globes (*crossed arrows*). With a lack of inflammatory disease in the paranasal sinuses one can assume that this is a fibrotic retractive process which is only associated with a scirrhous type of carcinoma; hence a radiologic diagnosis of scirrhous carcinoma of the breast or *possibility* scirrhous carcinoma of the pancreas or GI tract could be suggested.

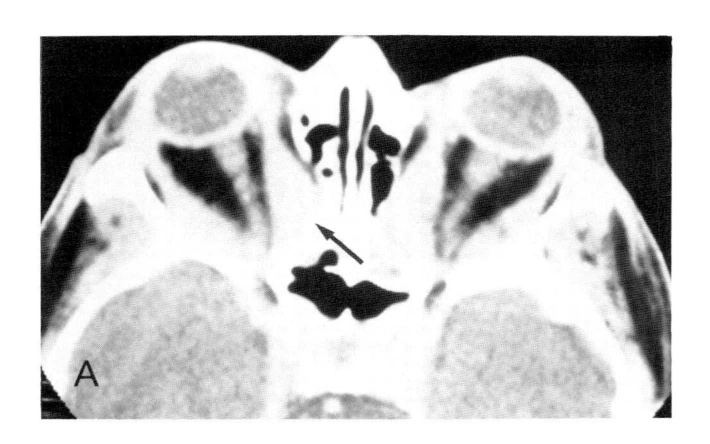

LESIONS OF THE EXTRA OCULAR MUSCLES

Plate 4.14. Endocrine *exophthalmous*.

Plate 4.14*A*. Massive enlargement of the medial rectus muscle is present bilaterally. The other extraocular muscles show enlargement involving the muscle bundles sparing of the tendonus insertions. The encroachment upon the optic nerve is maximal at the orbital apex and the muscle belly enlargement has produced deformity of the lamina papyracia bilaterally (*arrow*).

Intravenous contrast was given in this patient which demonstrates some scleral thickening anteriorly but the posterior margin of the globe and Tenon's capsule reflections do not show significant enhancement.

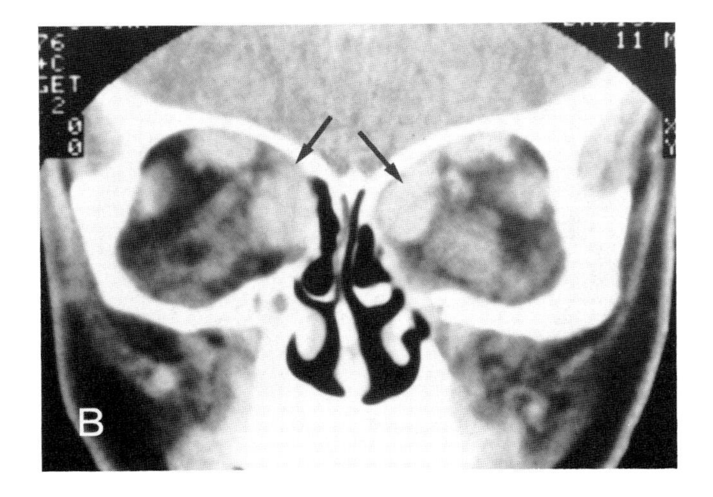

Plate 4.14*B*. Coronal scan demonstrates the massive amount of medial rectus thickening bilaterally (*arrows*). The remaining extraocular muscles are more involved on the right than on the left. The optic nerves are not encroached in the midportion of the orbit where this section was taken despite the massive degree of extraocular muscle enlargement.

This case demonstrates nicely why visual loss is due to optic nerve compression. Endocrine exophthalmous is a disease of the muscle bundles. The muscles occupy most of the orbit apex and hence this very narrow confined space is ideal for compressing the optic nerve.

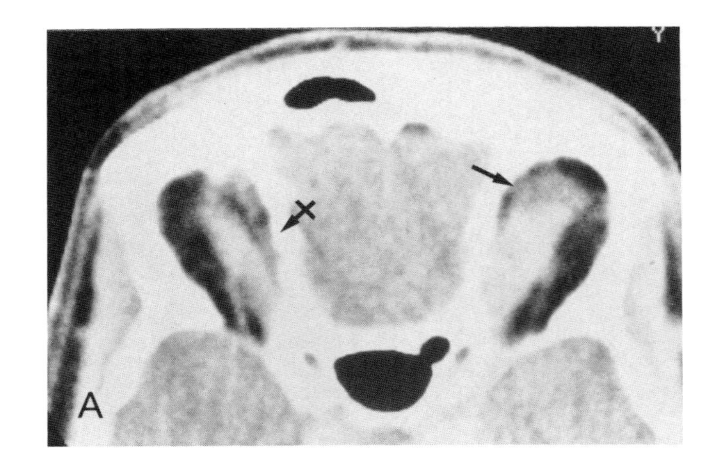

Plate 4.15. Granular cell myoblastoma.

Plate 4.15*A*. The; combined shadow of superior rectus and levator palpebral superiorus muscle show increase in density as well as bulk on the left (*arrow*) as compared to the normal muscle group on the right (*crossed arrows*).

Plate 4.15*B*. Coronal reformatted images confirm that the expansivity remains localized to the superior rectus muscle and levator palpebral superiorus on the left (*arrow*) as compared to a normal muscle grouping on the right (*crossed arrow*). All of the other extraocular muscles as visualized appear well within normal limits and there is no evidence of surrounding inflammatory disease.

Conceivably this type of muscle bundle enlargement could be due to myositis, metastatic tumor, vascular malformations, etc. As a matter of fact, the patient was followed for a period of 1 yr with a presumed diagnosis of myositis before exploration was performed and the diagnosis of granular cell myoblastoma was confirmed.

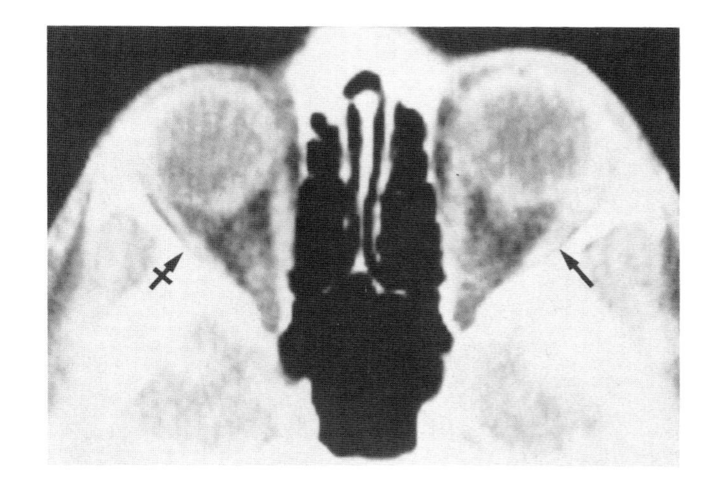

Plate 4.16. Myositis. Localized expansion of the lateral rectus muscle on the left is noted (*arrow*) as compared to a normal muscle bundle on the right (*crossed arrow*). The expansion of this muscle extends well beyond the muscular belly and into the tendonous insertion. There is a suggestion of some haziness to the posterior margin of the globe adjacent to the lateral rectus muscle tendon.

The fact that all of the other extraocular muscles were normal and there was no evidence of changes within the muscle cone fat placed this lesion in a differential diagnosis between myositis and pseudotumor. Since both of these diseases are probably on a autoimmune basis and represent spectra of the same entity it may well be that further differentiation is not warranted. Because this lesion seemed to be largely confined to the muscle and produced motility changes it was called myositis. The lesion responded promptly to steroids.

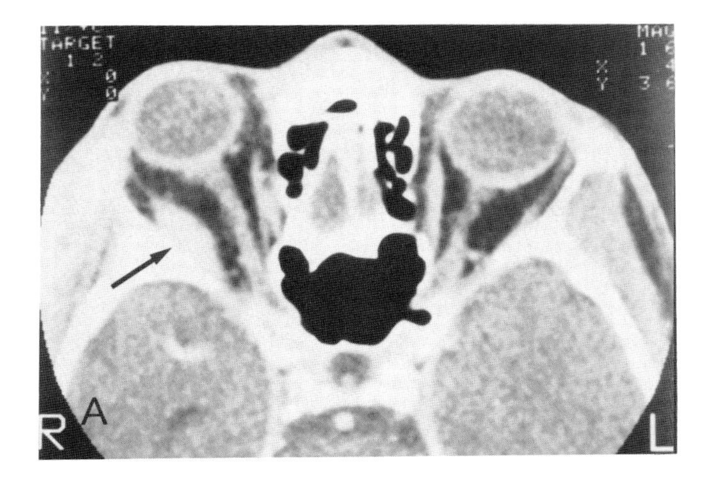

Plate 4.17. Lymphoma.

Plate 4.17*A*. The lateral rectus muscle is swollen (*arrow*) but the fat of the muscle cone region is well preserved. This localized disease could well be due to metastatic tumor, myositis, etc., but subsequent clinical course confirmed the diagnosis to be due to lymphoma infiltrations.

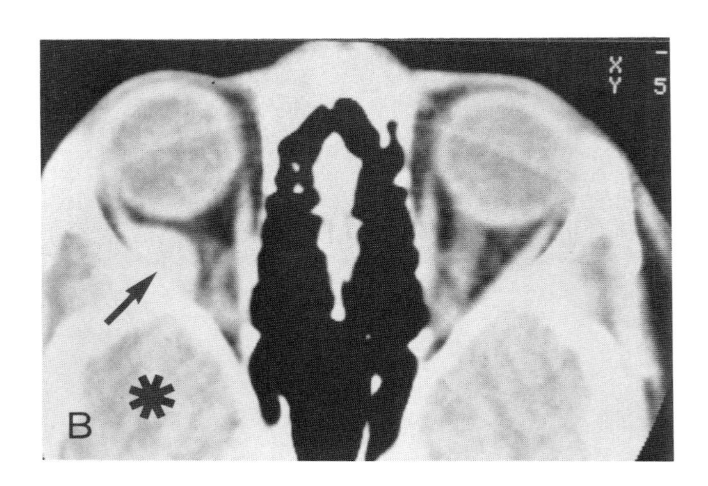

Plate 4.17*B*. Metastatic tumor. Scan with contrast (as denoted by *asterisk*) shows swelling of the lateral rectus muscle (*arrow*) almost identical in appearance to the patient shown in Plate 4.17*A*. There is a faint suggestion of some changes of the fat within the orbital apex. Again, isolated involvement of the lateral rectus muscle would be extremely unusual for endocrine exophthalmus and one must go through the differential of metastatic tumor, myositis, etc. The primary tumor was carcinoid.

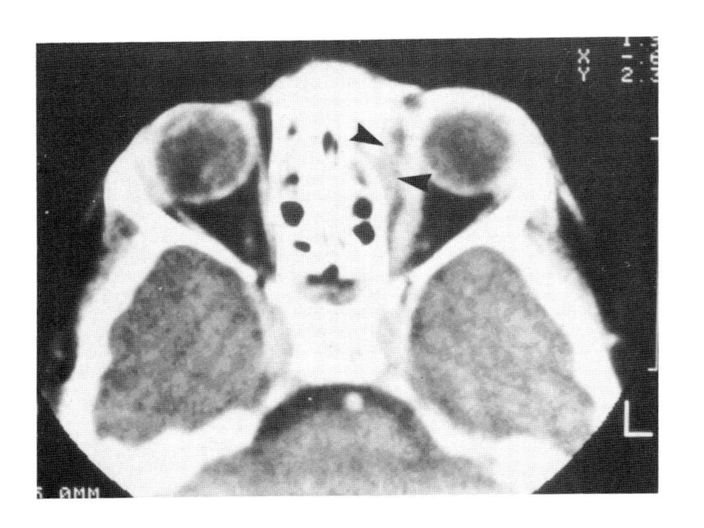

Plate 4.18. Inflammatory swelling of medial rectus. The medial rectus muscle is greatly enlarged and the reactive process extends anteriorly to involve the conjunctiva. The anterior and middle ethmoid air cells are opacified and the inflammatory process appears to follow the canal for the anterior ethmoid artery which is the defect in the medial wall of the orbit. A subperiosteal abscess is present within the medial wall of the orbit adjacent to the swollen muscle (*arrowheads*).

LESIONS OF THE LACRIMAL FOSSA

Plate 4.19. Mixed tumor lacrimal gland

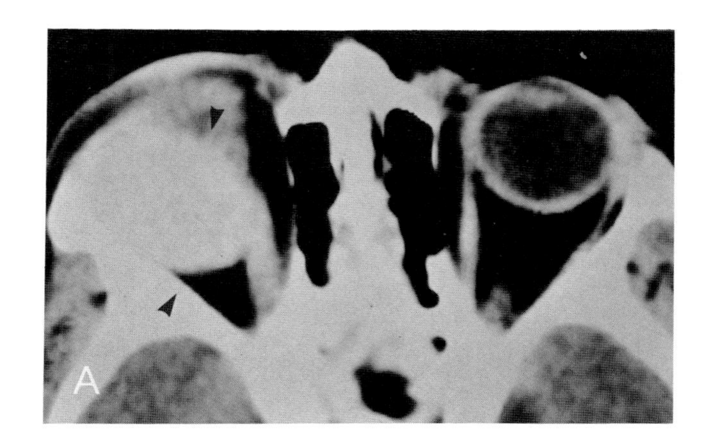

Plate 4.19*A*. A rounded mass is visible arising from the region of the lacrimal fossa (*arrowheads*). Other sections showed that the lacrimal fossa was eroded but no calcifications could be identified within the mass.

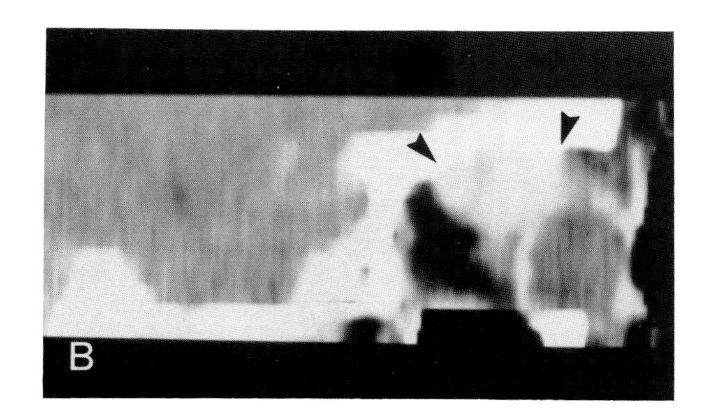

Plate 4.19*B*. Sagittal reconstructions show the rounded nature of the lesion arising from the lacrimal fossa (*arrowheads*). No infiltration into the muscle cone region or lateral wall of the orbit is present. The most likely diagnosis under these circumstances is a benign mixed tumor of the lacrimal gland. Small cell tumors and malignancies may remain this well confined. Pseudotumor also causes reactive changes in the tissues. The slow growing malignancies may remain in a rounded configuration for a considerable period of time so that one cannot be absolutely certain of the benign nature of the lesion from CT scanning.

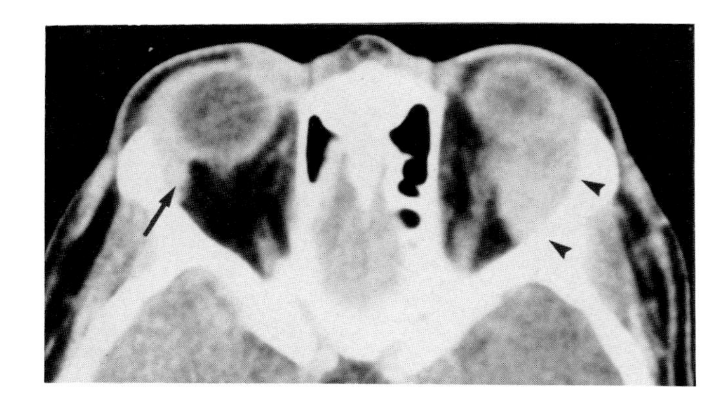

Plate 4.20. Lymphoma of orbit. An enhancing lesion is present on the left that arises from the region of the lacrimal fossa (*arrowheads*). Despite its rather extensive nature the margins remain fairly well confined. The opposite lacrimal gland on the right is also rather generous (*arrow*).

This bilateral process could be due to Sjögren's disease, pseudotumor, leukemia infiltrations, or possibly even sarcoid. The diagnosis requires total clinical evaluation of the patient and, in this case, cervical adenopathy proved to be the site of a definitive biopsy.

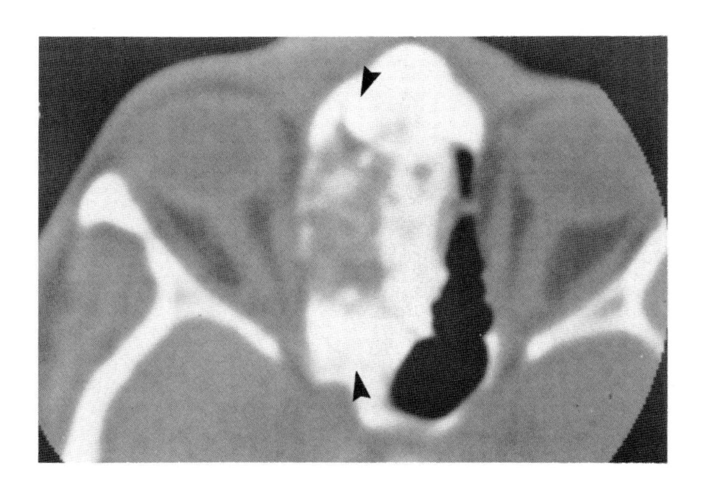

LESIONS OF THE BONY ORBIT

Plate 4.21. Fibrous dysplasia. Scan at bone window settings shows extensive involvement of the ethmoid regions extending back into the sphenoid sinus area (*arrowheads*). The extraocular muscles and optic nerves were normal on soft tissue settings. Note that the central portion of this process is of low density suggesting its fibrous nature.

There is a tendency for fibrous dysplasia to show the greatest amount of expansivity in the more immature fibrous elements. The more densely ossified areas are probably more mature and do not expand as much as the fibrous elements. This is an important differential to make in fibrous dysplasia when the process involves the region of the optic canal. One can suggest progressive disease if there is a lot of fibrous destructive process in this vital area.

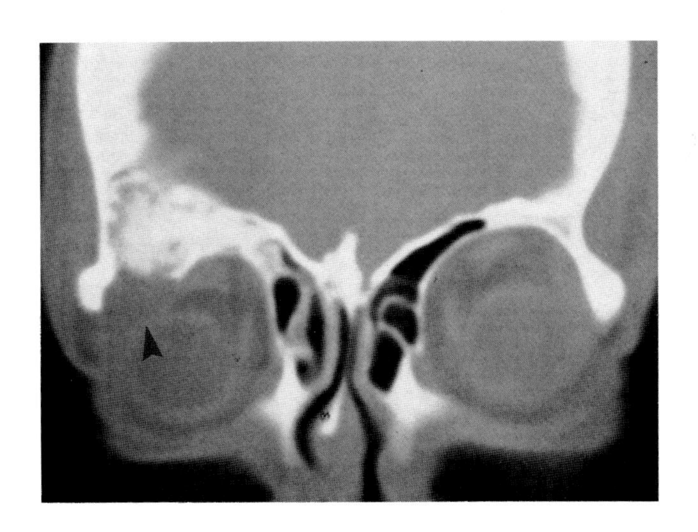

Plate 4.22. Fibrous osteoma. The fibrous osteoma expanded in the right frontal sinus and extended into the orbital roof. Like fibrous dysplasia, the more fibrous elements in this spectrum of disease show the greatest tendency for expansivity. Note that the globe is being displaced by noncalcified fibrous elements that have broken through the roof of the orbit on the right (*arrowhead*). The more dense portions of the fibrous osteoma are expanding within the sinus but not to the degree that the fibrous elements are expanding (8).

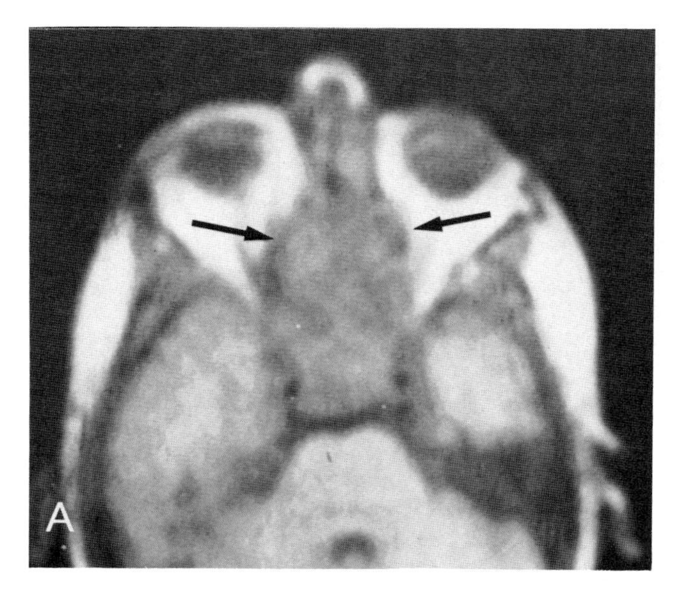

Plate 4.23. MRI of fibrous dysplasia.

Plate 4.23*A*. MRI scan shows the increased bulk to the posterior ethmoid air cells caused by fibrous dysplasia involving this region and extending into the sphenoid (*arrow*).

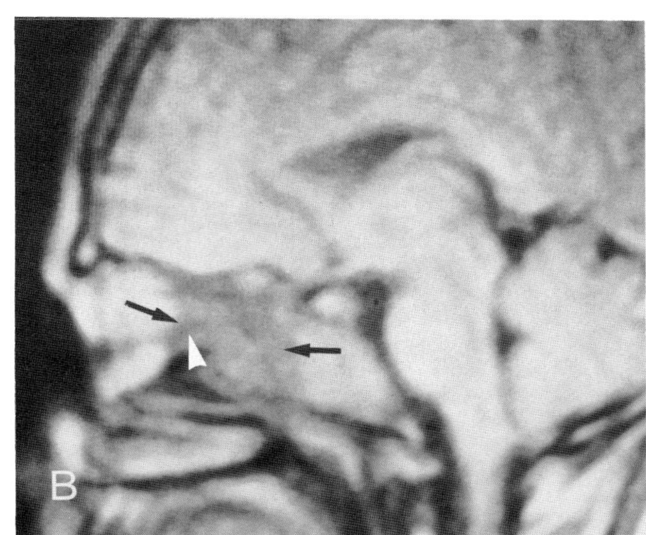

Plate 4.23*B*. Sagittal view shows the region of dysplastic bone (*arrows*) that extends cephalad to involve the planum sphenoidale. The signal is low by comparison to adjacent medullary bone but higher than cortical bone, apparently related to the fibrous tissue content. The optic nerve can be traced part of the way through the region of dysplastic bone (*arrowhead*). This particular patient had only minimal loss of visual acuity.

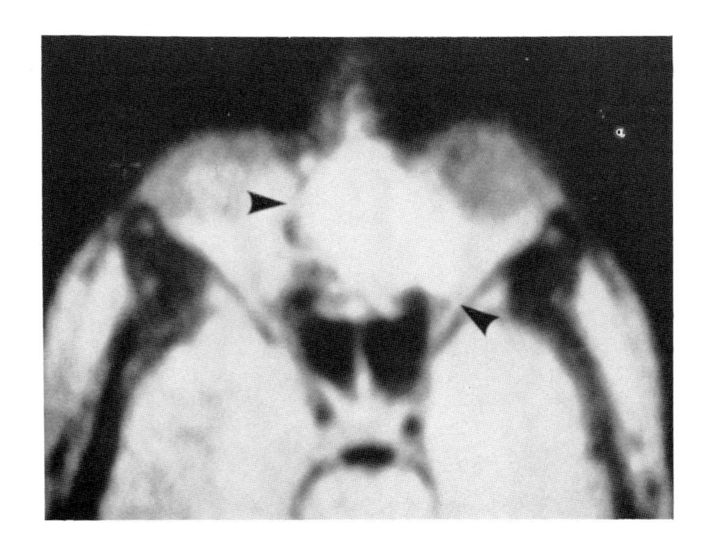

Plate 4.24. Esthesioneuroblastoma MRI. Axial scan shows bright signal arising from an esthesioneuroblastoma that has extended from the high nasal vault into the medial portions of the left orbit (*arrowheads*). There is a distinct signal difference between the fat of the orbit and the brighter signal that is coming from the esthesioneuroblastoma.

This signal difference is brought out be accentuating the T2 effects in spin echo pulsing sequences. The scan shown was delayed to a TE value of 110 msec. and a TR of 0.85 sec.

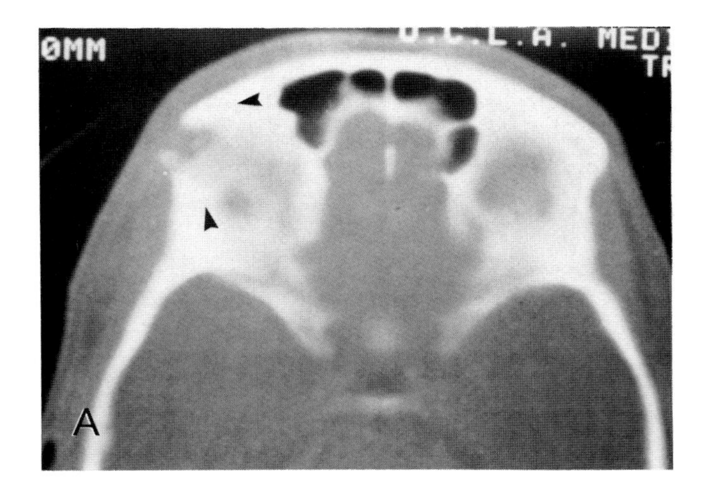

Plate 4.25. Eosinophilic granuloma.

Plate 4.25*A*. Axial scan at bone window settings show the defect in the roof and anterior margins of the right orbit. The lesion is almost completely destructive and has somewhat irregular margins (*arrowheads*). Two small calcifications are present within the lesion.

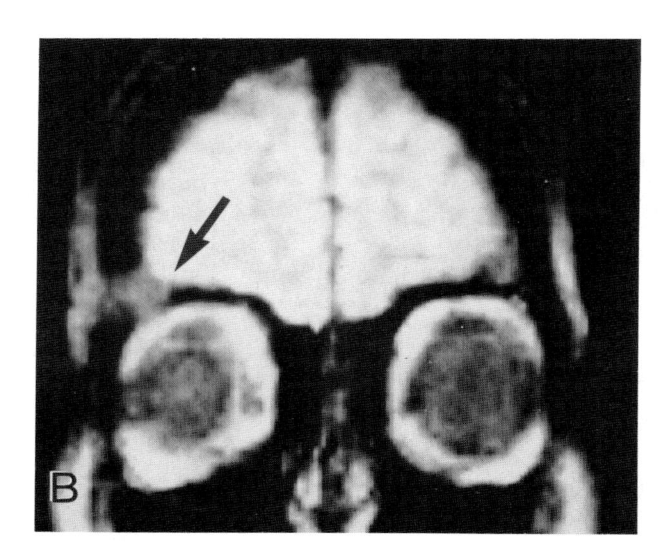

Plate 4.25*B*. Magnetic resonance imaging on this same patient in the coronal plane shows that an intermediate signal is present in the lesion (*arrow*). The lesion destroys the bone in the roof of the orbit and also extends into the base of the orbital process of the frontal bone. There is no evidence of extension of this tumor into the peripheral space of the orbit or intracranially.

By CT this could easily be a demoid, metastatic tumor, or any number of primary bone lesions. Biopsy was necessary to prove that this was an eosinophilic granuloma.

LESIONS OF THE GLOBE

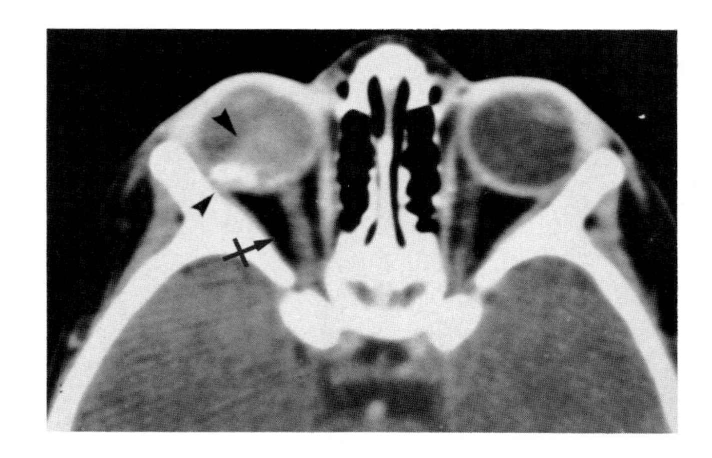

Plate 4.26. Retinoblastoma. This 14-month-old child with leukokoria (white pupillary reflex) shows a tumor mass in the posterior pole of the globe (*arrowheads*). The calcifications within this mass are quite dense as compared to the usual punctate type of calcifications seen in retinoblastoma. Other sections confirmed there was no extension into the optic nerve or outside the globe (*crossed arrow*).

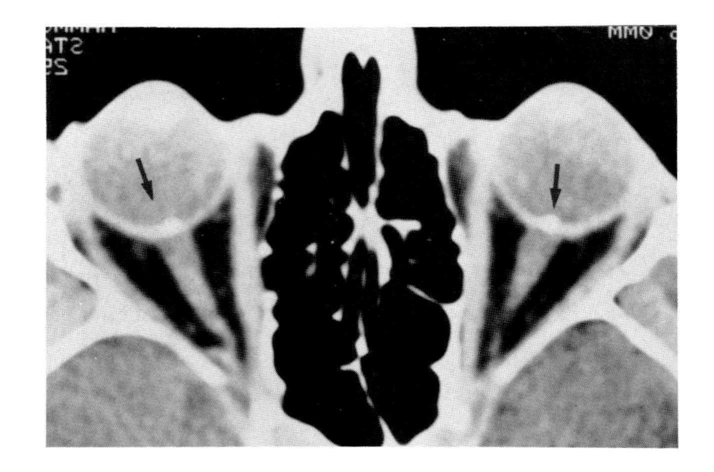

Plate 4.27. Drusen. Calcifications at the optic nerve disk (*arrows*) are characteristic of drusen. This benign condition is a degenerative process with hyaline deposits at the optic nerve heads. Slight diminution of visual acuity may be present. On fundascopic examination drusen closely resembles intracranial pressure and papilledema.

TRAUMA

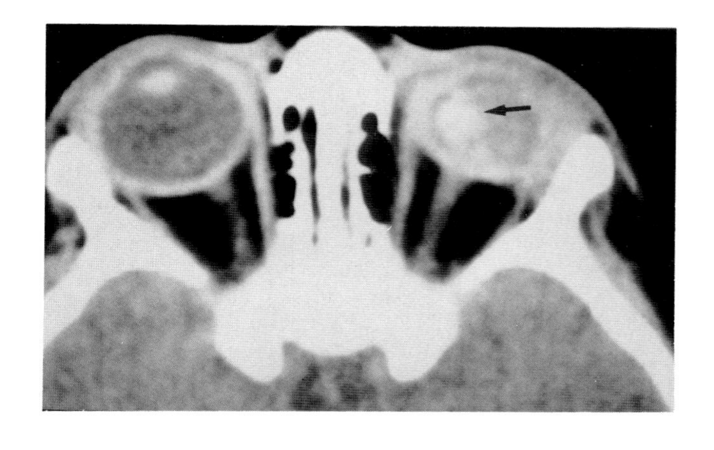

Plate 4.28. Calcified vitreous. Following trauma this globe has become shrunken and calcifications (*arrow*) are present within the vitreous secondary to old hemorrhage. Other sections showed that the lens was displaced inferiorly in the left eye.

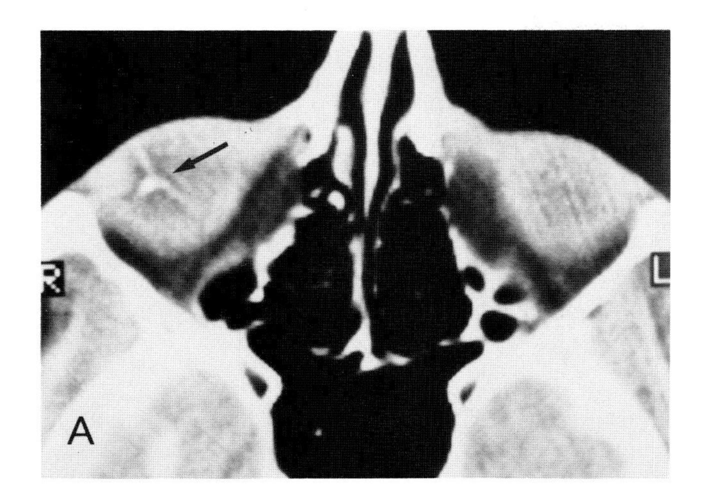

Plate 4.29 Foreign body.

Plate 4.29*A*. Axial sections show a foreign body (*arrow*) present on the right which appears to be at the level of the equator of the globe. Its vertical position cannot be clearly determined.

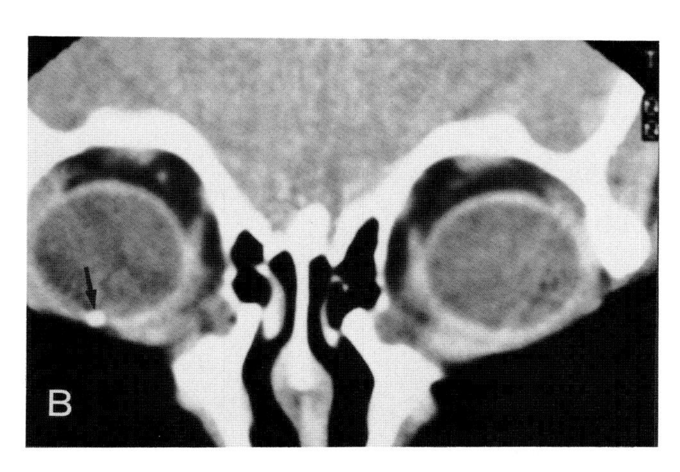

Plate 4.29*B*. Direct coronal scan confirms that the foreign body (*arrow*) is embedded in the sclera and is not floating free in the vitreous.

Chapter 5

Temporal Bone— Normal Anatomy, Pathology, Technique and Indication for Scanning

INTRODUCTION

Radiologic investigation of the temporal bone can be broken down into three main categories. Based on the clinical problems encountered they are: (*a*) inner ear studies for sensorineural hearing loss; (*b*) middle ear and mastoid air cells for conductive hearing loss; and (*c*) generalized studies for trauma, diffuse tumors, congenital lesions, and a variety of systemic diseases affecting the temporal bone (27). The degree of sophistication required from the radiologic studies will vary markedly depending on the basic underlying problem, philosophies of the referring physician and goals of patient management. For example, in a neurosurgical practice the patient presenting with a unilateral sensorineural hearing loss and corneal anesthesia would have to have a very large lesion in the pontocerebellar angle to account for involvement of nerves V and VIII. Transorbital views of the internal auditory canal would be diagnostic of an acoustic neuroma in a high percentage of cases. CT scanning of the temporal bone would be ordered to show the extent of the lesion for surgical planning and, in rare instances, for differential diagnosis. The otologic surgeon, on the other hand, usually encounters a patient with an acoustic neuroma at a much earlier stage. The plain film views of the internal auditory canal will be normal in 30 to 40% of the patients so that diagnosis will rely primarily on sophisticated CT or MRI techniques.

In this chapter emphasis will be placed primarily on computed tomography and to some extent MRI techniques. A brief introduction to conventional radiography together with some mention of pluridirectional tomography is mandatory in order to place the more modern techniques in perspective.

Technical advances are being made so rapidly in the field of CT scanning that any detailed text on methodology becomes outdated usually prior to printing (4, 16, 26, 29, 69, 73). Within this framework of limitations, we will try to present a combination of results using "off the shelf" hardware and some results obtainable with prototype models not available for general use.

Anatomy

The most commonly used CT display of the temporal bone is the axial projection. If one will make the effort to be thoroughly familiar with the anatomy as displayed in this projection, the vast majority of temporal bone problems can be completely evaluated. Multiple projections are possible by combining patient positioning with a tilting examining couch (65) but these unusual views are rarely needed.

When viewing the skull from above with the bony calvarium removed, one notices an X (Fig. 5.1). The upper limbs of the X are formed by the lateral walls of the orbit and the lower limbs of the X are formed by the posterior surface of the temporal bones. In the center of the X is the sella turcica (21). All of the structures of the **inner ear** orientate to the posterior

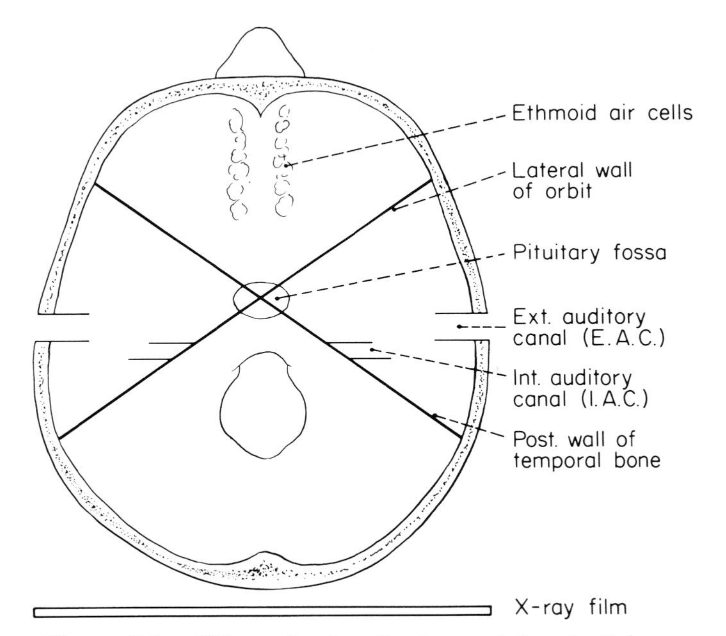

Figure 5.1. When viewing the base of the skull from above with the bony calvarium removed, an X is seen produced by the lateral walls of the orbit and posterior surfaces of the temporal bones. This X is very important for tensil strength of the base of the skull. All of the structures of the inner ear tend to align with the posterior surface of the temporal bone. The internal auditory canal and the external auditory canals would lie parallel to the plane of a piece of x-ray film placed behind the patient's head.

wall of the temporal bone or the lower limb of the X (Fig. 5.2). The axis of the cochlea with its 2¾ turns is perpendicular to the posterior wall of the temporal bone. The three semicircular canals are so arranged that the superior semicircular canal is perpendicular to the posterior wall, the posterior semicircular canal is parallel with the posterior wall, and the horizontal semicircular canal lies in the plane of the base of the skull. If one were to place a sheet of x-ray film behind the patient's head, the internal auditory canal would have its long axis parallel to the film and the external auditory canal would be approximately parallel to the film.

Continuing with this same concept, the **middle ear** structure and mastoid antrum all line up approximately on a 15° angle to the midsagittal plane. The combination of epitympanic recess, aditus ad antrum, and mastoid antrum also line up on this 15° angulation to the midsagittal plane (Fig. 5.3).

The facial nerve in its course is quite unique with the nerve arising from the lower pons to pass laterally and slightly forward in the pontocerebellar angle and enter the internal auditory canal (Fig. 5.4). Within the internal auditory canal the nerve lies in the anterior superior quadrant of the lumen. It exits through the lateral extremity of the canal to pass forward to the anterior genu which lies immediately above the cochlea. This is the labyrinthian portion of the facial nerve and is said to represent the narrowest portion of the bony canal. From the anterior genu the nerve courses posteriorly to dip beneath the horizontal semicircular canal and slightly above the oval window niche. This portion is called the horizontal portion (tympanic portion) of the facial nerve and it courses posteriorly also

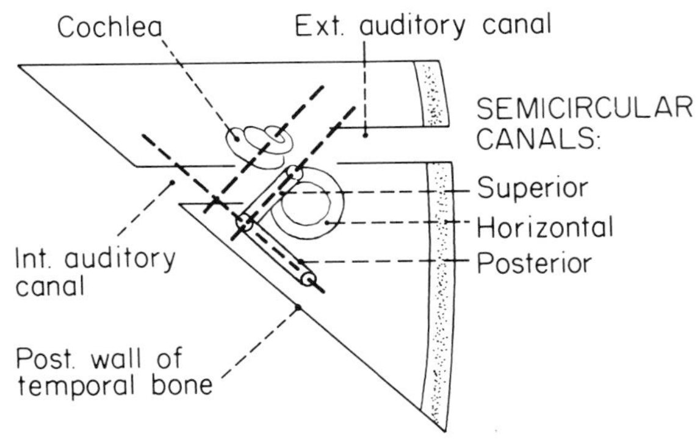

Figure 5.2. The temporal bone as viewed from above shows how simple the anatomy becomes if one keeps in mind that the inner structures take their alignment from the posterior surface of the temporal bone. The axis of the cochlea is perpendicular to the posterior wall. Of the three semicircular canals, the superior semicircular canal is also perpendicular to the posterior wall, while the posterior semicircular canal is parallel to the posterior wall. The horizontal semicircular canal lies in the plane of the paper.

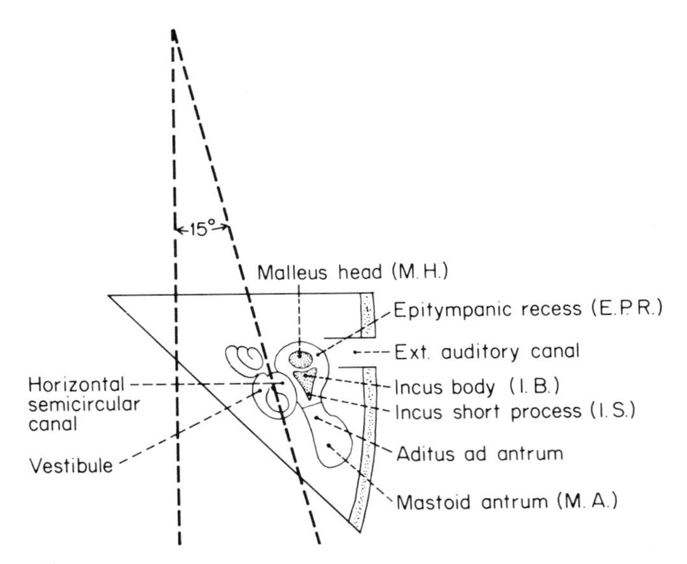

Figure 5.3. The middle ear structures align themselves on a plane that is 15° to the midsagittal plane. This relationship is easily appreciated on CT scanning in the base projection where one sees the epitympanic recess containing the head of the malleus and body of the incus lying immediately anterolateral to the horizontal semicircular canal. These two ossicles assume the configuration of an ice cream cone with the head of the malleus being the ice cream and the body of the incus being the cone. The short process of the incus is directed posterolaterally so that its tip lies in the aditus ad antrum. The aditus ad antrum communicates with the mastoid antrum so that all three chambers lie on the 15° angle plane. A distinct border is visible around the epitympanic recess but this margin is usually less distinct about the mastoid antrum. In the epitympanic recess, the ossicular chain should lie equidistant from all bony margins.

on an angle of approximately 15° to the midsagittal plane. At the level of the posterior limb of the horizontal semicircular canal, the facial nerve turns abruptly inferiorly and descends in a straight line to exit through the stylomastoid foramen. This terminal portion that is descending within the mastoid bone is called the descending portion of the facial nerve.

The remaining nerves in the internal auditory canal are the cochlea nerve lying in the anteroinferior quadrant of the lumen, and the superior and inferior vestibular nerves lying in their respective posterior quadrants of the internal auditory canal. The nervus intermedius accompanies the facial nerves in the anterosuperior quadrant and passes to the anterior genu and geniculate ganglion region.

The auditory artery, which may arise from the anteroinferior cerebellar artery, or as a separate branch of the basilar artery, enters the anterior portion of the pontocerebellar angle from in front of the pons and

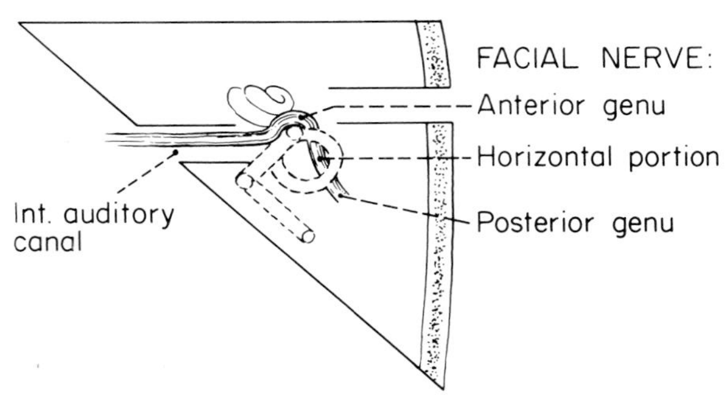

Figure 5.4. Facial nerve. The facial nerve lies in the anterior superior quadrant of the internal auditory canal. It exits from the internal auditory canal to pass above the cochlea where it forms an anterior genu within the geniculate ganglion. From the anterior genu, the nerve enters the tympanic cavity by passing beneath the horizontal semicircular canal and courses posterior. This portion is called the horizontal portion, or tympanic portion, of the facial nerve. As it reaches the posterior limb of the horizontal semicircular canal, it makes a bend interiorly which is called the posterior genu. The nerve then descends in the temporal (descending portion) to exit through the stylomastoid foramen.

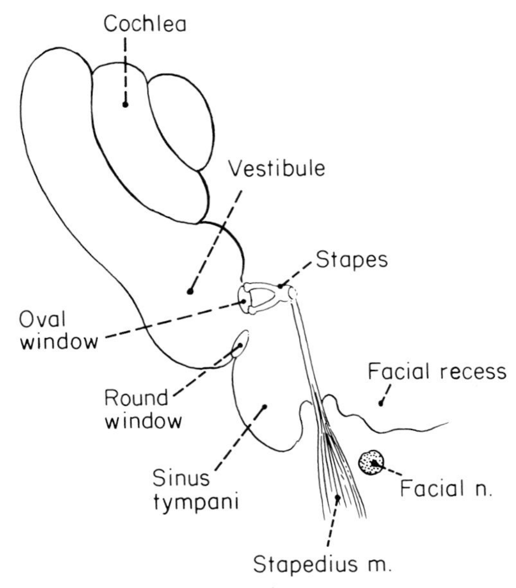

Figure 5.5. Sinus tympani. A very important recess lies in the posteroinferior medial aspect of the tympanic cavity and is called the tympanic sinus. This recess is hidden from view by the long process of the incus and stapes, and by the descending portion of the facial nerve. The margins of the round window form the anterior boundary of the sinus tympani, and the stapedius muscle and the facial nerve form its lateral boundary. The otic capsule lies medial and the extensions of the otic capsule which cover the posterior semicircular canal lie posterior. The sinus tympani is not to be confused with the facial recess which is an important surgical landmark, lying lateral to the descending portion of the facial nerve.

has a variable course within the internal auditory canal. The auditory artery has a characteristic loop that may be entirely within the pontocerebellar angle or within the internal auditory canal. The artery loops surround the nerves within the canal and, if arachnoidal reflections pass from the artery to the nerves, they may trap cerebrospinal fluid during air studies simulating an acoustic neuroma (34).

Sound is transmitted from the external canal through the ossicular chain to the oval window and fluid spaces of the inner ear. Amplification occurs because sound pressure is gathered over the broad area of the tympanic membrane and transferred to the small area of the footplate of the stapes. The sequence of sound transmission is: tympanic membrane to long process of malleus (manubrium), head of malleus, body of incus, long process of incus, stapes superstructure, footplate and, finally, to the inner ear. The long process of the incus lies posterior to the manubrium of the malleus. In the axial projection, their course is almost parallel. The long process of the incus ends as the poorly visualized lenticulaform process which articulates with the head of the stapes. As the name implies the stapes is "stirrup shaped" with the two crura forming an arch to insert in the footplate of the stapes. The crura measure between 0.1 and 0.3 mm in diameter and are extremely delicate. The footplate of the stapes is a biconcave disk witha a periphery of hyaline cartilage that sits in the oval window niche. Through the footplate, the sound is transmitted to the inner ear structures.

In the hypotympanum, along the posterior and inferior wall, lie some structures that are rarely visible in conventional radiographic techniques. This is the region of the sinus tympani (Fig. 5.5) (14). Its base is hidden from view through the external auditory canal because of the overlying tendon of the stapedius muscle and the stapes superstructure. The boundaries of the sinus tympani are as follows: medially and anteriorly lie the basal turn of the cochlea, otic capsule and round window. Posteriorly, are air cells that separate the sinus tympani from the sigmoid plate of the temporal bone. Laterally, a volcano-shaped mound of bone, the pyramidal eminence, combines with the canal for the descending portion of the facial nerve to form a boundary. This prominence contains the stapedius musle whose tendons pass directly anterior to insert into the neck of the stapes. At times, the bony separation between the facial nerve and stapedius muscle may be deficient. The inferior border of the sinus tympani is formed by a bony ridge (the subiculum) extending from the superior lip of the round window laterally to the facial canal (1).

A 1- to 3-mm plate of bone forms the roof (tegmen) of the middle ear. Thinning of the tegmen of the epitympanic recess and mastoid antrum is a common finding and, since 15% of normal patients have performation of this tegmen, it is difficult to be certain whether erosions are due to cholesteatoma or merely physiologic thinning. Nevertheless, in the clinical situation, absence of bony roof to the epitympanic recess and mastoid antrum is an ominous finding and may lead to brain abscess in the infected draining ear.

Anterior, in the middle ear cavity, the eustachian tube takes it origin from approximately the midportion rather than from the lower hypotympanum. Air cells of the petrous apex are alleged to drain directly into the eustachian tube as well as into the mastoid antrum.

Indications for Examination and Technique

The axial plane of scanning can be used in over 95% of the cases for studies of conductive hearing loss and for sensorineural hearing loss. The dividing line for the two types of hearing loss is the footplate of the stapes. Some alteration of approach is required when one is examining peripheral to the footplate of the stapes as opposed to central to the footplate of the stapes. The indications for study for conductive hearing loss involve all situations where detailed knowledge is not available for that portion of the anatomy between the footplate of the stapes and the tympanic membrane. The basic underlying pathology may be congenital malformations, trauma, chronic infection or tumors. In a similar fashion inner ear examinations cover not only the structures within the otic capsule but also the pathways for hearing and equilibrium in the central nervous system. As with other neuro CT scanning, intravenous contrast is essential for visualization of the cerebellar-pontine angle and brain. At the present time MRI examinations are showing considerable promise of examining the central nervous system as well as the contents of the internal auditory canal without the need of contrast agents. Unfortunately, MRI does not give details of bones, and especially the otic capsule, so that there will probably be a division of indications for study of MRI versus CT that divide at the lateral end of the internal auditory canal.

Our technique of scanning is based on the ability to achieve thin sections and examine either for soft tissues or bones using the same raw data. Details will be given for the GE 8800 and the reader with other equipment can modify the technique to suit his particular scanner. We scan at 1-mm intervals in the axial plane from the hypotympanum to the superior semicircular canal including both temporal bones in the field of scanning. 320 to 400 mas at 120 kV is used per section. Twelve adjacent sections will cover the field and permit adequate planar reformation when needed. Reformatted images are only used in one or two cases out of 100.

The scans are performed using conventional brain algorithms (not perspective bone scanning).

After viewing the first scan to see that the level is appropriate and assuring that both temporal bones are on the scan, the 12 sections are taken. The patient can leave the x-ray department and computer retargeting can take place as time is available.

Using "Re-View," scans are retargeted with bone algorithms so that each ear is viewed separately at 3.0 magnification. The data begin processing as computer time becomes available (usually over the next 45 min to 1 hr).

The scans are then photographed with each magnified temporal bone viewed with extended window (4000) and projected along the left and right hand margin of a 14 × 17 film. In the middle, the whole head is photographed at brain settings. By this combination of photography and image processing one can study the pontocerebellar angles, internal auditory canal and soft tissues of the middle ear at the same time as bony structures are being examined. One technique serves for both conductive and sensorineural hearing loss with the only difference being the addition of contrast for sensorineural hearing loss (Fig. 5.6).

If this initial technique fails to demonstrate pathology and there is a high suspect of acoustic neuroma, MRI or CT plus subarachnoid air can be used to study the contents of the internal auditory canal (4).

If MRI is used overlapping sections should be performed to get a section that is truly centered within the internal auditory canal, and sections should be taken in both the axial and coronal plane. The coronal display is superb for tracing the full course of the VIII nerve.

Air CT studies may be performed immediately following the conventional scan or the patient can be rescheduled at a future date after consultation with referring physicians. It has been our experience that patients who tend to have headaches or demonstrate significant sensitivity to symptoms are best examined on an inpatient basis for the postspinal headache which almost invariably occurs. Individuals who are more stoic and rarely have headaches do quite nicely with examination on an outpatient basis. It is surprising how alcoholics are seldom bothered by significant postspinal headache.

The spinal puncture and injection of air can either be performed with a patient prone or in the sitting position. We personally prefer the prone puncture using fluoroscopic control for the puncture and fluoroscopy during the actual injection of air. If the needle tip is kept in the center of view during fluoroscopy, air should be seen to bubble through the subarachnoid fluid and immediately flow away from the needle tip in a bubble. If air cannot be observed to bubble away from the needle tip and stays in the locale of the tip, in all probability there is a subdural injection and a second puncture at a different level is needed. Three to four cc of air or oxygen can be instilled into the lumbar subarachnoid space and the needle withdrawn. The patient can be left in the prone position for periods of 1 to 2 hr as needed waiting to get into the scanner

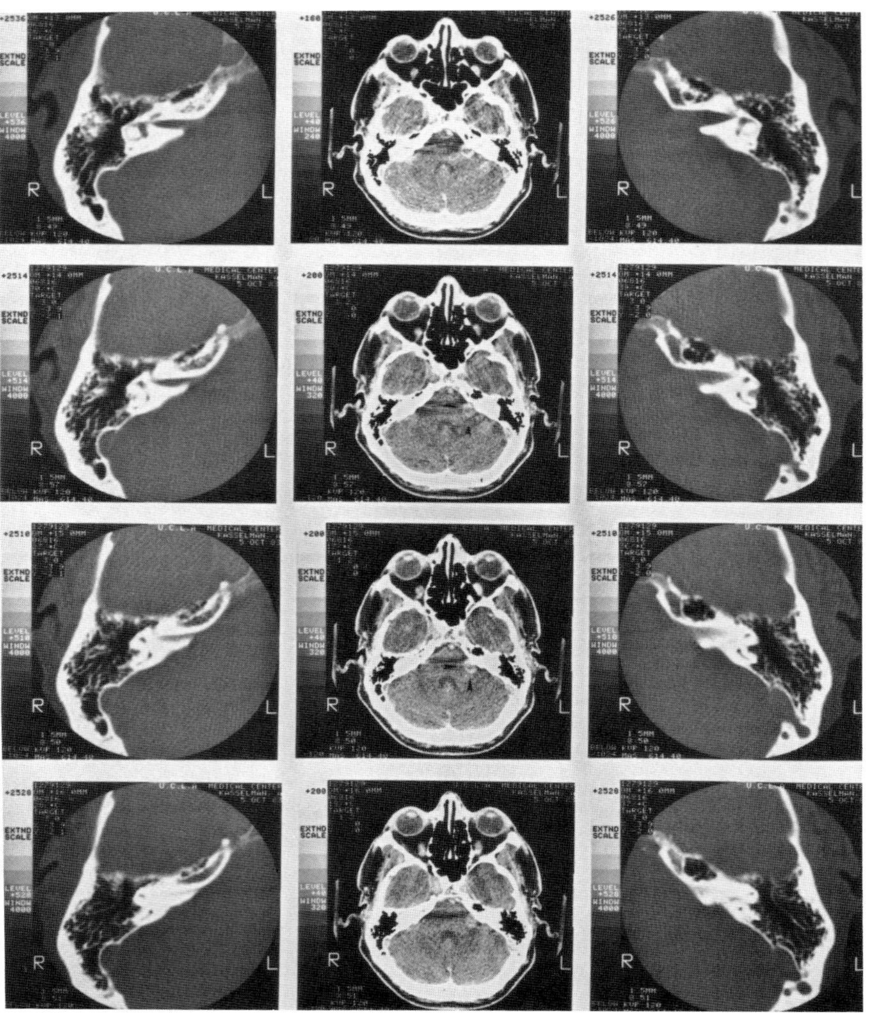

Figure 5.6. Photographic method. Using a 14 × 17 film format, 12 images are recorded on each film. The *center row* is photographed at brain settings while the images on either side represent the temporal bones photographed with bone window settings at expanded scales. Notice that in this patient with a small acoustic neuroma the tumor in the left pontocerebellar angle can be confirmed to be at the level of the internal auditory canal by comparing bone window settings with brain window settings. The radiologist giving the consult can confirm this type of finding using his viewing console but this is proof to the referring physician that such a relationship exists.

room. Once the scanning room becomes available, the patient is transferred to the examining couch in a prone position and then allowed to elevate his head and trunk with the suspect side highermost. The sagittal plane of the head is held almost parallel to the floor with only a slight tilt of the nose toward the floor in order to make the internal auditory canal assume a vertical position (Fig. 5.7). The patient will identify the rise of air from the lumbar region to the dorsal and cervical region by a slight local discomfort or pain. When he feels pain or pressure behind the ear the examiner can be assured that air has entered the pontocerebellar angle. The patient is returned into a lateral decubitus position and scans of the suspect internal auditory canal resumed. Scans should be taken at 1-mm intervals through the internal auditory canal and usually five sections will suffice. If air has not entered the canal the patient should be returned to the trunk elevated position with the suspect side again held high and the head lightly tapped or shaken in order to get the air to rise into the canal.

If the lumbar puncture is done in a sitting position the patient is instructed to tilt the head so that the suspect ear is uppermost and 3 to 4 cc of air injected. He is then placed in the lateral decubitus position and the needle is left in place in case air has not entered the pontocerebellar angle. Repositioning the air or reinjection can be performed as needed. In this sitting

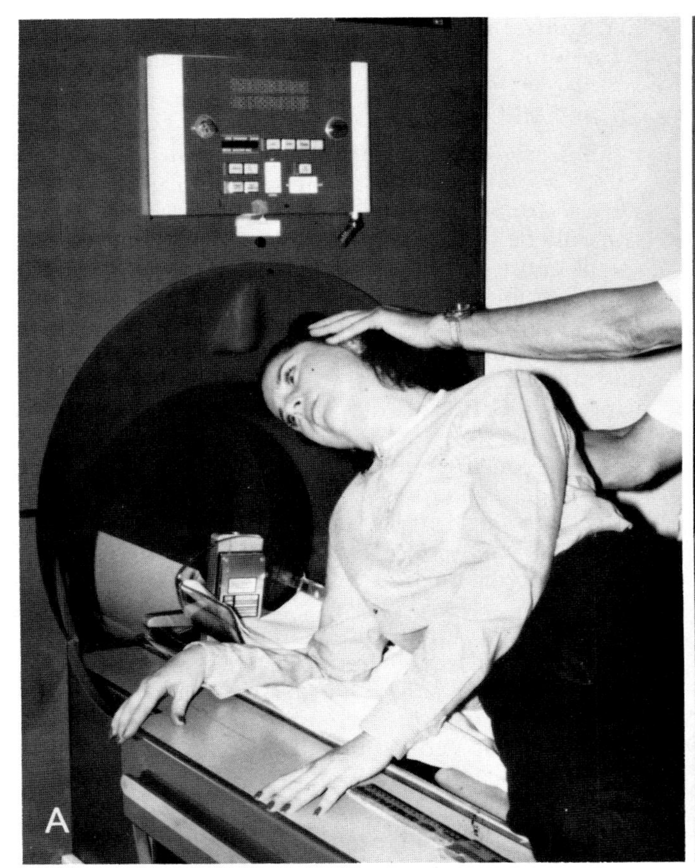

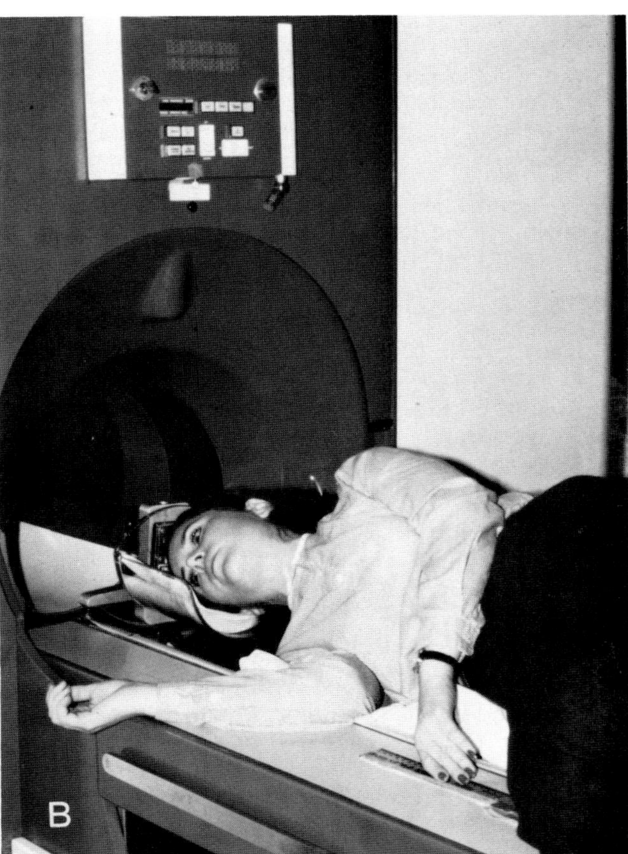

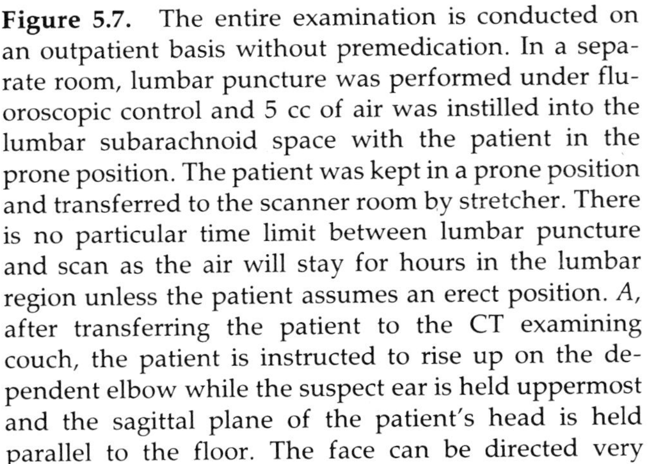
Figure 5.7. The entire examination is conducted on an outpatient basis without premedication. In a separate room, lumbar puncture was performed under fluoroscopic control and 5 cc of air was instilled into the lumbar subarachnoid space with the patient in the prone position. The patient was kept in a prone position and transferred to the scanner room by stretcher. There is no particular time limit between lumbar puncture and scan as the air will stay for hours in the lumbar region unless the patient assumes an erect position. *A*, after transferring the patient to the CT examining couch, the patient is instructed to rise up on the dependent elbow while the suspect ear is held uppermost and the sagittal plane of the patient's head is held parallel to the floor. The face can be directed very slightly toward the floor to aid in directing the air into the suspect pontocerebellar angle. Usually within 2 to 3 min, the patient will begin to experience a sensation of ache or discomfort, first in the cervical region and then in the occipital region of the suspect side. *B*, once the air has risen to the pontocerebellar angle the patient is placed in a lateral recumbant position with the chin extended. The initial scan is taken at a level 1 cm above the external auditory canal. Subsequent scans are taken at 1-mm intervals through the internal auditory canal. After completion of the suspect ear, the patient is rotated to the normal-ear-uppermost position by having the chin well extended and the body remaining recumbant.

position slightly more air will pass up over the convexity and additional air is frequently needed by this technique, as opposed to the prone position which is associated with a 98% success rate.

PATHOLOGY

A discussion of the pathology of the temporal bone is justifiably brief in order to concentrate on those topics that directly relate to CT and MRI scanning. A partial outline might be as follows:

- Congenital deformities
  - External canal
  - Middle ear
  - Inner ear
- Inflammatory disease
  - Type A reversible
    - Serous otitis
    - Glue ear
    - Chronic otomastoiditis with or without tympanic perforation
    - Malignant otitis externa
  - Type B irreversible without surgery

Tympanosclerosis
Retraction pocket with or without osteitis
Cholesteoma with or without polyps
Cholesteoma with complications
Trauma
Fracture
CSF fistula
Brain herniation
Tumors
Acoustic neuromas and inner ear tumors
Acoustic neuromas
Glomus jugulare tumors
Meningiomas
Facial nerve tumors
Middle ear tumors
External ear tumors
Systemic disease
Otosclerosis
Fibrous dysplasia
Paget's disease
Ménière's disease

CONGENITAL DEFORMITIES

External Ear

External canal obstructions can be either partial or complete and the obstructions may be membranous or bony. Once the external canal is obstructed, detailed knowledge of the underlying middle ear and inner ear becomes vital information for patient management. The medial end of the external canal may be separated from middle ear structures by a thick bony plate or a membranous soft tissue obstruction that is derived from remnants of the first and second embryologic branchial arches. Since the same two arches that form the external canal also form the ossicular chain, the dysplastic manifestation of canal development is usually extended to the ossicular chain. One might think of primoridal cartilage that normally would be used to build an ossicular chain being diverted for the bony plate of external canal atresia. Portions of the malleus, or the entire malleus and incus, may be incorporated into the bony plate. The ossicles may be fused to each other but separate from the bony plate. As a rule, the thicker the bony plate and ossicular deformity, the more atretic the middle ear cleft. The amount of middle ear cleft remaining also has a profound effect on whether conductive hearing can be reestablished. The width of the cavity and its relationship to the overlying temporal lobe can be evaluated by axial or coronal CT scanning.

The facial nerve in the normal patient descends posterior to the posterior external canal wall. With the lack of development of the external canal, abnormal positions of the facial nerve are quite common. The facial nerve may descend in an anterior direction from the tympanic cavity or may pass directly lateral from the anterior genu. Branching of the facial nerve may occur within the temporal bone. Accurate plotting of the course of the facial nerve requires high precision CT scanning with reconstruction capabilities.

CT and MRI Findings

CT study in the axial plane is the examination of choice for visualization of the external canal and middle ear structures. By taking sections at 1-mm intervals it should be possible to follow the facial canal throughout its entire length from the internal auditory canal to its exit from the stylomastoid foramen. The relationships of the facial nerve to the oval window and promintory are extremely important and CT may demonstrate the facial nerve canal lying within the oval window niche. The descending portion may begin shortly after the anterior genu. In this particular instance, reconstructed images for lateral projection and coronal projection may be of help in absolutely identifying the facial nerve after it leaves the middle ear cavity. Normally the descending portion lies immediately posterior to the pyramidal eminence and, more distally, the nerve is posterior to the thick cortex of the external canal. When there is failure of development of the external canal the facial nerve may align in a more anterior position. On axial CT the facial canal has a thicker wall than adjacent mastoid air cells. Its position can be verified by the reconstructed images or direct lateral CT if possible.

Middle Ear Deformities

OSSICLES

The ossicles may be fused or incompletely formed in external canal atresia. Abnormal articulations with the oval window are not at all uncommon and there may be abnormal thickness to the foot of the stapes lying within the oval window. Thickened footplate of the stapes is an important CT finding because there is frequently an associated abnormal communication between the perilymph of the inner ear with the subarachnoid space. If a surgeon operates for conductive hearing loss and removes the thickened footplate, a "gush" of fluid may start that can only be controlled by packing the middle ear with muscle and permanently destroying hearing in that ear (7, 21).

CSF FISTULA

Cerebrospinal fluid fistula in congenital abnormalities may occur through the footplate of the stapes or around the round window region. These openings are usually too small to be detected by any noninvasive radiologic means. On CT, one may see fluid in the middle ear cavity but this fluid cannot be differentiated from serous otitis. Metrizamide subarachnoid studies and CT scanning may confirm a communication to the middle ear cavity if meticulous technique is used but the exact site of the fistula is rarely demonstrated.

relates to the three layers that form the tympanic membrane; an outer squamous layer, a middle fibrous layer and an inner respiratory epithelial layer. Monomers form when the supporting collagenous fibers of Scarpa's membrane become destroyed and the tympanic membrane heals with only two layers of epithelium repairing the defect. If there is associated eustachian tube dysfunction and a negative pressure forms within the middle ear cavity, the epithelial layers cannot resist the negative pressure and are "sucked in" to the middle ear forming "retraction pockets." The inner lining of this retraction pocket continues to form new layers of cells. The collection of desquamated cells and epithelial lining becomes the cholesteatoma. Cholesteatomas have the ability to erode bone that is beyond the matter of mere pressure erosion and is probably related to enzyme content.

Since the most common site of perforation is in the pars flaccida (the posterosuperior quadrant of the tympanic membrane) one tends to find cholesteatomas associated with this region. They may extend into the mastoid antrum or into that portion of the middle ear cavity that lies lateral to the ossicular chain. The laminated collection of cellular debris of the cholesteatoma expands and produces erosion of the lateral wall of the epitympanic recess and of the margin of the posterior superior canal wall near the attachment of the tympanic membrane. This boney margin is known as the scutum. The expanding cholesteatomatous mass will conform to the bony margins of the cavity in which it lies to some extent. On the other hand the cholesteatoma acts like a rounded "keratoma" producing rounded indentation erosions of adjacent structures. The body of the incus and head of the malleus are quickly destroyed by cholesteatomas lying in the lateral portion of the epitympanic recess. After "pressure" erosions in the epitympanic recess they widen the aditus ad antrum and extend into the mastoid antrum. Lesions that gain access to the medial portions of the epitympanic recess and middle ear erode the bony covering of the facial canal and anterior portions of the lateral semicircular canal. Large lesions may even follow the facial nerve and anterior air cells to jump over the otic capsule and come to lie in the medial portions of the temporal bone.

Central perforations that give rise to cholesteatomas are frequently manifest by cholesteatomatous debris lying adjacent to the promintory and in the posterior hypotympanum. The lesions tend to displace the ossicular chain laterally and insinuate themselves around the stapes superstructure and oval window niche. Cholesteatomas that involve the region posterior to the round window are said to lie in the sinus tympani (Fig. 5.5) and present a particularly difficult problem in management. These cholesteatomatous materials cannot be removed without removal of the ossicular chain or by approaching the lesion from a retrofacial nerve approach. If cholesteatomatous deposits in the sinus tympani are overlooked or incompletely removed, then postoperation recurrence is inevitable.

As the cholesteatomas expand within the middle ear cavity they destroy the normal mucosal lining and occlude the opening of the eustachian tube. The cholesteatomatous masses interfere with aeration and drainage of the middle ear cavity and mastoid antrum accentuating the chronic inflammatory disease. Edematous changes take place within the epithelium of the external auditory canal and purulent drainage prevents healing of tympanic membrane perforations. Inflammatory polyps form from mucosa of the medial end of the external canal and are almost diagnostic of underlying cholesteatoma even when epithelial debris cannot be identified on otoscopic examination of the ear.

CONGENITAL CHOLESTEATOMA

Congenital cholesteatomas may well be related to epithelial rests that occur along the tract of migration of the primitive otocyst. In the embryo when the otocyst forms from ectoderm it migrates centrally to create the inner ear structures. Cells may be left behind in the mesenchyme giving rise to isolated epithelial elements. They usually become manifest in childhood but some do not make their appearance until the 2nd and 3rd decade. The lesions that involve the pontocerebellar angle are more prone to manifest late. In the young patient congenital cholesteatomas are particularly destructive in the middle ear cavity. They frequently invade over the top of the otic capsule to involve the petrous apex. At times congenital cholesteatomas may be limited to the petrous apex or even the external auditory canal.

CT and MRI Findings

Despite numerous attempts, no specific numbers have been found that positively identify cholesteatomatous material by either CT or MRI. On CT the density of these lesions is the same as granulation tissue or other inflammatory changes. With MRI the cholesteatomas create a bright spin echo image similar to high protein fluid or granulation tissue. The absence of bone shadows with MRI is so limiting that there seems to be little place at the present time for MRI in conductive hearing loss.

The CT diagnosis of cholesteatoma is based primarily on the following (68):

1. Ossicular erosion
2. Erosion of the lateral wall of the epitympanic recess
3. Erosions of the facial canal or horizontal semicircular canal
4. The presence of rounded soft tissue masses in the middle ear cavity or epitympanic recess
5. Identification of inflammatory polyps in the medial portion of the external auditory canal together with thickened tympanic membrane and chronic middle ear disease

All of the above findings are usually looked for in

the clinical setting of a draining ear with tympanic membrane perforation. When a normal tympanic membrane is encountered by the clinician and there is no history of inflammatory disease, the above listed CT changes are assumed to be on the basis of a congenital cholesteatoma. A congenital cholesteatoma may also be present in the petrous apex with or without continuity with the middle ear and mastoid antrum. The cholesteatomas and cholesterol granulomas in the petrous apex will produce an expansion with erosion of the posterior cortical surface of the temporal bone (15, 44). Cholesterol granulomas usually have a previous history of inflammatory disease but, from a practical point of view, they behave in a similar fashion and differentiation is a microscopic diagnosis. They may be associated with low density soft tissue masses in the cerebellopontine angle. Purely intracranial congenital cholesteatomas (epidermoid tumors) are frequently found in the cerebellopontine angle and have a lobulated border with the central portions of the tumor being of low density on CT. Tumors that are isodense with surrounding brain have been described. Usually the congenital cholesteatomas do not stain with intravenous contrast. A further description of this lesion will be given under pontocerebellar angle tumors.

Since the CT numbers of cholesteatomas in the middle ear cavity or mastoid antrum are nonspecific, a rather long differential diagnosis must be considered when comparing mass lesions in this region. Fluid in the middle ear associated with inflammatory disease may completely mask the presence of an underlying cholesteatoma. Malignant tumors or metastatic disease produces bone destruction of a more diffuse nature but early in the course of the illness may closely simulate local erosive changes of cholesteatomas. Malignant tumors tend to spread along mucosal surfaces and involve the external canal much earlier when compared to cholesteatomas. The metastatic tumors are more likely to involve periantral air cells as the primary site of deposit and produce destruction radiating from the point of origin.

Primary tumors of the middle ear clefts such as glomus tympanicum, ectopic jugular veins, and facial nerve tumors are usually identified clinically as a separate entity by the lack of chronic drainage.

OPERATIVE MANAGEMENT AND COMPLICATIONS OF CHOLESTEATOMA

The presence of cholesteatoma indicates irreversible disease and the need for surgical intervention. All aspects of surgical management of cholesteatoma are beyond the scope of this book. Only those facts that directly relate to CT evaluation will be discussed.

Some controversy exists whether cholesteatomas should be removed via radical mastoidectomy and leaving an open cavity, or whether a primary attempt at preserving conductive hearing should be made using a closed canal approach with primary tympanoplasty and limited mastoidectomy. The more radical approach insures a safe ear that can be easily inspected but sacrifices conductive hearing. The CT scanning prior to radical procedures is primarily to identify the facial nerve and its course. The presence of erosive changes along the facial nerve or erosions into the semicircular canal system may cause the surgeon to alter his approach and leave some cholesteatomatous material behind in order to lessen the chance of operative injury (47). The presence of disease lying throughout the sinus tympani or extending medial to the otic capsule also require altered approaches.

The same type of information is of value when planning the closed canal tympanoplasty but even more detailed information is desirous. At surgery the tympanic membrane is raised together with skin from the external canal and retained as a flap. The posterior and superior canal wall may be removed as necessary to gain surgical access and remove disease. Cartilage grafts can be used to restore structural support to the repaired tympanic membrane. By restoring the tympanic membrane, air conduction is preserved and the patient may lead a much more normal life with regard to exposure water in the external canal and freedom from drainage. A procedure that has been recommended is the "second look" operation after the initial surgery to manage any recurrent or residual disease. The incidence of recurrent disease is variously reported from 12 to 25%.

With present surgical techniques it is possible to perform complete removal of cholesteatoma that is located in the epitympanic recess. The ossicles are frequently diseased or destroyed by this type of cholesteatoma. The osteitis of ossicles involves primarily the body of the incus, long process of the incus and incudomalleal joint regions. Almost invariably the crura of the stapes are preserved with epitympanic recess cholesteatomas. Hearing can be restored very nicely by applying the restored tympanic membrane directly to the head and crura of the stapes or by using prosthetic materials or refashioned ossicles from the tympanic membrane to the footplate of the stapes.

When the cholesteatoma is secondary to a pars tensa perforation it is more likely to involve the middle ear cavity in a rather diffuse fashion and extend into the sinus tympani. The middle ear cavity cholesteatomas involve the long process of the incus and crura of the stapes to a much greater extent than the body of the incus or incudomalleal joint. If the crura of the stapes are destroyed and the cholesteatoma has involved the sinus tympani, much more radical management of the patient is required in order to prevent a very high recurrence rate or persistence of disease.

Quite naturally a surgeon is reluctant to perform a second surgical procedure when three-quarters of the examinations are going to be normal and the patients are clinically doing very well. The improved diagnostic capabilities of high resolution CT are ideally suited to excluding the latter group. In a recent review, Johnson et al. (32) reported very encouraging results among a group of 24 patients having a second look operation (Table 5.1). The true normal rate was 100% with regard to absence of cholesteatomas prior to the second look surgeries. When a well defined rounded mass could be seen on CT, the radiologic diagnosis was cholesteatoma and, again, these findings were confirmed at surgery. Diffuse soft tissue masses present the same problems in the postoperative patient as they do in the unoperated temporal bone and could not be differentiated. The only clue to the presence of a residual or recurrent cholesteatoma is the finding of new bone erosions or ossicular dislocations and destructions that were not present on previous studies. When new erosive changes were identified, the radiologic diagnosis of cholesteatoma was made and was invariably correct. When no erosions were present the accuracy of CT scanning was severely limited. Many patients who had postsurgical prosthetic materials or residual inflammatory disease could not be differentiated from true cholesteatomatous recurrences.

The clinical presentation of progressive hearing loss in the postoperative ear may be caused by lateralized grafts and ossicular fixation; both of which may simulate recurrent cholesteatoma. A lateralized graft occurs when the otologic surgeon repairs a tympanic membrane perforation by placing a graft on the external surface of the tympanic membrane. If for any reason the two layers of epithelium do not fuse completely and a space forms between the graft epithelium and the normal external epithelial air of the tympanic membrane, scarring with retraction will occur over a period of years in the postoperative period. The two layers will tend to separate around the periphery especially along the anterior margin of the tympanic membrane. No air will be found between the two layers but the overall thickness will be in the range of 1 to 2 mm. This separation will impede sound conduction and lead to progressive hearing loss. This thickened tympanic membrane will be indistinguishable from tympanosclerosis or other inflammatory disease unless one does a complete scan and demonstrates there is no evidence of residual or recurrent disease in the middle ear cavity or mastoid air cells. With a normal middle ear cleft and a thickened margin of the tympanic membrane one can safely make a CT diagnosis of lateralized graft.

Displacement of prosthetic ossicles or fixation of otogenous grafted ossicles to the margins of the middle ear cavity can be visualized on the axial scans. A close cooperation is necessary between the radiologist and the otologic surgeon so that he may know precisely what type of prosthetic material was used in order to verify whether displacement has occurred. Sometimes Teflon or plastic materials are used that are not radiopaque and their position cannot be verified. Occasionally there is sufficient air in the middle ear cavity to provide contrast and displaced prostheses can be identified as a "soft tissue" mass.

TABLE 5.1[a]
CT Findings in 21 Patients Who Had Repeat Surgery

CT appearance	No. of patients	Interpretation	Agreement with operative findings
			%
No significant soft-tissue mass	6	No recurrence	100
Mass with well defined margins	5	Cholesteatoma	100
Diffuse soft-tissue masses[b]			
No erosion	8	Cholesteatoma	25[c]
Erosion of the horizontal semicircular canal or facial canal	2	Cholesteatoma	100
Total	21		

[a] From Johnson DW, et al: Cholesteatomas of the temporal bone: role of CT. *Radiology* 148:733–737 1983.
[b] Scarring, fluid, surgically inserted material, etc.
[c] Two patients had progressive ossicular destruction, permitting an accurate diagnosis but cholesteatoma could not be differentiated from granulation tissue or scarring.

Brain Abscess

Brain abscesses secondary to inflammatory disease of the temporal bone are usually due to *Staphylococcus aureus*. The inflammatory process gains access to the brain either through defects in the dura or it starts as a phlebitis and follows the emissary veins. The temporal lobe is affected almost twice as frequently as the cerebellum (52, 70). This predilection for the temporal lobe may well be related to the physiologic thinning or even dehiscence of the tegmen that may take place in 15% of normal individuals.

The etiology as well as the method of diagnosis of brain abscesses has changed rather profoundly especially with the advent of CT scanning (54, 66). Only about 5% of all brain abscesses are related to the mastoid and middle ear cavity (58). At times the etiology of the abscess may be obscure. In the absence of evidence of acute mastoiditis or active cholesteatoma, one should make a careful search for an unsuspected congenital malformation that provides the pathway for inflammatory disease to pass from the middle ear cavity to the intracranial structures. Communications have been reported passing through the round window, the footplate of the stapes, or even accompanying the facial nerve to gain access to the internal auditory canal. Children with severely deformed cochleas of the Mondini type malformation are much more apt to have disruptions of the membranes of the inner ear and of communications between the middle ear cavity and subarachnoid space.

CT and MRI Findings

CT will demonstrate the abscess either in the temporal lobe or cerebellum as a low density center with ring enhancement with varying amounts of staining of adjacent brain. As the abscess begins to be controlled by body defenses the ring enhancement becomes more sharply defined and edema in adjacent brain diminishes. MRI is less able to distinguish between edema of the surrounding brain and the abscess cavity. With spin echo pulse sequence and prolongation of the TE value to range of 110 msec it may be possible to show the difference between edema and the liquid content of the cavity itself.

TRAUMA

Fractures

Trauma to the temporal bone can cause conductive hearing loss either through ossicular disruption or fluid within the middle ear cavity. Sensorineural hearing loss will accompany fractures through the otic capsule. Either type of hearing loss can be found from a fracture that begins in the parietal regions and extends the length of the temporal bone (longitudinal fracture) or from a fracture that occurs in the occipital region and crosses the long axis of the temporal bone (transverse fractures). Unfortunately, the fracture lines are not quite that simple except in the minority of cases (13, 30, 46). Usually, the fracture lines are quite jagged and difficult to detect without multiple projections.

OSSICULAR DISRUPTIONS

Ossicular disruptions most commonly involve the incus because it is only held in place by one ligament. The next most common injury occurs through the crura of the stapes, and the least most common is damage to the malleus either through crushing of its head or fracture through its neck (36, 66, 71).

FACIAL NERVE

The facial nerve may be involved by fractures either within the internal auditory canal or in its more peripheral portion from the geniculate ganglia to the stylomastoid foramen. An accurate determination of the level of fracture and any fragments within the canal plays a major role in patient management. Disruptions about the anterior genu or posterior genu constitute two of the most difficult areas to detect a fracture and represent some of the most common locations of injuries.

OTORRHEA

One of the complications associated with fractures extending into the middle ear cavity is cerebrospinal fluid otorrhea and herniation of the temporal lobe into the middle ear cavity (20). The mechanism of herniation is related to the fact that the mastoid is membranous bone and heals by endosteal union. Any tears of the dura or loss of cortical integrity heal quite slowly if at all. The meninges are able to pulsate and project into the middle ear cavity, thus creating a mass that interferes with conductive hearing. CSF communications to the middle ear may be suspected clinically by a persistence of fluid high in sugar content that otherwise closely mimics a serous otitis. It's amazing how frequently fluid levels are present in the sphenoid sinus with temporal bone CSF otorrhea.

CT and MRI Findings

The axial plane of presentation of CT has proven quite adequate for a majority of the trauma cases. The scans should be viewed in a nonmagnified mode as well as magnified so that the entire skull can be appraised a well as the temporal bone. As discussed under normal anatomy the X configuration at the base of the skull helps to explain the mechanism of trauma in many instances. If the force of impact "squeezes the skull" between the lateral margin of the orbit on one side and the opposite mastoid tip on the other, a compression type of injury in the mastoid region may occur. When a blow is applied to the mandible and transmitted through the external auditory canal, a sheering action takes place involving the more lateral portions of the temporal bone. A squeezing of the skull from bilateral application of pressure causes a sheering and compression force with disruption of the posterior table of the temporal bone and impaction fracture of the middle ear cleft. Blows high in the parietal region cause a longitudinal fracture down the long axis of the temporal bone. The mechanism of injury is quite interesting (23, 24). The parietal blow causes the skull to momentarily "indent" which is followed by "recoil." Lines of force radiate in a starlike pattern from the center of the blow with one of the fracture lines going down the temporal bone.

This same type of rebound phenomena can occur with occipital injuries, creating transverse fractures.

In patients with facial paralysis that have not been explained by routine axial views, either reformated images or special maneuvers should be performed to obtain additional views and, in particular, the lateral projection.

The CT diagnosis of ossicular disruptions is quite easy since the "ice cream cone" head of the malleus and body of the incus should lie centrally placed within the epitympanic recess. Displacements are readily recognized and usually the incus, when displaced, will fall well into the floor of the tympanic cavity. Fractures of the crura of the stapes are not quite so simple and indirect signs may provide the only clue to their presence. In the axial plane the manubrium of the malleus and long process of the incus should be seen as two parallel structures descending into the mesotympanum and directed toward the promintory. If they are no longer lying parallel or if their direction is anterior or

a rather abrupt change in the amount of pain and discharge (51). The association of development of carcinoma to chronic otitis media is so constant that some investigators have recommended routine pap smears (6). Cranial nerve palsies occur in approximately half of the patients, with the facial nerve being the most common nerve involved. These tumors have a tendency to spread medial to the otic capsule and involve the petrous apex. Once they have produced destruction in the petrous apex the tumor may cause paralysis of nerves V or VI as they extend intracranially. Diagnosis by radiologic means may be difficult since almost a third of the patients have already had radical mastoidectomies and only a disproportionate amount of pain in relationship to the degree of inflammation will provide the clue to perform biopsy for diagnosis.

CT and MRI Findings

CT again plays a dominant role in diagnosis because of the ability to identify bone destruction and bony canal enlargements. Possibly there may be a place for MRI in the future because of the reactive changes in adjacent bone and the ability to perform coronal scans when middle cranial fossa invasion has occurred. MRI can show the long T_2 signals of malignant lesions to good advantage.

One of the most important points of information required by the otologic cancer surgeon in the management of a middle ear cancer is the status of the carotid canal. Great care should be taken during scanning to include the entrance of the carotid artery into the skull base and the horizontal portion of the carotid canal to the level of its emergence into the cavernous sinus. This region must be free of disease for the conventional radical temporal bone surgery. The next important area to carefully evaluate is the lateral extremity of the internal auditory canal. Carcinomas of the middle ear cavity may directly invade the cochlea or semicircular canal system and gain access to the internal auditory canal by following the vestibular nerves or the cochlear nerve through the modiolus of the cochlea. Erosions of bone in these regions would indicate that the tumor mass gained access to the internal auditory canal and, hence, to the subarachnoid space. If the patient has a thin tegmen of the mastoid antrum and middle ear cavity, invasion may extend through the dura into the floor of the temporal fossa (middle cranial fossa). Coronal scanning or reformated images following intravenous contrast will demonstrate reactive changes in the adjacent brain. MRI is probably superior to CT for demonstrating this type of change since direct coronal scanning is no problem with MRI.

RHABDOMYOSARCOMA

Primary tumors of the middle ear cavity are rare lesions but show some specificity as to age groups. Rhabdomyosarcomas are the most common malignant neoplasm in children with the incidence of involvement of the ear being third to primary rhabdomyosarcoma in the orbit or nasopharynx (10). These lesions are highly malignant and the clinical course is rapidly fatal. Diagnosis is usually made late because of the similarity of the early lesions to inflammatory disease. Death usually occurs secondary to CNS involvement and modern treatment is directed toward removing the primary tumor and controlling the systemic nature of the disease with chemotherapy and radiation.

CT and MRI Findings

The rhabdomyosarcomas have the ability to rapidly invade and destroy bone. They produce soft tissue masses that extend intracranially as well as into the soft tissues overlying the mastoid process and temporal fossa. If intravenous contrast is used, enhancement occurs throughout the lesions in a mottled type of fashion.

HISTIOCYTOSIS

Histiocytosis may involve the middle ear cavity by extension from the matoid air cells or adjacent temporal bone. These lesions begin within the marrow cavity and extend to involve the inner and outer tables of the skull as well as the air spaces of the middle ear cavity. The lesions may be rapidly aggressive and destructive or there may be a slower growth pattern that allows some molding of the adjacent bony structures (38, 43, 55). Soft tissue involvement is common so that diagnosis is not purely that of a lesion confined to the bone. Involvement of the outer table of the skull to a greater extent than inner table is said to be relatively diagnostic of histiocytosis.

CT and MRI Findings

Usually the regions of multiple involvement seen with histiocytosis can be identified quite adequately with plain films of the skull and remaining skeletal system. The lesions are purely destructive with indistinct margins. The greater involvement of the outer table can be demonstrated quite nicely on axial CT scanning. The soft tissue masses stain to a greater extent than muscle following contrast.

METASTATIC TUMOR

Metastatic tumors of the temporal bone may occur from a wide variety of primary lesions with the most common being breast, lung, kidney, and prostate. The deposits are most likely to be present in the vicinity of the mastoid air cells and spread out from this central location. The clinical features are usually that of the primary tumor plus any local nerve or hearing structures that become involved. Pain is a prominent symptom due to bone involvement adjacent to dura.

CT and MRI Findings

The metastatic tumors are most commonly found in the periantral air cells and then spread to adjacent cavities and regions of cortical bone. They may occur in the medullary bone of the petrous apex or about the internal auditory canal but practically never originate within the otic capsule. The metastatic tumors show varying degrees of destruction and may even have productive changes if tumors originate in prostate, breast, thyroid, or colon.

SYSTEMIC DISEASE

The majority of systemic diseases affecting the vestibular apparatus and cochlear structures fail to produce radiologic changes of either the middle or inner ear. The bony proliferative and remodaling disorders associated with fibrous dysplasia, Paget's disease, osteopetrosis, or osteogenesis imperfecta all have characteristic appearance on conventional radiographic examinations. Similarly, the histiocytosis and storage diseases such as Gaucher's and Hurler's, when they affect the temporal bone, can be examined by conventional radiographic examinations or with CT in some instances.

CT and MRI Findings

Some specific entities that require CT are fibrous dysplasia when it involves the external auditory canal and mastoid portions of the temporal bone. The proliferative lesions cause narrowing and even obliteration of the external canal. Squamous epithelium becomes trapped in the medial portion of the external canal which then gives rise to cholesteatoma formation. The cholesteatoma has no possibility to spontaneously drain the exterior and may become infected by the eustachian tube. This type of lesion is potentially fatal unless managed by surgically enlarging the external auditory canal. The CT will identify the presence of cholesteatoma with destruction and expansion about the middle ear cavity.

Paget's disease, osteopetrosis, and osteogenesis imperfecta can all be identified on CT as they affect inner ear structures. The softening and bony proliferative changes of Paget's may cause encroachment upon the internal auditory canal. There may also be morphological distortions of the cochlea with destruction of the otic capsule and proliferative distortions of the lumen of the cochlea. The ossicular chain shows histological changes but these findings are not visible on CT.

External Auditory Canal

Lesions of the external canal do not require CT scanning unless they are related to more general processes of the temporal bone such as "malignant otitis," true malignancies or extensions of the middle ear cholesteatomas (see discussions under Middle Ear Tumors).

Benign Tumors

EXOTOSES

Benign exotoses are in reality a response to irritation as opposed to a true tumor. They are associated with cold water swimming and are readily diagnosed on clinical examination without the need of CT. Radiologic investigation is only performed to confirm the normal status of the middle ear structures which may be difficult to visualize on clinical examination due to the obscuring effect of the external canal exostoses.

CRISTOMAS

Cristomas are rests of normal salivary gland tissue that present as a tumor mass within the external auditory canal. They may develop tumors but generally are merely obstructing lesions that are diagnosed by clinical examination; radiologic investigation is rarely needed.

Malignant Tumors

Squamous cell carcinomas of the pino that invade the external auditory canal or those malignancies that arise within the external auditory canal constitute a severe problem of management. They produce tremendous amounts of swelling of the skin that usually obscures clinical examination of the middle ear cavity. Tumors may invade and destroy part of the bony external auditory canal. Their main access to vital structures is by destroying the tympanic membrane and invading into the middle ear.

CT and MRI Findings

MRI may show the high signal of these lesions and their changes late in the course of disease when brain invasion has occurred. In the earlier lesions that are amenable to surgery, fine detailed CT scanning is required to see the degree of bone destruction (see Carcinoma of the Middle Ear).

SUMMARY

The biggest "stumbling block" to CT of the temporal bone is learning the anatomy in the axial projection. Once this hurdle has been overcome the radiologist may provide an "in vivo" map of a wide variety of pathological processes affecting the temporal bone.

Magnetic resonance imaging will undoubtedly assume a dominant role for central nervous system imaging but the temporal bone will probably remain the domain of CT. This is indeed a fortunate arrangement because radiologic investigation of the temporal bone is being requested at an everaccelerating rate that is taxing the existing facilities. Hopefully, magnetic resonance imaging will free up time on our CT scanners to permit these requests to be fulfilled.

References

1. Anson BJ, Donaldson JA: *The Surgical Anatomy of the Temporal Bone and Ear.* Philadelphia, Saunders, 1967, pp 23, 28, 29.
2. Babin R, Hanafee WN: Anatomic and radiographic correlates in the middle ear. *Arch Otolaryngol* 101:474–477, 1975.
3. Batsakis JG: *Tumors of the Head and Neck. Clinical and*

Pathological Considerations, ed 2. Baltimore, Williams & Wilkins, 1979.

4. Bentson JR, Mancuso AA, Winter J, Hanafee WN: Combined gas cisternography and edge enhancement computed tomography of the internal auditory canal. *Radiology* 136:777–780, 1980.
5. Bergeron T: Personal communication, November 1978.
6. Bradley WH, Maxwell JH: Neoplasms of the middle ear and mastoid. Report of 54 cases. *Laryngoscope* 64:533–556, 1954.
7. Carlborg BIR, Farmer JC: Transmission of cerebrospinal fluid pressure via the cochlear aqueduct and endolymphatic sac. *Am J Otolaryngol* 4:273–282, 1983.
8. Chandler JR: Malignant otitis externa. *Laryngoscope* 78:1257–1295, 1968.
9. Chandler JR: Malignant external otitis: further considerations. *Ann Otol Rhinol Laryngol* 86:417–428, 1977.
10. Dehner LP, Chen KTK: Primary tumors of the external and middle ear. III. A clinicopathologic study of embryonal rhabdomyosarcoma. *Arch Otolaryngol* 104:399–403, 1978.
11. Delozier HL, Gacek RR, Dana ST: Intralabyrinthine schwannoma. *Ann Otol Rhinol Laryngol* 88:187–191, 1979.
12. Delozier HL, Parkins CW, Gacek RR: Mucocele of the petrous apex. *J Laryngol Otol* 93:177–180, 1979.
13. Dolan KD, Jacoby CG: Radiologic evaluation of temporal bone fractures. *Appl Radiol* 8:62–67, 1979.
14. Donaldson JA, Anson BJ, Warpeha RL, et al: The perils of the sinus tympani. *Trans. Pac Coast Otoophthalm Soc* 52:93–106, 1968.
15. Dubois PJ, Roub LW: Giant air cells of the petrous apex: tomographic features *Radiology* 129:103–109, 1978.
16. Ethier R, King DG, McLancon D, et al: Development of high resolution computed tomography of the spinal cord. *J Comput Assist Tomogr* 3:433–438, 1979.
17. Farrior JB, Hyams VJ, Benke RH, Farrior JB: Carcinoid apudoma arising in a glomus jugulare tumor: review of endocrine activity in glomus jugulare tumors. *Laryngoscope* 90:110–119, 1980.
18. Ferlito A: Histopathogenesis of tympanosclerosis. *J Laryngol Otol* 93:25–37, 1979.
19. Fisch U: Infratemporal fossa approach for glomus tumors of the temporal bone. *Ann Otol Rhinol Laryngol* 91:474–479, 1982.
20. Glasscock ME, Dickins JRE, Jackson CG, et al: Surgical management of brain tissue herniation into the middle ear and mastoid. *Laryngoscope* 89:1743–1754, 1979.
21. Goodhill V: *Ear Diseases, Deafness, and Dizziness*. Hagerstown, Md., Harper Row, 1979.
22. Graham M: The jugular bulb: its anatomic and clinical considerations in contemporary otology. *Laryngoscope* 87:105–125, 1977.
23. Gurdjian ES, Lissner HR: Deformation of the skull in head injury. A study with the stresscoat technique. *Surg Gynecol Obstet* 81:679–678, 1945.
24. Gurdjian ES, Lissner HR: Deformation of the skull in head injury studied by the "Stresscoat" technique, quantitative determinations. *Surg Gynecol Obstet* 83:219–233, 1946.
25. Hanafee WN, Bergstrom LV: Radiology of congenital deformities of the ear. *Head Neck Surg* 2:213–221, 1980.
26. Hanafee WN, Gussen R, Rand RW: Laminography of the mastoid in the basal projection. *AJR* 110:111–118, 1970.
27. Hanafee WN, Mancuso AA, Jenkins HA, et al: Computerized tomography scanning of the temporal bone. *Ann Otol Rhinol Laryngol* 88:721–728, 1979.
28. Hanafee WN, Mancuso AA, Winter J, et al: Edge enhancement computed tomography scanning of inflammatory lesions of the middle ear. *Radiology* 136:771–776, 1980.
29. Hanafee WN, Wilson G: Pontocerebellar angle tumors: newer diagnostic methods. *Arch Otolaryngol* 92:236–243, 1970.
30. Harwood-Nash DC: Fractures of the petrous and tympanic parts of the temporal bone in children: a tomographic study of 35 children. *AJR* 110:598–607, 1970.
31. Jenkins HA, Fisch U: Glomus tumors of the temporal region. *Arch Otolaryngol* 107:209–213, 1981
32. Johnson D, Voorhees, Lufkin RB, Hanafee WN, Canalis R: Cholesteatomas of the temporal bone: role of computed tomography. *Radiology* 148:733–737, 1983.
33. Kaseff LG, Nieberding PH, Shorago GW, et al: Fistula between the middle ear and subarachnoid space as a cause of recurrent meningitis: detection by means of thin-section, complex motion tomography. *Radiology* 135:105–108, 1980.
34. Khangure MS, Mojtahedi S: Air CT cisternography of anterior inferior cerebellar artery loop simulating an intracanalicular acoustic neuroma. *AJNR* 4:994–995, 1983.
35. Kieffer SA, Binet EF, Gold LHA: Angiographic diagnosis of intra- and extraaxial tumors in the cerebellopontine angle. *AJR* 124:207–209, 1975.
36. Lloyd GAS, du Boulay GH, Phelps PD, et al: The demonstration of the auditory ossicles by high resolution CT. *Neuroradiology* 18:243–248, 1979.
37. Lufkin RB, Barni JJ, Glenn W, Mancuso AA, Canalis RF, Hanafee WN: Comparison of computed tomography and pluridirectional tomography of the temporal bone. *Radiology* 143:715, 1982.
38. McCaffrey TV, McDonald TJ: Histiocytosis X of the ear and temporal bone: review of 22 cases. *Laryngoscope* 89:1735–1742, 1979.
39. Mendez G, Quencher RM, Post JD, Stokes NA: Malignant external otitis: a radiographic-clinical correlation. *AJR* 132:957–961, 1979.
40. Meyerhoff WL, Kim CS, Paparella MM: Pathology of chronic otitis media. *Ann Otol* 87:749–760, 1978.
41. Moller A, Hatam A, Olivecrona H: The differential diagnosis of pontine angle meningioma and acoustic neuroma with computed tomography. *Neuroradiology* 17:21–23, 1978.
42. Naidich TP, Lin JP, Leedsne, et al: Computed tomography in diagnosis of extraaxial posterior fossa masses. *Radiology* 120:333–339, 1976.
43. Ogura JH, Toomey JM, Setzen M, Sobel S: Malignant fibrous histiocytoma of the head and neck. *Laryngoscope* 90:1429–40, 1980.
44. Osborn AG, Parkin JL: Mucocele of the petrous temporal bone. *AJR* 132:680–681, 1979.
45. O'Sullivan TJ, Dickson RI, Blokmanis A, Roberts FJ, et al: The pathogenesis differential diagnosis, and treatment of malignant otitis externa. *J Otolaryngol* 7:297–303, 1973.
46. Potter GD: Trauma to the temporal bone. *Semin Roentgenol* 4:143–150, 1969.
47. Pratt LL: Complications associated with the surgical treatment of cholesteatoma. *Laryngoscope* 92:289–291, 1983.

48. Rice RP, Holman CD: Tumors of the glomus jugulare. *AJR* 89:1201–1208, 1963.
49. Schuknecht HF: *Pathology of the Ear.* Cambridge, Mass, Harvard University Press, 1974.
50. Shaffer KA, Houghton VM, Wilson CR: High resolution computed tomography of the temporal bone. *Radiology* 134:409–414, 1980.
51. Sinha PP, Aziz HI: Treatment of carcinoma of the middle ear. *Radiology* 126:485–487, 1978.
52. Som PM, Parisier SC, Wolf BS, Rose JS: Advantage of computerized tomography scan. Diagnosis of an otogenic cerebellar abscess. *Arch Otolaryngol* 104:542–543, 1978.
53. Som PM, Reede DL, Bergeron RT, Parisier SC, Shugar J, Cohen NL: Computed tomography of glomus tympanicum tumors. *J Comput Assist Tomogr* 7:14–17, 1983.
54. Stern J, Goldenberg M: Jugular bulb diverticula in medial petrous bone. *AJNR* 1:153–155, 1980.
55. Sweet RM, Kornblut AD, Hymans VJ: Eosinophilic granuloma in the temporal bone. *Laryngoscope* 89:1545–1552, 1979.
56. Sykora GF, Kaufman B, Katz RL: Congenital defects of the inner ear in association with meningitis. *Radiology* 135:379–382, 1980.
57. Thomsen J, Gyldensted C, Lester J: Computer tomography of cerebellopontine angle lesions. *Arch Otolaryngol* 103:65–69, 1977.
58. Thomsen J, Gyldensted C, Lester J, et al: Computerized tomography in otorhinolaryngology. *ORL J Otorhinolaryngol Relat Spec* 36:9–20, 1977.
59. Valavannis A, Schubiger D, Oguz M: High resolution CT investigation of nonchromaffin paragangliomas of the temporal bone. *AJNR* 4:516–519, 1983.
60. Valavanis A, Schubiger O, Wellauer J: Computed tomography of acoustic neuromas with emphasis on small tumor detectability. *Neuroradiology* 16:598–600, 1978.
61. Valvassori GE: Personal communication. Radiology of the Head and Neck Meeting. New York, May 1978.
62. Valvassori GE, Buckingham RA: Middle ear masses mimicking glomus tumors: radiographic and otoscopic recognition. *Trans Am Acad Ophthalmol Otolaryngol* 62:85–125, 1974.
63. Valvassori GE, Clemis JD: The large vestibular aqueduct syndrome. *Laryngoscope.* 88:723–728, 1978.
64. Valvasori GE, Mafee MF, Dobben: Computerized tomography of the temporal bone. *Laryngoscope* 92:562–565, 1982.
65. Van Waes PFGM, Zonneveld FW, Damsma P, Rabischong, J, Vignaud J: Direct or multiplanar CT of the petrous bone. Philips Medical Systems (represents scientific exhibit presented at 9th International Congress of Radiology in Otorhinolaryngology, June, 1982.
66. Vignaud J, Sultan A: Dislocations traumatiques de la chaine des osselets. *J Belge Radiol* 54:Fasc I, 203–208, 1971.
67. Virapongse C, Rothman SLF, Kier EL, et al: Computed tomographic anatomy of the temporal bone. *AJR* 139:739, 1982.
68. Voorhees RL, Johnson DW, Lufkin RB, Hanafee W, Canalis R: High resolution CT scanning for detection of cholesteatoma and complications in the postoperative ear. *Laryngoscope* 93:589–595, 1983.
69. Winter J: Edge enhancement of computed tomograms by digital unsharp masking. *Radiology* 135:234–235, 1980.
70. Wright JLW, Grimaldi PMGB: Otogenic intracranial complications. *J Laryngol Otol* 87:1085–1096, 1973.
71. Wright JW, Taylor CE, Bizal JA: Tomography and the vulnerable incus. *Ann Otol Rhinol Laryngol* 78:1–16, 1969.
72. Ylikoski J, Palva T, Collan Y: Eighth nerve in acoustic neuromas. *Arch Otolaryngol* 104:532–534, 1978.
73. Ziedes Des Plantes BE: Een Byzondere methode voor bet maken van roentgnfot's Van Schedelen. Inverrelkeom. *Med T Geneesk* 75:5219, 1931.

Chapter 5 Plates

Temporal Bone—
Normal Anatomy, Pathology, Technique and Indication for Scanning

The abbreviation used on Plates 5.1*B*–5.1*E* is: *SSC*, superior semicircular canal.

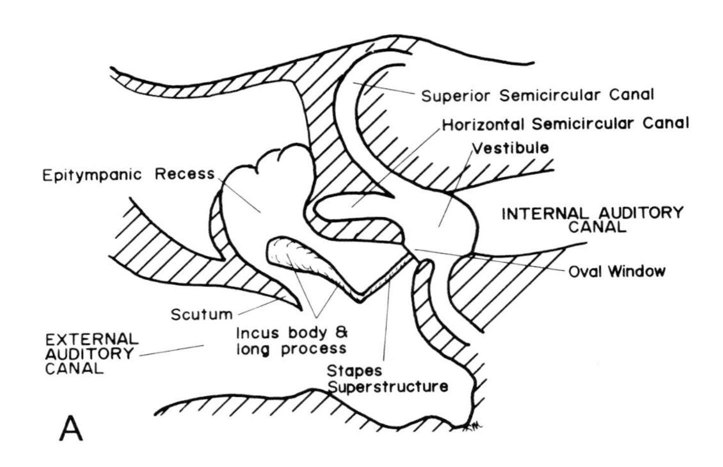

Plate 5.1*A*. Normal middle and inner ear anatomy at the level of the oval window as visualized in the Guillen projection. In this projection, with the head turned 15° toward the side under investigation, the medial wall of the middle ear cavity (promintory) lies parallel with the central beam of the x-ray tube. The body and long process of the incus form a V with the stapes superstructure in the middle ear cavity. The roof of the epitympanic recess appears further away from the ossicular chain than is present in the anatomical situation. This "roof" is actually the tegmen of the mastoid antrum which superimposes as a "ghost" shadow on the epitympanic recess, giving this cavity a more voluminous appearance.

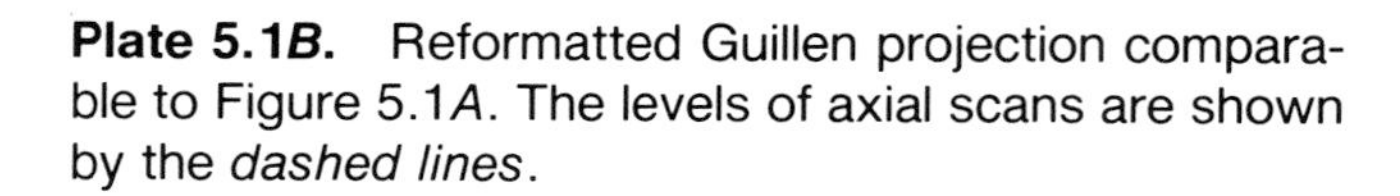
Plate 5.1*B*. Reformatted Guillen projection comparable to Figure 5.1*A*. The levels of axial scans are shown by the *dashed lines*.

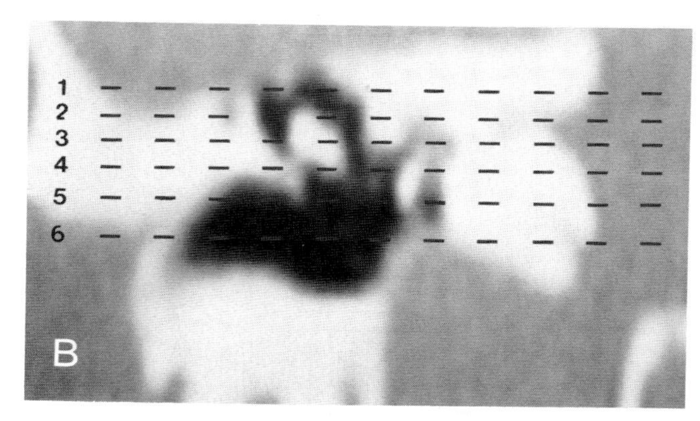

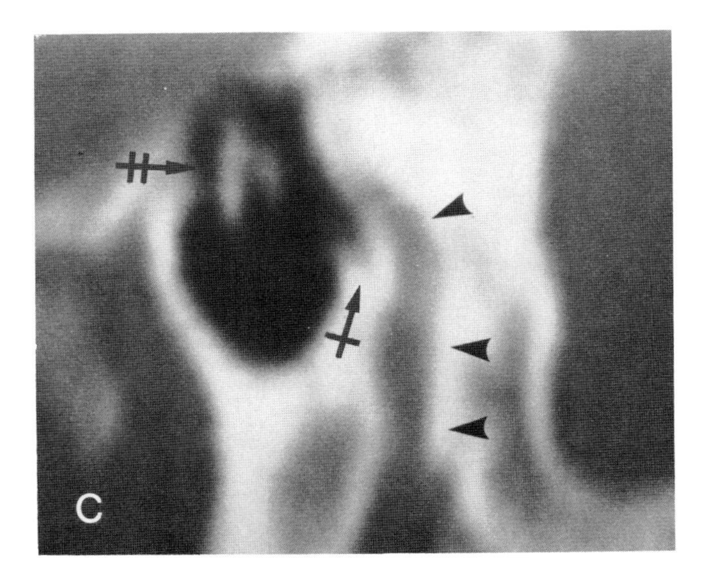

Plate 5.1*C*. Reformatted lateral view, or sagittal projection, to show the descending portion of the facial nerve and the ossicles in the epitympanic recess.

The facial nerve makes a curve and descends inferiorly immediately beneath the horizontal semicircular canals (*arrowheads*). The nerve exits from the stylomastoid foramen. Anterior to the descending portion of the facial nerve a volcano-shaped prominence of bone can be seen which houses the stapedius muscle. This mound is called the pyramidal eminence (*crossed arrows*).

The manubrium of the malleus and long process of the incus are clearly visualized (*double crossed arrow*) in the epitympanic recess.

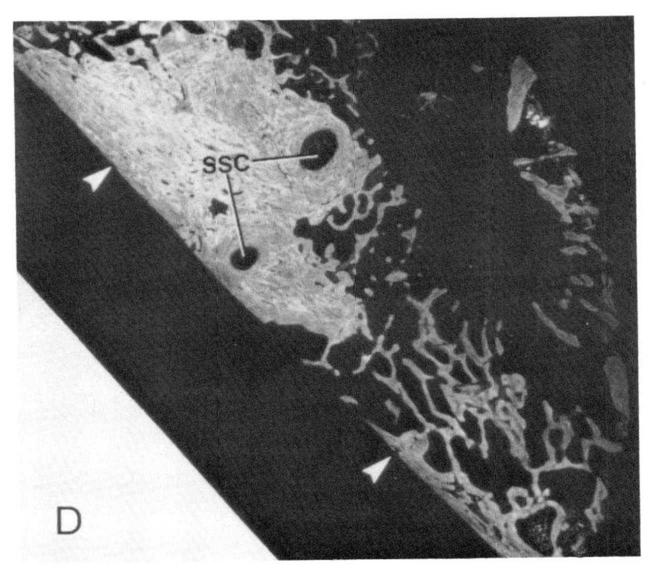

Level 1—highest level.

Plate 5.1*D*. Anatomical section 50-μm thick. *Arrowheads* indicate the posterior surface of the left temporal bone viewed from below as one would visualize an axial CT scan. *SSC* is the anterior and posterior limbs of the superior semicircular canal.

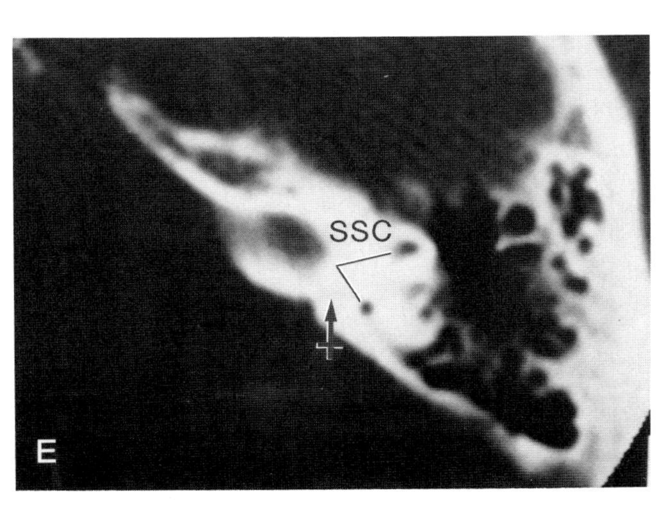

Plate 5.1*E*. Superiorly located axial section of the temporal bone. The superior semicircular canal is visualized as two "holes." The subarcuate vessels pass in the canal (*crossed arrow*) between the two limbs of the superior semicircular canal. This section and subsequent levels were taken using a GE 8800 scanner at 120 kV, 768 mas, 1.5-mm thick sections at 1-mm intervals.

The abbreviations used on Plates 5.2–5.4*B* are: *SSC*, superior semicircular canal; *PSC*, posterior semicircular canal; *CC*, crus communis; *V*, vestibule; *F*, facial nerve; *M*, head of the malleus: *I*. body of the incus; *HSC*, horizontal semicircular canal C_1, apical turn of cochlea; C_2, intermediate turn of the cochlea; C_3, basal turn of the cochlea; *MN*, modiolus nucleus; *LPI*, long process of incus; *MM*, manubrium of the malleus; *ED*, endolymphatic sac and duct; OVN, oval window niche; and T, tendon.

Level 2.

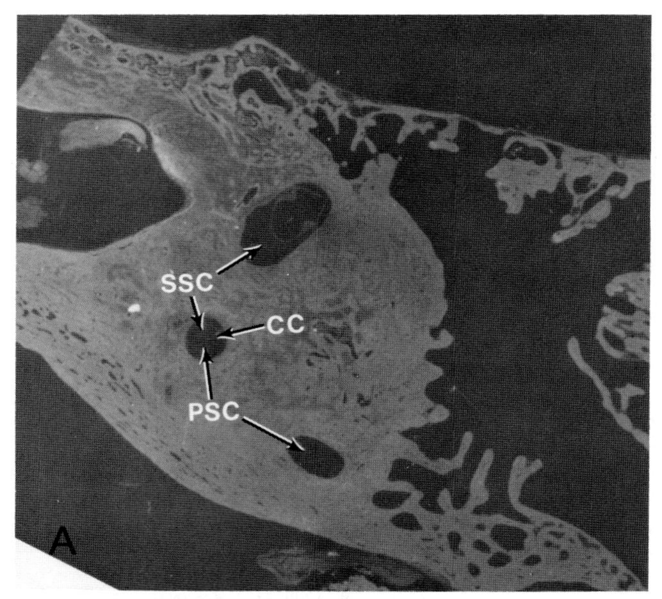

Plate 5.2*A*. Anatomic specimen.

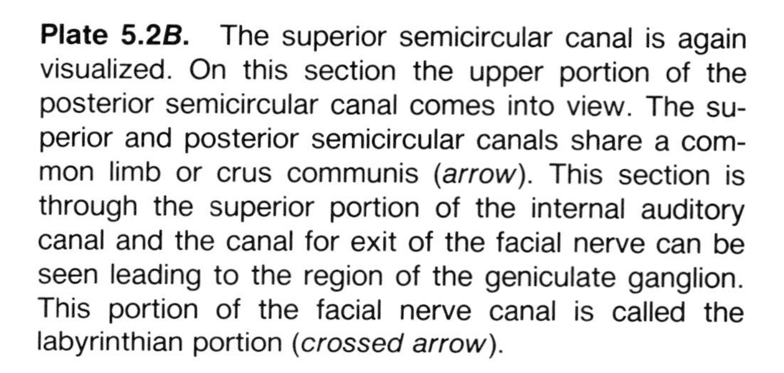

Plate 5.2*B*. The superior semicircular canal is again visualized. On this section the upper portion of the posterior semicircular canal comes into view. The superior and posterior semicircular canals share a common limb or crus communis (*arrow*). This section is through the superior portion of the internal auditory canal and the canal for exit of the facial nerve can be seen leading to the region of the geniculate ganglion. This portion of the facial nerve canal is called the labyrinthian portion (*crossed arrow*).

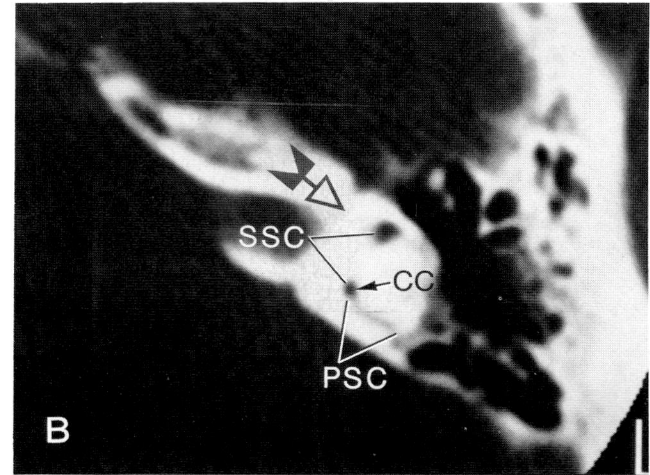

Level 3—horizontal semicircular canal level.

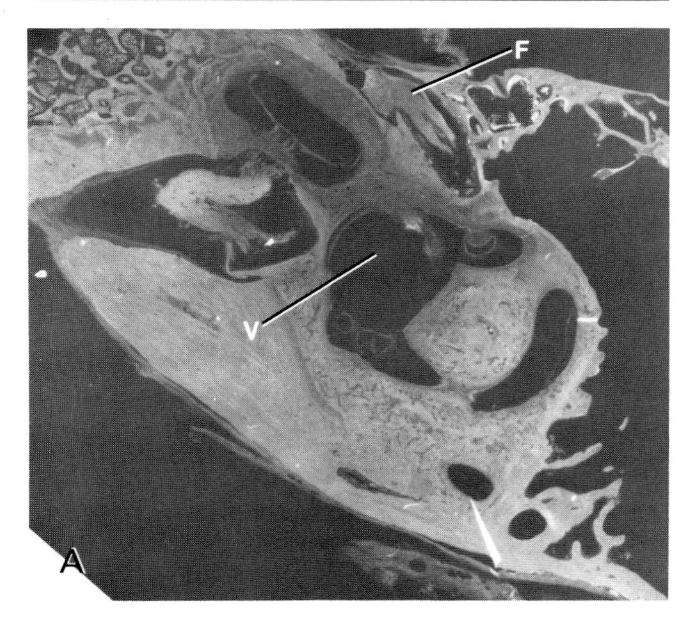

Plate 5.3*A*. Anatomical section shows the anterior genu of the facial nerve.

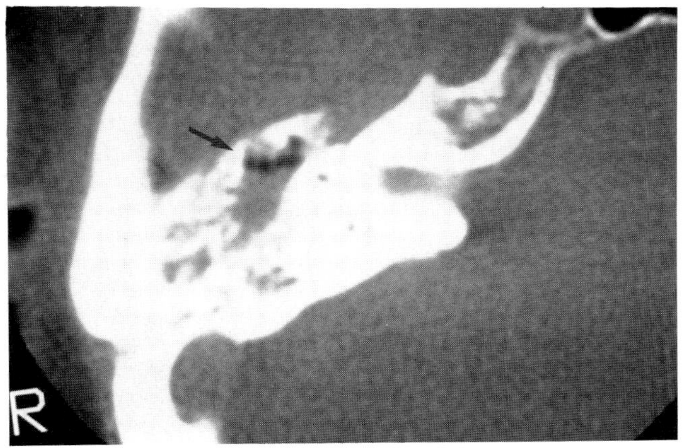

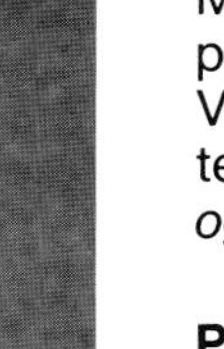

INFLAMMATORY DISEASE—REVERSIBLE

Many of these illustrations are from a joint project performed with Dr. Dexter Johnson: Johnson DW, Voorhees RL, Lufkin RB, et al: Cholesteatomas of the temporal bone: role of computed tomography. *Radiology* 148:733–736, 1983.

Plate 5.10. Section through the high epitympanic recess shows a distinct air fluid level (*arrow*) with the air rising to the anterior portion of the epitympanic recess because the patient is lying supine. All of the mastoid air cells are filled with fluid and there is no evidence of bone destruction.

One of the etiologies of serous otitis is nasopharyngeal carcinoma so the nasopharynx should be surveyed as part of the temporal bone examination.

Plate 5.11. Glue ear. This child with cleft palate and chronic ear infections was a good candidate for cholesteatoma. The normal appearing malleus and incus together with an intact lateral wall of the epitympanic recess would rule out any early mass or erosive changes.

Plate 5.11*A*. Scan with the patient supine (as indicated by *large arrow*) shows a normal appearing ossicular chain and margins to the epitympanic recess. The mastoid air cells are airless and only a small amount of air surrounds the ossicular chain. The tenacious fluid (*arrowheads*) layers over the oval window.

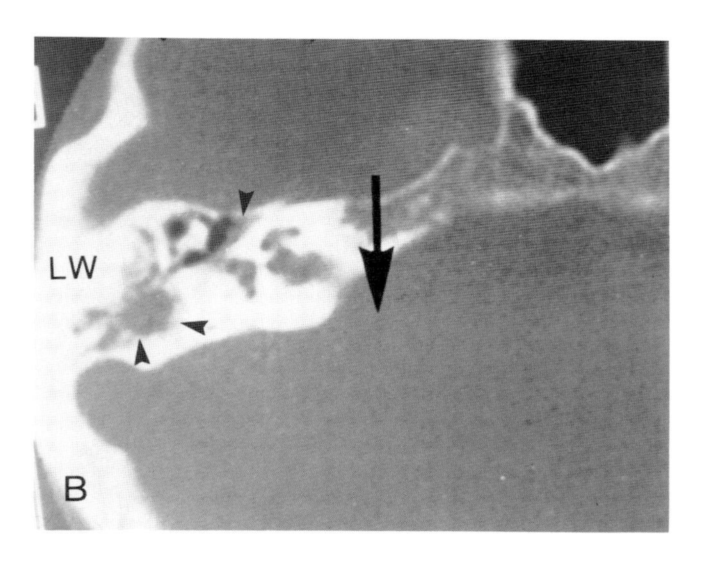

Plate 5.11*B*. Section through the same region with the patient prone (as indicated by *large arrow*) demonstrates that the air fluid interfaces (*arrowheads*) do not shift with changing patient position. The mastoid air cells stay opaque and the air remains trapped around the ossicles.

The abbreviations used on Plates 5.12*A*–5.15 are: *MA*, mastoid antrum; *JF*, jugular foramen; *CAL*, calcification; and *I*, incus.

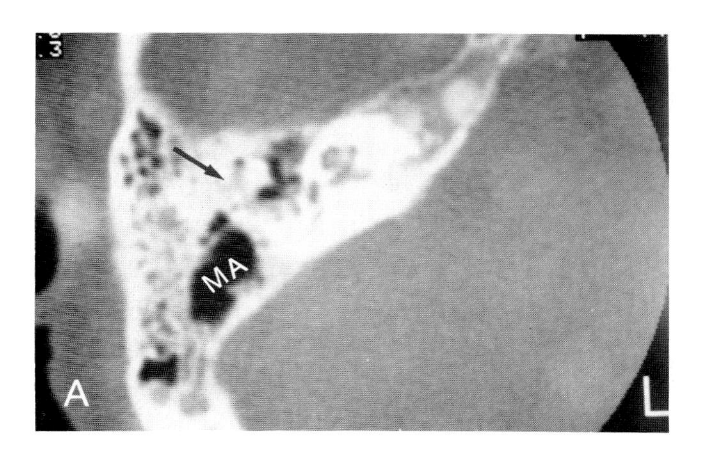

Plate 5.12. Chronic otitis media.

Plate 5.12*A*. Section through the midepitympanic shows that the ossicles are laterally displaced and soft tissue scarring connects the malleus and incus with the lateral wall of the epitympanic recess (*arrow*). The mastoid antrum is air containing but the periantral cells are largely filled with fluid. No bone erosions can be identified.

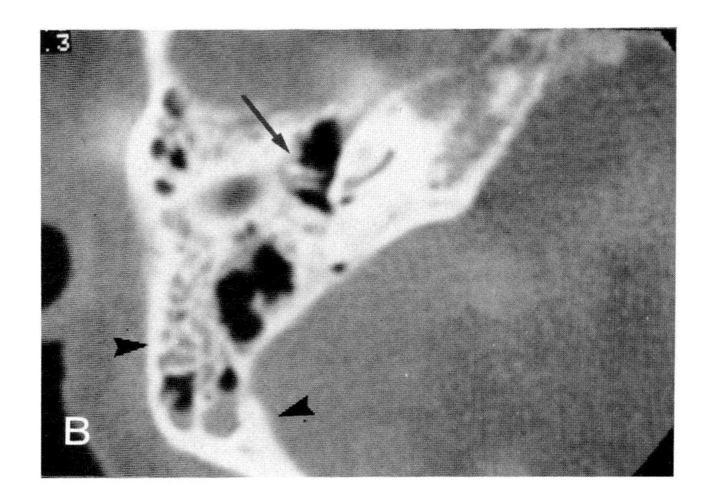

Plate 5.12*B*. Section through the mesotympanum shows that the manubrium of the malleus and long process of the incus are visualized lying parallel but much closer together than normal. They are also adjacent to a scarred and thickened tympanic membrane and tympanic annulus (*arrow*). The periantral cells are again shown to be largely fluid filled but multiple air fluid levels are visible (*arrowheads*).

The lack of erosive changes and deformity of the position of the ossicles would be consistent with chronic otitis media with retractive scarring.

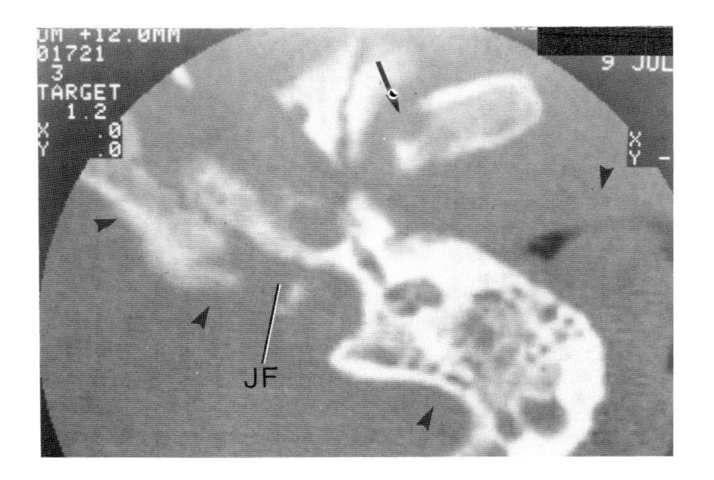

Plate 5.13. Malignant otitis externa. This acutely ill, uncontolled diabetic demonstrates soft tissue swelling occluding the external canal and opacification of the mastoid air cells. The inflammatory process extends along the under surface of the temporal bone from the external canal to the extreme petrous apex (*arrowheads*). Erosive changes involve the jugular foramen posteriorly and the temporomandibular joint anteriorly. The condyle of the mandible is eroded (*arrow*).

Despite the extent of this inflammatory process, the changes are still reversible with vigorous antibiotic therapy.

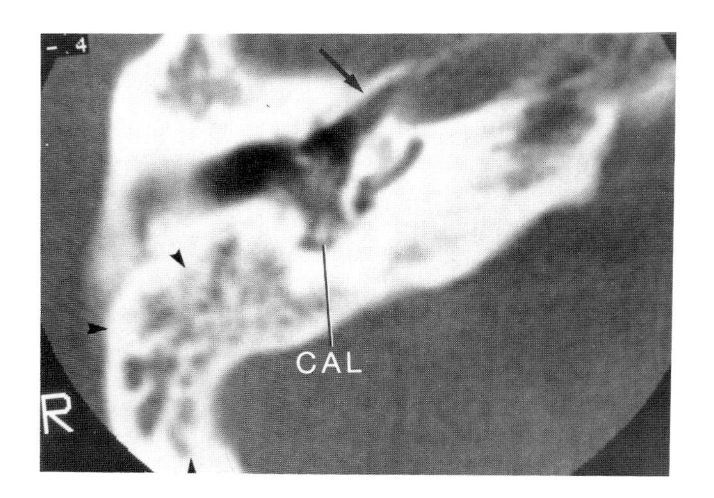

INFLAMMATORY DISEASE—IRREVERSIBLE

Plate 5.14. Tympanosclerosis. Section through the hypotympanum shows that inflammatory disease occludes the eustachian tube orifice (*arrow*). All of the mastoid air cells are opacified (*arrowheads*). In the posterior hypotympanum the scarring shows irregular densities and plaques of calcification lining the promintory and within the debris adjacent to the sinus tympani.

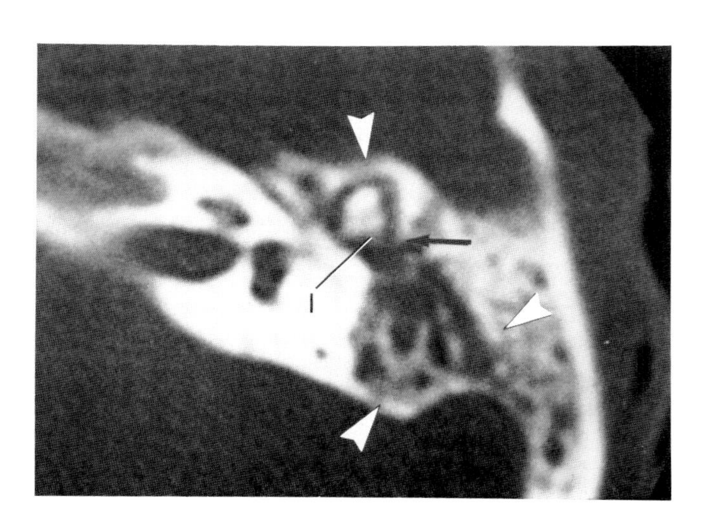

Plate 5.15. Early retraction pocket. In this patient with chronic inflammatory disease, a small retraction pocket could be identified from the pars flaccida region. Section through the epitympanic recess shows that the posterior portion of the lateral wall of the epitympanic recess is eroded (*arrow*). The posterior half of the body of the incus is also poorly mineralized indicating early erosion. The bony margins of the remainder of the epitympanic recess and mastoid antral region are reasonably well preserved. Some overall sclerosis of the mastoid air cell walls can be seen (*arrowheads*).

Retraction pockets are the beginning of cholesteatomas and have the ability to erode bone. They are formed from the healing phase of tympanic membrane perforations.

The abbreviations used on Plates 5.16*B*–5.16*D* are: *F*, facial nerve; *M*, malleus; *I*, incus; and *P*, polyp.

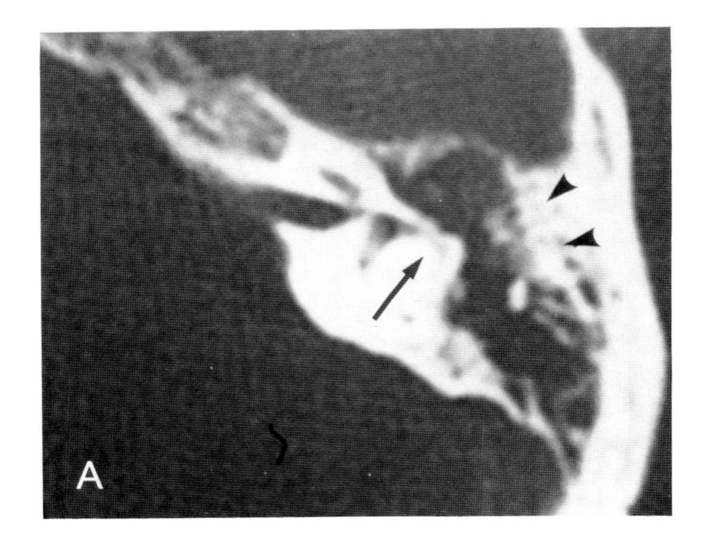

Plate 5.16. Cholesteatoma with polyps.

Plate 5.16*A*. Section through the epitympanic recess shows a total erosion of the ossicular chain and the lateral wall of the epitympanic recess. The aditus ad antrum region (*arrowheads*) that connects the epitympanic recess to the mastoid antrum is widened. The bony covering of the horizontal semicircular canal shows some erosion near the ampullary end (*arrow*) but the erosion is not completely through into the lumen of the canal. The periantral air cells are clouded but well preserved.

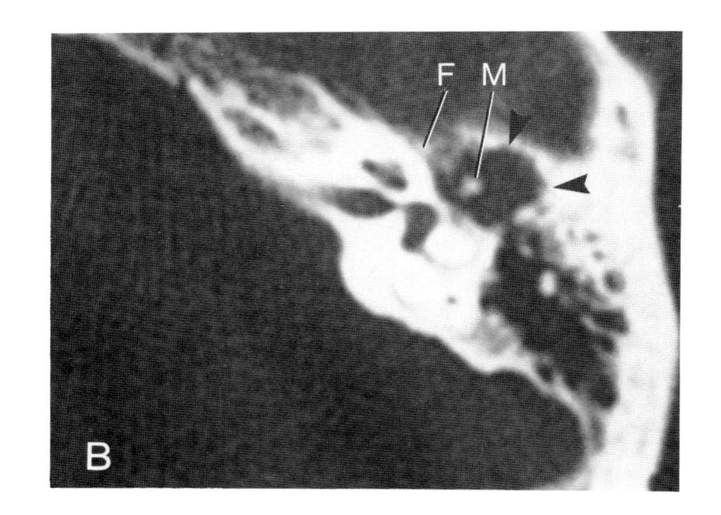

Plate 5.16*B*. Scan 1 mm inferior to 16*A* again shows enlargement of the epitympanic recess (*arrowheads*). A portion of the malleus neck is still preserved. On this section the facial nerve canal is seen along the medial wall of the tympanic cavity and has retained its bony covering.

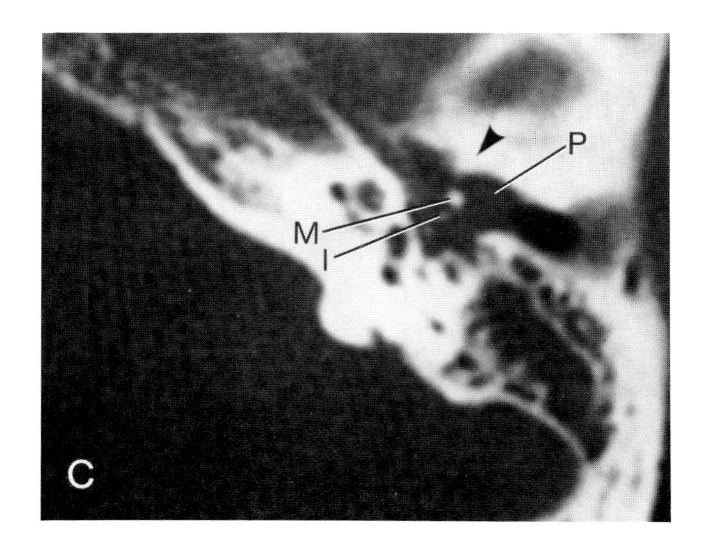

Plate 5.16*C*. Scan through the mesotympanum shows sclerosis of the margins of the tympanic cavity (*arrowhead*). Portions of manubrium of the malleus and long process of the incus can be seen in the soft tissue-filled middle ear cavity. The soft tissues that lie lateral to the ossicular remnants represent the attachment of the polyp to the roof of the external canal.

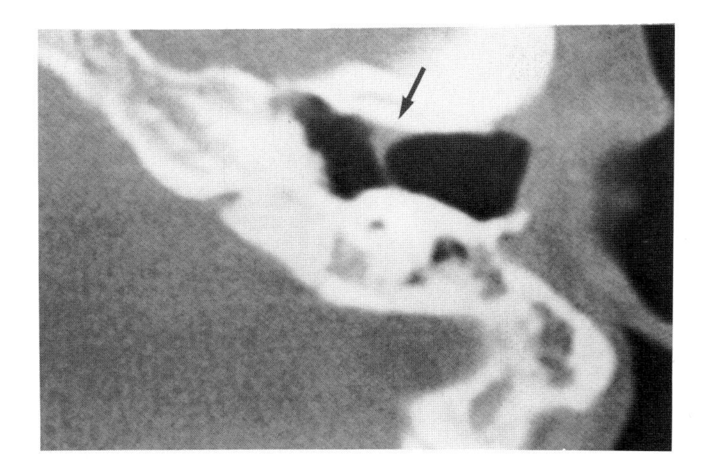

Plate 5.20. Lateralized graft. Because of progressive hearing loss in the postoperative period this patient was scanned. The mastoid cavity and middle ear cleft were entirely free of disease; however, at the level of the tympanic membrane one notices thickening and a characteristic wedge shape to the anterior margin (*arrow*). This increase in thickness and density of the tympanic membrane is related to separation of a graft from the remaining rim of normal tympanic membrane. The scarring that fills in the intervening space prevents good sound transmission and the separation increases with time.

Because the postoperative tympanic membrane is normally relatively opaque, the clinical diagnosis of lateral graft is not easy to make.

TRAUMA

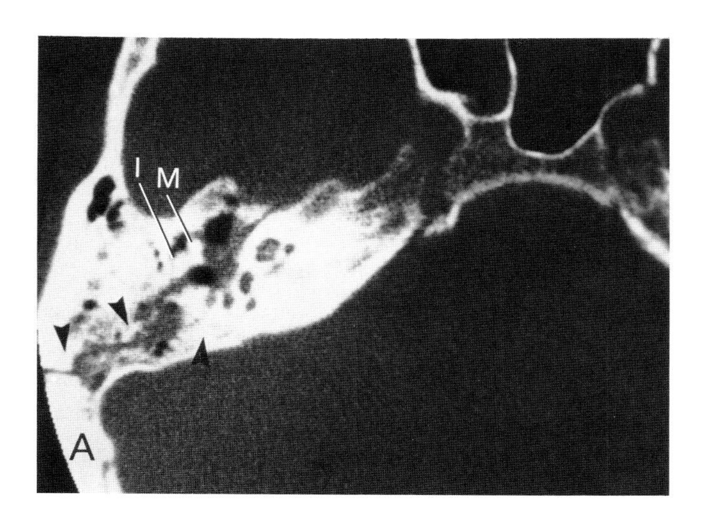

Plate 5.21*A*. Section through the epitympanic recess shows the longitudinal fracture as extended from the parietal region to involve the mastoid cavity and epitympanic recess (*arrowheads*). Both the malleus and incus are displaced laterally in the epitympanic recess.

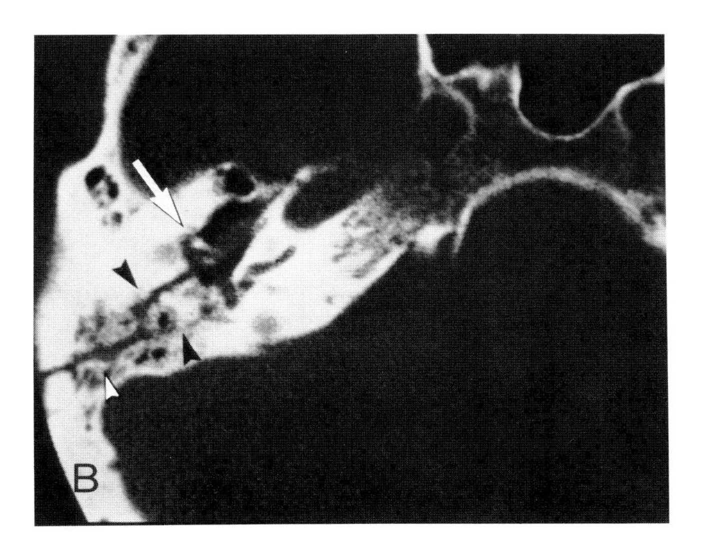

Plate 5.21*B*. The fracture is very jagged inferiorly and has one limb that goes through the middle ear cavity and another limb that goes posterior to the facial nerve and sinus tympani (*arrowheads*). The manubrium of the malleus and long process of the incus (*arrow*) are no longer parallel and are directed somewhat posterior, reflecting the displacement of the head and body of the incus seen in Plate 5.21*A*. The opacification of the mastoid air cells in this case was related to blood but could just as easily be related to cerebrospinal fluid.

Plate 5.22. Transverse fracture.

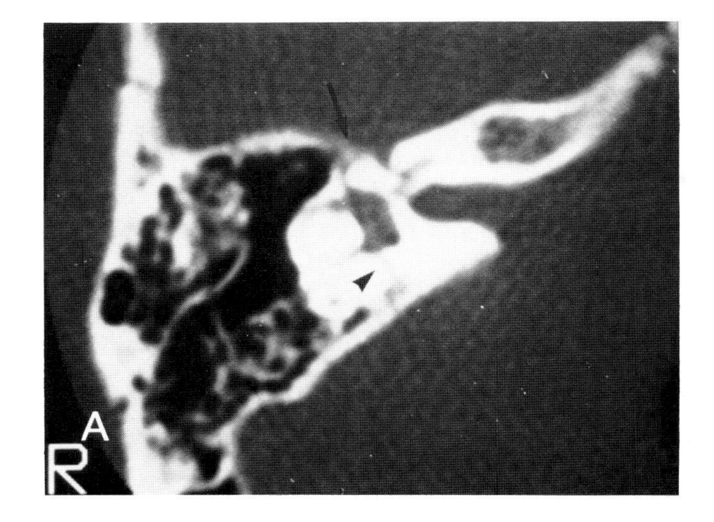

Plate 5.22*A*. Fracture through the superior semicircular canal region (*arrowhead*) involves the anterior limb but not the posterior limb of the superior semicircular canal. The fracture line goes into the region of the anterior genu of the facial nerve (*arrow*).

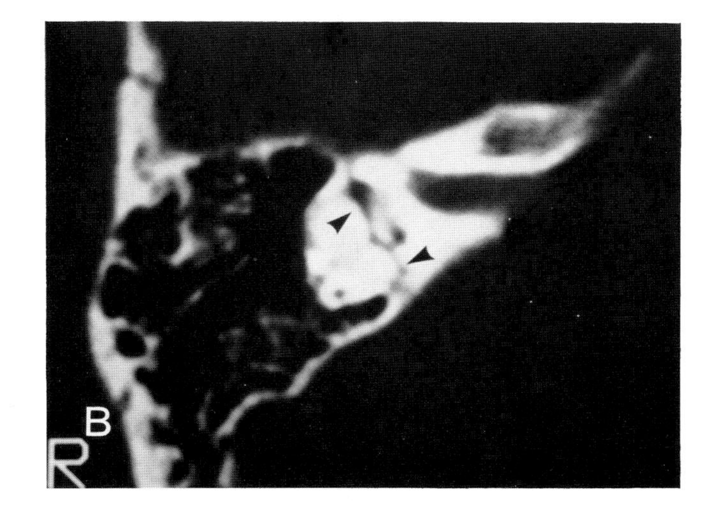

Plate 5.22*B*. The fracture line involves the vestibule (*arrowheads*) and distal anterior genu. The fracture fragments were separated but not displaced. This case demonstrates nicely how variable fracture lines can be. On one section they are faint and on a section 1 or 2 mm away the fracture lines are widely patent and clearly visible. The patient experienced total deafness but facial nerve function was intact.

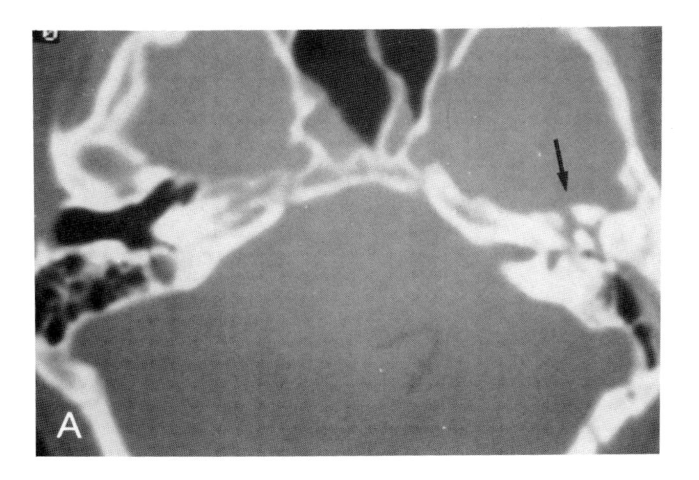

Plate 5.23. Fracture external canal. A variation of the transverse fracture is a force transmitted through the temporomandibular joint to the external canal and middle ear cavity.

Plate 5.23*A*. A comminution with a free fragment lying along the anterior wall of the epitympanic recess (*arrow*). The ossicles are displaced medially. Blood and cerebrospinal fluid are present in the epitympanic recess and mastoid cavity.

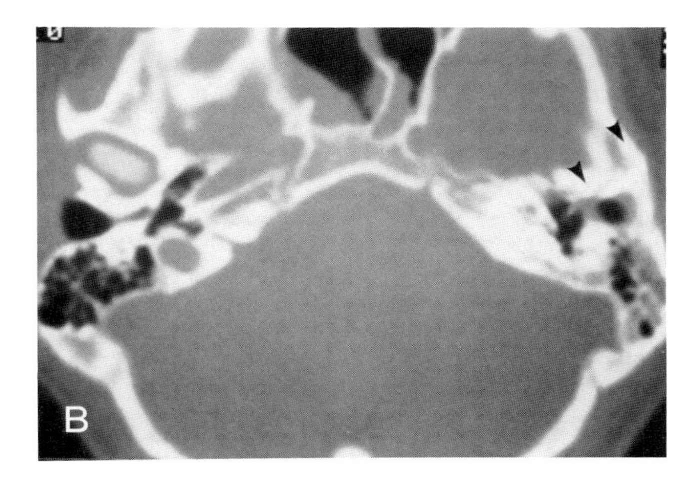

Plate 5.23*B*. The fracture extends from the temporomandibular joint into the hypotympanum (*arrowheads*) but not into the inner ear structures.

Plate 5.24. Combined acoustic neuroma and inflammatory disease.

Plate 5.24*A*. Scan following intravenous contrast viewed with brain window settings shows an acoustic neuroma (*arrowhead*) in the right pontocerebellar angle. The tumor can be identified arising from the internal auditory canal.

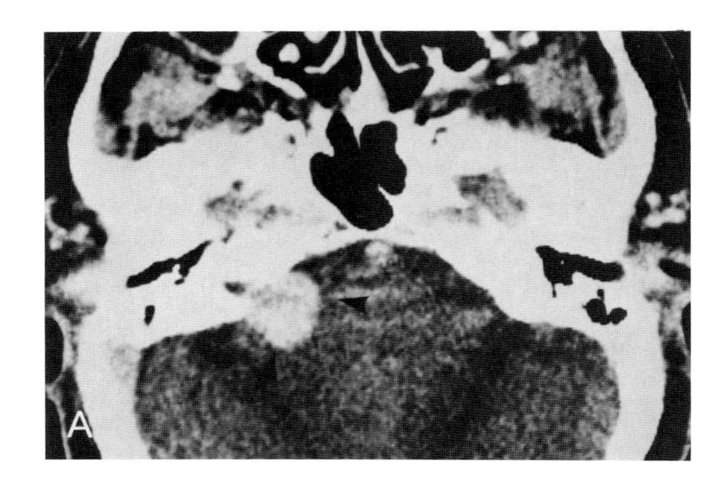

Plate 5.24*B*. The right temporal bone viewed with extended window and bone window settings shows a soft tissue density (*arrow*) lying along the medial wall of the middle ear cavity involving the oral window niche and the long process of the incus. The internal auditory canal is well visualized at this level but because of the extended scale and bone window settings we cannot identify the acoustic neuroma.

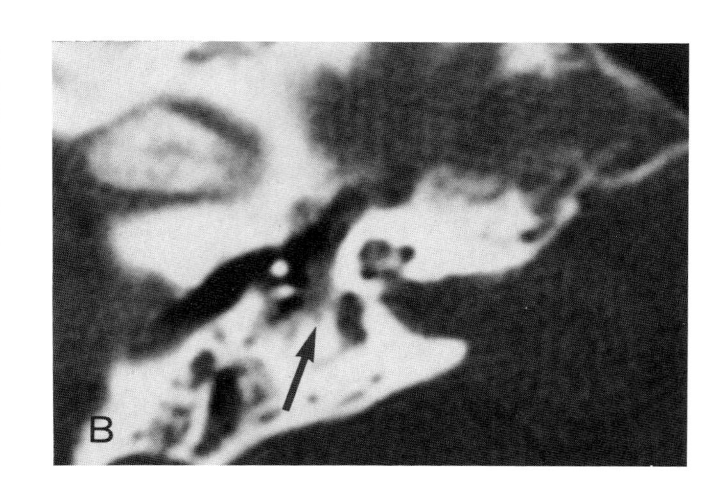

Plate 5.24*C*. Same patient. Sections near the inferior surface of the temporal bone using brain settings shows a density in the pontocerebellar angle (*crossed arrow*). This density closely resembles the acoustic neuroma but is not sharply outlined.

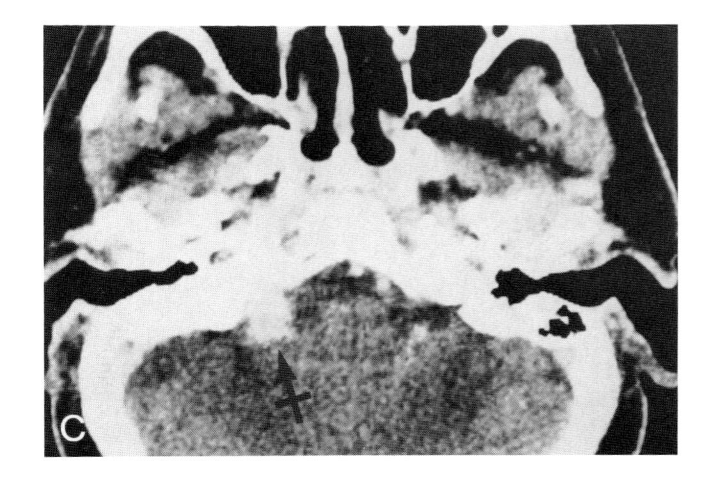

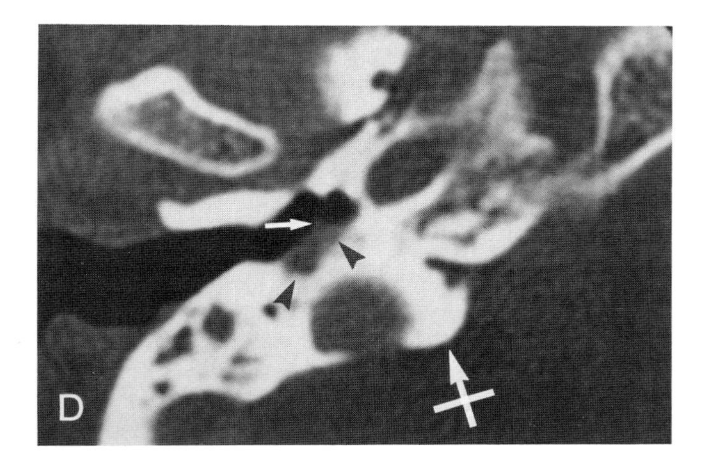

Plate 5.24*D*. Scan through the same level as Plate 5.24*C* using bone settings shows the jugular tuberosity (*crossed arrow*) is quite prominent in this patient. Partial volumning of the tuberosity makes it difficult to tell the inferior extent of the tumor. Note in the hypotympanum soft tissue density from inflammatory disease (*arrowheads*) that extends almost to the orifice to the eustachian tube. The tympanic membrane is retracted (*arrow*).

From the combined intracranial study and temporal bone study it is easy to understand how the patient presented with a combined sensorineural hearing loss (due to the tumor) and conductive hearing loss (secondary to inflammatory disease involving the long process of the incus and stapes superstructure).

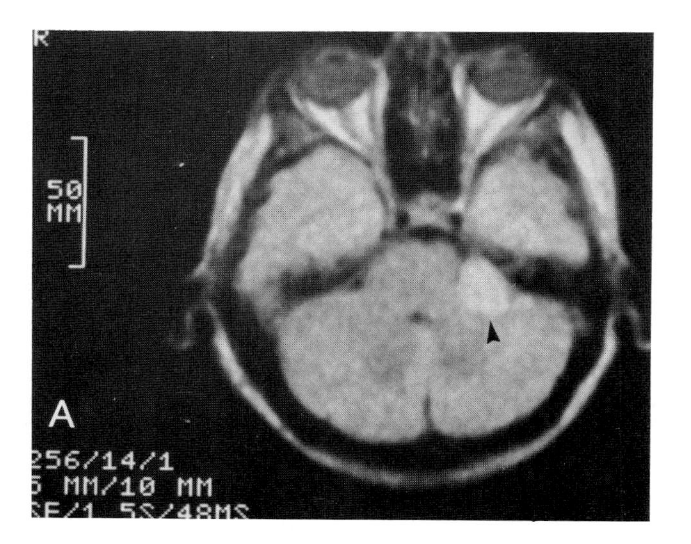

Plate 5.25. Acoustic neuroma—MRI study.

Plate 5.25*A*. This large tumor (*arrowhead*) produces distortion of the fourth ventricle and brainstem on the short TE of 48 msec. Its appearance is not unlike that of meningioma.

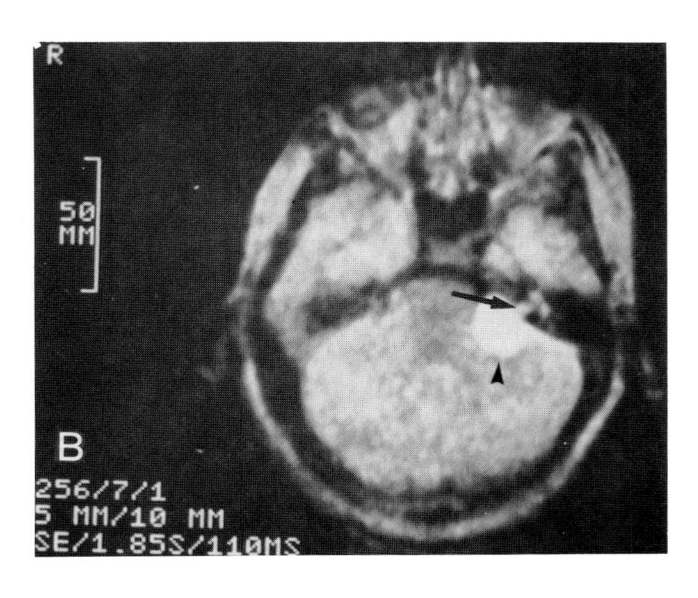

Plate 5.25*B*. With the long TE of 110 msec the acoustic neuroma is much brighter in signal (*arrowhead*) and extension of the tumor can be identified into the internal auditory canal (*arrow*). Occasionally low signal can be seen within portions of this type of tumor presumably related to blood vessels, necrosis or even calcification within the tumor.

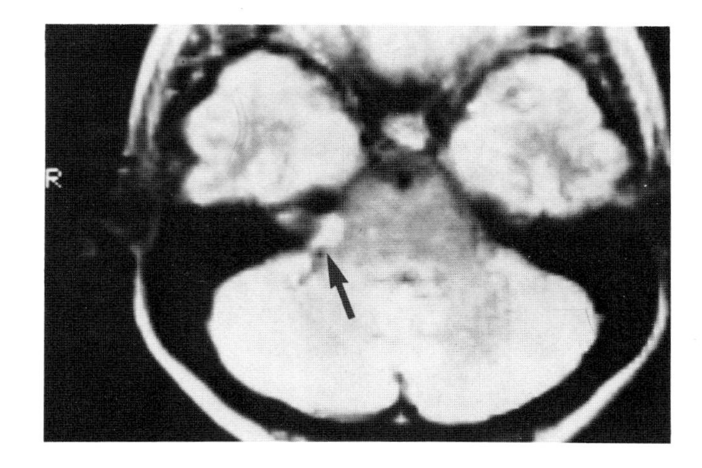

Plate 5.26. Acoustic neuroma—MRI study. This tumor of the CP angle merely bulges into the CP angle with the major bulk of the tumor lying within the internal auditory canal (*arrow*). (Courtesy of Dr. William Bradley, Huntington Research Foundation. Pasadena Calif.).

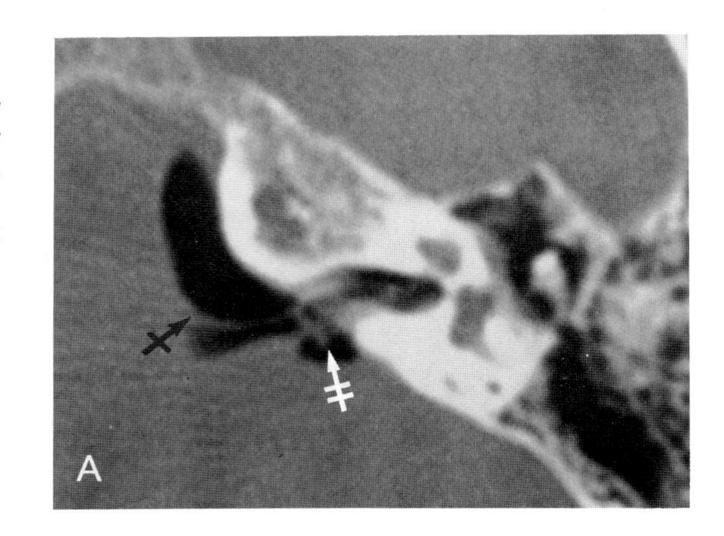

Plate 5.27. Normal air CPA study.

Plate 5.27*A*. The internal auditory artery (*crossed arrow*) characteristically makes a loop (*double crossed arrow*) adjacent to or around the VIII nerve. This loop may be outside the porus acusticus or within the canal. Note also that the internal auditory canal on this patient is not straight but has a distinct curve necessitating multiple 1-mm cuts to truly identify its contents.

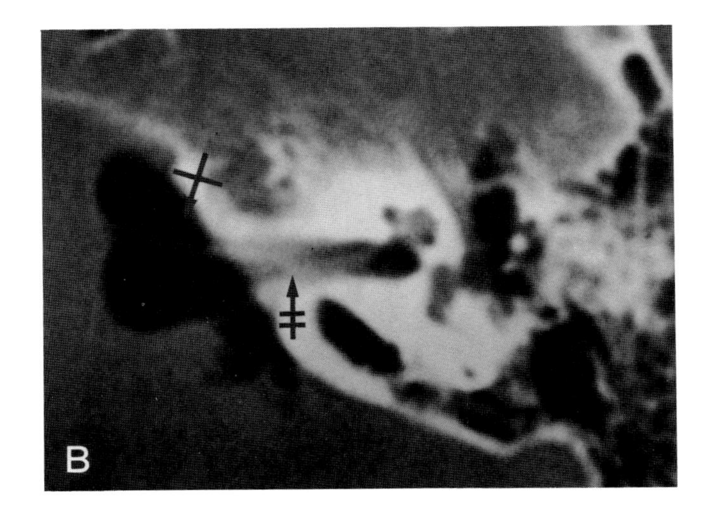

Plate 5.27*B*. The internal auditory artery (*crossed arrow*) in this patient has its loop (*double crossed arrow*) entirely within the internal auditory canal. If arachnoidal adhesions occur around the artery and trapped CSF, it may be extremely difficult to differentiate trapped CSF from intracanalicular tumor in a patient of this type.

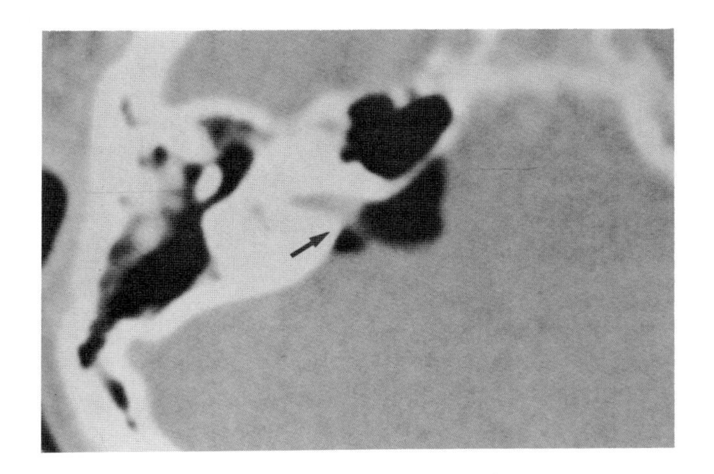

Plate 5.28. Intracanalicular tumors. Typical intracanalicular neuroma demonstrated with 3 cc of air maneuvered into the pontocerebellar angle. The tumor bulges into the pontocerebellar angle (*arrow*). The VIII nerve can be seen extending into the tumor.

In all probability these small intracanalicular tumors will become the domain of MRI because of ease of examination without lumbar puncture techniques.

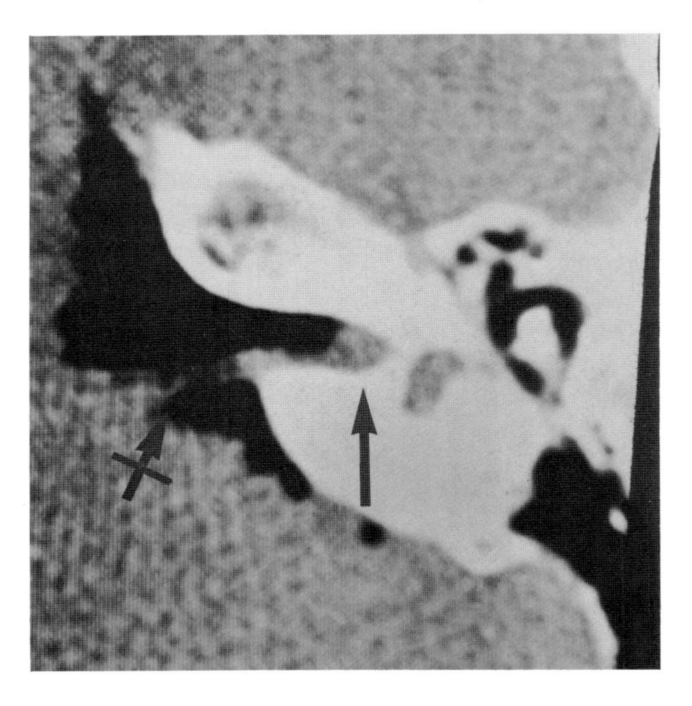

Plate 5.29. Intracanalicular tumor. CPA air study in this patient whose only symptom was tinnitus shows a normal appearing VIII nerve in the pontocerebellar angle (*crossed arrow*). In the extreme lateral portion of the canal, the tumor is seen with curved margins (*arrow*). If there is any question regarding the diagnosis in this type of case, the patient can be followed for a period of 6 to 9 months and examination repeated if symptoms progress.

The abbreviation used on Plate 5.30*B* is: *C*, cochlea.

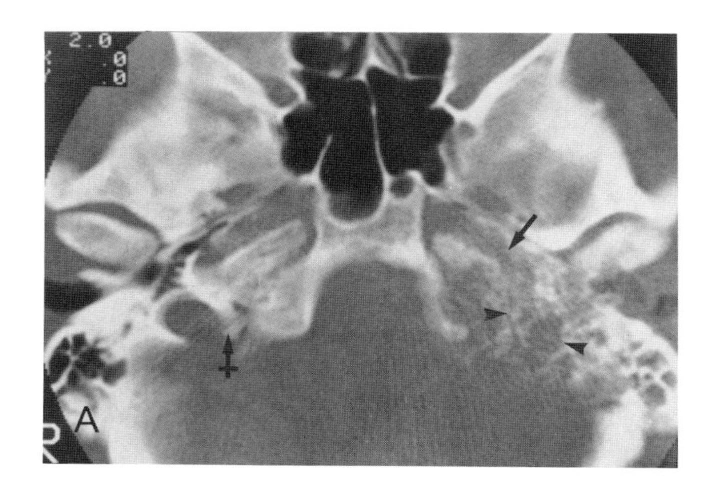

Plate 5.30. Glomus jugulare.

Plate 5.30*A*. Scan through the jugular fossa region shows total destruction of the jugular fossa region on the left involving both the vascular and neuro components of the foramen (*arrowheads*). On the right the jugular fossa is well preserved (*crossed arrow*). Exceedingly important information in this case is to note the status of the carotid canal. There is erosion of the lateral and entire posterior margin of the carotid canal (*arrow*).

The amount of intracranial extension into the posterior fossa and erosion about the carotid canal are important factors in determining operability of these lesions.

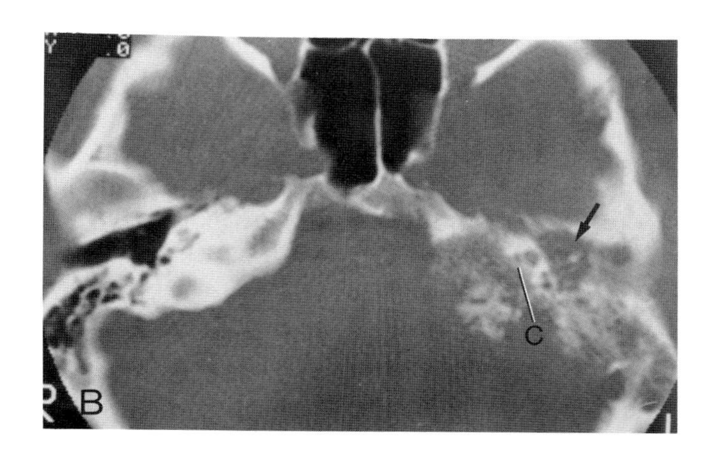

Plate 5.30*B*. Scan through the mesotympanum and mastoid air cell region shows the extensive bone destruction caused by this glomus jugulare tumor that extends all the way from the petrous apex throughout the mastoid air cells. The tumor has eroded portions of the cochlea and produced a soft tissue mass within the middle ear cavity (*arrow*).

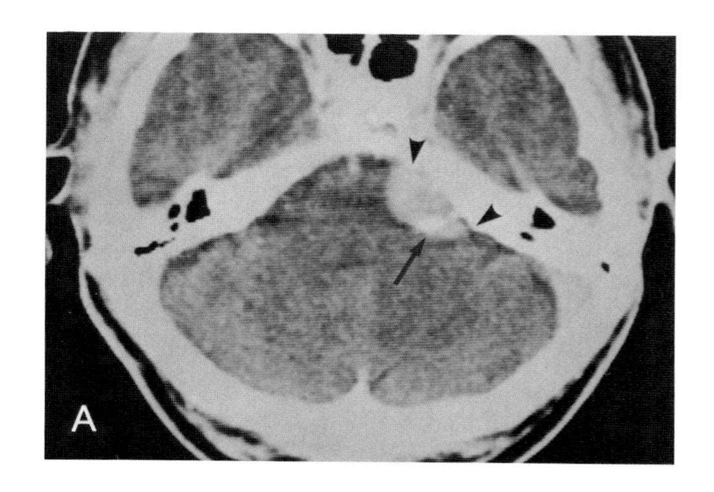

Plate 5.31. Meningioma.

Plate 5.31*A*. Meningiomas of the CPA characteristically have one margin that is flattened as the tumor creeps along the posterior surface of the temporal bone (*arrowheads*). Even though this tumor stains intensely on intravenous contrast, it is still posible to identify calcification (*arrow*) within the tumor.

Seeing a dense lesion in the CPA with a flattened margin on nonenhanced scanning would be virtually diagnostic of a meningioma.

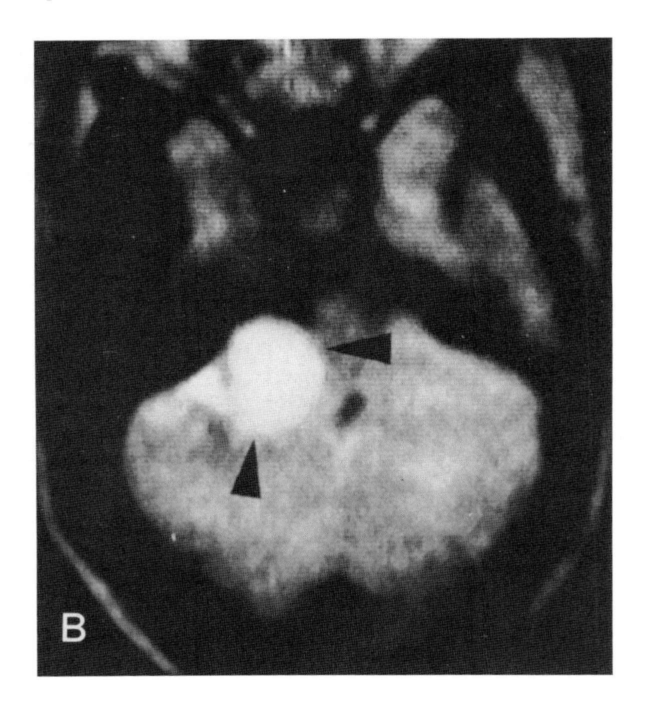

Plate 5.31*B*. Meningioma—MRI. In a different patient with a similar type of lesion one can see the intense stain produced by meningioma in the CPA (*arrowheads*). At the present time differential diagnosis of CPA from acoustic neuroma by MRI must rely on clinical findings and the fact that meningiomas rarely extend into the internal auditory canal to the extent that is found in acoustic neuromas.

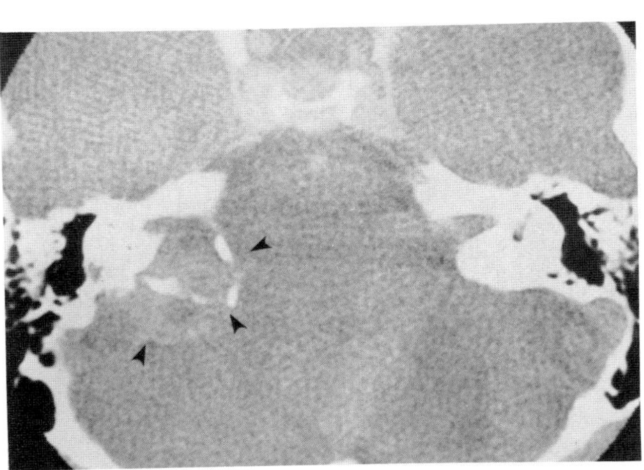

Plate 5.32. Primary cholesteatoma (epidermoid). Although the cells that give rise to epidermoids are undoubtedly present at birth, the lesions are extremely slow growing and take years to reach the size that cause clinical symptoms. The slow growth is evident in this patient by the degree of cortical expansion that has taken place in the pontocerebellar angle (*arrowheads*). The center of the tumor may appear isodense or hypodense by comparison to adjacent brain.

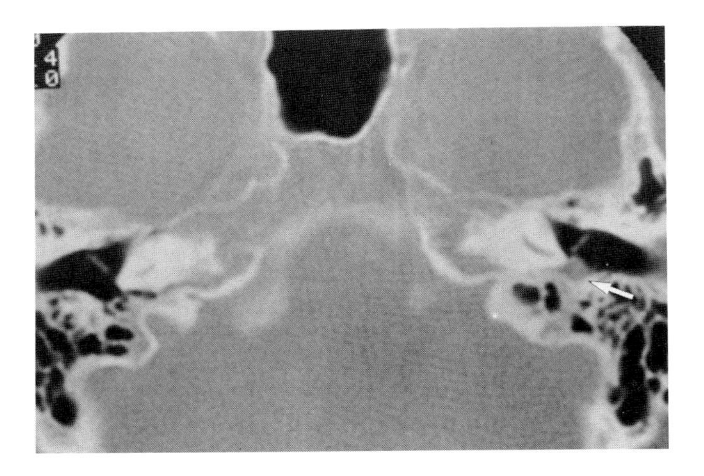

MIDDLE EAR TUMORS

Plate 5.33. Glomus tympanicum. The tumor mass (*arrow*) is well confined to the hypotympanum and bulges adjacent to the tympanic membrane. This particular lesion is more troublesome since it extends to involve the sinus tympani.

As with other glomus tumors it is important to see that there is no extension anteriorly into the region of the carotid canal or posteriorly into the intracranial cavity.

Plate 5.34. Squamous cell carcinoma.

Plate 5.34*A*. The tumor fills the mastoid antrum and epitympanic recess with soft tissue mass that is indistinguishable from chronic inflammatory disease (*arrowheads*). In this particular case bone destruction is seen along the posterior margin of the temporal bone and about the sigmoid sinus region (*arrow*). The right epitympanic recess is normal (*crossed arrow*).

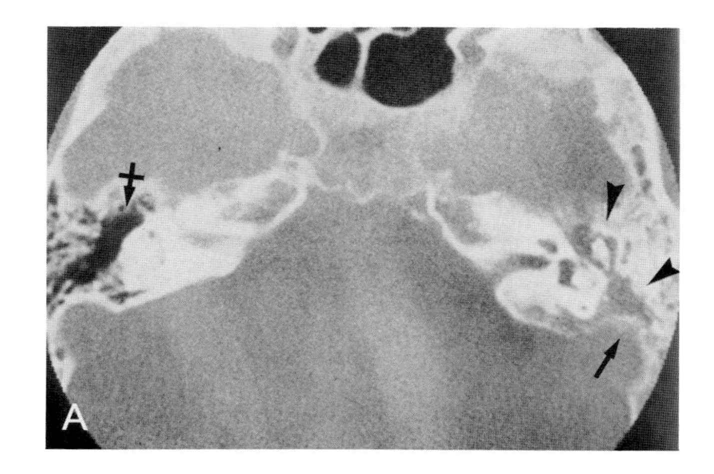

Plate 5.34*B*. This tumor causes destruction around the jugular foramen region (*arrowheads*) and throughout the middle ear cavity (*arrow*) not unlike a glomus jugulare. Tumor is extending into the external auditory canal and causes the eustachian tube orifice to be airless (*arrow with tail*). The right side is normal (*crossed arrow*).

Although the carotid canal remains normal the involvement of the eustachian tube is an ominous sign (see Plates 5.35*A* and *B*).

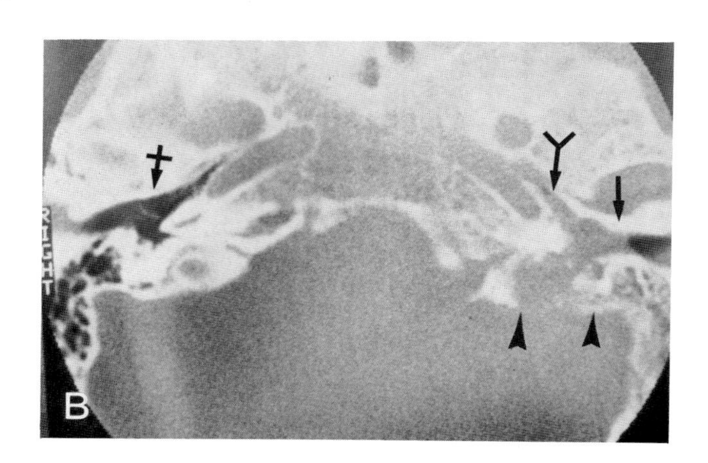

Plate 5.35. Carcinoma of the middle ear. (Courtesy of Dr. Rodney Million, University of Florida, Gainesville, Florida.)

Plate 5.35*A*. Tumor mass fills the middle ear cavity and causes destruction about the eustachian tube orifice (*arrow*).

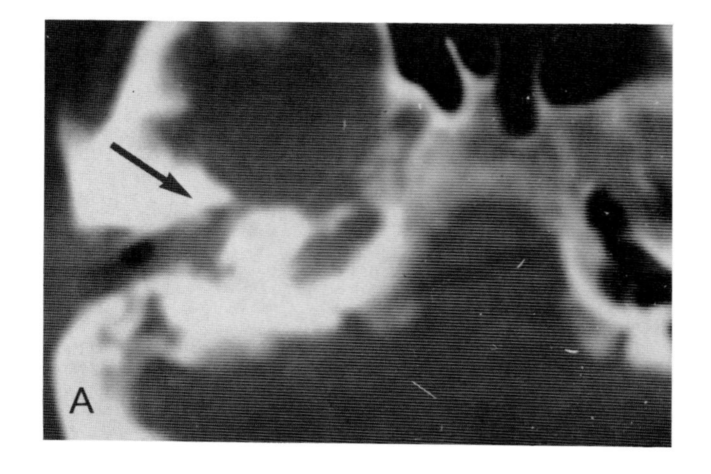

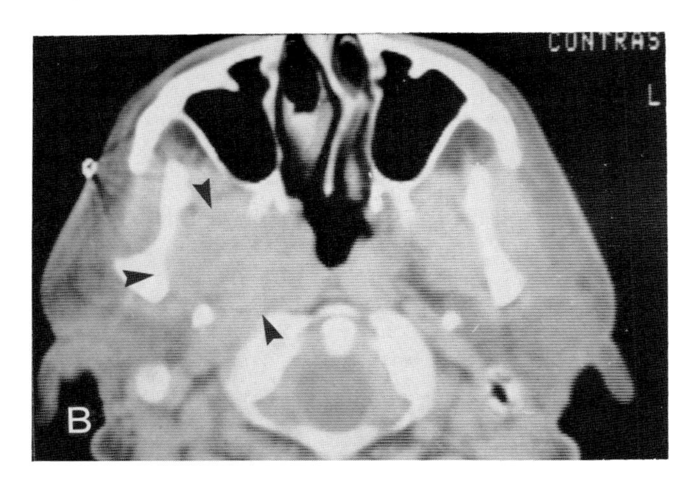

Plate 5.35*B*. The tumor has extended along the eustachian tube to cause obliteration of the fascial planes in the paranasopharyngeal space and in the infratemporal fossa (*arrowheads*).

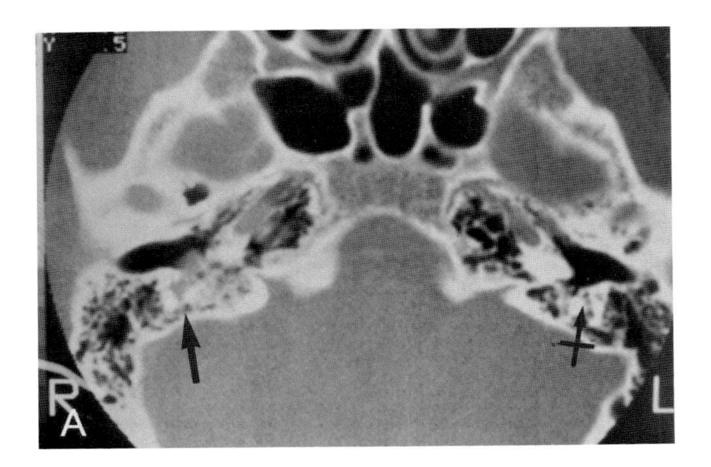

Plate 5.36. Facial neuroma.

Plate 5.36*A*. Scan through the hypotympanum shows a mass in the middle ear cavity that would not be unlike a glomus tympanicum or other small tumor. A clue to the diagnosis is the extension of the lesion along the facial nerve canal (*arrow*). Note normal facial canal on the left (*crossed arrow*).

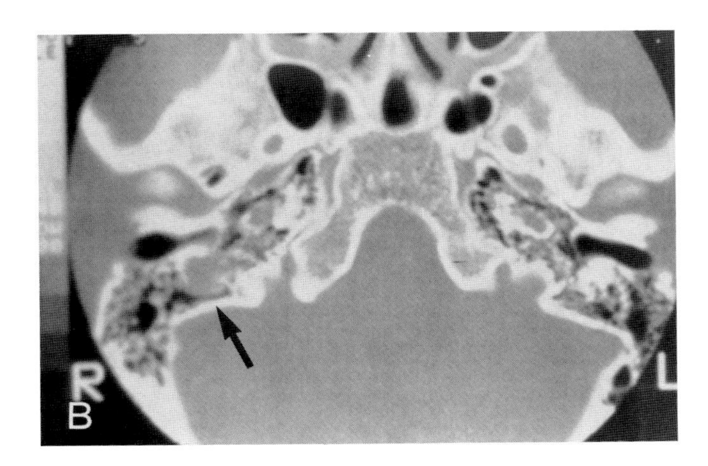

Plate 5.36*B*. The tumor appears larger as it extends toward the stylomastoid foramen (*arrow*).

This is an example of a plexiform type of neurofibroma. This particular tumor remains localized to the hypotympanum and descending facial canal. Neurofibromas may form large masses within the parotid gland and may extend throughout the deep neck muscles.

Plate 5.37. Metastatic tumor. Metastatic tumors produce a destructive process that usually centers about the mastoid air cells and extends throughout the cavities of the mastoid antrum and epitympanic recess (*arrowheads*). Characteristically, they produce cortical destruction rather than the usual type of sclerotic erosion seen with cholesteatomas.

This metastatic tumor was from the thyroid and its pattern of destruction would be indistinguishable from histiocytosis, rhabdomyosarcoma, or some of the other unusual tumors that can originate within the middle ear cavity.

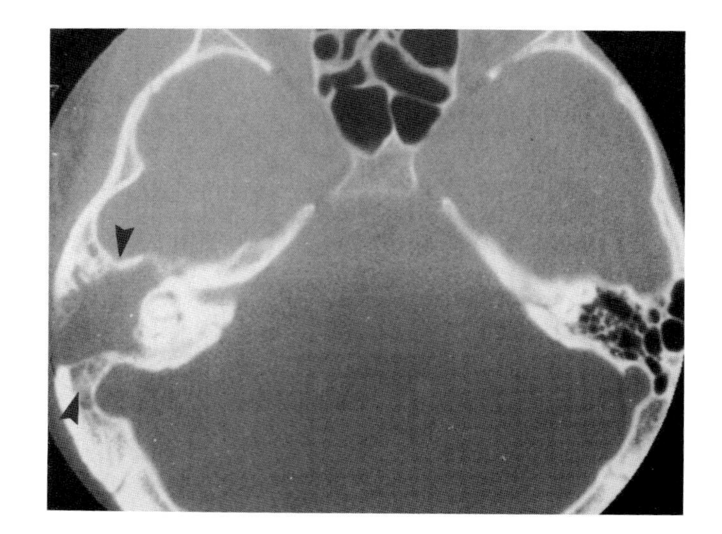

EXTERNAL EAR

Plate 5.38. Carcinoma.

Plate 5.38*A*. This tumor of the pina and external canal has invaded into the canal producing some bony destruction (*arrowheads*) but has failed to extend medially to the tympanic membrane. The middle ear cleft shows normal manubrium of the malleus and long process of the incus (*crossed arrow*).

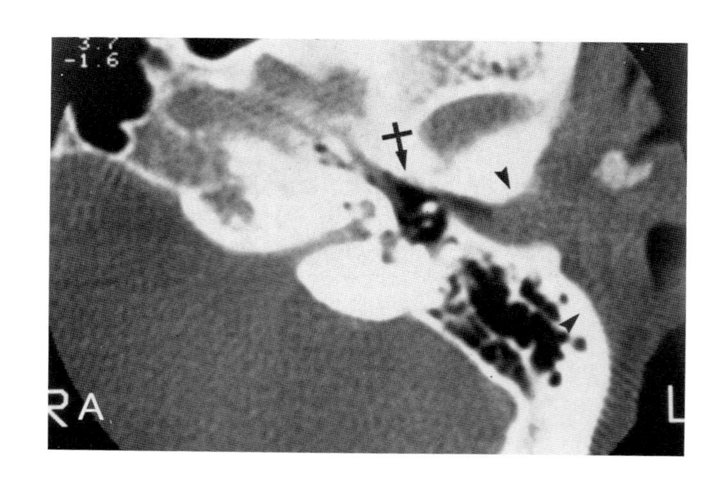

Plate 5.38*B*. Anteriorly, the tumor has destroyed the cartilagenous canal and invaded into the temporomandibular joint while, posteriorly, it has extended along the mastoid tip (*arrowhead*). The anterior portion of the hypotympanum (*crossed arrow*) gives reassurance that tumor has not extended into the eustachian tube or about the carotid canal, thus making radical temporal bone resection entirely feasible.

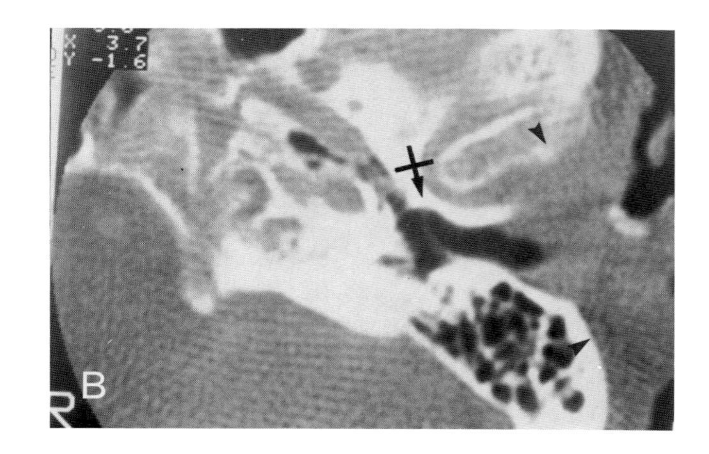

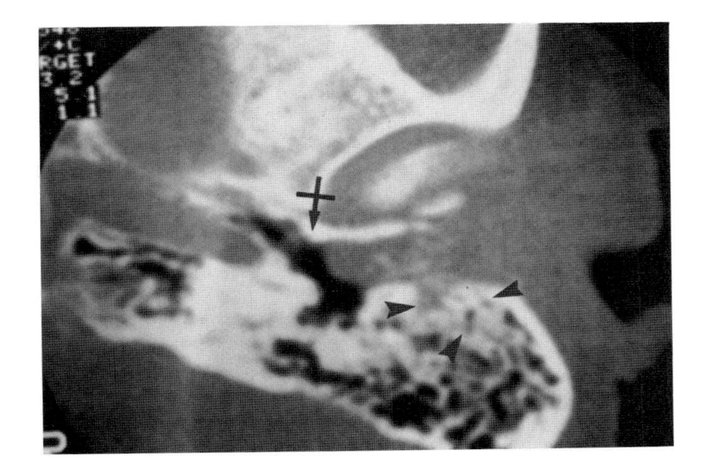

Plate 5.39. Primary cholesteatoma. The congenital cholesteatoma of the external canal has produced erosion of the posterior canal wall and invasion into the air cells low in the mastoid (*arrowheads*). The middle ear cleft is entirely normal (*crossed arrow*) as was the remaining epitympanic recess.

The absence of history of an ear infection plus a normal tympanic membrane are important diagnostic criteria of primary cholesteatomas. Unless secondarily infected they may show bulging masses that are similar in appearance to cristomas or other apocrine gland tumors.

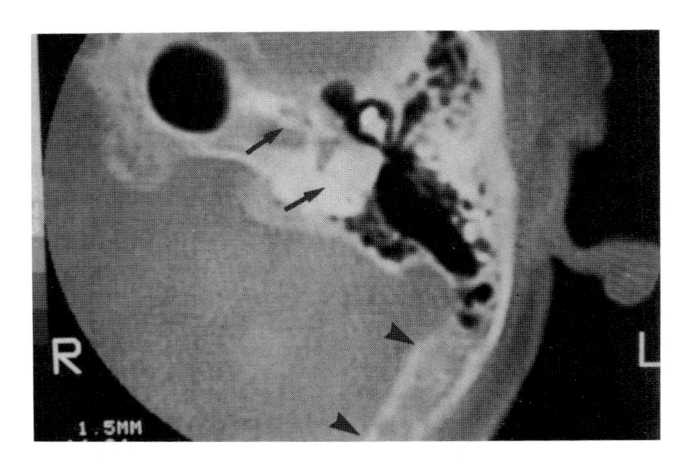

Plate 5.40. Paget's disease. The entire skull base has softened so that normal anatomic relationships are altered. The otic capsule surrounding the cochlea (*arrow*) and semicircular canals (*arrow*) shows patchy demineralization. The calvarium (*arrowheads*) shows typical widening of the diploic spaces.

Rarely is the sensorineural hearing loss of Paget's disease related to encroachment upon the internal auditory canals to a sufficient degree to compress the nerves.

Plate 5.41. Otosclerosis.

Plate 5.41*A*. Although the proliferative changes and calcific plaques of otosclerosis can occasionally be demonstrated, it is rarely necessary to perform CT scanning for the diagnosis. On this scan a plaque is seen adjacent to the basal turn of the cochlea (*arrow*).

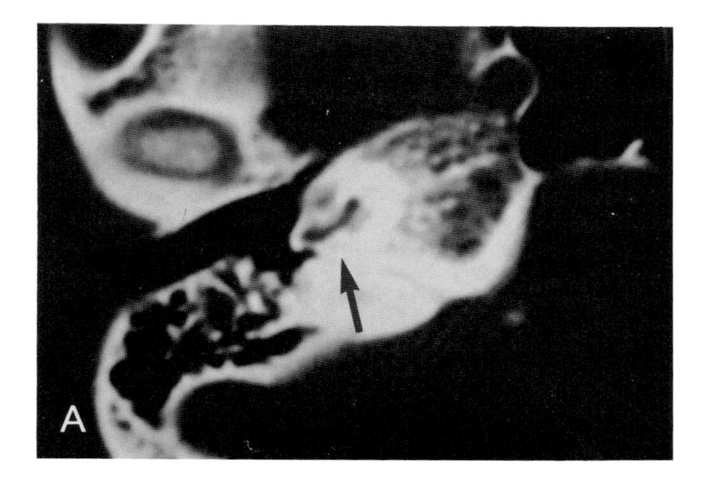

Plate 5.41*B*. In this patient the osseous spiral ligament in the cochlea appears much more prominent than usual (*arrow*). There is also thickening about the footplate of the stapes.

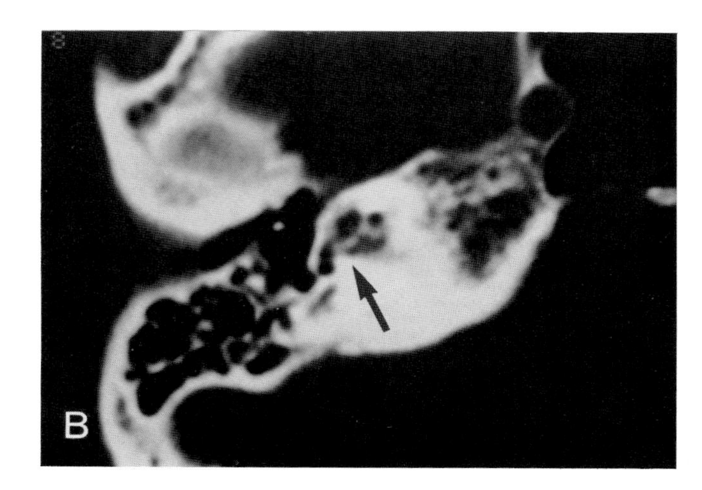

Salivary Gland

INTRODUCTION

Aspiration cytology has so altered the management of parotid lesions that all diagnostic modalities must be examined in the light of this examination (31). Radiology still combines with the clinical examination, the cytology examination and eventual therapeutic management to round out the total care of the patient by providing anatomical information. Occasionally histology of a lesion can even be suggested but, by and large, cytology has assumed the role of ultimate diagnosis. Within this framework we will now propose the role of radiology in salivary gland disease.

INDICATIONS FOR STUDY

Diffuse Parenchymal Disease

Diffuse parenchymal disease within the salivary gland is best studied by conventional film sialography (8). In many instances the patterns of ductal enlargement and parenchymal changes are pathonomonic of a particular disease entities such as Sjögren's, etc. The clinical impressions can be verified quickly and economically. Ductal stones in the parotid system are occasionally nonopaque and these filling defects can be identified. Similar information can be obtained from CT following IV contrast but the examination is more expensive and frequently not as specific. Submandibular gland stones are usually opaque and readily identifiable on intraoral films. Other pathology in the submandibular gland is not quite so clear cut as in the parotid gland and CT studies of the submandibular gland accompanied by intravenous contrast can be an effective method of studying for nonopaque ductal stones, diffuse disease, and differentiation of extrinsic from intrinsic pathology.

Tumors

The information necessary for management of a parotid lesion is:

1. Intrinsic versus extrinsic tumor.
2. Relationship to the facial nerve.
3. Benign versus malignant.

CIRCUMSCRIBED SUPERFICIAL LOBE TUMORS

There would seem to be very little justification for performing an expensive x-ray examination to confirm the presence of a tumor mass in the superficial lobe of the parotid gland which the patient himself probably discovered in the first place while shaving or applying facial cosmetics, the physician has already palpated, and the roentgen examination establishes with about an 85% accuracy rate. If the mass is in the superficial portions of the parotid gland and is well circumscribed, not stony hard, and freely movable, no additional anatomical information is needed. Aspiration cytology can easily be performed in a palpable lesion with the results giving the histology of the mass. The surgeon will then have all the information he needs for accurate preoperative informed consent and planning his surgical procedure which is a superficial parotidectomy.

ILL-DEFINED TUMORS—SUPERFICIAL OR DEEP

When a tumor mass has ill-defined borders on clinical examination, the cause may be due to a malignant infiltrating lesion, a localized inflammatory process, or that the examining physician is merely feeling the top of a deeply located tumor. Under these circumstances the clinician will not have information concerning the relationship of the facial nerve to the tumor nor can he or she be certain of whether the tumor has originated intrinsic or extrinsic from the salivary gland. CT performed with and without intravenous contrast enhancement will usually show the extent of the lesion. Its gross morphology may suggest the diagnosis but cytology with histological examination is essential for an accurate preoperative pathological diagnosis.

Relationship of Mass to Facial Nerve

During a superficial parotidectomy, the surgeon identifies the branches of the facial nerve in order to preserve its function and to develop the plane of dissection. He may do this by isolating the trunk of the facial nerve at its exit from the stylomastoid foramen or he may find branches of the facial nerve anterior to the parotid and trace them back to the main trunk in a retrograde fashion. Preoperatively, it is very helpful for the surgeon to know which approach to the facial nerve is necessary because the antegrade one is preferred (39). The surgeon will also want to know if it is necessary to remove the tip of the mastoid process in order to find the origin of the facial nerve at the stylomastoid foramen. Malignant tumors that have diffusely infiltrated the region of the plane of the facial nerve can be identified by CT sialography and advance notice can be given of an extremely difficult dissection or the necessity for sacrifice of the facial nerve.

If a tumor is located deep to the facial nerve (in the

deep lobe of the parotid), the management is changed from a superficial to total parotidectomy (43). Surgeons need to be aware of the circumstances preoperatively in order to plan the approach and to inform the patient of the added risk of morbidity to the facial nerve (34).

DEEP LOBE TUMORS

When a tumor mass can be palpated that is deep in location or bulges into the oral pharynx, more anatomical information is needed for surgical planning. CT study with intravenous contrast will generally show the confines of the lesion and whether it has originated from within the deep lobe of the parotid or is extrinsic to the parotid. Again, cytology will provide the ultimate pathological diagnosis for accurate surgical planning.

EXTRINSIC TUMORS

When a mass is near the borders of the parotid gland or submandibular gland the clinician may or may not be certain of the exact etiology or its extent. Cytology without CT may be sufficient under some circumstances because the pathological diagnosis may confirm the presence of a lesion such as a parapharyngeal cyst. On the other hand, if additional information is needed, CT with intravenous contrast should be added to the patient work-up.

Nondiagnostic Cytology Studies

CT with intravenous contrast will materially assist in identifying the areas for additional samplings or may even suggest the pathological diagnosis. Nondiagnostic cytology studies may be on the basis of inadequate material or the cellular material obtained was not specific for a known lesion. The diagnostic skills of pathologists vary considerably with regard to aspiration cytology. Many areas of the country do not have such skills available. Under such circumstances CT may contribute significantly to preoperative diagnosis in selected circumstances.

Retrograde Sialography and CT

Under some circumstances the diagnosis may still be obscure by all of the above examinations. The CT study can be performed by scanning following retrograde contrast injection to the ductal system. This is especially helpful when major ducts have been encased by tumor. The needle placement for aspiration cytology can be directed by CT control. Unusual tumor locations and other medical considerations may make intravenous contrast hazardous or inadequate.

MRI Examinations

With presently available MRI examinations parotid studies have been disappointing. The sections are quite thick and the bright signal originating from the parotid gland easily obscures a small lesion within the substance of the gland. Tumors rich in lymphatic tissue or inflammatory disease may give a similar bright signal to the normal parotid gland. For the present time CT is probably the examination of choice for tumor masses and plain film sialography for diffuse disease.

In the submandibular region MRI may play a more significant role because of its ability to demonstrate fascial planes of the neck, and the ease with which coronal sections can be performed better defines extent of disease.

TECHNIQUE OF CT SIALOGRAPHY

Parotid Masses

1. With the patient lying supine on the examining couch a skin mark is made over the center of the palpable mass. Depending on the tumor location, the chin is either extended or flexed to get any metal fillings in the teeth out of the plane of the tumor for CT scanning.
2. Prior to intravenous contrast, sections are taken at 5-mm thick through the center of the tumor, and an additional section is taken above and below the initial scan at 5-mm intervals.
3. Intravenous drip is started giving 150 ml of 60% contrast at a very rapid rate. During the final third of the contrast infusion the same three scans are repeated. Additional sections are taken either above or below the original sections so that the superior and inferior limits of the tumor mass have been clearly encompassed.

Submandibular CT with IV Contrast

The question to be answered in submandibular sialography is usually intrinsic versus extrinsic tumor. The submandibular gland is able to concentrate the iodine of the contrast material sufficiently in its cellular function so that ductal studies are rarely needed. Our technique may frequently be modified to conform to specific situations but in general is as follows:

1. A rapid intravenous drip of 150 ml of 60% contrast or 300 ml of 30% contrast is started with the patient lying supine on the examining couch with the chin well extended.
2. Scanning is begun 5 to 10 mm below the level of the hyoid bone using 5-mm-thick sections at 5-mm intervals. The first section is taken after approximately half to two-thirds of the IV infusion has been administered. The sections are conducted through the levels of the salivary gland up to the level of the tooth roots in the mandible.
3. If anything is suspicious as seen in the tongue base, tonsillar beds or retropharyngeal area, additional contrast is started and scanning continued until the questionable area has been completely evaluated.

Sialography and CT

When additional information is needed regarding the limits of a tumor mass, or in some special clinical situations, retrograde injection of the ductal system of either the parotid or submandibular gland can be added to the CT examination. Under most circumstances a conventional intravenous contrast study will already have been performed which would provide us with baseline studies of the tumor in question. If they are not available three noncontrast sections are taken through the tumor mass as outlined by intravenous contrast CT. The ductal system of the parotid or submandibular gland are cannulated with flexible catheters and needle combinations depending on the size of the orifice of the duct. For parotid lesions three cannula have proven useful (Fig. 6.1). Beginning with the larger bore and ending with a smaller cannula, the names of the needles are as follows:

1. Tapered tip Muller cannula
2. Lacrimal cannula manufactured by Becton-Dickinson
3. Rabinov sialography catheters manufactured by Cook and Company

In the submandibular gland either the "blue (0.30 mm outside diameter (OD)) or red (0.11 mm OD)" Rabinov sialography catheters prove ideal depending on the size of the duct orifice.

Sixty percent water-soluble contrast agents are used for both submandibular and parotid sialography and CT. A minute amount of contrast is injected for each section performed. Initially, a parotid gland may require 1 to 2 ml of contrast to fill the ductal system but subsequent injections will only require 0.1 to 0.3 ml of contrast to redistend the ductal system.

1. Sections are taken through the tumor using 5-mm-thick sections at 5-mm intervals.

2. The chin is either raised or lowered to get dental fillings out of the plane of the scan.

In the past ethiodized oil (Ethiodol) was used for CT retrograde sialography. Several patients have been restudied after 2- to 3-yr follow-up. Retained Ethiodol is still visible in some of the glands of these patients. Because of the possibility of granuloma formation we switched to water-soluble media.

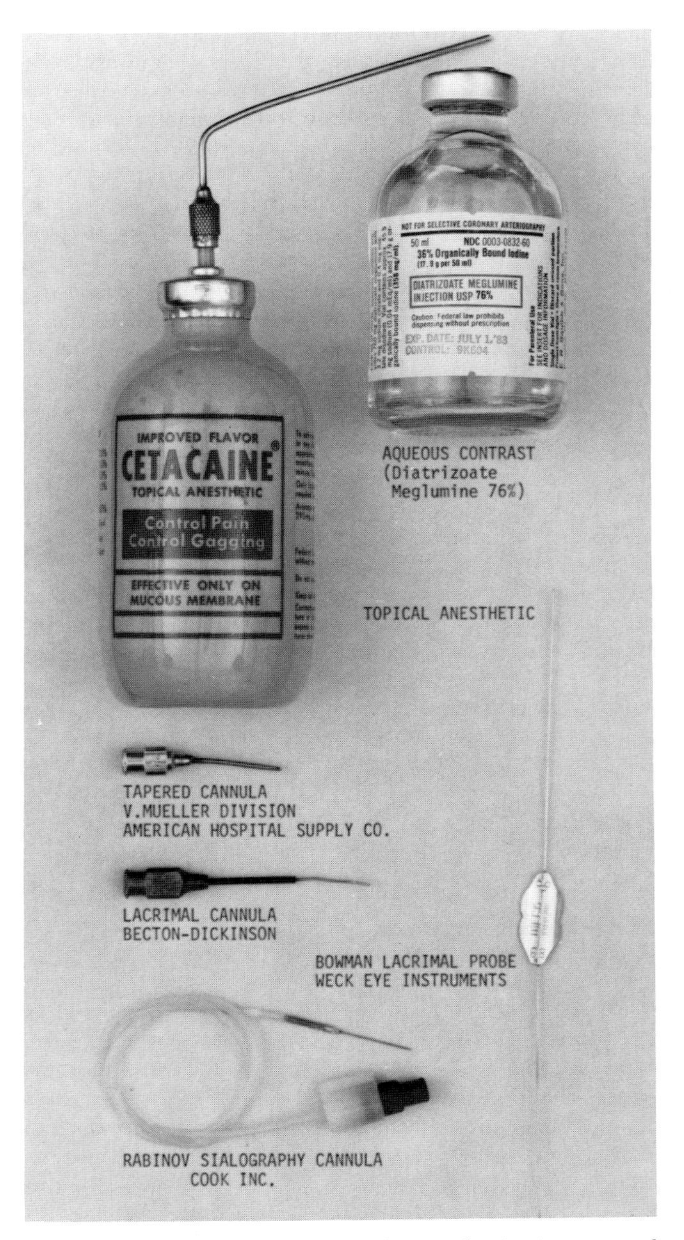

Figure 6.1. Equipment. Local anesthesia is sprayed on the intraoral opening of Stensen's duct. Water-soluble contrast material is used for contrast to insure a parenchymal staining. The exact concentration does not appear crucial and the less viscous material seemed to stay in the ducts for a longer period of time. An assortment of cannulas of various sizes and configurations can be quite helpful to assist in cannulating the very small ducts and to prevent reflux of contrast material into the oral cavity when the openings are patulous. At times, a Bowman lacrimal probe can be helpful in dilating stenotic orifices.

ANATOMY

Parotid Gland

Descriptions of the parotid gland, the largest of the salivary glands, are filled with myths and inaccuracies (18, 19). There are disagreements as to whether or not the parotid contains a distinct fascial capsule, although all agree that the deep cervical fascia penetrates completely through the gland in multiple areas producing lobes. The "lobes" are inconstant and therefore the various regions of the parotid glands are spoken of only as areas. Even the most commonly used differentiation of superficial and deep lobes is somewhat arbitrary and is created by the surgeon's scalpel as he traces the facial nerves.

The parotid gland molds itself to configuration of adjacent bony prominences and the muscle bundles giving rise to multiple indentations of the surface of the gland that may closely simulate extrinsic tumor masses on conventional sialograms. Most notable are those where the posteroinferior pole of the gland wraps around the tip of the mastoid process (and the sternocleidomastoid muscle) while the anteroinferior margin (sometimes spoken of as the tails of the parotid) arches around the angle of the mandible. The deep portion of the parotid takes on a curvilinear contour as it courses posterior to the neck of the mandible, passes deep to it, and comes to abut the parapharyngeal space. In extending deep to the mandible, the gland passes through a slight narrowing between the angle of the mandible anteriorly, the stylomandibular ligament and styloid process posteriorly, and the stylohyoid ligament inferiorly.

Stensen's duct is the excretory duct of the parotid gland. It passes anteriorly from the midportion of the gland to surround the masseter muscle then penetrates the buccinator muscle and fat pad, finally draining into the mouth opposite the second upper molar.

The arrangement of the facial nerve within the parotid gland is of major surgical concern. CT sialography can delineate the same anatomic landmarks that are used to isolate the nerve at surgery. The nerve exits the skull at the stylomastoid foramen and enters the posterior deep surface of the parotid gland (Fig. 6.2). The point of entry is immediately lateral to the styloid process, anterior (and slightly superior) to the posterior belly of the digastric, and medial to the tip of the mastoid process. The posterior belly of the digastric muscle can be located on CT scans by recognizing the typical anteromedial course of this well defined muscle bundle deep to the origin of the sternocleidomastoid muscle at the mastoid tip. Soon after entering the gland, the main facial nerve trunk divides. The number of nerves going through the substance of the gland are quite variable. They generally parallel the course of Stensen's duct and always lie lateral to the retromandibular vein. The nerve itself usually cannot be seen on CT scans.

The lymphatics of the parotid occur in three main groups: (*a*) a superficial group of lymphatics and nodes that drain from the temporal and frontal region of the scalp, the eyelids and anterior auricle: these empty into a posteriorly placed, superficial cervical chain of nodes; (*b*) a deep group of nodes lies within the substance of the gland and drains the eustachian tube, external auditory canal and deep portions of the face. These nodes may empty anteriorly into the subparotid nodes or posteriorly along the retromandibular vein into the deep cervical chain; and (*c*) a third group of deep nodal chains drains into the upper jugular chain which follows the course of the internal jugular vein. Enlargement of any of these nodes can be detected by CT when intravenous contrast material is used. The differential diagnosis of intrinsic from extrinsic masses which are due to deep cervical lymph nodes is more pressing when considering the submandibular gland than the parotid gland.

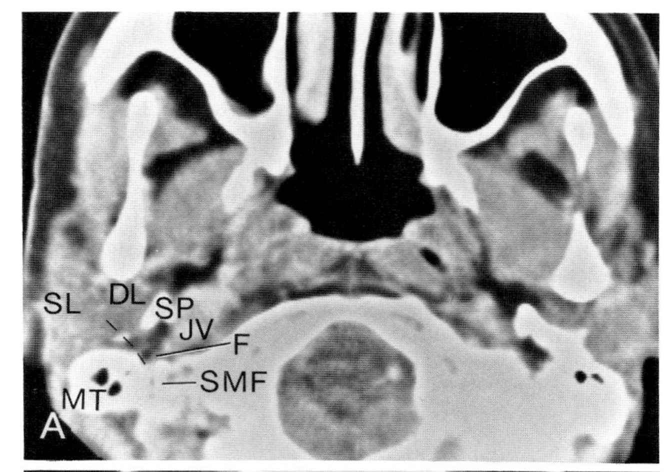

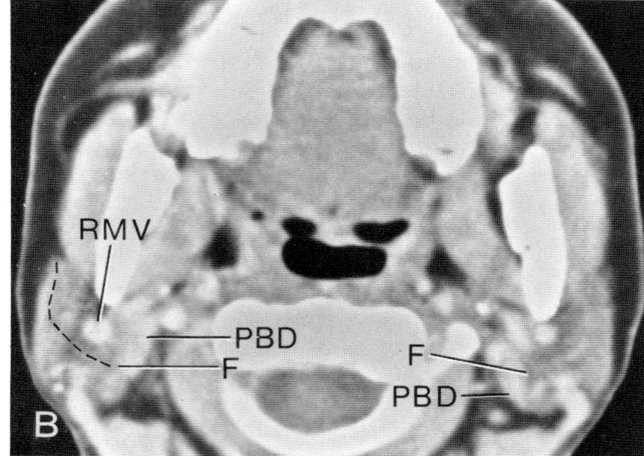

Figure 6.2. Course of the facial nerve. *A*, section through the mastoid tip (*MT*) is at the level of the stylomastoid foramen (*SMF*). The facial nerve (*F*) exits from the stylomastoid foramen and courses lateral and inferior (as shown by the *dashed line*). It lies lateral to the styloid process (*SP*) and jugular vein (*JV*). The main trunk of the facial nerve divides first into a postauricular branch and a muscular branch which supplies the posterior belly of the digastric. As the facial nerve courses anteriorly and inferiorly in the substance of the parotid gland it arbitarily divides the parotid into a superficial lobe (*SL*) and a deep lobe (*DL*). Within the substance of the parotid gland the nerve undergoes a plexiform arrangement (the pes anserinus). When it finally leaves the anterior border of the parotid gland there are approximately five divisions of the nerve. *B*, section through the midportion of the parotid gland shows the approximate course of the facial nerve (*dashed line*). On the patient's left, a branching shadow is visualized which may represent facial nerve giving off the branch to the posterior belly of the digastric (*PBD*) and then the nerve continues on anterolaterally. The nerve always lies lateral to the retromandibular vein (*RMV*).

The CT density of the superficial and intraparotid nodes is almost identical with that of normal parotid tissue. Following IV contrast, nodes that are present within the parotid gland will appear as low density regions as compared to normal parotid parenchyma which takes up intravenous contrast material. They are usually not quite as dense as tumor masses following contrast but this is not a reliable criteria of tumor versus lymph node. The intraparotid nodes may also be confused with the retromandibular vein which lies approximately 4 to 6 mm posterior to the neck and body of the mandible. Intravenous contrast during CT scanning will definitely identify the vein.

Submandibular Gland

The submandibular gland is less well defined by comparison to the parotid gland. It rests folded over the posterior free margin of the mylohyoid muscle. Approximately two-thirds of the gland lies inferior and superficial to this muscle while one-third rests on the superior surface of the mylohyoid muscle. The excretory duct originates from the anteromedial border of the gland and courses over the posterior surface of the mylohyoid muscle. It lies adjacent to the floor of the mouth until it reaches the region of the frenulum of the tongue where it finally exits through a small papilla. It may be joined by ducts from the sublingual glands near its opening. The main portion of the submandibular gland is indented by the lingual artery along its lateral border and the common facial vein along its posterolateral border. These vascular structures may closely simulate small tumor masses along the surface. Multiple lymph nodes lie on the surface of the submandibular gland draining the lips, submental nodes, anterior two-thirds of the tongue, and the nose. Because of the abundance of lymph nodes along the surface of this gland, differential diagnosis of extrinsic versus intrinsic tumors can be extremely difficult.

SALIVARY GLAND PATHOLOGY AND CT FINDINGS

Inflammatory Disease

As mentioned under Indications for Study, diffuse diseases of the salivary glands are best studied by conventional sialography rather than CT. Since localized masses may be found in some of these diseases at an advanced state, some mention of the CT findings will be discussed.

CHRONIC SIALADENITIS

Chronic sialadenitis is characterized by intermittent diffuse swelling of the gland and marked dilatation of the main duct. It may or may not be accompanied by stones. The duct develops secondary strictures and frequently assumes a sausage-link type of configuration. If a localized region of the gland becomes totally obstructed retained secretions may lead to a mucous-filled cyst.

CT Findings

The density of the gland is increased in an irregular fashion especially after the addition of intravenous contrast. The cystic lesions are then clearly visible as low density material with walls of varying thickness. The margins of these cavities are made up predominantly of compressed salivary tissue and fibrosis. The dilated Stensen's or Warthin's duct appears as a low density channel because they are filled with nonopacified saliva much like an obstructed kidney would have the ureter filled with nonopacified urine. Punctate calcifications in other areas of the gland assist in establishing the diagnosis of chronic sialadenititis. These calcifications may be too fine to be visualized on plain films.

GRANULOMATOUS DISEASE

Sarcoid is by far the most common granulomatous disease affecting the salivary glands. Tuberculosis is usually within the lymph nodes and may secondarily involve the gland. This is most common around the submandibular glands. Cat scratch fever and fungus infections are also part of the differential diagnosis of lymph node disease that secondarily invades the salivary glands.

CT Findings

In sarcoid the gland may show an increase in density with alternate areas of low density and high density scattered throughout the parenchyma. Fatty replacement of glandular tissue accounts for some of the low density regions, whereas the granulomatous changes cause an increase in density of the gland. Rarely do any of these lesions reach significant size so as to be mistaken for a tumor mass. Occasionally intraparotid nodes or the nodes lying along the surface of the submandibular gland enlarge to a significant extent and simulate new growth. When an intraparotid node is enlarged secondary to sarcoid, definitive diagnosis rests with pathology rather than the CT scan. As with most parotid lesions CT will merely tell the location of a mass and its relationships to the facial nerve. Histology rests with the pathologist and the total clinical presentation of the patient.

The nodes affected by tuberculosis or other necrotizing granulomatous diseases will usually show low density center and enhancing peripheral rims. The numbers of nodes affected by inflammatory disease is usually far in excess of the number of nodes that would be affected by a primary malignancy of the affected glands. This is probably the best clue that one can rely upon when evaluating CT study of chronic inflammatory disease as compared to malignant tumors.

AUTOIMMUNE DISEASE

The spectrum of autoimmune diseases ranges from Sjögren's to a variety of "benign lymphoepithelial lesions" (27). The autoimmune diseases are usually localized to the salivary and lacrimal glands. There may be an associated systemic component such as xerostomia, keratoconjunctivitis sicca, and rheumatoid arthritis (Sjögren's syndrome). When the rheumatoid arthritis is not present the syndrome is spoken of as a sicca syndrome which is merely a lesser penetrance of Sjögren's syndrome.

Recurrent parotitis of childhood is discussed with the autoimmune disease because the sialographic pattern is identical with that of adult Sjögren's. Since this disease is poorly understood and is characterized by recurrent bouts of acute parotitis which usually subsides at pubescence, autoimmune mechanisms are not an unreasonable speculation as the etiology of this disease.

CT Findings

In general the density of the salivary glands is slightly increased in autoimmune disease but punctate low density areas can be visualized throughout the gland. These low density regions correspond to the cystic-like changes that occur in the advanced stages of sicca syndromes of autoimmune origin. The main ducts are of normal or small caliber and are not seen as distinct structures. Occasionally one or two of the cysts become infected giving rise to localized masses which may be mistaken for tumors. By far the easiest method of diagnosing this group of diseases is by conventional sialography. If retrograde contrast injections are made for CT studies the gland becomes so dense with contrast material that little morphology is available for study.

Salivary Gland Tumors

GENERAL CLINICAL AND PATHOLOGIC BACKGROUND

Parotid tumors are said to constitute 70 to 80% of all tumors of the salivary glands (4, 32). Similarly, the likelihood of parotid tumors being benign is also approximately 80%. The situation is quite the reverse in the submandibular gland where approximately 50% of the tumors are malignant lesions. The relative incidence of parotid tumors is given in Table 6.1 taken from Batsakis (Ref 5, p 5).

"Perhaps no tissue in the body is capable of producing such a diverse histopathological expression than salivary tissue. This uniqueness may, in part, be due to the presence of the myoepithelial cell in the salivary glands; it is absent in the pancreas" (Ref 5, pp 1–75). Because of this diversity of tumor types, few observers gain sufficient experience to predict all of their various clinical manifestations. In the 1940s and early 1950s, pathologists and clinicians frequently referred to all salivary gland tumors as infiltrating or encapsulated types. Radiologic diagnostic techniques have not progressed to any major extent beyond this type of a gross description of morphology. In some respects, this limitation on refined diagnostic capabilities is not all that bad. The potential for complete surgical extirpation, metastatic spread and rapid growth is not always clear even from the microscopic pattern. Although tremendous advances have been made in understanding the behavior of these lesions by some pioneering patholog-

Table 6.1.[a]
Parotid Gland Tumors: Histological Diagnoses in Reported Series

Classification	Foote and Frazell (776 cases)	Bardwill (153 cases)	Eneroth (802 cases)	Lambert (83 cases)
Mixed tumors				
Benign	447(58%)	36(34%)	569(70.9%)	44(53%)
Malignant	46(6%)	34(22%)		1(1%)
Warthin's Tumor	50(6.5%)	5(3%)	41(5.1%)	16(19%)
Mucoepidermoid carcinoma				
Low grade	45(6%)	32(21%)	34(4.2%)	5(6%)
High grade	45(6%)			
Adenoid cystic carcinoma	15(2%)	13(8%)	19(2.4%)	2(2%)
Acinous cell carcinoma	21(3%)	8(5%)	36(4.5%)	3(5%)
Adenocarcinoma (miscellaneous)	32(4%)	16(11%)	17(2.1%)	
Oncocytic cell tumor	1(0.1%)	1(1%)	4(0.5%)	1(1%)
Squamous cell carcinoma	26(3%)	8(5%)	1(0.1%)	1(1%)
Miscellaneous				
Benign	3(0.4%)			
Malignant	0		15(1.8%)	5(6%)
Unclassified				
Benign	4(0.6%)			
Malignant	30(4%)			2(2%)

From JG Batsakis: *Tumors of the Head and Neck: Clinical and Pathological Considerations*, ed 2, Baltimore, Williams & Wilkins, 1979, p 22 (5).

ical studies such as Foote and Frazell (23) and others (20, 21), salivary gland tumors as a group are peculiar. This aspect is best summarized by Ackerman and Del Regato (1): "The usual tumor of salivary gland is a tumor in which the benign variant is less benign than the usual benign tumor and the malignant variant is less malignant than the usual malignant tumor." The cause of this biological behavior, expressions of local control, success of treatment, and ultimate prognosis cannot be expressed in 5 or 10 yr, but rather in 20 yr (Ref 5, pp 1–75).

The growth patterns of both benign and malignant lesions are not sufficiently characteristic to permit an accurate prediction of histology from CT scans. The aggressiveness of some benign tumors and the relative benignity of some malignant ones lend themselves well to a CT classification of general morphology that may have some bearing on predicting the eventful outcome for the patient. We are only suggesting the following categories so that we may help the surgeon by CT to further understand the relative aggressiveness of the lesions with which he is dealing:

1. Sharply circumscribed
2. Lobulated or slightly irregular but yet discrete
3. Diffuse and poorly defined

Table 6.2 lists the tumors that may fall under these various morphologic categories. Several lesions are listed under more than one morphologic group. This supports pathologic data indicating that a given histologic type ofparotid tumor may show different grades of aggressiveness (11, 41).

Benign Tumors

PLEOMORPHIC ADENOMAS

Pleomorphic adenomas are the most common neoplasm of the parotid gland. They constitute between 60 and 70% of all salivary gland tumors and occur most frequently superficial to the facial nerve (7). Women are affected slightly more frequently than men and, although the lesions are found in all age groups, it is most common in the 5th decade. The typical tumor is slow growing and rarely causes pain or paresthesias. The term "mixed tumor" is used because of the combination of epithelial and myoepithelial cells interspersed with a variety of mesenchymal products. The amount of stroma or myxoid material within the tumor is quite variable and calcifications are rare.

True malignant mixed tumors are exceedingly rare. They may be either malignant from the outset or may represent carcinoma arising in a previously benign pleomorphic adenoma. The true malignant mixed tumors may metastasize with both epithelial and stromal cells or they may metastasize with just the epithelial cells. The latter occurrence is more common in the carcinoma that arises in a pleomorphic adenoma that has been present for an extended period (10 to 15 years) (30).

CT Appearance

Benign mixed tumors are isodense with the normal parotid tissue prior to the intravenous injection of contrast material. Most of the lesions will stain with intravenous contrast and most of the tumors are more or less homogeneous in their staining characteristics. Low density centers with a thin contrast-enhancing rim are more in keeping with malignant lesions than a benign tumor.

When small pleomorphic adenomas are sharply outlined, as they reach 2 to 3 cm in diameter, the margins tend to be lobulated but the borders should be sharp. Lack of a sharp margin is suggestive of a more aggressive behavior of the lesion or associated inflammatory disease. This region should be chosen for aspiration cytology or microscopic examination following surgical extirpation. Unless the tumor is located in a region of the parotid gland remote from the facial nerve it is difficult to be absolutely certain whether some filaments of the nerve will be involved in the capsule of the tumor. A line drawn from the stylomastoid foramen to a position immediately lateral to the retromandibular vein will give a good approximation of where the facial nerve would lie. The course of the facial nerve is immediately cephalad to the posterior belly of the digastric. These landmarks should be included in the discussion of the location of tumor to assist the surgeon in planning his dissection.

By far one of the most important bits of information concerning facial nerve relationships is the position of a tumor mass with respect to the stylomastoid foramen. If there is not an adequate amount of normal parotid

Table 6.2.
CT Characteristics of Tumors

Rounded, Sharply Circumscribed	Lobulated or Irregular, but Sharply Demarcated	Irregular and Diffuse
Pleomorphic adenoma	Pleomorphic adenoma aggressive	Recurrent pleomorphic adenoma
Warthin's	Mucoepidermoid low grade	Mucoepidermoid—high grade
Acinous cell[a]		Adenocystic cancer
	Adenocarcinoma	Adenocarcinoma
	Oncocytoma[a]	Squamous cell cancer
Lipoma		Neuromas VII

[a] Rare tumors; experience is limited.

tissue adjacent to the stylomastoid foramen the surgeon may be required to remove the tip of the mastoid in order to get a clear margin around a tumor mass. He may also be forced to perform a retrograde dissection of the facial nerve rather than an antegrade dissection in establishing a cleavage plane between the deep lobes of the parotid and superficial lobe for superficial parotidectomy. At CT this border of normal gland parenchyma needs to be at least 3 to 4 mm from the stylomastoid foramen.

WARTHIN'S TUMOR

Warthin's tumor (cystadenolymphoma, papillary cystadenolymphomatosum) comprises 6 to 10% of all parotid neoplasms and characteristically does not produce facial nerve weakness or significant discomfort. The tumor is found mainly after the age of 40 with a male predominance of 5 to 1 over females (13). The age tendency is related to the oncocyte population of the gland which steadily increases after the age of 40. Histologically, Warthin's tumor is a cystic tumor composed of oncocytes in a lymphoid stroma. Because of the lack of symptoms, they frequently reach a large size by the time of surgical removal. If superficially located, they tend to be firm but if deep in the gland they may be described as soft and rubbery. The tumors are encapsulated but when they reach large size they tend to be lobulated in appearance. Necrosis within the tumor is rare. Warthin's tumor may be found in any location where normal parotid is intimately associated with lymphatic tissue. Accessory salivary tissue may harbor a Warthin's tumor and only rarely are they found in the submandibular gland.

Surgical removal is the treatment of choice and recurrences are difficult to document since Warthin's tumors are known to be multicentric in origin. Bilateral lesions are found in 10% of the patients (20, 38).

CT Findings

On the noninfused scans, Warthin's tumors are slightly increased in density as compared to normal gland parenchyma. The cystoidal areas are small and poorly outlined in contradistinction to cystoidal changes in chronic inflammatory disease which are more sharply defined. Multiple lesions may be present in the same gland as well as in bilateral configuration. When multiple tumor masses are present in the same gland it may be extremely difficult to differentiate this benign tumor from a malignant infiltrating tumor.

In the past it was thought that the affinity of these tumors for technetium was specific for a Warthin's tumor. The reliability of this sign has been seriously questioned (33) and patients with Warthin's tumors have been shown to exhibit cold spots on technetium scans, whereas some malignant tumors have been shown to have a high uptake of technetium.

LIPOMA

CT Findings

Lipomas of the salivary glands are readily recognized because of their sharp margins and low density centers. They may be located anywhere within the gland. Occasionally there will be a mixture of lymphangiomatous tissue within the lipoma in the younger patients which may raise the density of the lesion to isodense or even increase in density over the normal parotid tissue. The lymphangiomatous elements are not as sharply marginated as pure lipomas and create serious differential diagnostic problems.

Lipomas of the parotid gland can affect any age group. They are soft pliable lesions that may be difficult to palpate when deeply located. On clinical examination it may be extremely problematic to differentiate a lipoma from a branchial cleft cyst (see Chapter 8, The Neck). The branchial cleft cysts usually have an enhancing wall and are located in close proximity to the external auditory canal. There may even be a fistulous tract extending into the external auditory canal as seen by CT scanning.

OXYPHILLIC ADENOMAS (ONCOCYTOMAS)

These lesions are very uncommon and are only found over the age of 50. They are somewhat similar to Warthin's tumors in that they may show some affinity for technetium and they tend to be bilateral.

CT Findings

The oxyphilic adenomas (oncocytomas) are usually sharply marginated and are slightly increased in density as compared to normal parotid tissue. They may stain in homogeneous fashion much like the pleomorphic adenomas and require histology for differential diagnosis.

Malignant Tumors

The characteristics of malignant tumors as seen by CT are not specific and may be found in inflammatory disease and some of the benign tumors that are associated with ductal obstruction and inflammatory reactions. The most frequent CT findings are as follows:

1. Ill-defined areas of increased density
2. Central low density in a mass indicating necrosis
3. Mass with associated lymphadenopathy
4. Obstructed major ducts not associated with stones

One must constantly remember that small malignant lesions that are slow growing may have a sharply circumscribed margin much like any benign lesion. Equally important is the association of malignant tumors with other benign tumors. On multiple occasions we have encountered a malignant tumor within the center of a typical benign pleomorphic adenoma.

MUCOEPIDERMOID CARCINOMA

Mucoepidermoid carcinoma is the most common malignant tumor of the parotid gland and accounts for 6 to 9% of all parotid tumors (37). Foote and Frazell (22) pointed out that differentiation of mucoepidermoid carcinoma into a "benign" and malignant form was not warranted since metastases did occur from the "benign type." They recommended dividing the lesion into a high, intermediate and low degree of malignancy. This classification emphasizes the extremely indolent nature of some of the mucoepidermoid carcinomas. Their growth patterns are like that of the benign mixed tumor in that they may remain sharply circumscribed for a number of years. Lymph node metastases are only found late in the course of the disease. However, recurrences following surgical excision are not at all unusual. The more cellular types tend to be infiltrative and less sharply circumscribed and are prone to recur unless managed aggressively. Lymph node metastases are said to occur in 66% of the high grade mucoepidermoid carcinomas and distal metastases occur in approximately one-third of the patients (23, 24).

CT Findings

The mucoepidermoid carcinomas of the well differentiated variety are extremely slow growing and very late to metastasize. Their CT appearance is quite similar to the benign pleomorphic adenoma in that they are founded and sharply circumscribed or, occasionally, they may diffusely infiltrate like the highly malignant variety of mucoepidermoid carcinoma. They may have lobulations and may interfere with ductal filling. Again, the CT numbers of the malignant lesions are in no way specific enough to be characteristic of the histological diagnosis.

The more highly malignant mucoepidermoid carcinomas infiltrate and destroy glandular tissue. Major ducts may become encased and obstructed. The gland parenchyma may stain in an irregular fashion following intravenous contrast because of fat replacement of some of the parenchyma following obstruction of the ductal system. The pattern of diffuse margins persists throughout all levels of the tumor. As a rule calcifications are not present in this lesion or other malignant tumors. This is in contradistinction to chronic sialadenitis which also shows ill-defined margins but frequently contain calcium.

ADENOCARCINOMA, SQUAMOUS CELL CARCINOMA AND UNDIFFERENTIATED ANAPLASTIC CARCINOMA

These three lesions are discussed together since all are of about the same degree of aggressiveness and all are unusual tumors. They tend to infiltrate rather widely and produce ductal encasement and obstruction. All of the tumors invade adjacent musculature and metastasize to lymph nodes. Care must be taken in arriving at the diagnosis of primary lesions since metastases to intraparotid nodes or even distant metastases may closely mimic some of these squamous cell carcinomas.

CT Findings

The local findings in the parotid are nonspecific as with other malignant lesions. They merely show irregular borders and high density material. Careful searches should be extended to the nasopharynx and oropharynx to make certain that the lesion does not represent direct extension of another primary lesion. Metastasis to the parotid from squamous cell carcinomas of the skin in the temporoauricular region may look identical to anaplastic carcinoma.

ACINOUS TUMORS (SEROUS CELL) AND ONCOCYTIC CARCINOMAS

These two lesions are considered together because they are quite rare and generally exhibit a slow and circumscribed growth pattern. They may be lobulated or ill defined as the lesions become very large and more aggressive. Their location is nonspecific. Oncocytic carcinomas tend to be found in individuals over 70 yr of age (29). Malignancy was found in 2 of 21 oncocytomas reported by Conley (15).

The acinic cell tumors were primarily regarded as benign neoplasms until 1953 (12, 26). Although they only account for 1 to 2% of parotid neoplasms, their benign appearance causes considerable confusion. The tumor affects all ages and may grow quite slowly in a well circumscribed pattern. Metastases are said to occur late, with distant metastases only in the more aggressive varieties.

CT Findings

These rather rare tumors tend to be sharply circumscribed much like a pleomorphic adenoma. The acinic cell tumor is also circumscribed when encountered in children. There are no specific density ranges or patterns of necrosis that permit a definitive diagnosis (14, 29).

ADENOCYSTIC CARCINOMA

The adenocystic carcinoma is quite unique in its growth patterns and biological behavior. At one time, this tumor was classified as a benign lesion (cylindroma) but its persistence and distal metastases clearly identified it as one of the more relentless tumors of the salivary glands. The tumor may grow rapidly and metastasize to distant sites but, more frequently, it is very slow growing with distant metastases occurring 15 to 20 yr following surgical extirpation (16). Very early, the growth pattern may be sharply circumscribed but it generally permeates through the tissues in several areas and especially the perineural lymphatics of the facial nerve within the parotid gland. Pain or facial paralysis is said to occur in approximately one-third of the patients (Ref 5, pp 31–32). Microscopic evidence of perineural invasion was present in almost every instance from the experience of Batsakis (Ref 5, pp 31–32). Lymph node metastases occurs in approximately 15% of the patients and they are more likely to occur from direct invasion rather than by tumor cell embolization of the regional lymph nodes (2, 25).

CT Findings

The adenocystic carcinoma, like the mucoepidermoid carcinoma, tends to be rounded when very small. At the periphery of the tumor, afibroblastic response is excited which causes all ill-defined margin and compression of adjacent normal ductal systems. The density of contrast material surrounding the tumor may vary considerably.

One would hope that the characteristics of perineural invasion would provide the clue to definitive diagnosis but, unfortunately, this is not the case. The perineural invasion is microscopic in nature and not discernable by present CT techniques. Invasion about the stylomastoid foramen and into the descending portion of the facial nerve is also limited to a histological diagnosis until very late in the disease. One should not be tempted to call "funnel shaped enlargements of the stylomastoid foramen" either by CT or pluridirectional tomography unless there is a distinct asymmetry from the patient's normal side or the cortex of the facial canal has been demineralized.

LYMPHOMA

Lymphoma arising from within the parotid gland is exceedingly rare unless associated with Sjögren's syndrome (36). Lymphoproliferative orders are known to occur in autoimmune disease and many of these patients either start out or proceed to a full blown lymphomatous state (17). Atypical lymphomatous reactions (pseudolymphoma) may closely mimick the growth patterns of a true lymphoma of the parotid gland (6). They are generally accompanied by an enlargement of the preauricular (pretragal) lymph nodes that lie superficial to the parotid gland. The lymph nodes within the gland tend to occur in the posteroinferior third of the gland. True primary lymphoma of the parotid gland should begin in the gland parenchyma, however, differentiating parenchymal origin of lymphatic cells from origin within intraparotid lymph nodes may be very difficult if not impossible (3). In fact, the latter situation may be the histiogenesis of primary lymphoma (Ref 5, pp 49–50). Lymphomas that are primarily within the salivary glands have a much better prognosis if they occur in the absence of Sjögrens syndrome. Eleven of twelve such patients of primary salivary gland lymphomas without Sjögrens showed no evidence of recurrence 2 to 8 yr after treatment in a series reported by Nime et al. (35).

Patients with Sjögren's syndrome may proceed directly to a malignant lymphoreticular disorder or may first develop a pseudolymphomatous state and progress to a true lymphoma. There may or may not be an associated systemic autoimmune disease (rheumatoid arthritis). The extranodal origin of lymphoma in patients with Sjögren's disease is more likely to be in the lungs or abdomen.

Secondary involvement of the parotid gland by lymphoma is much more common. Tumor within the deep and superficial cervical lymph nodes may directly involve the gland as a tumor breaks out of the node capsule. In a clinical situation, little attention is paid to the parotid gland parenchymal involvement simply because the lymphadenopathy or systemic disease becomes the overwhelming problem. This type of parotid gland involvement is distinctly different from the inflammatory lesions of the salivary glands that occurs as a preterminal event with advanced lymphoma.

CT Findings

Primary lymphoma within the parotid gland is characterized by a very dense infiltration that totally replaces the gland parenchymal pattern. The margins are irregular and there are strands of tumor cells extending into adjacent normal parotid gland. There may or may not be associated lymphadenopathy in the surrounding nodes. The nodes are said to be of uniform density as opposed to low density centers found with metastatic squamous cell carcinoma. This distinction is not completely reliable. When one identifies the presence of large numbers of homogeneous staining pretragal and cervical nodes in association with a primary mass, lymphoma should be considered high on the list.

When lymphoma occurs secondary to Sjögren's disease, diagnosis is extremely difficult. Gland architecture is already modified by the Sjögren's disease and regions of increased density are unrecognizable until late in the course of the lymphoma. The diagnosis is made primarily by healthy clinical suspicion and early cytology studies rather than by radiologic study.

METASTATIC TUMORS

Secondary deposits within the parotid gland and in the paraparotid nodes most likely arise from the scalp or external ear. They also may originate from tumors about the orbit, eyelids, paranasal sinuses, or nasopharynx. The mucosal surface of the pharynx and tonsillar fossae may directly invade the deep portions of the parotid gland or may spread to the parotid by lymphatic extension. Melanoma is the most common tumor of the skin and appendages to involve the parotid gland. Almost invariably, there is enlargement of the preauricular node in addition to the other nodes either within gland substance or in the deep cervical chain. Mucosal tumors of the nasopharynx or pharyngeal cavity that extend to the parotid gland may totally replace the lymph nodes within the gland substance, making differentiation from primary tumors difficult.

Chapter 6 Plates

Salivary Gland

CONGENITAL LESIONS

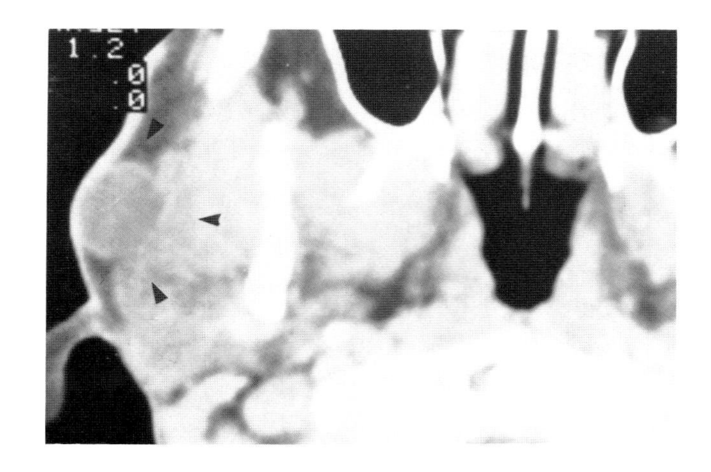

Plate 6.1. Branchial cleft cyst. Following intravenous contrast the mass lying superficial to the parotid gland shows an enhancing rim with central low density material (*arrowheads*). The very superficial nature of this "cystic-feeling" lesion suggested either adenopathy or some type of a cleft cyst on clinical examination. CT study was performed to show if there was any deep communications of the lesions.

STONES AND DIFFUSE DISEASE

Plate 6.2. Parotid duct stone.

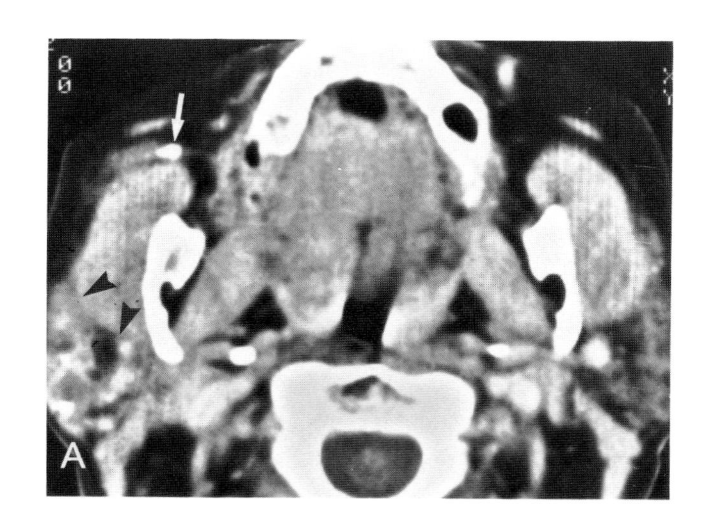

Plate 6.2*A*. CT with intravenous contrast shows a calcific density in the distal end of the parotid duct (*arrow*). This lesion is causing total obstruction of the ductal system and nonopacified secretions can be visualized within the gland as low density areas (*arrowheads*).

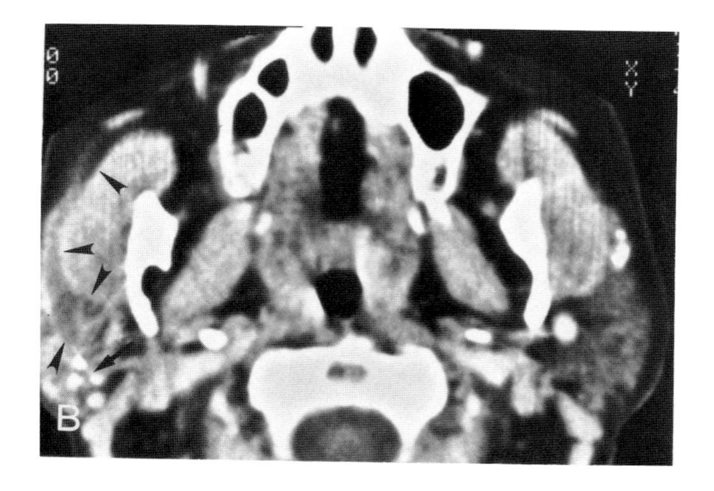

Plate 6.2*B*. Section taken further cephalad shows the entire parotid duct as a low density linear shadow (*arrowheads*) because the contrast material is not getting into the secretions. The wall of the parotid duct stains because of associated inflammatory changes. Calcifications are present in the posterior portion of the parotid gland (*arrow*). Calcifications can also be identified in the gland parenchyma on the opposite side.

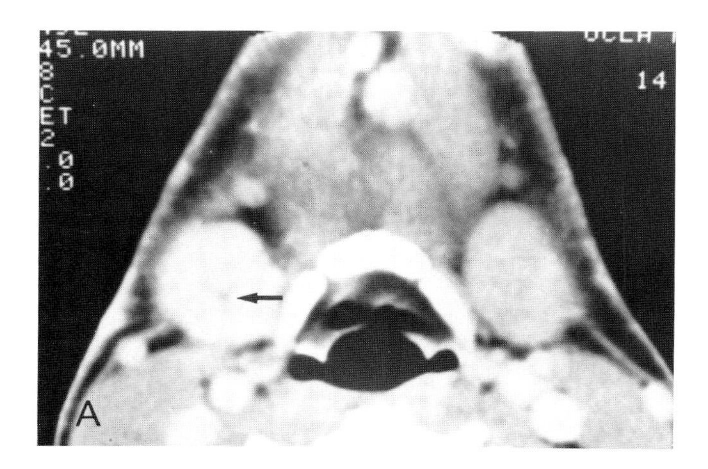

Plate 6.3. Submandibular stone.

Plate 6.3*A*. Following intravenous contrast there is a greater concentration of material in the right submandibular gland as compared to the left. Centrally, a lucent zone is identifiable which represents nonopacified secretions in the ductal system (*arrow*) because the iodine is normally secreted in saliva.

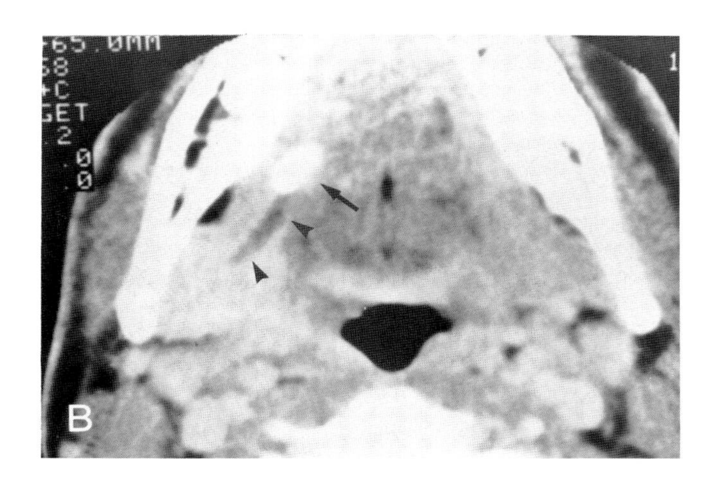

Plate 6.3*B*. Higher sections through that portion of the salivary gland above the mylohyoid muscle shows the nonopacified main duct (*arrowheads*) leading to a calculus (*arrow*). This patient had had a previous bout of submandibular swelling that was consistent with the stone. He had been symptom free for approximately 9 yr when persistent swelling reappeared. By history the swelling did not fluctuate, therefore, a CT study was performed.

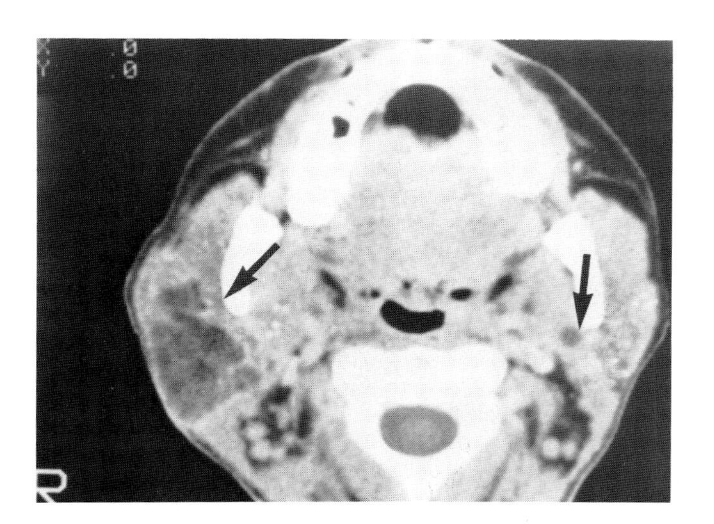

Plate 6.4. Sjögren's disease. The cystic changes present in the posterior portion of the right parotid gland was thought to represent an ill-defined tumor. CT study shows bilateral low density spaces (*arrows*) with an overall increase in density of the remaining parotid tissue. Punctate calcifications are present in scattered areas of the gland. This type of change is not diagnostic of Sjögren's disease. Differential diagnosis would include sarcoid and chronic sialadenitis which produce smaller, more evenly distributed, low density, lesions.

BENIGN TUMORS

Plate 6.5. Benign mixed adenoma.

Plate 6.5*A*. Scan prior to contrast shows a questionable density (*arrow*) adjacent to the masseter muscle but one cannot make a definite diagnosis of the tumor.

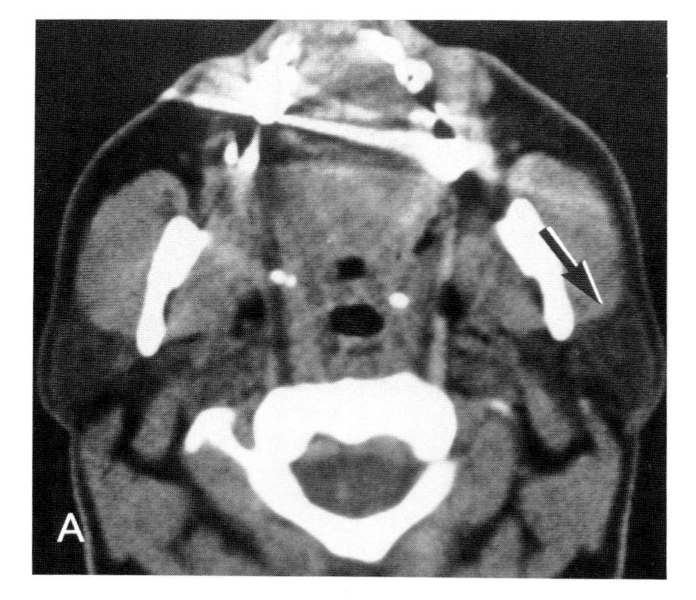

Plate 6.5*B*. Following intravenous contrast the tumor mass stains almost homogeneously (*arrowheads*). It is clearly visible adjacent to, and separate from, the masseter muscle.

This type of staining is seen in the vast majority of benign mixed adenomas but small malignancies with central necrosis can have an identical appearance. Some type of histology is essential for definitive diagnosis.

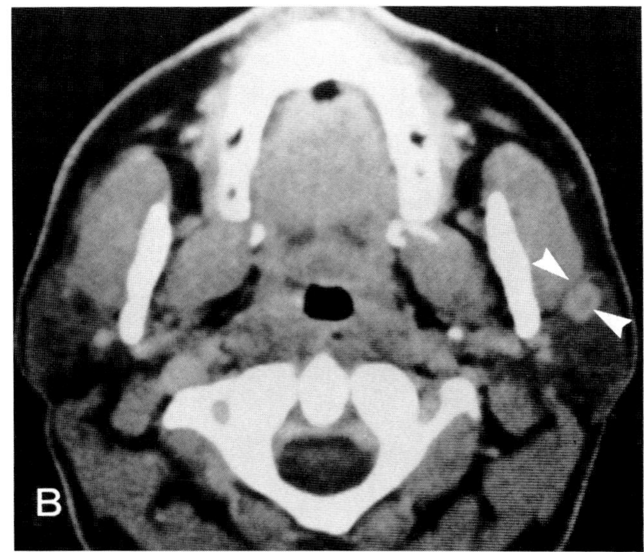

Plate 6.6. Intraparotid node. A circumscribed mass (*arrowhead*) can be seen in the superficial portion of the parotid gland on this CT sialogram. This study was performed at another institution and surgery was done without aspiration cytology. The CT appearance of this circumscribed mass is identical with that of a benign mixed tumor or a Warthin's tumor. (Courtesy of Dr. Daniel Johnson, New Orleans Radiology Group.)

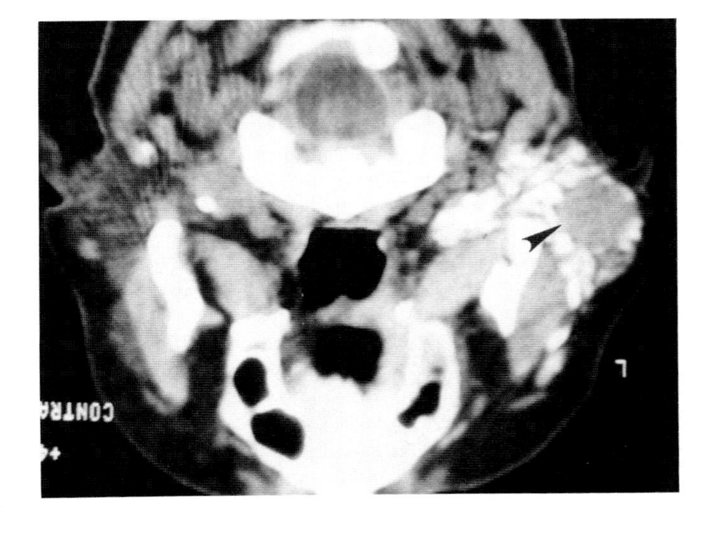

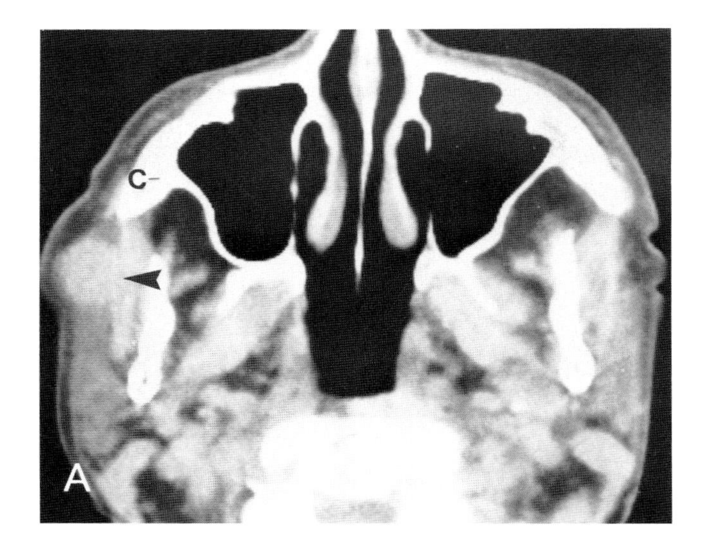

MALIGNANT TUMORS

Plate 6.7. Mucinous adenocarcinomas.

Plate 6.7*A*. Prior to contrast material (*C*–) a mass can be seen in the anterior portion of the right parotid gland (*arrowhead*). Because of its unusual location and history of rapid growth the CT study was performed.

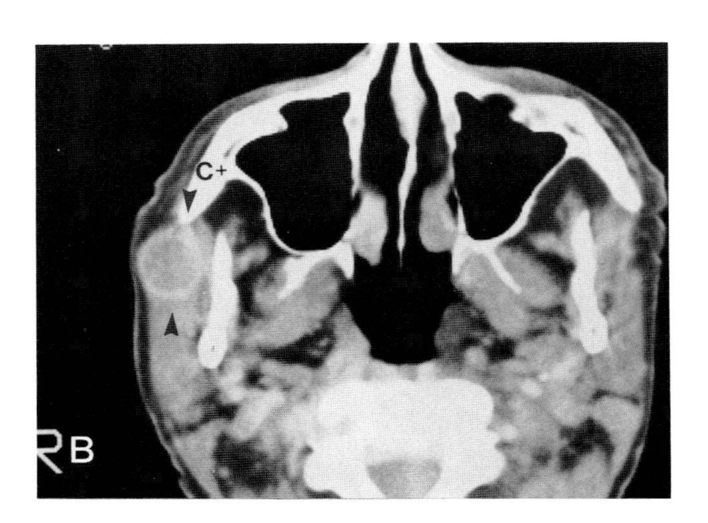

Plate 6.7*B*. Following the administration of intravenous contrast (*C*+) a rim type of enhancement is identified with low density center (*arrowheads*). This CT appearance could easily represent an anteriorly placed branchial cleft cyst or tumor with a large necrotic center. Aspiration cytology was necessary to give the histologic diagnosis of mucinous adenocarcinoma.

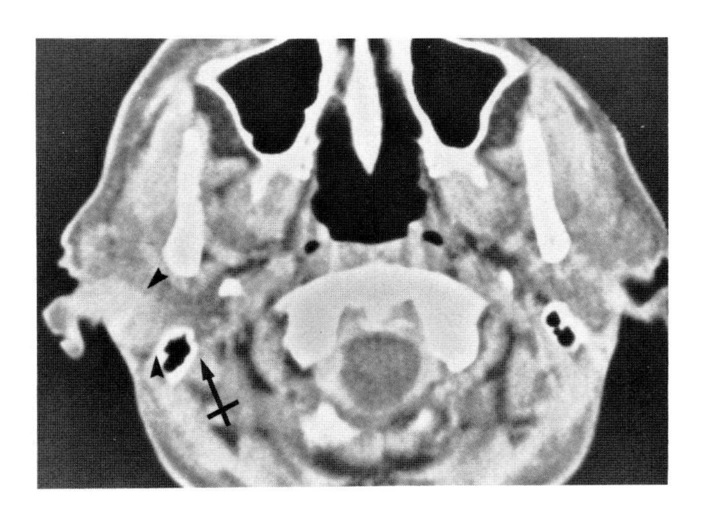

Plate 6.8. Mucoepidermoid carcinoma. The tumor mass (*arrowheads*) infiltrated posteriorly to the region of the mastoid tip. The deep portions of the tumor infiltrate close to the course of the facial nerve within the parotid gland, but the facial nerve is shown to be free of tumor at the level of the stylomastoid foramen (*crossed arrow*).

The surgeon still had to sacrifice the facial nerve but he knew that he had sufficient nerve outside of the mastoid for nerve grafting.

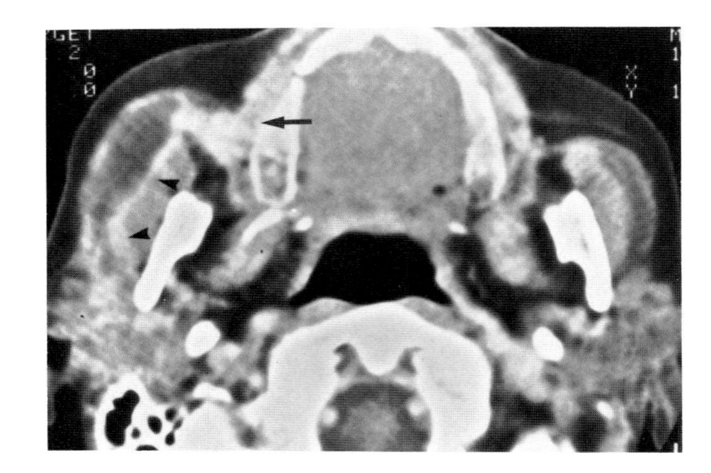

Plate 6.9. Squamous cell carcinomas of parotid duct. This patient had a history of long-standing chronic sialadenitis but swelling developed progressively on the right, associated with a mass near the opening of Stensen's duct.

CT with intravenous contrast shows an infiltrating mass of the distal Stensen's duct (*arrow*). More proximal and intraglandular portions of the ductal system are dilated (*arrowheads*) and filled with nonopaque parotid secretions. The entire gland parenchyma is dense and other sections show multiple areas of cystic dilatation of the ductal system.

Aspiration cytology of the distal duct confirmed the presence of a squamous cell carcinoma of the duct.

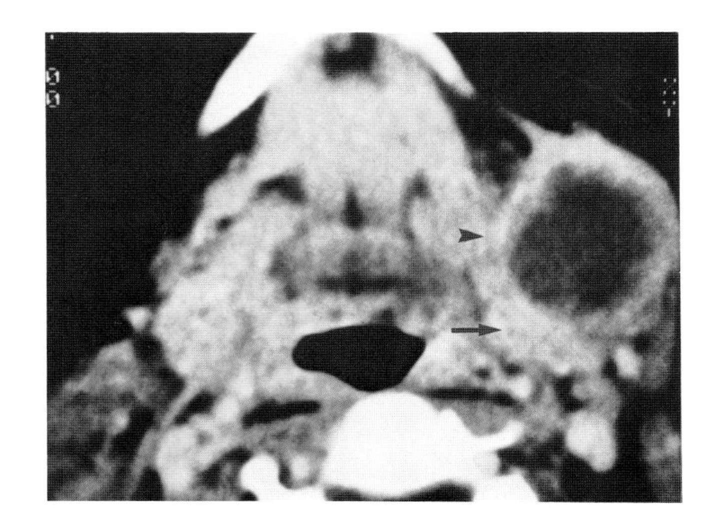

Plate 6.10. Anaplastic carcinoma submandibular gland. The tumor mass shows an enhancing border (*arrowhead*) and a low density center. The margins both internally and externally are somewhat irregular and the remaining submandibular gland tissue can be identified lying posteromedially (*arrow*). This type of lesion closely simulates a cyst and requires aspiration cytology for definitive diagnosis. The CT examination should survey the upper aerodigestive tract to rule out the presence of a primary lesion with the submandibular lesion representing a metastases.

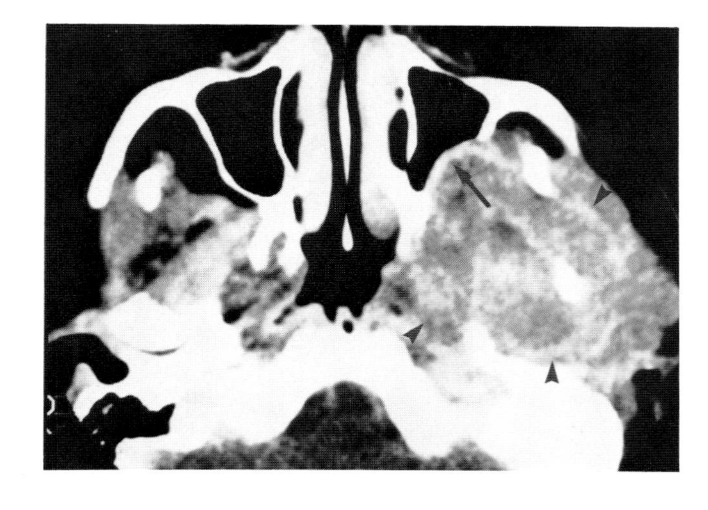

Plate 6.11. Massive diffuse involvement of the parotid can be seen (*arrowheads*) with the lesion extending into the infratemporal fossa and the paranasopharyngeal space, and indenting the posterior wall of the maxillary sinus (*arrow*). This type of massive tumor is almost diagnostic of lymphoma or some type of aggressive embryonic tumor such as rhabdomyosarcoma.

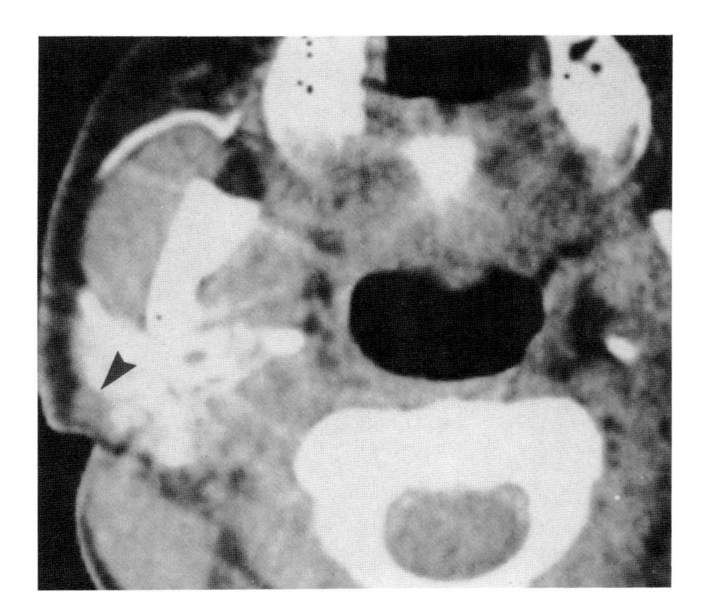

EXTRINSIC VERSUS INTRINSIC DISEASE

Plate 6.12. Pretragel lymph node This "mass" was extremely difficult to palpate and could not be seen by intravenous contrast. Because of the questionable nature of the lesion, positive contrast CT sialography was performed which shows a small mass on the surface of the parotid gland (*arrowhead*). The differential diagnosis was a small pretragal lymph node, versus a tumor, arising from the surface of the parotid gland. Under a trial of antibiotic therapy the node disappeared.

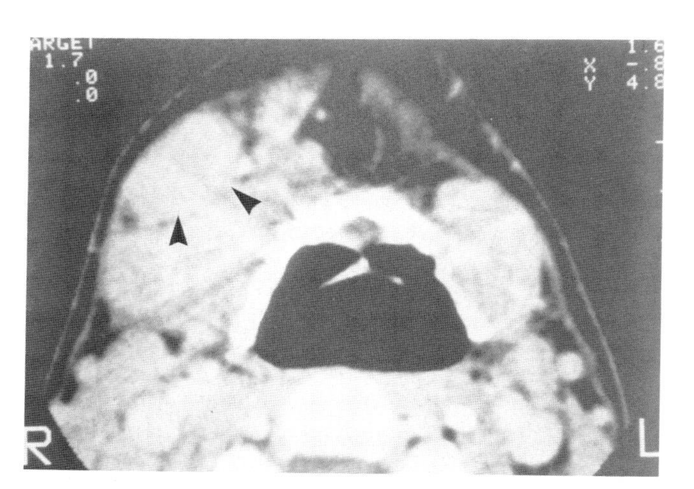

Plate 6.13. Extrinsic mass lymph nodes. In this patient with a known supraglottic carcinoma, intravenous contrast shows a node lying in the anterior jugular chain (arrowheads). The submandibular gland is clearly identified lying posterior to the node and being somewhat indented along its anterior margin.

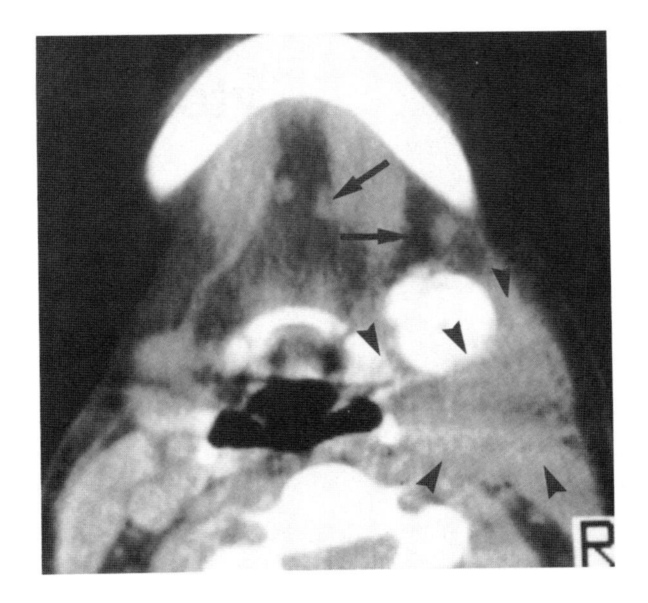

Plate 6.14. Extrinsic inflammatory disease. A diffuse mass is present (*arrowheads*) surrounding the region of the right submandibular gland which closely simulated aggressive malignancy arising from the gland itself. CT with intravenous contrast shows that the gland itself is reasonably intact and inflammatory process lies posterior to the gland. Some submental nodes are visible anterior to the gland (*arrows*).

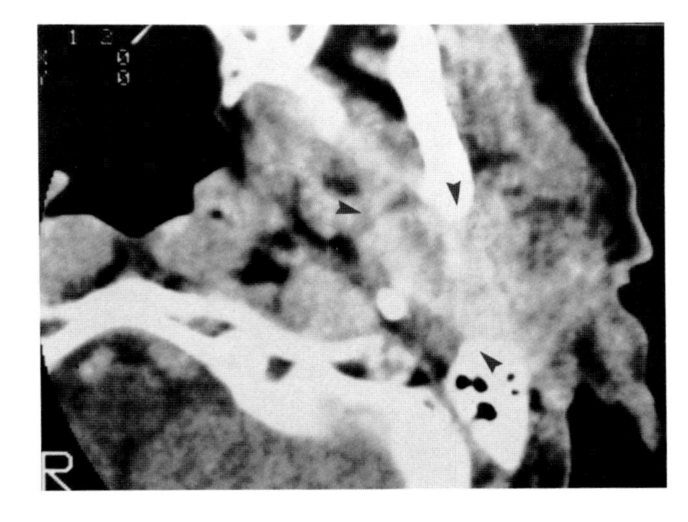

Plate 6.15. Carcinoma parotid deep lobe. On clinical examination ill-defined deep firmness was felt within the parotid gland. CT study shows replacement of the entire deep lobe of the parotid by an ill-defined mass. The low density medial border (*arrowheads*) indicates the tumor grows from within the parotid gland and is not from accessory salivary tissue in the nasopharynx. The tumor abuts the stylomastoid foramen making facial nerve involvement a certainty.

Chapter 7 Plates

Aspiration Cytology

Plate 7.1. Infratemporal rhabdomyosarcoma. This child was immune suppressed following 2 yr of chemotherapy for rhabdomyosarcoma of the nasopharynx. He developed trismus which was either on the basis of recurrent tumor or opportunistic infection.

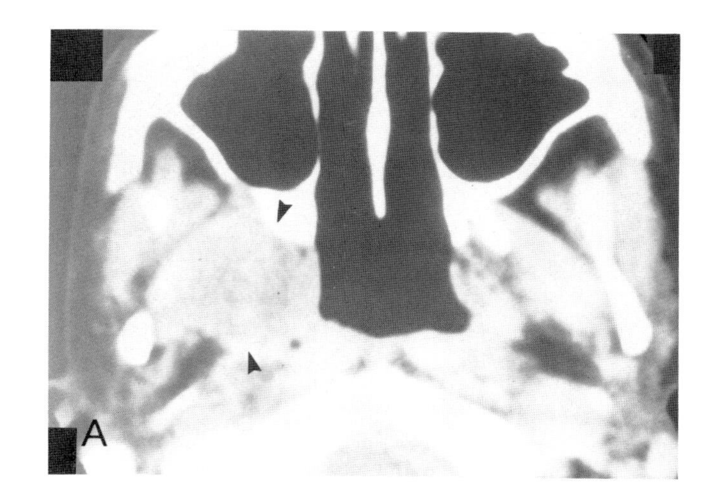

Plate 7.1*A*. Recurrent mass is visible invading the lateral pterygoid muscle and paranasopharyngeal space (*arrowheads*).

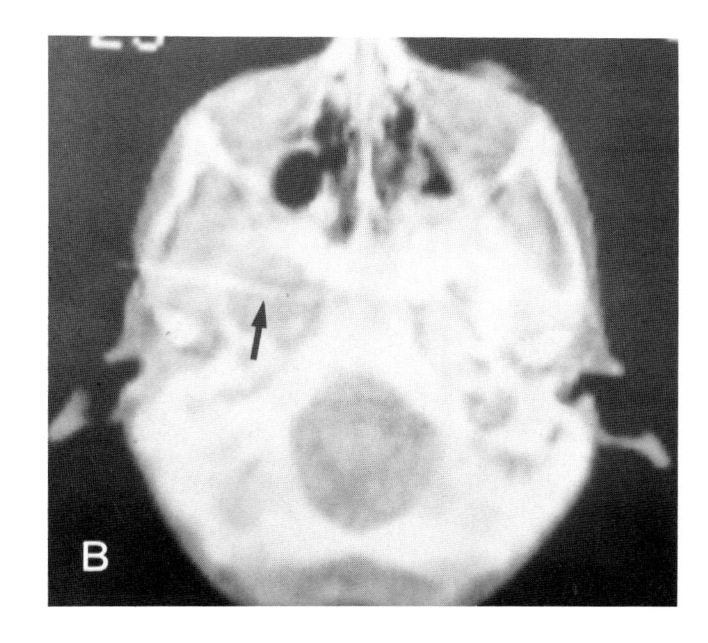

Plate 7.1*B*. Section shows a needle tip present in the mass in the infratemporal fossa (*arrow*). It is very important to see that the needle tip has ended and the "trailing edge" of the shadow of the metal confirms that the tip of the needle is truly within the plane of the section. This is extremely important if the needle has entered the mass from an oblique angle and does not lie entirely within the plane of the scan. In other words, if the needle is crossing the section at an angle, one may merely be making a section through the shaft of the needle and not through the tip of the needle.

tervening planes of high MR signal intensity defining the limits of the individual anatomic components. The airway is black due to its low signal. The internal portion of the cricoid has a strong signal where it is ossified and contains fatty marrow. Its ossified borders will have no signal and its cartilaginous portions show a signal slightly more intense than skeletal muscle. The signal from the musculature of the aerodigestive tract tends to be the same as that of surrounding skeletal musculature allowing these structures to be easily differentiated from surrounding fat. The signal intensity from the thyroid gland is slightly less intense than that of surrounding musculature (70). Detailed imaging of the normal thyroid gland and other neck structures will depend on more refined coil development than is presently available. Surface coils will produce the highest signal to noise. In preliminary work done at high field strengths (1.0–1.5 Tesla) with specially designed coils it is possible to visualize the gross internal architecture of the thyroid gland, the inferior thyroid artery and vein and perhaps even the normal parathyroid glands (18). This advanced technology obviously holds some promise for detailed imaging of this area which will rival that of both high resolution CT and realtime ultrasound.

Lateral Compartment

The investing layer of deep cervical fascia splits to envelop the sternocleidomastoid muscle (Fig. 8.1). This bulky muscle shifts from a primarily anterior position low in the neck to a lateral one in the midneck and a more posterior one in the upper neck. This reflects its course from its origins on the manubrium and clavicle to its insertions on the mastoid process and occipital bone (26, 36). This muscle is in reality two separate muscles and sometimes where one side is contracted the asymmetry can lead to the mistaken impression of a lateral neck mass on the side which is tensed. The omohyoid muscles run an oblique course deep to the sternocleidomastoid through the low neck and it is difficult to tell the two apart in most patients. This normal muscle should not be mistaken for adenopathy or other abnormalities (Fig. 8.3).

The carotid sheath lies within the fibroadipose tissue deep to the sternocleidomastoid muscle. The sheath in reality is not a fascia but a "dense feltwork of aerolar tissue" which surrounds the common and internal carotid arteries and vagus nerve (36). It is almost nonexistent over the jugular vein, thus allowing the vein to distend whenever necessary. The vagus nerve lies between the carotid and jugular and the internal jugular nodes surround the vein. Some authors would prefer to call this the carotid space. The terminology is less important than understanding the anatomy.

The lymph nodes of the neck are of prime importance in staging squamous cell carcinoma and other malignancies arising in this region. All of the nodal groups in the head and neck can be studied in a detailed manner with CT and MRI making an understanding of their normal anatomic distribution critical for interpretation of these studies. The deep cervical nodes lie in an envelope of fat predominantly within the lateral compartments. Other groups subsequently discussed do not; however, all of the important nodal groups in the head and neck region will be discussed at this juncture for the sake of continuity and completeness (Table 8.1).

On CT normal lymph nodes are typically of homogeneous density and have attenuation values at least 10–20 Hounsfield units (HU) less than the opacified blood vessels. Their tissue density is roughly equivalent to that of surrounding muscle. In surgical or anatomic specimens, nodes tend to be oval or elliptical but their shape on axial sections depends on their orientation to the transverse plane: if the long axis is oriented perpendicular to the plane of section they appear round; if it parallels the plane of section they look somewhat oblong. Larger nodes (1 cm) will sometimes appear slightly inhomogeneous. A peripheral rim of contrast enhancement is not normally present. Occasionally, fatty replacement may create eccentric lucencies at the hilum of normal-sized nodes. Accurate measurement of attenuation in the fatty foci may prove difficult because of their small size; however, visually their density appears identical to that of surrounding fat.

On spin echo MR images, the strong signal from the adipose tissue in the neck may hide normal-sized nodes. A short T_R pulse sequence will emphasize the contrast between the normal lymphoid tissue and surrounding muscle and fat. Normal nodes are routinely visible on such spin echo images made on MR units operating at field strengths of 0.15 Tesla or more. Surface coils are required for adequate imaging on most systems. Additional long T_R sequences with both early and delayed echos help define the extent of cervical adenopathy. A major advantage of MR in evaluating the lateral compartments is the low intensity of the signal from the flowing blood in the carotid artery and jugular vein. As a rule these vessels are easily distinguishable from nodes and other structures obviating the need for contrast enhancement; however some caution is necessary in this regard. Alterations in blood flow, either turbulence or velocity, can increase the signal intensity in these vessels (50). The flow rates and size of jugular veins, especially, may vary greatly from side to side and some variation from the "rule" must be expected due to this and the various manifestations of atherosclerotic disease which may affect the arterial side (50). In addition to seeing normal nodes and vessels on MRI, it is possible to visualize the vagus nerve as it lies between the carotid and jugular when high resolution signal acquisition techniques are employed.

The deep cervical nodes may be divided into three chains: internal jugular, spinal accessory, and transverse cervical chains (Fig. 8.3). The internal jugular

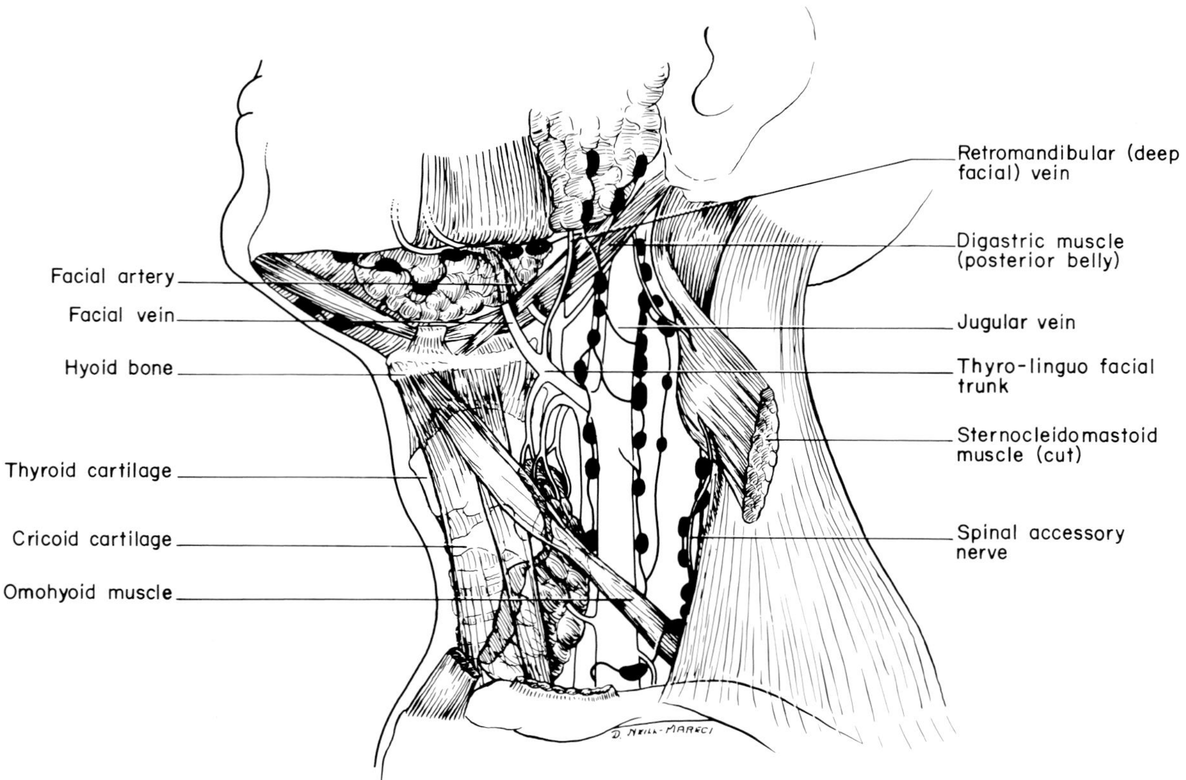

Figure 8.3. This diagram shows the major lymph node groups of the neck and the anatomic landmarks used to separate them into their various groups. The internal jugular nodes lie both anterior and posterior to the *jugular vein*. They may be divided into three groups: upper, from the *posterior belly* of the *digastric muscle* to the *hyoid bone* or, more properly, the junction of the *thyro-linguo facial trunk* with the jugular vein; middle, from the hyoid bone to the crossing of the *omohyoid muscle* and jugular vein; lower, from the *omohyoid muscles'* crossing with the jugular vein to the thoracic inlet. On CT the level of the *cricoid cartilage* may be used instead of the omohyoid muscle since the former is a landmark more easily seen on CT or MRI. Note how the internal jugular group and spinal accessory group are close together high in the neck and diverge as the vein and spinal accessory nerve separate lower in the neck. The submental, submandibular and periparotid nodes are also pictured.

group is the chief collecting chain. The spinal accessory chain diverges posteriorly from the nodes following the internal jugular vein and the transverse cervical group completes the triangle by bridging the gap between these two groups in the lower neck (just above the thoracic inlet (20, 59)). The inferior and medial point of this triangle marks the spot where the lymphatics empty into the venous system at the jugulosubclavian junction (20, 59).

The internal jugular nodes lie anterior, lateral and posterior to the jugular vein throughout its course in the neck (20, 59). They may be subdivided into anterior and lateral groups: the latter including those nodes lateral as well as posterior to the vein. There is a tendency shown both anatomically and by CT for the nodes in this chain to be the largest in the upper neck (Table 8.1) (20, 39, 40, 58, 59). This is particularly true of the nodes anterior to the vein. In the middle and lower neck the anterior nodes are very small and much less numerous so that few are normally visible on CT studies (39). The lateral division is easier to see in these lower sections since they are more numerous and lie in the more expansive fibrofatty space just posterior and lateral to the jugular vein (20, 39, 59).

Classically, the internal jugular chain is also divided into superior, middle and inferior groups (20, 59). The superior division begins at the posterior belly of the digastric muscle and by Rouvier's definition (59) extends caudally to the joining of the common facial vein and the internal jugular vein (Fig. 8.3). These nodes are also known as the digastric or jugulodigastric group. Subdigastric and tonsillar nodes are other synonyms. The highest of these may extend above the posterior belly of the digastric especially when pathologically enlarged (Fig. 8.4). We chose the hyoid bone as a landmark for the lower extent of the jugulodigastric nodes since the common facial vein-internal jugular bifurcation is not always seen on CT and the joining

Table 8.1
Size and Frequency of Visualization of Normal Cervical and Retropharyngeal Nodes[a]

Group	Seen in no. of patients	Size range	No. of Patiets with Nodes at Upper Limit of Range
		(mm)	
Occipital	0/30	—	—
Mastoid	0/30	—	—
Facial	0/30	—	—
Lingual	0/30	—	—
Parotid	7/30	3–5	1/7
Retropharyngeal			
Median	0/30	—	—
Lateral	20/30	3–7	2/20
Submental-submandibular	28/30	3–10	3/28
Internal jugular			
Superior	30/30	3–10	6/30
Middle	30/30	3–10	2/30
Inferior	30/30	3–5	5/30
Anterior jugular	0/30	—	—
Juxtavisceral-scalene			
Spinal accessory	28/30	3–5	5/28

[a] Reprinted with permission from Mancuso AA, et al: Computed tomography of cervical and retropharyngeal lymph nodes. Part 1. Normal anatomy. *Radiology* 148:709–716, 1983 (40).

of these two veins usually occurs about a centimeter or two below this level in the neck. Nodes in this region normally tend to be the largest of those found anywhere in the neck.

The middle internal jugular nodes, according to Rouvier (59), extend from the level of the thyroidal artery to the point where the omohyoid muscle crosses the jugular vein (Fig. 8.3). On CT we have elected to use the hyoid bone to separate the upper and middle groups and the level of the cricoid cartilage to separate the middle from the lower (39). Classically the lower division lies between the omohyoid muscle and the jugulosubclavian junction (59); the hyoid and cricoid are prominent CT and MR landmarks and there is little practical benefit derived from trying to be more precise.

The spinal accessory nodes follow the course of the XI cranial nerve for which they are named. The spinal accessory nerve courses under the posterior belly of the digastric about midway between the muscle's origin under the mastoid and its tendinous attachment to the hyoid (26, 36). The nerve then courses through dense fibroadipose tissue before it pierces the sternocleidomastoid muscle (Fig. 8.3). At this point the nodes of the internal jugular and spinal accessory groups blend together (20, 39, 59). The course of the nerve, and therefore nodes, then angles posterolaterally. The nodes of the spinal accessory chain are found lateral and posterior to the nerve; the latter basically coursing through the fat pad between the sternocleidomastoid and trapezius muscles. The nodes which lie deep to the sternocleidomastoid and just posterior to the jugular vein are anterior and medial to the nerve; these nodes are in the internal jugular, not the spinal accessory chain.

On axial CT images the normal nodes, ranging from 3–5 mm in diameter, are easily distinguished from the surrounding fat. The spinal accessory nodes also tend to be larger in the upper neck and in anatomic studies have been shown to range between 0.1 and 1.0 cm in size (20, 59). As few as 3 or as many as 20 may normally be present (20, 59). In the low neck the transverse cervical chain connects the spinal accessory and internal jugular nodes (20, 59). These nodes are located at and below the crossing of the omohyoid muscle and jugular vein. Frankly, it is quite difficult to differentiate lower internal jugular, spinal accessory and supraclavicular and transverse cervical nodes on axial images through this region. Practically, it is not important to do so; however, one group in this location is especially important—namely the scalene nodes. This group of nodes lies at the junction of the transverse cervical and internal jugular chains, just anterior to the scalene muscle (Fig. 8.3). They have been called the "Grand Central Station" of all lymphatics because they receive drainage from virtually every part of the body since they are immediately adjacent to the thoracic duct-subclavian vein junction (20, 59). Again, one cannot precisely identify a normal scalene node in this location on CT; however, if one understands the anatomic relationships of these nodes the presence of significant adenopathy in this region can be established with either CT or MRI.

There are a number of other important nodal groups in the head and neck region which must be considered in some detail. It is important for diagnostic imagers to become familiar with these nodes as well as those of the internal jugular chain since the regional nodal spread of head and neck primary and recurrent tumor varies considerably with its site of origin. All of these nodal groups can be evaluated by CT and MRI.

The parotid nodes may be intraglandular or extraglandular (20, 59). They are highly variable in size (1–5 mm), position and number (Table 8.1). When visible on CT studies, they are usually seen in the pretragal region or posterior to the retromandibular vein (posterior facial vein) near the latter's exit from the tail of the parotid (Fig. 8.3).

The submandibular nodes lie in front, adjacent to, and behind the gland (20, 59). Normal submandibular nodes are almost always visible during CT within the fat surrounding the gland (39). It might be difficult to distinguish an enlarged retroglandular node from one in the upper jugulodigastric group (Fig. 8.3). These nodes tend to be 3–10 mm in their largest dimension and oval or elliptical (Table 8.1).

The submental nodes lie in a paramedian position between the anterior bellies of the paired digastric muscles (20, 59). They are seen as elliptical densities

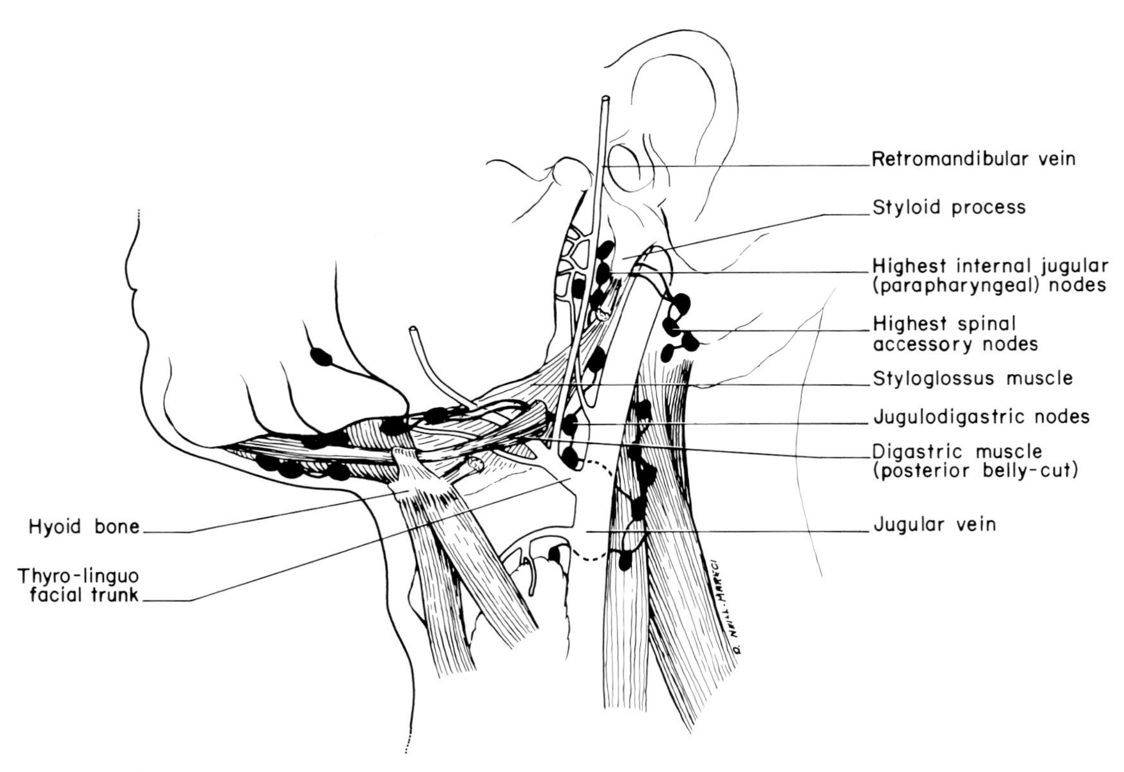

Figure 8.4. This diagram shows the important relationships of the deeper nodes in the internal jugular vein. Note that the *posterior belly* of the *digastric muscle* has been cut and the parotid gland has been removed. This reveals the *highest internal jugular nodes.* These nodes have also been called *parapharyngeal nodes* and they are sometimes included in the group of transitional nodes. The term transitional lymph nodes applies to these highest nodes in the spinal accessory and internal jugular groups which are so close together that they are sometimes difficult to differentiate from one another when enlarged. Also note that there are normally some communicating lymphatics between these groups.

within the fat between the floor of the mouth and hyoid bone.

The retropharyngeal nodes may be found from the C_1 level to the thoracic inlet but are most constant at the suprahyoid levels (20, 59). The medial group lies near the midline between the superior constrictor and prevertebral muscles (20, 59). These are in reality nodules located along the nasopharyngeal collecting trunk and are never seen in normals; they are occasionally seen when enlarged (39, 40). The lateral retropharyngeal group are true nodes and, even when normal, are frequently visible on CT scans at the C_1–C_2 levels (Table 8.1). They tend to be round, 3—5 mm soft tissue densities situated between internal carotid artery and longus colli and capitus muscles (see Chapter 11 Fig. 11.7).

The anterior cervical nodal group bears special consideration. Technically Rouvier intended that this include those nodes lying external to the deep cervical fascia (anterior jugular chain) and those lying deep in the neck around the thyroid trachea and esophagus (juxtavisceral group) (20, 50). The juxtavisceral nodes are usually not seen in normal CT studies or at least they cannot be reliably distinguished from small vessels. The juxtavisceral nodes have been subdivided by some authors into several groups and then even further subdivided and named for the particular structures they lie near or drain. We will consider only one of these groups, the nodes that lie along the course of the recurrent laryngeal nerve (Fig. 8.5). As discussed before, the recurrent nerve and inferior thyroid artery and vein run together in the fat pad between the thyroid gland and prevertebral muscles (longus colli). The normal juxtavisceral nodes cannot be distinguished from these neurovascular structures; however, when enlarged they may be detected by CT long before they are clinically palpable, mainly because of their deep location.

Posterior Compartment

The cervical spine is surrounded by groups of flexor and extensor muscles. The smaller flexor group lies anterior to the transverse processes of the cervical vertebrae and includes the longus capitis, rectus capitis, longus colli, and scalene group (Fig. 8.1). The larger extensor group lies posterior to the transverse processes. One need only know some of these muscles by name since they mark the position of more critical fluid collections without obvious loculation, especially the ones in the posterior neck. This apparent simple cystic appearance belies their true insinuating nature. Others have a mixed tissue density—cystic appearance especially on the face and in the tongue and floor of the mouth. Diagnostic imaging is aimed at showing its full extent and occasionally differentiate a hygroma from a first branchial cleft cyst. If excision is contemplated, those lymphangiomas showing a predominant cystic morphology should be much easier to remove than those with a more cavernous or angiomatous pattern (21).

Inflammatory Masses

Deep neck abscess, jugular thrombophlebitis and suppurative or malignant adenopathy are the three most common considerations entering into the differential diagnosis of an inflammatory lateral neck mass. These need to be differentiated from an infected branchial cleft cyst which is usually the primary diagnosis under consideration.

Deep neck abscesses may arise from many sites within the upper aerodigestive tract: breakdown of suppurative nodes or, in more recent years, due to neck injections made for intravenous substance abuse (15, 26, 45). Whatever the source, CT or MRI can be used to determine the likely cause and extent of the abscess. More importantly, CT can definitively distinguish neck abscess from cellulitis or jugular thrombophlebitis, and offers a high chance of differentiating all of these from malignant adenopathy (secondarily inflamed) or an infected branchial cleft cyst (15, 22, 26,45). Aspiration and biopsy may, however, be necessary under some clinical circumstances.

A deep neck abscess appears as a thick-walled cavity with a central cavity of relative low attenuation material. The wall is often crenated in appearance and enhances dramatically following IV contrast injection. The abscess may contain gas and the surrounding soft tissues show marked thickening. One must be careful to outline the full extent of the abscess because the surgical drainage depends on which neck compartments are infected (15, 26, 27, 45). Remember that there may be masticator, parapharyngeal, submandibular, or retropharyngeal space involvement depending on the point of origin. Cellulitis is easy to differentiate from abscess because it produces only diffuse swelling and obliteration of tissue planes without forming a frank, pus-containing cavity. A repeat CT may be necessary to be sure that a cellulitis has not "matured" into an abscess and CT can certainly be used to determine when an abscess is ripe for drainage (15).

Jugular thrombophlebitis will produce a characteristic CT appearance (1). On an infused study, the lumen of the internal jugular vein (and sometimes its tributaries) will not opacify but its wall will enhance. There may be diffuse soft tissue swelling around the vein which is the finding that may produce the mistaken impression on physical exam of an infected cyst, node or other mass. MRI may be expected to show lack of flow in the vein as well as surrounding morphologic changes similar to that of CT, but the relative use of these two modalities in this setting is yet to be worked out. Suppurative adenopathy in the child is usually very confined to a small area and may be impossible to distinguish from a limited abscess on the basis of imaging. In the adult, necrotic metastatic adenopathy potentially produces a similar dilemma but has never been a practical problem in our experience. The combination of needle biopsy, CT morphology, and clinical factors virtually assure that the clinician and imager are not taken by surprise.

Benign Neoplasms

The carotid body tumor is probably the most common benign neoplasm of the lateral compartment; it is the most common variety of paraganglioma in the head and neck (4). The clinical question being asked is usually whether a mass in the neck, near the carotid bifurcation, is a carotid body tumor, tortuous or aneurysmal bifurcation or other mass. If one wishes to do a screening type study the choices are CT, MRI, venous DSA, or ultrasound. Venous DSA or CT would probably be done by most people. Arteriography is the definitive study and indispensible if surgery is contemplated.

At CT paragangliomas enhance to the same degree as surrounding vasculature. They appear lobulated but well circumscribed as they lie between the internal and external carotid arteries. Cephalad growth may take the lesions to the skull base via the parapharyngeal space. On MRI their gross morphologic features are the same as those described for CT. On spin echo images, areas of both high and low signal intensity are visible probably due to differential flow rates within the tumor (50).

Other benign tumors of the lateral neck are uncommon. Neurilemomas or neurofibromas usually arise from the vagus or, rarely, the spinal accessory nerve where they present as lateral neck masses. A tumor of vagal origin will present in the anterior triangle while the spinal accessory variety presents beneath the sternocleidomastoid muscle or in the posterior triangle. If the mass is a neurilemoma (schwannoma) it may shows areas of cystic degeneration or calcification on CT (48). Schwannomas tend to have a thick capsule and appear hypovascular. Plexiform neurofibromas are also hypovascular but do not appear encapsulated and usually have lower attenuation than surrounding vessels. Internally, neurofibromas do not usually manifest the retrogressive cystic or calcific changes which may be seen in neuromas (4, 48, 65).

Lipomas are easily recognized on CT by their characteristic low attenuation. The fatty growths may appear localized at CT but in reality be very insinuating in their growth pattern. On MRI they will be seen as lesions of high signal intensity with, perhaps, better depiction of internal septae on high resolution studies. Dermoids (benign cystic teratomas) tend to occur near the midline and in the floor of the mouth. Where they are more laterally situated (in the submandibular space) they may mimic other lateral compartment masses (28). CT or MRI will show typically fat and fluid content with CT being more sensitive for calcifications. Other benign tumors are extremely rare (Table 8.2) and may arise from mesenchymal elements or accessory (ectopic) salivary tissue.

Nodal Metastases

Metastatic cervical adenopathy is our most common reason for performing CT of the neck. Before CT and ultrasound there was no practical means of imaging the cervical nodes. Either CT or MRI will remain the imaging examination of choice for staging cervical nodes for the foreseeable future, simply because either study can visualize all groups of interest. The more complex issue of which will provide more accurate data about the status of cervical nodes is yet to be determined.

Cervical metastatic disease is an extremely important issue in the diagnosis and management of head and neck malignancies. The vast majority of malignant head and neck neoplasms are squamous cell carcinomas of the upper aerodigestive tract. The incidence of nodal metastasis is quite high in these patients depending on the size, site and biologic aggressiveness of the lesion. As a general rule, the presence of histologically confirmed metastases halves the 5-yr survival for any given primary (4). Specific rates will be discussed in detail in subsequent chapters. Overall, the survival rates for patients with an upper aerodigestive tract carcinoma and nodes at the time of presentation is 30% (6). When patients present with a squamous cell cervical metastasis from an unknown primary, presumably in the upper aerodigestive tract, 5-yr survival is halved to 15%. Adeno- or undifferentiated carcinoma metastatis to the neck carry 5 and 9.4% survival rates (12).

Earlier in this chapter we considered the size range and morphology of normal nodes in the head and neck region. The criteria for abnormal nodes are presented in Table 8.4. The cutoff between normal and probable metastatic nodes is taken as 1.5 cm. Choosing a size criterion for this purpose immediately introduces a source of error (39, 40, 42, 58, 60). The limitations here are: (*a*) false positives due to benign, reactive nodes larger than 1.5 cm; and (*b*) false negatives due to microscopic or small macroscopic deposits in normal-size nodes. These are unavoidable. The rationale for the size criteria is based in large part on our experience and that of McGavran et al. (44), the latter reporting that necks with nodes 1.5 cm or larger are very likely to contain tumor, even if the largest node does not. Since most necks normally do not contain nodes larger than 1.0 cm, one could lower the size criteria for "positivity" or call those nodes in the 1.1–1.5 range suspicious (9, 20, 39, 40, 59). This would likely raise the false positive rate and lower the false negative rate, but have little impact on approach to treatment.

When tumor emboli enter the periphery of the node they become trapped and will grow to replace the normal tissue (20). Squamous cell carcinoma has a distinct tendency to become necrotic sometimes before the metastatic focus has grown to 1 cm (20, 37, 40, 72).

Table 8.4
Criteria Used for Staging Necks Modified from AJC[a] Guidelines[b]

N_0	Nodes less than 15 mm in size and of homogeneous density
N_1	Nodes 15–29 mm in size in largest diameter or node of any size with evidence of necrosis clearly demonstrated
N_2	Single homolateral node 3–6 mm in size; more than one positive homolateral node 15 mm or greater, or less than 15 mm with necrosis; conglomerate homolateral nodal mass 3–6 mm in size
N_3[c]	Homolateral nodal mass greater than 6 cm, bilateral nodes or contralateral nodes

CRITERIA USED FOR PREDICTING CAPSULAR RUPTURE AND EXTRANODAL EXTENSION

Intranodal tumor

Well circumscribed mass with a distinct interface between it and surrounding fat

Extranodal tumor

Ill-defined staining margin without clear distinction between it and surrounding fat

Evidence of edema or thickening of surrounding fibroadipose tissue or muscles

Fixation

Combination of extranodal characteristics and loss of plane between mass and structure in question (e.g. carotid, sternocleidomastoid muscle). Especially if mass surrounds structure in question

Adherent or abutting

Combination of intranodal characteristics and loss of plane between mass and structure in question

[a] From American Joint Committee for Cancer Staging and End-Results Reporting. *Manual for Staging of Cancer 1983.*
[b] Reprinted with permission from Mancuso AA, et al: Computed tomography of cervical and retropharangeal lymph nodes. Part II. Pathology. *Radiology* 148:716–723, 1983 (41).
[c] In bilateral adenopathy each neck should be staged separately.

spaces. In the oropharynx there seems to be almost the opposite. The tonsillar regions (lingual, faucial) appear relatively thinner than in normals suggesting involution of lymphoid tissue at these locales. The parapharyngeal spaces appear relatively larger because of this superficial atrophy and will look normal if there was no prior involvement by tumor.

The major salivary glands will show excessive contrast enhancement for varying periods (up to 3–6 months) postradiation. Following this, involution with fatty replacement is not uncommon and correlates with the postradiation "sicca syndrome" these patients experience.

If a particular site has been involved with tumor the findings will vary. In general, though, bulky masses are uncommon following successful radiotherapy. Some minor alteration in the tumor bed with fibrous strands is to be expected but in general the lesions should disappear. If the disease was very extensive, such as a nodal mass with a lot of extranodal spread, then a remaining residual fibrotic mass is more likely. With CT there is really no way to tell from a single study whether a residual mass is tumor, scar or both (23, 67). It is essential to obtain a baseline scan and followup at appropriate intervals (23, 67). This need not be done in all patients but patients at high risk for recurrence should be selected and followed for 2 yr or as the clinical situation warrants (23). Some risk factors would include: (*a*) advanced disease—stage III–IV, (*b*) extranodal spread, (*c*) close surgical margins, (*d*) persistent postradiation edema, and (*e*) persistent pain at primary or nodal sites or other symptoms referable to the head and neck region. With MRI, the signal from tumor may differ enough from that of fibrosis to allow more precise diagnosis than is possible with CT; this will be the subject of intense study over the next several years. Preliminary data does show that the signal from tumor decreases during radiation therapy and tends to approach that of skeletal muscle. This is most striking on T_2-weighted sequences.

Surgery alters the anatomy a good deal more than radiation but, again, the remaining anatomy has an extraordinary ability to return to a near normal appearance. In a classical radical neck dissection, the sternocleidomastoid muscle, jugular vein and submandibular will be sacrificed. In a Bocca (conservation) neck dissection, all of the normal structures will be left except the lymph node bearing fibrofatty tissue surrounding the carotid sheath. If the procedure is not complicated there will be very little "residual scar"; however, if there has been a wound infection or other complication, thickening of the deep tissue planes may be extensive. Again, a baseline study is necessary if follow-up is contemplated.

Modern surgical techniques sometimes require myocutaneous flaps for closing and reepithelializing large surgical defects. The muscle is in large part denervated during preparation and rotation of the flap so that on CT it will appear as a largely fatty density laced with strands of residual atrophic muscle. The appearance is unmistakable. There will often be some scar at the margins of the flap so that a baseline study is essential for accurate follow-up and early detection of residual tumor (64).

The vast majority of primary and recurrent head and neck cancers can be confirmed by routine clinical biopsy procedures. Sometimes the disease is in relatively inaccessible areas (e.g. retropharyngeal nodes) or beneath areas of extensive scarring. CT-directed needle biopsy of virtually any lesion is safe and easy to accomplish. The approaches to the infratemporal fossa and retropharyngeal areas are discussed in detail in Chapter 7. CT directed biopsy may be integrated with baseline and follow-up scans to provide a comprehensive strategy to detect and confirm recurrent tumor as early as possible (17).

References

1. Allgayer B, Reiser M, Ries G: Computed tomographic demonstration of venous thrombosis of different etiologies. *Eur J Radiol* 1:204–206, 1981.
2. Ballantyne AJ: Significance of retropharyngeal nodes in cancer of the head and neck. *Am J Surg* 108:500, 1964.
3. Bashur B, Ellis K, Gold RP: Computed tomography of intra-thoracic goiters. *AJR* 140:455–460, 1983.
4. Batsakis JG: *Tumors of the Head and Neck: Clinical and Pathologic Considerations*, ed 2. Baltimore, Williams & Wilkins, 1979.
5. Beahrs O: Surgical anatomy and technique of radical neck dissection. *Surg Clin North Am* 57:633–700, 1977.
6. Brady JV: The present status of treatment of cervical metastases from carcinoma arising in the head and neck region. *Am J Surg* 111:56, 1971.
7. Bocca E: Conservative neck dissection. *Laryngoscope* 85:1511–1515, 1975.
8. Brandenburg JH, Lee CYS: The eleventh nerve in radical neck surgery. *Laryngoscope* 91:1851, 1981.
9. Cachin Y, Sancho-Garnier H, Micheau C, Marandas P: Nodal metastasis from carcinomas of the oropharynx. *Otolaryngol Clin North Am* 12:145–154, 1979.
10. Cocke EW: Benign cartilagenous tumors of the larynx. *Ann Otol Rhinol Laryngol* 72:167, 1962.
11. DeSanto LW, Holt JJ, Beahrs OH, O'Fallon WM: Neck dissection: is it worthwhile? *Laryngoscope* 92:502–509, 1982.
12. Didolkar MS, Fanous N, Elias EG, Moore RH: Metastatic carcinomas from occult primary tumors. A study of 254 patients. *Ann Surg* 186:625, 1977.
13. Doppman JL, Brennan MF, Koehler JO, Marx SJ: Computed tomography for parathyroid localization. *J Comput Assist Tomogr* 1:30–36, 1977.
14. Doppman JL, Krudy AG, Marx SJ, Saxe A, et al: Aspiration of enlarged parathyroid glands for parathyroid assay. *Radiology* 148:31–35, 1983.
15. Endicott JN, Nelson RJ, Saraceno CA: Diagnosis and management decisions in infections of the deep fascial spaces of the head and neck utilizing computerized tomography. *Laryngoscope* 92:630–633, 1982.
16. Fermont DC: Malignant cervical lymph adenopathy due

to an unknown primary. *Clin Pathol* 31:355–358, 1980.
17. Gatenby RA, Mulhern CB, Strawrtz J: CT-guided percutaneous biopsies of the head and neck masses. *Radiology* 146:717–719, 1983.
18. General Electric Corporation Technical Exhibit: Chicago, Radiologic Society of North America Scientific Assembly, 1983.
19. Glazer HS, Aronberg DJ, Lee JKT, Sagel SS: Extralaryngeal causes of vocal cord paresis: CT evaluation. *AJR* 141:527–531, 1983.
20. Haagensen CD, Feind CR, Herter FT, Slanetz CA, Weinberg JA: *The Lymphatics in Cancer*. Philadelphia, Saunders, 1972.
21. Harkins GA, Sabiston DC: Lymphangioma in infancy and childhood. *Surgery* 47:811, 1960.
22. Harnsberger HR, Mancuso AA, Byrd S, Muraki AS, Johnson L, Hanafee WN: Branchial cleft anomalies and their mimics: the role of CT. *Radiology* 152:739–748, 1984.
23. Harnsberger HR, Mancuso AA, Muraki AS, Parkin JL: The upper aerodigestive tract and neck: CT evaluation of recurrent tumor. *Radiology* 149:503–510, 1983.
24. Hawkins DB, Jacobsen BE, Klatt EC: Cysts of the thyroglossal duct. *Laryngoscope* 92:1254–1258, 1982.
25. Himalstein MR: Branchial cysts and fistulas. *Ear Nose Throat J* 59:47–54, 1980.
26. Hollinshead WH: *Anatomy for Surgeons: Volume 1: The Head and Neck*, ed 2. Hagerstown, Md, Harper Rowe, 1968.
27. Holt CR, McManus K, Newman RK, et al: Computed tomography in the diagnosis of deep neck infections. *Arch Otolaryngol* 108:693–696, 1982.
28. Hunter TB, Paptanus SH, Chernin MM, Coulthard SW: Dermoid cyst of the floor of the mouth: CT appearance. AJR 141:1239–1240, 1983.
29. Jesse RH: The philosophy of treatment of neck nodes. *Ear Nose Throat J* 56:58, 1977.
30. Jesse RH, Fletcher GH: Treatment of the neck in patients with squamous cell carcinoma of the head and neck. *Cancer* 39:868–872, 1979.
31. Kalnins IK, Leonard AJ, Sako K, Razack MS, Shedd DP: Correlation between prognosis and degree of lymph node involvement in carcinoma of the oral cavity. *Am J Surg* 134–450, 1977.
32. Kalovidouris A, Mancuso AA, Sarti DA: Static grey-scale parathyroid ultrasonography: is high resolution real-time technique required? *Clin Radiol* 34:385–393, 1983.
33. Kalovidouris A, Mancuso AA, Dillon W: A CT-clinical approach to patients with symptoms related to the Vth, VIIth, IXth–XIIth cranial nerves and cervical sympathetics. *Radiology* 151:671–676, 1984.
34. Krudy AG, Doppman JL, Brennan MF, Marx ST, et al: The detection of mediastinal parathyroid glands by computed tomography, selective arteriography and venous sampling. *Radiology* 140:739–744, 1981.
35. Krudy AG, Doppman JL, Miller DL, Marx SJ: Work in Progress: abnormal parathyroid glands. *Radiology* 148:23–29, 1983.
36. Last RJ: *Anatomy Regional and Applied*, ed 5. Edinburgh, Churchill-Livingston, 1972.
37. Lenz M, Bahren W, Hasse ST, Ranzinger G, Wierschin W: Computer tomographie zervikaler lymphknotenmetastasen bei Malignomen des Kopf-Heb-Bereichs. ROFO 139:281–284, 1983.
38. Mancuso AA, Hanafee WN: Elusive head and neck carcinomas beneath intact mucosa. *Laryngoscope* 93:133–139, 1983.
39. Mancuso AA, Maceri D, Rice D, Hanafee WN: CT of cervical lymph node cancer. *AJR* 136:381–385, 1981.
40. Mancuso AA, Harnsberger HR, Muraki AS, Stevens MH: Computed tomography of cervical and retropharyngeal lymph nodes: normal anatomy, variants of normal and applications in staging head and neck cancer. Part 1: Normal anatomy. *Radiology* 148:709–714, 1983.
41. Mancuso AA, Harnsberger HR, Muraki AS, Stevens MH: Computed tomography of cervical and retropharyngeal lymph nodes: normal anatomy, variants of normal and applications in staging head and neck cancer. Part II: Pathology. *Radiology* 148:715–723, 1983.
42. Martin H, Morfit HM: Cervical node metastasis as the first symptom of cancer. *Surg Gynecol Obstet* 76:133–159, 1944.
43. McGuirt WF, McCabe BF: Significance of node biopsy before definitive treatment of cervical metastatic carcinomas. *Laryngoscope* 88:594–597, 1978.
44. McGavran MH, Bauer WC, Ogura JH: The incidence of cervical lymph node metastases from epidermoid carcinoma of the larynx and the relationship to certain characteristics of the primary tumor. *Cancer* 14:55–66, 1961.
45. Mehar GI, Colley DP, Clark RA, et al: Computed tomographic demonstration of cervical abscess and jugular vein thrombosis: a complication of intravenous abuse in the neck. *Arch Otolaryngol* 107:313–315, 1981.
46. Meyers DS, Templer J, Davis WE, et al: Aspiration cytology for diagnosis of head and neck masses. *Trans Am Acad Ophthomol Otolaryngol* 86:650–655, 1978.
47. Miller D, Ervin T, Weichselbaum R, Fabian RL: The differential diagnosis of the mass in the neck. A fresh look. *Laryngoscope* 91:140–145, 1981.
48. Miller EM, Norman D: The role of computed tomography in the evaluation of neck masses. *Radiology* 133:145–149, 1979.
49. Million RR, Cassisi NJ: *Management of Head and Neck Cancer: A Multidisciplinary Approach*. Philadelphia, Lippincott, 1984.
50. Mills CM, Brandt-Zawadski M, Brooks LE, et al: Nuclear magnetic resonance: principles of blood flow imaging. *AJNR* 4:1161–1166, 1983.
51. Moss WT, Brand WN, Battifora H: *Radiation Oncology—Rational, Technique, Results*, ed 5. St. Louis, Mosby, 1979.
52. Muraki AS, Mancuso AA, Harnsberger HR: Metastatic cervical adenopathy from tumors of unknown origin: the role of CT. *Radiology* 152:749–753, 1984.
53. Nahum AM, Bone RC, Davidson TM: The case for elective prophylactic neck dissection. *Laryngoscope* 87:588–599, 1977.
54. Nystrom JS, Weiner JM, Wolf RM, et al: Identifying the primary site in metastatic cancer of unknown origin. *JAMA* 241:381–383, 1979.
55. Picus D, Balfe DM, Koehler RE, Roper CL, et al: Computed tomography in the staging of esophageal carcinoma. *Radiology* 146:433–438, 1983.
56. Reading CC, Charboneau JW, James EM, Karsell PR, Purnell DC, Grant CS, vanHeerden JA: High resolution parathyroid sonography. *AJR* 139:539–546, 1982.
57. Reed GF, Rabuzzi DD: Neck dissection. *Otolaryngol Clin North Am* 547–563, 1969.
58. Reede DL, Whelan MA, Bergeron RT: Computed tomog-

raphy of the infrahyoid neck—parts I and II. *Radiology* 145:389–402, 1982.

59. Rouvier H: *Anatomy of the Human Lymphatic System,* ed 1. Ann Arbor, Mich, Edwards Brothers 1938, pp 1–82.
60. Sako K, Pradier RN, Marchetta FC, Pickren JW: Fallibility of palpation in the diagnosis of metastasis to cervical nodes. *Surg Gynecol Obstet* 118:989, 1964.
61. Sample WF, Mitchell SP, Bledsoe RC: Parathyroid ultrasonography. *Radiology* 127:485–490, 1978.
62. Schaefer SD, Merkel M, Diehl J, Maravilla K, Anderson R: Computed tomographic assessment of squamous cell carcinoma of oral and pharyngeal cavities. *Arch Otolaryngol* 108:688–692, 1982.
63. Simeone JF, Mueller PR, Ferucci JT, vanSonnenberg E, Wang Chiu-An, Hall DA, Wittenberg J: High resolution real-time sonography of the parathyroid. *Radiology* 141:745–751, 1981.
64. Solbiati L, Montali G, Groce F, Belloti E, Giangrande A, Ravetto D: Parathyroid tumors detected by fine needle aspiration biopsy under ultrasonic guidance. *Radiology* 148:793–797, 1983.
65. Som PM, Biller HF: Computed tomography of the neck in the postoperative patient: radical neck dissection and the myocutaneous flap. *Radiology* 148:157–160, 1983.
66. Som PM, Shugar JM, Biller HF: The early detection of antral malignancy in the postmaxillectomy patient. *Radiology* 143:509–512, 1982.
67. Stark DD, Moss AA, Gooding GAW, Clark OH: Parathyroid scanning by computed tomography. *Radiology* 148:297–299, 1983.
68. Templar J, Perry MC, Davis WE: Metastatic cervical adenocarcinoma from unknown primary tumor. *Arch Otolaryngol* 107:45–47, 1981.
69. VonBahren W, Haase ST, Lenz M, Ranzinger G, Wierschin W: Computer tomographie zervikaler lymphknotenmetastasen bei Malignomen des Kopf-Heb-Bereichs. *ROFO* 139:281–284, 1983.
70. Ward PH, Fredrickson JM, Strandjord NM, Valvasson GE: Laryngeal and pharyngeal pouches: surgical approach and the use of cinefluorographic and other radiologic techniques as a diagnostic aid. *Laryngoscope* 73:564, 1963.
71. Wilkinson EJ, Hause L: Probability in lymph node sectioning. *Cancer* 33:1269, 1974.
72. Work WP: Newer concepts of first branchial cleft defects. *Laryngoscope* 87:1581–1593.

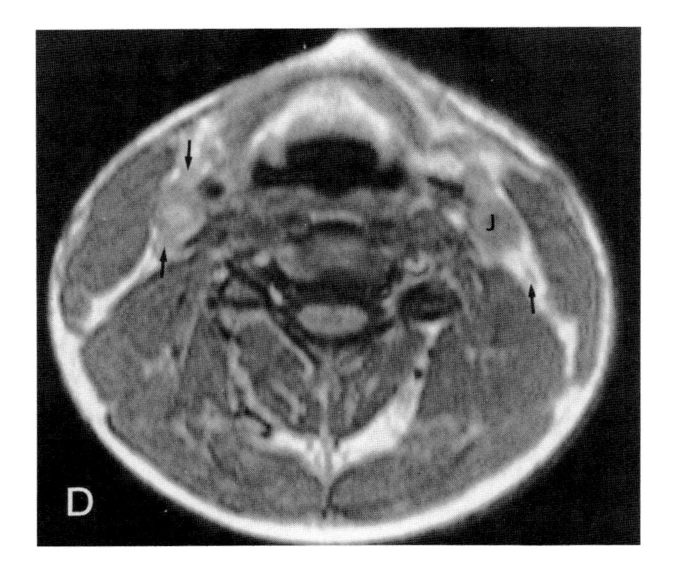

Plate 8.3*D*. The midneck and the midsupraglottic larynx. The nodes seen around the carotid sheath are normal in size. On an MRI it is difficult to be certain whether the rounded structures (*arrows*) in front of the carotid sheath are lymph nodes or merely small veins in which blood is flowing slowly. Note that the signal intensity in these structures is approximately equivalent to that of the left internal jugular vein. This is a potential pitfall in staging cervical lymph node disease with MRI.

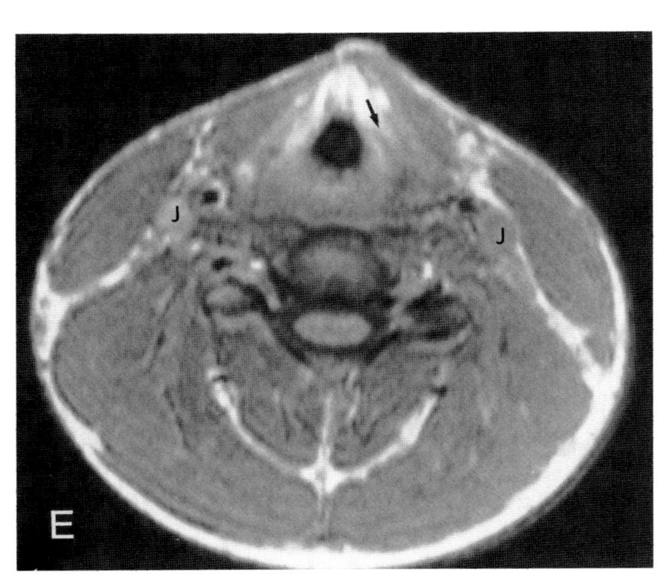

Plate 8.3*E*. The midneck at the level of the false vocal cords and therefore still in the region of the middle internal jugular group of nodes. Small nodes can lie posterior to the jugular vein. In prior sections as well as this one, the fatty envelope separating the muscle bundles within the neck have been very well seen despite the fact that this patient has very little fat compared to musculature. The superior contrast resolution of MRI makes the evaluation of patients with thin necks easier than it is with CT. The fat within the paralaryngeal space is visible as well (*arrow*). Normal laryngeal anatomy will be discussed in detail in Chapter 9. The neural elements are quite easily seen; these will be detailed in Plate 8.12.

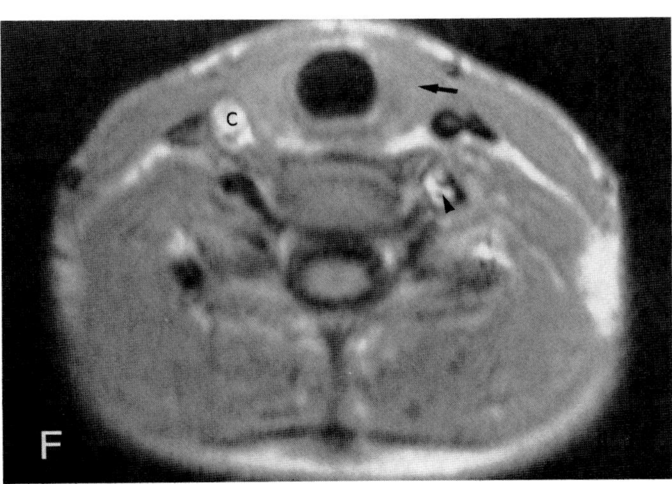

Plate 8.3*F*. The low neck at the level of the first tracheal ring. The thyroid gland has a slightly shorter T_1 than the surrounding muscle giving it a slightly more intense signal on this pulse sequence (*arrow*). The carotid artery on the right side as well as the vertebral artery (*arrowhead*) on the left side are brighter than the jugular veins. We would expect the flow in these vessels to be somewhat faster than that in the veins. The arteries are brighter because, during interleaved multislice acquisition, protons perturbed in one slice may be "read out" in another.

The abbreviations used in Plates 8.4–8.6 are: *J*, jugular vein; *C*, carotid artery; *LC*, longus capitus muscles; *V*, retromandibular vein; *PG*, parotid gland; and *SMG*, submandibular gland.

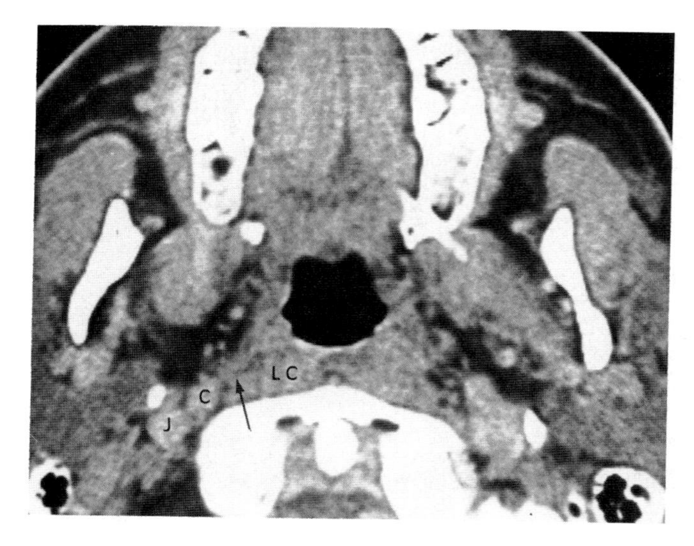

Plate 8.4. Before CT it was fairly difficult to evaluate the retropharyngeal nodal group. The lateral retropharyngeal nodes (*arrow*) normally lie between the longus colli and longus capitus muscles and the structures of the carotid sheath. Normal densities, presumably lymph nodes, are routinely visible in this location and may measure up to 5 or 7 mm in size. Nodes over 1 cm in size in this region are considered abnormal with the exception of those seen in young children. These nodes do involute with increasing age but are commonly present in adults.

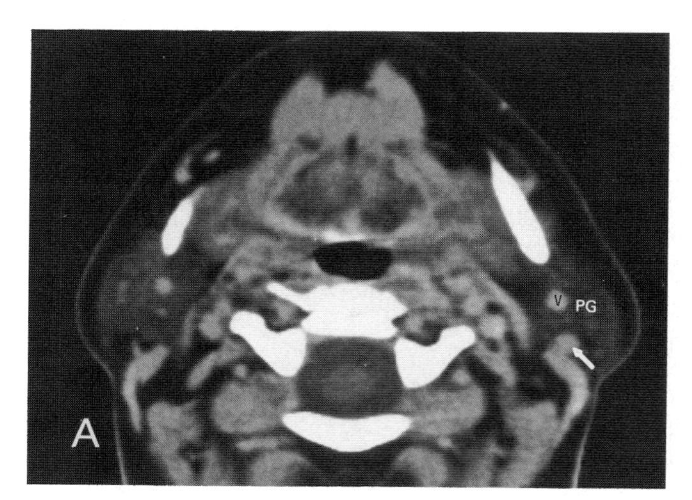

Plate 8.5. Intra- and periparotid nodes are normally visible on CT studies although they are seen much less frequently than those of the internal jugular chain. They tend to be approximately 3–5 mm in size when present and they occur in farily characteristic locations. One usual location is the pretragal region not pictured in the next two plates.

Plate 8.5*A*. A slightly off-axis view through the parotid gland. The retromandibular vein is a common landmark within the substance of the parotid gland. A normal periparotid node (*arrow*) lies between the substance of the gland and the sternocleidomastoid muscle.

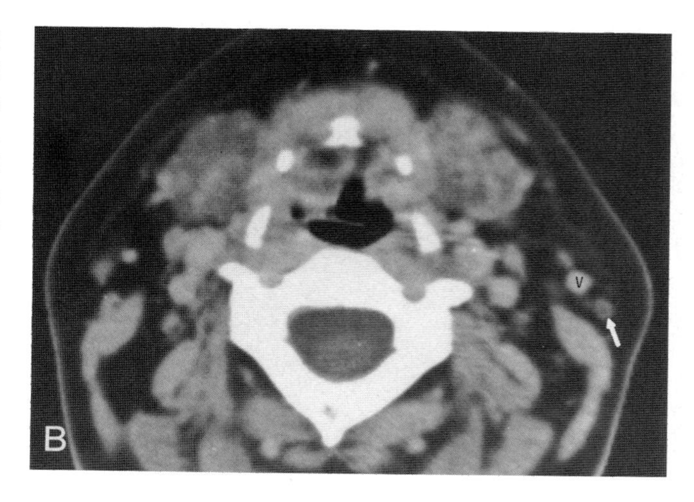

Plate 8.5*B*. The section made approximately 2 cm lower shows the retomandibular vein near its exit from the parotid gland. The nonenhancing periparotid node (*arrow*) is immediately adjacent to the tail of the parotid. This is the most common location of normal periparotid nodes as seen on CT studies. These nodes may also be found in the pretragal region and within the substance of the gland along the course of the retromandibular vein.

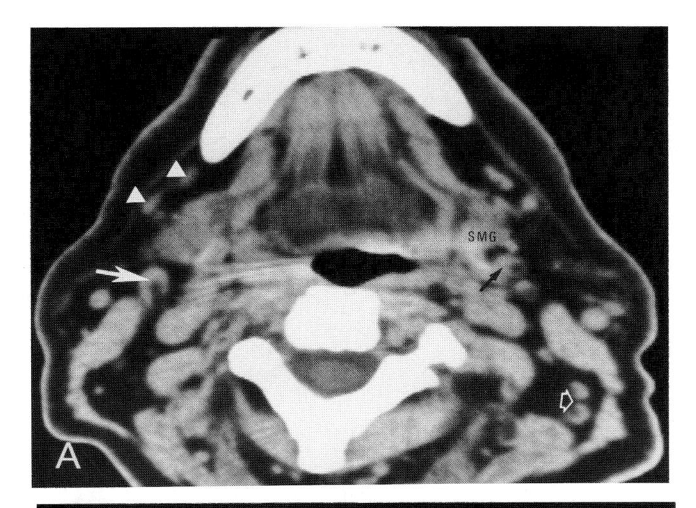

Plate 8.6. The following three sections are made through the submandibular and submental regions. Normal lymph nodes are commonly seen in this area and they may appear fairly large at times. Reactive nodes are very common in this region due to the high incidence of periodontal and tooth infections.

Plate 8.6*A*. Normal submandibular nodes can be seen within the submandibular space both in front (*white arrowheads*) and behind (*white arrow*) the submandibular gland. Nodes may also appear to be wrapped within the substance of the gland (*black arrow*). Nodes which have been enlarged may involute, sometimes with fatty replacement within their hila (*white and black arrows*). This is a normal variant and should not be mistaken for a necrotic metastatic focus within a normal-sized node. Similar foci are present in the upper spinal accessory nodes on the left side (*open arrowhead*). This also illustrates the proximity of all the nodal groups high in the neck. Sometimes it is difficult to be sure whether enlarged nodes in this region are from the submandibular group, internal jugular group or spinal accessory group, or any combination of the three.

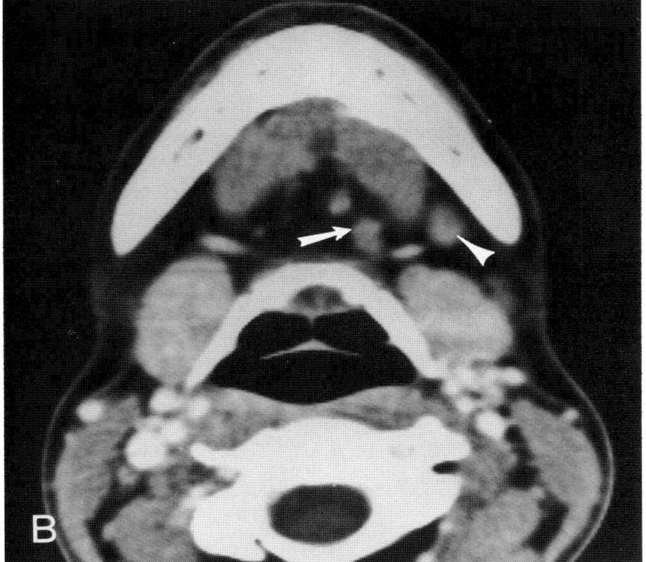

Plate 8.6*B*. The section through the lower submandibular and submental spaces shows nodes in the submental triangle or suprahyoid portion of the neck (*arrow*). A 1-cm submandibular node (*arrowhead*) is also present.

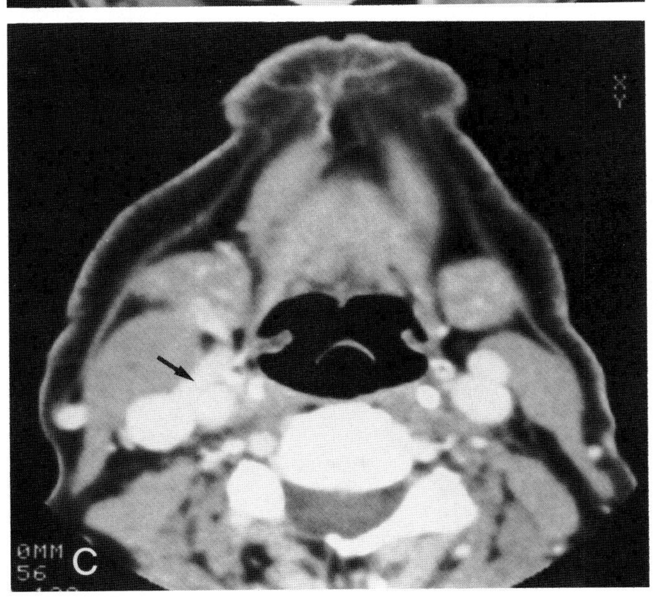

Plate 8.6*C*. This section illustrates the degree of variation that can be seen from patient to patient. Virtually no nodes are visible in the submandibular or submental region in this patient and the internal jugular nodes are also very small. The junction of the common facial vein and the internal jugular vein (*arrow*) is, technically, the inferior boundary between the upper (jugulodigastric) and middle internal jugular groups.

The abbreviations used in Plates 8.7–8.9 are: *ScM*, scalene muscle group; *J*, jugular vein; *CA*, *CCA*, carotid artery; *LC, longus colli muscle; IHS*, infrahyoid strap muscles; *C*, cricoid cartilage; *ES*, esophagus, *SCM*, sternocleidomastoid muscle; *Tr*, trachea; and *TG*, thyroid gland.

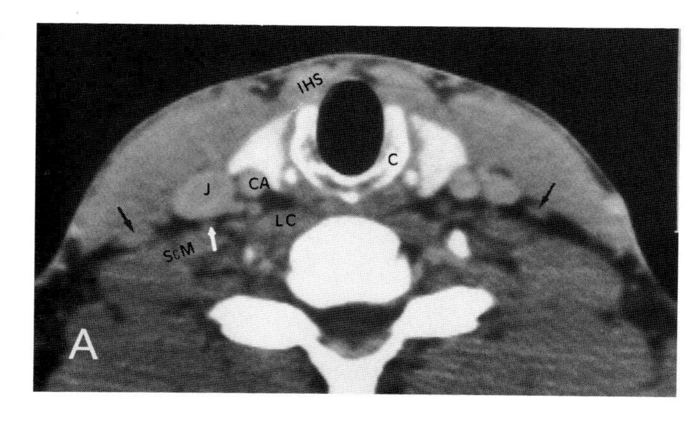

Plate 8.7. Sections through the low neck will show the anatomy of the thyroid gland and thoracic inlet with remarkable clarity. The proper visualization of all of the structures in this region requires thin section studies and usually intravenous contrast injection. The anatomy varies considerably as one proceeds from the thyroid bed to the thoracic inlet.

Plate 8.7*A*. The thyroid gland itself is unmistakable as it lies immediately adjacent to the cricoid cartilage. The high attenuation of the two thyroid lobes is due to their iodine content which varies from patient to patient. The infrahyoid strap muscles create soft tissue density just anterior to the infraglottic larynx at this level. The parathyroid regions are located between the carotid sheath, the longus colli muscle and the thyroid gland. Normal lymph nodes may be seen in this location. The spinal accessory and lower internal jugular groups lie lateral to the jugular vein (*black arrows*) and occasionally a normal scalene node may be seen (*white arrow*) just anterior to the scalene muscle group.

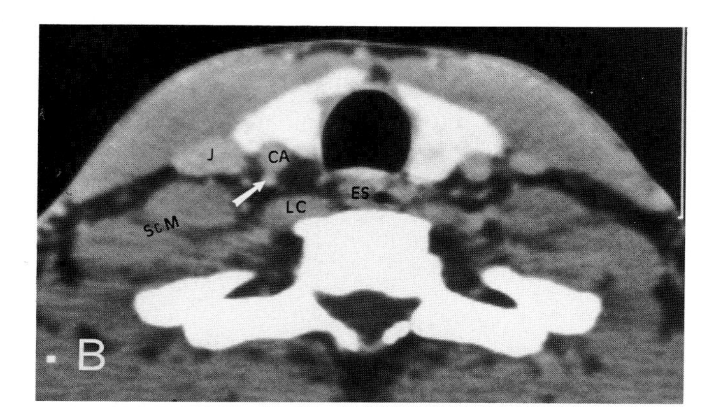

Plate 8.7*B*. Once below the cricoid cartilage, the anatomy changes. The structure of the cricopharyngeus muscle gives way to the typical elliptical soft tissue density of the esophagus. The jugular vein lies lateral rather than posterolateral to the carotid artery. The minor neurovascular bundle (*arrow*), which includes the recurrent laryngeal nerve and inferior thyroid artery and vein, lies in the fat behind and medial to the carotid artery.

Plate 8.7*C*. At the thoracic inlet the lung apices become visible. The esophagus continues to lie behind the trachea slightly to the left of midline. The longus colli muscles tend to become somewhat smaller as does the scalene group as it narrows to tendons which insert on the first rib. The vessels of the minor neurovascular bundle (*arrow*) are still visible. The phrenic nerve lies just anterior to the anterior scalene muscle and the cervical sympathetics ascend just anterior and lateral to the transverse processes.

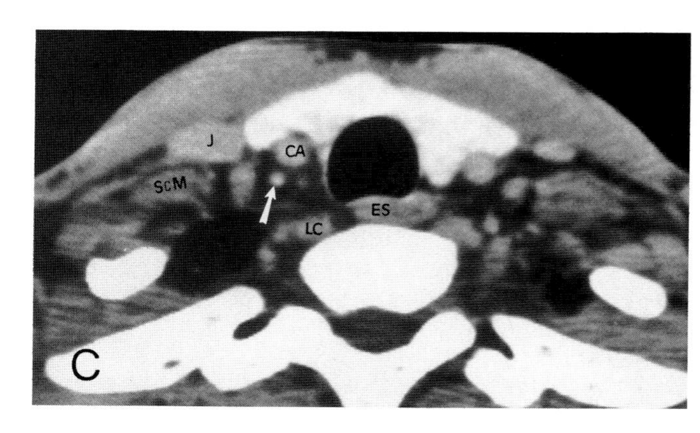

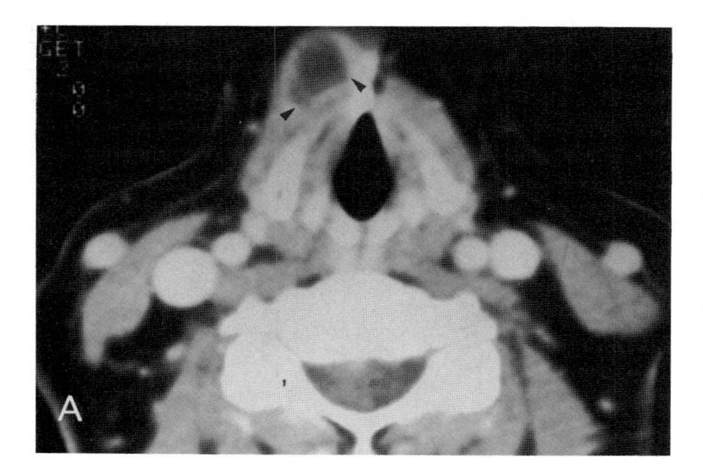

Plate 8.20. These are three axial CT sections of a patient with a thyroglossal duct cyst.

Plate 8.20*A*. At the level of the true vocal cords, a cystic mass (*arrowheads*) lies in a paramedian position just anterior to the infrahyoid strap muscles. Thyroglossal duct cysts are usually in, or just off, the midline. Occasionally, they can be slightly more lateral and then might be confused for masses arising from the lateral compartment.

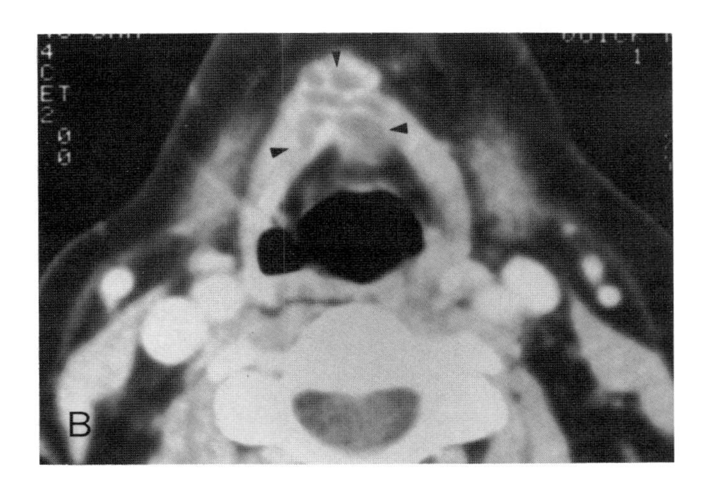

Plate 8.20*B*. At the midsupraglottic level, the multiloculated thyroglossal duct cyst occupies the midline (*arrowheads*). It is wedged in between the infrahyoid strap muscles at the level of the thyrohyoid membrane. Some extension into the preepiglottic space is evident.

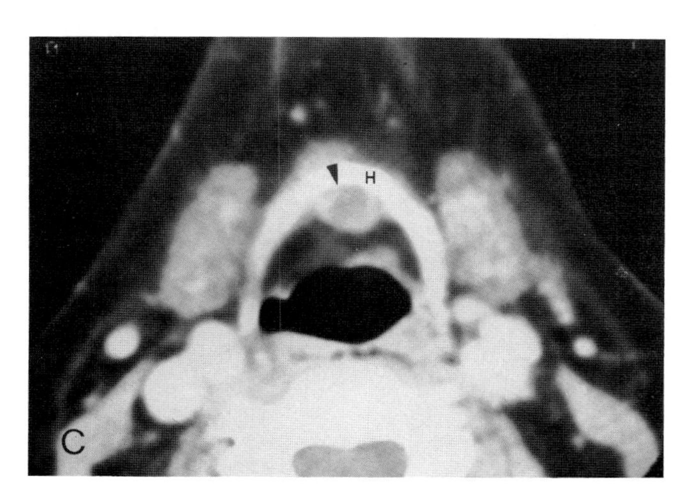

Plate 8.20*C*. At the level of the hyoid bone the cyst (*arrowhead*) lies posterior to the body of the hyoid bone within the upper preepiglottic space. This is a very typical location and is the reason that the middle segment or body of the hyoid bone must be removed during surgical excision of these cysts.

The abbreviations used in Plates 8.21–8.24 are: *H*, hyoid bone; *SCM*, sternocleidomastoid muscles; *CCA*, carotid artery; *Tr*, trachea; *Cy*, cyst; and *LC*, longus colli muscle.

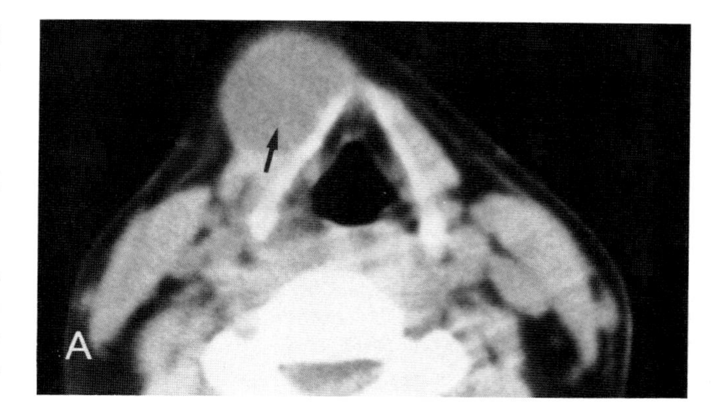

Plate 8.21. A different patient with a thyroglossal duct cyst.

Plate 8.21*A*. A section at the level of the false cords shows the large cystic mass (*arrow*) in the paramedian position. This lesion could be mistaken on physical exam for an external laryngocele or mass arising from the lateral compartment.

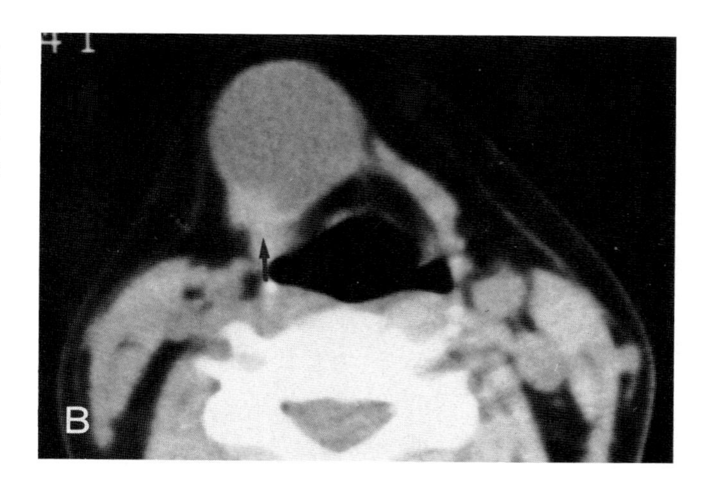

Plate 8.21*B*. A section through the midsupraglottic larynx shows the thyroglossal duct cyst to be somewhat larger and there is displacement of the infrahyoid strap muscles posteriorly and laterally (*arrow*). (This case appears courtesy of Daniel H. Johnson, M.D., New Orleans Radiological Group, Inc.)

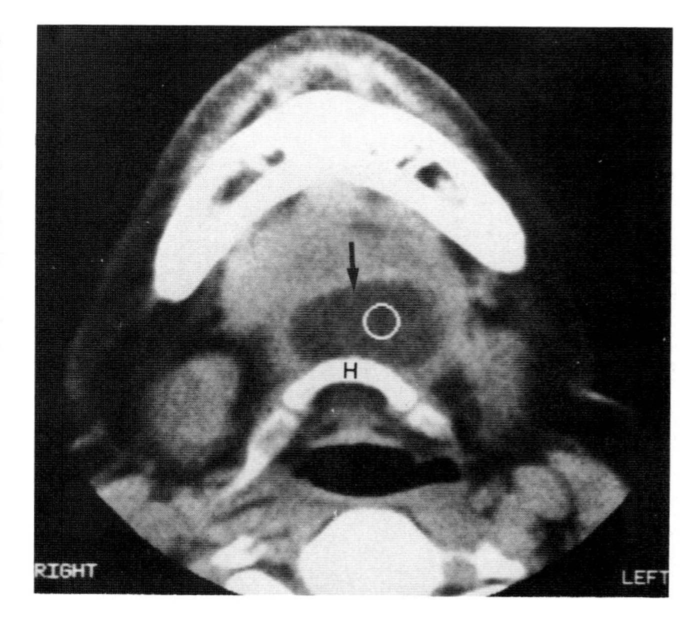

Plate 8.22. A thyroglossal cyst in another patient. The lower sections showed the typical course of the cyst in the low neck. Portions of thyroglossal duct cysts often extend into the tongue base although they are usually in the forms of small tracts. Because of this, for these lesions a core of tissue is normally taken in the tongue base to the foramen cecum during surgery. This patient demonstrates an unusually large cystic extension (*arrow*) into the tongue base. This was of obvious value in preoperative planning since the tongue base is a common site of recurrence if the lesions are not properly excised. (This case appears courtesy of G. Gamsu, M.D., Department of Radiology, University of California at San Francisco.)

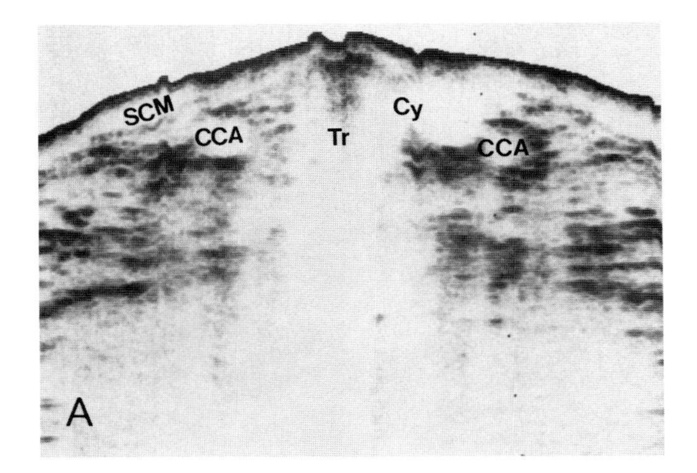

Plates 8.23 and 8.24. Most often, thyroid ultrasound is done to determine whether a palpable, "cold," thyroid nodule is cystic or solid.

Plate 8.23*A*. On the transverse image there is an anechoic region located between the trachea and carotid artery. It has a very sharply circumscribed back wall and there is evidence of enhanced through-transmission of sound. This appears to be a simple cyst.

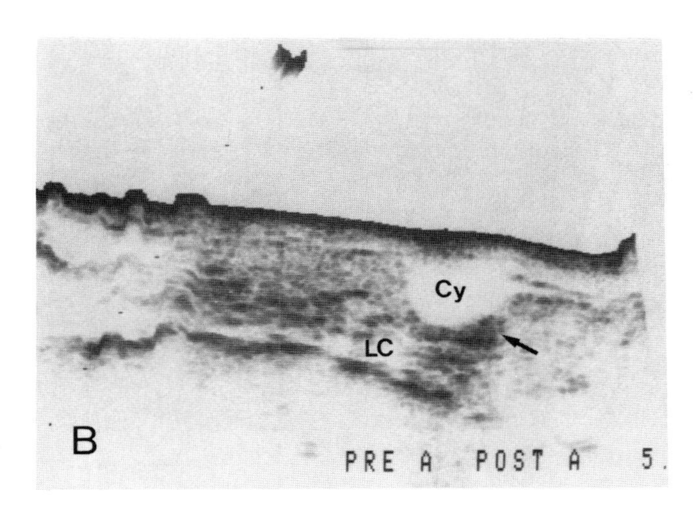

Plate 8.23*B*. The presence of the mass is confirmed on the longitudinal view. The cyst lies anterior to the longus colli muscle. Its cystic nature is confirmed by its strictly anechoic appearance, sharpness of its contours, and enhancement of through-transmission (*arrow*).

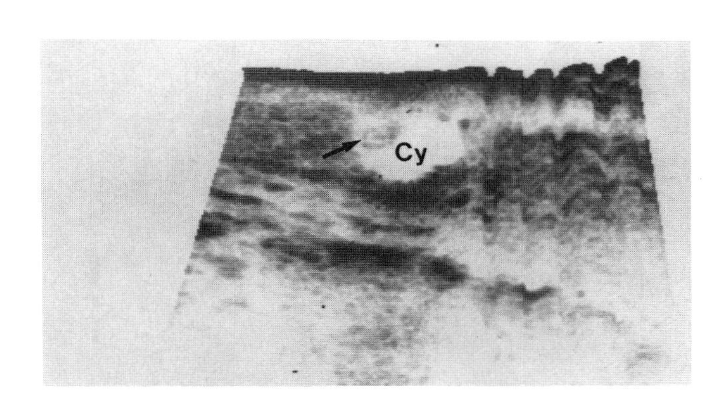

Plate 8.24. This is primarily a cystic mass in the thyroid gland; however, a focal nodular area of increased echogenicity projects from its wall (*arrow*). This takes the cyst out of the simple category and places it into the complex, primarily cystic category. The latter classification may be due to hemorrhagic cysts, necrotic adenomas or carcinoma. The appearance of these lesions is nonspecific and requires further evaluation.

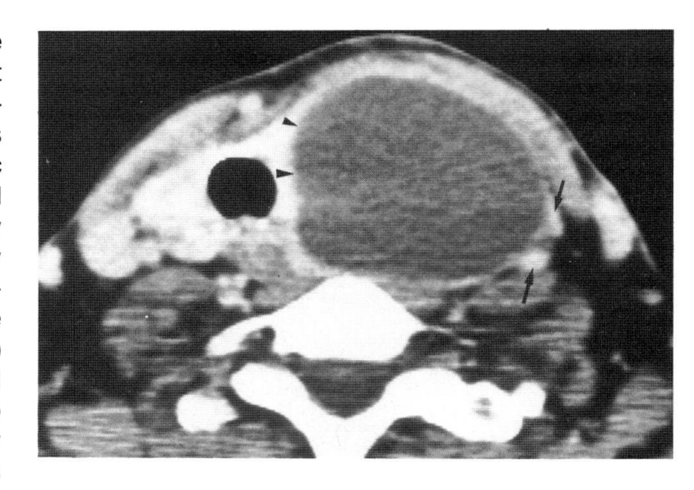

Plate 8.25. Computed tomography is of limited value in evaluating lesions of the thyroid gland. This patient presented with a large lower neck mass and was suspected of having a branchial cleft cyst. CT study was done to aid in the differential diagnosis. The large cystic lesion is obvious and could be arising either in the visceral or lateral compartments of the neck. The carotid artery and jugular vein on the left side are displaced far laterally (*arrows*). The trachea and, in fact, entire visceral compartment is displaced to the right. The left lobe of the thyroid gland is stretched over the mass (*arrowheads*) suggesting that the mass arose within the thyroid gland rather than outside of it. This is somewhat analagous to the "beak" sign seen in renal cysts. At surgery a very large thyroid cyst was confirmed. (This case appears courtesy of H. R. Harnsberger, M.D., Department of Radiology, University of Utah Medical School.)

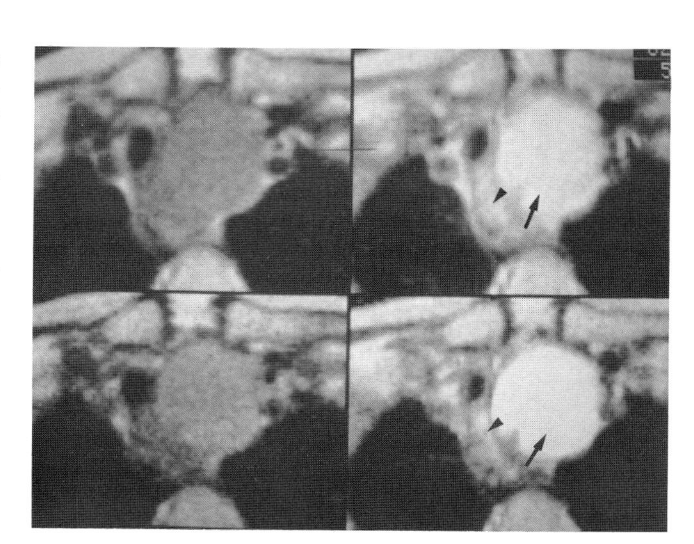

Plate 8.26. Magnetic resonance images made at the thoracic inlet show a large goiter. The T_2-weighted images (*above* and *below on the right*) show areas of differing T_2. The large, very will circumscribed cyst (*arrow*) has a T_2 which is somewhat longer than the other portions of the enlarged thyroid gland (*arrowhead*). It is likely that MRI will have a limited role in imaging pathology within the thyroid gland, perhaps somewhat similar to that of CT. Ultrasound is a good deal less expensive and quite good for determining the cystic versus solid nature of solitary, cold nodules. This is the single most frequent indication for imaging of the thyroid gland. (This case is courtesy of Drs. D. Stark and G. Gamsu, Department of Radiology, University of California at San Francisco.)

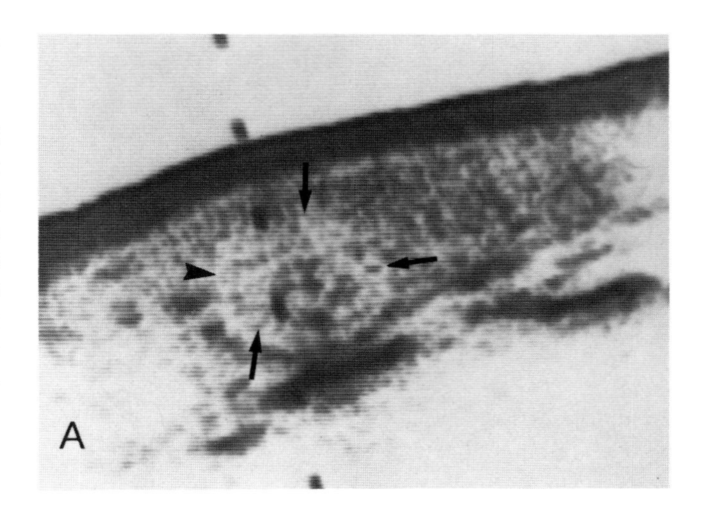

Plate 8.27. A variety of purely solid lesions as seen by ultrasound.

Plate 8.27*A*. A longitudinal scan shows a 1.3-cm solid thyroid nodule (*arrows*). The texture change between the normal thyroid gland and the mass is dramatic. Along one of its lateral margins there is a suggestion of a surrounding "halo." This was a solitary, cold nodule. The odds are overwhelming that it is a benign adenoma. The surrounding "halo" (*arrowhead*) strongly suggests that it is benign; however, there is no way to definitely tell that this is a benign lesion by any ultrasound criteria. The definitive diagnosis requires histologic proof.

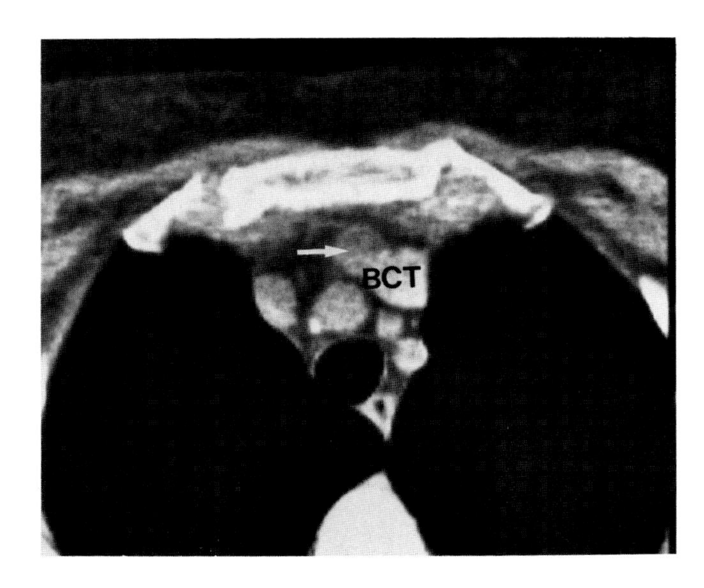

Plate 8.37. A CT scan from a patient who had clinical and biochemical evidence of a parathyroid adenoma. Ultrasound examination of the neck revealed no evidence of a parathyroid adenoma in the usual locations. The CT examination showed a small intramediastinal mass (*arrow*) just anterior to the left brachiocephalic trunk. Surgery confirmed an ectopic parathyroid adenoma.

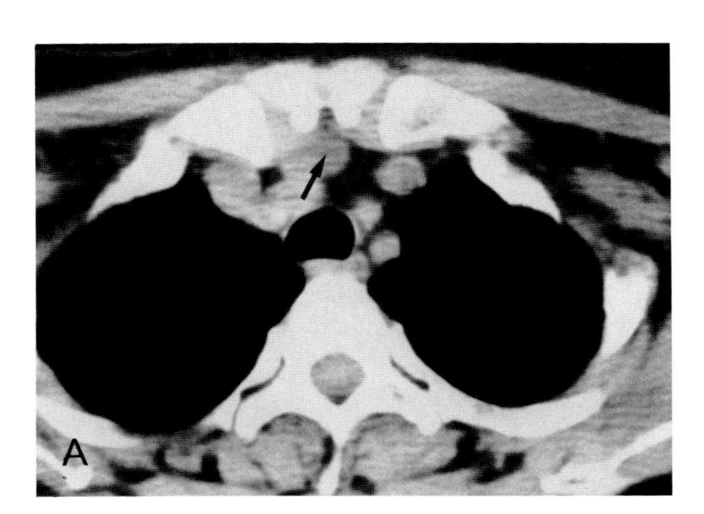

Plate 8.38. A CT and MR examination in a patient with hyperparathyroidism.

Plate 8.38*A*. The CT study at thoracic aorta strongly suggests a superior mediastinal parathyroid adenoma (*arrow*).

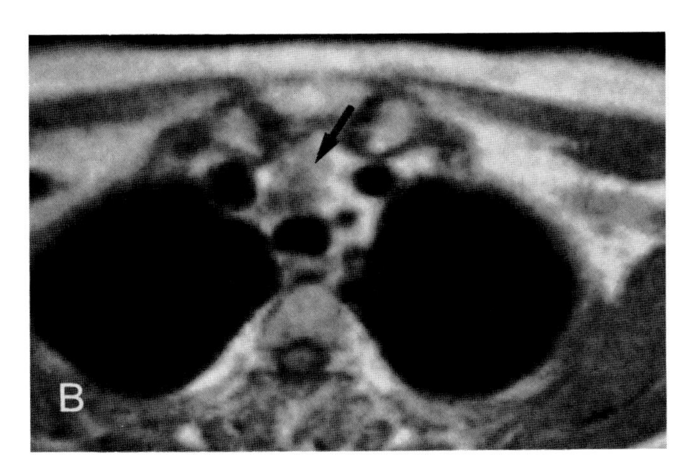

Plate 8.38*B*. A T_1-weighted sequence (1000/28) shows the ectopic parathyroid adenoma to best advantage since the signal from the adenoma (*arrow*) is nicely contrasted from that of the surrounding mediastinal fat. (This case appears courtesy of Drs. D. Stark and G. Gamsu, Department of Radiology, University of California at San Francisco.)

The abbreviations used in Plates 8.39–8.43 are: *SCM*, sternocleidomastoid muscle; *J*, *JV*, jugular vein; *CA*, carotid artery; *SMG*, submandibular gland; and *PS*, parapharyngeal space.

LATERAL COMPARTMENT

Plates 8.39–8.43. Both CT and ultrasound can be used to evaluate cystic neck masses. Ultrasound often adds little to the clinical impression; it may confirm the cystic nature of the mass but it rarely shows its entire extent and relationship to the other structures within the neck. CT better defines the anatomy relative to the pathology and may provide data which will markedly alter surgical or other therapeutic approaches. MRI should prove roughly equivalent to CT in these regards.

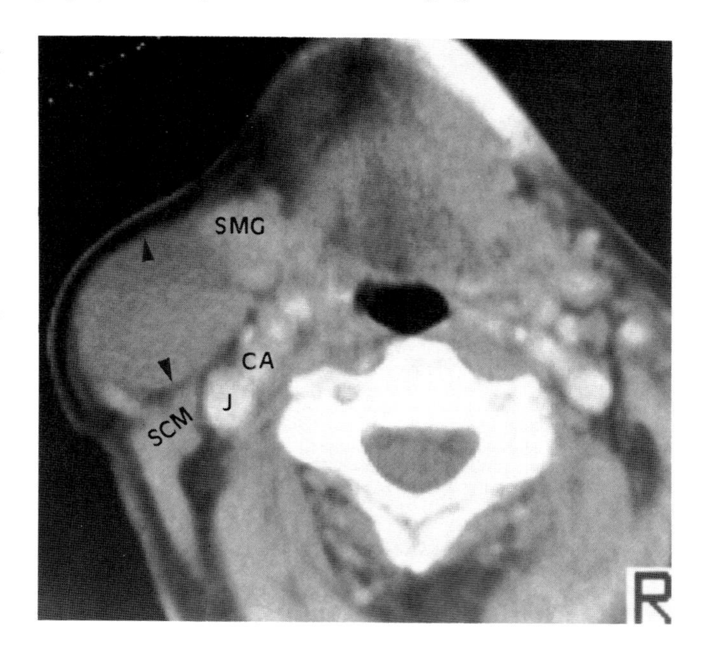

Plate 8.39. A well circumscribed mass is present in the lateral neck; its attenuation value is nearly equivalent to water. The appearance and location (*arrowheads*) of this cystic mass is virtually pathognomonic for a second branchial cleft cyst. The submandibular gland is pushed anteriorly and medially while the sternocleidomastoid muscle is displaced posteriorly. The carotid artery and jugular vein are sometimes displaced medially. The mass itself has a very thin wall (*arrowheads*) which really cannot even be appreciated on the CT study. This was compatible with the lack of history of prior infection.

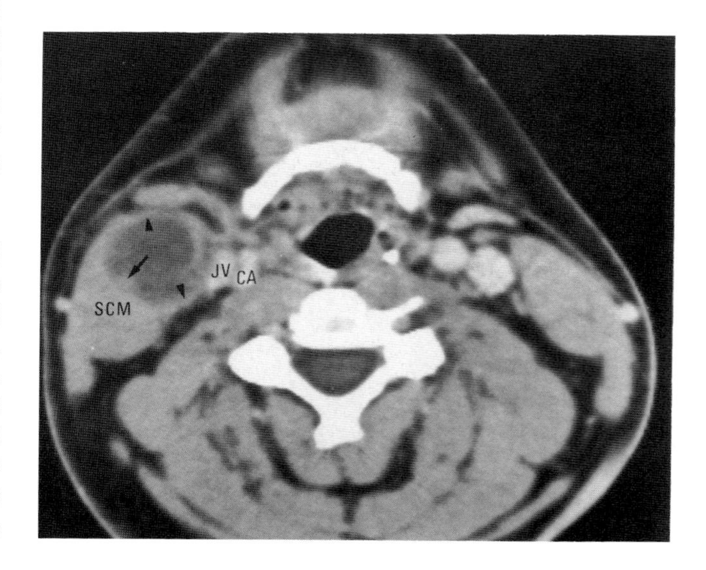

Plate 8.40. This branchial cleft cyst has a thickened (*arrowheads*) and sometimes nodular (*arrow*) wall. The typical displacement of surrounding structures is noted and described in detail in Plate 8.39. This patient had a history of numerous previous infections within the cyst. The nodule within the wall should *not* strongly suggest the possibility of carcinoma arising within a branchial cleft cyst. That situation rarely, if ever, occurs. If squamous cell carcinoma is present in a cystic neck mass then this is much more likely to be a metastatic node which has become necrotic.

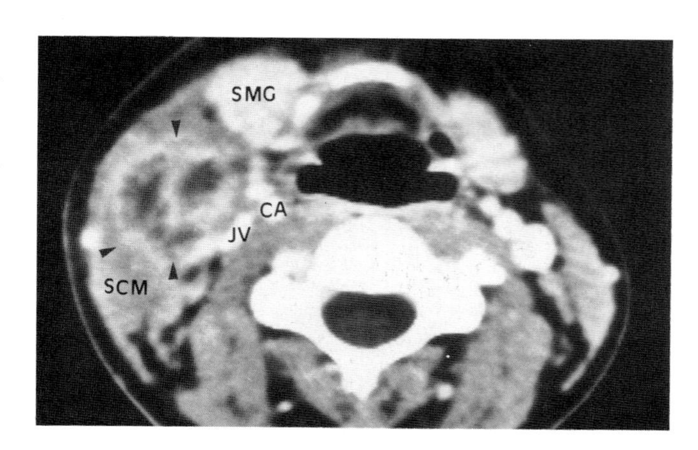

Plate 8.41. This branchial cleft cyst again shows typical displacement of surrounding structures which was described in Plate 8.39. The mass itself shows a thick, indistinct, enhancing wall (*arrowheads*). It is obviously multiloculated and the surrounding soft tissues appear indurated. It is predictable from this CT study that this mass will be adherent to the sternocleidomastoid muscle, submandibular gland, and jugular vein. This was the case at surgery. Some surgeons appreciate this sort of morphologic information prior to neck exploration.

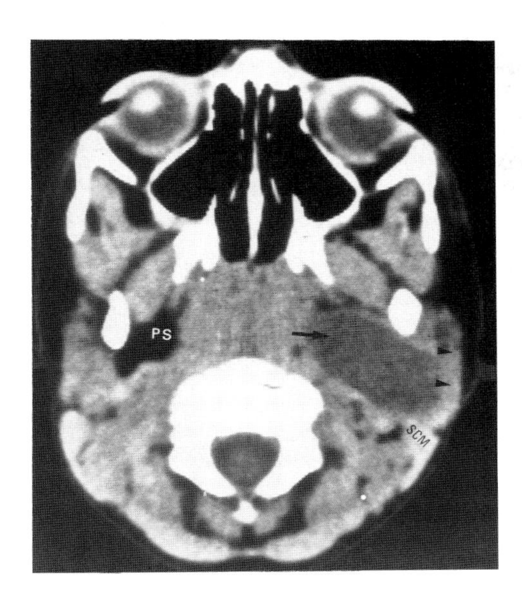

Plate 8.42. This infant was diagnosed as having a first branchial cleft cyst manifested mainly by swelling in the parotid region. The cystic mass (*arrow*) obviously filled the parapharyngeal space. The sternocleidomastoid muscle is displaced posteriorly and the parotid gland is compressed and pushed laterally (*arrowheads*). From the CT study this is obviously not a first branchial cleft cyst but a parapharyngeal extension of a second branchial cleft cyst. The CT information altered the surgical approach to this lesion and forewarned the head and neck surgeon that the facial nerve would be very close to the skin surface within the compressed parotid gland.

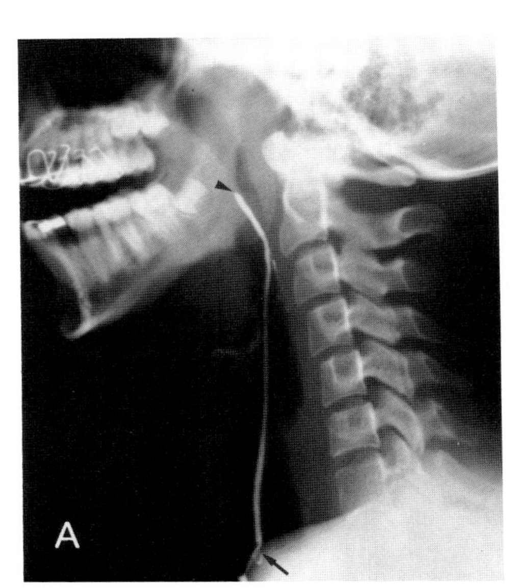

Plate 8.43. Branchial anomalies may be classified as cysts, sinuses, or fistulae. CT or MRI are best used to image the cystic neck masses because they show the full extent of the lesion better and may change the therapeutic approach or even the diagnosis. Fistulae and sinuses are better studied by conventional radiographic techniques.

Plate 8.43*A*. A patient with a fistula draining anterolaterally in the low neck. The fistulous track was injected in the low neck (*arrow*) and filled cephalad to a point in approximately the midtonsillar fossa (*arrowhead*).

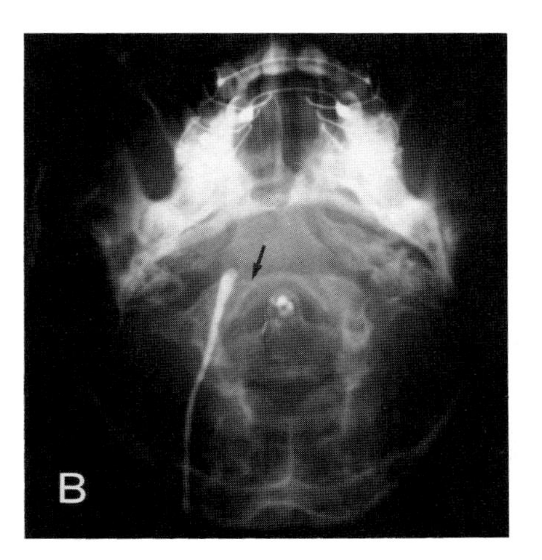

Plate 8.43*B*. The frontal view shows the upper margin of the fistula and a small tract draining from the fistula into the oropharynx. This study is certainly more informative than CT or MRI would have been in this patient.

The abbreviations used in Plates 8.44–8.48 are: *PG*, parotid gland; and *SCM*, sternocleidomastoid muscle.

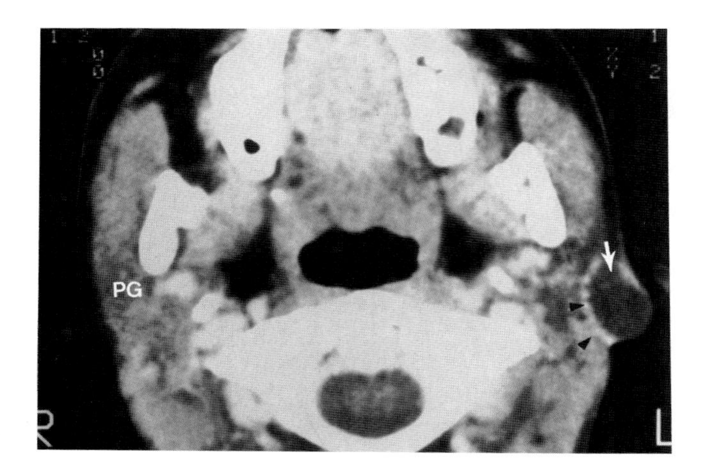

Plate 8.44. First branchial cleft cysts present as parotid or periparotid masses. They are frequently associated with a sinus tract which extends back to, and sometimes drains into, the external auditory canal. This cyst (*arrow*) is seen within the substance of the parotid gland. The patient had undergone several prior attempts at excision and the peripheral enhancement (*arrowheads*) was due to many episodes of prior inflammation. Histories of inflammation and inadequate incision are unfortunately common in these patients. The cyst and its fistulous track should be completely excised the first time so that repeated operations do not threaten facial nerve function.

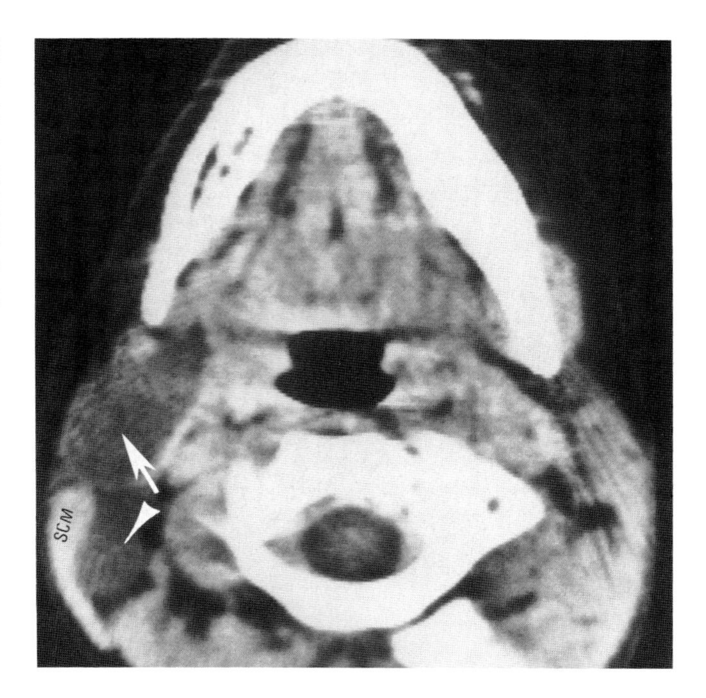

Plate 8.45. Another cystic mass in the upper neck and on higher sections near the tail of the parotid. Components extending into the anterolateral portion of the neck (*arrow*) and behind the sternocleidomastoid muscle (*arrowhead*) are visible. Occasionally a cystic hygroma can be mistaken for a cystic mass arising from the branchial apparatus. However, the diagnosis is usually suspected clinically. The insinuating growth pattern of this cystic hygroma is clearly different from that of branchial cleft cysts and characteristic of cystic hygromas.

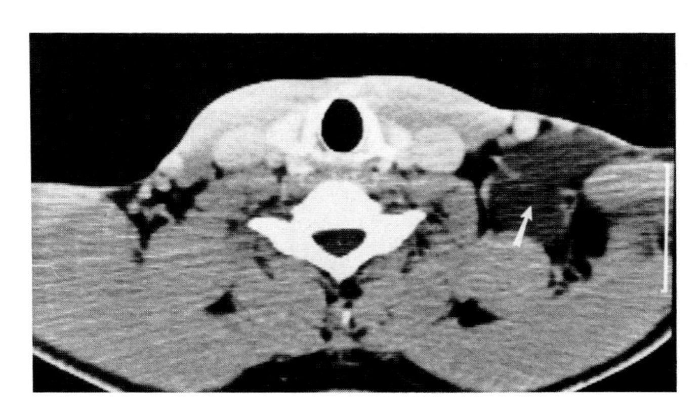

Plate 8.46. The cervicothoracic junction is another common place for cystic hygromas (*arrow*). The mass can be seen insinuating between the small vessels in the supraclavicular fossa and posterior triangle region. Lymphangiomas and hemangiomas show a similar growth pattern though their gross appearance on CT studies is quite different.

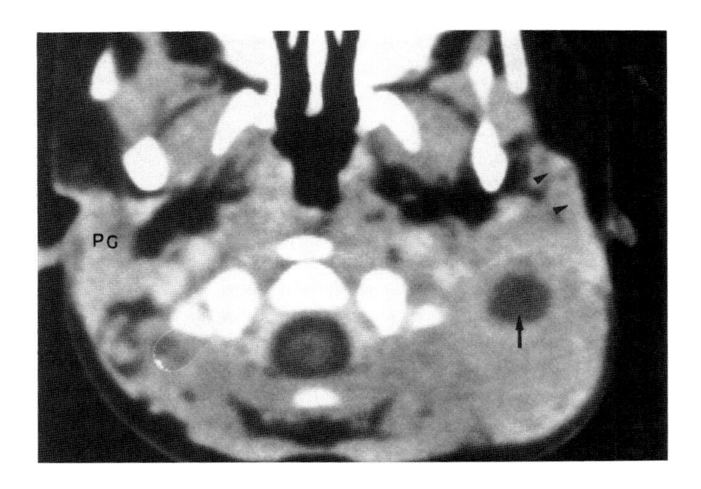

Plate 8.47. Masses presenting in the posterior triange are most often related to enlarged lymph nodes. This young child had an inflammatory mass in the posterior neck. CT revealed a cystic component (*arrow*) within a large area of induration in the posterolateral aspect of the neck. The parotid gland on the abnormal side (*arrowheads*) appears to be adherent to the mass. This was due to suppurative adenopathy which resolved on antibiotic therapy. We have only observed necrosis due to suppurative adenopathy in the pediatric age group. Other inflammatory conditions such as tuberculosis can produce necrosis in lymph nodes of the neck.

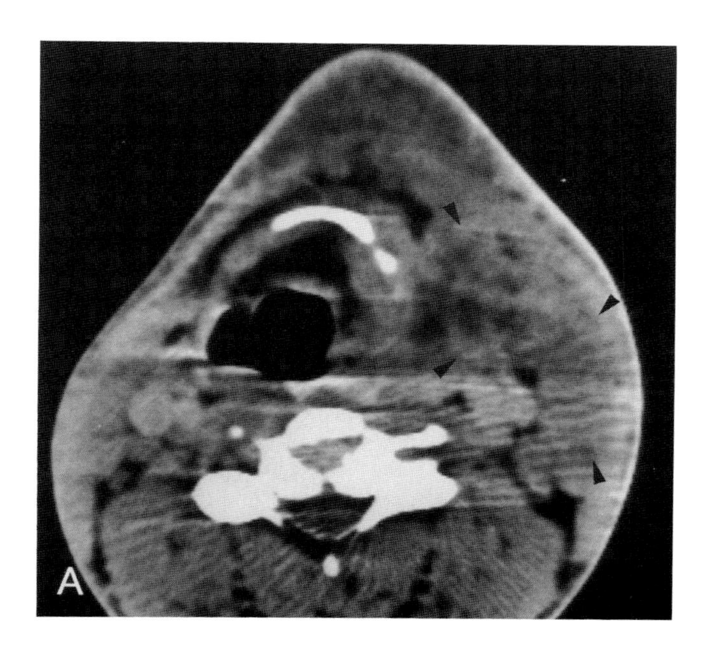

Plate 8.48. Two sections done on a patient who is a known intravenous drug abuser. The patient presented with a diffusely swollen neck and there was some question of whether this was cellulitis, a deep neck abscess or an inflamed branchial cleft cyst since the history of I.V. drug abuse was not apparent to the referring clinician at presentation.

Plate 8.48*A*. A section at the hyoid level shows diffuse soft tissue swelling within the lateral compartment (*arrowheads*). Some areas of nonconfluent low attentuation are present. Differentiation of cellulitis and abscess at this level might be considered difficult.

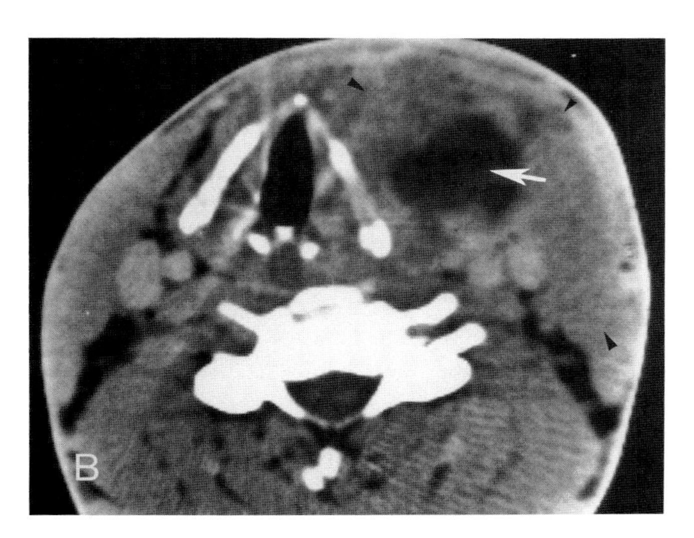

Plate 8.48*B*. Lower in the neck there is still a tremendous amount of swelling in the lateral compartment (*arrowheads*). An obvious large area of low attenuation is present (*arrow*). This represented a deep neck abscess. CT was pivotal in helping the clinician decide that surgical drainage was necessary. The distinction between cellulitis and abcess in the neck is one that is sometimes difficult to make on strictly clinical grounds.

The abbreviations used in Plates 8.49–8.52 are: *SMG*, submandibular gland; *H*, hyoid bone; *SCM*, sternocleidomastoid muscle; *Tr*, trachea, *TG*, left lobe; and *CA*, carotid artery.

Plate 8.49. This patient has an obvious deep neck abscess and the case illustrates how CT can be used to determine the full extent of the abscess and, therefore, help plan the surgical drainage procedure.

Plate 8.49*A*. The abscess containing gas (*arrow*) is visible in the suprahyoid neck just posterior to the submandibular gland.

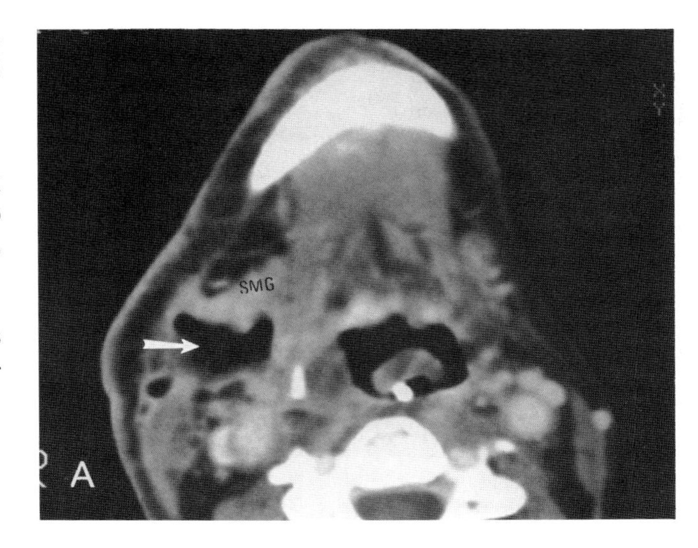

Plate 8.49*B*. A section at the level of the hyoid bone shows the abscess "pointing" from the deep neck through the thyrohyoid membrane (*arrow*). There is a suggestion that the infectious process is headed toward the retropharyngeal space (*arrowhead*) at this point.

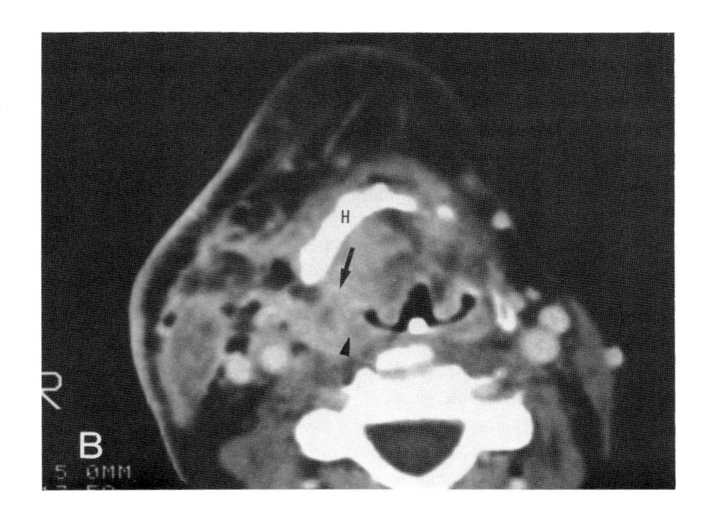

Plate 8.49*C*. A section slightly lower again shows the marked swelling of the skin, subcutaneous tissues and sternocleidomastoid muscle. The abscess (*arrow*) "points" through the thyrohyoid membrane. At this level it is clear that it is also extending into the retropharyngeal space (*arrowhead*). There was some delay in the diagnosis of this patient's neck abscess. CT study showed that the mass was extending into the retropharyngeal space and, therefore, would soon enter the mediastinum if not drained. All CT findings were confirmed at surgery which was done immediately following the CT study.

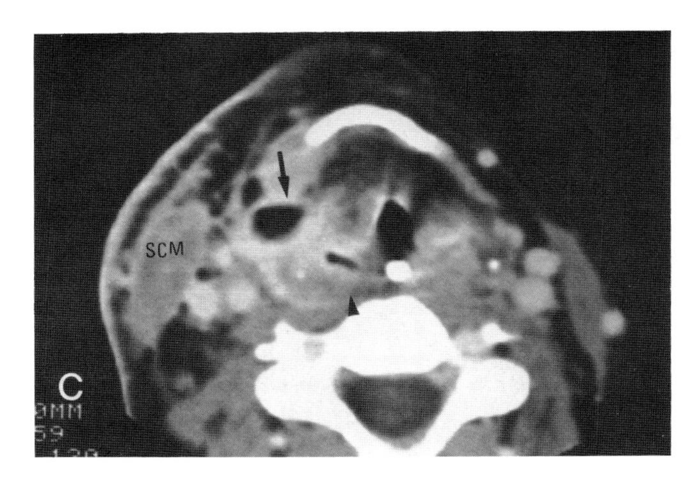

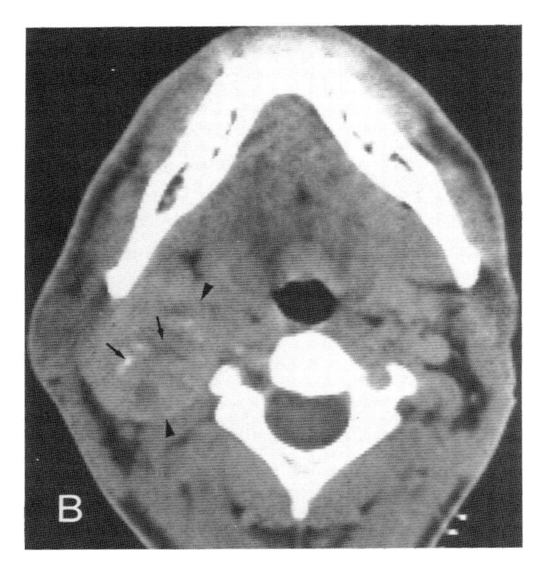

Plate 8.57*B*. Again the very well circumscribed nature of the mass is apparent (*arrowheads*). Punctate areas of calcification are present and, again, there are zones of low attenuation within the mass (*arrows*). This certainly is not an enlarged lymph node. The position is good for a branchial cleft cyst; however, the internal morphology is atypical. This spectrum of findings could probably only be produced by a neuroma or a dermoid tumor. A lymph node due to granulomatous infection could conceivably have the same appearance. This turned out to be a neuroma.

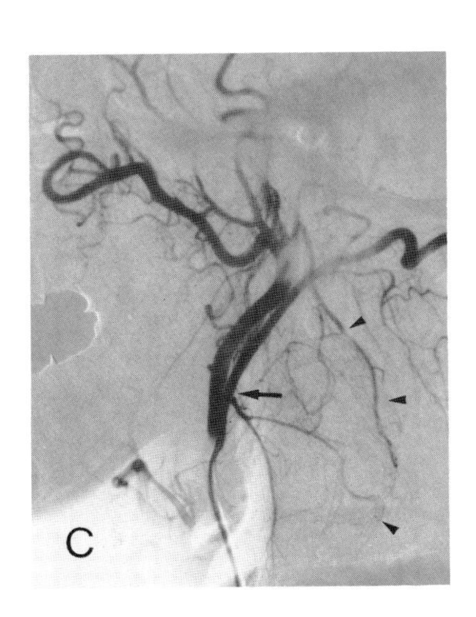

Plate 8.57*C*. Neuromas, even though they may enhance fairly intensely on CT studies, have a relatively avascular appearance on angiography. Here the displacement of the proximal external carotid branches (*arrow*) and the course of more distal feeding vessels (*arrowheads*) shows the mass is relatively hypovascular. This is mainly of importance in differentiating neuromas from paragangliomas when this difference is not apparent on CT studies. (This case appears courtesy of G. Gamsu, M.D., Department of Radiology, University of California at San Francisco.)

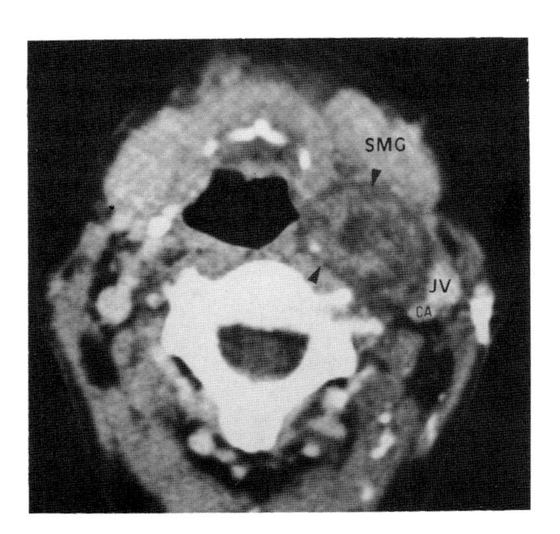

Plate 8.58. Neuromas are an unusual cause of masses within the lateral compartment. This 40-yr-old female presented with a 2-month history of a lateral neck mass. Angiography showed a relatively hypovascular mass splitting the internal and external artery branches. The clinical diagnosis was branchial cleft cyst versus tumor. CT revealed a somewhat inhomogeneous soft tissue mass (*arrowheads*) displacing the submandibular gland anteriorly and the carotid artery and jugular vein posterolaterally. This is obviously not a branchial cleft cyst. The mass was excised and pathology revealed a ganglioneuroma. (This case appears courtesy of W. Dillon, M.D., Department of Radiology, University of California at San Francisco.)

The abbreviations used in Plates 8.59–8.63 are: *JV*, jugular vein; and *SMG*, submandibular gland.

Plate 8.59. CT is of some value in confirming the clinical impression of a carotid body tumor. This is especially true when surgical removal is not anticipated and the clinician only wants to confirm his clinical impression.

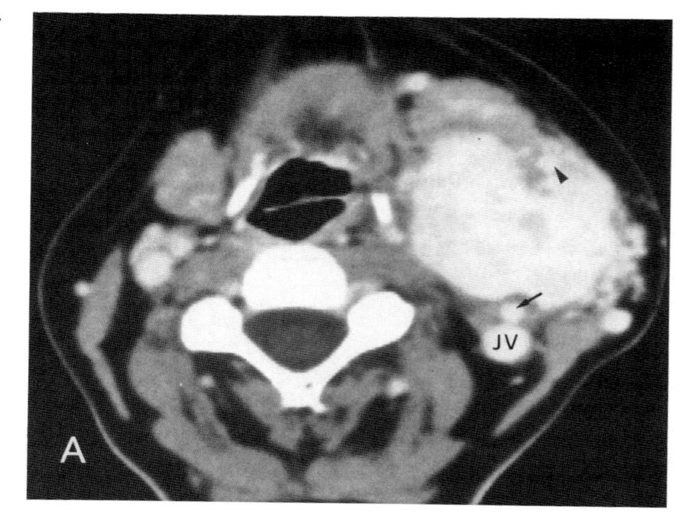

Plate 8.59*A*. This shows the typical, intense staining pattern of a carotid body tumor (and paraganglioma at any site). There is posterior displacement of the carotid artery (*arrow*) and probably anterior displacement of the external carotid (*arrowhead*), although this could be a large draining vein or some other vessel.

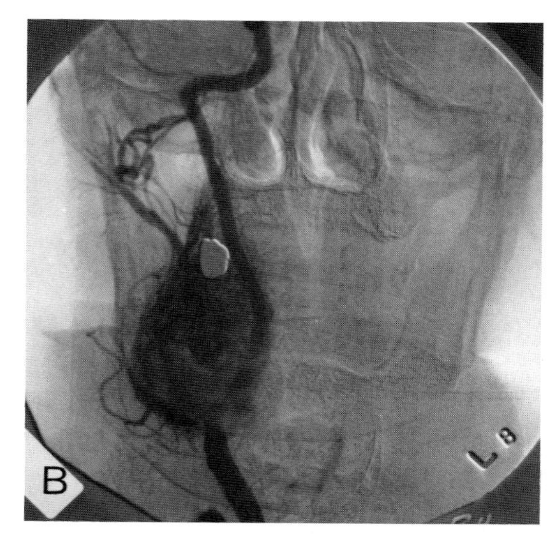

Plate 8.59*B*. A common carotid injection in another patient shows a typical vascular pattern of a paraganglioma at the carotid bifurcation. Contrast this with the appearance of the neuroma in Plate 8.57.

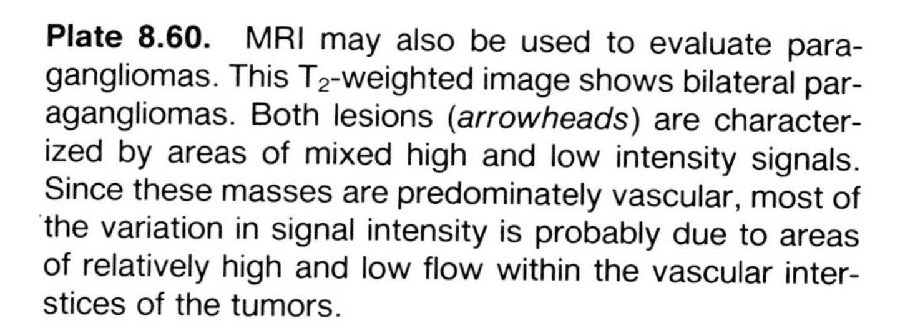

Plate 8.60. MRI may also be used to evaluate paragangliomas. This T_2-weighted image shows bilateral paragangliomas. Both lesions (*arrowheads*) are characterized by areas of mixed high and low intensity signals. Since these masses are predominately vascular, most of the variation in signal intensity is probably due to areas of relatively high and low flow within the vascular interstices of the tumors.

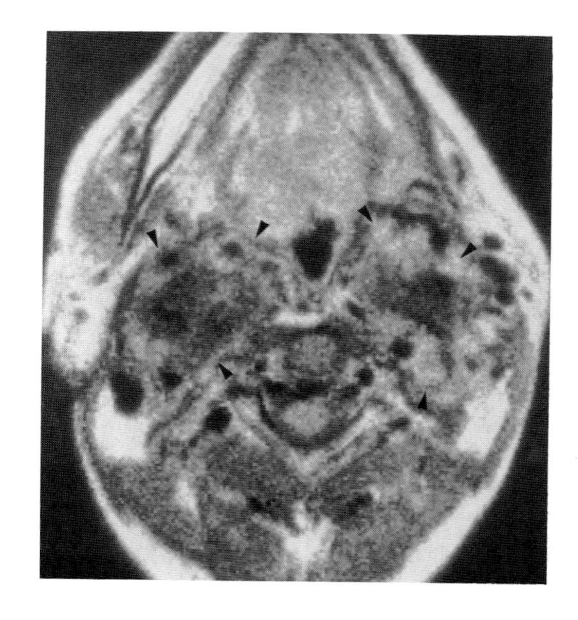

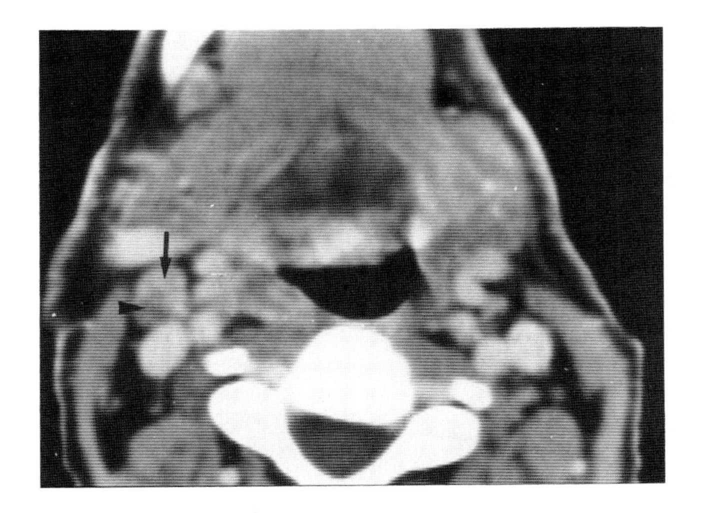

Plates 8.61–8.73. CT may be used to detect and stage cervical lymph node disease. It is complementary to the clinical examination and can provide information that is extremely helpful in planning the therapeutic approach to the neck.

Plate 8.61. In the untreated neck CT will upstage a clinically N_0 neck to N_1 in approximately 5–6% of the time. This is usually based on the visualization of necrotic metastatic foci within normal-sized nodes. This patient had a 15-mm node in the midjugulodigastric region on the right (*arrow*). This was positive by CT size criteria alone; however, a low attenuation metastatic deposit (*arrowhead*) could be seen within the node. This neck was staged N_1 clinically so that CT and the clinical exam were in agreement.

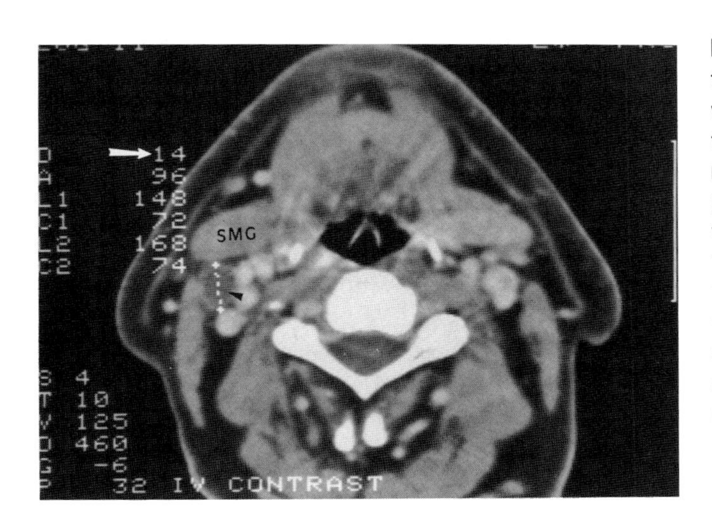

Plate 8.62. Most of the difficulty in physical examination, in untreated patients, comes in those individuals with short, fat necks. This patient had a carcinoma of the tongue base. The neck was staged N_0 clinically. CT revealed a 14-mm node (*arrowhead*) which was completely necrotic. One can see the thickness of the intervening soft tissues and the fact that the node lies deep to the submandibular gland. The actual measurement of the node in millimeters is shown by the *white arrow* which corresponds to the distance between the two cursor points. In this particular patient there was a second necrotic node so that the neck was in reality N_{2B} rather than N_0.

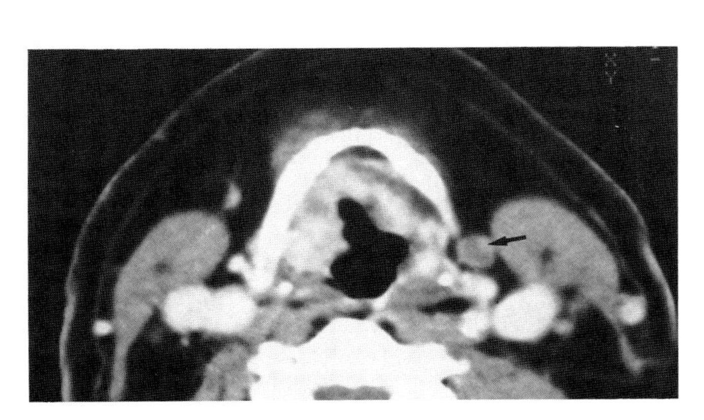

Plate 8.63. Most necrotic nodes will be larger than 1.5 cm in diameter; however, nodes as small as 8 mm may contain low attenuation foci due to metastatic disease. These will only be seen if one pays very careful attention to CT technique in the examination of these patients. This patient had an infiltrating carcinoma of the supraglottic laryngx and a nonpalpable, 8-mm lymph node on the left (*arrow*). Radical neck dissection confirmed the presence of the node and it was filled with a small metastatic deposit.

The abbreviations used in Plates 8.64–8.65 are: *SMG*, submandibular gland; *N*, nodes; *JV*, jugular vein; and *CA*, carotid artery.

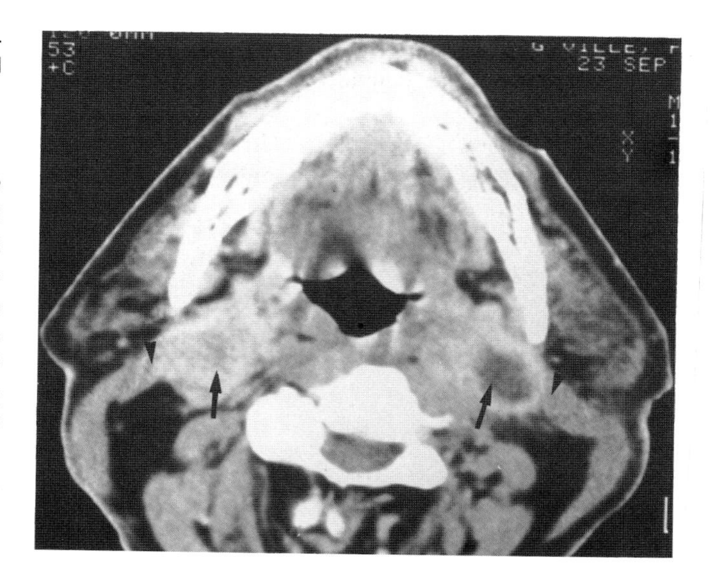

Plate 8.64. The physical examination becomes more difficult after a patient has had surgery and/or radiation therapy. This patient had been radiated and the neck examination was negative although limited by postradiation fibrosis. There was an 8- to 10-mm ulcer on the posterior pharyngeal wall but no other evidence of recurrent disease. CT revealed the unsuspected finding of two large necrotic metastases (*arrows*) in the highest internal jugular nodes. Note that these are at the level of the posterior belly of the digastric muscle (*arrowheads)* which are abnormally thickened as well. The highest internal jugular nodes (parapharyngeal) are normally not palpable unless massively enlarged and then one cannot be certain of the exact upper limit of their extent.

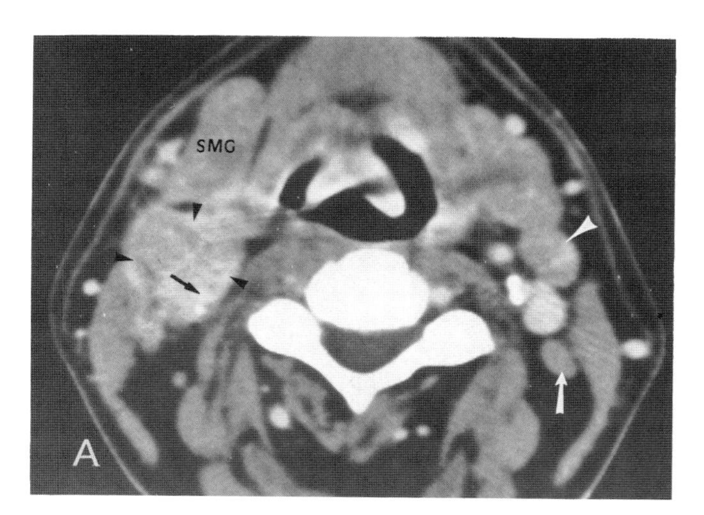

Plate 8.65. The clinical evaluation of extranodal disease is open to a wide range of subjective interpretation based on the relative degree of fixation of palpable nodal masses. CT can add some objectivity to this evaluation but has its limitations in this regard.

Plate 8.65*A*. A patient with pyriform sinus carcinoma had a bilateral positive neck. The neck mass on the right was characterized as being partially mobile. CT showed a fairly diffuse mass (black *arrowheads*) with complete obliteration of the tissue planes surrounding the carotid artery (black *arrow*). Such findings virtually always indicate carotid fixation. In the opposite neck, the anterior interior jugular node (white arrowhead) was palpable but the physical examination failed to detect posterior internal jugular nodes which also proved to be positive (white arrow).

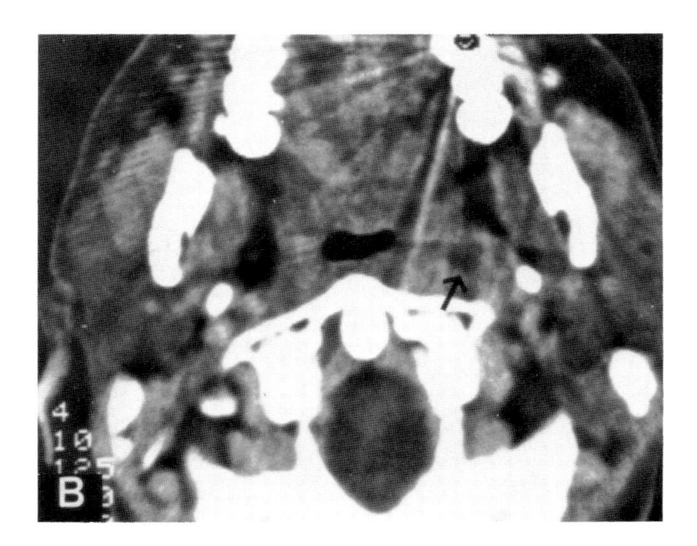

Plate 8.69*B*. An enlarged, necrotic retropharyngeal node is present on the left side (*arrow*). This pattern of adenopathy virtually insures that the primary tumor is in the nasopharynx.

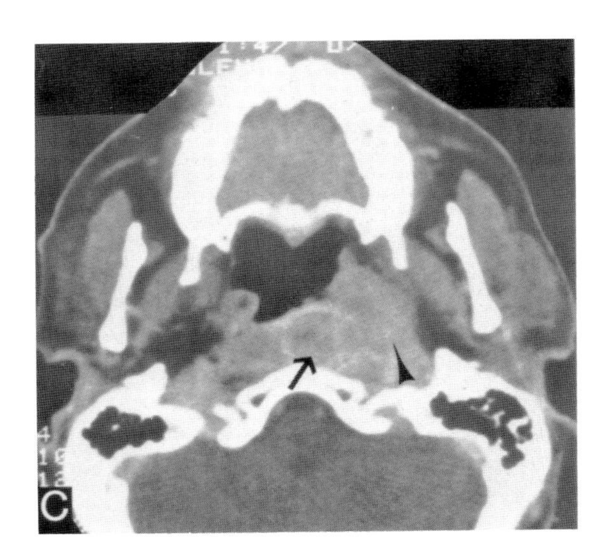

Plate 8.69*C*. The deeply infiltrating carcinoma (*arrowhead*) produced no visible mucosal abnormality at panendoscopy. Biopsy at this site confirmed poorly differentiated squamous cell carcinoma. An enlarged medial retropharyngeal node was also present (*arrow*).

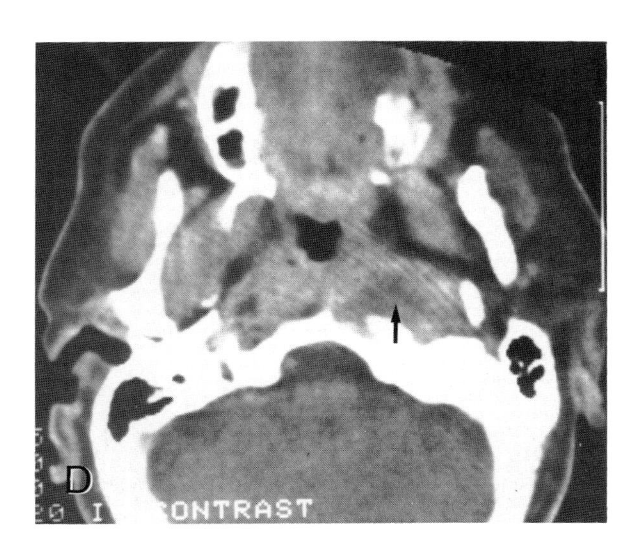

Plate 8.69*D*. This is an example of an enlarged retropharyngeal node (*arrow*) on a patient who had prior surgery and radiation for pyriform sinus carcinoma. The patient presented with a suboccipital neck pain of uncertain etiology. Retropharyngeal lymph nodes in this position are generally considered to be a sign of incurable disease when due to primary tumors outside the nasopharynx. When these are due to squamous cell carcinoma arising outside the nasopharynx it is usually in the setting of recurrent tumor (after lymphatic drainage has been altered by therapy) or when the tumors have, themselves, spread to the nasal cavity, hard palate or nasopharynx.

The abbreviations used in Plates 8.70–8.72 are: *N*, lymph node; *SMG*, submandibular gland; and *H*, hyoid bone.

Plate 8.70. Masses in the submandibular space will arise either from the submandibular gland or lymph nodes. Other masses are quite unusual. CT can usually make the distinction of a glandular or extraglandular mass with ease.

Plate 8.70*A*. This patient presented with an enlarged mass in the submandibular region. This was an enlarged, reactive lymph node which was clearly separate from the submandibular gland. Smaller, normal submandibular lymph nodes were present on the other side (*arrows*).

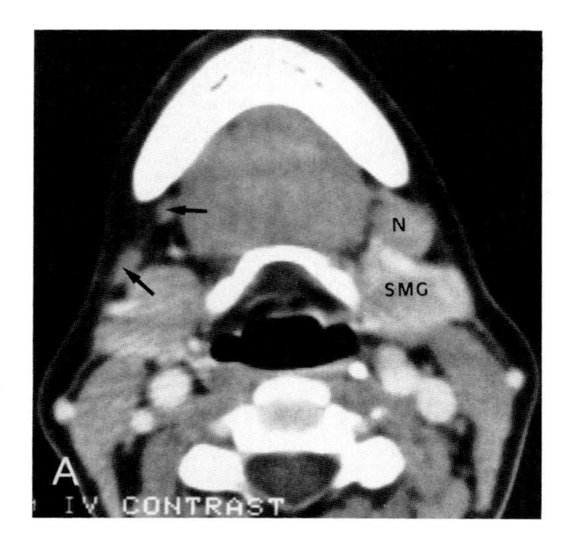

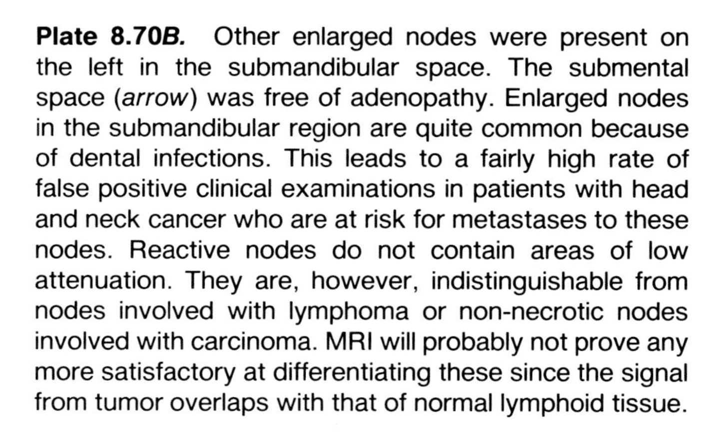

Plate 8.70*B*. Other enlarged nodes were present on the left in the submandibular space. The submental space (*arrow*) was free of adenopathy. Enlarged nodes in the submandibular region are quite common because of dental infections. This leads to a fairly high rate of false positive clinical examinations in patients with head and neck cancer who are at risk for metastases to these nodes. Reactive nodes do not contain areas of low attenuation. They are, however, indistinguishable from nodes involved with lymphoma or non-necrotic nodes involved with carcinoma. MRI will probably not prove any more satisfactory at differentiating these since the signal from tumor overlaps with that of normal lymphoid tissue.

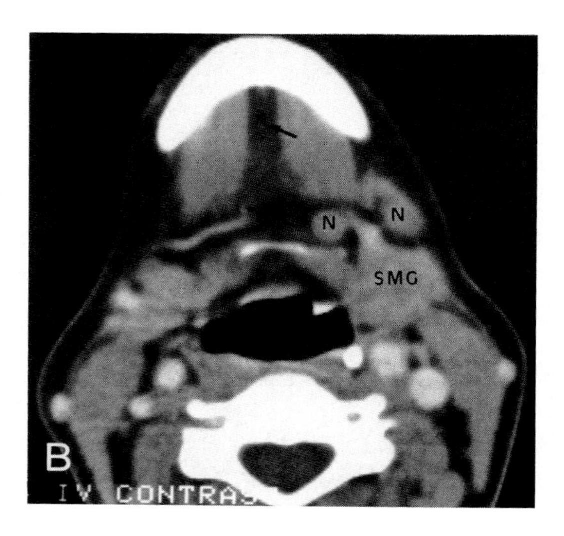

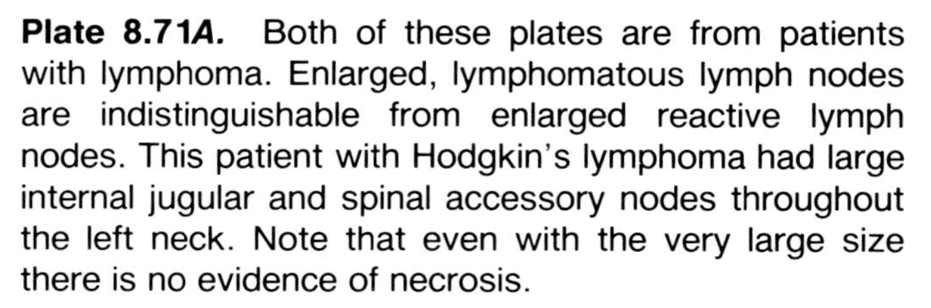

Plate 8.71*A*. Both of these plates are from patients with lymphoma. Enlarged, lymphomatous lymph nodes are indistinguishable from enlarged reactive lymph nodes. This patient with Hodgkin's lymphoma had large internal jugular and spinal accessory nodes throughout the left neck. Note that even with the very large size there is no evidence of necrosis.

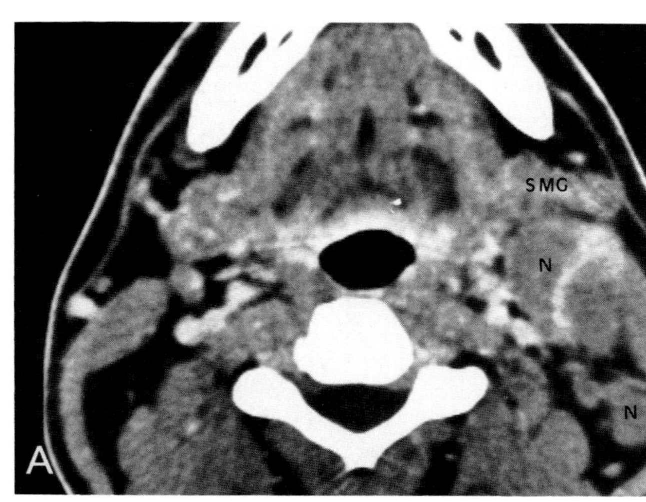

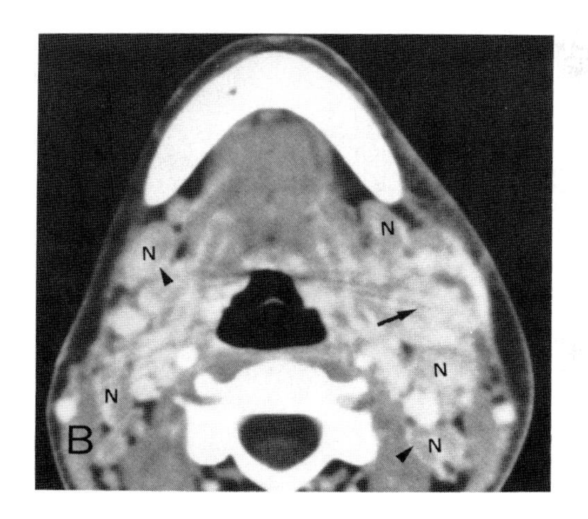

Plate 8.71*B*. This patient has a diffuse histiocytic lymphoma involving Waldeyer's ring and the neck. Numerous enlarged lymph nodes are present in the submandibular, internal jugular and spinal accessory chains. It is difficult to distinguish the nodes from the submandibular gland (*arrow*). Note that many of the nodes have a thin rim of surrounding contrast enhancement. This may be seen in large reactive nodes or lymphomatous nodes and it is a nonspecific finding. This may be related to hyperemia of the lymph node capsules.

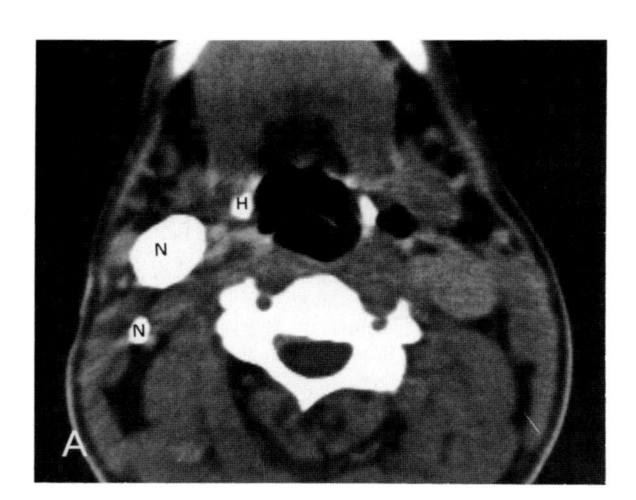

Plate 8.72. This patient presented with a neck mass of unknown etiology. This was believed to be due to a carcinoma somewhere in the upper aerodigestive tract. CT was done as part of the initial work-up after office endoscopy was negative. Fine needle aspiration biopsy was not done.

Plate 8.72*A*. This was photographed at a fairly wide window setting to show the very dense calcification in the jugular digastric and posterior internal jugular nodes. These should not be confused with the normal hyoid bone calcification.

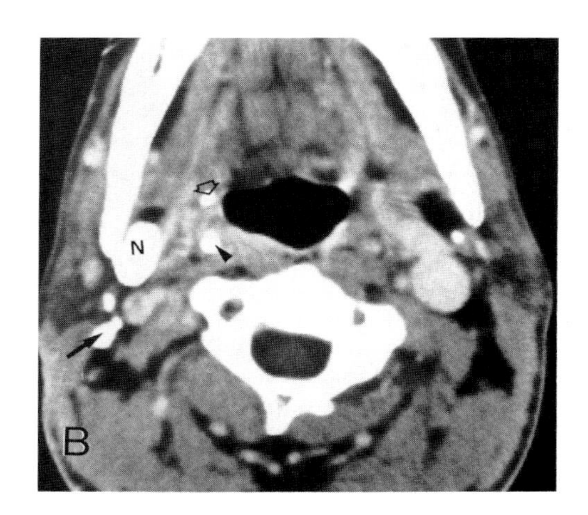

Plate 8.72*B*. Shows calcification of a submandibular node, a retropharyngeal node (*arrowhead*) and upper neck nodes (*arrow*) either in the spinal accessory group or posterior internal jugular group. There was also calcification in the tonsillar fossa (*open arrowhead*). After the CT study, a more complete history was obtained and the patient had apparently had tuberculosis. PPD was positive and these were presumed to be calcified nodes due to prior tuberculosis cervical adenitis.

The abbreviations used in Plates 8.73–8.74 are: *ScM*, scalene muscles; *JV*, jugular vein; *CA*, carotid artery; *E*, esophagus; and *C*, carotid.

Plate 8.73. The scalene, juxtavisceral (recurrent laryngeal) nodes may be involved from primary sites below the clavicles. If the primary site is above the clavicle, then it is quite likely to be the thyroid gland, subglottic space or cervical esophagus.

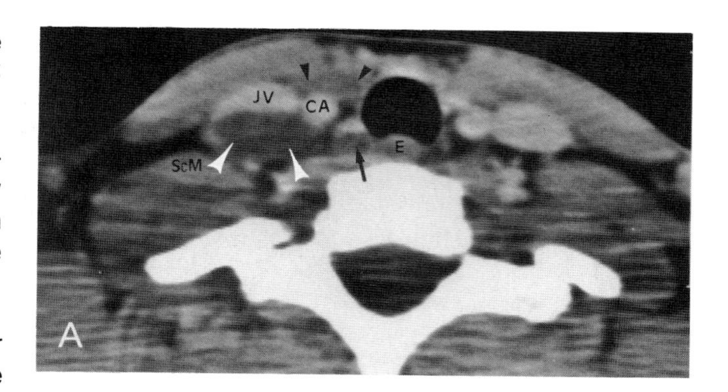

Plate 8.73*A*. A patient with prior history of surgery for papillary follicular carcinoma of the thyroid gland. The patient had no specific symptoms and the referring clinician did not palpate any residual disease. Because of the vague symptoms, a CT study was done. A low attenuation mass (*white arrowheads*) was seen between the scalene muscle and the carotid sheath structures. A smaller, juxtavisceral lymph node (*arrow*) in the tracheoesophageal groove was also enlarged. The study also suggested recurrent tumor in the resected thyroid bed (*black arrowheads*). Surgery confirmed all of the CT findings.

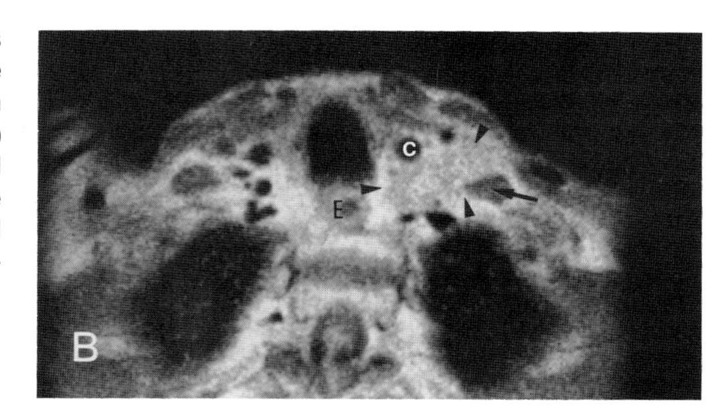

Plate 8.73*B*. MRI may also show enlarged lymph nodes in the same area (*arrowheads*). At this pulse sequence (2000/28) the tumor-bearing nodes tend to blend in with the surrounding fat. The carotid, scalene muscle (*arrow*) and esophagus, all have different signal intensities and are easily distinguished from the fat and the lymph node mass. (This illustration is courtesy of Drs. D. Stark and G. Gamsu, Department of Radiology, University of California at San Francisco.)

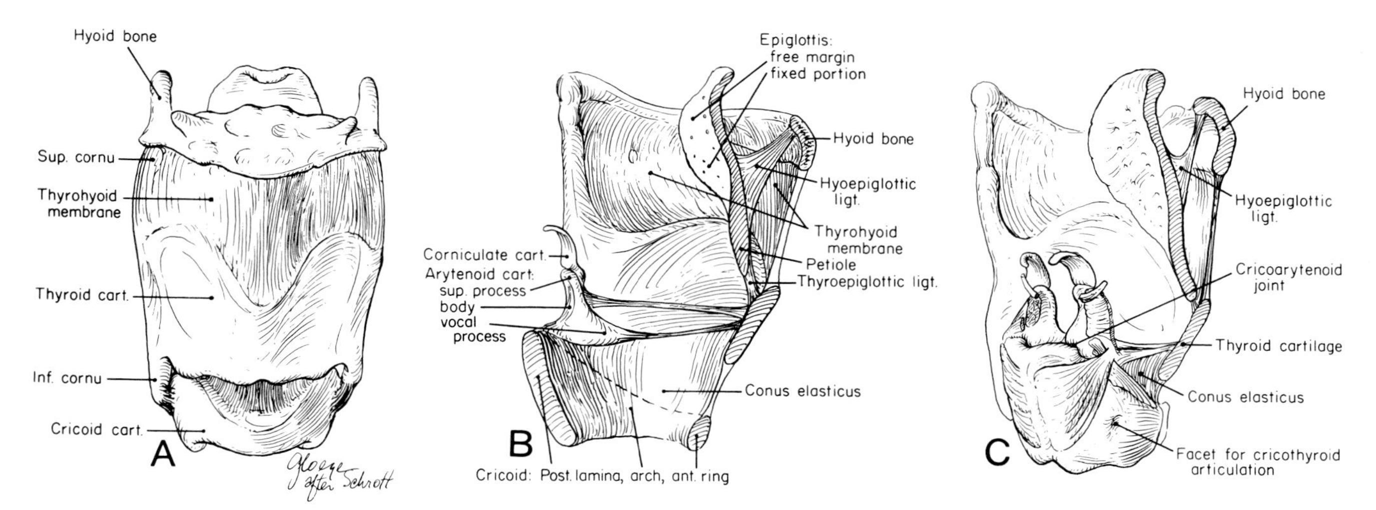

Figure 9.2. Frontal *A*, lateral *B*, and oblique posterior *C* views of the laryngeal skeleton. The basic components are the thyroid, cricoid, arytenoid, and epiglottic cartilages and hyoid bone which are held together by numerous membranes and ligaments named for their attachments. The *conus elasticus* is one of the most important membranes. Note how it attaches to the sloping margin of the cricoid (*B* and *C*).

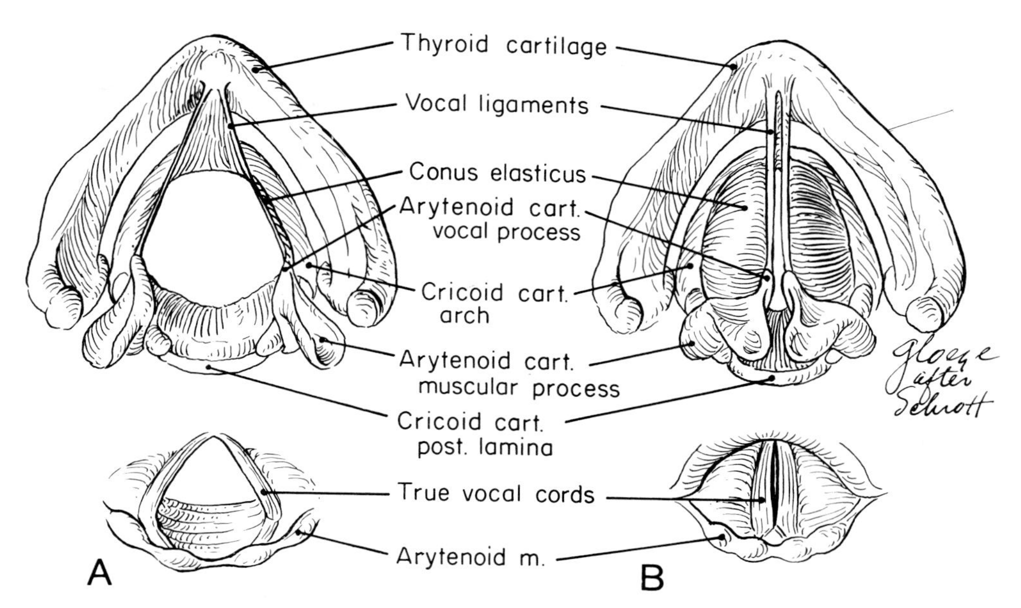

Figure 9.3. Views from above of the cricothyroid complex made during quiet respiration (*A*) and phonation (*B*) with accompanying endoscopic views. Note how the arytenoids glide medially on the facets of the cricoarytenoid joint during phonation. The medial movement also moves them superior because they ride up the slope of the cricoid lamina (*see* Fig. 1C). The vocal processes of the cricoid also rotate medially during phonation. The origin of the conus elasticus at the vocal ligament and its sweep inferiorly and laterally to insert on the cricoid arch creates the morphology of the undersurface of the true cords as seen on CT scans.

of the laminae anterior to the laryngeal soft tissues is an unmistakable landmark at the level of the true cords (Fig. 9.1). Infraglottically, the anterior contour of the fused laminae appears more rounded as the thyroid cartilage lies anterior and lateral to the cricoid (Fig. 9.1). The inferior cornua of the thyroid laminae articulate with the cricoid posteriorly. The cricothyroid joint allows the cricoid and thyroid to be tilted upon one another so that the distance between their opposed anterior margins may be varied (30, 37). Supraglottically, the appearance of the thyroid cartilage changes again. About 1 cm above the glottis, the thyroid notch appears between the two laminae (Figs. 9.1, 9.2). The superior cornua of the thyroid laminae are visible posteriorly from about the mid- to upper laryngeal vestibule (1–2 cm above the true cords). The cornua attach to the hyoid by the lateral thyrohyoid ligament which may contain the tiny calcified triticeal cartilages (30, 37). The thyrohyoid ligament is actually the posterior free margin of the thyrohyoid membrane. The thyrohyoid membrane is not separable from the density of the infrahyoid strap muscles in most patients. This muscle group includes the sternothyroid, sternohyoid, thyrohyoid, and omohyoid. It creates a tissue-density band lying external to the thyroid laminae and paralleling their oblique position in the neck. Above the superior margins of the thyroid cartilage, these muscles maintain this same orientation to their insertion on the hyoid (Fig. 9.1).

The degree and extent of the thyroid cartilage calcification and ossification varies. In general, they are more calcified (ossified) in males and with increasing age. The cartilages of most of our teenage patients have been mineralized enough for diagnostic purposes. The laminae tend to be symmetric in given individuals.

The morphology of the laminae is similar to the cricoid in that there tends to be a surface, "cortical," type of calcification with a more lucent central space. If the cartilage is ossified, this may be a true medullary cavity. Focal areas of ossification are common. The laminae are thinnest in the paramedian region at the glottic level. At this location the internal and external calcified surfaces of the laminae may merge and obliterate the central, low density, space.

On MR images the margins of an ossified thyroid cartilage will, like the cricoid, produce little signal. Fat within the ossified portions of the laminae will have a high signal. Foci of dense ossification and mineralized cartilage have much lower signal intensities. The medium intensity signal of the infrahyoid strap muscles is clearly demarcated from the adjacent fat and laryngeal skeleton (41, 79).

Epiglottis

The epiglottis forms the anterior wall of the laryngeal vestibule (Figs. 9.1, 9.2). Its free portion projects superiorly just posterior to the valleculae; this is sometimes called the suprahyoid portion of the epiglottis. The body or fixed portion of the epiglottis tapers rapidly below the level of the hyoid. Its inferior tip, the petiole, attaches to the thyroid laminae via the thyroepiglottic ligament, at a point in the midline just below the thyroid notch. The top of the infrahyoid epiglottis is attached to the hyoid by the hyoepiglottic ligament.

The epiglottis is elastic cartilage and, as such, only occasionally displays significant calcification. Normally there are small perforations within the cartilage (30, 37); these cannot be seen on imaging studies. Calcification, when it occurs, is plaquelike and lies just deep to the mucosa of the supraglottic larynx.

Inferiorly, near the petiole, the epiglottis is very narrow in its lateral dimension, becoming broader superiorly. The body of the epiglottis is broadest at the hyoid bone level where the hyoepiglottic ligament courses anteriorly within the preepiglottic space deep to the cartilage (Fig. 9.2, 9.4).

Arytenoids

In quiet respiration, with the true vocal cords abducted, these pyramidal-shaped cartilages lie mainly laterally atop the cricoid lamina. The arytenoid's muscular process projects laterally toward the thyroid lamina; sometimes they almost appear to touch. The vocal process points anteriorly to where it joins the vocal ligament; this is the best landmark for identifying the true vocal cords (Fig. 9.1, 9.3). At this level the top of the cricoid is still visible between the arytenoids. About 5–10 mm cephalad, the superior process of the arytenoid may still be visible, although this is the level of the false vocal cords. The cricoid is no longer seen. The slightly more cephalad corniculate cartilages are inseparable from the superior processes of the arytenoids on axial CT scans. Rarely, the tiny cuneiform cartilages are seen slightly anterior to the superior processes within the substance of the aryepiglottic folds. The arytenoids are usually visible on CT studies made at scan speeds under 10 sec. Suspended respiration may show them abducted or adducted depending on whether the patient valsalvas while breath holding. Voluntary maneuvers such as valsalva, or modified valsalva, produce varying degrees of adduction of the true vocal cords. When the cords are adducted, the arytenoids slide medially and slightly superiorly along the sloping shoulders of the cricoid. They also rotate so that the vocal process points somewhat medially toward the narrowed rima glottidis (Fig. 9.3).

The arytenoids are very often densely and homogeneously calcified. In fact, on CT, the arytenoids often appear to be the most radiodense structure in the laryngeal skeleton. The signal intensity on MR images again varies with the relative degree of calcification, ossification and fat content.

Hyoid Bone

The hyoid bone has three basic parts. The body is a bar of bone lying anterior to the supraglottic larynx. The two greater horns project posterolaterally. Often, a zone of lucency, representing a fibrous connection, is present between the body and greater horns (30, 37).

The hyoid is suspended in the neck between the supra- and infrahyoid strap muscles (30, 37). It is also attached to the epiglottis by the hyoepiglottic ligament, to the thyroid cartilage by the thyrohyoid membrane, and to the styloid process of the temporal bone by the stylohyoid ligament (Fig. 9.2).

The hyoid is a very useful landmark. It generally indicates the transition from fixed to free portion of the epiglottis and, therefore, the upper aspect of the preepiglottic space (Fig. 9.1). It also marks the lower level of the jugulodigastric group of nodes.

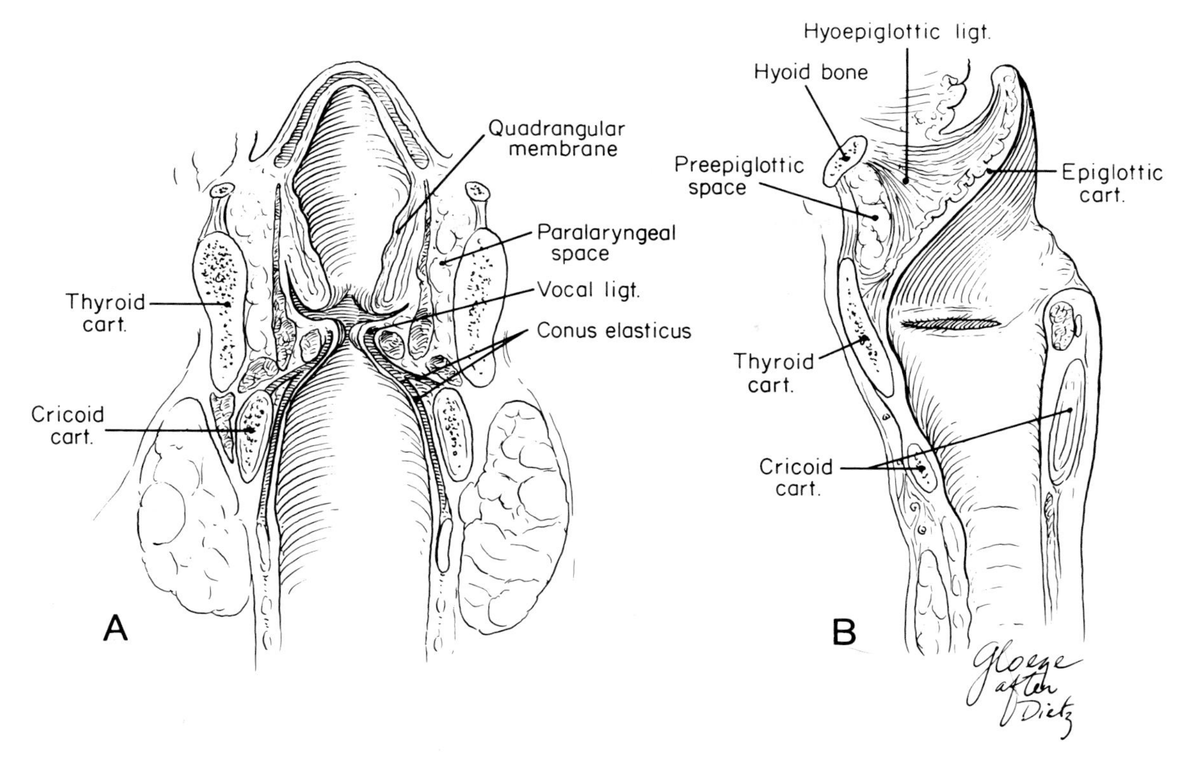

Figure 9.4. Deep tissue planes of the larynx. *A*, coronal section through the midplane of the true cords shows that the paralaryngeal space is composed of the tissues (mainly fibrofatty) between the laryngeal skeletal and all of the paired mucosal structures of the larynx. Note how the conus elasticus extends from the vocal ligament to the cricoid arches. *B*, midsagittal section of the larynx emphasizes the boundaries and components of the preepiglottic space.

DEEP TISSUE PLANES—AN OVERVIEW

To paraphrase Tucker (81, 82), clinically, the paralaryngeal space is best thought of as that part of the larynx containing all the paired structures of the larynx not normally seen by the examiner. If one extends this definition to include the paired deep spaces of the infraglottis and the unpaired preepiglottic space, then one of the major contributions of CT and MRI in laryngeal diagnosis becomes obvious (Fig. 9.4). CT and MRI show the deep planes while the clinical examination, including laryngoscopy and biopsy, evaluates the mucosa. The studies are complimentary and each yields important diagnostic information.

The conus elasticus and, to a much lesser extent, the quadrangular membrane deserve special attention with regard to the deep laryngeal anatomy (Fig. 9.5). It is this elastic tissue that separates the various deep compartments (11, 68, 69, 79, 82). They are not seen as distinct structures on CT or current MR images but the conus elasticus' ability to limit and direct the growth of tumors is seen. High resolution MRI may be able to differentiate the conus from adjacent fat and muscle.

The conus elasticus (cricothyroid membrane) attaches to the upper margin of the cricoid and the free upper border thickens to become the vocal ligament (Figs. 9.2–9.5). The conus separates the deep planes of the infraglottis into an anterior wedge which is intimately associated with the cricothyroid ligament (an anterior midline thickening of the cricothyroid membrane) and two lateral compartments which are in continuity with the true vocal cords above (69, 81) (Figs. 9.3, 9.4). The supraglottic larynx is similarly divided by the less well developed quadrangular membrane. The quadrangular membrane attaches to the borders of the epiglottis and sweeps back to the arytenoid cartilages in the aryepiglottic folds (Fig. 9.5*A*). The preepiglottic space now becomes the anterior wedge separating the two lateral paralaryngeal (paraglottic) spaces (Fig. 9.5*B*). The paralaryngeal spaces are also continuous with the soft tissue of the neck via areas of inherent weakness in the laryngeal skeleton; its connecting membranes.

REGIONAL ANATOMY

The following anatomic description uses the laryngeal skeleton for basic orientation in order to maintain

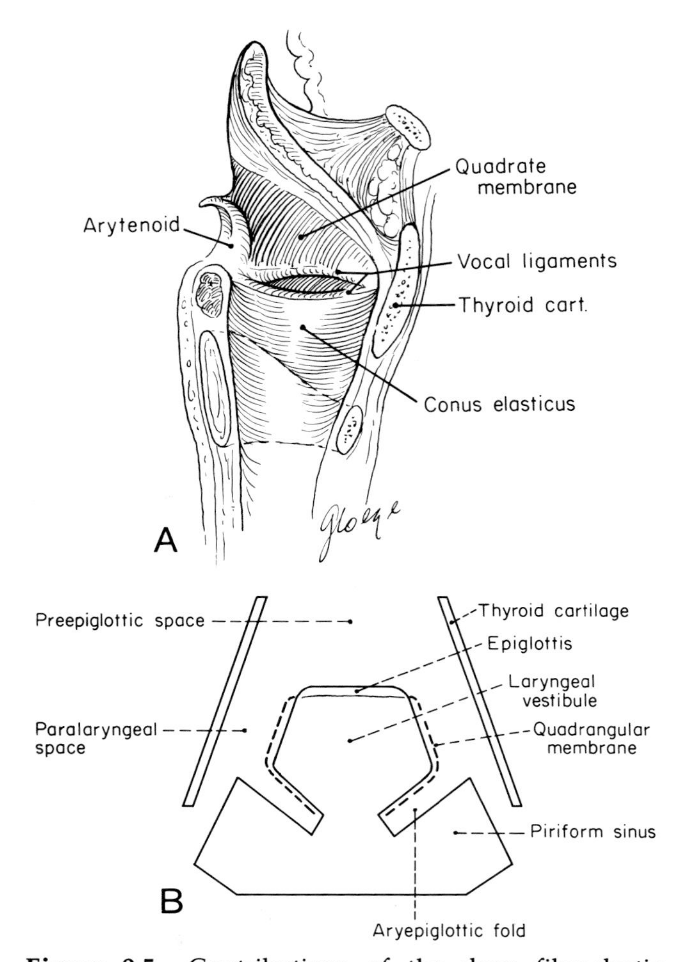

Figure 9.5. Contributions of the deep fibroelastic membranes to the structure of the larynx. *A*, the conus elasticus is a very tough membrane which extends from the vocal ligament to the cricoid (*light dashed lines*) where it splits; an outer portion of it then attaches to the cricoid's superior margin and the inner portion is reflected medially to lie just deep to the mucosa (*see* Fig. 9.4*A*). The quadrate membrane is relatively flimsy and lies deep to the walls of the laryngeal vestibule and within the aryepiglottic fold. *B*, schematic axial view of the midsupraglottis shows the relationship of the quadrangular membrane (*dark dashed lines*) to the deep tissue planes. More importantly, it emphasizes the continuity of the paralaryngeal and preepiglottic spaces and their intimate contact with the laryngeal skeleton and hypopharynx. The aryepiglottic fold is the boundary (transition zone) from endolarynx to hypopharynx (pyriform sinus).

continuity of thought between this and the previous section. Recall that the deep tissue planes simply lie between the superficial mucosal landmarks and the laryngeal skeleton and an accurate overall perspective is assured.

Glottic Region

SUPERFICIAL AND CARTILAGINOUS LANDMARKS

When studying this region in the axial plane it is best to separate it into three zones. From inferior to superior, we refer to these zones as (*a*) the undersurface of the true vocal cords, (*b*) the midplane of the true cords, and (*c*) a transition zone between the true and false cords at the level of the laryngeal ventricle. Such thinking helps the inexperienced interpreter overcome much of the insecurity stemming from not seeing all of the vocal cord in one section or even not knowing where the true cords are. Sections less than 5 mm thick, multiplanar reformations and sections done during suspended respiration or modified valsalva can help sort out subtle pathologic changes in these areas. Direct coronal and sagittal MR images will prove useful in selected circumstances. The following description is of a larynx in quiet respiration since this is obligatory on MRI, and commonly used at CT. More recently I have been using suspended respiration as a routine to reduce motion artifacts. This produces a marked improvement in image quality but makes it necessary to become familiar with the variation it produces in the anatomic display.

In the midplane of the true vocal cords, the arytenoids sit atop the lateral aspect of the cricoid lamina (Figs. 9.2, 9.3). The cricoid is still very prominent posteriorly (Fig. 9.1). The vocal processes point anteriorly, slightly medially, and are attached to the vocal ligament. The vocal ligaments continue anteriorly, forming the free edges of the true vocal cords. The ligaments join in the midline anteriorly at the inner surface of the thyroid laminae creating the anterior commissure of the rima glottidis. At this point, the V-shaped orientation of cartilage's angle is visible anteriorly and the paramedian regions of the laminae are at their thinnest (Fig. 9.1). The anterior commissure lies about 1 cm above the inferior margin of the thyroid cartilage. With the true cords abducted, no tissue density should be visible at the anterior commissure. At this point, the true cord mucosa is essentially lying on perichondrium (33, 65, 82). With adduction, the true vocal cords are pressed together making detection of subtle pathologic changes at the anterior commissure impossible.

The posterior commissure lies between the arytenoids on the anterior surface of the cricoid lamina. With the true vocal cords abducted, there should normally be no visible thickening of the soft tissues here. If the vocal cords are adducted, the soft tissue in this region may become heaped up and subject to misinterpretation as a tissue mass; this should not be interpreted as pathology without checking the comparable sections made in quiet respiration. This "thickening" is probably due to contraction of the interarytenoideus muscle.

We refer to the area 5 mm above the midplane of

the true vocal cords as the transition zone between true and false cords or the ventricular level. On axial sections the laryngeal ventricle is occasionally seen as a distinct air-filled space but it is not a reliable landmark. Also, if one keeps the flexible approach this transition zone concept offers, trying to interpret minor variations caused by scans made slightly oblique to the central axis of the larynx will be less frustrating. Ideally, axial sections should be perpendicular to the central axis of the larynx. This can be assured by doing preliminary projection radiographs and angling the gantry appropriately (Fig. 9.6). With MRI multiplanar imaging will be used in questionable cases.

In the transition zone the top of the cricoid cartilage is just barely or no longer visible, depending on the section thickness (Fig. 9.1). The thyroid cartilage is usually still thin in the paramedian regions. If the thyroid notch is a deep one, the separation between laminae may become apparent on sections in the transition zone. The superior process of the arytenoid and, more superiorly, the corniculate cartilages are seen posteriorly. Remember not to call this the true cord level just because the arytenoids are visible. The cartilages eventually lead to the base of the aryepiglottic fold; a prominent soft tissue landmark at this level and above. The inferior portion of the pyriform sinus lies lateral to the aryepiglottic fold. Scans done in modified valsalva accentuate these anatomic relationships and may make it easier to interpret abnormalities in this region.

Just above the anterior commissure some normal thickening of the soft tissues is present (Fig. 9.1). This is due in part to the insertion of the thyroepiglottic ligament (Fig. 9.2). The appearance is in marked contrast to that of the anterior commissure of the true vocal cords where no more than 1 mm of tissue thickness is normally present.

The undersurface of the true vocal cords extends about 5 mm below the midplane of the true cords. As others have done (65, 81), we exclude this portion of the vocal cords from the laryngeal infraglottis. At this level the cricoid ring is C-shaped, open anteriorly, and bridges the airway between posterior margins of the thyroid laminase (Fig. 9.1). The thyroid laminae have a rounded anterior contour and a fairly uniform thickness. The free margins of the true cords are roughly concentric to the curve of the ipsilateral thyroid laminae. The cords taper from posterior to anterior and normally no tissue thickness is present in the midline anteriorly or posteriorly. The true cords almost appear to originate posteriorly from the cricoid; however, this is not the case. The arrangement of the deep space anatomy, specifically the conus elasticus, in this region will explain the true cord's appearance.

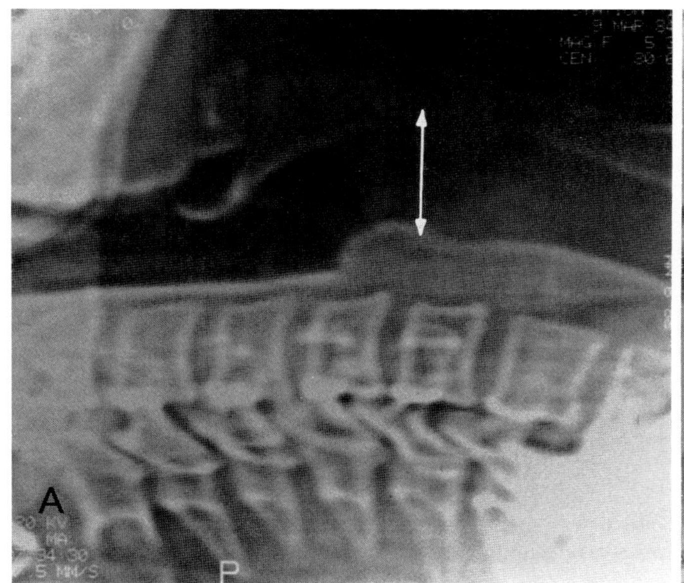

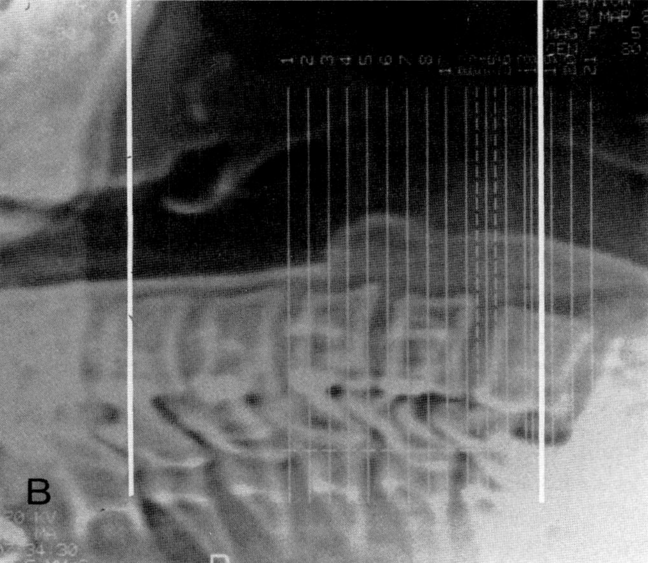

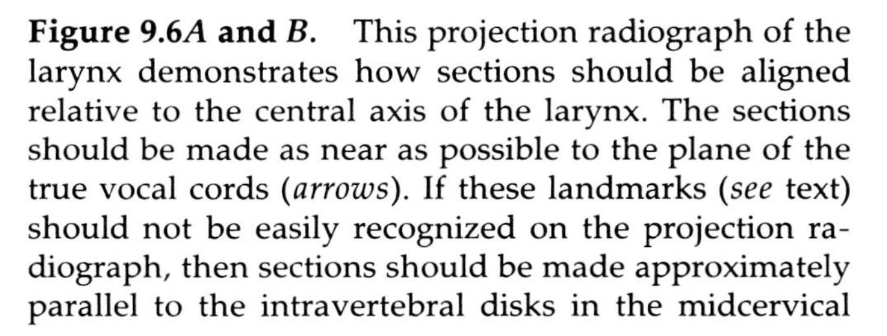

Figure 9.6*A* and *B*. This projection radiograph of the larynx demonstrates how sections should be aligned relative to the central axis of the larynx. The sections should be made as near as possible to the plane of the true vocal cords (*arrows*). If these landmarks (*see* text) should not be easily recognized on the projection radiograph, then sections should be made approximately parallel to the intravertebral disks in the midcervical spine. It so happens that the plane at the intravertebral disk is usually the same as that for the true vocal cords. A complete study of the larynx usually requires sectioning from approximately the angle of the mandible to the bottom of the cricoid cartilage or upper trachea (*large white lines*). Overlapping sections at the true and false cord level and in the infraglottic larynx are sometimes desirable.

DEEP TISSUE PLANES

At the midplane of the true vocal cords the paralaryngeal space is a very narrow band of tissue between the vocal cord and the thyroid lamina. Most often it is not visible on axial images. When seen on CT it appears as a very thin, low density zone between the muscle-density vocal cord and the calcified thyroid lamina. On high resolution MRI it appears as a thin band of increased signal intensity between the much lower signals from the thyroid lamina and vocal cord musculature. This narrow inferior extension of the paralaryngeal space is seen routinely on whole organ sections but is not visible frequently enough on CT to be of practical benefit. It is most important to understand that this space provides a conduit for submucosal spread of tumor to the infraglottis.

Moving from the true cord through the level of the laryngeal ventricle there is an abrupt change in the tissue deep to the mucosal surfaces. The muscle tissue which occupies nearly the full volume of the true cord thins out quite a bit and is mixed with the fibrofatty tissue which predominates in the paralaryngeal space at this level and above. At CT, the space is of low attenuation, on MR images it has a very high signal. The laryngeal ventricle separates the true and false cords but is rarely visible on axial images. The reverse "E" maneuver so useful during laryngography to distend the ventricles is not practical on CT and no maneuver is practical on MRI.

The base of the aryepiglottic folds will be visible at the true-to-false transition zone (ventricular level). The folds are most easily appreciated when the pyriform sinuses are distended with air and thereby separated from the lateral pharyngeal wall. The fatty tissue within the paralaryngeal space extends posteriorly into the aryepiglottic folds making them low attenuation or high signal intensity on CT and MR, respectively.

The transition between the glottis and infraglottis is the undersurface of the true cords (11, 81). There are no reliable density changes that allow differentiation of the paralaryngeal space from the undersurface of the true vocal cords at this level but it should be recalled that the cords are normally a maximum of 5 mm thick. They also taper dramatically from posterior to anterior commissure, where they are approximately 1 mm thick (11). Therefore, as one sections from the midplane of the true cords through their undersurface and on the infraglottis, the largest volume of paralaryngeal space will be anterolateral (11, 81, 82). The important contribution of the conus elasticus to this arrangement will be considered in the next section.

Infraglottic Larynx

CARTILAGINOUS LANDMARKS

Below the undersurface of the true cords, the cricoid becomes the dominant cartilaginous landmark (Fig. 9.1). High in the infraglottis, the cricoid's sloping superior shoulders have given off extensions (anterior arch) which, like arms, begin to reach anteriorly and embrace the airway along its posterolateral margins (Figs. 9.1, 9.3). The thyroid cartilage has become more rounded anteriorly and overlaps the cricoid laterally while it alone still protects the airway anteriorly (Fig. 9.1). In the midinfraglottis, the arches of the cricoid reach more anteriorly until it is almost U-shaped. The inferior cornua of the thyroid cartilage and the cricothyroid joining come into view posteriorly. Anteriorly, the thyroid cartilage is fading from view to be replaced by the density of the cricothyroid ligament (membrane) and associated anterior neck muscles. In the low subglottis, the arches of the cricoid finally join anteriorly forming a complete ring (Figs. 9.2, 9.3). Posteriorly, the rounded inferior cornua of the thyroid cartilage articulates with the cricoid; the thyroid laminae are no longer visible.

DEEP TISSUE PLANES

The glottis and infraglottis are a continuum. The conus elasticus weds these two regions while separating the deep planes of the infraglottis into an anterior and two lateral compartments. The lateral compartments are continuous with the undersurface of the true cord and the paralaryngeal space above, and may be considered continuous across the midline posteriorly. The anterior compartment is wedge shaped and lies in the midline (68, 81, 82). This is analogous to the supraglottic arrangement of the deep tissue planes where the preepiglottic space is the anterior wedge-shaped compartment and the paralaryngeal spaces lie laterally on either side. In both the infraglottic and supraglottic regions, the lateral and anterior compartments are continuous with one another.

The conus elasticus arises along the upper margin of the cricoid and sweeps upward to the free margin of the true vocal cord where it thickens and forms the vocal ligament (Figs. 9.2–9.5). The vocal ligament runs from the anterior commissure to the vocal process of the arytenoids. It is not visible on CT or MRI as a distinct structure. High resolution MRI may be the first study to show us this structure in vivo. Even though the conus is not seen one can appreciate its affect on the anatomy of the paralaryngeal space. Its strong attachment along the cricoid encourages the tissue beneath the under surface of the true cord to conform to the space between it and the thyroid cartilage. In the high infraglottic region, this arrangement produces the typically symmetric tissue density bordering the airway anterolaterally (Fig. 9.1). As the arches of the cricoid reach more anteriorly, this space becomes increasingly confined until there is no space left at all. The conus is so adherent to the cricoid that whenever the cricoid is visible, there should be no tissue thickness between its inner surface and the airway. If tissue is present, it is pathologic.

An anterior thickening of the conus, the cricothyroid ligament and its lateral extensions, the cricothyroid

institutional bias. All treatment plans require an accurate account of the extent of the tumor.

Surgically there are several options (16, 17, 53, 54, 58–61, 63, 76, 78):

1. Laser excision or cordectomy for limited lesions usually in poor operative candidates or limited post radiation recurrence.
2. Total laryngectomy with or without neck dissection.
3. Partial laryngectomy with conservation of glottic function: sphincteric, phonation and respiration (Fig. 9.8):
   a. Vertical hemilaryngectomy
   b. Supraglottic laryngectomy
   c. Extended partial laryngectomy for more advanced lesions.

The decision between a partial and total laryngectomy can be difficult and requires very precise localization of the tumor if recurrences are to be avoided while maintaining glottic function. As Ogura and Henneman state, "This is precision surgery . . ." (61). Specifics of how CT or MRI helps make this decision are considered in the following sections.

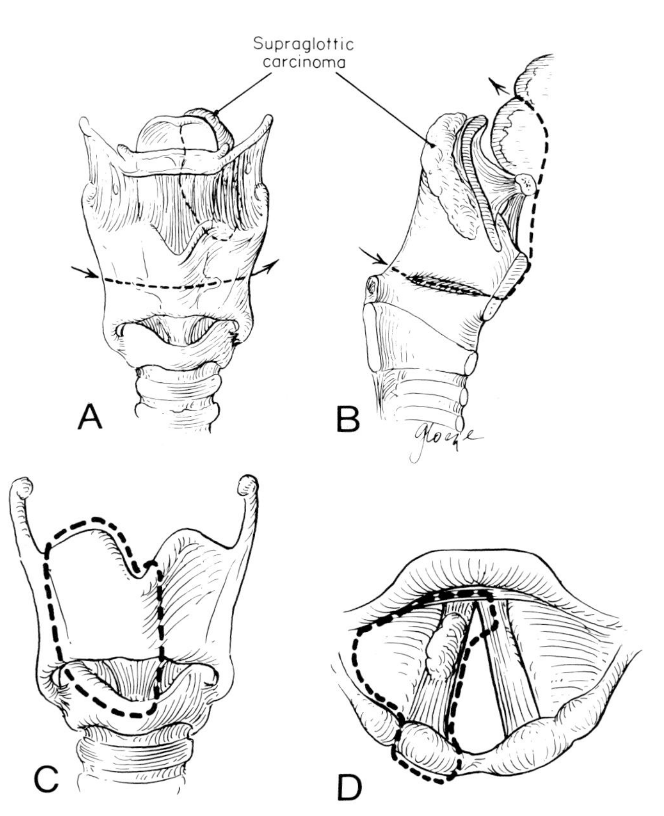

Figure 9.8. Typical margins for supraglottic laryngectomy are indicated by *dashed lines* and *arrows*. *A*, anterior view. *B*, note how close the margin passes to the anterior commissure. *C*, usual extent of resection in vertical hemilaryngectomy. *D*, a laryngoscopic view indicating that the arytenoid, anterior commissure and anterior 30% of contralateral true cord may be sacrificed in a vertical hemilaryngectomy.

The radiation oncologist also needs to know where the tumor is located before he can offer his help. Extensive spread to the preepiglottic space, cartilage invasion, deeply infiltrating aryepiglottic fold lesions, and significant infraglottic extension are only some of the factors which might limit the effectiveness of radiotherapy (4, 53, 54, 56, 83).

WHERE DOES MODERN IMAGING FIT IN?

CT complements direct laryngoscopy and biopsy better than any other radiologic exam (3, 21, 26, 44–51, 72) for the following reasons:

1. It reveals deep extensions not visible by any other means
2. It detects relatively subtle cartilage invasion
3. It demonstrates nodal metastases not evident clinically

As a result of these factors CT has helped avoid partial laryngectomy that surely would have failed and has increased the confidence of surgeons that their planned conservation procedure would succeed. CT has its pitfalls and shortcomings. In the following sections, these and strengths of CT relevant to treating laryngeal cancer will be discussed. MRI may improve on CT in some or all of these regards.

Mucosal and Minimal Lesions

The best way to look at the mucosa of the larynx and hypopharynx is with a mirror or endoscope (Fig. 9.7). CT does not show mucosal detail. Small mucosal lesions are often undetectable at CT. They may cause asymmetries of the larynx which fall within the range of normal variation. Most of these occur around the true and false cords. Motion artifacts makes interpretation of minor changes even more equivocal. If one true cord is weakened or paralyzed, it might appear "thickened" and be mistaken for tumor. These surface asymmetries may be due to tumor, fibrosis, edema, hemorrhage, or inflammation. It is not possible to make a tissue diagnosis of structural abnormalities seen on CT appearance or density values alone. One must be constantly mindful that CT does not "diagnose cancer" but it used to show the extent of disease. While MRI shows some promise of increased specificity it will still suffer some limitations in this regard. From preliminary work it seems likely that tumor and fibrosis may differ in signal (52). Overlap of signal intensities of reactive edema and inflammatory lesions is to be expected.

In learning to interpret any imaging study, a physician must learn what findings have the potential to alter the prognosis or therapeutic approach. Close communication between the radiologist and otolaryngologist, who already has visualized the laryngeal mucosal surfaces, will usually be enough to decide whether these minor imaging findings are significant. In general, mucosal lesions will produce only some slight bulging of the surfaces bordering the airway. Sometimes the scan will look entirely normal but the clinician may declare that there is a tumor present. The radiologist should not be discouraged or misled by the paucity or absence of findings; limited findings indicate a small lesion amenable to partial laryngectomy or radiation cure in a large percentage of cases.

If there is extensive motion artifact, one should not overread surface irregularities but, on the other hand, one should not ignore valuable information that might still be present. Laryngeal CT scans are rarely nondiagnostic at scan speeds of 10 sec or less. In almost all circumstances it is possible to "read through" the artifacts to give the clinician information regarding deep extension of tumors. The problem of significant motion artifact has been virtually eliminated with faster scanners and the use of suspended respiration.

PATTERNS OF SPREAD

The American Joint Committee on Staging of Cancer recognizes three regions of origin for laryngeal tumors: glottic, subglottic, and supraglottic and stages the lesions T_1 through T_4 (2). Glottic tumors have growth patterns which are much different than supraglottic lesions and the very rare primary subglottic tumors (11, 65, 69, 81). It is important to consider both routes when staging and diagnosing laryngeal tumors. Before discussing regional spread patterns of laryngeal carcinoma we will consider how distortion of the laryngeal skeleton may aid or hinder interpretation of imaging studies.

Laryngeal Skeleton Distortion

Distortions of the thyroid, cricoid, and arytenoid cartilages and hyoid bone are common in diseased larynges (46). The distortions must be recognized because:

1. The laryngeal skeletal components are major anatomic landmarks and a disturbance of their usual relationship might confuse an inexperienced interpreter
2. They may herald cartilage invasion or exolaryngeal extension by tumor which might otherwise go undetected
3. Following radiation therapy, they may indicate chondronecrosis which could lead to progressive airway obstruction
4. Distortions resulting from old, unremembered, trauma may present as mass lesions on clinical examination of the larynx or confuse clinical staging of malignancies (28).

The tendencies for cartilage invasion by tumor will be discussed subsequently.

Thyroid Cartilage Distortions

Even though the thyroid is fairly rigid and suspended within the neck by strong fibrous and muscular attachments, we have observed a high incidence of these distortions in diseased larynges. In a detailed review of our first 66 CT scans, 29 had some distortion of the thyroid cartilage (46). With several years more experience it seems that the true incidence of distortions is lower but the principles are still important.

The thyroid cartilages unite in the midline at the angle. On either side of the angle there is an area of relative weakness. This paramedian region responds much as a wishbone does when one pushes or pulls on either or both of the thyroid laminae. Several types of abnormal patterns may be observed.

Inward Buckling and Infracture. These are seen when the paramedian regions are weakened by tumor invasion or perichondritis and chondronecrosis and the ipsilateral thyroid lamina assumes an inwardly directed convexity (Fig. 9.9). Minimal, symmetric inward bowing is a normal variant; rarely, it can be quite dramatic and symmetric. Anytime abnormal bowing is present in conjunction with tumor at or near the anterior commissure, cartilage invasion must be suspected. If the destructive process is advanced, actual infracturing may occur (Fig. 9.9).

Tethering and Spreading. Bulky posteriorly placed tumors may push the ipsilateral thyroid lamina outward resulting in tethering in the paramedian region (Fig. 9.9). We have observed this phenomenon in benign and malignant lesions. Any extensive spreading force combined with a destructive process at the anterior commissure can lead to an actual infracture. Sometimes a bulky tumor may fill the larynx displacing both thyroid laminae laterally to produce a symmetric "spread" appearance (Fig. 9.9).

Cricothyroid Distortions

A line drawn through the cricothyroid joint along the axis of the superior and inferior cornua will be approximately parallel to the vertical axis of the body (Fig. 9.10). If the cricothyroid joint is disrupted, or the integrity of the cricothyroid membrane lost, this cornual axis usually tilts anteriorly. The inferior margin of the thyroid cartilage will then approach the anterior arch of the cricoid. On axial CT sections through the infraglottis, one will then see "too much of" the cricoid and thyroid on the same section (Fig. 9.11). We have observed this anterior tilting in several different clinical settings including:

1. Anterior subglottic extension of tumor
2. Posterior paralaryngeal extension of tumor in the cricothyroid space
3. Postradiation fibrosis
4. Postradiation perichondritis and chondronecrosis
5. Blunt trauma

We have not observed posterior tilting relative to the vertical axis. If the neck is not slightly hyperextended, some approximation of the cricoid and thyroid might mimic a pathologic distortion.

On axial sections, the midsagittal plane of the patient

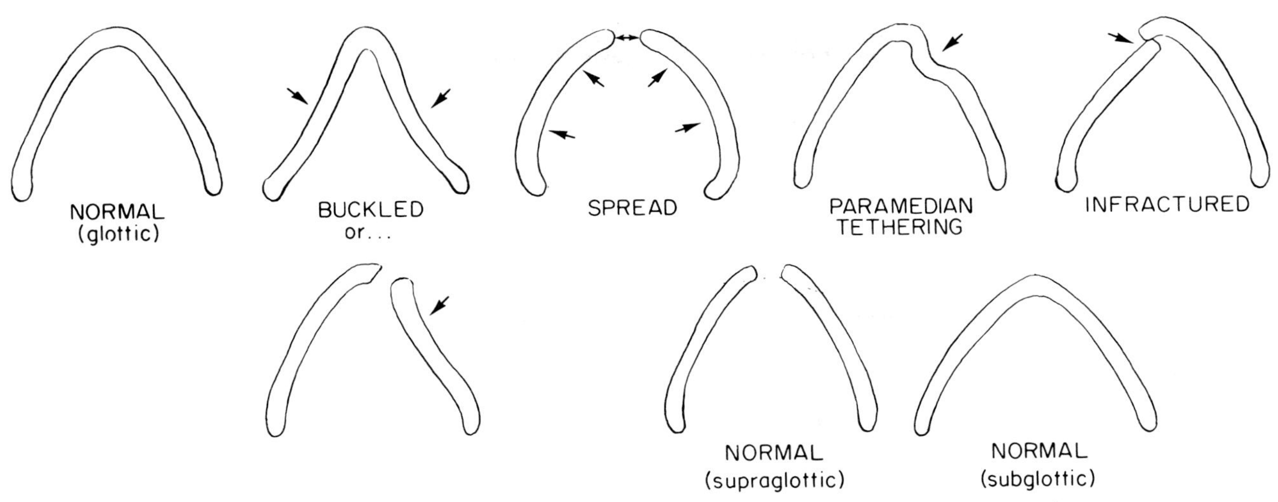

Figure 9.9. Various distortions (*arrows*) of the thyroid laminae are compared to the normal glottic, infraglottic and supraglottic contours. Distortions may be due to laryngeal tumor, trauma and inflammatory lesions or may be secondary to lesions arising from other structures in the neck.

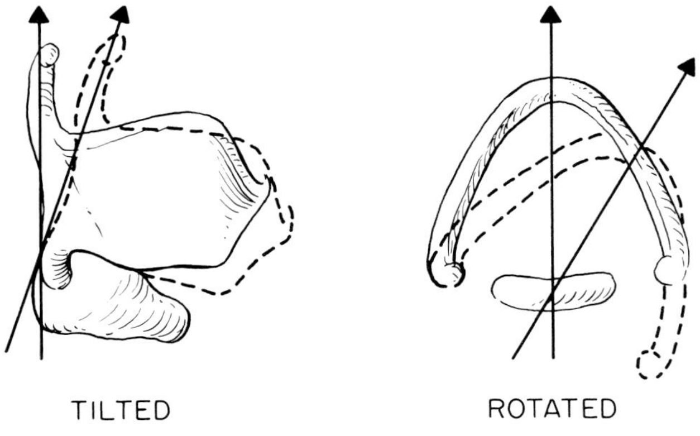

Figure 9.10. *Dashed line* superimpositions represent tilting of the cornual axis and rotation of the thyroid laminae relative to the cricoid. Etiologies for such distortions are varied.

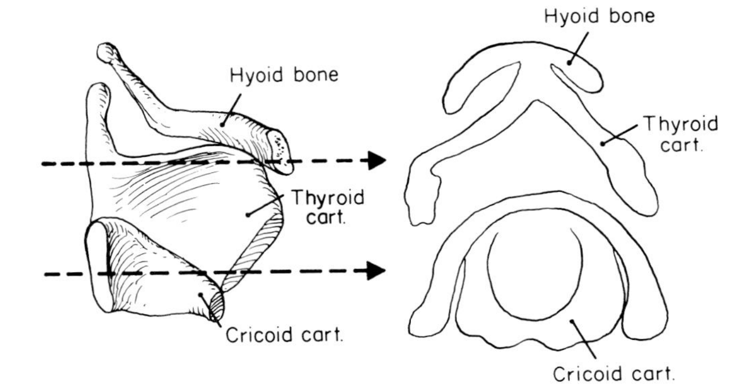

Figure 9.11. When the integrity of the connecting membranes (thyrohyoid and cricothyroid) of the larynx is upset, the cartilages and hyoid bone "telescope" (*dashed lines and arrows*) upon one another producing the distorted CT images of the laryngeal skeleton pictured on the *right.*

should pass through the midline of the larynx, bisect the cricoid, and divide the thyroid laminae into two equal halves (Fig. 9.10). The thyroid cartilage may rotate clockwise or counterclockwise relative to this plane. Such rotation is usually due to pathology in the paralaryngeal space extending to the cricothyroid space and perhaps disrupting the cricothyroid joint. This also has been observed in a variety of clinical situations including tumor, postsurgical changes, postradiation therapy, and following trauma.

Thyrohyoid Distortions

These may be due to displacement of the thyroid cartilage and/or hyoid bone or loss of integrity of the thyrohyoid membrane. Bulky tumors or benign masses (e.g. laryngocele) may spread superiorly in the preepiglottic or paralaryngeal spaces and separate the hyoid bone and thyroid cartilage. On axial images the appearance will usually be that of a tilted hyoid.

Fibrosis of the thyrohyoid membrane may pull the thyroid cartilage and hyoid together. Tumor invading this membrane may cause them to collapse one upon the other. On axial images one then observes the thyroid cartilage and hyoid on the same section (Fig. 9.11). This only occurs normally if the neck is markedly flexed. A valsalva maneuver may also bring the two together, mimicking pathology.

Arytenoid Distortions

The arytenoid may be displaced or dislocated for many reasons. Usually it is displaced anteriorly, medially, and superiorly by pathology involving the paralaryngeal space or cricoarytenoid joint. The arytenoid also moves in the same direction during phonation or when a cord is paralyzed. Depending upon the clinical situation, it might be difficult to distinguish a paralyzed cord from a dislocated arytenoid. Sometimes one arytenoid may be slightly more superior than the other and this confuses interpretation of exactly where the true vocal cord level is unless other anatomic landmarks, such as the top of the cricoid and anterior commissure, are analyzed in addition to the pathology causing the actual arytenoid displacement. Occasionally, the arytenoids will dislocate posteriorly into the hypopharynx. In these instances they lie posterior to the cricoid.

GLOTTIC REGION

T_1 glottic lesions are confined to the true cords with the mobility remaining normal. T_{1A} lesions involve one cord, T_{1B} lesions both. These tumors may be treated by radiation, laser, cordectomy, or vertical partial laryngectomy. All methods of treatment show high cure rates (16, 53). Radiation is usually chosen because of the superior voice quality following therapy.

CT scan of these lesions may look normal or show true cord thickening. The widening of the true cord may appear focal or diffuse. Most of the time one can predict the amount of true cord involvement in thirds, e.g. anterior one-third involved, posterior two-thirds involved. Diffuse or predominantly posterior thickening, near but not involving the vocal process of the arytenoid, may indicate a tumor more amenable to surgery than radiation therapy (78). CT density differences between nonenhancing tumor and muscle are not large enough to allow one to predict invasion of the true cord musculature in most instances. Bulky tumors do not necessarily infiltrate deeply but may be mainly exophytic. MRI signals from tumor and muscle differ so much that the distinction between an exophytic and deeply infiltrating lesion is easier. Improved coil design should improve spatial as well as contrast resolution in the neck.

True cord tumors may spread to involve the mucosa over the arytenoids or at the anterior commissure. When the true cords show limited mobility but are not truly fixed, the tumors usually fall into the T_2 category. Strictly defined T_2 lesions have spread to either the supraglottic or subglottic region and show normal or only impaired mobility, of the TVC (2). With CT, impaired mobility may be implied in scans by diminished motion artifacts on the involved side as well as the typical paramedian position of the cord and more medially situated arytenoid (on scans done in quiet respiration). CT and MRI provide an exquisite view of the anterior commissure and its relationship to the supraglottic and subglottic compartments. Since the mucosa is basically adherent to the perichondrium at the anterior commissure (11, 66, 68), any more than 1 mm of tissue thickness is considered abnormal.

Once true vocal cord lesions reach the anterior commissure, they may spread infraglottically, extend to the contralateral cord, and invade the thyroid cartilage (Fig. 9.12) (34, 67). Sometimes tumors are so bulky that they bulge across the midline and clinically either appear to extend across the commissure or hide it from view. Axial sections are ideal for sorting out such problems because a cleavage plane may be visible between the tumor, anterior commissure, and contralateral true vocal cord if the tumor has not extended across the commissure. If biopsy confirms greater than 30% involvement of the contralateral true vocal cord, partial laryngectomy is usually contraindicated unless accompanied by a permanent tracheostomy (Fig. 9.8) (7). In advanced lesions, almost circumferential spread at the true cord level may make the larynx look very symmetric although grossly abnormal. The amount of anterior midline thickening gives the clue to the true situation.

Tumor at the anterior commissure may follow the central tendon of the vocal ligaments into the thyroid cartilage (34, 66). This is very difficult to detect by clinical exam and may be by CT. One must not overread an area of demineralization in the paramedian region as definite cartilage invasion (Fig. 9.12). Focal asymmetric areas of diminished density may be a normal variant in this locale. If there is tumor, an obvious difference between the two sides and evidence of weakening of the laryngeal skeleton indicate invasion. MRI holds some promise for greater accuracy in diagnosing early cartilage invasion since the signal from tumor should differ dramatically from cartilage regardless of its state of mineralization. Detection of subtle cartilage invasion by MRI will require high resolution thin section (5-mm) techniques; this necessitates optimal coil design and/or high field strengths.

Glottic tumors may extend posteriorly causing thickening of the tissues around or in between the arytenoids. Any thickening of the tissue overlying the cricoid, or medial to the arytenoid, must be interpreted as abnormal (Fig. 9.12). Sometimes it will be tumor; at other times it may be reactive edema or an intense fibroblastic response termed "pseudosarcomatous" reaction (6). Structural abnormalities will always have some explanation; it may not always be tumor nor will it always be obvious before the final pathology report is made. Keep in mind that posterior commissure (interarytenoid) spread of tumor is very unusual except in advanced lesions.

Large T_1 and T_2 glottic lesions may be treated by surgery or radiation (7, 16, 39, 53). If surgery is chosen, it will often be by some modification of the classical vertical hemilaryngectomy since this operation was originally designed for tumors that went only to the vocal process of the arytenoids (7, 53, 58, 78). Spread to the false cord and significant subglottic spread (see next section) may obviate vertical hemilaryngectomy (7, 53, 58, 78). Further extension may fix the vocal cord producing a T_3 lesion; this definitely worsens the prog-

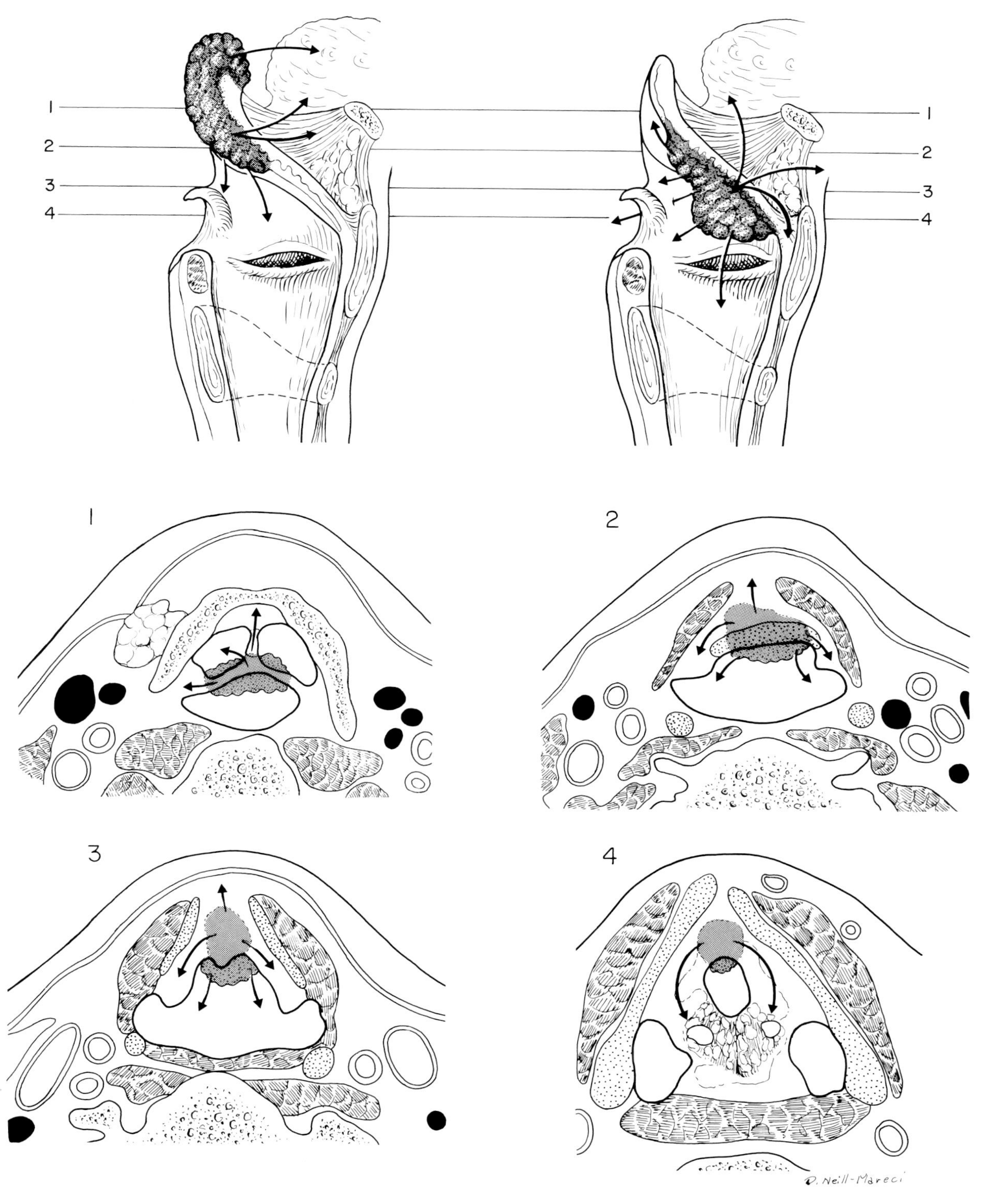

Figure 9.13. Patterns of spread (*curved arrows*) of anterior supraglottic (epiglottic) carcinoma. Epiglottic carcinoma may involve the supra- or infrahyoid portion of the epiglottis. The two whole organ diagrams at the *top* of the page indicate the level of the following axial sections; they also show the possible routes of spread of cancer arising on either part of the epiglottis. The following four sections illustrate patterns of spread as they may appear on CT images through the supraglottic larynx.

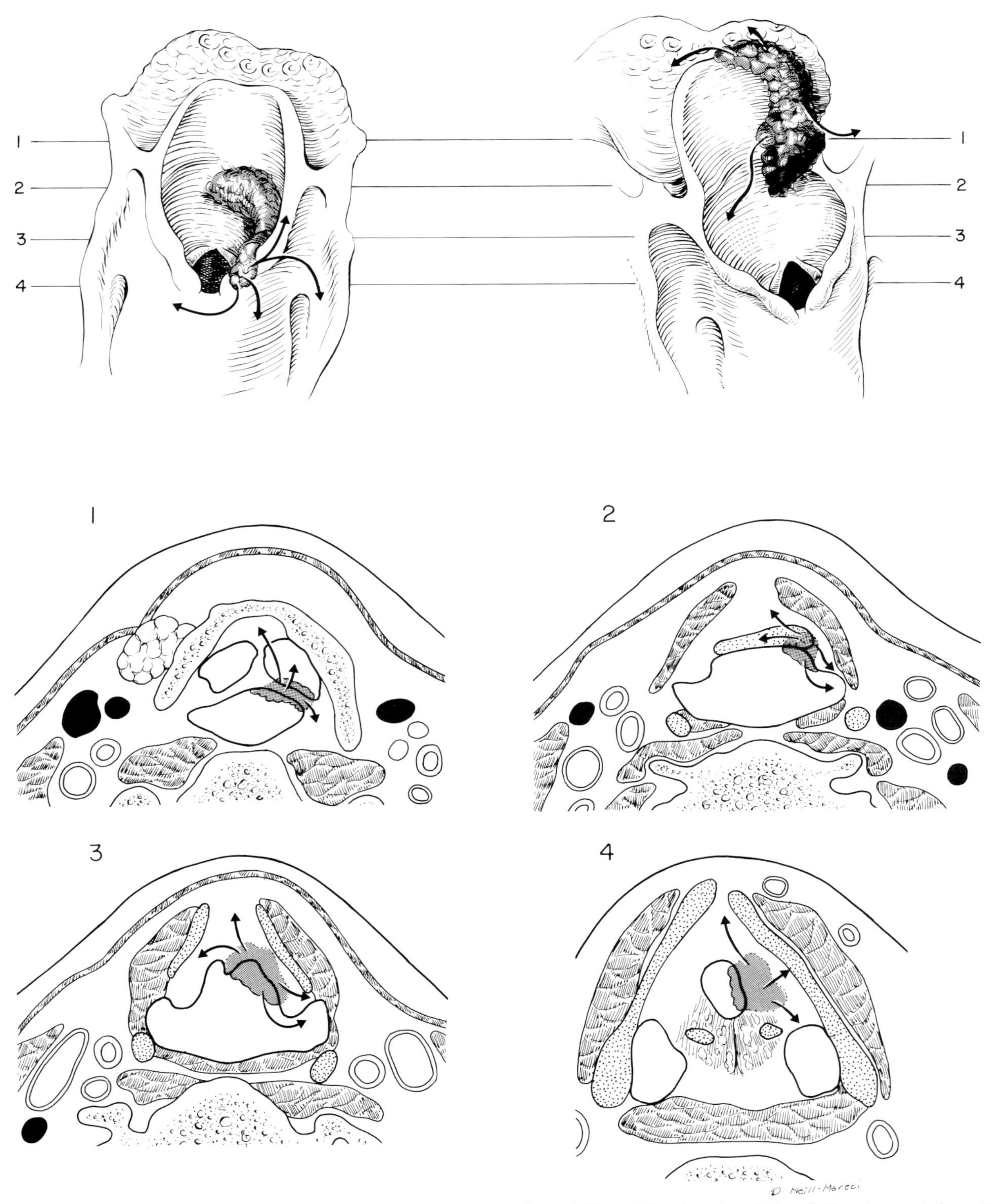

Figure 9.14. Patterns of spread (*curved arrows*) of marginal supraglottic carcinoma. The two whole organ diagrams at the *top* of the figure illustrate the gross patterns of spread of these lesions. Compare this to Figure 9.13 to recognize the differences and similarities between these lesions and those which arise more in the midline. The levels of the following axial sections are indicated on the whole organ diagrams; the axial diagrams show patterns of spread as they might appear on axial MR or CT images through the supraglottic larynx.

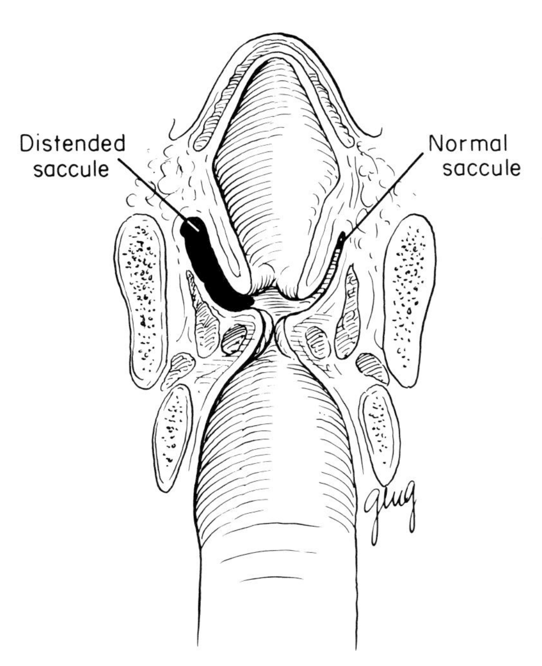

Figure 9.17. Pathogenesis of cysts of the saccule (appendix) of the laryngeal ventricle. Saccule may become enlarged and distended with air in patients who play wind instruments, and glass blowers, etc. If the outlet of the saccule to the ventricle is obstructed (usually by scarring), secretions build up and may progress (*see* Fig. 9.18).

cele) presenting as a mass in the neck (Fig. 9.18). When confined to the larynx it is called an internal laryngocele. Chronic inflammation may narrow the neck of the laryngocele either at its origin or where it passes through the thyrohyoid membrane (13, 24, 80). Fluid then may accumulate within the dilated saccule producing a cystic mass which interferes with cord function. CT is virtually diagnostic of fluid-filled laryngoceles (24). The well circumscribed fluid density mass extends superiorly in the paralaryngeal space from the false cord level (Fig. 9.18). The ipsilateral thyroid lamina is displaced. Both internal and external components may be imaged (23, 24). Differentiation from a second branchial cleft cyst may be difficult from simple neck exam, although laryngoscopy will usually clarify the situation. A major pitfall to remember is that a small cancer near the origin of the saccule may obstruct its neck producing a fluid-filled laryngocele that looks exactly like one secondary to a benign etiology (13, 80).

NEOPLASTIC

Benign neoplasms of the larynx are unusual lesions. The overwhelming majority are papillomas (76), many of which occur in the pediatric age range. Adenomas and mesenchymal tumors make up the remainder of the group except for chondromas. Chondromas of the laryngeal cartilages are rare; there are about 160 reported cases (6, 15). The radiographic diagnosis would probably be clear from routine "soft tissue" views or xeroradiographs. If clinical circumstances require more precise diagnostic information or one suspects a subtle lesion, CT provides an unmatchable view of the laryngeal skeleton which should reveal the typical pattern of punctate and ring-shaped choncroid matrix ossification.

We have seen two paragangliomas of the larynx. One produced only localized thickening of the aryepiglottic fold; the other looked like an internal laryngocele containing high attenuation fluid.

Hemangiomas of the adult variety are difficult to distinguish from angiomatoid speaker's nodules or granulation tissue; these cases should not come to imaging. The infantile type are true angiomatous malformations seen mainly in the infraglottic larynx. They are usually self-limited lesions and the diagnosis is made easily by endoscopy (76).

Rhabdomyomas, lipomas, neuromas, and fibromas may occur. All are rare lesions in the larynx (6, 76).

Trauma

The larynx may be injured accidentally or iatrogenically in a variety of ways. The CT appearance of injured larynges are as varied as the circumstances that

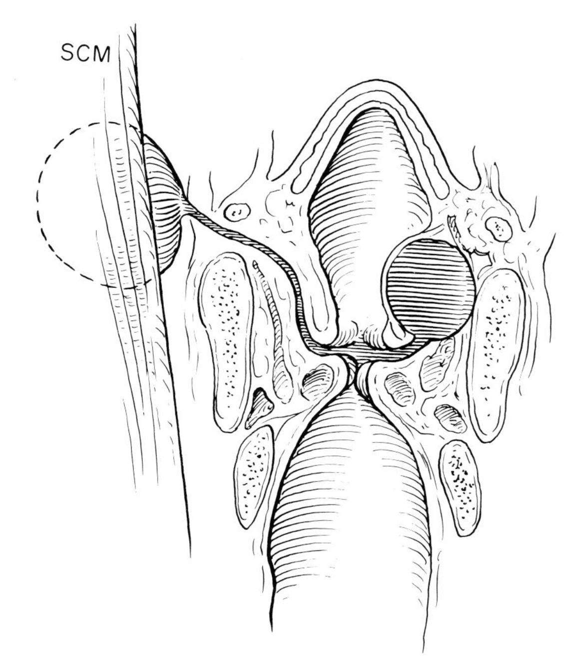

Figure 9.18. When a cyst of the saccule or an air-filled saccule enlarges, it may stay within the paralaryngeal space (internal laryngocele) or extend through the thyrohyoid membrane to present as a mass anterior and medial to the sternocleidomastoid muscle (*SCM*) (external laryngocele). Cysts are usually totally internal or combined internal and external.

lead to these pathologic distortions. The role of CT is to show the extent of injury to the laryngeal skeleton and tissues. This unique means of visualizing the injured larynx has advanced our understanding of how the larynx resonds to various types of trauma (28, 48).

BLUNT AND PENETRATING TRAUMA

Clinical Background

Laryngotracheal complex injuries may compromise critical glottic functions. Airway maintenance is the primary concern. In addition, preserving a normal voice and assuring swallowing without aspiration are always the goals of therapy (1, 18, 55, 62, 64). In acute injuries, the examination of the larynx must be quick, accurate, and as free of added morbidity as possible. In this regard CT represents a major advance for the laryngologist. Even following severe head and neck injury, the larynx may be safely and conveniently studied at the same time as the brain and facial structures without moving the patient who needs only to lie down and breathe quietly during the exam.

Physical examination may reveal signs of serious laryngeal injury. Sometimes these may be discounted in a patient who has other extensive head and face injuries unless life-threatening airway obstruction is present (1, 18, 55, 62). Also, swelling of the anterior cervical soft tissues might obscure fairly gross distortions of the normal laryngeal contour from the palpating hand. If, for these or other reasons, the extent of laryngeal injury is not appreciated acutely, the resulting delay can lead to a laryngeal stenosis which may prove more difficult to manage at a later time. Such delay also may lead to a less satisfactory therapeutic result (1, 18, 55, 62).

After reestablishing the airway and treating other life-threatening injuries, the diagnostic work-up of the traumatized laryngotracheal complex must go forward without delay. Laryngoscopy and bronchoscopy may be necessary to identify mucosal tears, cartilaginous fragments, false passages, or tracheal rupture (2, 18, 55, 62). Sometimes soft tissue swelling is so extensive that much of the larynx cannot be visualized at laryngoscopy. The remaining work-up is radiographic.

When the airway is adequate and swallowing is not disturbed, conservative management is possible if further work-up shows no mucosal tears or displacement of the laryngeal cartilages. These injuries are usually contusions of the laryngeal soft tissues which do not require surgery (55, 62).

Good functional results are obtained if acute laryngeal fractures are reduced within 3–7 days postinjury (18, 55, 62). Delay in surgery leads to scarring and chronic laryngeal stenosis. Surgical indications include (62, 64):

1. Progressive airway obstruction with cervical emphysema
2. Clinical or x-ray evidence of displaced fractures of the thyroid, cricoid, or epiglottic cartilages
3. Endoscopic evidence of avulsed soft tissue or mucosal laceration
4. Evidence of false passages
5. Dislocation of arytenoids

Ogura and others (62) have devised the following classification for injuries of the laryngotracheal complex. Even though many injuries involve one or more of these sites, the classification is useful for discussing diagnosis and management of blunt trauma and we will adhere to this system in the following sections.

A. Acute
   1. Soft tissue
   2. Infraglottic
   3. Glottic
   4. Supraglottic
B. Chronic stenosis

CT Diagnosis of Acute Injuries

In blunt trauma to the laryngotracheal complex, the offending force is usually directed from anterior to posterior. Typical are impacts with dashboard or fist and "clothesline" type of force; in each case the larynx is compressed between the offending agent and cervical spine (Fig. 9.19). The actual extent of injury will vary with the amount and vector of the applied force and the impacting area and contour of the object. Other factors which may influence the almost infinite variety of findings include (Fig. 9.20):

1. The level of the larynx that was the central point of injury
2. The size of arthritic spurs along the anterior aspect of the cervical spine
3. The tensions exerted by the intrinsic laryngeal musculature and the supra- and infrahyoid strap muscles.

Soft Tissue Injuries

The bleeding and edema that results from blunt trauma spreads through the deep planes of the larynx much like tumor. Infraglottically, the extensions are directed circumferentially by the conus elasticus manifested on CT as thickening of the tissues between the inner border of the cricoid and the airway. At and above the glottis, blood and edema tend to spread in a craniocaudad direction within the laryngeal spaces. Since the paralaryngeal and preepiglottic spaces are in continuity, spreading hematoma may cross the midline within the latter. Because these spaces contain variable amounts of fat at, and above, the false cords, their CT density will increase when averaged with that of fluid and blood.

The all important effects of this bleeding and edema is the extent to which they narrow the airway. The axial projection of CT gives true cross-sectional area of the airway that can be compared directly to clinical examination. It does not tend to overestimate the size of an irregularly shaped airway as conventional radiography may (Fig. 9.21). Accutely, swelling within the supraglottic larynx or hypopharynx may prevent a

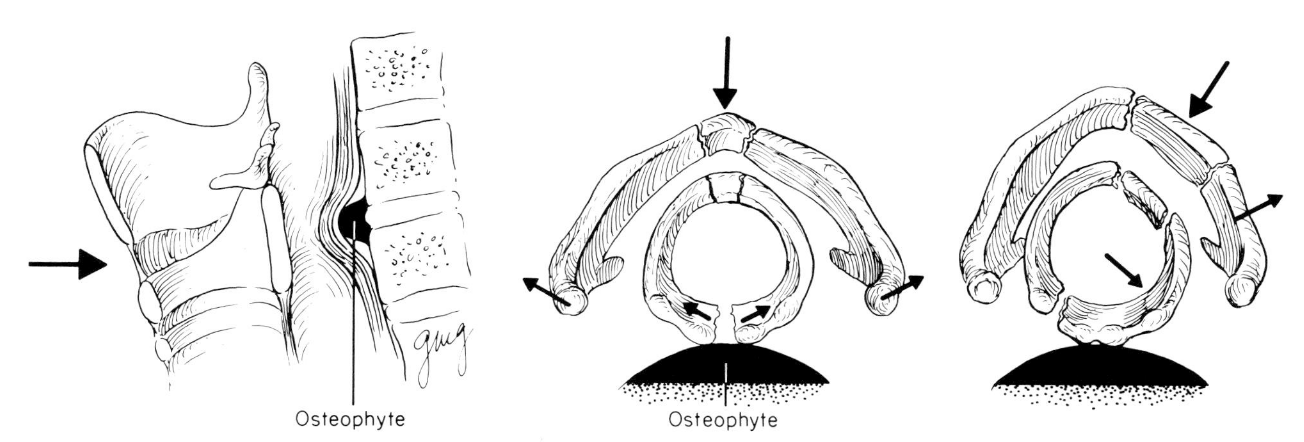

Figure 9.19. Mechanisms and patterns of cricothyroid complex fractures due to blunt laryngeal trauma. A compressive force (*large arrows*) squeezes the larynx against the cervical spine. The patterns of cricoid and thyroid cartilage fractures (*small arrows*) vary depending on the vector and velocity of the force, level of maximum impact, status of the cervical spine, and "elasticity" of the laryngeal skeleton (*see* text).

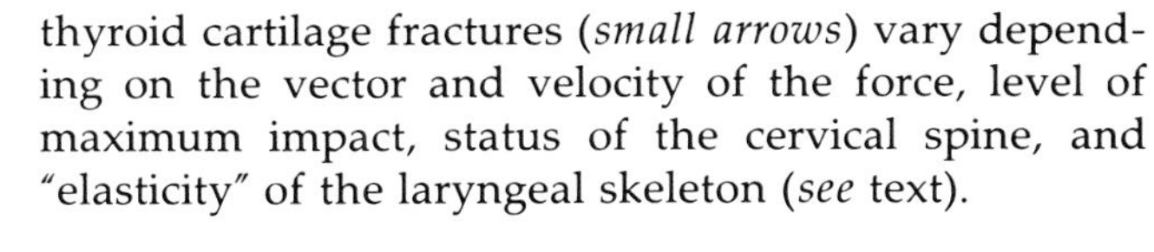

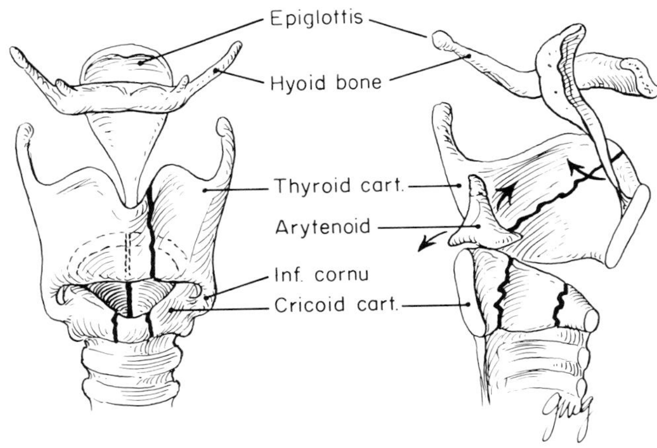

Figure 9.20. Overall perspective of laryngeal skeletal fractures and dislocations. Many combinations are possible. The thyroid cartilage may suffer longitudinal paramedian, transverse, or comminuted fractures. The cricoid always breaks in two places, the posterior laminar component going unrecognized clinically. The epiglottis may be avulsed posteriorly and superiorly (*top right arrow*) and the arytenoid dislocated either anteriorly or posteriorly (*lower two arrows*).

complete clinical examination of the lower airway during laryngoscopy; in such cases, airway appraisal by CT is even more important.

Dysfunction of the true vocal cord may result simply from the effect of the mass due to bleeding and edema. CT allows us to rule out other causes of limited true cord function such as cricoarytenoid joint dislocation or other laryngeal skeletal injury.

The extent of mucosal injury must still be established by laryngoscopy. The exact mechanism of tears is unclear but mucosal avulsions are quite common. In almost all cases, this combination of laryngoscopy and CT will provide all the information necessary for management. Laryngoscopy shows the mucosal detail, glottic function, and size of the airway. CT shows the status of the laryngeal skeleton deep planes and confirms the clinical appraisal of airway size. Laryngography, or water-soluble contrast swallow, may be held in reserve to demonstrate an otherwise undetectable false passage.

Infraglottic Injuries

Infraglottic injuries (including laryngotracheal separation) are mainly manifested by fracture of the cricoid cartilage (62) and the resultant airway narrowing due to displacement of cartilaginous fragments and associated hematoma and edema. The anterior cricoid may be comminuted and posteriorly displaced. Such findings are occasionally difficult to detect on CT scans, being indicated only by an absence of the anterior ring, but they are relatively easy to predict clinically because the cricoid prominence is lost or depressed (62).

CT displays the rest of the cartilage which is basically hidden from clinical examination. The cricoid, being a ring, will fracture in two places. If the force is applied directly anteroposterior, both the anterior arch and the posterior lamina may fracture near the midline and the two halves of the cartilages will "spring" apart (Figs. 9.19, 9.20). If the applied force is from the side, the cricoid will be compressed obliquely against the cervical spine; the arch fracture will then be more lateral in location and ipsilateral to the applied force. The posterior lamina fracture will also be off midline and either on the same side or side opposite to the applied force (Figs. 9.19, 9.20).

The springing apart of the cricoid appears to widen the airway posteriorly, while the acute airway narrowing caused by bleeding and edema is contained mainly laterally and anteriorly in the compartments formed by the attachments of the conus elasticus. The posterior spreading may serve a protective function by fortuitously widening an otherwise narrowed airway.

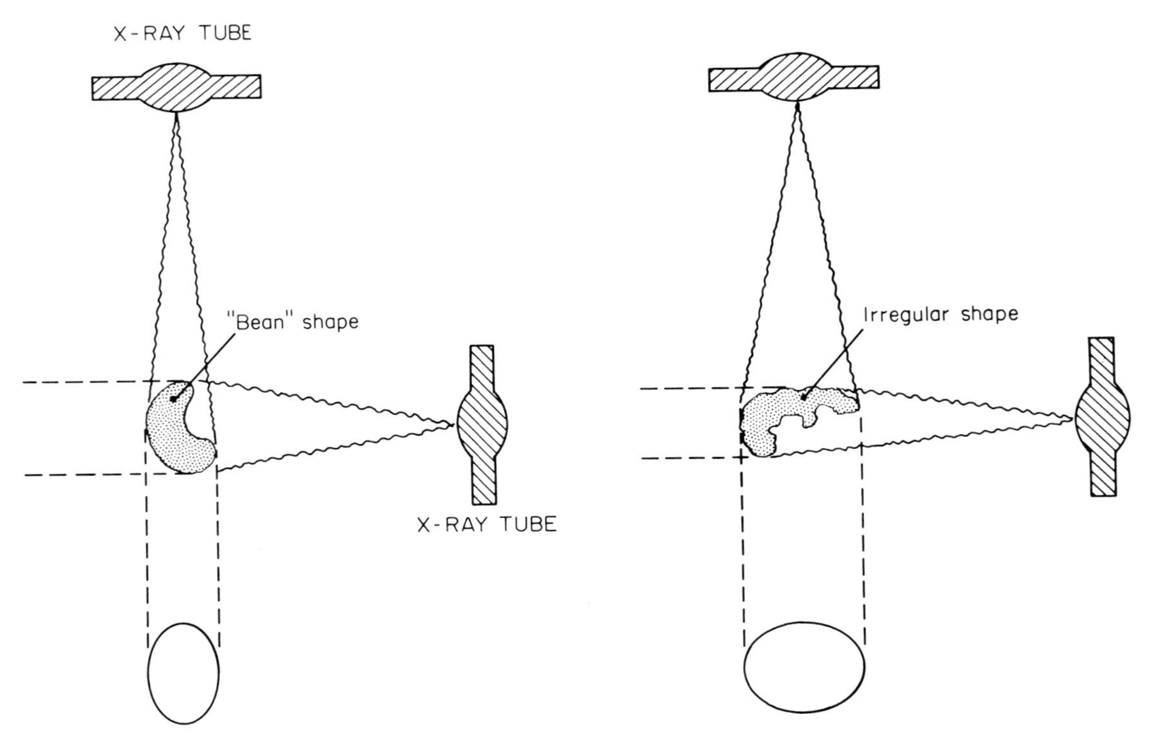

Figure 9.21. Conventional anteroposterior and lateral views may give a false impression of cross-sectional area of the airway. CT correlates better with the clinical examination findings because of its axial sections.

Fractures and dislocations affecting the cricothyroid joint may contribute to dysfunction of the true vocal cords. If this joint is disrupted the cricothyroid muscle may not be able to properly lengthen the vocal cords in preparation for phonation. Also, deformation of the cricothyroid joint may herald injury to the recurrent laryngeal nerve which passes closely posterior to the joint capsule.

The status of the cricoid cartilage is one of the most important factors in planning the repair of an injured larynx and predicting the probability of successful management (1, 18, 55, 62, 64). The cricoid is the foundation of the larynx and CT demonstrates the extent of cricoid injury more precisely than any other diagnostic method (48, 43).

The cricoid is most often severely injured in cases of massive trauma with laryngotracheal separation. When the point of separation is between the cricoid and trachea, the trachea recoils toward the mediastinum and the larynx is pulled upward. The cricoid rotates so that the lamina is oriented oblique to its normal position in the neck with the lamina facing inferiorly rather than posteriorly. This can cause a confusing CT image in that almost all of the lamina will be imaged in one or two axial sections. If the major separation is between the cricoid and thyroid (i.e. rupture of the cricothyroid membrane) the cricoid will be pulled down with the trachea so that the cricoid lamina will then lie oblique to its normal axis, only this time the lamina faces superiorly. Fractures of the cricoid lamina and arches may further complicate the CT appearance and there is almost always gross disruption of cricothyroid joints. If these injuries go unrepaired, the fractures heal with marked deformity which results in cricothyroid and cricoarytenoid joint dysfunction and infraglottic stenosis. The massive cricothyroid joint injuries also herald the possibility of crushing injuries to the recurrent laryngeal nerve which runs so close to the joint capsule.

Glottic Injuries

The thyroid cartilage, being opened posteriorly, will be splayed out when compressed against the cervical spine. This force causes tethering in the paramedian regions resulting in a vertical paramedian fracture (Figs. 9.19, 9.20). The effect is much like breaking a wishbone. A more laterally directed force can produce both paramedian and horizontal fractures. The laminae may be comminuted with displacement of fracture fragments. Widely dispersed fragments provide no support for the true or false cords. Also, the anterior commissure may be disrupted. Glottic and paraglottic soft tissue swelling is often a predominant finding. The soft tissue and cartilaginous injury can combine to produce marked rotational distortions of the cricothyroid complex. On routine soft tissue views or xeroradiography, these distortions are difficult to interpret and can lead to false impressions of arytenoid dislocation.

The arytenoid cartilages do not fracture as a rule but

they often dislocate. When dislocated, the arytenoids may lie more anteriorly and superiorly than normal with foreshortening of the ipsilateral vocal cord (Fig. 9.20). The vocal cord also will assume a paramedian position sometimes making it difficult to distinguish a dislocation from a cord paresis (70). The arytenoids also may be avulsed superiorly and posteriorly into the pharynx (70) (Fig. 9.20). When the cricothyroid joint is subluxed, the aryepiglottic fold will usually be medially situated, thus widening the ipsilaterally pyriform sinus.

Extensive soft tissue injury alone may cause glottic dysfunction, especially if it involves the regions of the anterior or posterior commissures (62). There may be associated cricoid cartilage fractures or fractures-dislocation of the cricothyroid joint.

Supraglottic Injuries

In supraglottic injuries, the soft tissue changes seen on CT scans may indicate significant cartilage injury. The epiglottis may become avulsed from its attachment to the thyroid cartilage (Fig. 9.20). The concomitant laceration of the thyroepiglottic ligament may produce hemorrhage into the inferior part of the preepiglottic space causing its density to increase while separating the epiglottis posteriorly. Because epiglottic cartilage itself is not imaged on CT, the possibility of its avulsion is only inferred from the amount of separation between the base of the epiglottis and thyroid cartilages. With the present techniques, extensive mucosal hemorrhage may appear identical to epiglottic avulsion. Associated transverse or vertical fractures of the thyroid cartilages near the anterior commissure increases the index of suspicion for epiglottic avulsion.

The hyoid bone is more often displaced by laryngeal and exolaryngeal hemorrhage and edema than it is fractured. The hyoid's normal segmented appearance should not be mistaken for fractures.

CHRONIC LARYNGEAL STENOSIS DUE TO BLUNT TRAUMA—CT DIAGNOSIS

Clinical Background

Chronic stenosis of the larynx often follows delayed treatment or inaccurate evaluation of an acutely injured larynx (1, 18, 55, 62, 64). Scarring due to intubation will be discussed under "postintubation injuries." The distorted laryngeal skeleton will become progressively more abnormal as granulation tissue evolves to scar. The resulting soft tissue "webs" and retraction lead to structural shifts which narrow the airway and cause glottic dysfunction. The likelihood of adequate return to normal function following surgery is greatly reduced in this setting. These injuries should be diagnosed and treated in the acute phase for optimal results (1, 18, 55, 62, 64).

Clinically, chronic stenosis is classified as: (*a*) supraglottic and glottic, (*b*) infraglottic and glottic, and (*c*) mixed. Overall, the information necessary for adequate repair includes:

1. The status of the thyroid cartilage, including the extent of fracture and location of displaced fragments
2. The possibility of avulsion of the supraglottic laryngeal structures
3. The status of the glottic level especially with regard to webs at the anterior or posterior commissure and the possibility of arytenoid dislocation
4. The extent of injury to the cricoid cartilage, an especially important factor in laryngotracheal disruption.

A useful classification of CT findings due to chronic injuries is: (*a*) mucosal or submucosal, (*b*) deep distortion, and (*c*) mixed. The actual appearance of the tissues within these categories will vary with the elapsed time between the injury and the CT examination.

Mucosal and Submucosal Lesions

Granulation tissue within the supraglottic larynx in the subacute phase of an injury (10–30 days post-trauma) looks the same as hemorrhage and edema does in the acute phase. At the glottic and infraglottic levels, we have observed nearly complete airway occlusion by a fairly homogeneous soft tissue mass which appears to be of lower density than the nearby neck musculature. At surgery, this has proven to be granulation tissue. The reasons for its lower density than other tissues is not clearly evident but it is probably due to edema.

With time, granulation tissue will organize and contract. During this process it still remains lower in density than normal muscle tissue. Soft tissue webs thus formed present a sharply contoured thickening of the soft tissues adjacent to the airway. Fibrotic webs are most easily recognized at the anterior and posterior commissures. In the infraglottis, a circumferential rind of tissue may narrow the airway. The encroachment is usually sharply defined and symmetric but may be asymmetric and irregular depending on the nature and extent of the original insult. Occasionally, small herniations (false diverticuli) of the airway may be seen extending into the separation between fragmented portions of the laryngeal cartilages.

Deep Distortion

In the chronic phase of an injury, the deep tissue planes may progressively increase in density due to infiltration by granulation tissue. With complete resolution, the normal low density appearance of the supraglottic paralaryngeal and preepiglottic spaces usually returns even when progressive scarring or retraction has altered laryngeal contours.

Laryngeal skeletal fractures appear acutely as simple lucencies of "off sets" in the ossified or calcified cartilages. With time, minimally displaced fractures heal without deformity. In more severe, chronic, cases distortions of the laryngeal skeleton will be out of proportion to the soft tissue changes visible on CT scans; the soft tissue changes having passed through the more acute phases of healing and gone on to scarring and retraction. Cartilage fractures may be completely or partially healed; the only clue to their presence may be nonunion or malunion which presents on scans as free fragments or deformed cartilage contours, respectively. The arytenoid cartilages may be dislocated or held fixed in a paramedian position by fibrotic webs at the posterior commissure or surrounding the cricoarytenoid joint itself. It may be difficult to distinguish a paramedian cord due to arytenoid dislocation, nerve damage or fibrosis by CT alone.

"OCCULT" TRAUMA

Occasionally, CT of the larynx shows considerable distortion with findings that look exactly like those following blunt trauma; however, the patient does not recall the incident (28). The findings include webs, marked irregularity of the thyroid cartilages due to healed fractures and distortions of the laryngeal skeleton.

These patients sometimes present with voice fatigue, hoarseness and pain in the throat, which lead to clinical evaluation of the larynx. The distortions caused by the forgotten trauma may mimic mass lesions at laryngoscopy even though no mucosal abnormalities are seen. Two patients with such symptoms and a "mass" seen at laryngoscopy have been surgically explored in spite of a CT scan that indicated that no tumor was present (28). The surgery confirmed that multiple old fractures of the thyroid cartilage with resultant deformity of its contour had produced the paralaryngeal mass. No tumor was present. In other patients, who repeatedly deny a history of trauma at the time of clinical or CT examination, such a history has been forthcoming only after conversations with friends or relatives. Others have had a history of alcohol abuse or participation in violent sports (e.g. wrestling, football), although no specific traumatic event stands out in their minds. There is no clear correlation between the patients' presenting complaints and the old injuries. But it is possible that the anatomic distortion caused misuse of the glottis and the pattern of voice fatigue and hoarseness is the result. Treatment is conservative (28).

IATROGENIC TRAUMA

Postintubation

The vast majority of chronic laryngeal and subglottic stenoses are related to prolonged intubation. The endotracheal tube may denude the mucosa of the glottic or infraglottic region either due to pressure necrosis or chronic irritation. As the denuded cords reepithelialize, webs form which may compromise the airway and limit vocal mobility. The infraglottic region heals with proliferation of scar tissue over a variable extent that reduces the lumen of the airway. CT demonstrates the diameter of the residual airway in the transglottic and infraglottic regions well but shoulder artifacts may interfere with accurate delineation of the trachea at the thoracic inlet. Such artifacts also may render reformatted images suboptimal for interpretation. The vertical extent of such stenosis must be shown accurately in order to plan a surgical resection. Conventional tomography or xeroradiography may be used to better advantage than CT at giving this type of measurement if significant artifacts are present. The multiplanar capability of MRI may prove useful in this situation and will certainly be of higher resolution than reformatted CT images.

Aside from scarring, glottic dysfunction following endotracheal intubation may be related to paresis of the vocal cords or cricoarytenoid joint dislocation (70). In both cases the affected cord or cords will be paramedian in position during quiet respiration. There may be a lack of motion artifact on the affected side. The arytenoids will often appear to be asymmetric. When truly dislocated, the affected arytenoid tends to be more anterior and medial than the normal one and the ipsilateral true cord appears foreshortened. Subtle subluxation of the cricoarytenoid joint is indistinguishable from vocal cord paresis on CT scans. Adjunctive findings in cricoarytenoid dislocation or vocal cord paralysis include medial displacement of the aryepiglottic fold with resultant opening of the ipsilateral pyriform sinus.

Surgery for Benign Disease

CT provides valuable information in patients who have complaints of persistent pain, hoarseness, or with symptoms related to aspiration or airway encroachment following surgery for benign disease, e.g. repair of traumatic injuries or Teflon injections for glottic incompetence. Since Teflon is of higher density than soft tissues and does not cast artifacts, CT can precisely localize the injection sites. Misplaced injections can be recognized and may account for otherwise inexplicable postoperative symptoms.

Following repair of a traumatized larynx, CT may reveal previously unrecognized, and therefore untreated injuries, that explain the patient's persistent complaints. This is especially true of the cricoid cartilage where unsuspected fracture may occur posteriorly on the side contralateral to the impact. Inadequately reduced or misaligned fracture fragments are depicted often with as much clarity as postreduction views of long bones (out of plaster). In the postoperative setting, the extent of infections which may have progressed to chondritis and partial collapse of the laryngeal skeleton may be evident. This cartilaginous collapse along with the marked soft tissue swelling produces severe airway encroachment.

Hyoid interpositions or other types of grafts may

become denuded of mucosa and/or infected. CT has confirmed clinical suspicions of such occurrences.

References

1. Alonso WA, Pratt LL, Zollinger WK, Ogura JH: Complications of laryngotracheal disruption. *Laryngoscope* 84:1276–1290, 1974.
2. American Joint Committee on Staging of Cancer: Staging head and neck sites and of melanoma, Chicago, AJR, 1980.
3. Archer CR, Yeager VL, Friedman WH, et al: Computed tomography of the larynx. *J Comput Assist Tomogr* 2:404–411, 1978.
4. Baclesse F: Carcinoma of the larynx. *Br J Radiol* (Suppl 3) 1–68, 1949.
5. Ballantyne AJ: Significance of retropharyngeal nodes in cancer of the head and neck. *Am J Surg* 108:500, 1964.
6. Batsakis JG: *Tumors of the Head and Neck: Clinical and Pathological Considerations*, ed 2. Baltimore, Williams & Wilkins, 1979.
7. Biller HF, Ogura JF, Pratt LL: Hemilaryngectomy for T_2 glottic. *Arch Otolaryngol* 93:238–243, 1971.
8. Bridger MW, Jahn AF, Van Nostrand AWP: Laryngeal rheumatoid arthritis. *Laryngoscope* 90:296–303, 1980.
9. Browning GG: Clinical value of selective serial sectioning of laryngectomy specimens. *Proc R Soc Med* 69:417–419, 1976.
10. Browning GG, Busuttil A, Mclay A: An improved method of reporting on laryngectomy specimens. *J Pathol* 119:101–106, 1976.
11. Bryce PD: The laryngeal subglottis. *J Laryngol Otol* 89:667–685, 1975.
12. Camnitz PS, Biggers WP, Fischer ND: Avoidance of early complications following radical neck dissection. *Laryngoscope* 89:1553–1562, 1979.
13. Canalis RF, Maxwell DS, Hemenway WG: Laryngocele—an updated review. *J Otolaryngol* 6:191–199, 1977.
14. Chandler JR: Radiation fibrosis and necrosis of the larynx. *Ann Otol Rhinol Laryngol* 88:509–514, 1979.
15. Cocke EW: Benign cartilaginous tumors of the larynx. *Anal Otol Rhinol Laryngol* 72:167, 1962.
16. Cocke EW, Wang CC: Part 1. Cancer of the larynx selecting optimum treatment. *CA* 26:194–200, 1976.
17. Cocke EW: Cancer of the larynx: surgery. *CA* 26:201–211, 1976.
18. Cohn AM, Larson DL: Laryngeal injury: a critical review. *Arch Otolaryngol* 102:166–170, 1976.
19. Fisch: *Lymphography of the Cervical Lymphatic System.* Philadelphia, Saunders, 1958.
20. Friedman I: *Centennial Conferences on Laryngeal Cancer, Workshop No. 7, Toronto*, Alterti PW, Bryce PD, (eds): New York, Appleton Century Crofts, 1974.
21. Gamsu G, Webb WR, Shallit JB, Moss AA: Computed tomography in carcinoma of the larynx and pyriform sinus—the value of phonation CT. *AJR* 136:577–584, 1981.
22. Gatenby RA, Mulhern CB, Richter MP, Moldofsky Rj: CT-guided biopsy for the detection and staging of tumors of the head and neck. *AJNR* 5:287–289, 1984.
23. Giovanniello J, Grieco RV, Bartone NF: Laryngocele. *AJR* 108:825–829, 1979.
24. Glazer HS, Mauro MA, Aronberg DJ, et al: Computed tomography of laryngocoeles. *AJR* 140:549–552, 1983.
25. Goodrich WA, Lenz M: Laryngeal chondronecrosis following roentgen therapy. *AJR* 60:22–28, 1948.
26. Gregor RT, Lloyd GAS, Michaels L: Computed tomography of the larynx: a clinical and pathologic study. *Head Neck Surg* 3:284–296, 1981.
27. Haagensen DC, Feind CR, Herter FP, et al: The lymphatics in cancer, ed 1. Philadelphia, Saunders, 1972, pp 60–208.
28. Hanson DG, Mancuso AA, Hanafee WN: Pseudo mass lesions due to occult trauma of the larynx. *Laryngoscope* 92:1249–1253, 1982.
29. Harnsberger HR, Mancuso AA, Muraki AS: The upper aero-digestive tract and neck: CT evaluation of recurrent tumors. *Radiology* 149:503–509, 1983.
30. Hollinshead WH: *Anatomy for Surgeons: vol 1, The Head and Neck*, ed 2. Hagerstown, Md, Harper & Row, 1968.
31. Keene M, Harwood AR, Bryce DP, et al: Histopathological study of radionecrosis in laryngeal carcinoma. *Laryngoscope* 92:173–180, 1982.
32. Kernan JD: The pathology of carcinoma of the larynx studied in serial sections. *Trans Am Acad Ophthalmol and Otolaryngol* 55:10–21, 1950.
33. Kirchner JA: Two hundred laryngeal cancers: patterns of growth and spread as seen in serial section. *Laryngoscope* 87:474–482, 1977.
34. Kirchner JA, Som ML: The anterior commissure technique of partial laryngectomy: clinical and laboratory observations. *Laryngoscope* 85:1308–1317, 1975.
35. Klein R, Fletcher GH: Evaluation of the clinical usefulness of roentgenologic findings in squamous cell carcinomas of the larynx. *AJR* 92:43–54, 1964.
36. Larsson S, Mancuso AA, Hoover L, Hanafee WN: Differentiation of pyriform sinus cancer from supraglottic laryngeal cancer by computed tomography. *Radiology* 141:427–432, 1981.
37. Last RJ: *Anatomy, Regional and Applied*, ed 5. Edinburgh, Churchill Livingston, 1972.
38. Lawry GV, Finerman ML, Hanafee WN, Mancuso, AA, et al: Laryngeal involvement in rheumatoid arthritis: a clinical laryngoscopic and computerized tomographic study. *Arthritis Rheum* 27:873–882, 1984.
39. Lederman M: Place of radiotherapy in treatment of cancer of the larynx. *Br Med J* 5240:1639–1646, 1961.
40. Lehmann QH, Fletcher GH: Contribution of the laryngogram to the management of malignant laryngeal tumors. *Radiology* 83:486–500, 1964.
41. Lufkin RB, Larsson SG, Hanafee WN: Work in progress: NMR of the larynx and tongue base. *Radiology* 148:173–175, 1983.
42. McGavran MH, Bauer WC, Ogura JH: The incidence of cervical lymph node metastases from epidermoid carcinoma of the larynx and their relationship to certain characteristics of the primary tumor. *Cancer* 14:55–66, 1961.
43. Maceri D, Mancuso AA, Bahna M, et al: Value of computerized axial tomography in severe laryngeal injury. *Arch Otolaryngol* 108:449–551, 1982.
44. Mafee MF, Schild JA, Valvassori GE, Capek V: Computed tomography of the larynx: correlation with anatomic and pathologic studies in cases of laryngeal carcinoma. *Radiology* 147:123–128, 1983.
45. Mancuso AA, Hanafee WN, Juilliard GJF, et al: The role of computed tomography in the management of cancer of the larynx. *Radiology* 124:243–244, 1977.

46. Mancuso AA, Calcaterra TC, Hanafee WN: Computed tomography of the larynx. *Radiol Clin North Am* 16:195–208, 1978.
47. Mancuso AA, Hanafee WN: A comparative evaluation of computed tomography and laryngograpy. *Radiology* 133:131–138, 1979.
48. Mancuso AA, Hanafee WN: Computed tomography of the injured larynx. *Radiology* 133:139–144, 1979.
49. Mancuso AA, Tamakawa Y, Hanafee WN: CT of the fixed vocal cord. *AJR* 135:429–434, 1980.
50. Mancuso AA, Maceri D, Rice D, Hanafee WN: CT of cervical lymph node cancer. *AJR* 136:381–385, 1981.
51. Mancuso AA, Hanafee WN: Elusive head and neck cancers beneath intact mucosa. *Laryngoscope* 93:133–139, 1983.
52. Mancuso AA, Fitzsimmons J, et al: MRI of the upper pharynx and neck: variations of normal and possible applications in detecting and staging malignant tumors. (Presented at the RSNA Scientific Assembly, November 1984, Washington, D.C.) (Submitted for publication.)
53. Million RR, Cassisi NJ: Larynx. In *Management of Head and Neck Cancer: A Multidisciplinary Approach.* Philadelphia, Lippincott, 1984, chap 19.
54. Millian RR, Cassisi NJ:Hypopharynx: Pharyngeal Walls, Pyriform Sinus and Postcricoid Pharynx. In *Management of Head and Neck Cancer: A Multidisciplinary Approach.* Philadelphia, Lippincott, (in press), chap 21.
55. Morgenstein KM: Treatment of the fractured larynx. *Arch Otolaryngol* 101:157–159, 1975.
56. Moss WT, Brand WN, Battifora H: Radiation Oncology, ed 5. St. Louis, Mosby, 1979.
57. Muraki AS, Mancuso AA, Harnsberger HR: Metastatic cervical adenopathy from tumors of unknown origin: the role of CT. *Radiology* 154:749–753, 1984.
58. Ogura JH, Biller HF: Conservation surgery in cancer of the head and neck. *Otolaryngol Clin North Am* 2:641–665, 1969.
59. Ogura JH, Biller: Reconstruction of the larynx following blunt trauma. *Ann Otol Rhinol Laryngol* 80:492–506, 1971.
60. Ogura JH, Dedo HH: Glottic reconstruction following sub-total glottic-supraglottic laryngectomy. *Laryngoscope* 75:865–878, 1965.
61. Ogura JH, Henneman H: Conservation surgery of the larynx and hypopharynx—selection of patients and results. *Can J Otolaryngol* 2:11–16, 1973.
62. Ogura JH, Henneman H, Spector GJ: Larynog-tracheal trauma: diagnosis and treatment. *Can J Otolaryngol* 2:112–118, 1973.
63. Ogura JH, Sessions DG, Spector GJ: Conservation surgery for epidermoid carcinoma of the supraglottic larynx. *Laryngoscope* 85:1808–1815, 1975.
64. Ogura JH, Powers WE: Functional restitution of traumatic stenosis of the larynx and pharynx. *Laryngoscope* 74:1081–1110, 1964.
65. Oloffsson J, van Nostrand AWP: Growth and spread of laryngeal and hypopharyngeal carcinoma with reflections on the effect of preoperative irradiation. 139 cases studied by whole organ serial sectioning. *Acta Otolaryngol.* (Suppl 308):1–84, 1973.
66. Oloffsson J: Specific features of laryngeal carcinoma involving the anterior commissure and subglottic region. *Can J Otolaryngol* 4:618–630, 1975.
67. Oloffsson J. Sokjer H: Radiology and laryngoscopy for the diagnosis of laryngeal carcinoma. *Acta Radiol* (Diagn) 18 Fasc 4:449–476, July 1977.
68. Pressman JJ: Submucosal compartmentation of the larynx. *Anal Oto Rhinol Laryngol* 77:165–172, 1956.
69. Pressman JJ, Simon M, Morell C: Anatomic studies related to the dissemination of cancer of the larynx. *Am Acad Ophthalmol Otolaryngol Trans* 64:628–638, 1960.
70. Quick CA, Merwin GE: Arytenoid dislocation. *Arch Otolaryngol* 104:267–270, 1978.
71. Rideout DF, Poon PY: Radiologic studies of larynx after radiotherapy for carcinoma. *J Can Assoc Radiol* 28:182–186, 1977.
72. Rouvier H: *Anatomy of the Human Lymphatic System*, ed 1. Ann Arbor, Mich, Edwards Brothers, 1938, pp 1–82.
73. Sagel SS, Aufderheide JF, Aronbert DJ, Stanley RJ, Archer C: High resolution computed tomography in the staging of carcinoma of the larynx. *Laryngoscope* 91:292–300, 1981.
74. Salmon LFW: Chronic laryngitis. In Ballantyne, J, Groves J (eds): *Scott-Brown's Diseases of the Ear, Nose and Throat*, ed 4. London, Butterworths, 1979, vol 4, pp 333–373.
75. Schaefer SD, Merkel M, Diehl J, et al: Computed tomographic assessment of squamous cell carcinoma of oral and pharyngeal cavities. *Arch Otolaryngol* 108:688–692, 1982.
76. Shaw H: Tumors of the Larynx. In Ballantyne J, Groves J (eds): *Scott-Brown's Disease of the Ear, Nose and Throat*, ed 4. London, Butterworths, 1979, vol 4, pp 375–457.
77. Silverman PM, Johnson GA, Korobkin M: High resolution sagittal and coronal reformatted CT images of the larynx. *AJR* 140:819–822, 1983.
78. Som ML: Cordal cancer with extension to vocal process. *Laryngoscope* 85:1298–1307, 1975.
79. Stark DD, Moss AA, Gamsu G, et al: Magnetic resonance imaging of the neck. Part I: normal anatomy. *Radiology* 150:447–454, 1984.
80. Stell PM, Maran AGD: Laryngocele. *J Laryngol Otol* 89:915–923, 1975.
81. Tucker GF, Jr: The anatomy of laryngeal cancer. *Can J Otolaryngol* 3:417–427, 1974.
82. Tucker GF, Smith HR: A histological demonstration of the development of laryngeal connective tissue compartments. *Am Acad Opthalmol Otolaryngol Trans* 66:308–318, 1962.
83. Wang CC: Part III: Cancer of the larynx; radiation therapy. *CA* 26:212–218, 1976.
84. Ward PH, Berci G, Calcaterra TC: New insights into the causes of postoperative aspiration following conservation surgery of the larynx. *Ann Otol Rhinol Laryngol* 86:724–731, 1977.
85. Ward PH, Hanafee Wn, Mancuso AA, Shallit J, Berci G: Evaluation of computerized tomography, cinelaryngoscopy, and laryngography in determining the extent of laryngeal disease. *Ann Otol Rhinol Laryngol* 88:454–462, 1979.
86. Ward PH, Calcaterra TC, Kagan AR: The enigma of postradiation edema and recurrent or residual carcinoma of the larynx. *Laryngoscope* 85:522–529, 1975.
87. Yeager VL, Archer CR: Anatomical routes for cancer invasion of laryngeal cartilages. *Laryngoscope* 92:449–452, 1982.

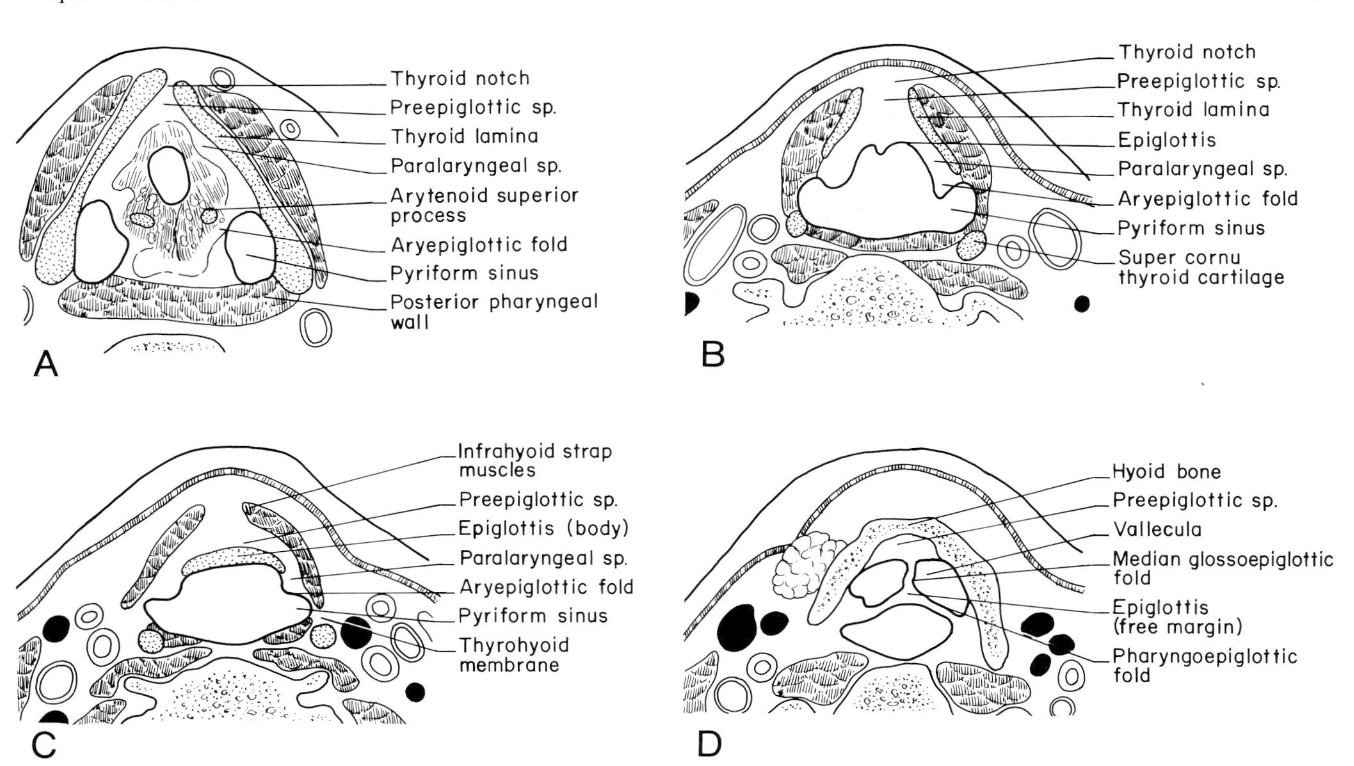

Plate 9.7*A–D*. normal supraglottis and pyriform sinuses

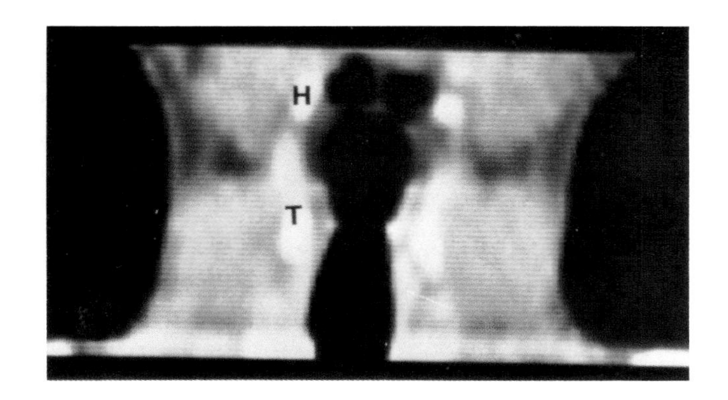

Plate 9.8. The relationship between the hyoid bone and thyroid lamina is seen in a different perspective in the midcoronal place. While theoretically useful for showing the extent of a true cord lesion relative to the subglottic space and false vocal cords, reformations are usually not of sufficient quality to justify their use in most tumors of the larynx. The coronal plane may be more useful in MRI where direct coronal sectioning is possible.

The abbreviations used on Plates 9.9–9.10 are: *PES,* preepiglottic space; *H,* hyoid bone; *PS,* paralaryngeal space; *PYS,* pyriform sinus; *A,* arytenoids; *APP,* appendix of the laryngeal ventricle; *T,* thyroid lamina; *SC,* superior cornual of thyroid cartilage; and *LV,* laryngeal vestibule; *AEF,* aryepiglottic fold; *E,* epiglottis.

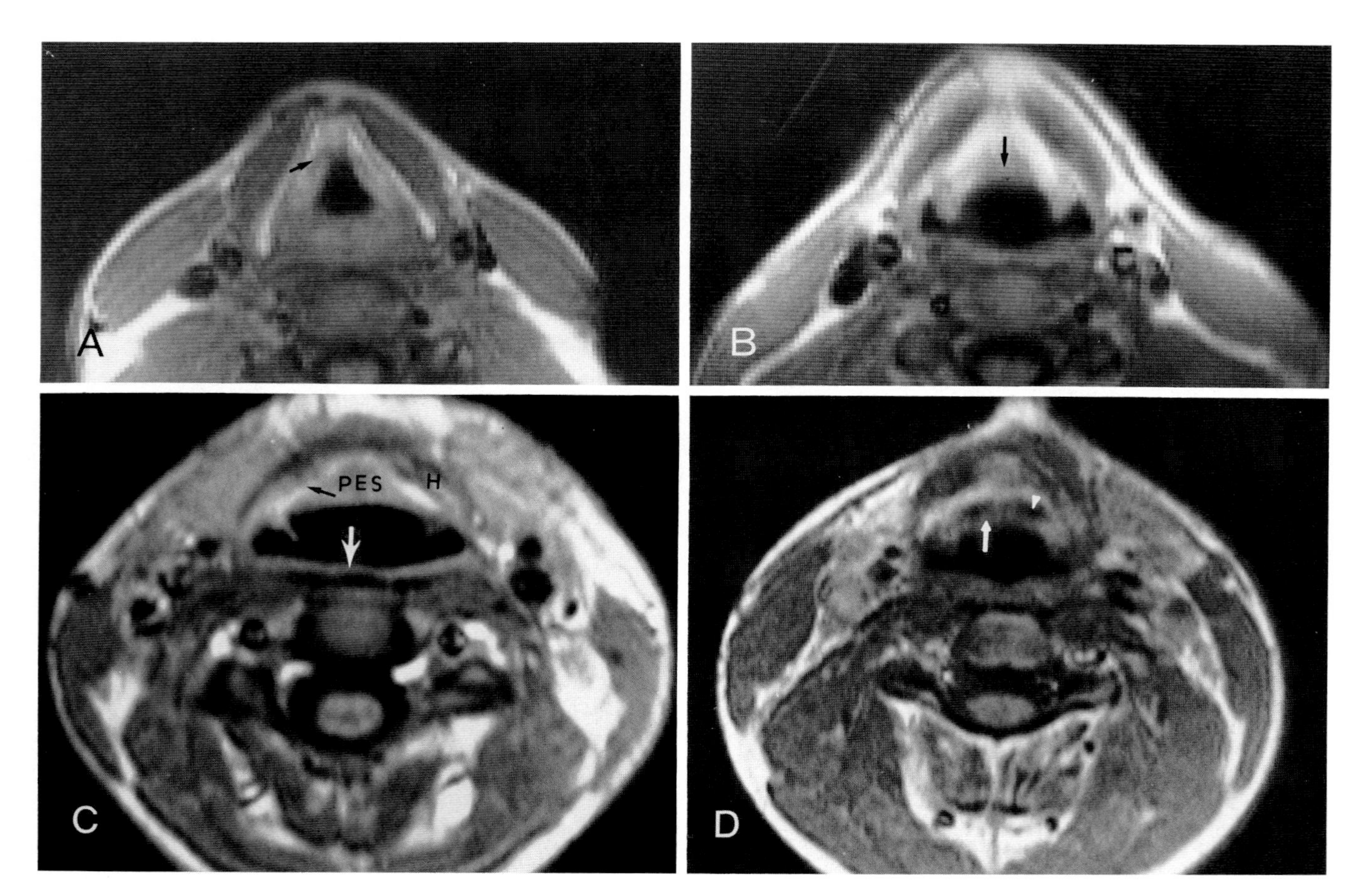

Plate 9.9. MRI images made with a spin echo 500/30 technique and utilizing a surface coil. The sections in plates *A* and *B* were made on an 0.15 Tesla resistive unit and the slice thickness is approximately 1 cm. Those in figures *C* and *D* were made on an 0.3 Tesla permanent unit and the slice thickness is approximately 4 mm.

Plate 9.9*A.* This section is just above the false cord level. The signal intensity of the paralaryngeal space (*arrow*) is mixed due to the content of both fat and muscle. The high intensity signal of the thyroid lamina is due to its fat content. This was described previously.

Plate 9.9*B.* A section made in the low supraglottic larynx. The fat in the preepiglottic space produces a relatively high signal intensity. Notice that just deep to the laryngeal surface of the epiglottis (*arrow*) there is normally an area of somewhat lower signal intensity due to the presence of the epiglottic cartilage, glandular tissue and ligamentous structures depending on the level of the section. This is very well illustrated in Plates 9.10 *A–D.*

Plate 9.9*C.* Section through high supraglottis. The intensity of preepiglottic space is decreased relative to the paralaryngeal space (*black arrow*) for reasons discussed previously. The hyoid bone is visible and the full thickness of the posterior pharyngeal wall (*white arrow*) is easy to appreciate.

Plate 9.9*D.* Section through the uppermost supraglottic larynx shows the free margin of the epiglottis (*arrow*) and vallacullae (*arrowhead*).

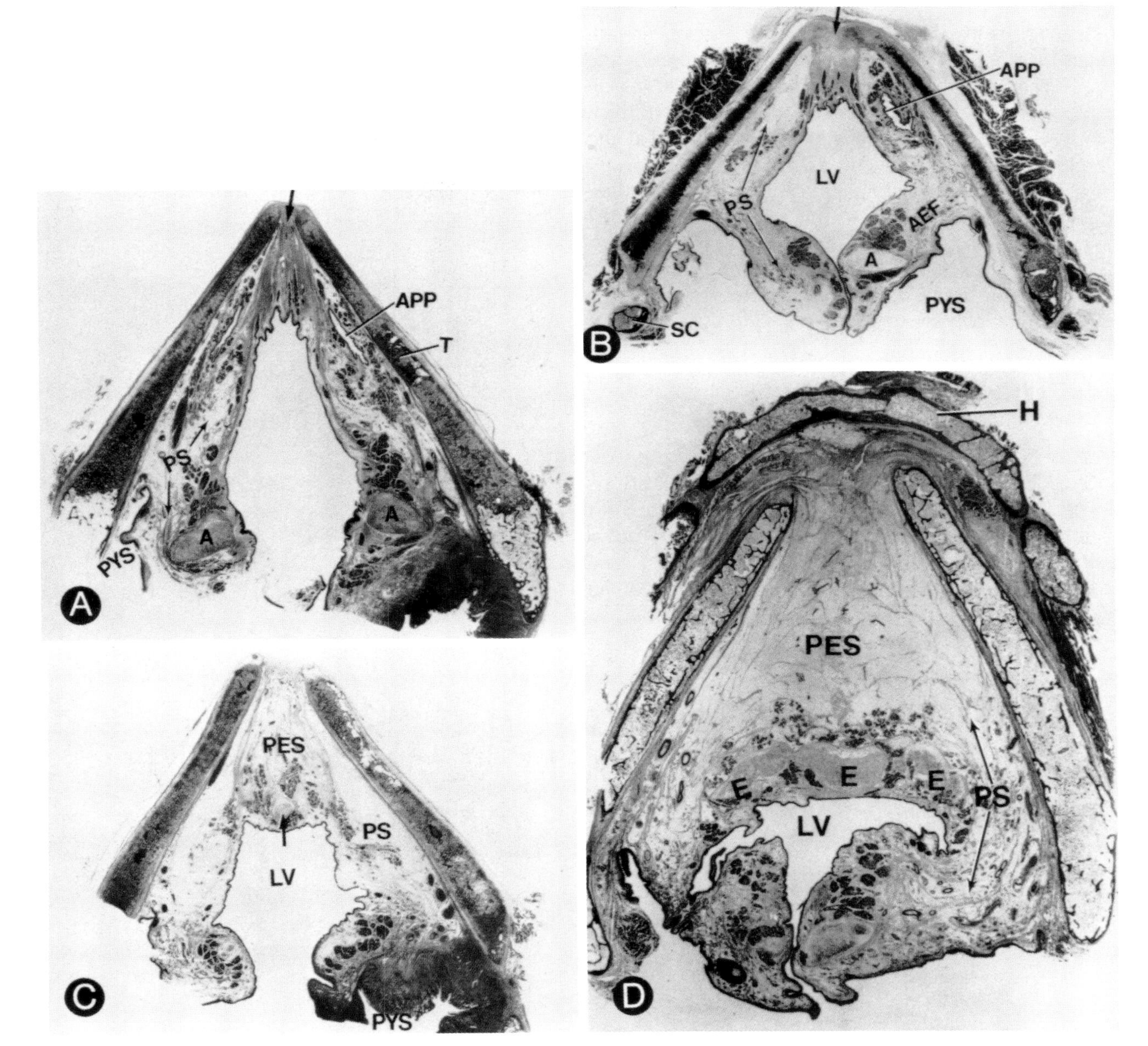

Plate 9.10*A*. The small gap between the thyroid lamina anteriorly (*arrow*) is the thyroid notch; a characteristic of the supraglottic larynx. Posteriorly, the superior processes of the arytenoids are seen within the base of the aryepiglottic fold. The tissue within the false cords has changed from the primarily muscular content of the true cord level to mainly fibrofatty content at the false cord level. This is the paralaryngeal space.

Plate 9.10*B*. The thyroid notch (*arrow*) marks the supraglottic larynx. Pyriform sinuses are distended and the fat within the paralaryngeal space and the aryepiglottic fold is obvious and explains the high signal intensity seen on MR images in these structures.

Plate 9.10*C*. Most of the preepiglottic space is fibrofatty in nature but note the presence of glandular tissue and the petiole of the epiglottis (*arrow*) just deep to the mucosa. This creates the lower signal intensity seen in this region on the MR images.

Plate 9.10*D*. The hyoid is not usually visible at this level but as an artifact of preparation it indicates that the integrity of the thyrohyoid membrane has been lost. The preepiglottic space is predominantly fatty although the epiglottis will create some diminished signal just deep to the airway. This is seen on CT as an area of increased density. The fatty preepiglottic and paralaryngeal spaces are in continuity in the supraglottic larynx.

The abbreviations used on Plates 9.14–9.16 are: *C,* cricoid and *PES,* preepiglottic space.

Plate 9.11. MR offers the advantage of simple multiplanar imaging. This may be useful in the supraglottic larynx. This is a midsagittal spin echo 500/30 image through the neck using a surface coil. It is 1 cm thick and was done on 0.15 Tesla unit. The fat within the preepiglottic space produces a high signal. The body of the epiglottis (*arrow*) causes the signal to be somewhat diminished just deep to the airway. The anterior commissure is easily visualized (*arrowhead*). The cricoid is seen by virtue of its fat content (*curved arrow*). Note the direct relationship between the preepiglottic space and the tongue base.

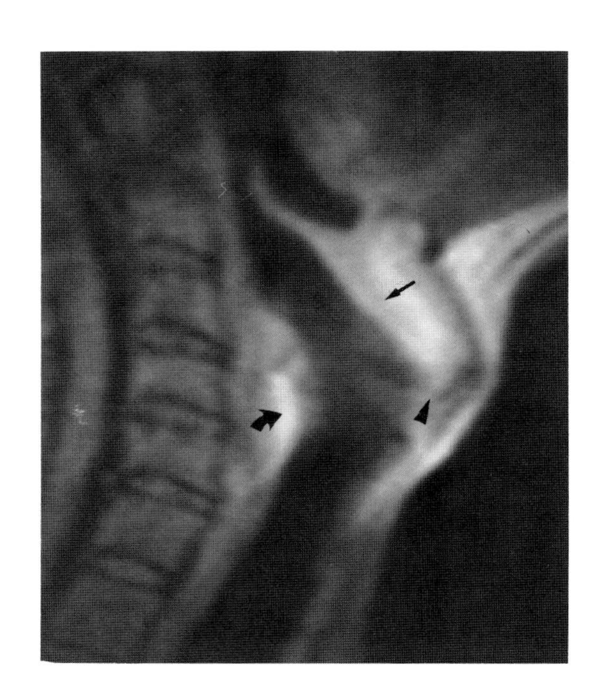

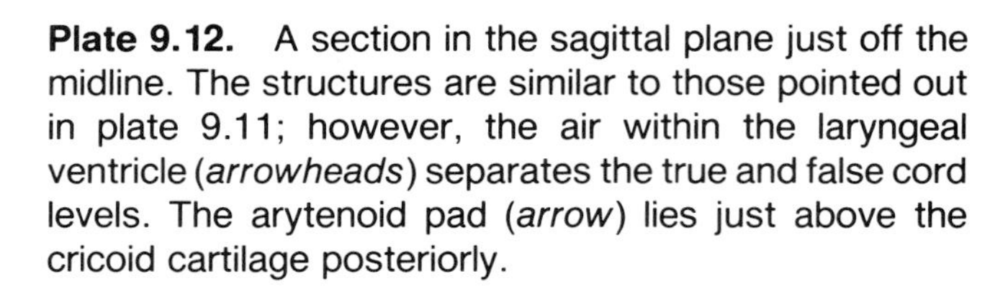

Plate 9.12. A section in the sagittal plane just off the midline. The structures are similar to those pointed out in plate 9.11; however, the air within the laryngeal ventricle (*arrowheads*) separates the true and false cord levels. The arytenoid pad (*arrow*) lies just above the cricoid cartilage posteriorly.

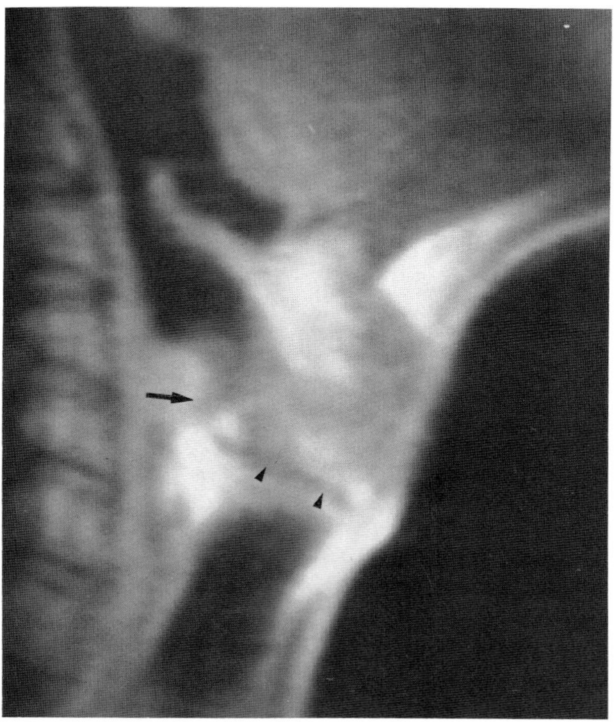

Plate 9.13. This CT section of the mid- to lower supraglottic larynx shows the position of the three coronal MRI sections in Plates 9.14–9.16. As you compare the two, note how the fat spaces seen as low attenuation on CT compare with the fat-containing tissues as seen on MRI where they will have relatively high intensity.

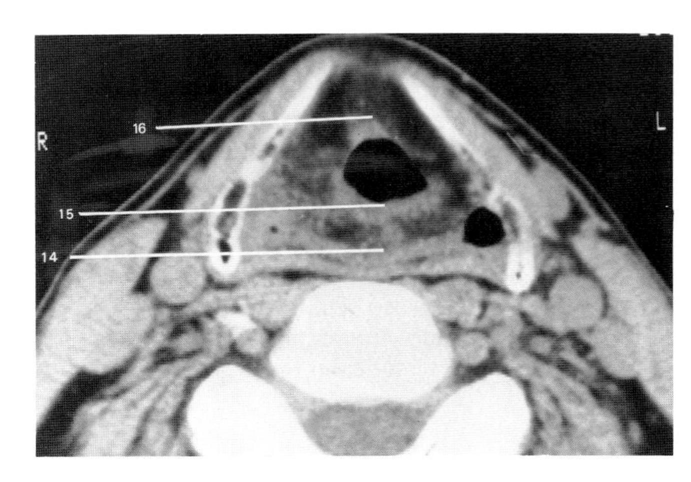

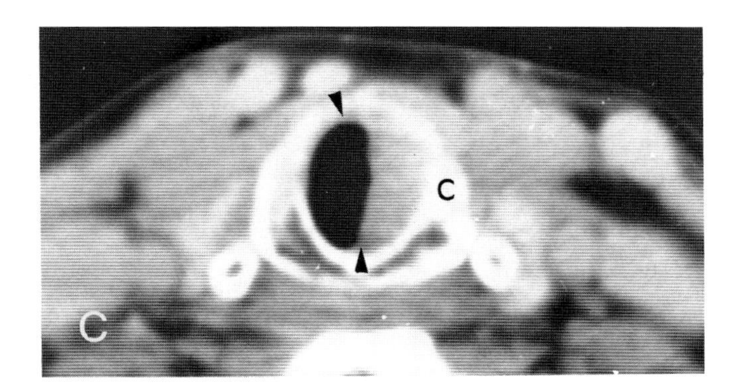

Plate 9.26*C*. In the low subglottis a circumferential but the predominantly left-sided pattern of tumor spread is seen. The tumor has clearly extended across the midline both anteriorly and posteriorly (*arrowheads*). It does not appear to be eroding the cricoid cartilage. Most of this spread was submucosal as it was not appreciated on direct laryngoscopy.

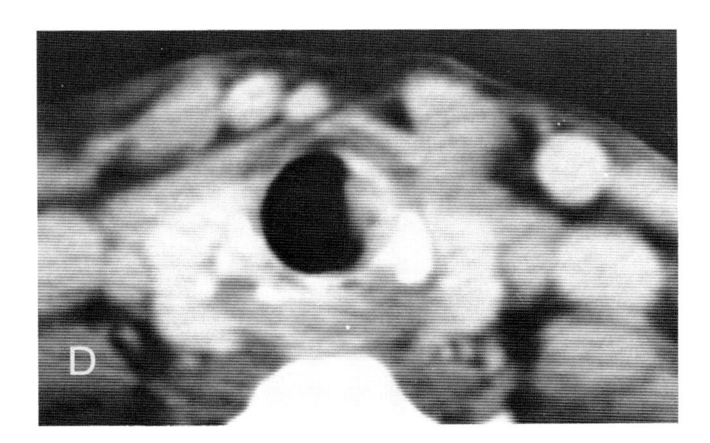

Plate 9.26*D*. Gross tumor is still visible at the lower border of the cricoid cartilage. Because of the CT findings, the patient had a total laryngectomy which confirmed the extent of the tumor as seen on CT.

The abbreviations used on Plate 9.27 are: *C*, cricoid; *A*, arytenoid; and *T*, thyroid cartilage.

Plate 9.27. This patient had numerous previous biopsies which all returned pathologic diagnoses of dysplasia and atypia but no definite evidence of cancer. Progressive hoarseness and the appearance of a lymph node led to repeat endoscopy which produced a positive biopsy.

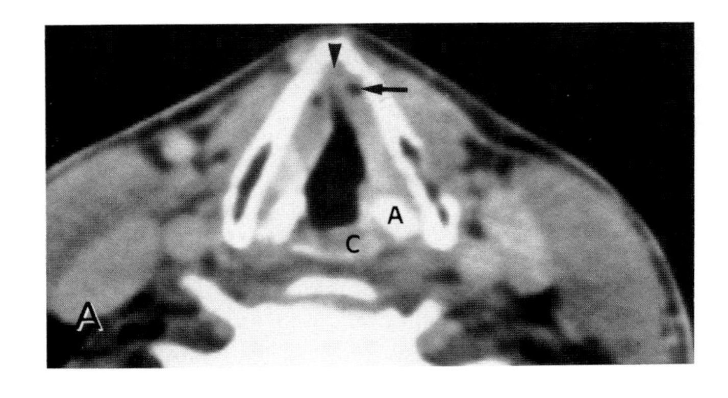

Plate 9.27*A*. The arytenoid and cricoid mark the midplane of the true vocal cords. Although the cords appear slightly asymmetric, there is little evidence of tumor at this location. The anterior commissure is slightly thickened (*arrowhead*). Incidentally, the appendices of the laryngeal ventricles are visible (*arrow*) which indicates that the scan is slightly oblique to the plane of the true vocal cords.

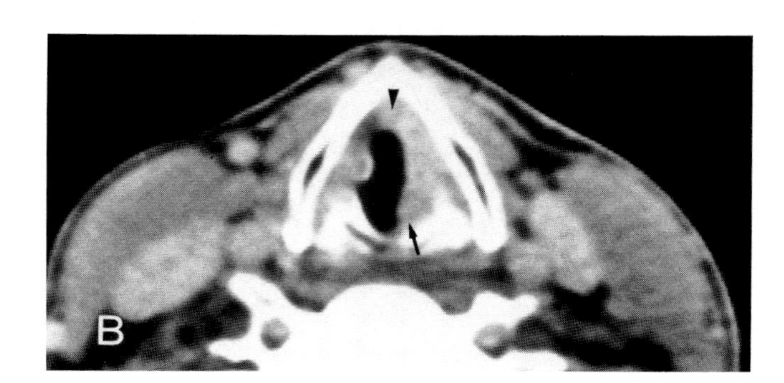

Plate 9.27*B*. A section through the undersurface of the true vocal cords shows the tumor growing across the anterior (*arrowhead*) and posterior (*arrow*) commissures.

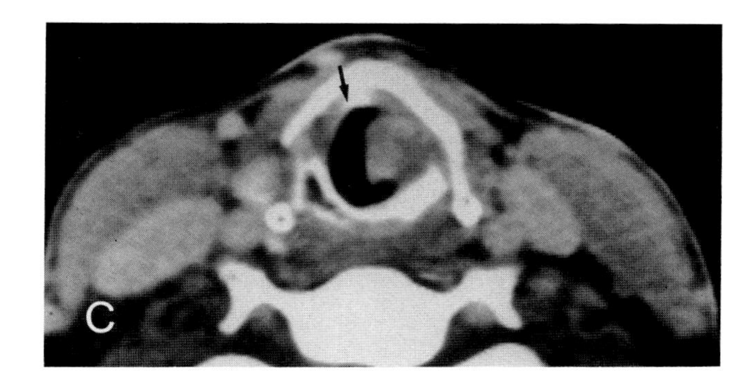

Plate 9.27*C*. In the midsubglottis a circumferential pattern of tumor growth typical of subglottic spread is present. The predominant mass is left-sided but a rind of tissue surrounds the airway (*arrow*).

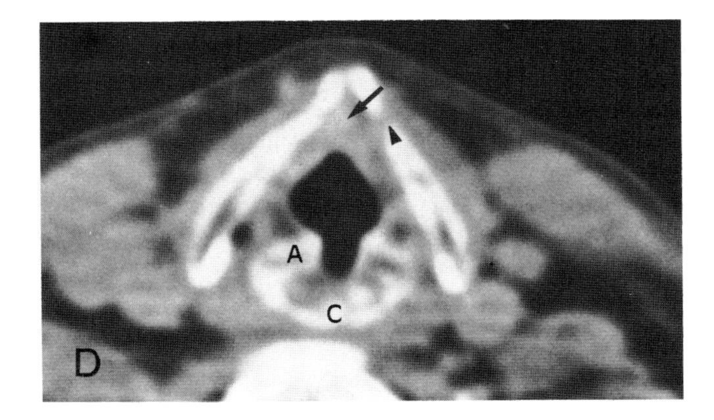

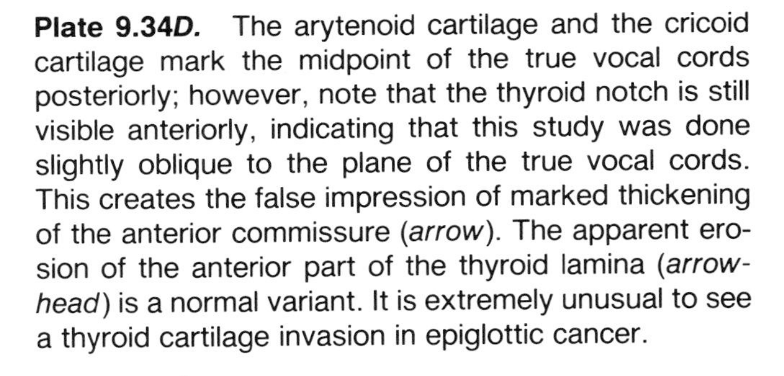

Plate 9.34*D*. The arytenoid cartilage and the cricoid cartilage mark the midpoint of the true vocal cords posteriorly; however, note that the thyroid notch is still visible anteriorly, indicating that this study was done slightly oblique to the plane of the true vocal cords. This creates the false impression of marked thickening of the anterior commissure (*arrow*). The apparent erosion of the anterior part of the thyroid lamina (*arrowhead*) is a normal variant. It is extremely unusual to see a thyroid cartilage invasion in epiglottic cancer.

Tumor was within millimeters of the anterior commissure; however, the obliquity of the sections through the midpoint of the true vocal cords gave the false impression that the anterior commissure was grossly involved. Such obliquity in sectioning should be avoided at all costs in studies where the status of the anterior commissure is in question, and should be avoided in general in CT of the larynx.

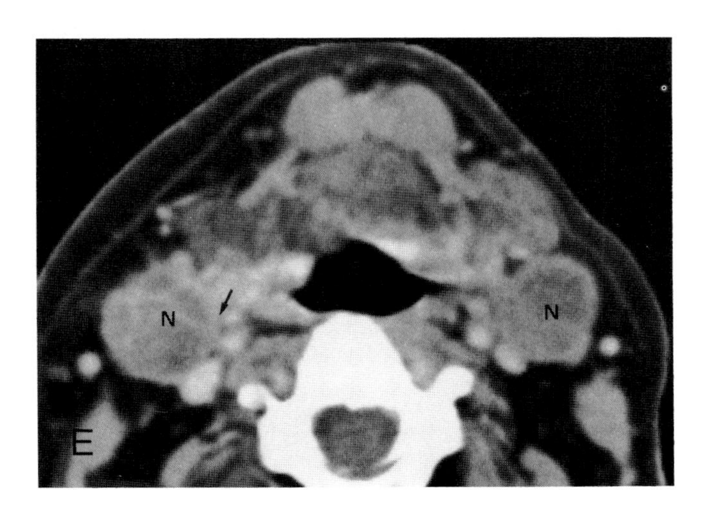

Plate 9.34*E*. This patient had large bilateral nodes which were easily palpable. The nodes were described clinically as being reduced in mobility raising some question of fixation to deep structures. The nodel masses are obvious and their borders are very well circumscribed. A clear cleavage plane can be seen between the carotid vessels and the nodal mass (*arrow*). This indicates that there is certainly no evidence of carotid fixation and really no findings to support extranodal spread of tumor. The CT findings were confirmed at surgery.

The abbreviation used on Plate 9.35 is: *A,* arytenoid.

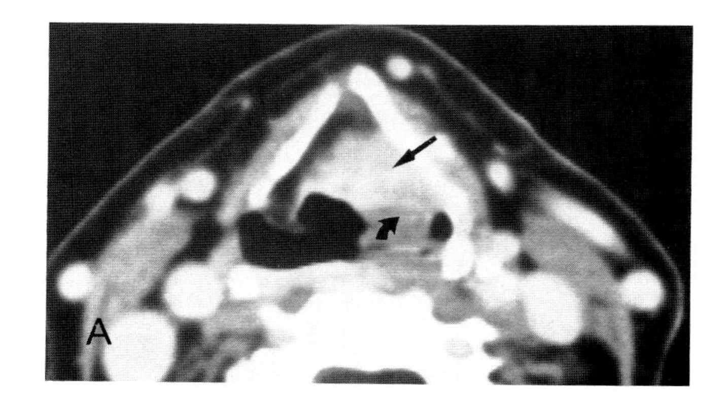

Plate 9.35. This patient had a large marginal supraglottic carcinoma. The study was done in suspended respiration to reduce the motion artifacts. Contrast the appearance of this lesion with the two prior cases which were examples of anterior supraglottic or epiglottic carcinoma.

Plate 9.35*A*. A predominantly lateral, deeply infiltrating, mass is present with much of its spread being in the paralaryngeal space (*arrow*). The tumor clearly is involving parts of the preepiglottic space and extending across the midline. Typical of marginal supraglottic lesions, the aryepiglottic fold (*curved arrow*) is quite thickened as the tumor grows down this toward the arytenoids.

Plate 9.35*B*. A section through the upper false cords shows that the entire paralaryngeal space is obliterated on the left side compared to the right. The laryngeal cartilages are intact. This scan looks a little odd because the false vocal cords have been pressed together during breath-holding.

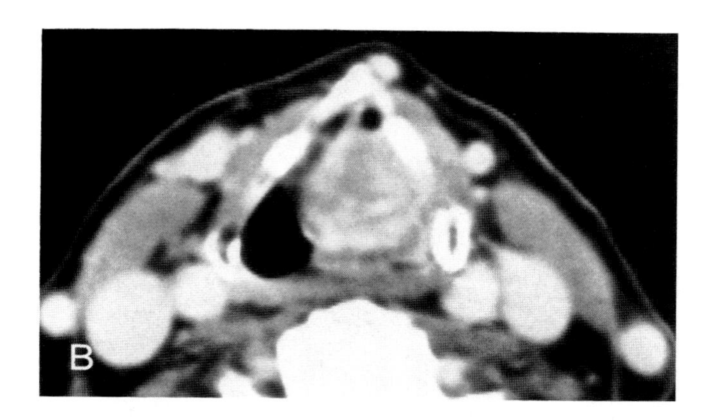

Plate 9.35*C*. At the level of the false cords the arytenoid on the side of the tumor is dispalced superiorly and perhaps somewhat anteriorly. The tumor has extended down the aryepiglottic fold to surround the aytenoid cartilage (*arrowheads*). Compare the normal fatty density within the right aryepiglottic fold with the infiltrated base of the aryepiglottic fold on the left side. This pattern of spread of marginal supraglottic lesions down to the arytenoid and sometimes cricoarytenoid joint is quite characteristic of these lesions. Carcinomas involving the medial wall of the pyriform sinus and sometimes transglottic carcinomas have a similar growth pattern. Epiglottic cancers tend to do this only when they are very large and exhibit a circumferential growth pattern.

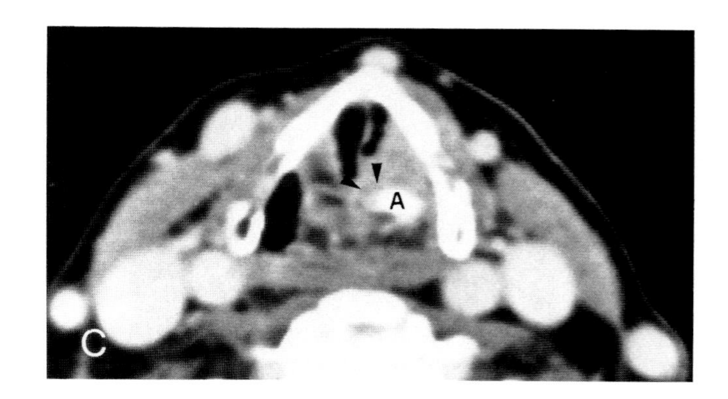

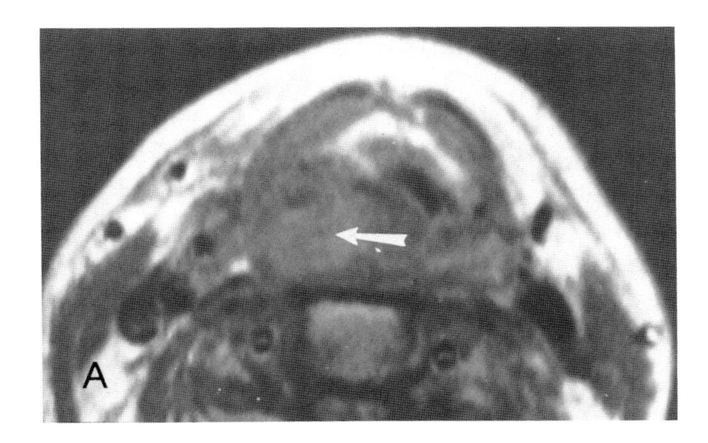

Plate 9.39. Spin echo 500/28 pulse sequence done on a patient with a large supraglottic mass.

Plate 9.39*A*. The mass has nearly the same signal intensity as surrounding muscle. In the center (*arrow*) its signal intensity is higher due to necrosis and hemorrhage or highly proteinacous fluid.

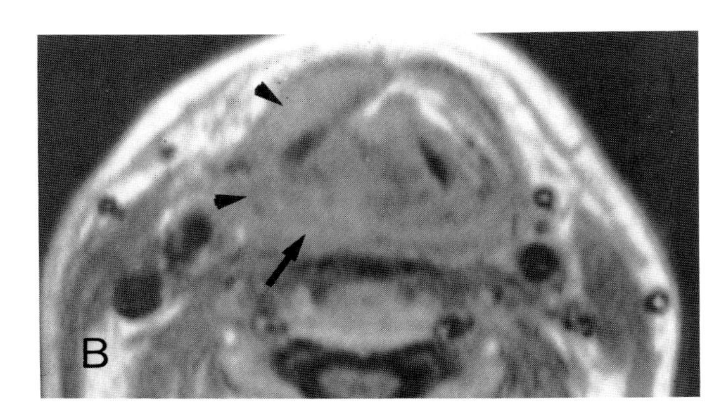

Plate 9.39*B*. The tumor spreads along the posterior pharyngeal wall (*arrow*) and beyond the laryngeal skeleton (*arrowheads*); behavior typical of pyriform sinus carcinoma (see Plates 9.40–9.46).

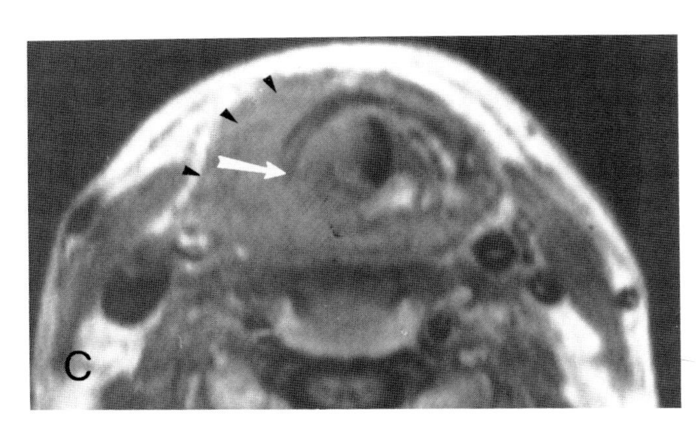

Plate 9.39*C*. The tumor has destroyed the thyroid cartilage (*arrow*) and spread into the soft tissues of the neck (*arrowheads*). The MRI clearly shows this to be a pyriform sinus and not a supraglottic carcinoma; the images are much like those of CT and MR *may* come to replace CT in the evaluation of laryngeal cancer.

Plate 9.40. This patient has a relatively small pyriform sinus carcinoma. In *A* and *B*, the sections are made during quiet respiration. In *C* and *D*, they are made during breath-holding. It is extremely important to do high quality CT if one is going to contribute to the care of patients with carcinoma of the larynx. This illustrates the importance of careful attention to detail and technique. The sections in *A* and *B* are virtually uninterpretable except for the obvious presence of a lymph node mass. In *C* and *D*, CT is of value in staging the fairly limited pyriform sinus carcinoma.

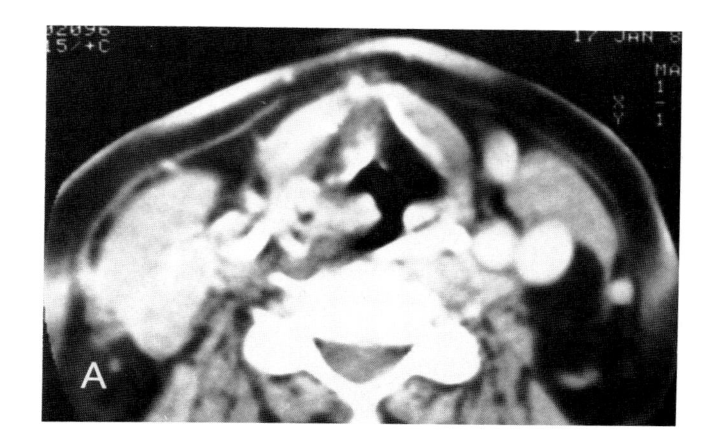

Plate 9.40*A*. This is done in the low supraglottic larynx and shows motion artifacts which render interpretation of the larynx basically impossible. There is a lymph node mass present on the right side but its margin cannot be adequately evaluated.

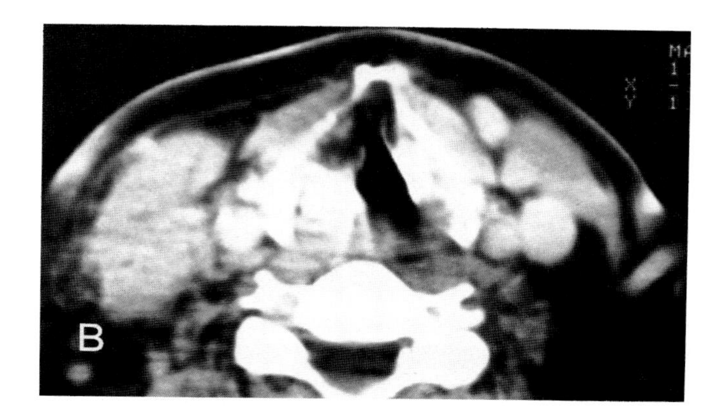

Plate 9.40*B*. This selection is made at the level of the false vocal cords and, again, the endolaryngeal structures are uninterpretable except to say that there is no gross mass present. The lymph node mass is fairly well seen on the right side and its margins are poorly circumscribed.

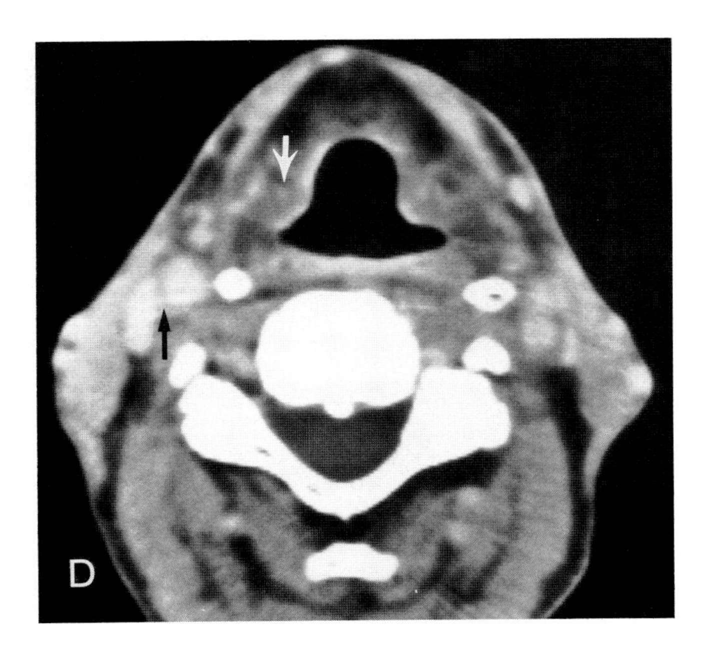

Plate 9.49*D*. A section taken in the midsubglottic larynx shows generalized increased density in the paralaryngeal space (*white arrow*) This is present bilaterally and is quite symmetric. Again the planes around the carotid sheaths (*black arrow*) are less distinct than usual although they are not completely obliterated.

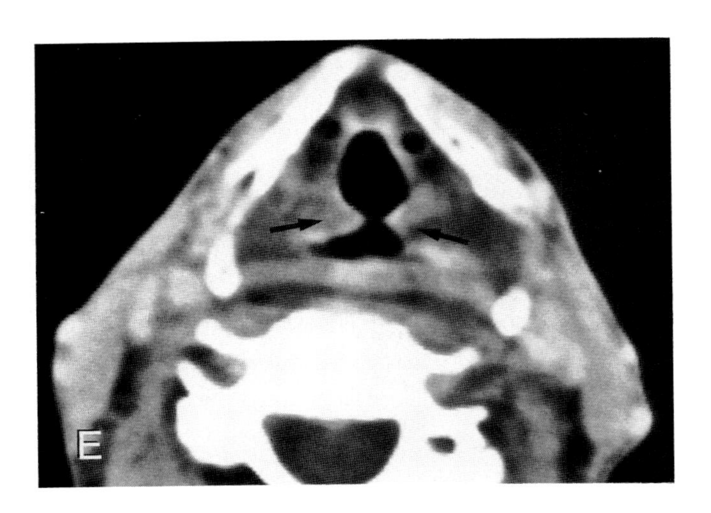

Plate 9.49*E*. A section through the false cords shows the aryepiglottic folds to be nearly opposed in the midline (*arrows*). This scan was done during quiet respiration. Again there is a generalized increased density in the normally fatty paralaryngeal spaces, especially posterolaterally. There is slight thickening of the posterior pharyngeal wall.

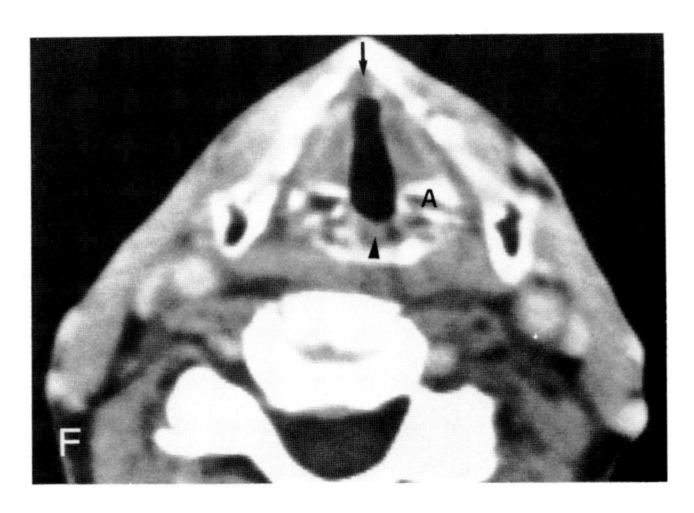

Plate 9.49*F*. At the midplane of the true cords, both cords are in a paramedian position. This is somewhat unusual for a scan done in quiet respiration. Also note the lack of any motion artifact; this is quite unusual in a scan done with a 10-sec scan time. Note that the arytenoid cartilages are rotated and displaced somewhat medially. Many of the findings at this level and above are due to the fibrosis at the anterior (*arrow*) and posterior (*arrowhead*) commissures. These are so-called anteroposterior commissure webs. The presence of these findings at the true cord explains why the aryepiglottic folds are pulled somewhat medially on other sections. The increased density of the deep tissue planes may be edema and/or fibrosis.

The abbreviations used on Plates 9.50 *A–D* are: *A*, arytenoid; *E*, epiglottic cartilage; and *N*, node.

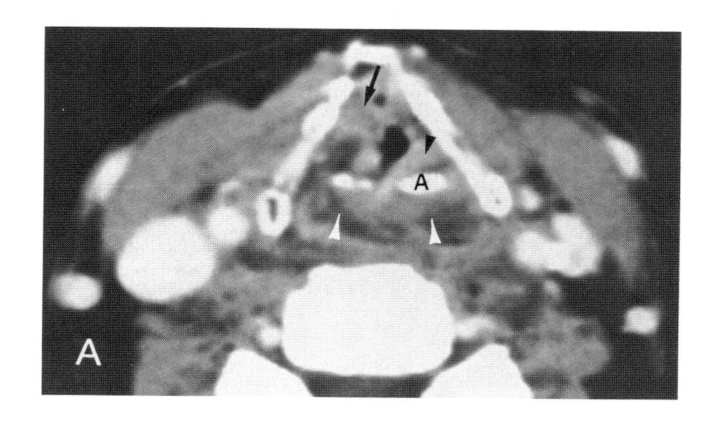

Plate 9.50. The problem of whether swelling within the larynx represents postradiation edema or recurrent tumor is often a difficult diagnostic dilemma. CT has its limitations in this regard, as does laryngoscopy and even biopsy. In fact, in this setting, it is often wise to avoid biopsy if possible because of the increased risk of infection or aggravation of the reaction within the larynx.

This patient had radiation therapy for a supraglottic carcinoma. There was persistent edema several months after therapy but no tumor could be seen. CT was done in order to help decide whether or not to biopsy the larynx.

Plate 9.50*A*. At the level of the false cords the left arytenoid was at a slightly different level than the right. There was some slight increase in density just anterior to the left arytenoid (*black arrowhead*) and some thickening of the posterior pharyngeal wall (*white arrowheads*). All of these were fairly nonspecific findings as was the slight thickening in the anterior portion of the false cords (*arrow*).

Plate 9.50*B*. In the low supraglottic larynx, just above the false cords, there was some diffuse swelling in the larynx; however, there was an area of asymmetric infiltration within the paralaryngeal space (*arrows*).

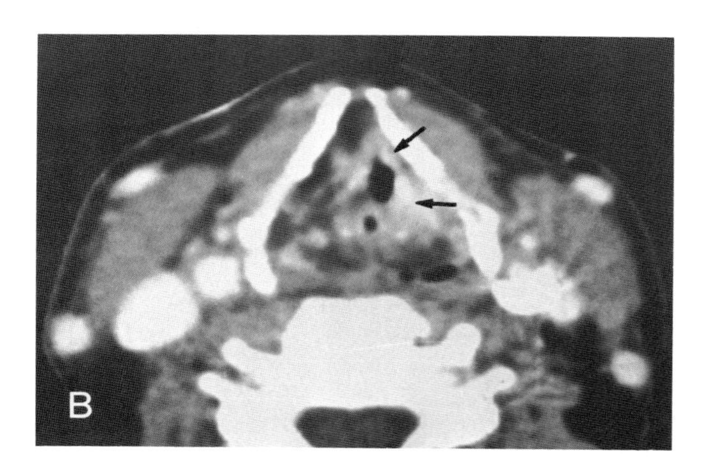

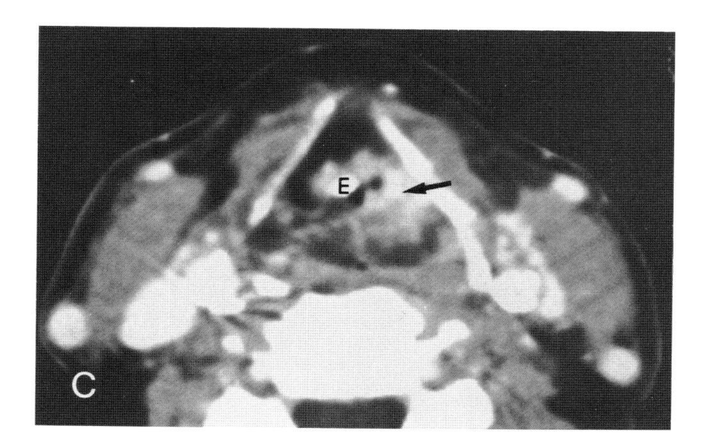

Plate 9.50*C*. In the midsupraglottic larynx, the epiglottic cartilage happened to be calcified in this patient. Again the entire larynx appeared somewhat swollen but there was a focal area of increased density within the paralaryngeal space on the left and extending to the aryepiglottic fold (*arrow*). The focal mass in the larynx was taken as strong evidence of recurrent tumor and biopsy was encouraged. The mucosa in this area appeared normal with the exception of the diffuse edema noted everywhere in the larynx. The biopsy returned tumor.

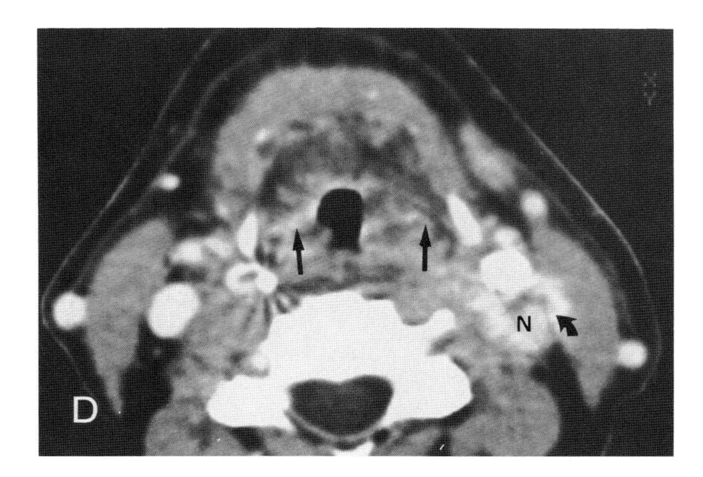

Plate 9.50*D*. In the same patient, a section made through the uppermost supraglottic larynx again showed diffuse swelling of the endolaryngeal soft tissues (*arrows*). At this higher level, no focal abnormalities were seen in the larynx. However, there was an abnormal node present in the left neck. This node was necrotic and there was evidence of indistinct periphery (*curved arrow*) suggesting extranodal spread of tumor. This node could not be felt clinically because of the postradiation induration in the soft tissues of the neck. The presence of a nodal metastases with extranodal spread of tumor was confirmed at surgery.

The abbreviations used on Plate 9.51 are: *T*, thyroid cartilage; *C*, cricoid; and *A*, arytenoid.

Plate 9.51. The diagnosis of recurrent tumor following surgery can also be a difficult diagnostic dilemma. Partial laryngeal surgery, especially vertical hemilaryngectomy, considerably alters the anatomical appearance of the larynx. CT may sometimes help in detecting recurrent tumor in these patients before it is obvious on the physical examination.

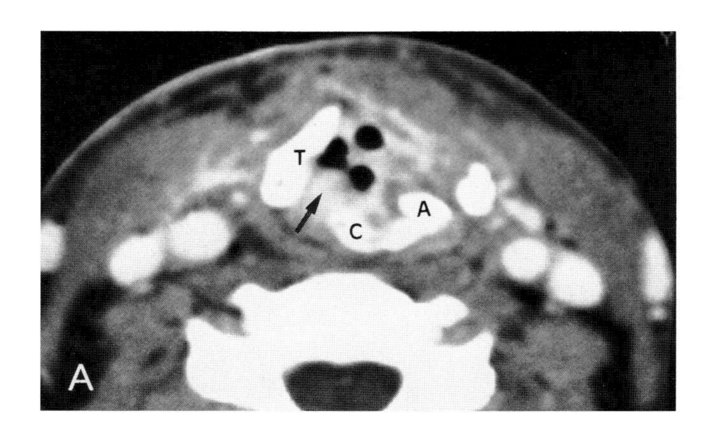

Plate 9.51*A*. This patient had a vertical hemilaryngectomy which was followed by radiation therapy for a glottic carcinoma. This therapy was chosen because the patient refused total laryngectomy. The intent was to salvage with radical surgery if this combined treatment plan failed. A remnant of the thyroid cartilage is present on the right side. The cricoid and arytenoid cartilages mark the approximate level of the reconstructed glottis. The bulging area on the right side (*arrow*) was considered somewhat suggestive of recurrent tumor.

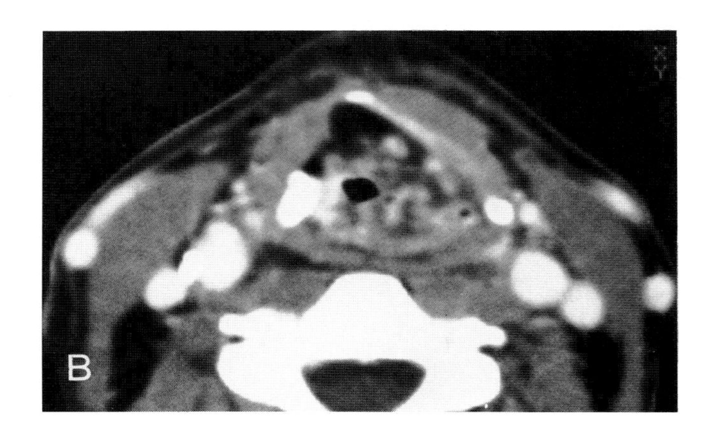

Plate 9.51*B*. A section just above this showed findings compatible with diffuse edema but no definite evidence of tumor.

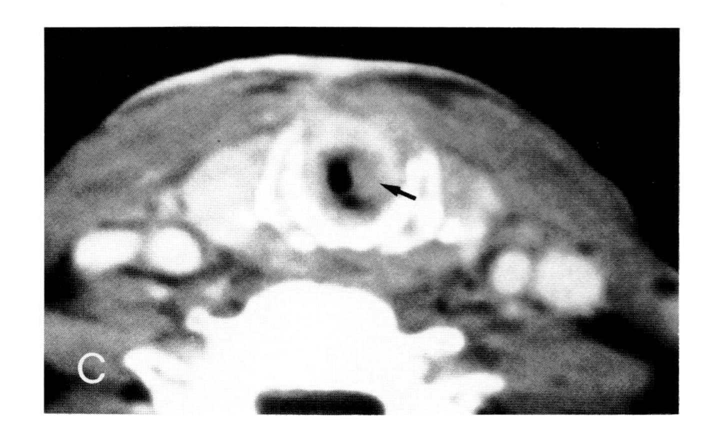

Plate 9.51*C*. The sections through the infraglottic larynx were the most disturbing of all. These showed a rind of abnormal tissue surrounding the entire subglottic region (*arrow*). This was taken as very strong evidence of persistent tumor. The patient was lost to follow-up for some time after this study after he was told that total laryngectomy might be necessary following a repeat endoscopy and biopsy. He did have a tracheostomy tube in place at this time because of persistent swelling in the larynx and difficulty breathing.

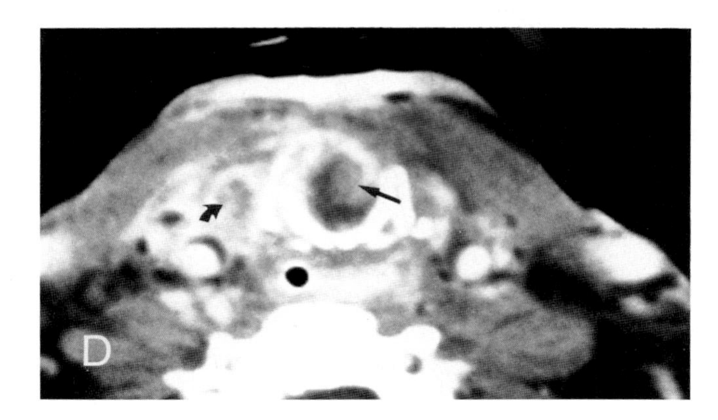

Plate 9.51*D*. Returning approximately 2–3 months later, the subglottic airway was completely occluded by what we believed to be a combination of tumor and inflammatory tissue (*arrow*). In addition, there was an obvious mass lesion adjacent to the laryngeal skeleton on the right side (*curved arrow*). This was thought to be enlarged lymph node or perhaps direct or perineural extension of tumor outside of the larynx. The patient submitted himself to surgery at this time and recurrent tumor was found within the larynx confirming the CT findings. Cords of tumor were also found within the neck.

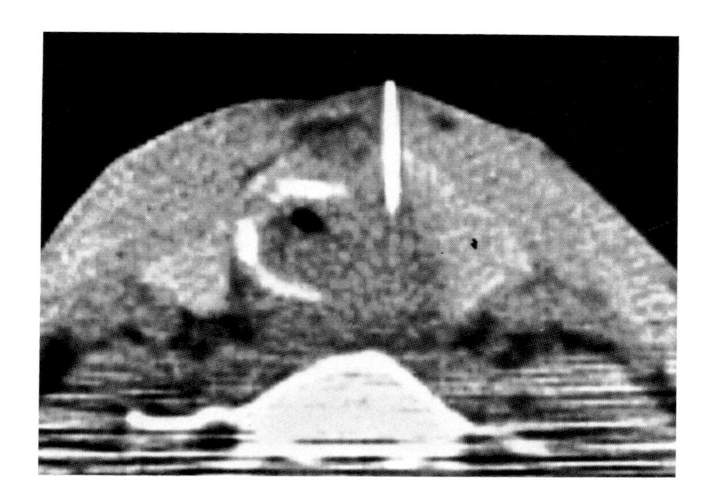

Plate 9.52. This patient had had a vertical hemilaryngectomy for carcinoma of the true vocal cords. He had some hoarseness and direct laryngoscopy showed edema but no definite evidence of mucosal tumor. CT revealed a large subglottic and exolaryngeal tissue mass. A CT-directed biopsy was able to confirm the diagnosis of recurrent tissue. Needle biopsy must, of course, be used with some caution when the mass being biopsied is immediately adjacent to the airway.

The abbreviations used on Plate 9.53 are: *T*, thyroid cartilage; *C*, cricoid; and *A*, arytenoid.

Plate 9.53. Following radiation therapy the larynx may have a varied appearance. It might be entirely normal or show some evidence of swelling or obvious recurrent tumor. Occasionally there will be gross rearrangement of the laryngeal skeleton and related soft tissues due either to perichondritis or chondronecrosis. Predominantly fibrotic changes were illustrated previously.

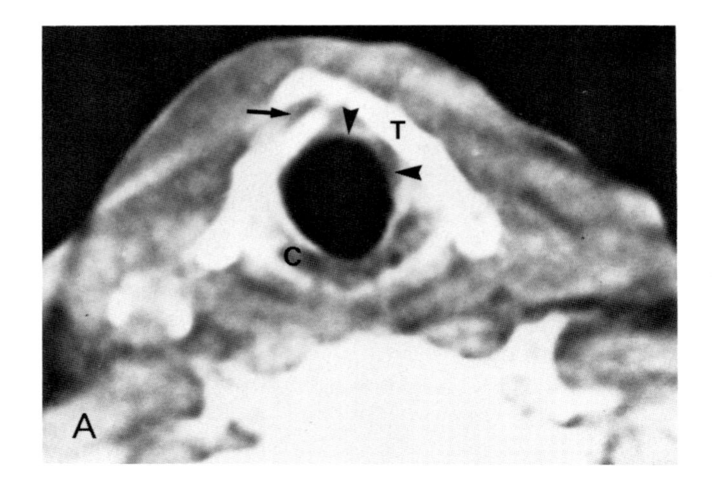

Plate 9.53*A*. At the midinfraglottic there is a paramedial fracture (*arrow*) of the thyroid cartilages. The cricoid is intact and there is some soft tissue swelling (*arrowheads*) in the infraglottic larynx.

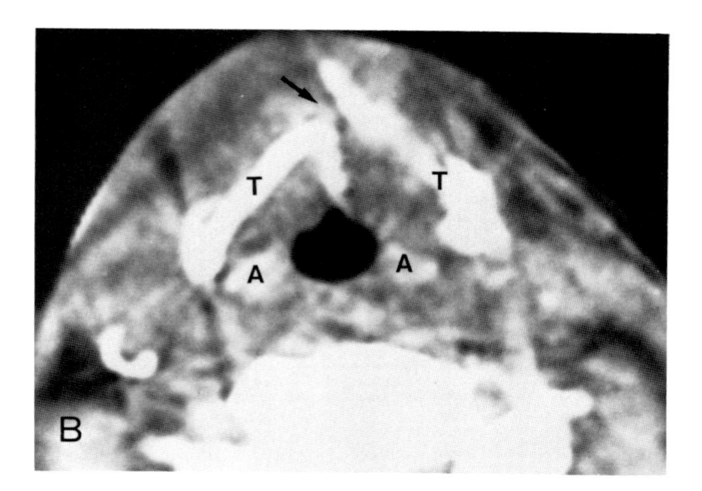

Plate 9.53*B*. At the glottic level there is marked comminution and infracturing (*arrow*) of the thyroid laminae. The soft tissues of the larynx are markedly swollen and there is no way to distinguish tumor from edema or the combination of the two.

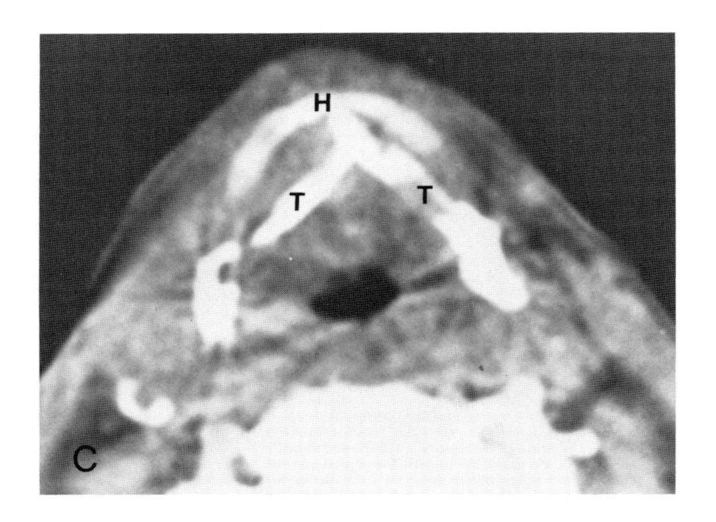

Plate 9.53*C*. In the midsupraglottis the hyoid bone and the thyroid cartilages are both seen in the same section indicating that the integrity of the thyrohyoid membrane has been interrupted and the two parts of the skeleton have collapsed together. The marked soft tissue swelling could be due to edema, inflammatory reaction and tumor, or all three of these. This patient eventually developed a draining cutaneous fistula and required total laryngectomy for the management of this severe chondronecrosis. No tumor was found in the laryngeal specimen.

Plate 9.54. This patient was treated primarily with radiation therapy for a supraglottic carcinoma. Following radiation, a radical neck section was done on the right to remove residual lymph node disease. She returned with severe and marked soft tissue swelling in the larynx and was tracheostomy dependent.

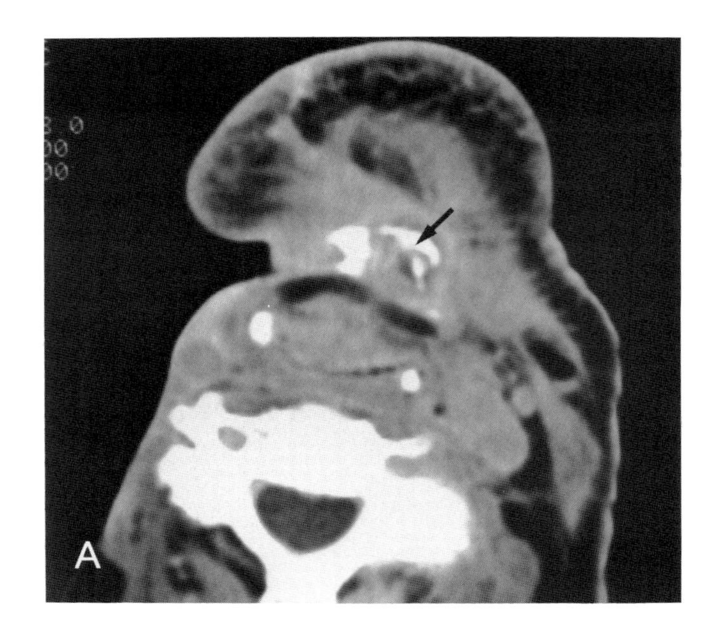

Plate 9.54*A*. The section through the high supraglottic larynx shows the effects of radical neck dissection. The extreme thickening of the skin and subcutaneous soft tissues is typical of that seen in patients who have had very high doses of radiation therapy. This thickening is inflammatory for reasons described subsequently. The hyoid bone is markedly abnormal in appearance at this level (*arrow*) and the larynx is obviously airless.

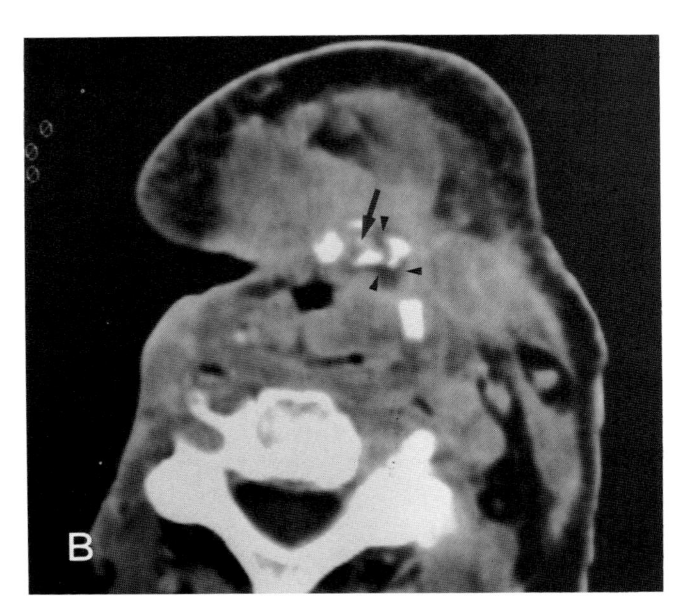

Plate 9.54*B*. A section slightly lower shows a fragmented hyoid bone (*arrow*) with a peripherally enhancing fluid collection around the fragments (*arrowheads*).

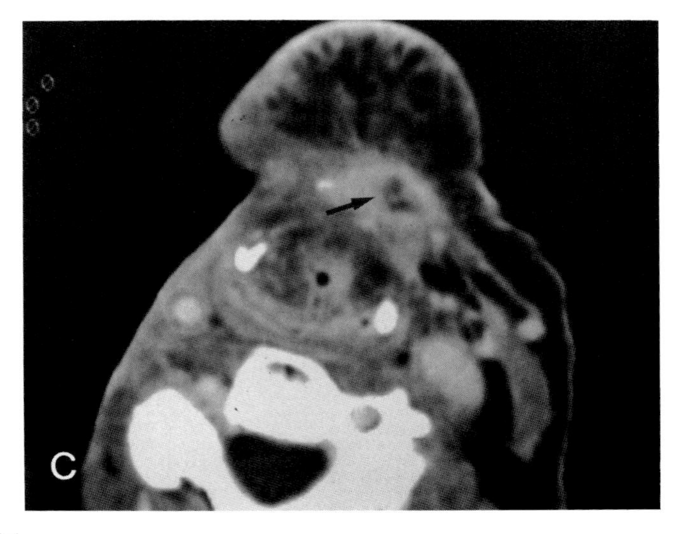

Plate 9.54*C*. A section through the midsupraglottic larynx again shows the diffuse edema within the endolarynx. There is an obvious peripherally enhancing fluid collection along the course of the thyrohyoid membrane (*arrow*).

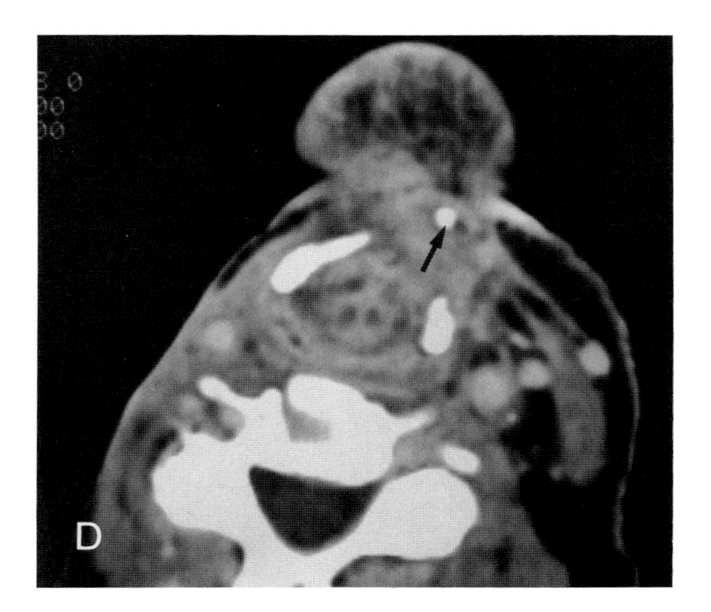

Plate 9.54*D*. At the level of the false vocal cords diffuse edema within the larynx causes it to appear airless. Only a small portion of the anterior aspect of the thyroid laminae is left (*arrow*) and this appears to be surrounded by a small, again peripherally enhancing, fluid collection. This is probably part of the thyroid lamina which is behaving like a sequestrum.

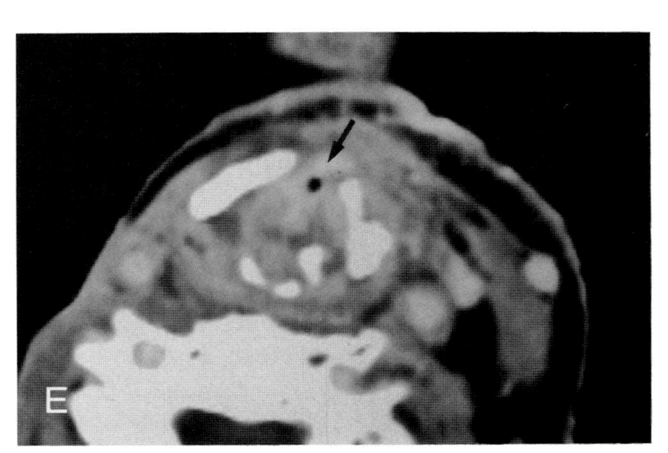

Plate 9.54*E*. At the level of the true vocal cords, again, the anterior thirds of the thyroid laminae are not visible. There is excessive staining of the soft tissues at the area of the anterior commissure (*arrow*).

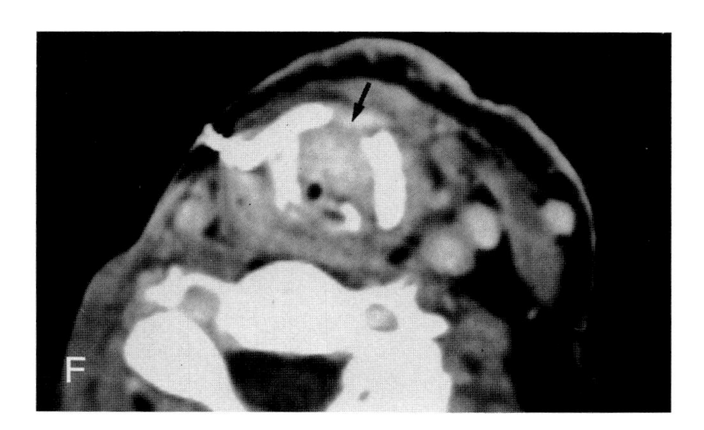

Plate 9.54*F*. This is done at the level of the undersurface of the true cords and shows excessive enhancement of all the laryngeal soft tissues. Again, this is perhaps slightly more obvious in the region just below the anterior commissure (*arrow*).

This findings in all of these sections are indicative of severe chondronecrosis predominantly anteriorly within the larynx. The hyoid bone and probably portions of the thyroid lamina are acting as sequestra in what is basically an osteomyelitis of the laryngeal skeleton. These findings were confirmed at surgery.

BENIGN LESIONS—INFLAMMATION

Plate 9.55. Section from a patient with a nonspecific granulomatous infection of the larynx.

Plate 9.55*A*. Both aryepiglottic folds are swollen (*arrows*).

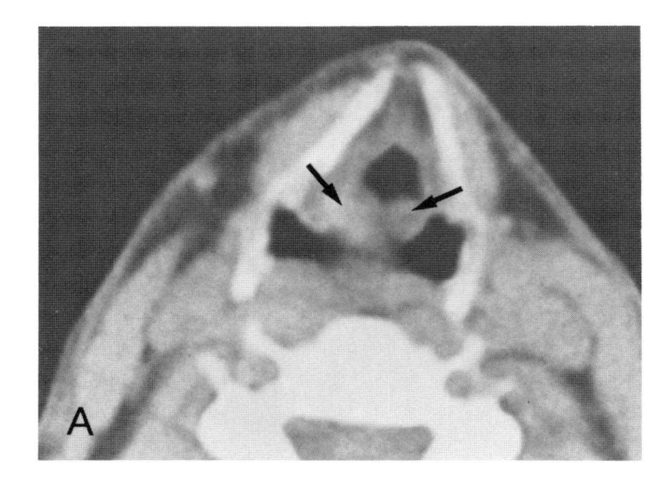

Plate 9.55*B*. An infiltrating process increases the density of the paralaryngeal space (*arrow*). Compare this with the normal paralaryngeal space (*curved arrow*). The nature of the chronic granulomatous infection in this case was "nonspecific" and was believed to be atypical for contact granulomas. The case illustrates that infiltrating processes in the deep tissue planes of the larynx cannot be differentiated by their CT appearance. The paralaryngeal space density could just as well have been due to tumor or tuberculosis.

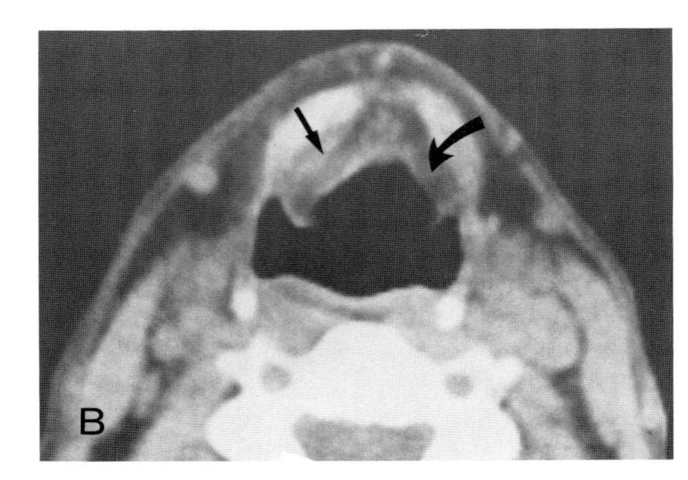

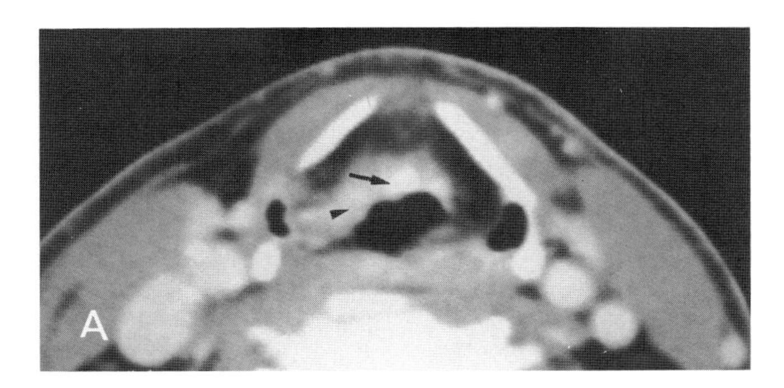

Plate 9.56. Patient with an infiltrating supraglottic mass.

Plate 9.56*A*. In the midsupraglottis there is a mass involving the epiglottis (*arrow*) and aryepiglottic fold (*arrowhead*).

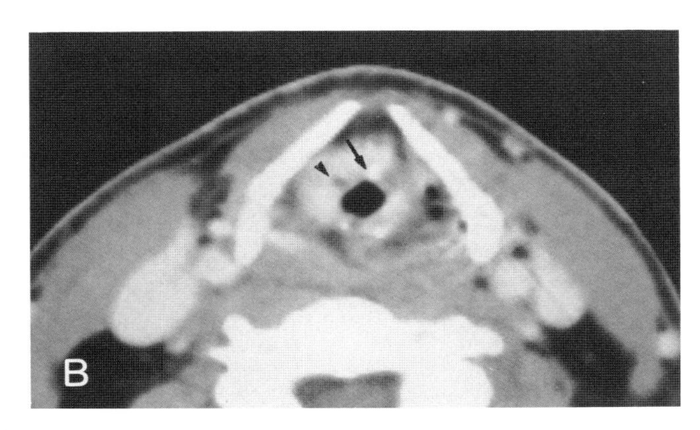

Plate 9.56*B*. At the false cord level the mass shows considerable deep spread within the paralaryngeal space (*arrowhead*) and goes on to involve the petiole of the epiglottis (*arrow*). The patient had laryngeal tuberculosis. Inflammatory and neoplastic masses may have identical appearances on CT, and tumor and inflammation often coexist.

The abbreviations used in Plates 9.57–9.58 are: *C*, cricoid cartilage; *A*, arytenoid; and *T*, thyroid cartilage.

Plate 9.57. This patient presented with signs and symptoms strongly suggestive of laryngeal cancer. A biopsy showed a large mass within the larynx and biopsy did return squamous cell carcinoma. A CT study was done to stage the extent of the lesion. At endoscopy this lesion did not appear to be ulcerative or particularly inflamed. In fact, the endoscopist believed that a good deal of the lesion might be submucosal in origin.

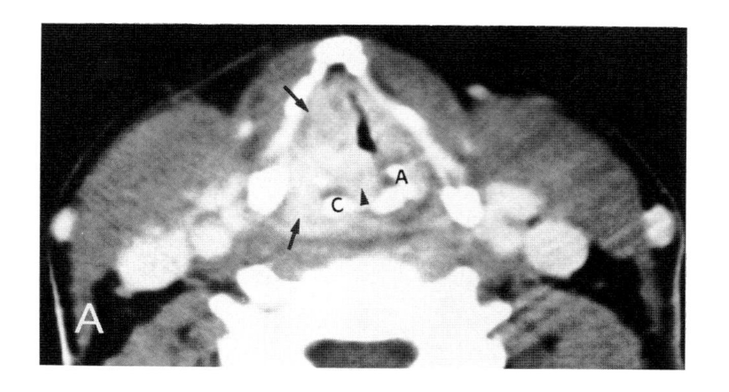

Plate 9.57*A*. The arytenoid and top of the cricoid cartilage mark the level of the true cords. There is a very large mass present which extends into the cricothyroid space and postcricoid region (*arrows*). In addition, the mass appeared to cross into the interarytenoid region (*arrowhead*). The extreme amount of enhancement suggested that if there was tumor there must at least be a large component of inflammation associated with it; however, this made little sense since no ulceration was visible clinically.

Plate 9.57*B*. This section was made on the same date through the infraglottic larynx and showed extreme soft tissue swelling and, in addition, a large defect in the lamina of the cricoid cartilage (*arrow*). The findings of marked enhancement and a large midline defect in the cricoid cartilage were considered quite unusual for a laryngeal cancer. It was suggested that this might be an inflammatory lesion or an infected tumor. For this reason the patient was placed on antibiotics and restudied approximately 10 days later. The results of this repeat study are seen in the following two plates.

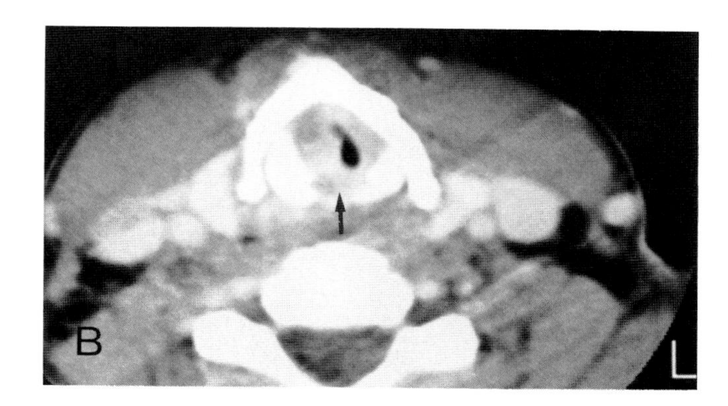

Plate 9.57*C*. The level of this section corresponds with that in plate 9.57*A*. There has been a dramatic interval decrease in the soft tissue swelling at the level of the true vocal cords. The deep tissue planes are still abnormal but are returning to normal.

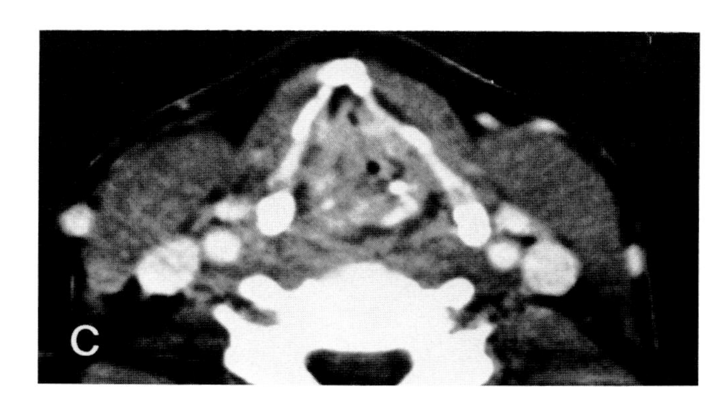

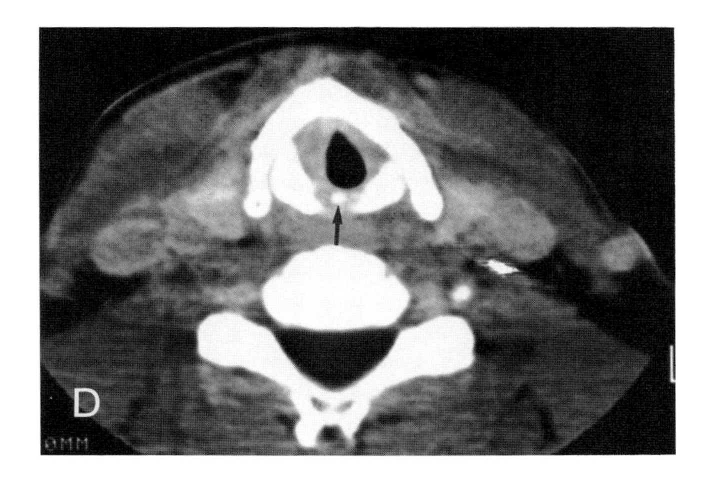

Plate 9.57*D*. While there is still persistent soft tissue swelling in the infraglottic larynx there has been marked decrease in the amount of enhancement in the two examinations. In addition, a small fragment of cricoid cartilage is visible in the defect within the posterior cricoid lamina (*arrow*). A sequestrum such as this would be highly unusual in an area of cartilage destruction due to tumor. A repeat endoscopy showed that this patient, in fact, had laryngeal syphillis coincident with a rather limited tumor of the true vocal cords.

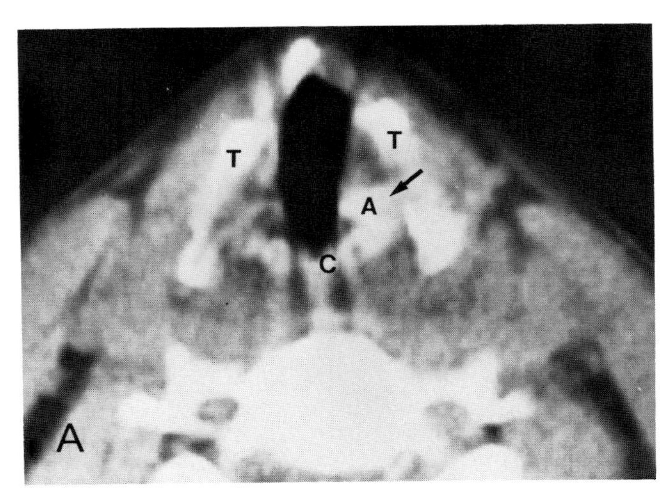

Plate 9.58. A patient with rheumatoid arthritis who presented with hoarseness and presumed rheumatoid involvement of the cricoarytenoid joint.

Plate 9.58*A*. At the glottic level one arytenoid is more medial and anterior than the other (*arrow*). There is also a lack of motion artifact on the involved side on this scan with an EMI CT 5005 unit.

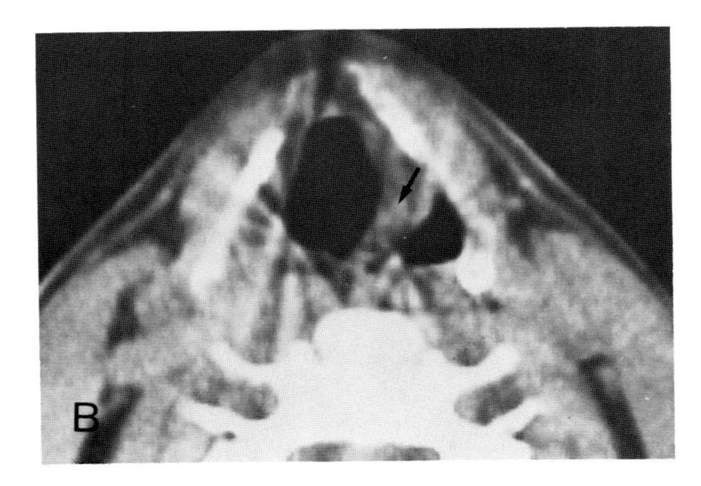

Plate 9.58*B*. While the involved cord was in the paramedian position in the prior section, the ipsilateral aryepiglottic fold is also medially displaced (*arrow*) and the pyriform sinus is opened. This finding has been observed in a number of patients with rheumatoid arthritis and hoarseness and may represent subluxation of the cricoarytenoid joint. The findings are somewhat nonspecific and cannot be used to predict the activity of the disease. The findings are identical to those seen in a paralyzed or paretic vocal cord. Subtle erosion of the arytenoid or cricoid cartilage may be visible on more modern scanners.

The abbreviations used in Plates 9.59–9.60 are: *PyS*, pyriform sinuses; *L*, laryngocele; and *V*, vallecula.

BENIGN MASSES

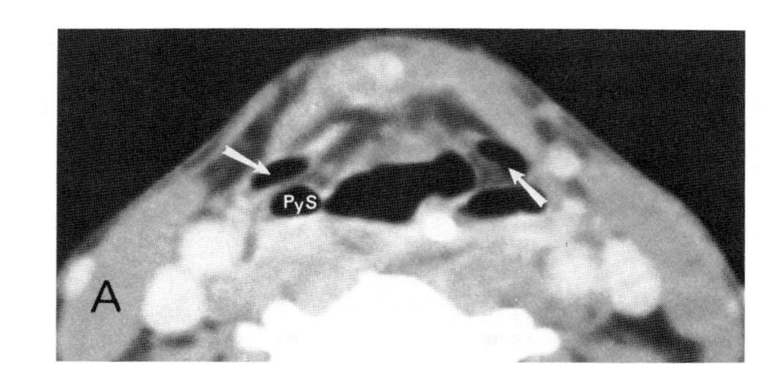

Plate 9.59*A*. This section through the midsupraglottic larynx demonstrates bilateral air-filled laryngoceles (*arrows*). This represents accumulations of air within enlarged appendices of the laryngeal ventricles. These typically lie more anterior and medial than the pyriform sinuses but as one proceeds more superiorly in the larynx they come to lie in a slightly more lateral position.

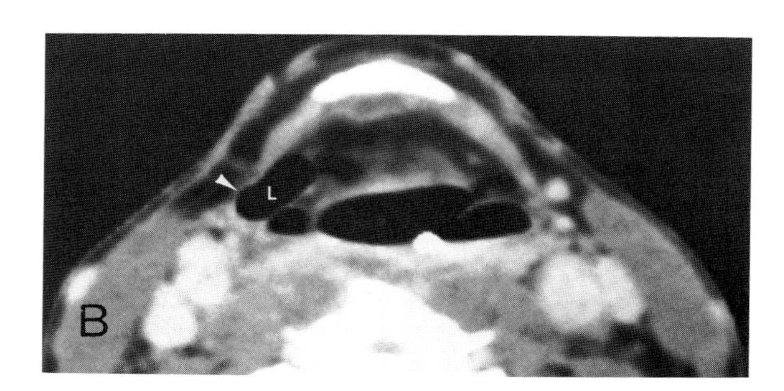

Plate 9.59*B*. A section through the upper supraglottic larynx shows the internal laryngocele as it produces a slight bulge in the thyrohyoid membrane (*arrowhead*).

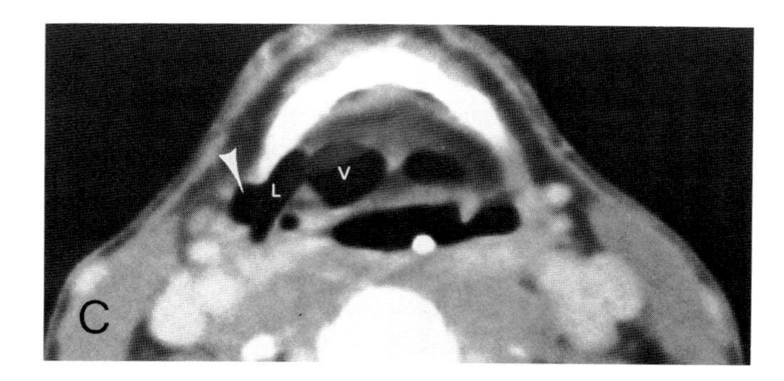

Plate 9.59*C*. The laryngocele has actually extended (*arrowhead*) slightly beyond the thyrohyoid membrane at this level and has technically become both an internal and external laryngocele. On the previous sections, only, the internal component was demonstrated. Although this appears to be in continuity with the vallecula a thin wall of mucosa could be seen between the two at wider window settings. If this truly was continuous with the vallecula it could be mistaken for a pharyngocele, but then it would not have its lower component within the paralaryngeal space which identifies it as a laryngocele.

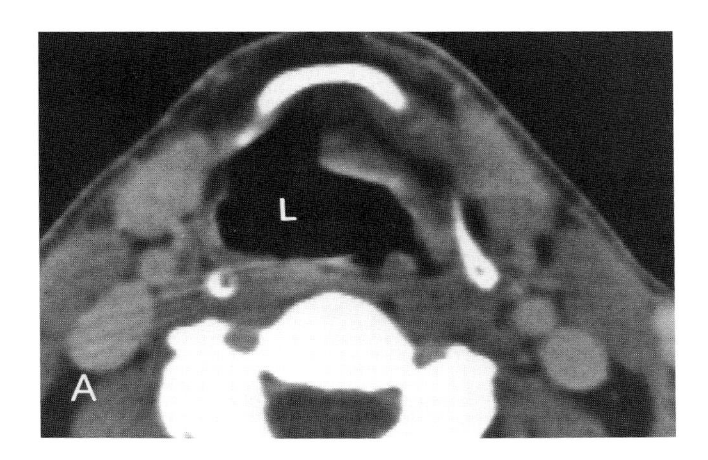

Plate 9.60. This man presented with acute and severe airway obstruction requiring immediate tracheostomy. Following tracheostomy he had a CT study of his larynx.

Plate 9.60*A*. The images seen here and below caused some confusion at initial interpretation. The residents doing the study wondered why a man with such a large airway would have any trouble breathing. In fact, this section through the supraglottic larynx shows that the entire airway is obliterated by a large, air filled laryngocele.

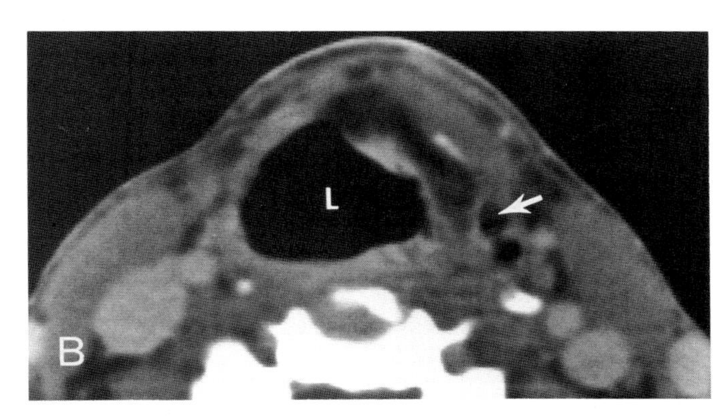

Plate 9.60*B*. The true nature of the mass becomes a little more apparent at the midsupraglottic level where the laryngocele is somewhat more eccentric and the contralateral anatomy appears more normal. Some air within the opposite pyriform sinus is visible (*arrow*).

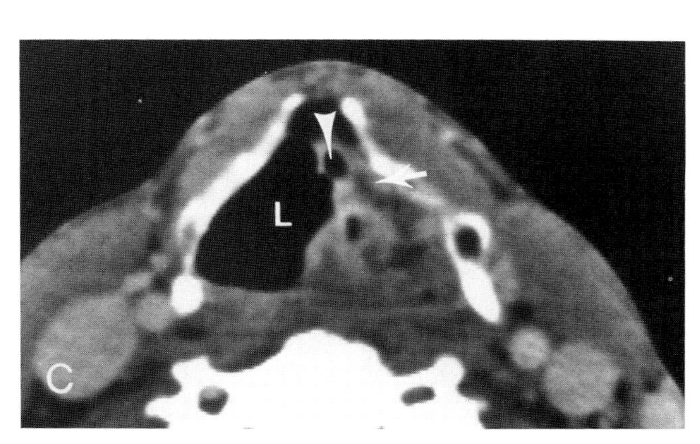

Plate 9.60*C*. At the false cord level a small amount of airway is visible (*arrowhead*) between the laryngocele and the compressed left false vocal cord (*arrow*). Careful sectioning was done through the level of the true and false cords to exclude an obstructing tumor. No tumor was found. During the operation to decompress the laryngocele, the laryngeal ventricle, true cord and false cord were carefully explored and there was no evidence of an obstructing lesion.

The abbreviation used in Plate 9.61 is: *SC*, superior cornu.

Plate 9.61. Cysts of the saccule (appendix) of the laryngeal ventricle may present as either laryngeal or neck masses. When they present as a neck mass they are sometimes difficult to distinguish from branchial cleft cysts unless a careful examination of the endolarynx is performed. CT is capable of making a definitive diagnosis of this unusual lesion and can shown its endo- and exolaryngeal extent.

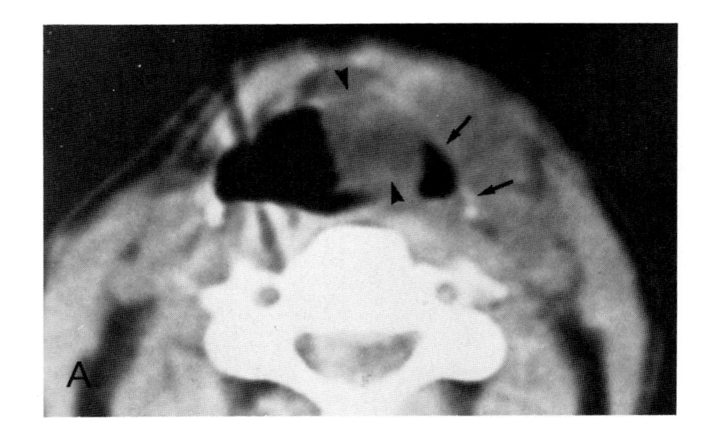

Plate 9.61*A*. This patient presented with hoarseness, a neck mass, and a submucosal mass which was producing marked distortion of the larynx. A submucosal tumor was suspected clinically. The thyroid lamina (*arrows*) was poorly mineralized in this elderly female. An obvious fluid-density mass obliterates the parapharyngeal space and widens the aryepiglottic fold (*arrowheads*).

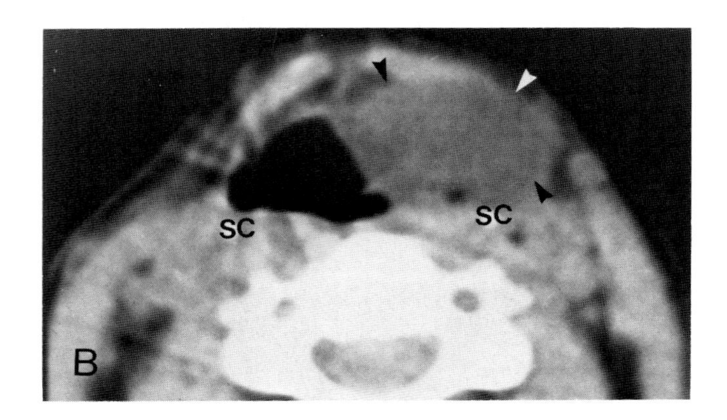

Plate 9.61*B*. A very well circumscribed mass extends from the paralaryngeal space through the region of the thyrohyoid membrane and into the neck (*arrowheads*). Note that the mass is of higher attenuation than fat and lower than surrounding musculature; the actual CT numbers indicated that it was fluid density. The scan is virtually diagnostic of a cyst of the laryngeal saccule (i.e. a fluid-filled internal and external laryngocele).

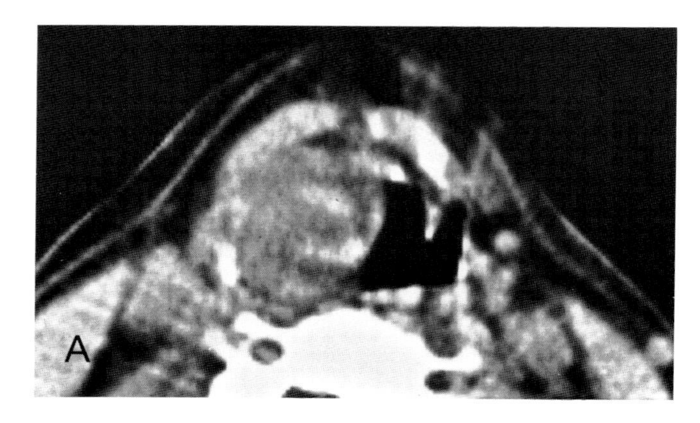

Plate 9.62 *A* and *B*. These are sections obtained in two different patients in the midsupraglottic larynx. Both patients presented with hoarseness and at laryngoscopy had submucosal masses within the larynx. The mass in part *A* was an internal laryngocele while that in part *B* was a paraganglioma. The paraganglioma has a slightly increased attenuation on this noncontrast study. This is a nonspecific finding since highly proteinaceous fluid can have a similar appearance. A contrast-enhanced scan would have shown considerable difference between the two lesions; however, this was not commonly done in the earlier days of neck CT.

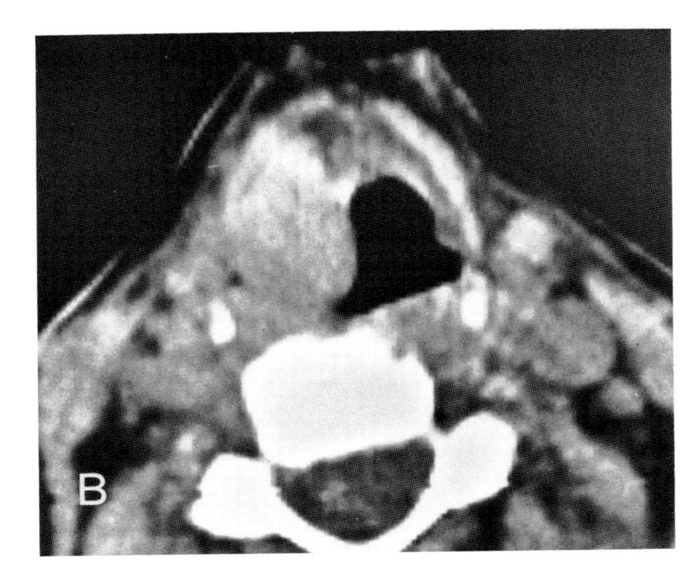

Plate 9.62*B*.

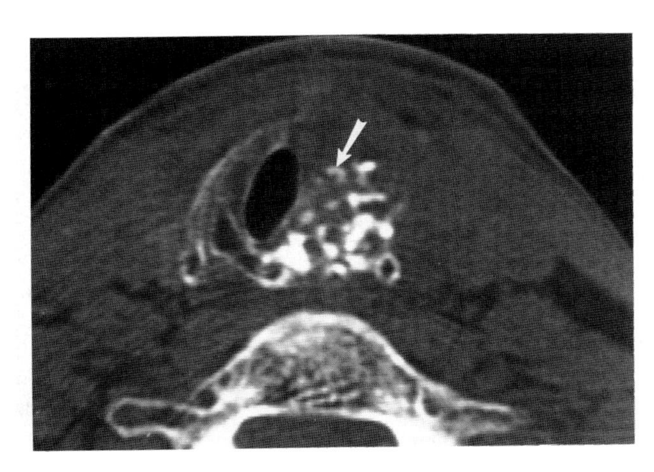

Plate 9.63. Benign tumors of the laryngeal skeleton are unusual and almost always turn out to be enchondromas. This section done through the level of the undersurface of the true cords shows the typical calcified matrix of an enchondroma. The ring-like pattern (*arrow*) is unmistakable. (Case courtesy of Radiology Group, Phoenix Baptist Hospital, Phoenix, Ariz.)

The abbreviations used in Plates 9.64–9.65 are: *C*, cricoid; *T*, thyroid laminae; *O*, osteophyte; *A*, arytenoid; and *IC*, inferior cornua of the thyroid cartilage.

BLUNT TRAUMA

Plate 9.64. This series of scans of an elderly man who was strangled; the force was applied in a direct anterior to posterior manner which pulled the larynx against the patient's large cervical spine osteophytes.

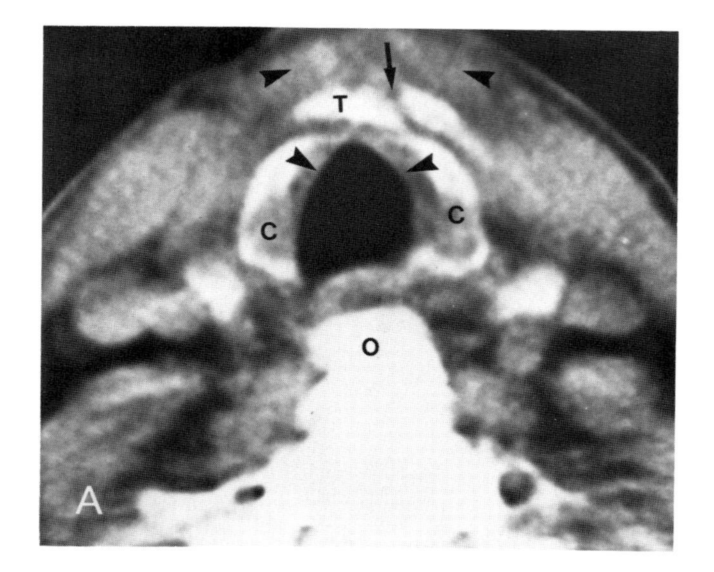

Plate 9.64*A*. The fractured (*arrow*) thyroid laminae are anterior to the lowest portion of the cricoid ring indicating collapse or rupture of the cricothyroid membrane. Soft tissue swelling is seen both in the infraglottic region and the soft tissues of the neck (*arrowheads*). A large osteophyte is visible posteriorly.

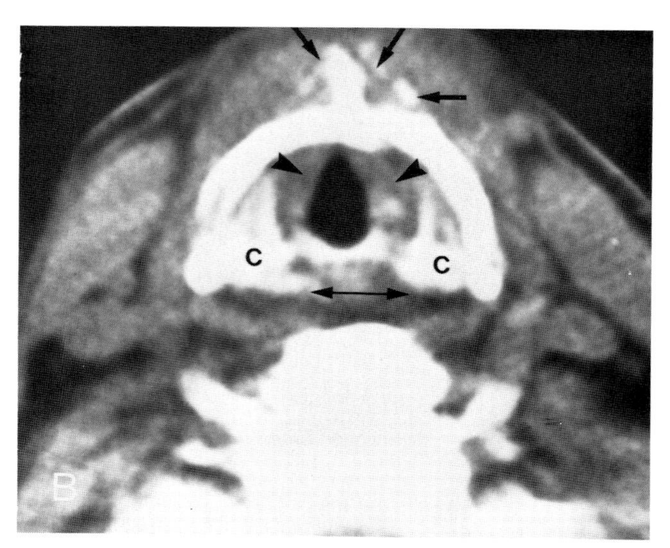

Plate 9.64*B*. Extensive cricothyroid complex injuries are seen at the midinfraglottic level. Fragments of the comminuted thyroid cartilage lie anteriorly (*arrows*). The cricoid cartilage is split posteriorly and its major fragments are distracted (*double-headed arrow*). Smaller cricoid cartilage fragments border the airway posteriorly and associated edema and hemorrhage narrow the airway (*arrowheads*).

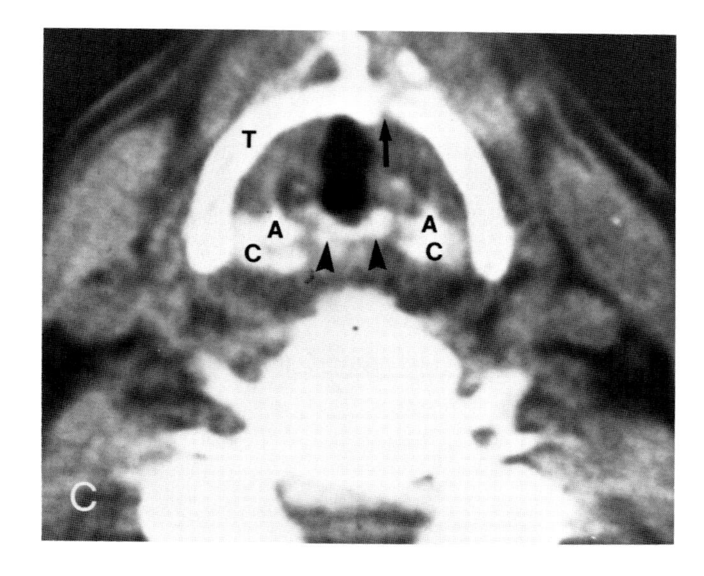

Plate 9.64*C*. There is a right paramedian fracture of the thyroid cartilage (*arrow*). The fragments of the cricoid cartilage are widely separated posteriorly and each arytenoid has stayed with its respective half of the cricoid cartilage. Small fragments of the cricoid border the airway posteriorly (*arrowheads*). There is a marked amount of swelling in the true vocal cords due to hemorrhage and edema. The patient remained aphonic many months after the injury due in large part to the wide separation of the arytenoids secondary to the fracture and dislocation of the cricoid cartilage. Even if the arytenoids were fully mobile they would not be able to oppose one another in the midline.

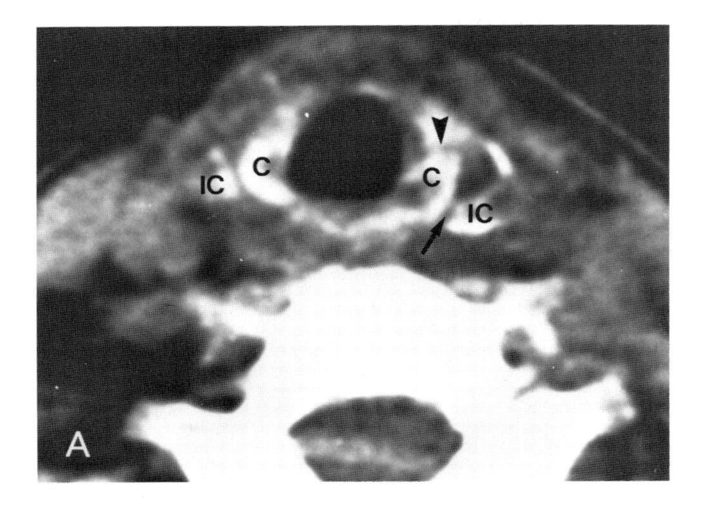

Plate 9.65. An acute injury sustained in a car accident when the left side of the patient's larynx was pressed against a dashboard. The force in this case is applied obliquely from the left side.

Plate 9.65*A*. There is a slightly displaced fracture (*arrowhead*) of the cricoid arch. Related soft tissue swelling borders the airway. The cricothyroid joint is grossly dislocated (*arrow*).

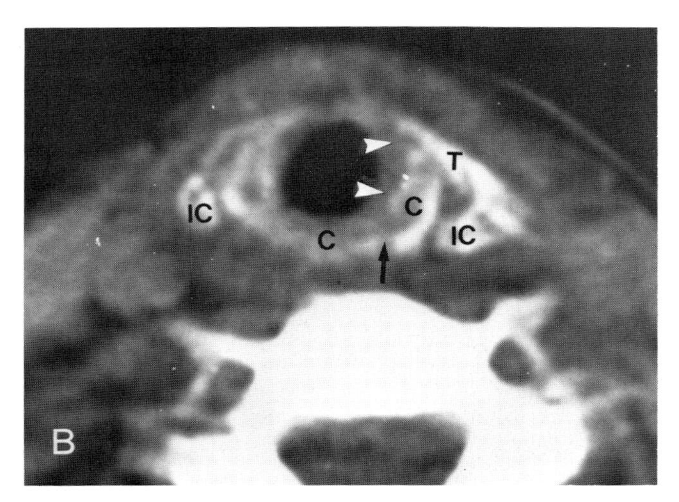

Plate 9.65*B*. A second fracture of the cricoid ring is seen more posteriorly at the junction of the lamina and arch (*arrow*). The fracture fragment is laterally displaced and there is related soft tissue swelling surrounding the airway (*arrowheads*). The gross disruption of the cricothyroid joint is again obvious and might herald damage to the recurrent laryngeal nerve which runs very close to this joint.

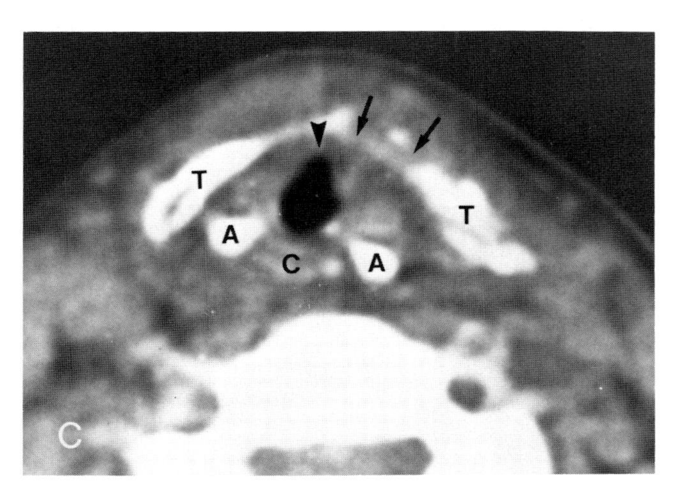

Plate 9.65*C*. At the midplane of the true vocal cords, the thyroid laminae appeared to be spread apart and the left thyroid lamina shows evidence of comminuted but minimally displaced fractures (*arrows*). A left paralaryngeal hematoma separates the thyroid lamina from the cricoarytenoid complex. The anterior commissure is swollen (*arrowhead*). The relationship between the cricoid and arytenoid cartilages are normal indicating that there is no cricoarytenoid joint dislocation.

The abbreviations used in Plate 9.66 are: *IC*, inferior cornua of the thyroid cartilage; *C*, cricoid cartilage; *T*, thyroid cartilage; *A*, arytenoid; *PES*, preepiglottic space; and *E*, epiglottis.

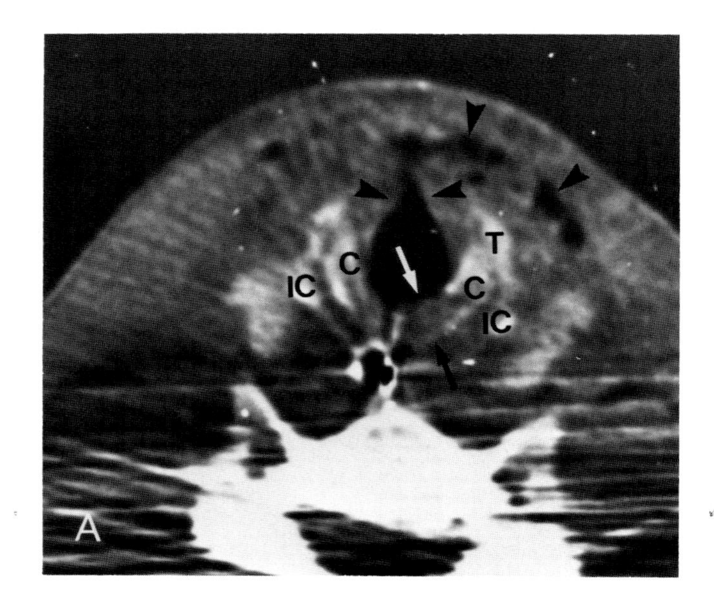

Plate 9.66. Pre- and postoperative views of a young man who sustained a clothesline type injury while riding a motorcycle. Parts *A–C* are of suboptimal quality because they were done under circumstances of extreme emergency and due to artifacts produced by the nasogastric tube. Physical examination of the patient's larynx was nearly impossible because the supraglottic part of the airway was, according to the surgeon, "so macerated and swollen I couldn't see anything below that level."

Plate 9.66*A*. In the mid- to low infraglottic level there was a marked amount of subcutaneous emphysema and soft tissue swelling (*arrowheads*). Posteriorly, a focal area of soft tissue swelling suggests a fracture (*arrows*) in the cricoid cartilage. There was no definite evidence of a displaced fracture of the cricoid ring.

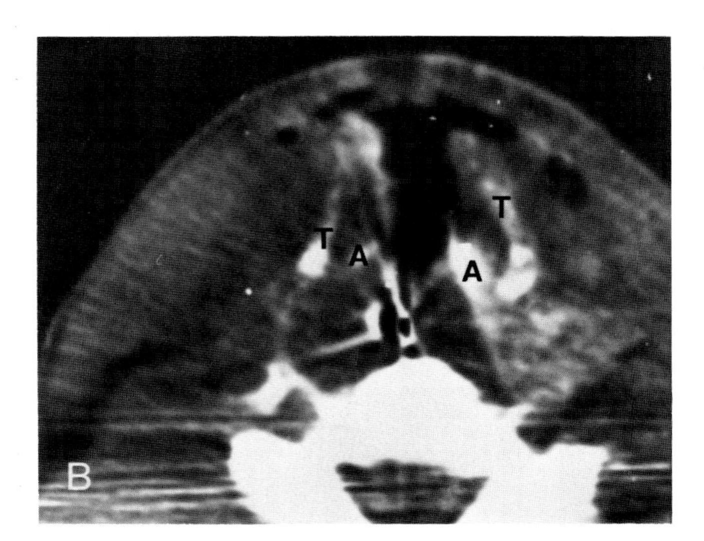

Plate 9.66*B*. The larynx has been split open anteriorly and is in continuity with the soft tissues of the anterior neck. The thyroid laminae are widely separated. The arytenoids are only grossly visualized posteriorly.

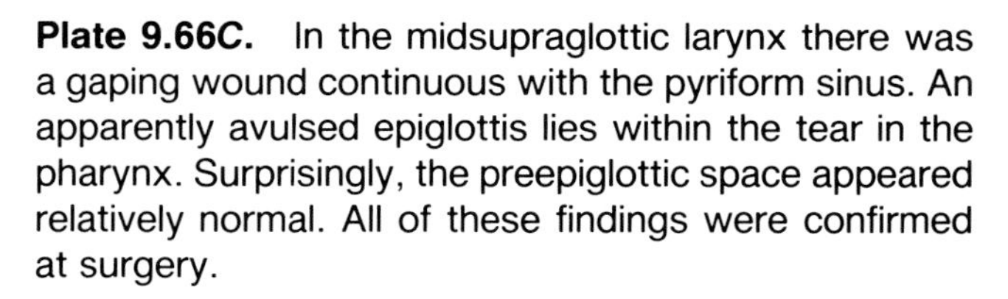

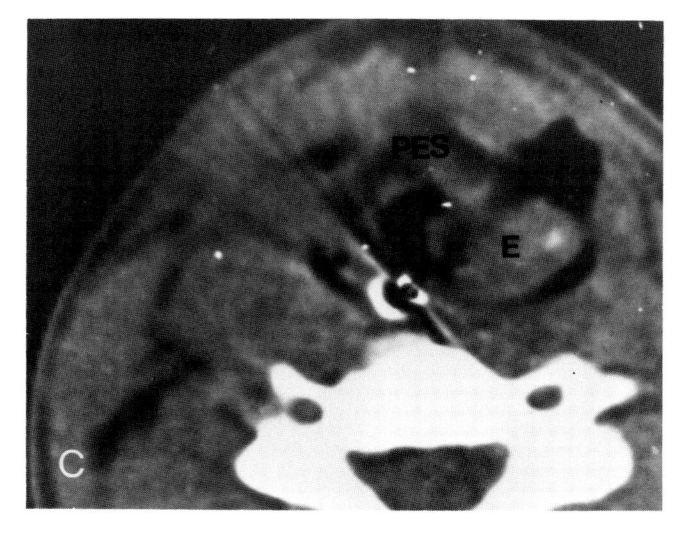

Plate 9.66*C*. In the midsupraglottic larynx there was a gaping wound continuous with the pyriform sinus. An apparently avulsed epiglottis lies within the tear in the pharynx. Surprisingly, the preepiglottic space appeared relatively normal. All of these findings were confirmed at surgery.

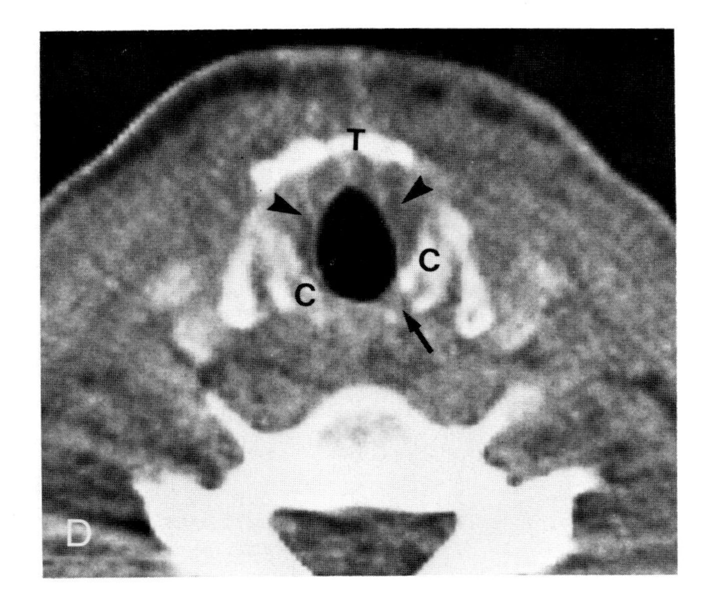

Plates 9.66 *D* and *E* are postoperative views of this patient.

Plate 9.66*D*. This section is made of approximately the same level as that seen in 9.66*A*. The thyroid laminae now bridge the airway anteriorly as is normally the case. The amount of thyroid visible this low in the larynx indicates that some of the length of the crico-thyroid membrane has been lost either due to the injury or scarring. The fracture (*arrow*) of the cricoid lamina is minimally displaced and was confirmed at surgery. There is a fairly marked amount of abnormal soft tissue surrounding the airway (*arrowheads*). This represents organization of the previous edema and hemorrhage. Despite this persistent swelling, glottic function was excellent.

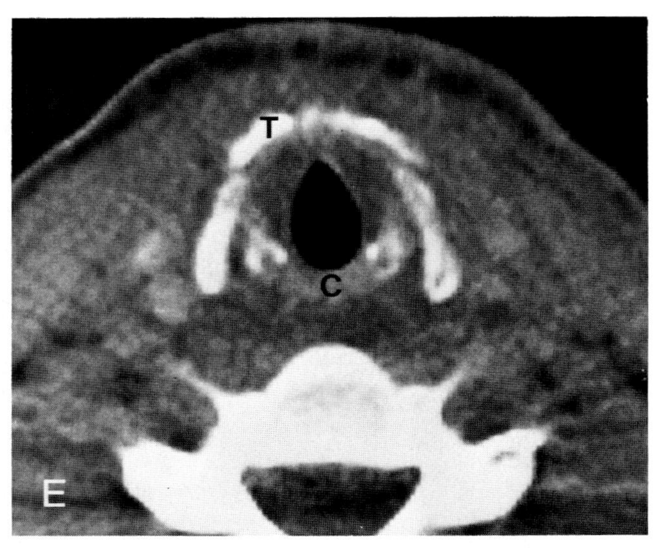

Plate 9.66*E*. A section made through the undersurface of the true cords and shows the cricoid cartilage is intact and the reduced fragments of the thyroid cartilage are in anatomic alignment. The surgeon can certainly judge postreduction alignment of the larynx much the same way the orthopaedic surgeon might do with conventional films in long bone fractures.

The abbreviations used in Plate 9.67 are: *C*, cricoid cartilage and *IC*, inferior cornua of the thyroid cartilage.

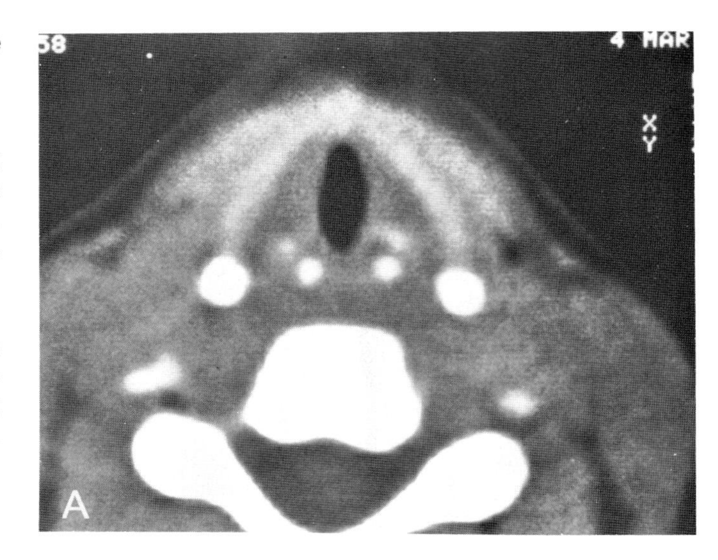

Plate 9.67. This young female was in an auto accident and sustained life-threatening injuries. The injury to the larynx went unnoticed until several weeks posttrauma. (This case courtesy of J. Knickerbocker, M.D., Vancouver, British Columbia.)

Plate 9.67*A*. A section at the glottic level shows a relatively normal appearance of all of the laryngel skeleton although there is a distinct lack of motion artifact which is uncommon on scans of 10 sec duration. This suggests that some scarring might be present.

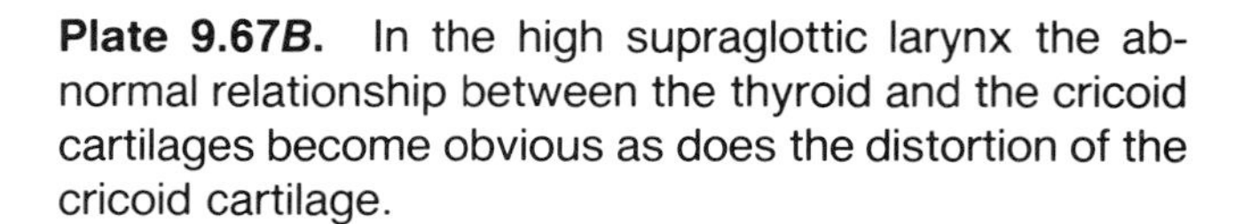

Plate 9.67*B*. In the high supraglottic larynx the abnormal relationship between the thyroid and the cricoid cartilages become obvious as does the distortion of the cricoid cartilage.

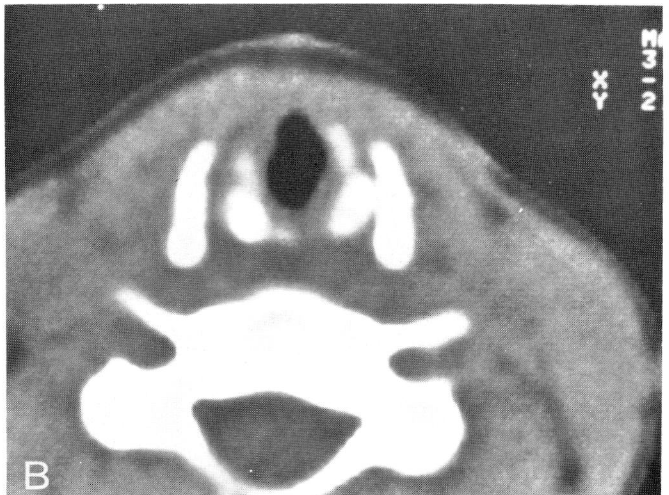

Plate 9.67*C*. In the midsupraglottic larynx there is evidence of marked disruption of the posterior aspect of the cricoid cartilage and a definitely abnormal relationship between the thyroid cartilage and the cricoid.

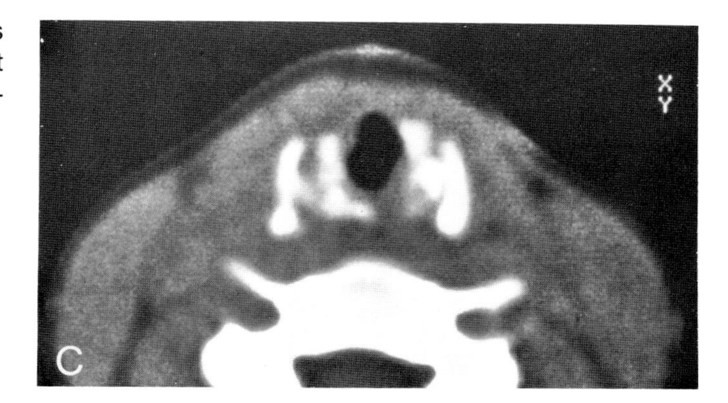

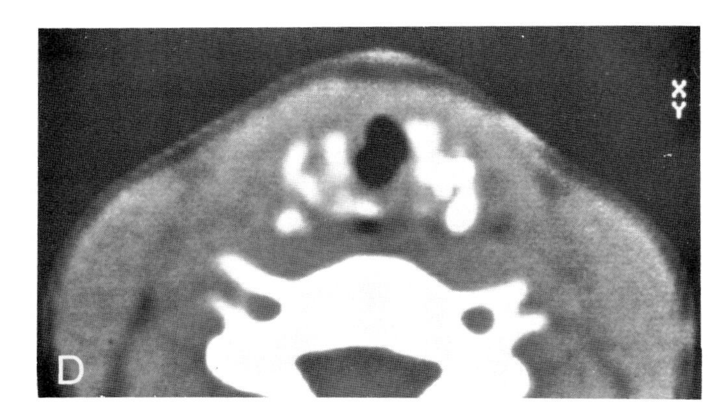

Plate 9.67*D*. Further evidence of a comminuted fracture of the cricoid cartilage. Several fragments are present and there is obviously related soft tissue swelling in the infraglottic larynx on this and other sections.

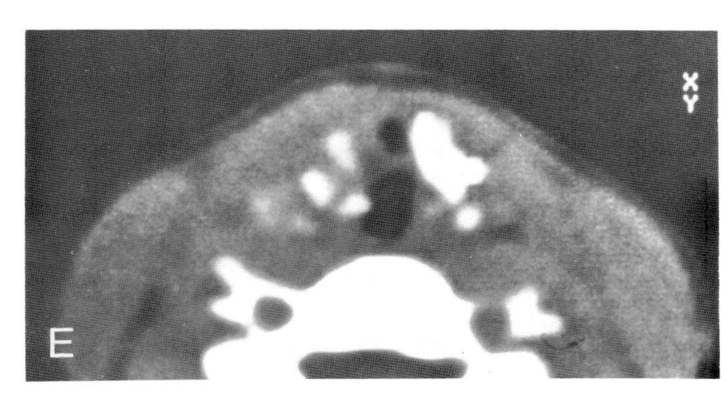

Plate 9.67*E*. It is perhaps this section which emphasizes the severe amount of injury to the cricothyroid complex. The cricoid cartilage is markedly rotated. The relationship between inferior cornua of the thyroid cartilage and the cricoid is completely disrupted. Such findings are usually indicative of severe cricothyroid complex injury and often cricothyroid or cricotracheal separation.

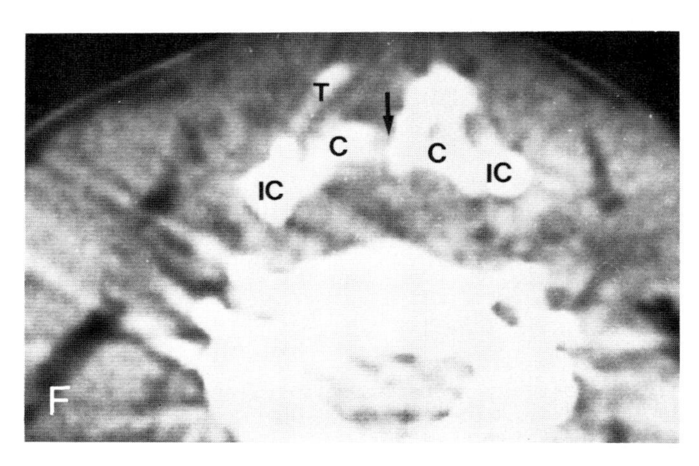

Plate 9.67*F*. A different patient who also sustained severe injury to the cricothyroid complex. This patient had clear-cut evidence of laryngotracheal separation. The cricoid cartilage is split posteriorly (*arrow*) on this and several other sections. The cricoid is so severely rotated that the posterior lamina is oriented into the axial plane and what is visualized in this section is basically that part of the cricoid cartilage which usually faces the cervical spine. The relationships between the inferior cornua of the thyroid cartilage and the cricoid cartilage are grossly abnormal indicating disruption of the cricothyroid joints. Laryngotracheal separation is a very serious injury that most often results in death. In this case the patient survived but the prognosis for good glottic function in the long-term is guarded.

The abbreviations used on Plates 9.68–9.71 are: *T*, thyroid laminae; *A*, arytenoid, *HC*, cornua of the hyoid; and *HB*, body of the hyoid.

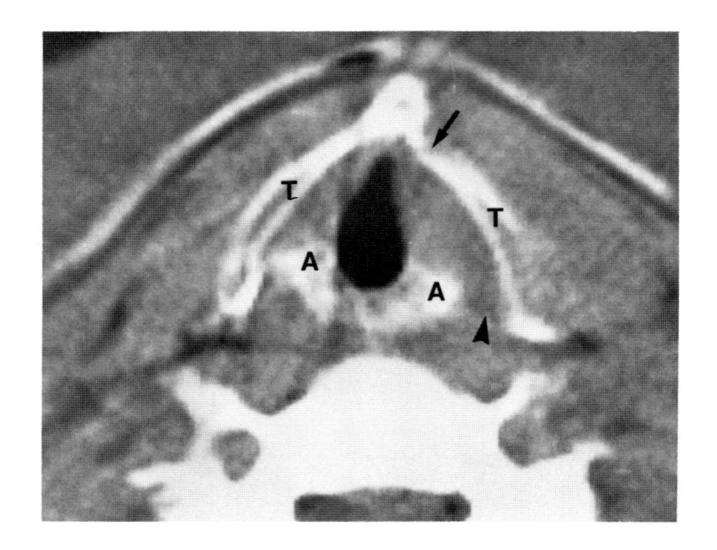

Plate 9.68. A nondisplaced paramedian fracture of the thyroid lamina is present in this patient who was involved in a fist fight. The fracture (*arrow*) and paraglottic hematoma account for the widening of the cricothyroid space (*arrowhead*). There is also slight swelling at the anterior commissure. There is no evidence of arytenoid dislocation. This patient recovered full glottic function with conservative management. In the past it has been a surgical dictum that all laryngeal fractures should be explored. This probably is not true in nondisplaced laryngeal fractures which are discovered during CT. We have experienced several cases where such fractures were seen and all patients recovered full glottic function without surgical intervention.

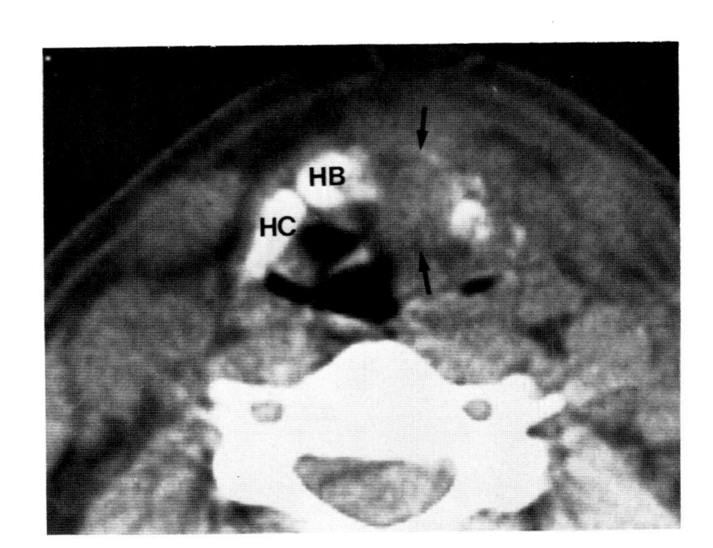

Plate 9.69. A fracture of the hyoid bone was sustained in a dashboard-vs.-neck injury. The *arrows* indicate the area of fracture and related soft tissue swelling. Great care must be taken in evaluating hyoid bone as it often appears somewhat asymmetric on various sections. The body of the hyoid and the cornua of the hyoid normally make up separate parts of this bone and the junctions of these should not be overinterpreted as fractures. Isolated fractures of the hyoid bone are of little clinical importance.

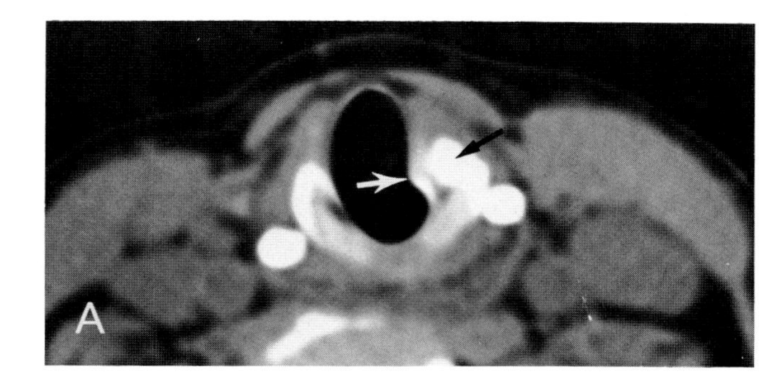

Plate 9.70. This patient presented with hoarseness and submucosal mass noted in physical examination. CT revealed the cause of both findings.

Plate 9.70*A*. A section through the mid- to low infraglottis shows a healed fracture of the cricoid cartilage (*black arrow*) which is causing distortion of the mucosal surface (*white arrow*).

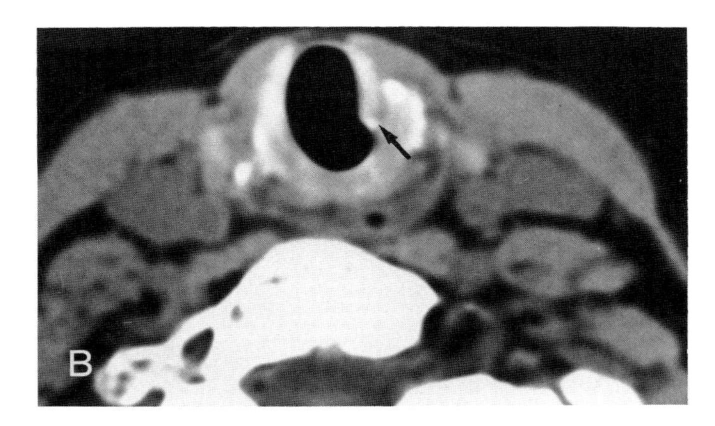

Plate 9.70*B*. A section in the low subglottis shows the displaced fracture of the cricoid arch (*arrow*). Despite this fairly dramatic fracture the patient had good glottic function.

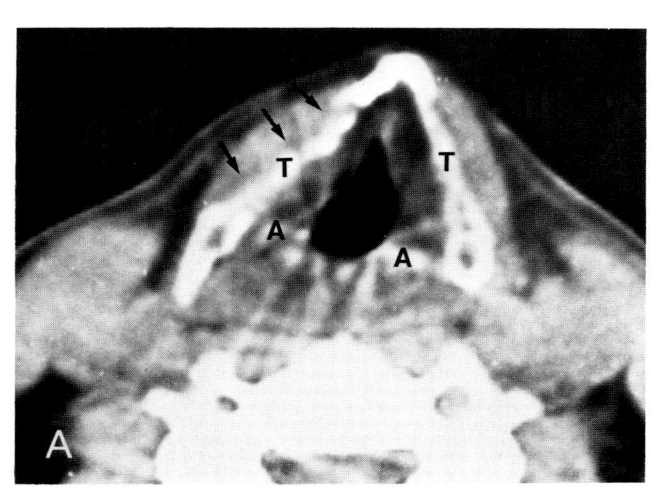

Plate 9.71. CT has proven to be a unique tool capable of delineating the nature and extent of laryngeal skeletal abnormalities from old trauma that sometimes leads to the erroneous clinical impression of a significant laryngeal mass lesion. The patient in the following illustrations presented with neck pain and slight hoarseness, and the clinician saw a "submucosal mass" on the right side at endoscopy.

Plate 9.71*A*. A section at the false cord level shows old, healed fractures (*arrows*) of the thyroid lamina. There is also some inbuckling of the thyroid cartilage. The arytenoids are in normal position.

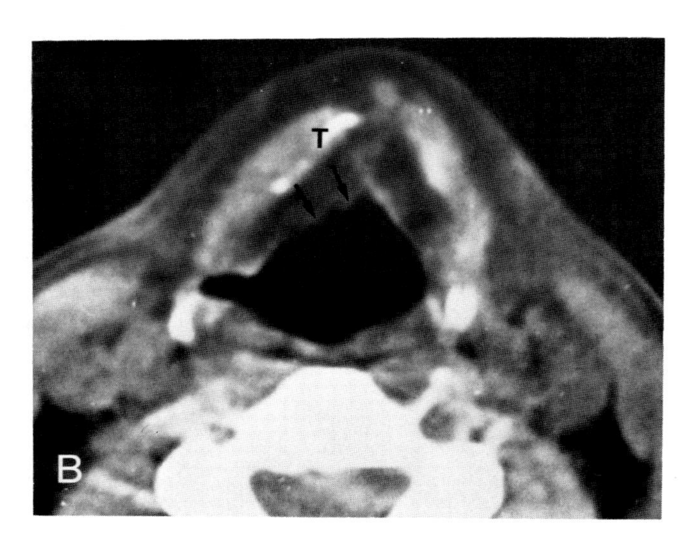

Plate 9.71*B*. In the midsupraglottic larynx, the laryngeal anatomy continues to be asymmetric. The thyroid lamina appears somewhat more medial in position on the right compared to the left. There is also an obvious bulge in the mucosa (*arrows*) which was producing the "pseudomass" effect noted at endoscopy. Open exploration of the larynx revealed multiple healed fractures of the thyroid cartilage. Upon further questioning the patient remembered a prior traumatic episode in which he had signs and symptoms of a laryngeal injury but had not sought medical attention.

The abbreviation used on Plate 9.72 is: *A*, arytenoid.

Plate 9.72. CT may be used to evaluate glottic, subglottic and tracheal stenosis. Most often this is caused by prolonged intubation. The role of CT in this regard may be thought of as complementary to that of positive contrast studies of the trachea. CT may be used to show the length and airway narrowing but lacks some of the dynamic information that positive contrast tracheography may provide in certain instances.

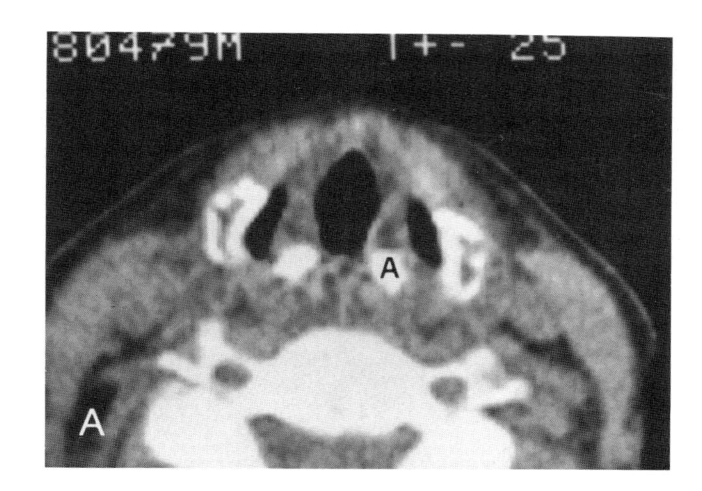

Plate 9.72*A*. A patient who had glottic and subglottic stenosis following a prolonged intubation. At the level of the false cords the arytenoids are drawn medially along with the aryepiglottic folds. This situation is similar to that seen in the fibrotic larynx due to radiation therapy.

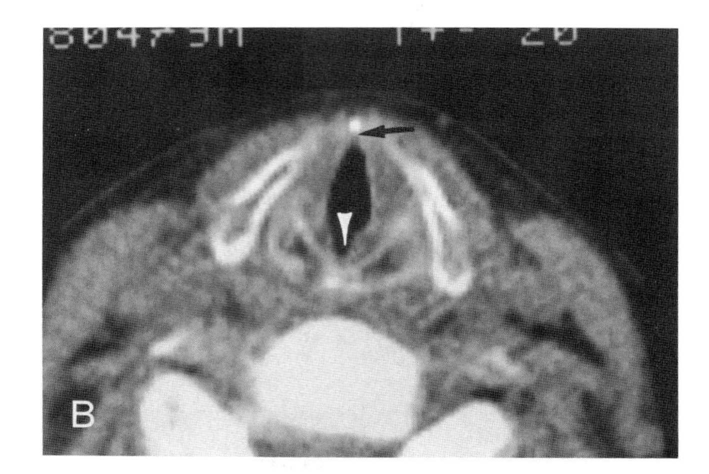

Plate 9.72*B*. At the undersurface of the true cords small amounts of tissue thickening are seen at the anterior (*arrow*) and posterior (*arrowhead*) commissures. These are small, fibrotic webs at these locations.

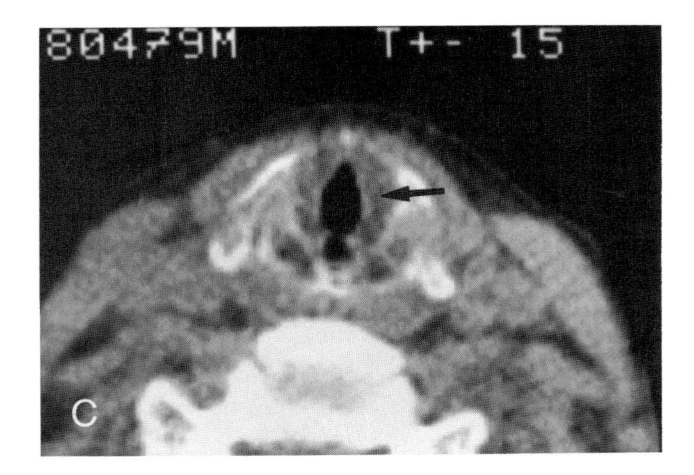

Plate 9.72*C*. In the midinfraglottic larynx a rind of abnormal tissue surrounds the airway and narrows it in both its anteroposterior and transverse dimensions (*arrow*). This tissue thickening could represent granulation tissue or fibrotic tissue depending on the stage of the disease. If this were an acutely traumatized patient this could represent tissue swelling and hemorrhage as well.

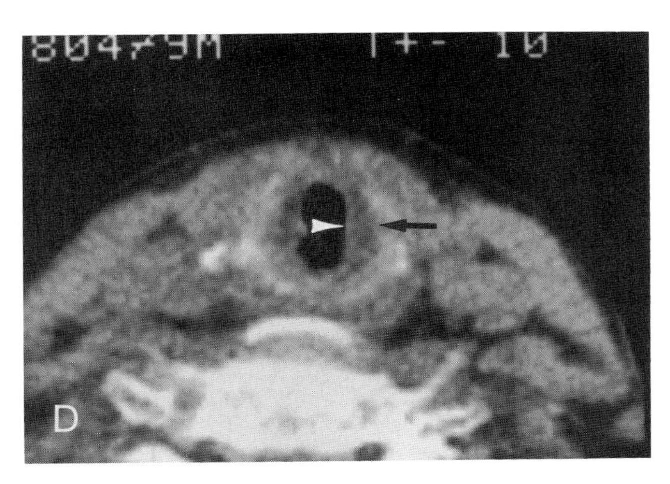

Plate 9.72*D*. In the low subglottic level the rind of tissue still surrounds the airway producing a moderate amount of subglottic stenosis. All of the tissue *between the white arrowhead and black arrow* is abnormal. The tissue seen in chronic subglottic stenosis tends to have a lower attenuation than tumor and usually will not enhance if intravenous contrast is given. If this is active inflammatory or granulation tissue considerable enhancement may be seen and, in the proper setting, this tissue might not be distinguishable from tumor or other inflammatory changes except by biopsy.

The abbreviations used on Plate 9.73 are: *A*, arytenoid; *C*, cricoid; and *T*, thyroid lamina.

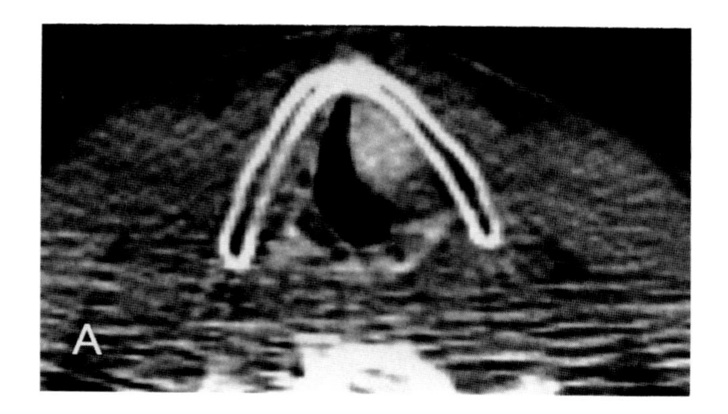

Plate 9.73. Patients with true vocal cord paresis sometimes have Teflon injected into the weakened true vocal cord in order to fix it in the midline. This allows the functioning contralateral cord to oppose it in adduction and create a competent glottic.

Plate 9.73*A*. The section is at the undersurface of the true vocal cords and the high attenuation of the Teflon is deposited in the anterior two-thirds of the left true vocal cord. Note how the cord has come to, and perhaps slightly across, the midline.

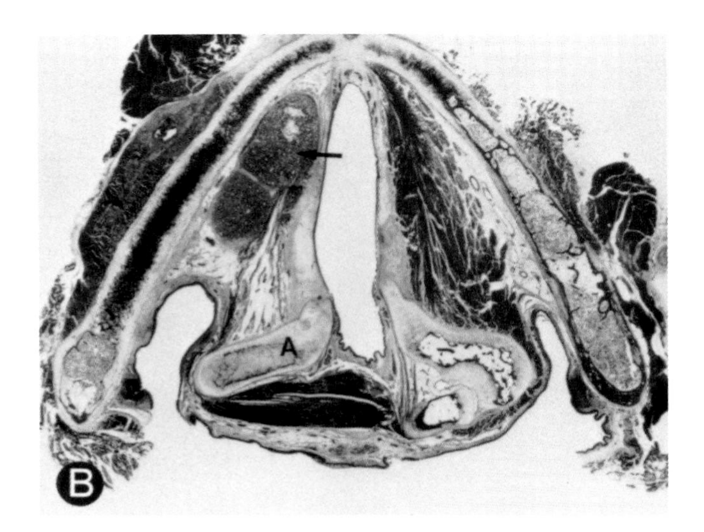

Plate 9.73*B*. This axial whole organ section is taken through the midplane of the true cords and demonstrates the appearance of a Teflon injection in the cord (*arrow*). Note the extreme atrophy of the muscles within the paralyzed cord compared to the normal side.

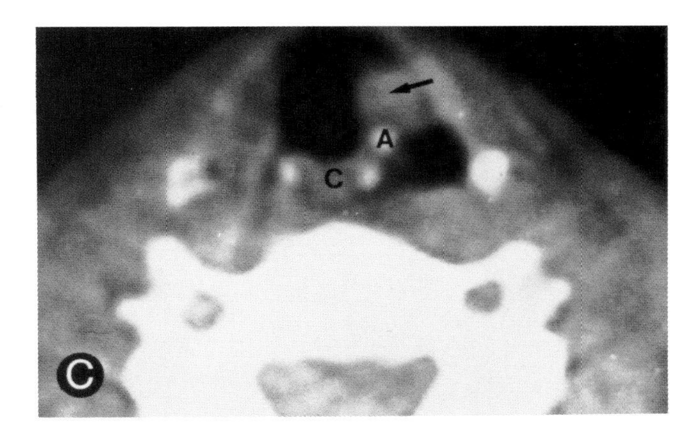

Plate 9.73*C*. This is a different patient who also had Teflon injected into the vocal cord. The patient had an idiopathic paresis of the left true cord. Approximately 2 yr after the Teflon injection the patient developed progressive hoarseness and pain in the larynx. A CT was done to search for some laryngeal abnormality to explain the complaints. The arytenoid and cricoid cartilages mark the plane of the true vocal cords. The Teflon injection thickens the posterior third of the true vocal cord and base of the aryepiglottic fold on the left (*arrow*).

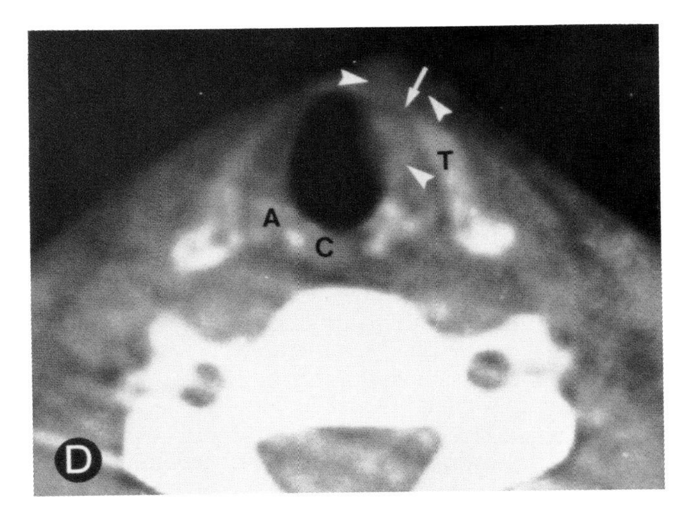

Plate 9.73*D*. A scan slightly lower shows increased density in the anterior two-thirds of the left true cord (*arrowheads*). However, some of the high density material can be seen working its way through the anterior portion of the thyroid lamina. This also appeared to be extending into the soft tissues of the neck (*arrow*). This obviously does not represent a common use of CT but does show how it can sometimes contribute information in unusual clinical circumstances.

Chapter 10

Oral Cavity and Oropharynx Including Tongue Base, Floor of the Mouth and Mandible

Improvements in CT technology since 1979–1980 have made routine, high quality, studies of this region possible. Recently, several groups have shown that CT contributes very valuable information to the management of squamous cell carcinoma of the oropharynx (6, 30, 32). MR images are equally impressive and may surpass the informational content of those made by CT (5, 7–9, 23, 25). Whatever the eventual outcome of this comparison, it is now very clear that diagnostic imagers have very powerful tools to evaluate pathology in this region. It is incumbent on us to become contributing members of the team caring for patients with squamous cell carcinoma and other lesions of the oropharynx, oral cavity, and related structures.

NORMAL ANATOMY AND VARIANTS

Oral Cavity

The oral cavity includes the hard palate, floor of the mouth, gingivobuccal sulci, retromolar trigone and anterior two-thirds (oral portion) of the tongue. The roof of the mouth is the hard palate. The soft palate is in the oropharynx (1, 16, 20). The hard palate has contributions from three bones: premaxilla, maxilla and palatine bones (Fig. 10.1). Its main mass is made by the palatal process of the maxilla (16, 20). The hard palate is covered by tightly adherent mucoperiosteum which becomes discontinuous only over the tooth sockets within the maxillary alveolar ridge (Fig. 10.2) (16, 20). The maxillary tuberosity is an important surgical landmark (27, 28); it is a fairly prominent bulge just behind the third molar socket (Fig. 10.1) of the alveolar process of the maxilla. The pterygoid plates lie just posterior and superior to the tuberosity with the upper buccinator space and lower infratemporal fossa being just lateral.

The upper gingivobuccal sulcus is a recess which lies between the cheek and the maxillary alveolar ridge (Fig. 10.2). The sulcus is, therefore, composed of mucoperiosteum on its medial surface and the mucosa reflecting over the buccinator muscle. The buccinator muscle forms muscular substance of the cheek. It arises on the posterior aspect of the maxillary alveolar ridge, upper medial surface of the mandible at the junction of the ramus and body and along the pterygomandibular raphe (Figs. 10.1, 10.2) (16, 20). The muscle then sweeps forward to interdigitate with the orbicularis oris at the angle of the mouth. The lower gingivobuccal sulcus lies between the mandibular alveolar ridge and cheek. On axial images the buccinator is seen bridging the gap between the posterior maxilla and mandible with the fatty buccinator space lying just laterally (Fig. 10.2). The parotid duct traverses the buccinator space and pierces the muscle just prior to its opening onto the buccal mucosa opposite the upper second molar (Figs 10.1, 10.2). The buccinator space is continuous with the infratemporal fossa above.

Like the maxilla and hard palate above, the mandible is the most obvious bony landmark of the oral cavity. Axial images prove ideal for tomographic study of the mandible and floor of the mouth.

The lingual surface (inner table) of the mandible serves as a primary attachment for the extrinsic tongue muscles and those forming the floor of the mouth. The myloid line is the origin of the mylohyoid muscle (Fig. 10.1). This important landmark along with the extrinsic tongue musculature separates the floor of the mouth from the neck (16, 20). The extrinsic tongue muscles include the genioglossus, geniohyoid, hyoglossus and styloglossus; in between these muscles and the mylohyoid lie important surgical spaces and related neurovascular structures.

The mylohyoid slopes downward from its attachment on the mandible to form a median raphe; its posterior fibers attach to the hyoid (Figs. 10.1, 10.2). The genioglossus and geniohyoid muscles arise from the genial tubercles on the lingual surface of the mandible and are separated by the fatty midline lingual septum. The sublingual space lies between these paired paramedian muscle bundles and the mylohyoid. The sublingual space may be considered a division of the submandibular space, but they are in fact separated by the mylohyoid muscle. The submandibular (submaxillary) space is cradled between but lies mainly below the mandible and mylohyoid muscle (Figs. 10.3, 10.4). The submandibular gland has a small component which extends over the free (posterior) edge of mylohyoid muscle to lie in the sublingual space (Fig. 10.1). The submandibular gland is, therefore, partly in the floor of the mouth and partly in the suprahyoid neck (16, 20).

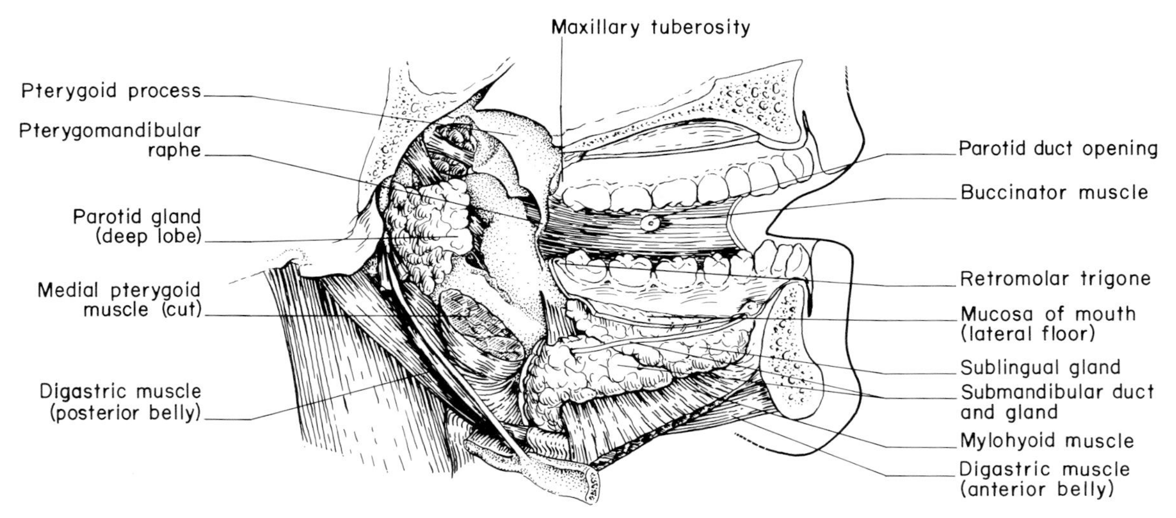

Figure 10.1. This sagittal view of the oral cavity and oropharynx provides a graphic representation of the intricacy of the anatomy and the important relationships that exist between these two regions. In particular, note how the pterygomandibular raphe provides a direct pathway for spread of tumor from the retromolar trigone to the region of the maxillary tuberosity and buccinator space. The complex interrelationship between these and other areas explains many of the patterns of spread seen on CT and MRI studies.

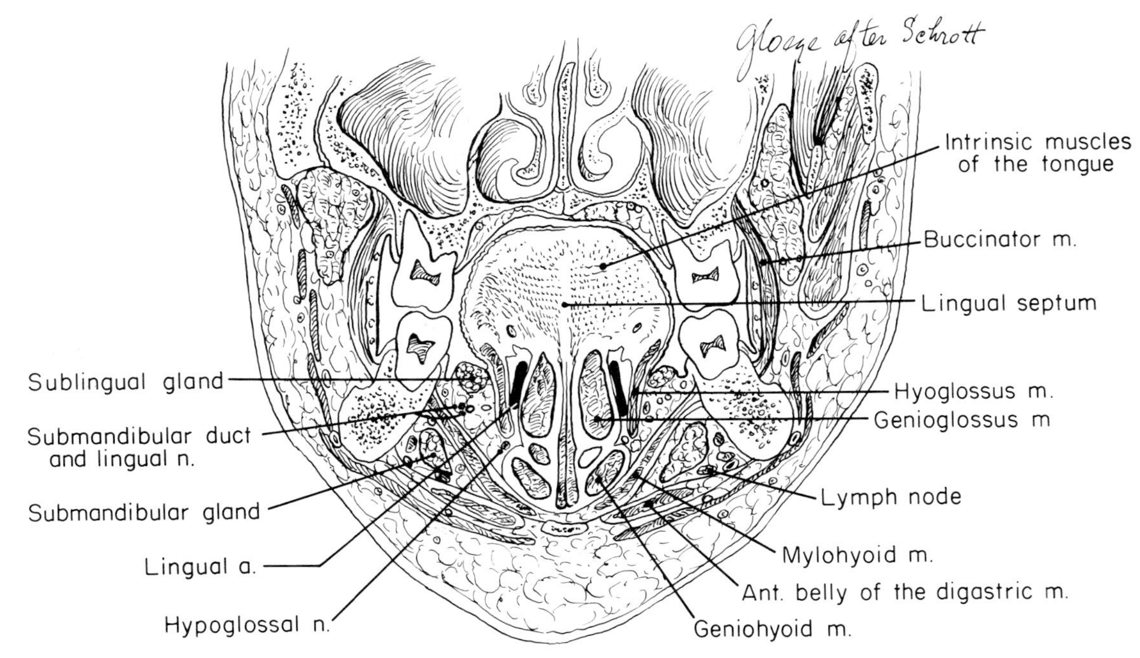

Figure 10.2. Coronal sections through the midface, oral cavity and floor of the mouth. The mylohyoid muscles form the floor of the mouth; below these lie the suprahyoid portion of the neck. The various spaces of the floor of the mouth are presented in Figures 10.3 and 10.4. Carefully note the relationship of the neurovascular and glandular structures to the extrinsic tongue musculature.

On axial images the lingual septum normally lies in the midline. The sublingual and submandibular spaces and intervening muscle bundles are always visible and symmetric allowing for differences caused by slightly skewed positioning (Table 10.1). Obliteration of these planes should be considered pathologic (24, 30). The paired anterior bellies of the digastric muscle lie below the mylohyoid in the suprahyoid neck. These two prominent muscle bundles diverge from their origin on the lower, inner surface of the mandible to their tendinous sling on the hyoid (Figs. 10.1, 10.2). The posterior belly of the digastric then continues on to its

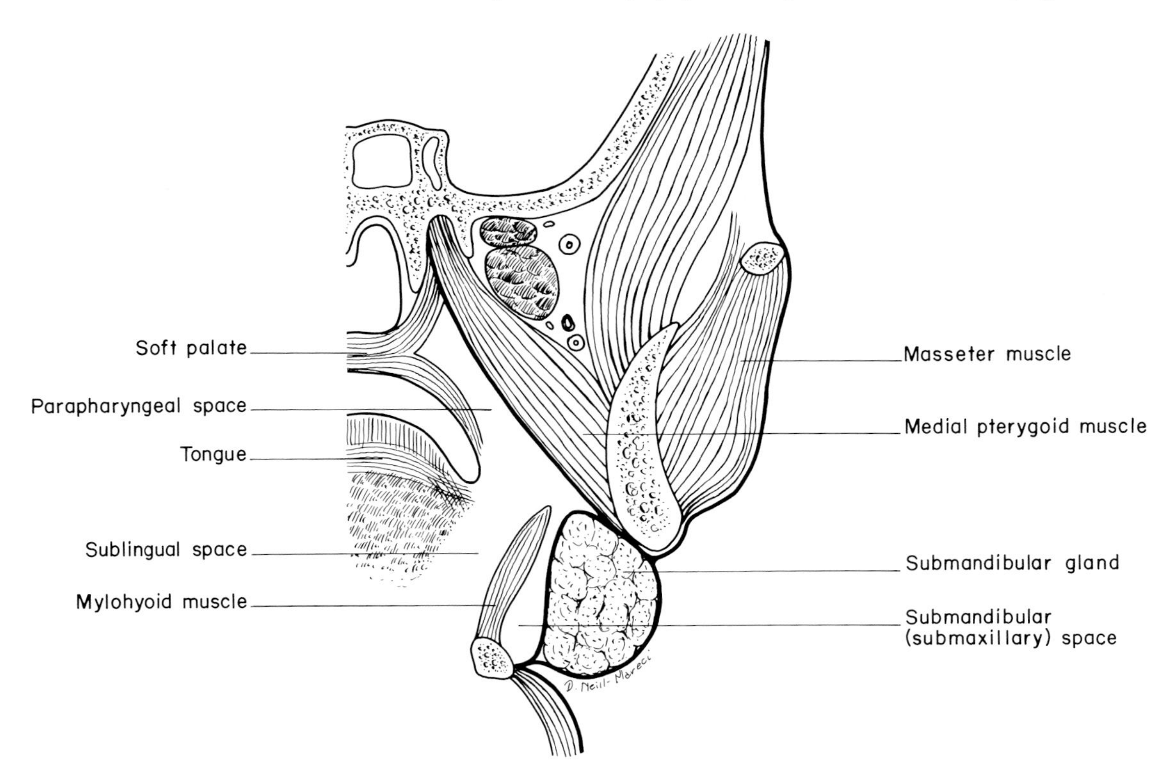

Figure 10.3. The *heavy black line* represents the superficial layer of cervical fascia. Note how it splits around the mandible to run medial to the medial pterygoid muscle and lateral to the masseter and temporalis muscle. The cervical fascia at this point limits the masticator space. The parapharyngeal space is an exceedingly important one in CT and MRI diagnosis and lies medial to the reflection of the cervical fascia. The sublingual space is medial to the mylohyoid muscle while the submandibular or submaxillary space lies lateral and inferiorly within the suprahyoid neck.

insertion on the medial aspect of the mastoid process. The submental space lies between the anterior bellies of the digastrics and below the mylohyoid; it contains fat and lymph nodes. These muscles may appear asymmetric on axial sections due to skewing of the mandible relative to the plane of section (30). Visualization of the submental space depends on the amount of fat it contains and the plane of section.

The oral portion (anterior two-thirds) of the tongue ends posteriorly at the circumvallate papillae. The base of the tongue (root, posterior third) is in the oropharynx. The oral tongue is loosely attached to the floor of the mouth so that it can move freely. The submandibular ducts terminate at papillae which lie in the floor of the mouth just anterior and lateral to the frenulum of the tongue (Figs. 10.1, 10.2). The anterior two-thirds of the tongue has a fairly nondescript, homogeneous appearance on CT compared to the dramatic appearance of the multiple muscles and spaces in and below the floor of the mouth. However, on sagittal MR images the medium to low intensity signal of the intrinsic tongue musculature is interwoven with the high intensity of surrounding fat producing a layered appearance in the oral tongue.

The retromolar trigone is a small triangular mucosal surface behind the third molar covering the ascending ramus of the mandible; it is continuous with the maxillary tuberosity above (Fig. 10.1). Its deep relationships put it in contact with structures of the oral cavity, oropharynx, nasopharynx, and floor of the mouth, placing it at the crossroads between each of these regions. The pterygomandibular raphe lies beneath the mucosa of the retromolar trigone. The raphe attaches to the hamulus of the lateral pterygoid plate (Fig. 10.1) and the posterior myloid ridge of the mandible and serves as the insertion of the buccinator, orbicularis oris and superior constrictor muscles (16, 20). The pterygomandibular space lies just posterior and medial to the raphe. This small triangular fat pad lies between the medial pterygoid muscle and the mandible; it is continuous with the parapharyngeal space and contains the lingual nerve. The pterygomandibular space is normally visible and fairly symmetric in appearance on axial CT or MR images.

Oropharynx

The oropharynx, as defined by the American Joint Committee on Cancer Staging, extends from the anterior tonsillar pillars to its inferior limit at the pharyngoepiglottic folds (1). It includes the soft palate supe-

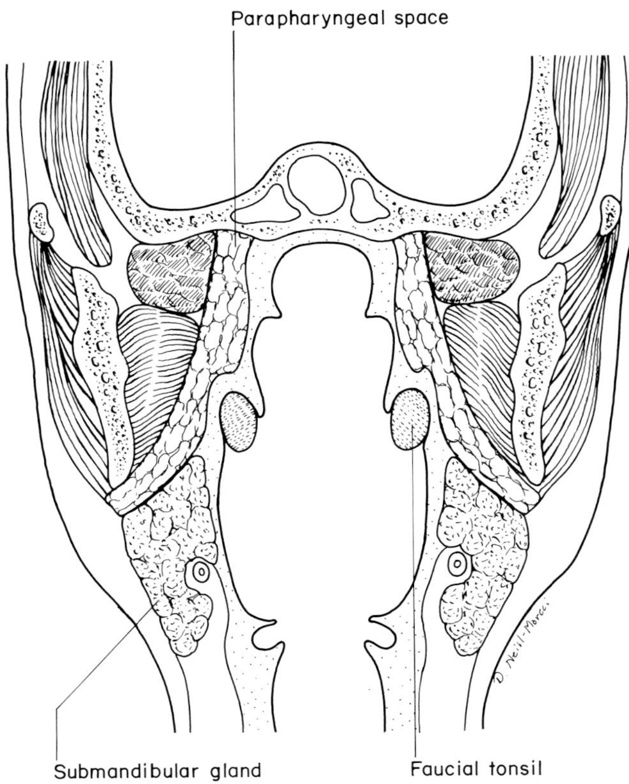

Figure 10.4. The parapharyngeal space is an exceedingly useful deep tissue plane for evaluating the nasopharynx and upper oropharynx. It is shown here as a fat-containing space which extends from the skull base to the floor of the mouth. Note how the space shifts laterally away from the mucosal surfaces of the upper aerodigestive tract (*lightly stipled area*) near the lower pole of the faucial tonsils. This bit of anatomy is exceedingly important in making subtle diagnoses in the oropharynx.

riorly. Its faucial pillars (anterior and posterior tonsillar pillars) surround the palatine tonsils and merge to continue inferiorly as the glossotonsillar (glossopharyngeal) sulci. The tongue base (posterior third) begins at the level of the circumvallate papillae and extends to the valleculae (16, 20). The mucosa of the valleculae and that of the lingual (anterior) surface of the epiglottis is in the oropharynx (16, 20).

The anatomy of the oropharynx as seen on CT or MRI should be divided into superficial and deep components. Superficial structures are formed by mucosal reflections over lymphatic tissue (lingual and faucial tonsils) and muscle bundles which surround this part of the upper aerodigestive tract. The deep components include the parapharyngeal space, carotid sheath, deep musculature, and other intervening tissue spaces described in detail in the preceding section on the floor of the mouth and subsequently, the most important factor to keep in mind is that the superficial structures have a fairly, and sometimes highly, variable appearance; whereas the deep anatomy is much more constant in appearance both between patients and from side to side individual patients (30).

Faucial (Tonsillar) Pillars and Palatine (Faucial) Tonsils

The anterior tonsillar pillar is formed by a mucosal reflection over the palatoglossus muscle (Fig. 10.5) (16, 20). The muscle arises on the soft palate and interdigitates with one of the deep extrinsic tongue muscles, the styloglossus. The posterior tonsillar pillar is a mucosal reflection over the palatopharyngeus muscle (Fig. 10.5). This muscle arises from both the hard and soft palate and runs deep to the glossotonsillar sulcus before it inserts on the thyroid cartilage (16, 20). The faucial (palatine) tonsil lies between the tonsillar pillars; it is the only component of Waldeyer's ring that has a capsule (16, 20). The capsule is a thin, specialized portion of the pharyngobasilar fascia which separates the tonsil from the superior constrictor muscles. On CT

Table 10.1
Analysis of 35[a] CT Scans of Oropharynx and Floor of the Mouth[b]

	Present	Symmetric	Asymmetric
A. Superficial structures			
Faucil tonsil—tonsillar pillars	30	14	16[c]
Glossopalatine sulci	35	21	14[c]
Lingual tonsils	35	5	30
Valleculae	35	13	22
Pharyngoepiglottic folds	35	15	20
B. Deep tissue spaces			
Parapharyngeal space (upper-faucial tonsil level)	30	14	16
Plane between submandibular gland and glossopalatine sulci	0	—	—
Planes surrouding carotid sheath	35	NA[d]	NA
Sublingual space	35	33	2
Lingual septum	35	35[e]	0
C. Deep musculature			
Genihyoid—genioglossus		35	0
Styloglossus—hyoglossus		31	4
Mylohyoid		30	5
Digastric—anterior belly		29	6
Digastric—posterior belly		30	5

[a] Only 30 paitents had complete scans through the upper oropharynx.
[b] Reproduced with permission of Radiology from AS Muraki, AAH Mancuso, and HR Harnsberger: CT of the oropharynx. . . *Radiology* 148:725–731, 1983 (30).
[c] Transverse dimension of pharyngeal walls differs by 3 mm or more.
[d] Due to variability of surrounding muscles and vessels.
[e] Not deviated to one side.

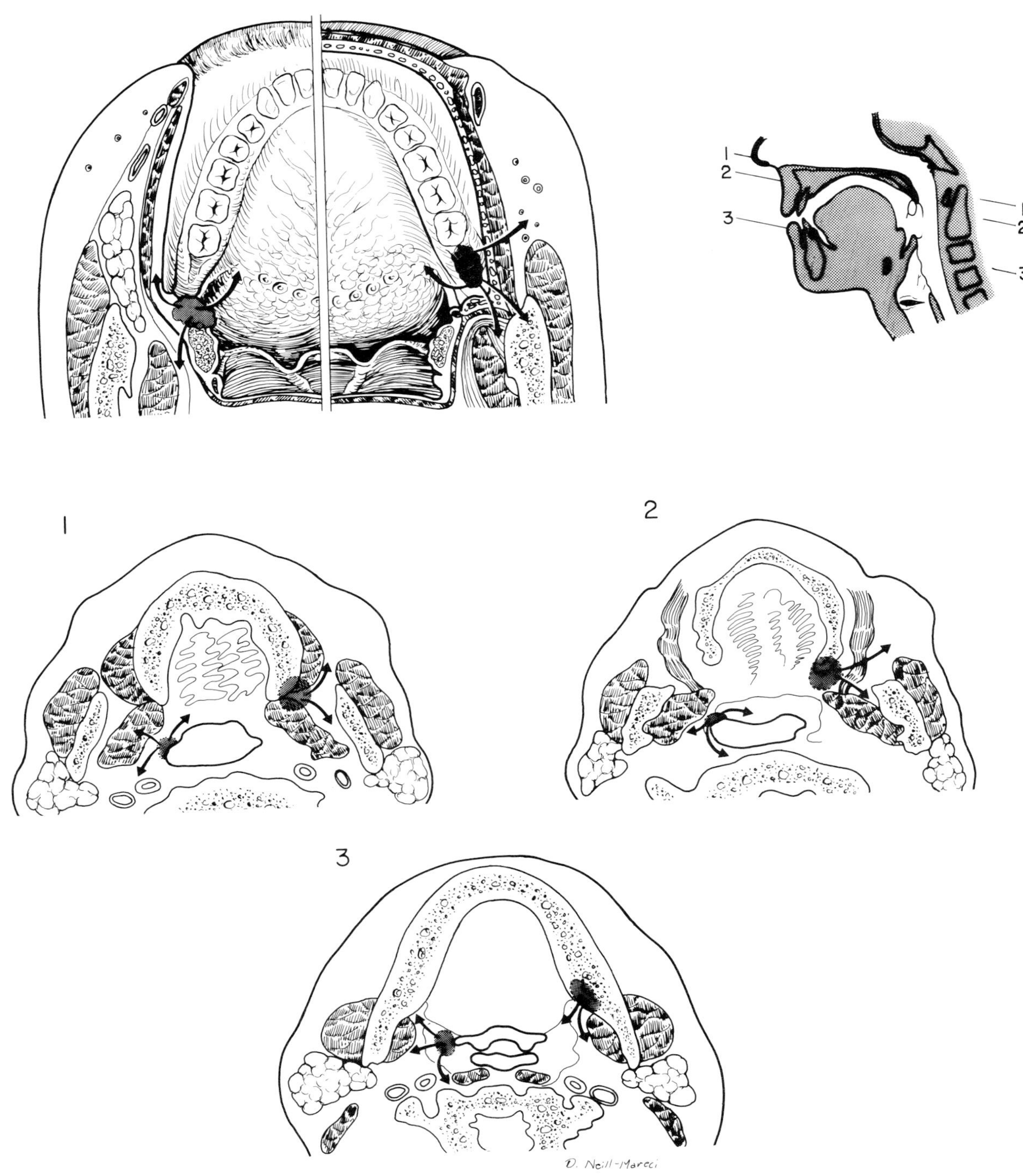

Figure 10.5. Patterns of spread of retromolar and tonsillar cancer. The level of the large anatomic diagram is slighly higher on the left than on the right. An anterior tonsillar pillar lesion is pictured on the *left* side of the diagram and its corresponding routes of spread on axial images are shown on the *left hand side* of the three following axial line diagrams. A retromolar lesion is shown on the *right hand side* of the anatomic diagram and its pattern of spread on axial images is shown on the right hand side of the line diagrams. The level of each line diagram corresponds approximately to that shown in the schematic drawing in the upper right hand corner.

images the tonsillar tissue density is inseparable from that of muscle. On MR images the signal from the tonsil is clearly separated from the parapharyngeal space by the lower intensity signal of its capsule and the superior constrictor muscle (7–9, 25).

The amount of tonsillar tissue varies widely between patients and in individual patients. The thickness of the tonsil plus the pharyngeal wall may vary 3–6 mm from side to side in a CT study on any given individual (Table 10.1). The parapharyngeal space lies deep to the tonsillar fossa and is always present in normals at the level of the upper palatine tonsils (upper to midoropharynx) (Fig. 10.4) (30). The area of the parapharyngeal space appears to be inversely proportional to the amount of residual lymphoid tissue in the fossa and the size of surrounding musculature. Complete loss of the parapharyngeal space at this level indicates pathology (30). More caudal axial sections will show that the parapharyngeal space disappears as the submandibular glands come into view (Fig. 10.4). In fact, the submandibular space is an anterior continuation of the parapharyngeal in the suprahyoid neck but the bulk of the submandibular gland normally effaces the fat plane between the gland and pharyngeal wall (Figs. 10.3, 10.4) (16, 20). Lack of visualization of the parapharyngeal space deep to the glossotonsillar sulcus (in the low oropharynx) is a *normal* finding (Table 10.1) (30). The styloglossus and stylopharyngeus muscles and stylohyoid ligament are intimately related to the lateral wall of the pharynx at this point (Fig. 10.1).

The overall CT density of palatine tonsillar tissue is equivalent to that of surrounding muscle. Some enhancement with I.V. contrast is not unusual and is presumed due to the chronic inflammation which is often present in patients who smoke, drink and have poor oral hygiene. Focal areas of low attenuation probably due to old infection and abscess are not uncommonly present. Focal areas of calcification, probably dystrophic, are another common variant of normal. The MRI signal of tonsillar tissue on spin echo images tends to be much more intense than muscle especially on late echo T_2-weighted images (25). The signal intensity may approach that of surrounding fat if the lymphoid tissue is very hypertrophic (25).

Tongue Base

The circumvallate papillae separate the base of the tongue from the anterior two-thirds (oral portion) (16, 20). The mucosa of the tongue base is continuous with that of the lower tonsillar fossae and together they form two laterally placed gutters; the glossotonsillar (glossopalatine) sulci. Submucosal lymphoid nodules of the lingual tonsil raise ridges beneath the mucosa of the tongue base. Like the palatine tonsils, the lingual tonsils vary a great deal in size and appearance between individuals. In a given patient the tonsillar tissue is likely to appear asymmetric; this will produce asymmetry of the airway, the valleculae, and in the deeper intrinsic muscle mass of the tongue (Table 10.1) (30). In fact, the lingual tonsil often appears to "penetrate" the tongue base for 3–5 mm on axial sections. This tissue may show fairly marked enlargement and a contrast enhancement presumably due to chronic inflammation. On spin echo MR images the signal of lingual tonsillar tissue appears to be higher in intensity than that of surrounding pharyngeal musculature; it may be higher, lower or near equal to that of the intrinsic muscle mass of the tongue base depending on the fat content of the tongue. Tissue contrast between this lymphoid tissue and surrounding muscles is again maximum on T_2-weighted spin echo images.

The musculature of the tongue base is continuous with that of the oral tongue. The intrinsic muscles make up the core of the tongue base (root of the tongue) and their several bundles interdigitate with each other and the extrinsic musculature. At CT the intrinsic muscle mass usually appears as a low attenuation region surrounded by the interwoven fibers of the hyoglossus and styloglossus muscles. The other extrinsic tongue muscles were discussed in conjunction with the floor of the mouth. Coronal, and especially sagittal, MR images show the intrinsic tongue muscle to best advantage.

The lingual artery penetrates the tongue base medial to the hyoglossus and about 1–2 cm above the hyoid level (Fig. 10.2) (16, 20). The hypoglossal nerve and lingual vein lie lateral to the hyoglossus muscle and in an axial plane at roughly the same level as that of lingual artery's main trunk (Fig. 10.2) (16, 20). The hypoglossal nerve courses deep to the posterior belly of the digastric just before it penetrates the tongue base (16, 20). Portions of these vessels can be seen on MRI by virtue of their low signal (due to the flow-void phenomenum) or on contrast enhanced CT studies.

Soft Palate

The soft palate is the roof of the oropharynx and it separates the oropharynx from the nasopharynx. It has important deep communications with the oral tongue via the palatoglossus muscle and with the nasopharynx via the tensor and levator palati muscles (21, 22).

The soft palate lies in the axial plane and for about one-third of its extent posterior to its junction with the hard palate. The junction of the soft palate and anterior tonsillar pillars can be appreciated on axial sections through the level of the maxillary alveolar ridge and lower pterygoid plates. More distally, the palate lies oblique to the axial plane and is, therefore, not well visualized with axial sections. Midline and paramedian sagittal MR images show the internal structure of the soft palate. Its signal intensity approaches that of surrounding fat and lymphoid tissue, presumably due to its content of lymphatic, glandular and fatty tissue (16, 20). On CT sections the soft palate tends to be of an attenuation intermediate to that of muscle and fat.

Technique of Examination

CT

POSITIONING AND RESPIRATION

The patient should lie supine and, as a rule, the neck should be mildly extended to facilitate avoiding fillings. The only drawback to extending the neck is that patients might become uncomfortable quicker and increased motion artifacts may result. The head should rest in a molded plastic holder and be immobilized. The chin should be secured by tape, or better yet, a chin strap. The patients are instructed not to move, cough, swallow, or move their jaw during scanning. Scans may be done during quiet respiration or, if there are significant motion artifacts, during suspended respiration. We use suspended respiration as a routine. If there is excessive motion of the tongue the patient can hold the tip of the tongue between his teeth. All removable bridgework and false teeth should be taken out of the mouth.

TECHNICAL FACTORS—PLANNING THE STUDY

A localizing lateral projection radiograph is obtained. This will aid in planning how to avoid fillings and show whether the patient is positioned correctly. The latter is best judged by the overlap of mandibular rami and angles. The preferred plane of section is roughly parallel to the inferior border of the mandible or hard palate (Fig. 10.6). Ten to 20° angling of the gantry will usually suffice to avoid fillings. The angle is craniad above the maxillary teeth and caudally below. Semiaxial (or semicoronal) section may be necessary to avoid fillings when the lesions are of the tonsillar pillars and fossae; in these cases one may hyperextend the neck and angle the gantry 20° caudally so that the plane of section is almost parallel to the mandibular condyle. Such attention to positioning will virtually always produce artifact-free, detailed images of the tonsillar pillars, fossae and retromolar trigone areas.

Sections should be 4–5 mm thick and done contiguously. The data should be obtained, if possible, with a convolution filter that allows one to view bone detail without sacrificing too much contrast resolution. Most CT systems have various approaches to this end and they should be exploited because bone invasion is often a critical issue in management decisions. Sections 1.5–3 mm thick might be useful at times for detecting areas of subtle bone destruction. Technique factors should be set to allow for maximum, or near maximum, spatial and contrast resolution. Zoom reconstructed images will improve spatial resolution and make the study more pleasing aesthetically.

Sections should begin at the hard palate or slightly above (C_1 level) and continue to the hyoid bone as a routine. In cases of carcinoma this will cover the primary site as well as nodal groups at highest risk. It is wise to include the middle internal jugular nodes in known cancer cases and, if nodes are positive elsewhere in the neck, the low neck nodes should also be surveyed (perhaps leaving 5-mm intervals between sections).

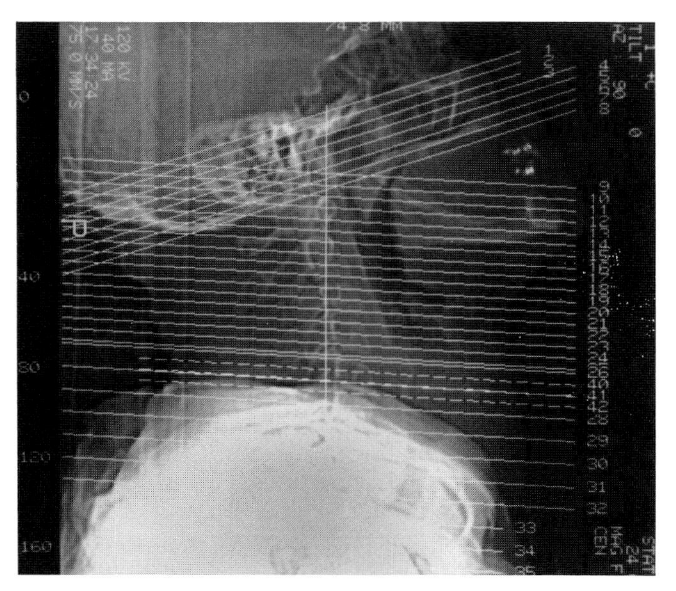

Figure 10.6. This projection radiograph shows how one might approach a basic study of the oropharynx. Within the nasopharynx the plane of section should stay roughly parallel to the hard palate. Within the oropharynx it should lie about parallel to the ramus of the mandible; however, some allowances must be made to avoid interference from dental fillings. At and below the level of the hyoid bone, the section should be about parallel to the intervertebral disks so that they will approximate the plane of the true vocal cords lower in the neck. This is less important when the larynx is not being examined.

Sagittal and coronal reformations add little to the CT study of this region. Direct coronals may prove useful in lesions involving the hard palate and tonsillar regions, especially if they need to be done to avoid artifacts due to dental work.

CONTRAST INFUSION

Intravenous contrast should be used in all patients with malignant tumors and the vast majority of those with benign lesions. Those with benign mandibular lesions, trauma or temporomandibular joint problems usually can be done without contrast infusion. The method of combined bolus and rapid drip infusion we use was described in detail in Chapter 8. Contrast infusion is particularly valuable to ulceroinfiltrative squamous cell carcinoma (the most common lesion we are asked to study) because these tumors will often be inflamed and incite an inflammatory response at their borders with normal tissue; this makes staging of the primary tumor easier and more precise. Contrast infusion is indispensible for accurate staging of the cervical nodes as described in Chapter 8.

MRI

We image this region with spin echo pulse sequences. These are obtained via a multislice technique with the option of obtaining simultaneous images at multiple T_Es depending on the T_R which is selected. In general we do at least a T_1-weighted sequence for general anatomic display (T_R 500 msec; T_E 30 msec) and a T_2-weighted sequence which maximizes tissue contrast (T_R 2000 msec; T_E 30, 60, 90, 120). Sections should be 1 cm or less in thickness with the smallest interscan intervals possible. The optimal section thickness for studying this area is probably about 4–6 mm. Sections in this thickness range are now possible on 0.15 Tesla units even with their reduced signal-to-noise ratios. Sections 7–8 mm thick of excellent quality can be done on 0.15 and 0.3–0.5 Tesla units and are adequate for studying the oropharynx. The high field strength units (1.5 Tesla and larger) produce excellent images with sections as thin as 3 mm. Improvements in coil design and system optimization, however, are leading to rapid improvements at lower field strengths and these instruments are fully capable of studying the soft tissue extent of tumor in the pharynx. Axial sections are most useful but one should consider coronal imaging in tonsillar, retromolar trigone, floor of the mouth, and hard and soft palate tumors. Sagittal imaging may be useful in staging carcinomas of the oral tongue and tongue base.

INDICATIONS FOR STUDY

CT is used most frequently in this region for pretreatment staging of squamous cell carcinoma of the oropharynx and oral cavity and related cervical adenopathy. It contributes most to the evaluation of lesions which are T_2 or larger or show evidence of deep infiltration. Mucosal lesions less than 2 cm in size with clinically negative necks should not be studied with CT or MRI. An ulceroinfiltrative growth pattern characterized by surrounding induration and possible fixation of deep structures are some indicators of when CT and MRI may contribute clinically important data. More specific indications vary with the primary site and will be discussed further in the Pathology section of this chapter. The staging of cervical lymph node metastases was discussed in Chapter 8. CT has proven value in the search for occult primary tumors of the tongue base and faucial tonsils in patients presenting with ear pain of uncertain etiology or enlarged cervical node (*see* Chapter 8) (18, 24, 29, 30). CT is also of proven value for detection of recurrent carcinoma in this region and may do so in the absence of mucosal disease (15, 24, 26, 33). It may also be used to follow patients at high risk for recurrent primary site or nodal disease following therapy.

MRI may prove valuable in oncologic imaging of this region as well. In this and other regions of the body, it tends to show the borders between tumor and normal tissue with greater clarity than CT without the need for infusion of contrast medium. If the signal from posttreatment scarring differs enough from tumor, MRI may prove very useful in following patients at significant risk for recurrence (25).

CT currently remains the imaging examination of choice for evaluating parapharyngeal masses; MRI could prove more specific and come to replace CT for this purpose. This is true for its use in the evaluation of the following benign processes as well: suspected benign mass lesions or cysts of the floor of the mouth, mandibular vs submandibular gland vs nodal masses, tumors of the mandible, lingual thyroid, abscess vs cellulitis of the masticator, and parapharyngeal or submandibular spaces. CT may also be used for investigating temporomandibular joint dysfunction although MRI will likely replace it and most arthrography. The role of CT and MRI relative to other imaging will be discussed under specific disease categories in the following sections.

PATHOLOGY

Malignant Neoplasms

HISTOLOGY

Cancer of the oral cavity and oropharynx accounts for about 5% of all malignant tumors in the United States (4, 27, 28). About 90% of these are squamous cell carcinomas (4). Some authors suggest that the histologic grade tends to increase as the primary site moves from anterior (oral cavity) to posterior (tongue base) (4). The histologic grade does have a bearing on prognosis and therapy but the extent of the primary and status of the cervical nodes are of more tangible importance in actually planning management.

Verrucous carcinoma, generally considered to be a "low-grade" squamous cell carcinoma, is a particularly interesting variant in that cure rates are usually much higher than those of its more invasive counterparts (4, 11, 27). These probably make up 5% of oral carcinoma (4, 11). Their warty, exophytic appearance may sometimes hide a more deeply infiltrating component from the unwary. In fact, these may be confused with *benign* neoplasms of squamous cell origin. They most commonly arise along the gingivobuccal sulcus, the alveolar mucosa of the mandible and in the retromolar trigone region (4, 11, 27).

Malignancies other than squamous cell carcinoma are infrequent. Adenoidcystic carcinoma may arise from minor salivary glands or accessory salivary tissue and is notorious for its infiltrative and perineural patterns. Perineural growth of other carcinomas is unusual at presentation but not uncommon if tumor recurs (10). The clinical course of adenoidcystic carcinoma is usually one of multiple local recurrences over a protracted period (5–15 yr) with the patient eventually succumbing to the tumor (4, 27). Adenocarcinoma most commonly arises from the soft palate, probably because of its increased density of glandular tissue (4, 28). Lymphoma may involve the oropharyngeal portion of Waldeyer's ring. Melanoma may also occur. Sarcomas and tumors of the mandible, odontogenic and nonodontogenic, round out the group of less common tumors arising in or around the oropharynx and oral cavity. Rhabdomyosarcoma is the most frequent malignant tumor of the region encountered in childhood.

PATTERNS OF SPREAD

Squamous cell carcinoma of the upper aerodigestive tract grows mucosally, submucosally and by deeper pathways. Prior to the CT era, Ballantyne (2) and Lederman (21, 22) described how deep spread occurs along preexisting anatomic pathways of least resistance in the upper aerodigestive tract. In the early 1970s these tendencies were confirmed by whole organ sectioning specimens removed from patients with laryngeal and hypopharyngeal cancer (*see* Chapter 9). It is more difficult to get such precise anatomic correlation in the oropharynx and oral cavity but the accumulating surgical, clinical and CT experience (6, 15, 18, 24, 29, 30, 32) supports the basic observations of Ballantyne (2) and Lederman (21, 22). In fact, by studying CT scans of tumors at various stages, one can piece together the natural history of oral cavity and oropharyngeal cancer. Combining all of the prior knowledge and that now available from CT, the basic modes of deep spread which occur include the following:

1. *Along muscle bundles and surrounding fascial planes.* Once tumors burrow through the mucosa and submucosa and more superficial muscles of the pharyngeal wall they reach deeper muscles by growing within the parapharyngeal space (Fig. 10.4). Both muscles of the pharyngeal wall and those surrounding the parapharyngeal space have attachments to the skull base, hyoid bone, mandible, and other muscular structures of the upper aerodigestive tract. Tumors are free to spread within the loose fibrofatty tissue planes surrounding the muscles as well as along the muscles themselves. Such growth seems to be directed mainly along the muscle bundles from origin to insertion (or vice versa). The neurovascular bundles and ducts coursing through the deep tissue spaces may also serve as conduits for spread of tumor. These tendencies are of critical importance to the diagnostic imager; with CT and MRI the imager has the tools to detect and quantitate growth which may occur beyond the clinician's ability to detect its full extent even with modern endoscopic technique and good palpation. For example, a tumor of the faucial tonsil may penetrate the mucosa, the superior pharyngeal constrictor and its fascia, spreading to the soft palate and nasopharynx via the palatoglossus or palatopharyngeus and the levator palati, respectively. Much of this spread may occur beneath normal looking mucosa (15, 24, 29, 30). Caudal spread along the palatopharyngeal muscle could direct growth toward the tongue base, floor of the mouth and lateral pharyngeal wall. All of the oropharynx and oral cavity are interconnected by not only contiguous mucosal surfaces, but also by the deep planes and musculature which are vital to proper function. Some authors lump carcinoma of the oropharynx into one group, citing the tendency for this region to behave as a "tumor field." In fact, it does—not only because of its response to the toxic effects of tobacco and alcohol but also because of these anatomic relationships.
2. *Spread along periosteal surfaces.* Cancers spreading along the deep muscle bundles and fascial planes and spaces will eventually reach the bones. Growth continues along the periosteal surface until a pathway in the bone presents itself. This may be a neurovascular channel or an area of natural weakness such as the tooth sockets of the mandible and maxillary alveolar ridges. Bone destruction may also occur directly at the point of attachment.
3. *Spread within marrow space.* Once tumor enters a bone it may continue spreading within its marrow cavity. In the head and neck region this pattern is most easily appreciated in the mandible.
4. *Perineural spread.* Tumor may grow along neurovascular bundles or actually within the perineural space. True perineural growth may be seen in the first presentation of squamous cell carcinoma; it is more commonly seen in the setting of recurrent disease (10). This may explain the high incidence of pain as an early and major complaint of patients with recurrent tumor. Perineural spread is a characteristic of adenoidcystic carcinoma.
5. *The status of regional lymph nodes* is a major determinant of treatment and prognosis in squamous cell carcinoma of the oral cavity and oropharynx. The presence of a cervical metastasis essentially halves the patient's chances of survival (4, 27, 28). This is a trend that seems to hold regardless of the stage or histologic grade of the primary. Since the preferred site and incidence of nodal disease varies so much with that of the primary we will consider those tendencies along with local spread patterns and related clinical and treatment considerations by site.

ORAL CAVITY

Gingivobuccal Sulcus, Hard Palate, and Retromolar Trigone

Lesions arising on the mucosa along the mandibular and maxillary alveolar ridges spread to secondarily involve the hard palate, soft palate, buccal mucosa, floor of the mouth or underlying bone, depending on their site of origin (Fig. 10.7) (2).

Lesions of the upper gum spread superiorly to the hard palate and eventually the inferior nasal cavity and nasal septum in very advanced lesions. Invasion of the maxillary antrum is also a late finding and usually related to continued growth after direct bone invasion or penetration of an open tooth socket in the maxillary alveolar ridge (2, 27). Posteriorly and laterally the lesion may penetrate the buccinator and go on to involve the massater and pterygoid muscles with subsequent spread to the mandible (2, 27, 30).

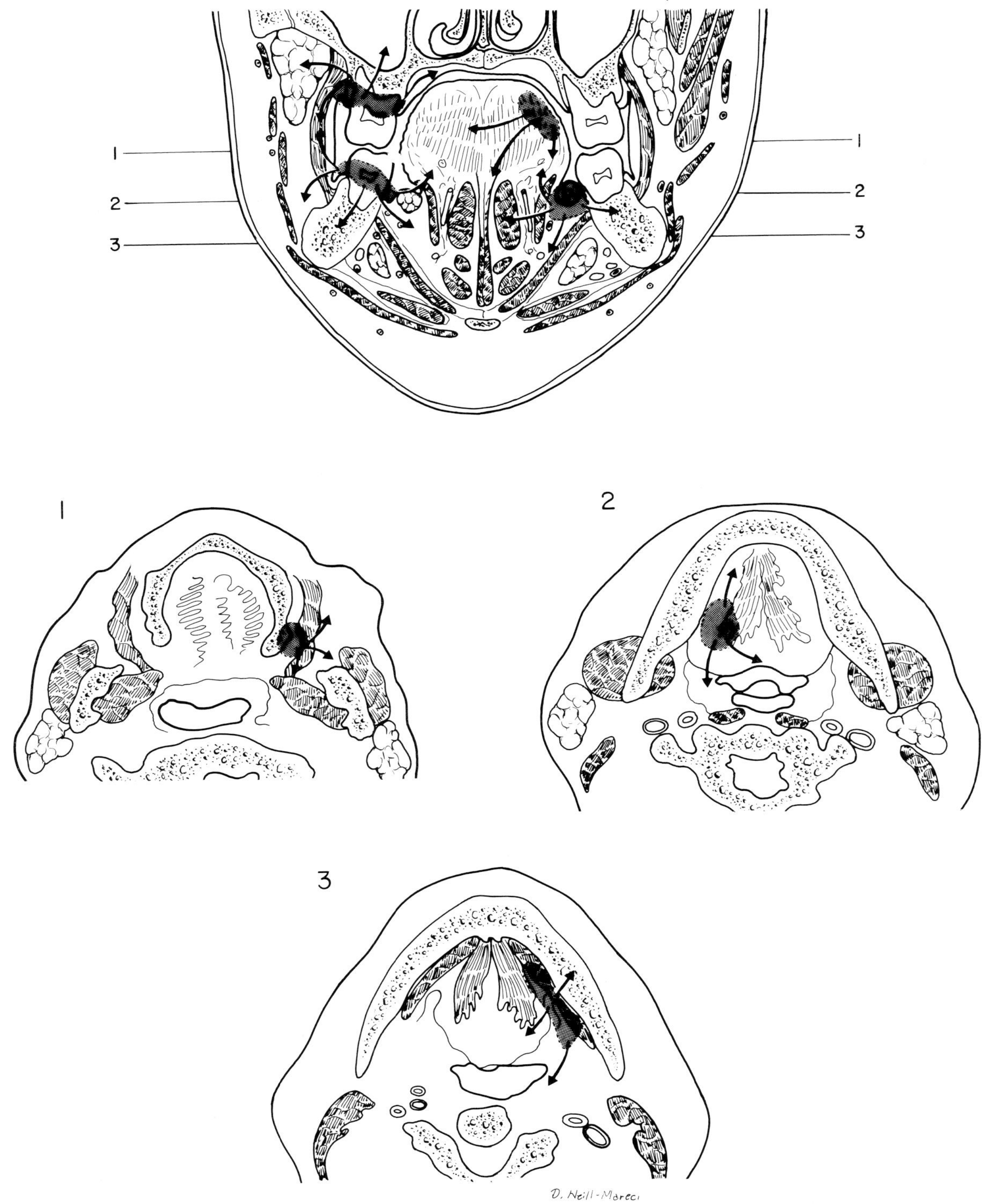

Figure 10.7. Patterns of spread carcinoma arising within the oral cavity. Each of the *shaded areas* represents possible sites of primary tumors within the oral cavity and oral tongue. The *arrows* spreading out from these tumors show areas of likely involvement in more advanced lesions. Involvement of many of these lesions will critically alter therapeutic decisions and the functional deficit resulting from therapy. Pictured here are patterns of spread for upper and lower gingivobuccal sulcus, lateral floor of the mouth and oral tongue lesions. The three axial sections correspond to the levels indicated on the coronal section.

Lesions of the lower gum invade periosteum, adjacent buccal mucosa and floor of the mouth (Fig. 10.7). A slow growing tumor will sometimes produce a saucerized area of remodeling in the mandible before frank invasion occurs (27). Lateral and posterior spread, as in the upper gum, takes the lesion to the buccinator space where it can reach the massater muscle, pterygomandibular raphe and medial pterygoid muscle (2, 27, 30). From this point the spread pattern will be much like that of a primary retromolar trigone carcinoma.

Spread of squamous cell carcinoma of the retromolar trigone is intermediate between the pattern of anterior tonsillar pillar and gingivobuccal sulcus cancer. Some authors include this oral cavity lesion in discussions of anterior tonsillar pillar carcinoma; however, since the anatomical and treatment considerations are more akin to gingivobuccal sulcus tumors we include it here. The retromolar trigone is small and spread to adjacent areas occurs early. The anterior tonsillar pillar and adjacent buccal mucosa may be involved (Fig. 10.5). Deep, posterior spread characteristically involves the tiny pterygomandibular space and medial pterygoid muscle (Fig. 10.5) (2). Lateral spread involves the buccinator space and muscle as well as the pterygomandibular raphe. Cephalad mucosal or deep spread takes the tumor to the maxilla and it may destroy the lower pterygoid plates and maxillary tuberosity while extending into the lower reaches of the infratemporal fossa (2, 15). Spread to the mandible is early but frank destruction is usually a late finding in this area of very dense cortical bone (27).

CT and MRI are ideally suited to showing the deep extent of all of these lesions. The planes surrounding the gingivobuccal sulci and retromolar trigone are quite constant and obliteration of these planes follows the patterns described for each site. These studies are particularly useful to show unsuspected spread to the lower part of the infratemporal fossa (15). The CT views of the pterygoid plates and lingual (medial) surface of the mandible are exceptional; however, orthopantomography and dental occlusion films are still necessary for complete evaluation of the mandible.

The submandibular and jugulodigastric nodes are the first order drainage sites (14, 27, 31). Thirty to 40% of patients will have positive nodes at presentation (4, 27). The risk for subclinical disease is about 25% (4, 27). The retropharyngeal nodes may be involved if the tumor has spread to the hard palate, soft palate or nasal cavity (3, 14, 15, 26, 30, 31).

Lesions of this region are generally treated with surgery. Options include: (*a*) local excision taking a cuff of periosteum or bone, (*b*) rim resection taking the tumor and rim of adjacent mandibular alveolar ridge, and (*c*) excision including partial mandibulectomy. The decision is based mainly on extent of periosteal or bone involvement (27). Superficially spreading lesions covering a large area of mucosa, such as those spreading to the tonsillar pillar, soft palate or buccal mucosa, often receive primary radiation therapy but bone invasion requires partial mandibulectomy (27). If surgery is not feasible then radiation is sometimes used as the initial therapy reserving surgery for salvage (27). In advanced lesions postoperative radiation is usually added to the surgical management (27).

FLOOR OF THE MOUTH

Ninety percent of the squamous cell carcinomas of the floor of the mouth arise within 2 cm of the anterior midline (27). Mucosal penetration and early deep spread is a frequent finding (27). These patterns are particularly well visualized on axial sections, although coronal MR images may prove useful. Spread here is primarily within the areolar tissue of the sublingual space (Fig. 10.7). Tumor may penetrate the sublingual glandular tissue and spread posterior along the course of the submandibular gland duct and lingual artery and genioglossus into the root of the tongue (2, 27). More commonly it spreads laterally to involve the mylohyoid muscle to its attachment along the lingual surface of the mandible (Fig. 10.1). The periosteum at this point forms an effective barrier and frank mandibular invasion is a late finding, usually occurring after tumor has grown up over the mandibular alveolar ridge (2, 27).

The submandibular and jugulodigastric nodes are first order drainage sites (14, 31). There is a 2% incidence submental node involvement (27). Since this usually is a midline lesion the patients are at high risk for bilateral nodal spread. Overall, 30% of patients with T_1 lesions will have clinically positive necks at presentation; however, 30% of these will turn out to be false positive physical findings. This reflects the frequency of enlarged, reactive submandibular nodes in these patients who often have very poor dentition (27). T_2 and T_3 tumors have a 65 and 71% rate of histologically confirmed nodal metastases.

Diagnostic imaging of the primary site should include CT or MRI in any lesion if deep infiltration is suspected from the clinical examination. Superficially spreading lesions are common in this area and needless imaging should be avoided. Inflamed and/or obstructed sublingual salivary glands can mimic tumor. Obstruction of the submandibular duct can cause enlargement, abnormal enhancement and even frank suppurative necrosis in the submandibular gland; this should not be mistaken for abnormal nodes. CT does well at showing bony invasion along the lingual surface of the mandible but orthopantomography and occlusal views may still be required for complete evaluation of the mandible. Imaging of the lateral floor of the mouth lesions seems to contribute useful information more often than studies of those in the anterior midline. The cervical nodes should be studied at least to the midcervical region. Skip from the submandibular to the lower deep cervical nodes (juguloomohyoid) is possible but quite unlikely (14, 27, 31).

Surgery or radiation will produce similar cure rates in T_1 and early T_2 lesions (27). Lesions attached to the periosteum will usually receive a rim resection (27). Mandibular tori raise the risk of bone exposure with interstitial implants and these patients are usually treated by rim resection (27). Involvement of the tongue usually indicates radiation because of the better functional results (27). Lateral lesions are usually handled surgically (27). Larger lesions (T_2 and T_3) are usually managed with a combination of interstitial and external irradiation because of the major cosmetic deformity and functional disability produced by surgery (27). Periosteal involvement is not a contraindication to radiation but raises the risk of bone exposure and necrosis (27). If rim resection is possible it will be followed with radiation. Tumors which are T_4 due to bone invasion only are occasionally cured by combined surgery and radiation. Extensive tumors with fixation of the tongue, massive neck disease or spread to the submental skin are palliated (27).

ORAL (ANTERIOR TWO/THIRDS) TONGUE

Tumors of the anterior portion of the oral tongue tend to present early and are usually superficial at presentation (4, 27). Those of its midportion may be more deeply infiltrating and tend to spread to the floor of the mouth (27). The lesions of the posterior oral tongue behave more like cancer of the tongue base; they are more invasive. There are no anatomic barriers to restrict a deeply infiltrating cancer of the tongue. Spread occurs freely along the intrinsic and extrinsic muscles. Extension to the tonsillar pillar and soft palate via the palatoglossus muscle is possible (2, 27). Posterior inferior extension via the glossopharyngeus to the glossotonsillar sulcus may occur. Spread to the sublingual space can carry the tumor to the mylohyoid and geniohyoid with all of the implications of spread in the floor of the mouth that were discussed in the preceding section (Fig. 10.7). Growth along the intrinsic musculature can carry it to the tongue base.

First-order nodes are the submandibular and jugulodigastric groups (14, 27, 31): 30–40% of patients stage T_1 or T_2 have clinically positive nodes at presentation (4, 27). This rate doubles to 72% in T_3 and T_4 lesions (4, 27). Both necks are at high risk because the lesions often involve and cross the midline and because of the crossed lymphatic drainage of the tongue. (4, 14, 27, 31).

CT or MRI may be done for lesions staged T_2 or larger, or for a lesion of any size which shows evidence of deep infiltration. Any lesion of the posterior portion of the oral tongue should probably be studied because of their tendency for deep invasion. Coronal and sagittal MR images will augment the axial sections by showing the full extent of these relative to the intrinsic and extrinsic musculature to the tongue base (23). Orthopantomography is still a useful complement to the axial CT and MRI view of the mandible in cases where mandibular invasion is suspected clinically or by the imaging studies. Nodes should always be staged to at least the midcervical region. The retropharyngeal nodes are not at significant risk unless the lesion has spread to the soft palate, nasopharynx or posterior pharyngeal wall, or if the patient is being studied for recurrent disease.

The interpretive pitfalls mentioned in the discussion of the floor of the mouth may also apply in the oral tongue. The reader is referred to the anatomic and pathologic sections on the tongue base and floor of the mouth for a review of normal variations and abnormalities that can lead to mistakes in defining the deep extent of malignant lesions.

Operation and radiation yield equal local control rates for T_1 and T_2 lesions of the oral tongue (27). The choice is based on expected functional and cosmetic loss and the patient's desires. Wide local excision is suitable for well demarcated T_1 lesions. Radiation is usually chosen for large T_1 and T_2 lesions to preserve speech and swallowing (27). In the moderately advanced T_2 and T_3 lesions, superficial lesions may be handled with radiation and deep ones usually recieve combined therapy (27). CT and MR imaging are most valuable in deeply infiltrating lesions. Deep projections of the primary tumor are sometimes difficult to recognize and the likelihood of cutting across tumor are greater than for other sites in the head and neck (27). The preoperative map available from imaging can help the surgeon plan a glossectomy that will yield adequate margins. T_4 lesions are treated for cure either with total glossectomy/laryngectomy or total glossectomy with myocutaneous flap reconstruction combined with postoperative radiation. Most patients with such advanced lesions are treated palliatively (27).

OROPHARYNX

Tonsillar Area and Soft Palate

Early tumors may only show superficial thickening which is accepted as a normal variant in this area (30). Tumors of the anterior tonsillar pillar frequently spread deeply via the palatoglossus muscle to the soft palate above and posterolateral portion of the oral tongue below. Their pattern of deep spread differs from lesions of the retromolar trigone until they extend anteriorly to the mucosa of the trigone (Fig. 10.5). From the retromolar trigone (RMT), lesions may go on to the buccinator space, along the pterygomandibular raphe and to the mandible. The posterior edge of the hard palate may be invaded but destruction of the maxilla is usually a late finding. Extension from the soft palate to the nasopharynx and skull base via the parapharyngeal space and tensor and levator palati muscles is also a late phenomenum (3, 21, 22, 28). Such spread often is considered resectable until it reaches the eustachian tube orifice.

Lesions of the tonsillar fossa begin either within remnants of the palatine tonsil or from the mucosa in the recess between the two pillars. Deep infiltration will obliterate the underlying parapharyngeal space and the tumors are then free to spread to the level of the nasopharynx and skull base (Fig. 10.5). As these lesions advance they also tend to grow posteriorly and inferiorly along, and deep to, the glossotonsillar sulcus. They may spread to the tongue base (styloglossus, glossopharyngeus) the pharyngeal wall (glossopharyngeus), soft palate (palatopharyngeus), and directly to the carotid sheath (Fig. 10.5). Advanced lesions may also invade the mandible and spread beyond the pharyngoepiglottic fold to the pyriform sinus.

Discrete tumors of the posterior tonsillar pillars are unusual (28). Their patterns of spread are as those just described for tonsillar cancers once they reach the posterior tonsillar pillar.

Cancers of the soft palate virtually always arise on its oral surface (4, 28). Deep spread to the nasopharynx is along the tensor or levator palati muscles and parapharyngeal space and may continue cephalad to the skull base (2, 21, 22). Caudal spread is first to the tonsillar pillars and then on to the oral tongue (palatoglossus) and tongue base or pharyngeal wall (palatopharyngeus). The tumors may also spread forward and destroy the hard palate.

Anterior tonsillar pillar lesions have about a 45% risk of clinically positive nodes at presentation (4, 28). The submandibular and jugulodigastric nodes are first echelon drainage sites (4, 31). Patients with tonsillar fossa and posterior tonsillar pillar cancers run a 74% chance of having clinically positive nodes at presentation (4, 28). The jugulodigastric, postglandular submandibular, junctional, and spinal accessory nodes are all at high risk (4, 14, 28, 31). Anterior pillar lesions spread to contralateral nodes about 5% of the time increasing to 11% in tonsillar lesions (28). The risk for subclinical disease is 10–15% and 50–60% respectively (4, 28).

About 35–50% of patients with carcinoma of the soft palate will have positive nodes at presentation (4, 28). First order nodes are mainly the jugulodigastric (15, 31). The retropharyngeal, junctional, spinal accessory, and submandibular nodes are also at risk (3, 14, 28, 31). Sixteen percent of patients will have bilateral nodes (4, 28).

CT should be used to stage the primary site and nodal groups at risk in patients with cancers 2–4 cm (T_2) or larger. Any size lesion with findings suggesting deep spread should also be studied. Recall that the palatine tonsil is a preferred site of origin for occult carcinoma and this area should be studied in detail in patients presenting with cervical metastases of uncertain etiology (*see* Chapter 8). Studies of the primary should always include the area from the midnasopharynx to midsupraglottic larynx and be extended, if necessary, anywhere from the skull base to the thoracic inlet.

The MRI signal from tumor overlaps with that of surrounding lymphoid tissue (25). This means that diagnosis of the presence and extent of lesions will be based mainly on morphologic criteria as it is with CT. MRI's rendering of the anatomy may prove superior to that of CT, and eventually make MRI the examination of first choice. The issue of bone invasion will have to be evaluated very carefully on CT and MRI comparative studies before MRI can be thought of as completely replacing CT. Coronal CT sections should be done whenever subtle destruction of the skull base or hard palate is suspected but cannot be confirmed on axial CT or MRI images. If the mandible is at risk and CT or MRI studies are inconclusive, orthopantomography is indicated. Dental occlusal views should not be necessary in lesions this far posterior.

The N work-up should include imaging of the retropharyngeal to midcervical nodes in detail and survey sections through the low neck. All but the lower midjugular and lower jugular nodes will be included in the routine study of the primary site as described in the prior paragraph. The M work-up routinely includes only a chest x-ray.

Radiation is the usual choice for tonsillar region cancer except for the occasional small lesion that can be handled by wide local excision or tonsillectomy (28). Surgical resection usually requires removal of the mandible, tonsil, both pillars, and perhaps part of the tongue, and the functional loss is not justified with the high success rate of radiation (28). T_2 and T_3 lesions have a 20–50% chance of failure even when adequate doses are applied (28). These may be salvaged with surgery. Lesions staged T_4 only because of mandibular invasion receive surgery followed by radiation (28). CT may help avoid unnecessarily morbid surgical procedures in advanced lesions by showing deep spreads which lie beyond the limits of curative resection (6, 15, 24, 30, 32). Radical irradiation in these patients may control approximately 25% of such lesions (28). Lesions of the soft palate are usually irradiated with surgery reserved for salvage of radiation failures (28).

BASE OF THE TONGUE

Cancer of the tongue base may cause ear pain, a sense of fullness in the back of the throat, and/or a mild case of dysphagia. All of these symptoms may be vague and sometimes considered "functional" by referring clinicians. These complaints are important to the diagnostic imager because we are often asked to perform conventional studies such as barium swallows on such patients who come from physicians who may not be aware that the signs and symptoms can be related to tongue base cancer. Under the proper circumstances, we should take the initiative as consultants and suggest a CT or MRI as well as referral to a head and neck surgeon. Cancer of the tongue base is notorious for its tendency to infiltrate deeply, sometimes without any sign of mucosal disease; this is, in part, because it arises in the deep epithelial lined crypts below the circumvallate papillae. Examination of the tongue base requires palpation which may be impossible for even an experienced clinician if the patient is unable to cooperate.

Cancers of the tongue base tend to begin laterally and remain on one side until advanced. Coexisting, normal tonsillar tissue must not be mistaken for tumor spreading across the midline; this distinction sometimes requires biopsy. These malignancies grow within the intrinsic muscle mass at the root of the tongue and often surround the glossopharyngeal sulcus (Fig. 10.8). Tumors may grow inferiorly and burrow into the pre-epiglottic space. Superfically, they may spill over the pharyngoepiglottic fold from the vallecula and into the pyriform sinus. Spreading anteriorly, they may infiltrate the posterior edge of the mylohyoid muscle and grow along the muscle and within the sublingual space to involve the anterior floor of the mouth. They may also spread freely along both intrinsic and extrinsic muscles to the anterior two-thirds of the tongue. Tumors may spread laterally from the lower glossotonsillar sulcus directly into the soft tissues of the neck and the carotid sheath (Fig. 10.8). Mandibular destruction is unusual in all but the most advanced or recurrent tumors. Tumor usually reaches the mandible by spread along the pterygoid or mylohyoid muscles. A favored site of invasion is at the entry of the neurovascular bundle on the medial aspect of the mandible.

About 75% of patients will have clinically positive necks at presentation; 30% will have bilateral nodes (4, 28). The jugulodigastric nodes are first order drainage sites (14, 31). Submandibular nodes are at risk in lesions which involve the floor of the mouth. Spinal accessory nodes are at small but definite risk and are included in treatment plans (28). Even in a clinically negative neck the chances are 50–60% that an occult metastasis will be present (4, 28).

CT should be done as part of the T work-up for all tongue base cancers. MRI is potentially accurate for staging if the signal from tumor can be distinguished from that of the usually extensive lingual tonsillar tissue surrounding the tongue base and lower glossotonsillar sulci. Our studies indicated that the signal from tumor does not differ enough from that of normal lymphoid tissue to allow this distinction (7–9, 25). The area from the hard palate (C_1) to the midsupraglottic larynx should be studied in detail. This will also include the areas of major concern in the N work-up but the remainder of the mid- and lower cervical nodes should be surveyed. A chest x-ray will usually suffice for the M work-up.

The key to the use of CT (or MRI) lies in its ability to help decide whether curative resection is possible. Total glossectomy is considered an unacceptable alternative in many institutions because of its extreme morbidity and small chance of cure in such extensive lesions (28). Partial glossectomy requires preservation of one lingual artery and hypoglossal nerve, and CT can virtually always determine whether these structures can be saved while maintaining an adequate margin of resection (6, 15, 19, 24, 30, 32). CT may also help in subtle decisions of whether partial glossectomy with or without laryngectomy should be done. Occult spread to the lateral soft tissues of the neck may indicate a lesion beyond the limits of curative resection and radical radiation may be done to attempt cure. In some institutions the choices of therapy are less complex since radiation is used as the primary therapy with surgery reserved for radiation failures (28).

RECURRENT DISEASE

Recurrent carcinoma of the oral cavity and oropharynx will declare itself within 2 yr of treatment (4, 28). Longer delays are less common, probably occurring in only 5–10% of patients who eventually do recur (28). Recurrence after 2 yr may well be a second primary, rather than lack of control at the primary site.

Patients with a primary site recurrence will almost always complain of local pain. Referred pain to the ear is a common symptom in tongue base or glossotonsillar sulcus lesions. Trismus is a late sign and usually indicates gross involvement of the musculature within the infratemporal fossa. Trismus is probably most commonly present in recurrent tonsillar and retromolar trigone lesions since these lesions have a predilection for spread along the pterygoid muscle groups (2, 15).

Physical examination of patients who have been treated by radiation or surgery for cancer in this region is sometimes difficult. These patients are at risk for recurrence in the neck as well as at the primary site, especially if the primary site is not controlled (4, 28). Scarring and induration following treatment limit even the experienced examiner's evaluation of the neck (15, 28). Inaccessible nodes such as the retropharyngeal and highest internal jugular groups are at increased risk due to altered lymphatic drainage or spread of tumor to the nasopharynx (2, 3, 13, 15, 26). In some cases, recurrence at the primary site may produce small areas of local ulceration or induration but provide little other evidence of their true extent on inspection or endoscopic exam; this is due to a tendency for recurrent tumor to go "underground" and spread primarily along the deep pathways described previously (2, 15, 24, 28). Imaging becomes very important under these circumstances.

CT has proven extremely valuable in both the detection and quantitation of recurrent cancer of the oral cavity and oropharynx (15, 24, 28, 33). When recurrent tumor is detected by CT the patients have always been symptomatic. To date there are no prospective studies demonstrating that CT can detect asymptomatic recurrences. A patient may have symptoms without CT evidence of recurrence. Follow-up CT studies may reveal tumor recurrence suggesting that the original symptoms were in part because of perineural spread. In such cases the patient's symptoms will be unremitting and progressive. In other patients with persistent

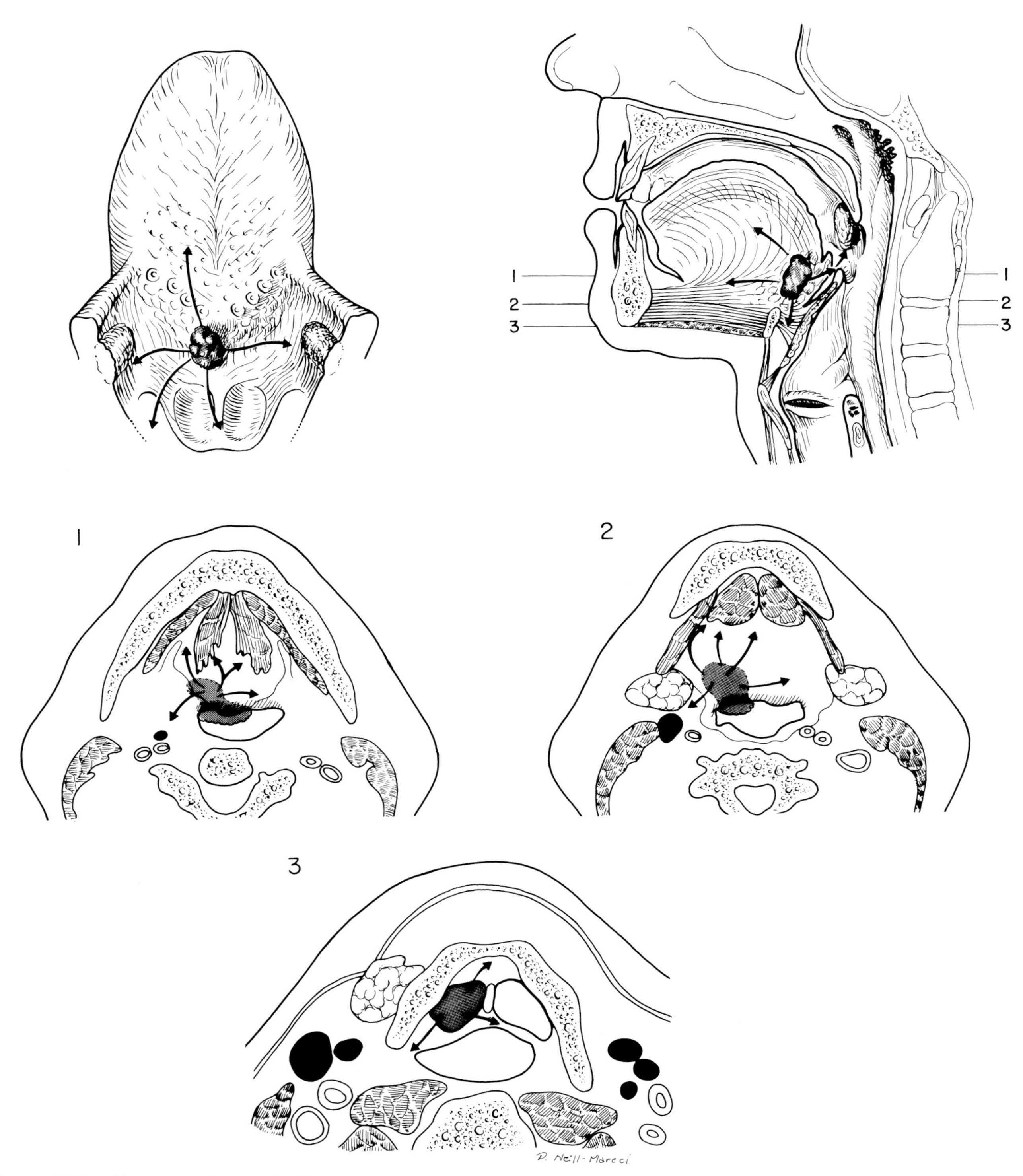

Figure 10.8. Patterns of spread of tongue base cancer. The superficial and deep spread patterns are indicated by the whole organ and sagittal diagrams of the tongue and related structures. The axial sections *below* correspond to the levels indicated on the sagittal drawing. These patterns are discussed in detail in the text.

pain CT studies may also prove negative and the cause of the pain or other findings remains unknown. Possible reasons include perineural spread without a gross mass lesion, soft tissue necrosis, low grade osteonecrosis, scarring of pterygoid muscles and resultant temporomandibular joint dysfunction, scarring and retraction of neck musculature with shoulder dysfunction, and neck pain (28).

An unequivocal diagnosis of recurrence can often be made on a single CT study; however, residual scarring can sometimes produce persistent obliteration of deep tissue planes. Usually the deep anatomy returns to a near-normal appearance following successful radiation therapy and surgical defects appear very "clean-cut" (i.e. appropriate anatomy is missing but residual anatomy looks normal). If patients are at high risk for recurrence we and others suggest a baseline study at 6–12 weeks following the completion of therapy (15, 33). Follow-up scans are done at 3-month intervals twice and 6-month intervals twice. To date, salvage of primary site failures has a poor success rate (28). Earlier detection by follow-up CT or MRI may improve the patient's chances for long-term survival. There is a chance that MRI will be able to differentiate the signal of fibrotic residue of treatment from that of tumor (25). If a baseline study is done, residual fibrosis should have a signal about equivalent to muscle. Increasing signal or persistently high signal on T_2-weighted spin echo images may indicate recurrent or persistent tumor, respectively.

Earlier in this chapter patterns of deep spread were discussed for each of the anatomic regions within the oral cavity and oropharynx. When tumor recurs at these sites, it has even more potential to spread anywhere from the skull base to the hypopharynx; however, there are certain predilections for local spread depending on the primary site and these are basically the same as those discussed previously. The retropharyngeal nodes must always be studied in patients with known or suspected recurrence as well as all of the cervical nodes (3, 13, 15, 26).

MANDIBULAR MALIGNANCIES

In most practices, squamous cell carcinoma arising in the oral cavity or oropharynx will be the most common malignancy affecting the jaw. Many other tissue types may arise in, or secondarily involve, the mandible. All are unusual and they are most logically considered from the viewpoint of their tissue of origin. The hematopoietic tissue within the marrow cavity can produce a myeloma, plasmacytoma, or malignant lymphoma. Chondrosarcoma, osteogenic sarcoma, and fibrosarcoma all occur in the mandible although their frequency in the facial skeleton is much lower than that in the rest of the body (4). Angiosarcoma, neurogenic sarcoma, Ewing's sarcoma, malignant fibrous histiocytoma, and malignant giant cell tumor may also occur (4). Malignant varieties of odontogenic tumors, namely amieloblastoma do occur but they are rare.

The imaging of these tumors is fairly straightforward. Orthopantomography and plain films are usually done as part of the initial evaluation. A question then sometimes arises as to the full extent or origin of the lesion. CT at present is the best tomographic study of the mandible for bone detail and is usually most informative in the axial plane. Soft tissue extent of lesions is equally well shown by CT and MRI, while MRI will probably be more sensitive to spread of tumor within the marrow cavity of the mandible.

Benign Masses

The most common benign tumors affecting the oral cavity and oropharynx are squamous cell papillomas (4). These do not come to imaging. The majority of benign masses that do present for diagnostic imaging do so because they produce bulges under normal appearing mucosa or in the neck. Other presenting complaints may include dysphagia (odynophagia). Airway obstruction is a late finding.

Benign tumors of the mandible include odontogenic cysts and the benign variants of the malignant tumors discussed previously. CT or MRI may help differentiate odontogenic cysts from amieloblastoma based on the density or signal from the contents within the expanded shell of mandibular cortex.

Cystic hydromas, lymphangiomas and hemangiomas involve the tongue and floor of the mouth but present predominantly as neck masses. Neck schwannomas or neurofibromas may extend along the most peripheral course of the cranial nerves and involve the tongue, floor of the mouth and mandible. The full extent of these lesions is sometimes appreciated only by studying the patient in detail from the skull base to the final destination of these nerves. Benign masses in the lateral floor of the mouth, submandibular space, or suprahyoid neck are usually related to nodes or the submandibular gland; however, other lesions must be considered. Teratomas in this region usually occur in the midline and may lie either above or below the mylohyoid and geniohyoid (4). They may appear multiloculated or contain fat and calcific densities as well as fluid. About 1.6% of all dermoids occur in this area (4). Plugging of the sublingual gland ducts can lead to the backup, retention, and possibly leakage of mucous secretions. The resulting collection of fluid can produce the cystic mass in the floor of the mouth and/or submandibular space known as a rannula (4). On CT, these appear as thin-walled fluid collections. Treatment is surgical excision.

Thyroglossal duct cysts will be in midline and at the level of the hyoid bone. These were discussed in detail in Chapter 8. Lingual thyroid tissue produces a typically high attenuation mass within the intrinsic musculature. Nuclide studies must be done to be sure functioning thyroid is present in the neck before symptomatic ectopic thyroid tissue is removed from the tongue base (4).

A number of lesions may present as parapharyngeal

masses at the level of the oropharynx including: benign lesions of accessory salivary tissue within the parapharyngeal space (usually benign mixed tumors), paragangliomas, schwannomas and, rarely, chordomas and lipomas. These are discussed in detail in the Chapter 11 on the nasopharynx. The tumors of accessory salivary tissue, usually benign mixed tumors, may occur anywhere but are especially common near the junction of the hard and soft palates.

Inflammatory Lesions

The diagnosis of a parapharyngeal or masticator space infection is most often a straightforward clinical problem. Sometimes it is not clear whether pain and induration is due to cellulitis or abscess (12, 17). Also, deep spread of pus may be beyond detectable limits by physical examination just as occurs in tumors. CT has proven very effective in differentiating cellulitis from abscess and in showing the full extent of abscess cavities so that adequate surgical drainage is assured (12, 17).

Masticator space abscesses are usually the result of dental infections. At CT the fluid collection will be located between the muscles of mastication and the mandible and typically show a thick, intense rim of enhancement. They often contain gas. The surrounding soft tissues will be grossly edematous, characterized by enlargement of the muscles and thickening and increased density of surrounding fat planes.

Parapharyngeal abscesses produce similar gross morphologic features. The most common etiology for a parapharyngeal abscess remains direct spread of a peritonsillar abscess. Penetrating trauma or spread of masticator space abscess are other common causes. Once in the parapharyngeal space pus may spread to the skull base, or involve the cranial nerves in the upper carotid sheath. Caudal spread can involve the submandibular space or spill over the posterior aspect of the mylohyoid muscle and into the deep neck compartments. Penetration into the retropharyngeal space and spread to the mediastinum is possible but unusual in this era of antibiotics and sophisticated diagnostic imaging.

Infections in the anterior floor of the mouth or mandible may spread to the submental space. Abscesses will assume their typical appearance while cellulitis will cause diffuse obliteration of tissue planes with no clear-cut fluid collection. The intense cellulitis seen in Ludwig's angina is a good example of this. Patients with Ludwig's angina, however, must be decompressed surgically since the intense swelling in the confined spaces of the floor of the mouth can lead to severe airway obstruction.

Inflammatory lesions such as those described above are always in a state of evolution. Sometimes a patient may be studied before a cellulitis has "matured" to a frank collection of pus. Also the diffuse inflammation and swelling surrounding early abscess may mask the fluid collection. If the diagnosis of cellulitis is made by CT and the patient does not respond to appropriate antibiotic therapy, then one should not hesitate to restudy the patient to be sure of whether or not an abscess cavity has formed.

References

1. American Joint Committee on Staging of Cancer: *Staging of Head and Neck Sites, and of Melanoma 1980.* Chicago AJC, 1980.
2. Ballantyne AJ: Routes of spread. In Fletcher GH, MacComb WS (eds): *Radiation Therapy in the Management of Cancers of the Oral Cavity and Oropharynx.* Springfield, Ill, Charles C Thomas, 1962, chap 3.
3. Ballantyne AJ: Significance of retropharyngeal nodes in cancer of the head and neck. *Am J Surg* 108:500, 1964.
4. Batskasis JG: *Tumors of the Head and Neck; Clinical and Pathological Considerations,* ed 2. Baltimore, Williams & Wilkins, 1979.
5. Bryan RN, Ford JJ, Schneiders NJ: *NMR Parameters of Nasopharyngeal Soft Tissues.* Presented at the 16th Annual Meeting, American Society of Head and Neck Radiology, San Antonio, May 1984.
6. Byrd SE, Schoen PJ, Gill G, et al: Computed tomography of palatine tonsillar carcinoma. *J Comput Assist Tomgr* 7:976–982, 1983.
7. Dillon WP, Mills CM, Brant-Zawadski M, et al: *NMR Imaging of Head and Neck Pathology.* Presented at the 69th Scientific Assembly Radiologic Society of North America, Chicago, November 1983.
8. Dillon W, Woolf M, Engelstand B: *Magnetic Resonance Imaging of Cervical Adenopathy-Optimal Imaging Parameters.* Presented at the 16th Annual Meeting American Society of Head and Neck Radiology, San Antonio, May 1984.
9. Dillon WP: *Clinical Head and Neck Magnetic Resonance Imaging with a 0.35 Tesla unit.* Presented at the 16th Annual Meeting American Society of Head and Neck Radiology, San Antonio, May 1984.
10. Dodd GD, Dolan PA, Ballantyne AJ, et al: The dissemination of tumors of the head and neck via the cranial nerves. *Radiol Clin North Am* 8:445–462, 1970.
11. Duckworth R: Verrucous carcinoma presenting as mandibular osteomyelitis. *Br J Surg* 49:332, 1961.
12. Endicott JN, Nelson RJ, Saraceno CA: Diagnosis and management decisions in infections of the deep fascial spaces of the head and neck utilizing computerized tomography. *Laryngoscope* 92:630–633, 1982.
13. Fisch U: *Lymphography of the Cervical Lymphatic System.* Philadelphia, Saunders, 1968.
14. Haagensen DC, Feind CR, Herter FP, et al: *The Lymphatics in Cancer,* ed 1. Philadelphia, Saunders, 1972, pp 60–208.
15. Harnsberger HR, Mancuso AA, Muraki AS: The upper aerodigestive tract and neck: CT evaluation of recurrent tumors. *Radiology* 149:503–509, 1983.
16. Hollinshead WH: *Anatomy for Surgeons*: vol 1, *The Head and Neck,* ed 2. Hagerstown, Md, Harper & Row, 1968.
17. Holt GR, McManus K, Newman RK: Computed tomography in the diagnosis of deep neck infection. *Arch Otolaryngol* 108:693–696, 1982.
18. Kalovidouris A, Mancuso AA, Dillon WP: A CT-clinical

approach to patients with symptoms related to the V, VII, IX-XII and cervical sympathetics. *Radiology* 151:671–676, 1984.

19. Larsson S, Mancuso AA, Hoover L, Hanafee WN: Differentiation of pyriform sinus cancer from supraglottic laryngeal cancer by computed tomography. *Radiology* 141:427–432, 1981.
20. Last RJ: Anatomy Regional and Applied, ed 5. Edinburgh, Churchill Livingston, 1972.
21. Lederman M: *Cancer of the Nasopharynx: Its Natural History and Treatment.* Springfield, Ill, Charles C Thomas, 1961.
22. Lederman M: Cancer of the pharynx. *J Laryngol* 81:151, 1967.
23. Lufkin RB, Larsson SG, Hanafee WN: Work in progress: NMR of the larynx and tongue base. *Radiology* 148:173–175, 1983.
24. Mancuso AA, Hanafee WN: Elusive head and neck cancers beneath intact mucosa. *Laryngoscope* 93:133–139, 1983.
25. Mancuso AA, Fitzsimmons J, Mareci T, Million R, Cassisi N: MRI of the upper pharynx and neck: variations of normal and possible applications in detecting and staging malignant tumors, Parts I and II—Normal anatomy and pathology (submitted for publication).
26. Mancuso AA, Harnsberger HR, Muraki AS: Computed tomography of cervical and retropharyngeal lymph nodes: normal anatomy, variants of normal and applications in staging head and neck cancer, Part II: Pathology. *Radiology* 148:715–723, 1983.
27. Million RR, Cassisi NJ: Oral Cavity. In Million RR, Cassisi NJ (eds): *Management of Head and Neck Cancer: A Multidisciplinary Approach.* Philadelphia, Lippincott, 1984, chap 8.
28. Million RR, Cassisi NJ: Oropharynx. In Million RR, Cassisi NJ (eds): *Management of Head and Neck Cancer: A Multidisciplinary Approach.* Philadelphia, Lippincott, 1984.
29. Muraki AS, Mancuso AA, Harnsberger HR: Metastatic cervical adenopathy from tumors of unknown origin: the role of CT. *Radiology* 152:749–753, 1984.
30. Muraki AS, Mancuso AA, Harnsberger HR: CT of the oropharynx, tongue base and floor of the mouth: normal anatomy and range of variations and applications in staging carcinoma. *Radiology* 148:725–731, 1983.
31. Rouvier H: Anatomy of the Human Lymphatic System, ed 1. Ann Arbor Mich, Edwards Brothers, 1938, pp 1–82.
32. Schaefer SD, Merkel M, Diehl J, et al: Computed tomographic assessment of squamous cell carcinoma of oral and pharyngeal cavities. *Arch Otolaryngol* 108:688–692, 1982.
33. Som P, Shugar J, Biller H: The early detection of antral malignancy in the postmaxillectomy patient. *Radiology* 143:509–512, 1982.

Chapter 10 Plates

Oral Cavity and Oropharynx Including Tongue Base, Floor of the Month and Mandible

NORMAL ANATOMY

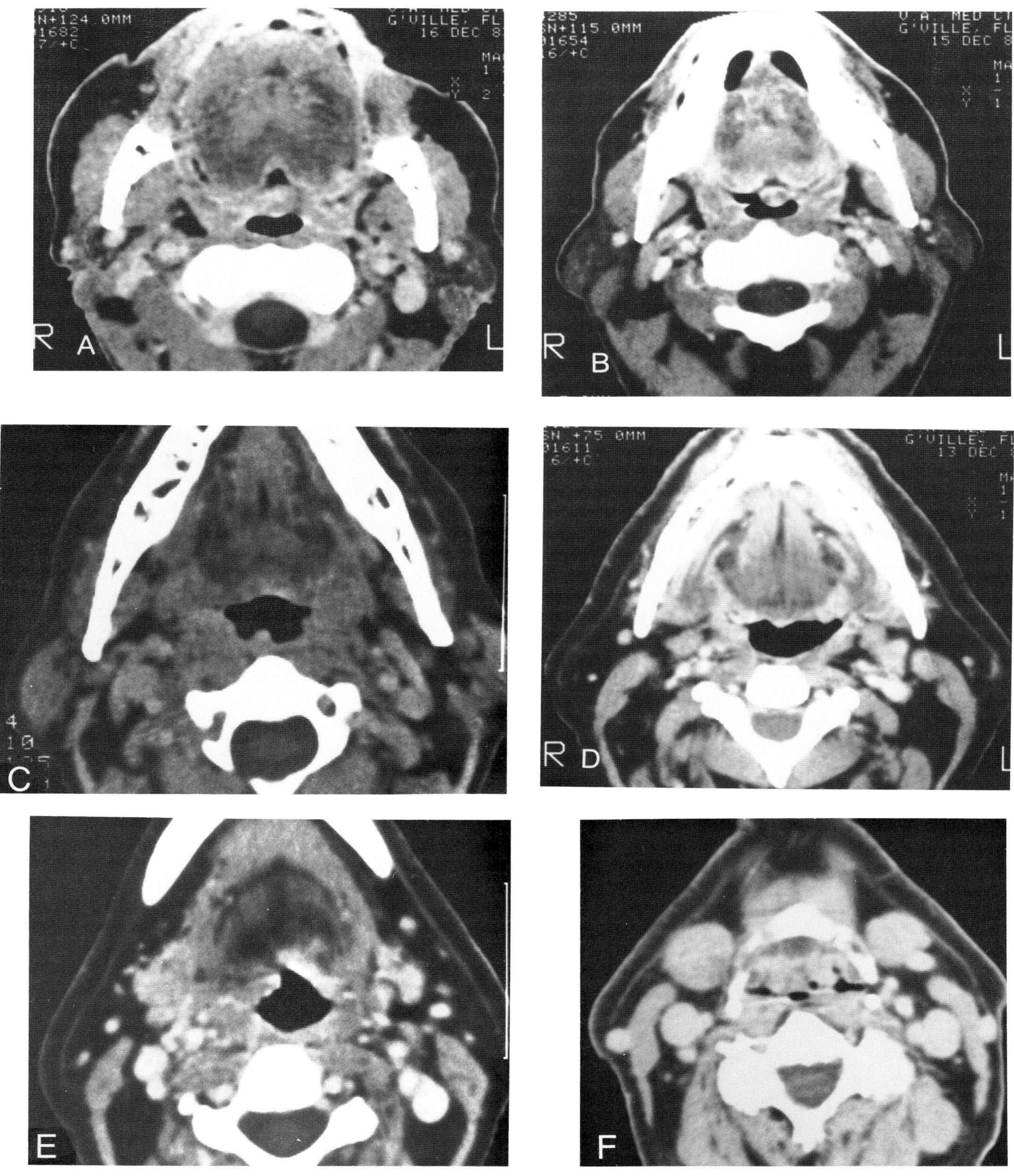

Plate 10.1 A–F.

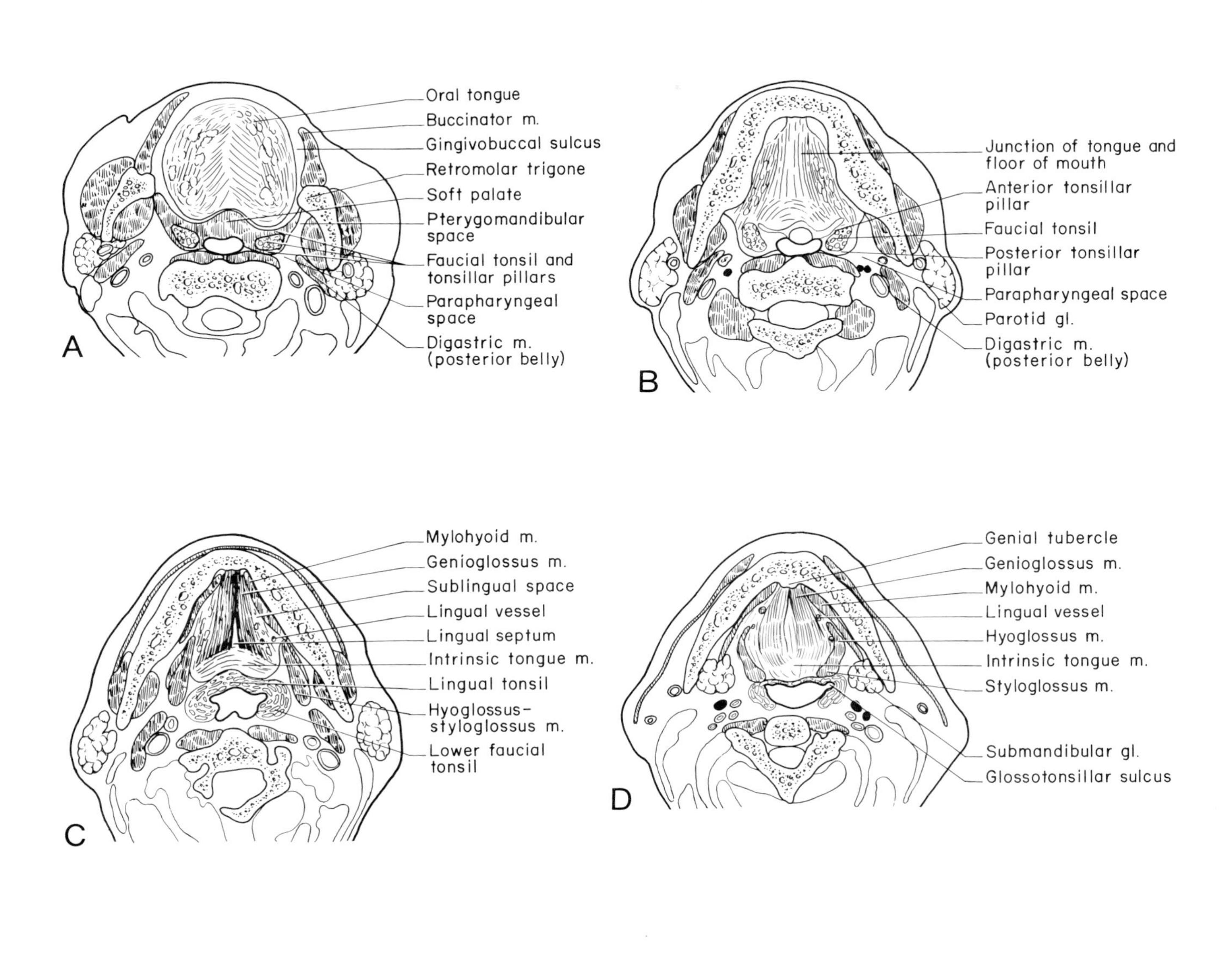
Oral tongue
Buccinator m.
Gingivobuccal sulcus
Retromolar trigone
Soft palate
Pterygomandibular space
Faucial tonsil and tonsillar pillars
Parapharyngeal space
Digastric m. (posterior belly)
A
Junction of tongue and floor of mouth
Anterior tonsillar pillar
Faucial tonsil
Posterior tonsillar pillar
Parapharyngeal space
Parotid gl.
Digastric m. (posterior belly)
B
Mylohyoid m.
Genioglossus m.
Sublingual space
Lingual vessel
Lingual septum
Intrinsic tongue m.
Lingual tonsil
Hyoglossus-styloglossus m.
Lower faucial tonsil
C
Genial tubercle
Genioglossus m.
Mylohyoid m.
Lingual vessel
Hyoglossus m.
Intrinsic tongue m.
Styloglossus m.
Submandibular gl.
Glossotonsillar sulcus
D

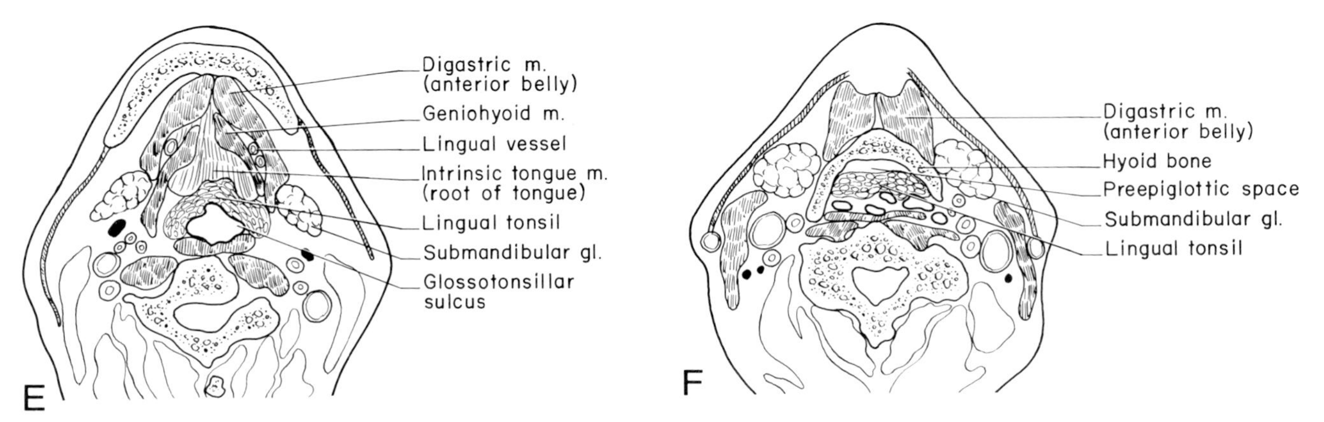
Digastric m. (anterior belly)
Geniohyoid m.
Lingual vessel
Intrinsic tongue m. (root of tongue)
Lingual tonsil
Submandibular gl.
Glossotonsillar sulcus
E
Digastric m. (anterior belly)
Hyoid bone
Preepiglottic space
Submandibular gl.
Lingual tonsil
F

Plate 10.2 A–F.

Plate 10.3. These are axial MRI sections done on a 0.15 Tesla resistive unit. A custom-designed surface coil was used. The anatomic detail in all these images is roughly equivalent to that of CT although, of course, the conrast resolution of different structures noted is based on entirely different physical principles.

Plate 10.3*A*. A spin echo 500/30 image shows the anatomy in the upper oropharynx. The parapharyngeal space (*arrow*) appears bright because of its fat content. At this pulse sequence the tissue within the tonsillar fossa (*arrowhead*) does not produce a dramatic contrast with surrounding muscle tissue.

For the specific anatomy in this and the following Plates refer to the line diagram in Plate 10.2.

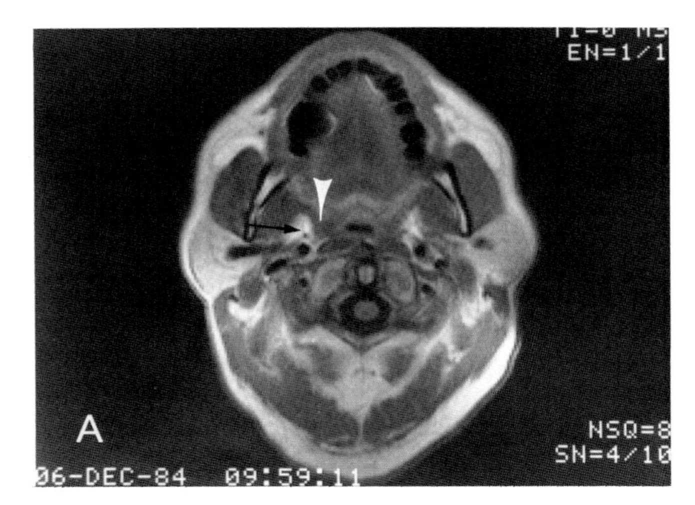

Plate 10.3*B*. This spin echo 500/30 image was obtained through the floor of the mouth. The high intensity signal (*black arrow*) between the genioglossus and mylohyoid muscles is due to fat and glandular tissue within the sublingual space. The uncinate process of the submandibular gland projects into the floor of the mouth over the edge of the mylohyoid muscle (*white arrow*). At this pulse sequence there is very little contrast between the faucial tonsillar tissue (*arrowhead*) and surrounding musculature.

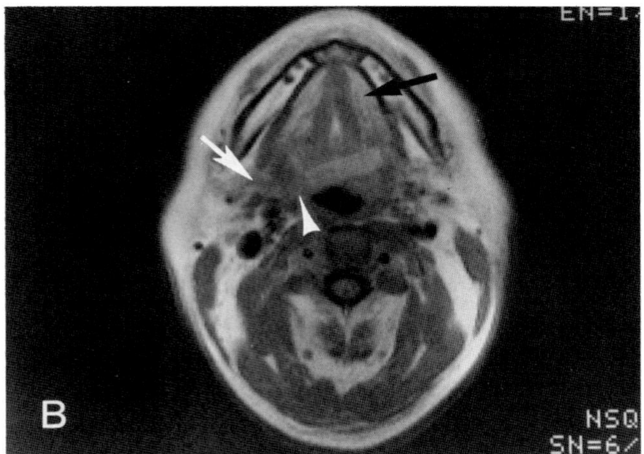

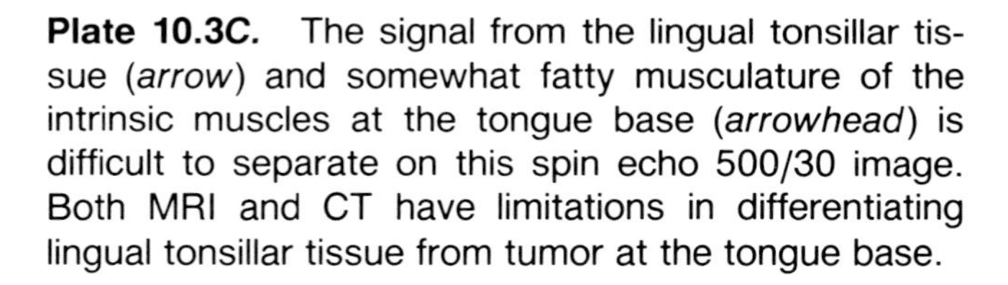

Plate 10.3*C*. The signal from the lingual tonsillar tissue (*arrow*) and somewhat fatty musculature of the intrinsic muscles at the tongue base (*arrowhead*) is difficult to separate on this spin echo 500/30 image. Both MRI and CT have limitations in differentiating lingual tonsillar tissue from tumor at the tongue base.

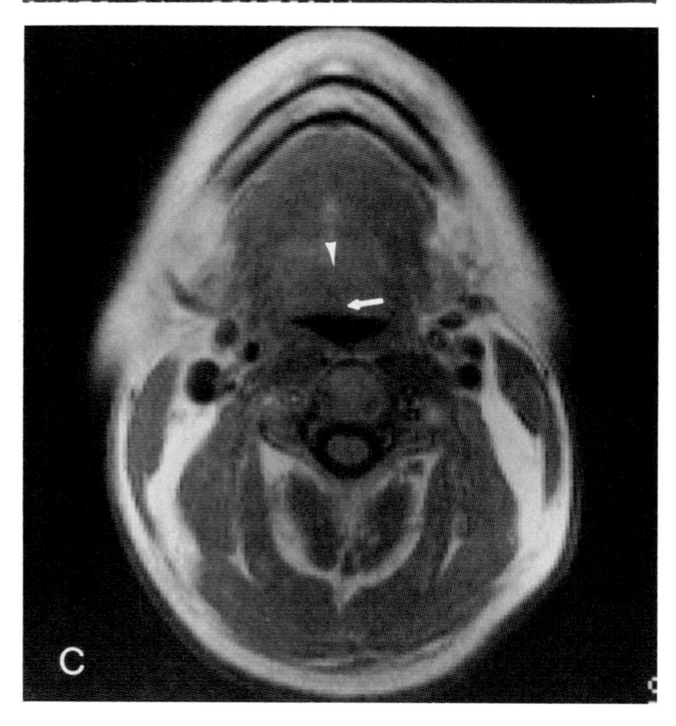

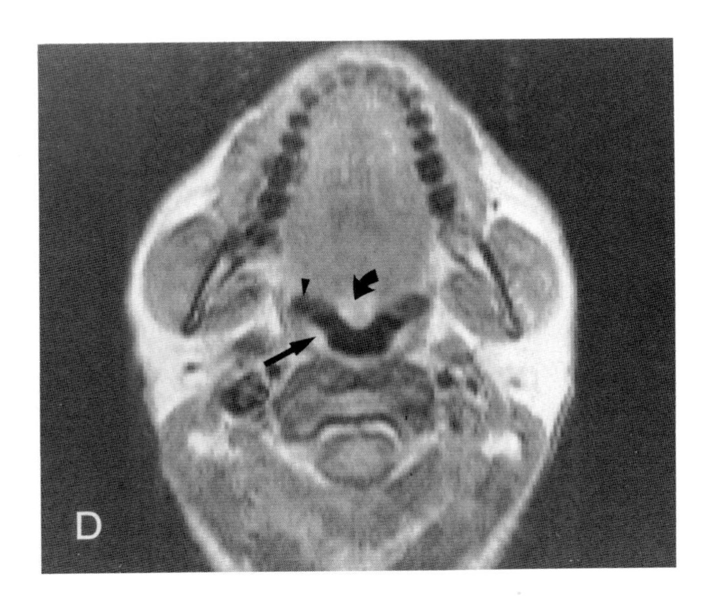

Plate 10.3*D*. The following three images are part of a multislice multiecho acquisition using a T_2-weighted sequence. These are specifically spin echo 2500/90 images. At the level of the faucial tonsils the residual tonsillar tissue along the posterior and lateral pharyngeal wall (*arrow*) has an obviously higher signal than the musculature of the pharynx at this level. This person had a tonsillectomy explaining the residual fibrosis (*arrowhead*) in the more anterior part of the tonsillar bed. The uvula of the soft palate and, in fact, most of the soft palate has a high signal intensity much like that of lymphoid tissue (*curved arrow*); this is probably due to its high content of glandular tissue.

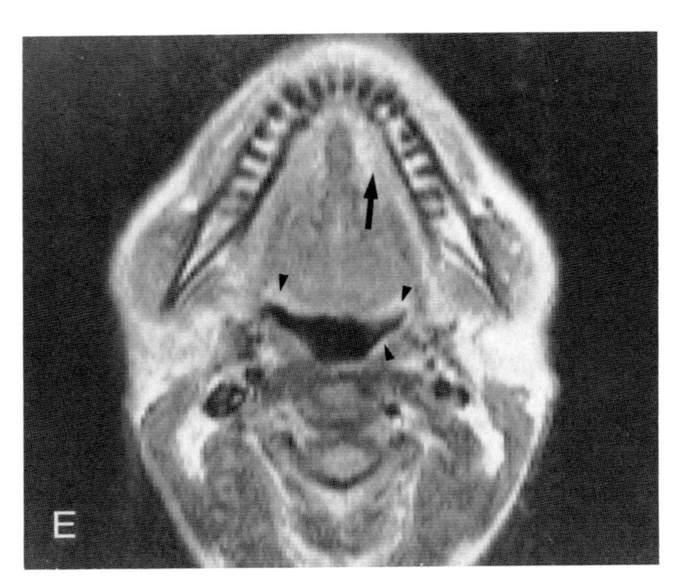

Plate 10.3*E*. A section made just below the lower poles of the faucial tonsils through the upper floor of the mouth. A smaller amount of lymphoid tissue lines the tongue base and posterolateral wall of the oropharynx at this level (*arrowheads*). The intermediate intensity signal within the sublingual space (*arrow*) is due to its mixture of fat and salivary gland tissue.

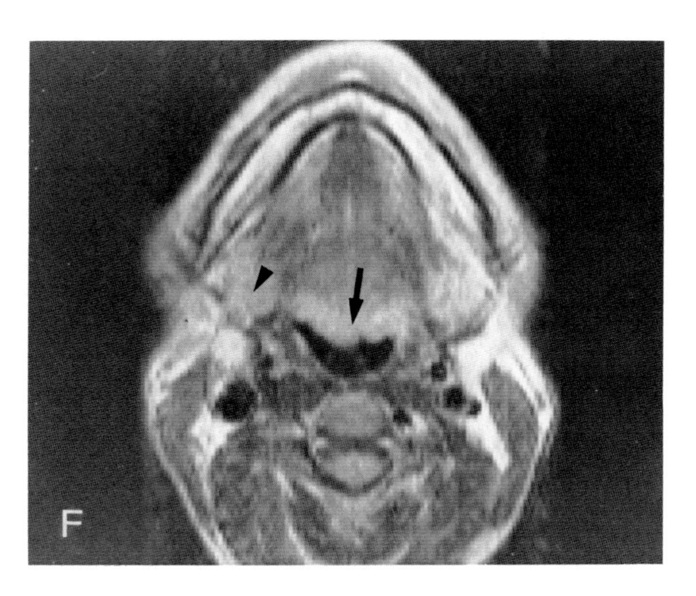

Plate 10.3*F*. A section made through the tongue base shows the relatively high signal intensity of the lingual tonsils (*arrow*). Unfortunately, the signal intensity of squamous cell carcinoma and lymphoma (as well as other tissue types) overlap broadly with those of lymphoid tissue regardless of the pulse sequences used. This means that, like CT, we will have to see some evidence of deep infiltration to be sure whether these masses within the oropharynx represent tonsillar tissue or a more aggressive process. This makes MRI, like CT, complementary to the physical examination with the same limitations in determining the mucosal extent of disease as CT suffers. Note that the signal of the lingual tonsil does not differ very much from that of the submandibular gland (*arrowhead*); the signal from the parotid and salivary glands vary a great deal depending on their relative content of fat and glandular tissue. The appearance of these major salivary glands will vary greatly with age on both MR and CT images.

The abbreviations used on Plates 10.5 and 10.6 are: *GH*, geniohyoid muscle; *D*, digastric muscle; *GG*, genioglossus muscle; and *SMG*, submandibular gland.

Plate 10.4. This plate emphasizes normal variations of tonsillar tissue as seen on both CT and MRI.

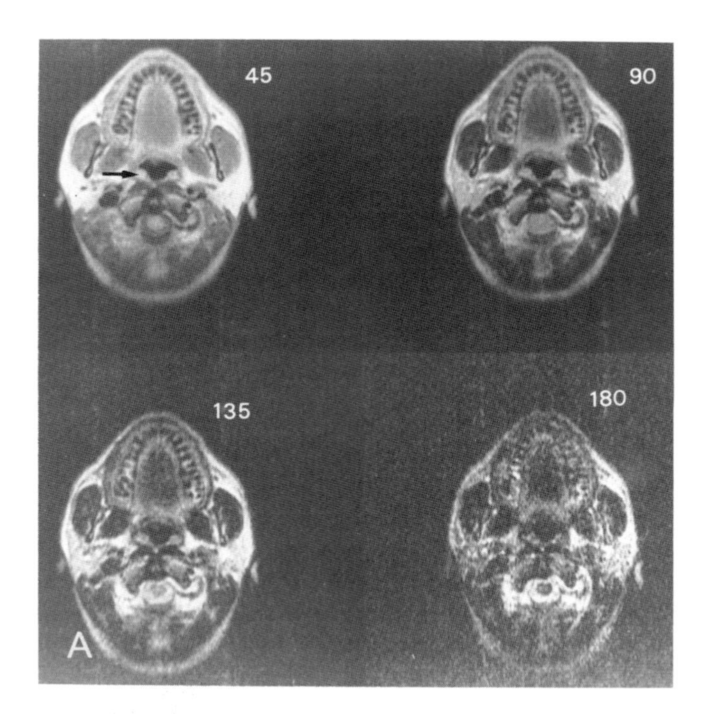

Plate 10.4*A*. This group of four images represents one section of 10 acquired during a multislice multiecho acquisition. The TR was 2500 msec, and the echo delay times were *45*, *90*, *135*, and *180* msec. These show a progressive increase in the contrast between the tonsillar tissue (*arrow*) and surrounding musculature. Unfortunately, the same trends are seen for the tumors which commonly involve the upper aerodigestive tract so that we must still depend on evidence of deep extension to be sure that a mass in these locations represents a malignancy.

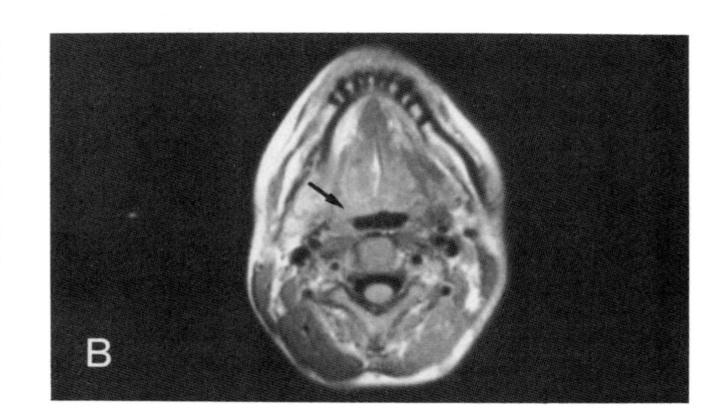

Plate 10.4*B*. A spin echo image from a 1.5 Tesla device. This image shows slightly better contrast and spatial resolution of the tonsillar tissue adjacent to the tongue base (*arrow*). Comparison with images in plate 10.3 A–C will show that effective imaging of the upper aerodigestive tract and neck can be done at a broad range of field strengths with good results. (This image is reproduced courtesy of General Electric Corporation.)

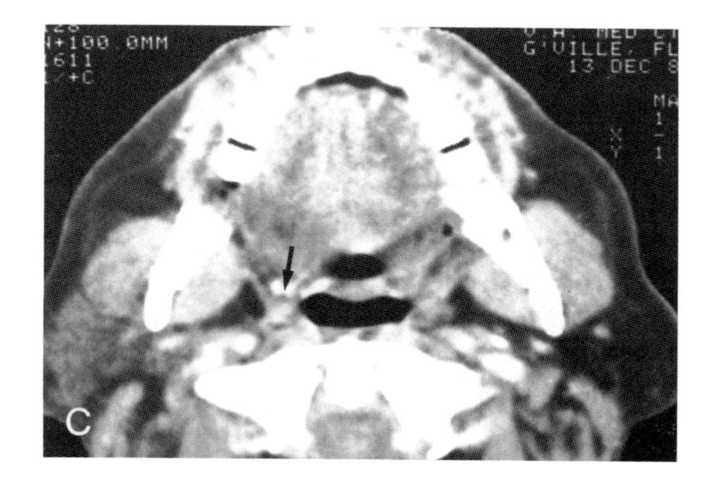

Plate 10.4*C*. Occasionally, punctate or globular foci of calcification will be seen in the faucial (*arrow*) and, more rarely, in the lingual tonsillar region. These are a common normal variant and probably represent dystrophic calcification in areas of prior inflammation.

Focal areas of low attenuation presumably related to small fluid collections are also seen on occasion in the faucial tonsils. This is illustrated in Plate 10.1*B*.

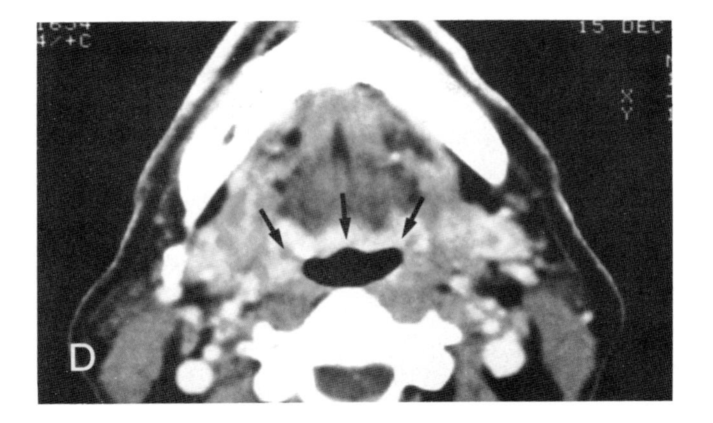

Plate 10.4*D*. Diffuse and fairly intense enhancement is often seen within the tonsillar tissue of the oropharynx. It is more common to see this in the region of the lingual tonsils and the lower glossotonsillar sulci as this figure illustrates (*arrows*) although it may also be seen in the faucial tonsils (*see* Plates 10.1 *A* and *D*). Such enhancement may be seen during an upper respiratory tract infection; however, it is more usual to see this in chronic smokers and people with poor oral hygiene in whom it is most likely a result of chronic low grade infections due to these repeated insults to the mucosa.

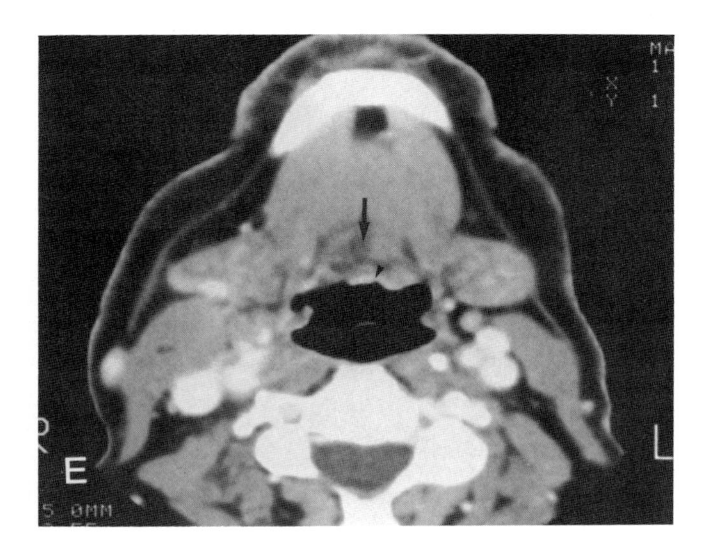

Plate 10.4*E*. The amount of lymphoid tissue anywhere in the oropharynx shows great variation from patient to patient. In most of the prior illustrations the patients had a significant amount of lymphoid tissue present. In this patient there is virtually no lymphoid tissue at the tongue base. There is, perhaps, a small amount immediately adjacent to the airway (*arrowhead*). The normal, slightly low, attenuation of the intrinsic muscle at the very bottom of the root of the tongue (*arrow*) is due to the fat content of this portion of the tongue.

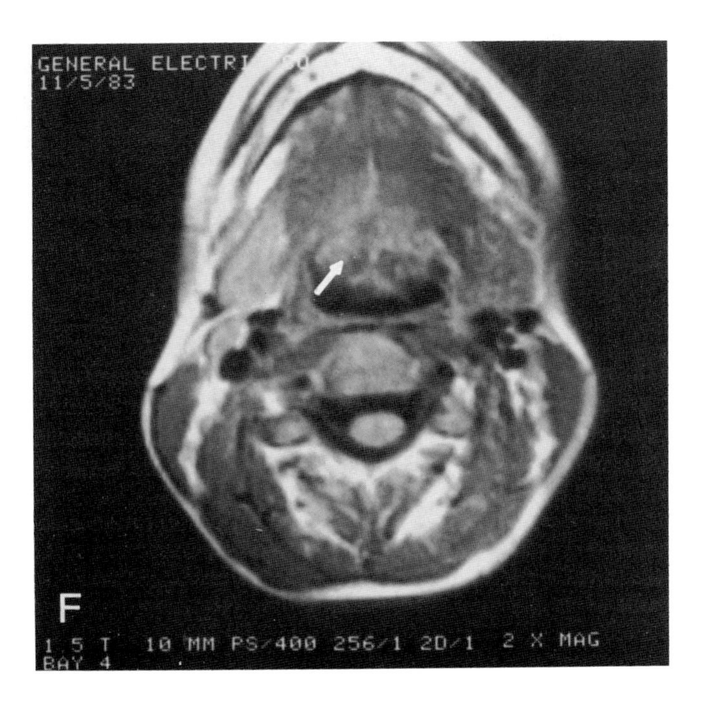

Plate 10.4*F*. MRI done at a field strength of 1.5 Tesla shows relatively sparse lingual tissue in this patient. This should be compared to Plates 10.3 *C* and *F*. Note that the small amount of tonsillar tissue (*arrow*) is very near in intensity to the fat-containing intrinsic muscles at the tongue base. Also note that good MR imaging of the upper aerodigestive tract may be done at a variety of field strengths with roughly comparable image quality. Images the quality of those seen in Plate 10.3 may be obtained with acquisition times between 6 and 7 min; higher field strength acquisition time can be lowered. A full discussion of the relative advantages of different field strengths is beyond the scope of this text. (This image is reproduced courtesy of General Electric Corporation.)

The abbreviations used on Plates 10.5 and 10.6 are: *GH*, geniohyoid muscle; *D*, digastric muscle; *GG*, genioglossus muscle; and *SMG*, submandibular gland.

Plate 10.5. One of the advantages of MR imaging is its multiplanar capabilities. Any part of the anatomy may be studied in virtually any plane although the axial and true coronal and sagittal planes are the most frequently used. The sagittal plane provides some advantages in tongue lesions.

Plate 10.5*A*. A midline sagittal section through the tongue base. The vallecula is filled with lymphoid tissue (*arrowhead*) and it is somewhat difficult to see a definite line of demarcation between the lymphoid tissue and intrinsic muscle mass at the base of the tongue. It is relatively easy to see the fat which lies between these various layers of intrinsic muscles (*arrow*). The fact that the tongue base is contiguous with the preepiglottic space just behind the hyoid bone (*curved arrow*) is an important consideration in the spread of epiglottic as well as tongue base cancers.

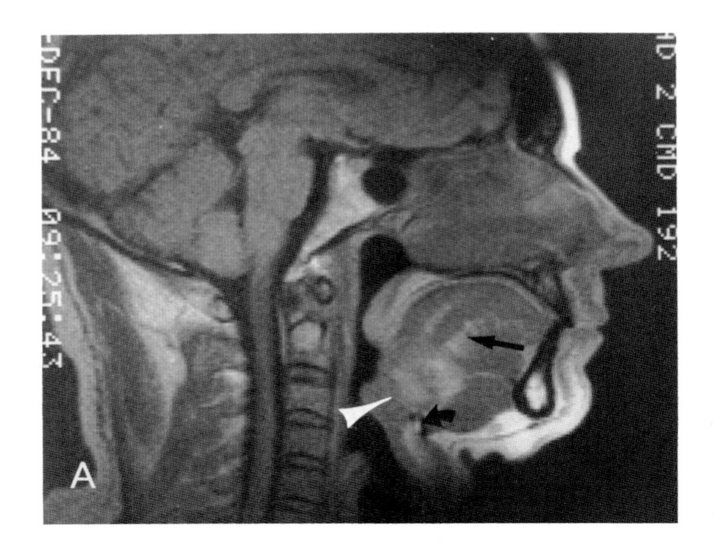

Plate 10.5*B*. A parasagittal section done with a spin echo 500/30 pulse sequence. Again, the tongue base region bulges somewhat as we move laterally (*arrow*) and it is difficult to be sure how much of the tissue is related to a large lingual tonsil and how much to prominent intrinsic muscles of the tongue. The branching lingual vessels are also visualized (*arrowheads*). The bulky geniohyoid muscle is a prominent landmark in the floor of the mouth. The anterior belly of the digastric muscle arises just inferior to the geniohyoid.

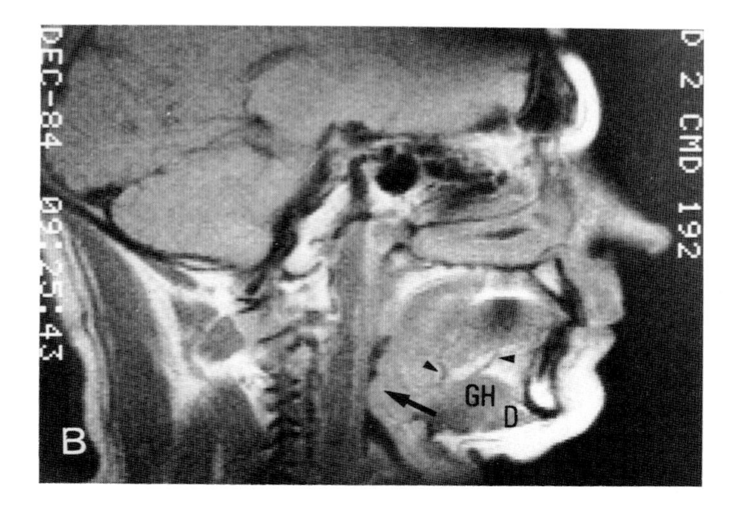

Plate 10.5*C*. A spin echo 2000/60 pulse sequence reveals that a good deal of the prominence at the tongue base is related to the intrinsic musculature (*arrow*) with which this more T_2-weighted acquisition contrasts better than with the fat within the tongue base and the relatively thin strip of lymphoid tissue on the surface of the tongue base (*arrowhead*). Both the geniohyoid and genioglossus muscles form a good deal of the bulk of the floor of the mouth and tongue. Note how the genioglossus fans out anteriorly and superiorly.

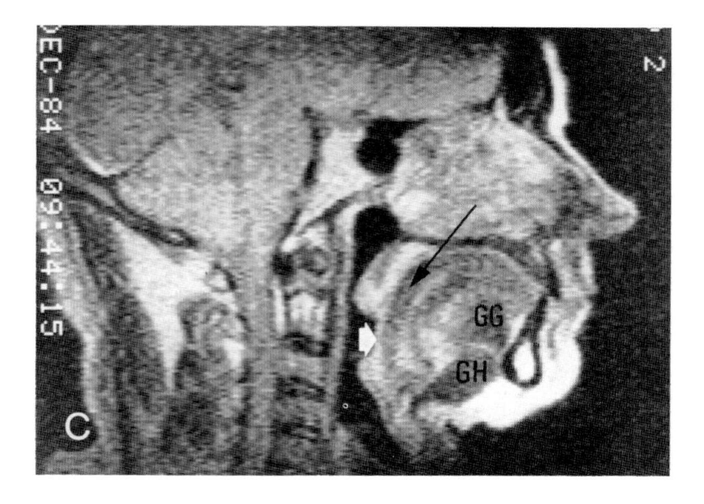

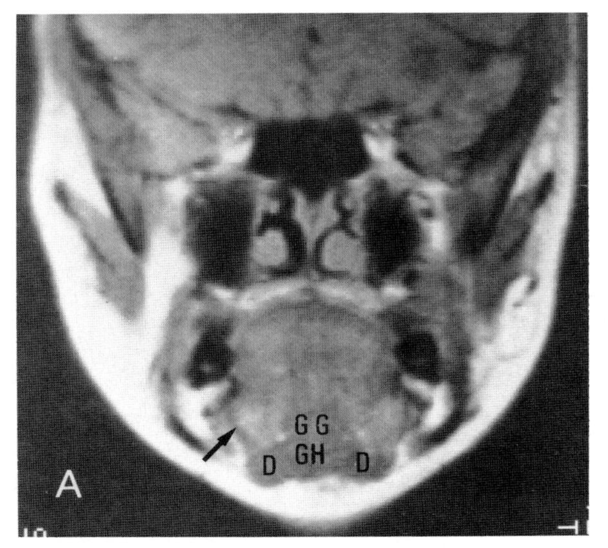

Plate 10.6. Coronal sections can be used to advantage to study both the tongue and floor of the mouth. While axial sections are best suited to the overall study of the tongue and floor of the mouth lesions, the sagittal plane is probably the best supplement to tongue base lesions and the coronal plane best for more anteriorly placed oral tongue and floor of the mouth lesions.

Plate 10.6*A*. A section made relatively anteriorly using a spin echo 500/30 pulse sequence. The anterior edge of the mylohyoid is barely visible (*arrow*). The geniohyoid and genioglossus muscles form the main bulk of the tongue and floor of the mouth in the midline. The anterior bellies of the digastric muscles are seen more inferiorly; since these lie below the plane of the mylohyoid muscle they are really in the suprahyoid neck rather than the floor of the mouth.

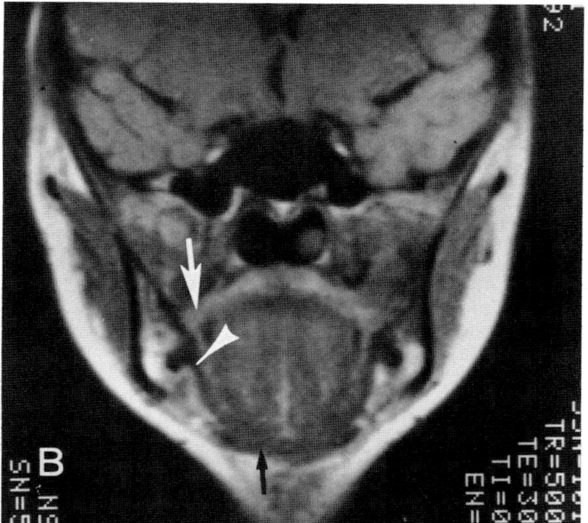

Plate 10.6*B*. A more posterior section gives a very nice view of the oblique course of the mylohyoid muscle as it arises along the myloid ridge of the mandible (*arrowhead*) and runs inferiorly and laterally to attach on the hyoid bone (*black arrow*). The lingual nerve runs through the pterygomandibular space (*white arrow*). Note that this space is occupied by an abnormal mass on the opposite side on this and the following plate. This will be discussed in detail in Plate 10.11.

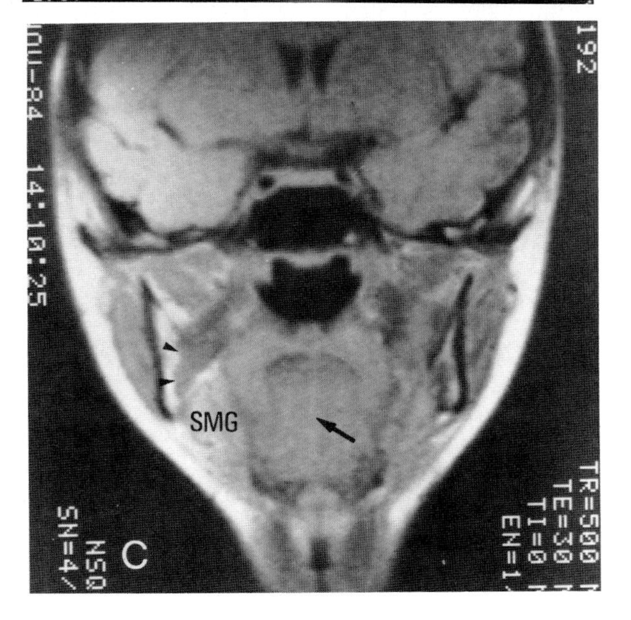

Plate 10.6*C*. An even more posterior section which shows the symmetry of the more posterior part of the oral tongue. The lingual septum is predominantly fatty and lies in the midline (*arrow*). The submandibular gland occupies most of the submandibular space at this point. Note the insertion of the medial pterygoid muscle on the mandible (*arrowheads*).

PATHOLOGY—MALIGNANT TUMORS

Plate 10.7. Masses in the gingivobuccal sulcus are usually limited lesions which do not require CT evaluation. Oftentimes physical examination and orthopantomography augmented by occlusal views will show its relationship to the mandible and the mucosal extent will be apparent clinically. In larger lesions it is sometimes helpful to do either MR or CT imaging of these lesions to be sure of their deep extent. This is especially true when planning operations such as rim resections of the mandible or other variations of partial mandibulectomy.

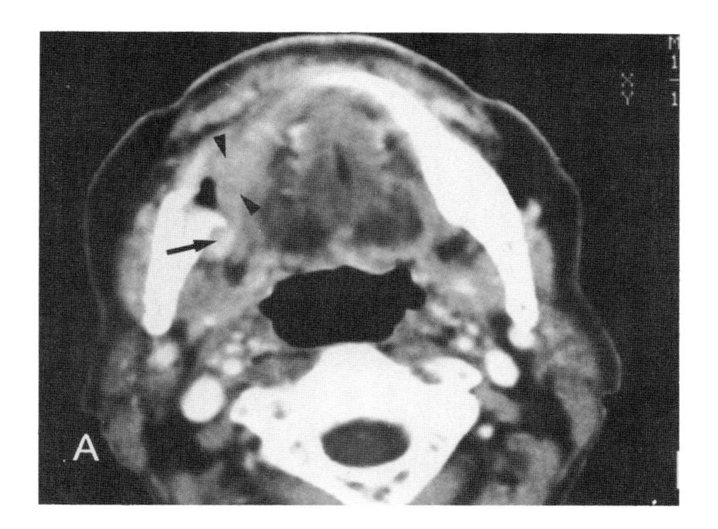

Plate 10.7*A*. A patient with a somewhat large gingivobuccal sulcus lesion. The soft tissue extent of the mass was apparent clinically. On CT the mass (*arrowheads*) spills over into the pterygomandibular space (*arrow*). This is very close to the lingual nerve.

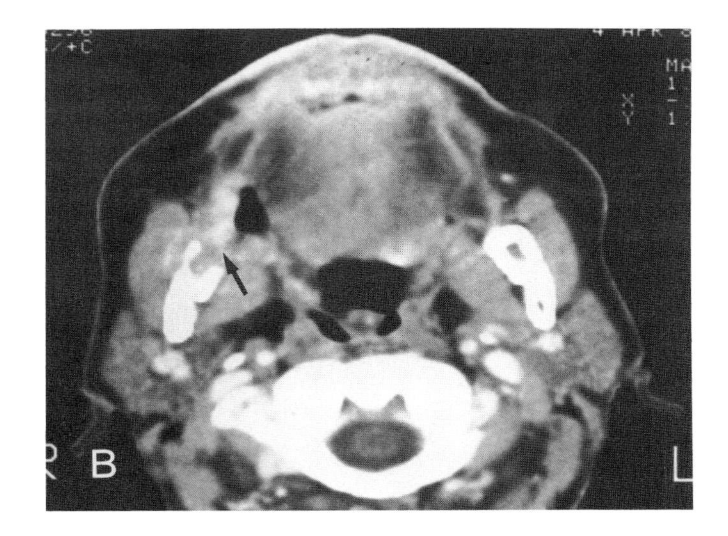

Plate 10.7*B*. A section slightly higher shows that the lesion erodes a large portion of the mandibular alveolar ridge. On the prior section, and this one, the lesion has spilled over into the area usually involved in retromolar trigone lesions. At this point (*arrow*) the mass has access to the pterygomandibular raphe.

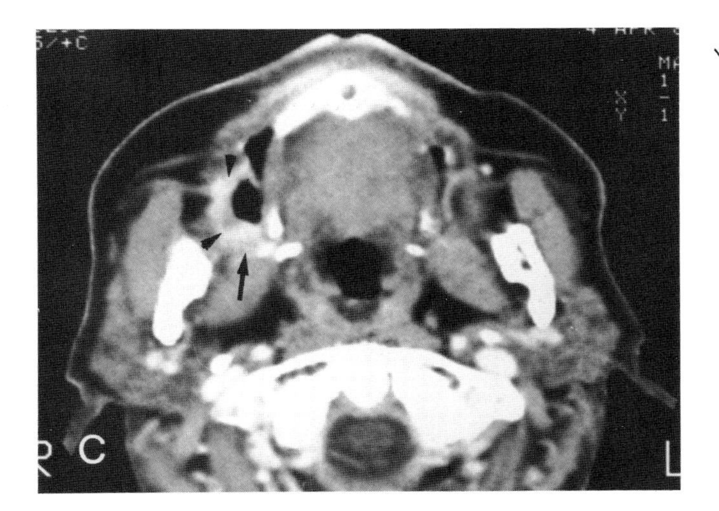

Plate 10.7*C*. The lesion extends directly into the lower buccinator space (*arrowheads*) and along the course of the pterygomandibular raphe (*arrow*).

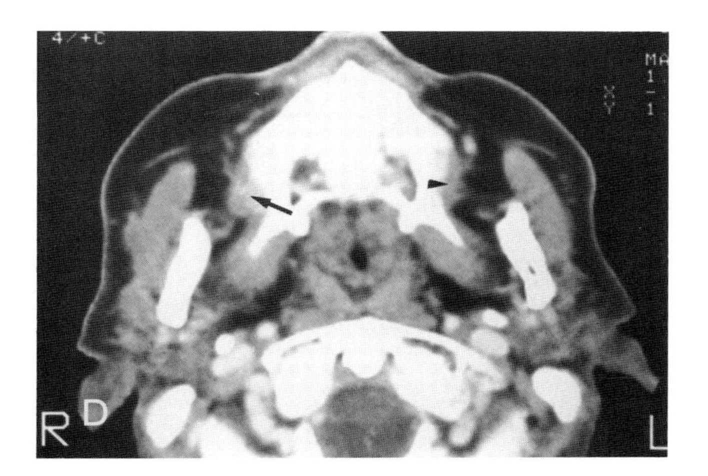

Plate 10.7*D*. The top of the mass can be seen lying just above the level of the maxillary tubercle (*arrow*) within the buccinator space. The normal attachment of the buccinator muscle (*arrowhead*) is seen on the opposite side.

Plate 10.8. In parts *A–H* we have the opportunity to observe the natural history of a retromolar trigone carcinoma. *A–C* shows the study done when the patient originally presented. At that time, the patient refused the therapy then returned several months later at which time the study illustrated in Plates *D–H* was obtained. There was no therapeutic intervention at any time between the two studies.

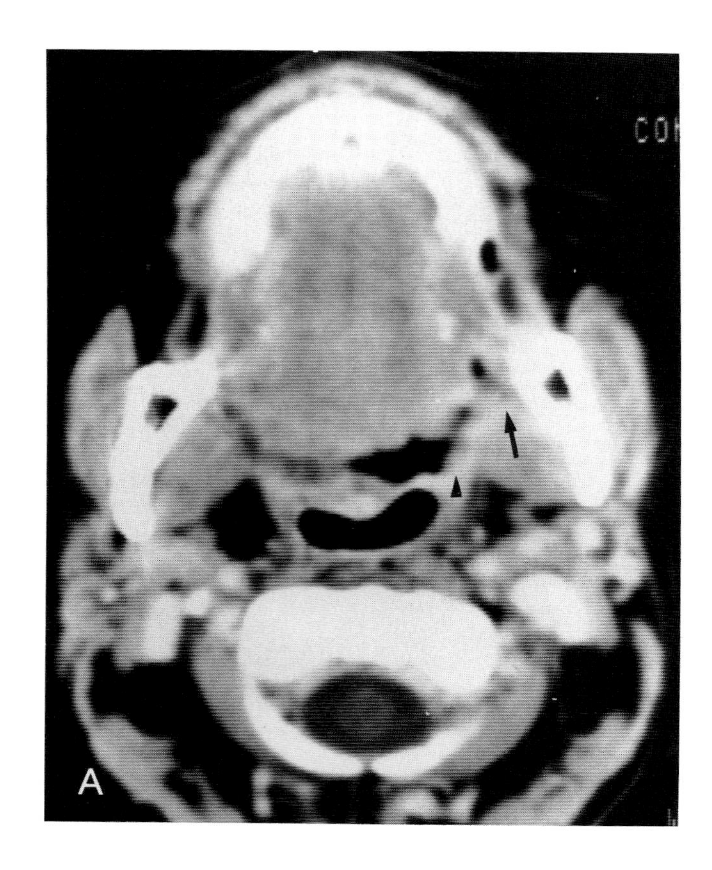

Plate 10.8*A*. A section through the region of the retromolar trigone shows an infiltrating mass obliterating the pterygomandibular space (*arrow*); this should be compared to the normal opposite side. There is also some thickening of the anterior tonsillar pillar region (*arrowhead*).

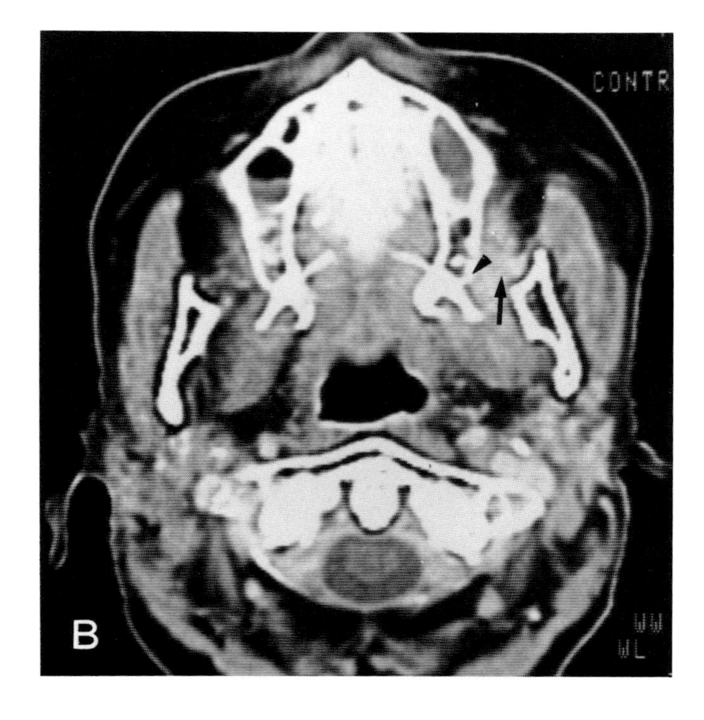

Plate 10.8*B*. This section through the low nasopharynx shows the clinically occult spread of this deeply infiltrating lesion along the pterygomandibular raphe and into the buccinator space (*arrow*) adjacent to the maxillary tubercle (*arrowhead*).

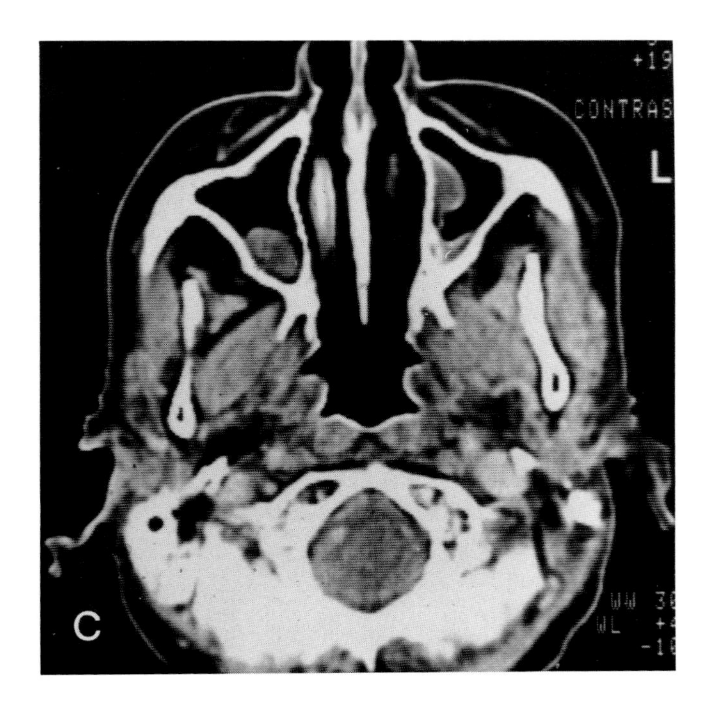

Plate 10.8*C*. A section made through the midnasopharynx shows no definite abnormalities at this time aside from some incidental mucoperiosteal thickening in the maxillary sinuses. This is included for comparison with plate 10.8*H*. Note that at this time the deep soft tissue planes of the nasopharynx appear normal and symmetric bilaterally.

Plate 10.8*D*. This is the first of the sections taken on the follow-up study some months after the patient presented initially. This section is comparable to the level in plate 10.8*A*. The mass in the retromolar trigone region has now become frankly ulcerative; air fills a good deal of the ulcer (*white arrow*). There is continued obliteration of the pterygomandibular space (*black arrow*); this should be compared to the normal opposite side. The spread back into the anterior tonsillar pillar and tonsillar fossa is more extensive than on the prior study.

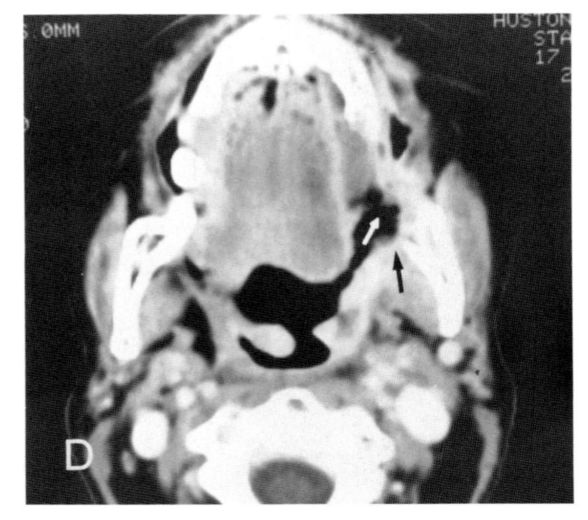

Plate 10.8*E*. A section about 1 cm lower than the prior one showing that the tumor has spread into the oral portion of the tongue (*arrow*) and along the mylohyoid muscle (*arrowheads*) into the floor of the mouth. The marked enhancement of the tumor is due to the ulceration and secondary inflammation.

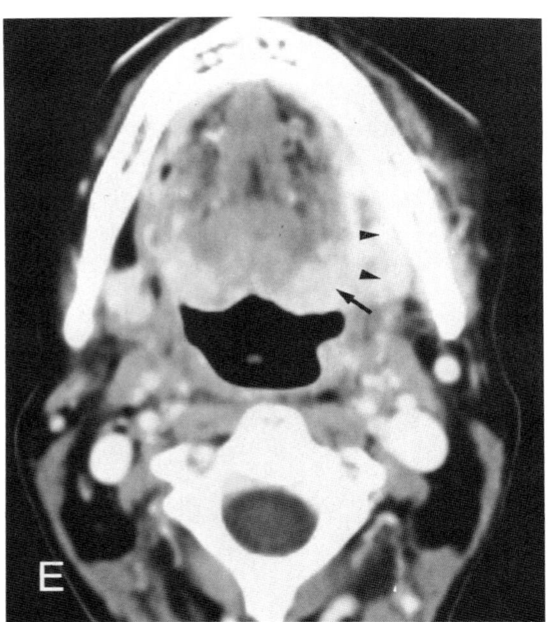

Plate 10.8*F*. A section about the same level as part *B*. The large ulcerative mass involving the junction of the tonsillar pillar and soft palate is obvious. There is tumor surrounding (*arrowheads*) the region of the maxillary tubercle (*arrow*).

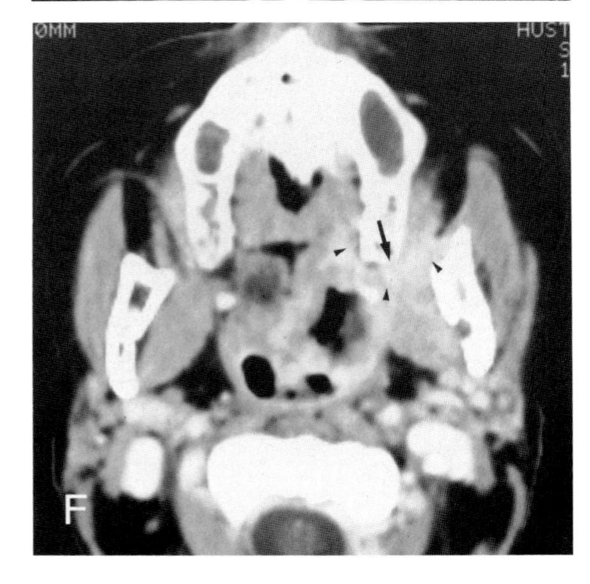

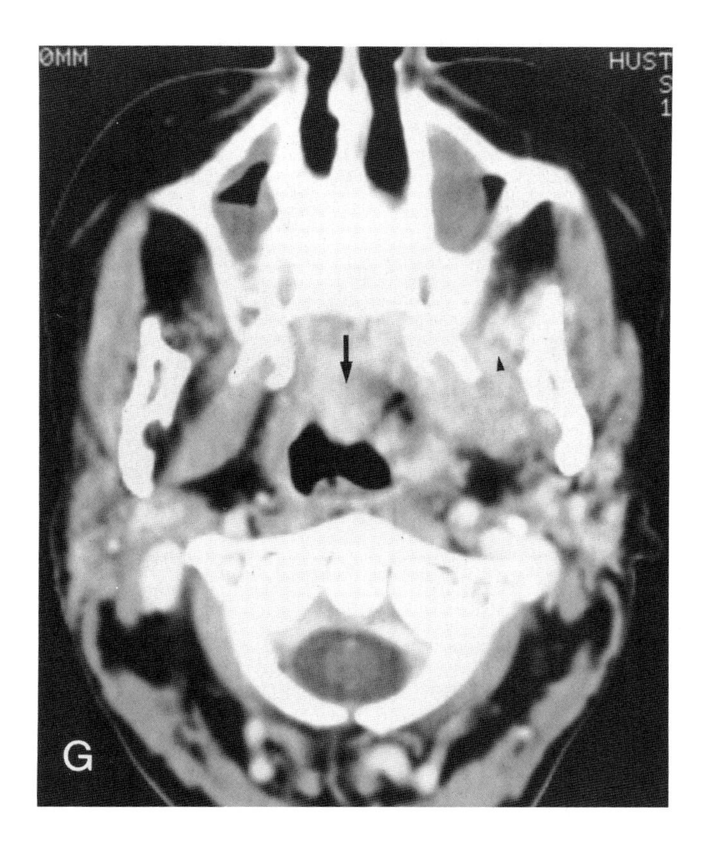

Plate 10.8*G*. The ulcerative mass involves nearly the entire soft palate (*arrow*). Since the mass has access to the soft palate it may spread cephalad along the tissue compartment around the levator palati muscles into the nasopharynx. The continued spread of tumor cephalad from the upper buccinator space into the lower part of the infratemporal fossa is also apparent on this section (*arrowhead*).

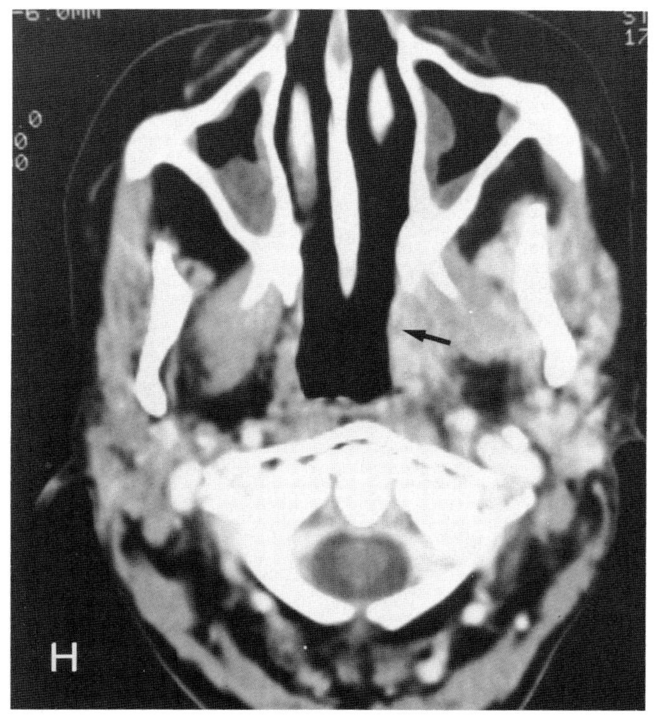

Plate 10.8*H*. This should be compared to part *C*. A definite infiltrating mass is present within the parapharyngeal space and well up into the nasopharynx (*arrow*). This was not visible on the endoscopic examination. This spread took place along the tensor and levator palati muscles and therefore was deep to the mucosal surfaces within the nasopharynx.

In summary, this case shows the natural history of untreated retromolar trigone carcinoma. The tumor grew along the pterygomandibular raphe into the low buccinator space and on to the infratemporal fossa. It extended up the anterior tonsillar pillar to involve the soft palate and from there grew along the palatal muscles into the nasopharynx. Inferiorly, it infiltrated the small pterygomandibular space and from there spread down to the mylohyoid muscle where it grew along the deeper aspects of the floor of the mouth. It also reached the floor of the mouth directly along the lower gingival surface. Its growth downward in the anterior tonsillar pillar took it to the junction of the pillar and the free margin of the oral tongue.

Plate 10.9. Most squamous cell carcinomas of the oral tongue are superficial, fairly localized, lesions when first discovered. These do *not* require imaging for complete evaluation. When the lesions become larger and more deeply infiltrative it is often difficult to be sure of the deep margins from physical examination alone. Even at surgery it may be difficult to be certain that an adequate margin of resection has been obtained when operating in the oral portion and base of the tongue. Both MRI and CT can be used to show the deep extension of these lesions and their spread to involve contiguous sites.

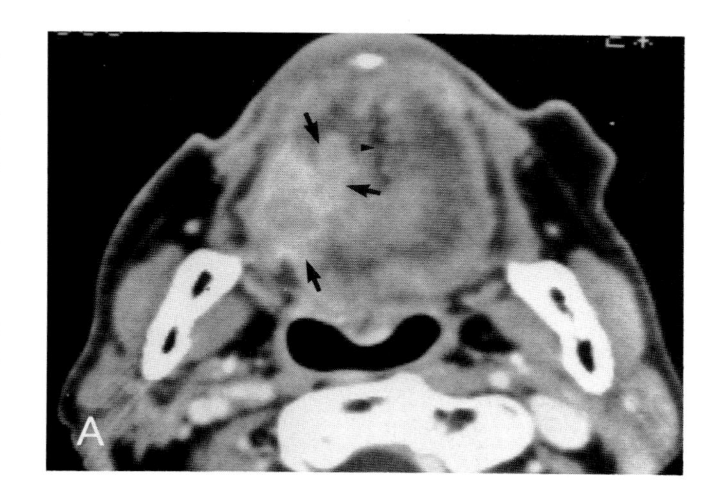

This patient had a T_2 carcinoma of the oral tongue with some evidence of ulceration and induration.

Plate 10.9*A*. A section through the oral tongue shows the precise deep margins of the lesion (*arrows*). The lesion extends to within millimeters of the midline lingual septum (*arrowhead*).

Plate 10.9*B*. This mass turned out to be more deeply infiltrating than it originally appeared. More inferiorly on this section through the floor of the mouth, the mass involves the anterior and right side of the floor of the mouth fairly extensively (*arrows*). It also lateralized to the region of the mylohyoid muscle on the right side, although there was no evidence of frank mandibular invasion (*arrowhead*).

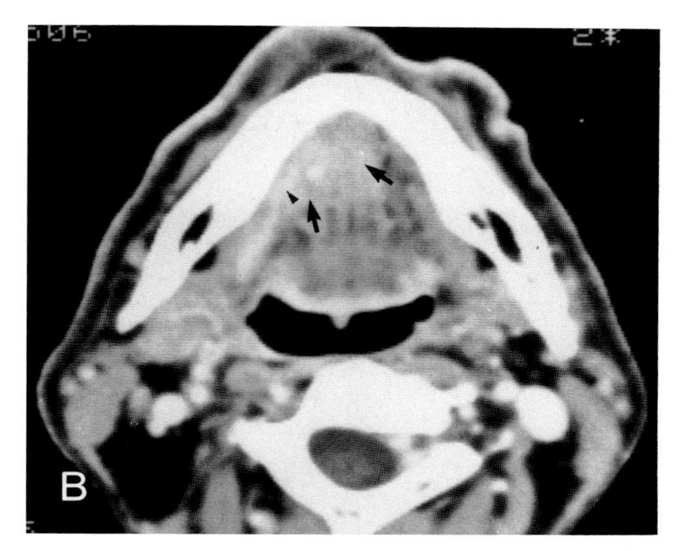

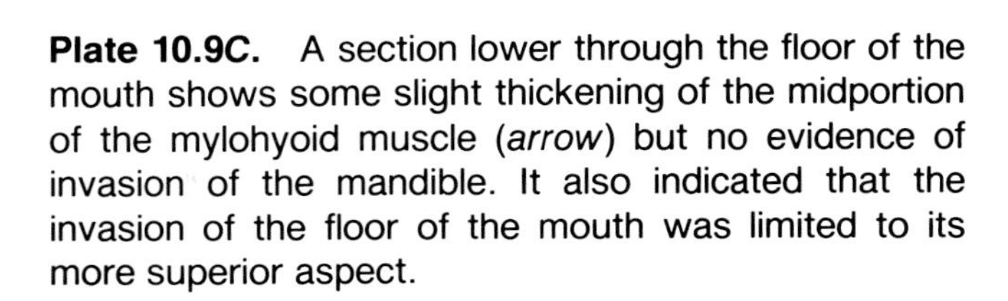
Plate 10.9*C*. A section lower through the floor of the mouth shows some slight thickening of the midportion of the mylohyoid muscle (*arrow*) but no evidence of invasion of the mandible. It also indicated that the invasion of the floor of the mouth was limited to its more superior aspect.

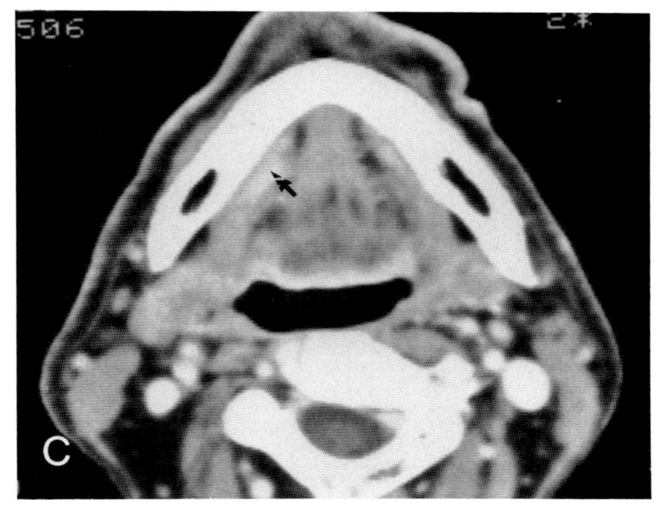

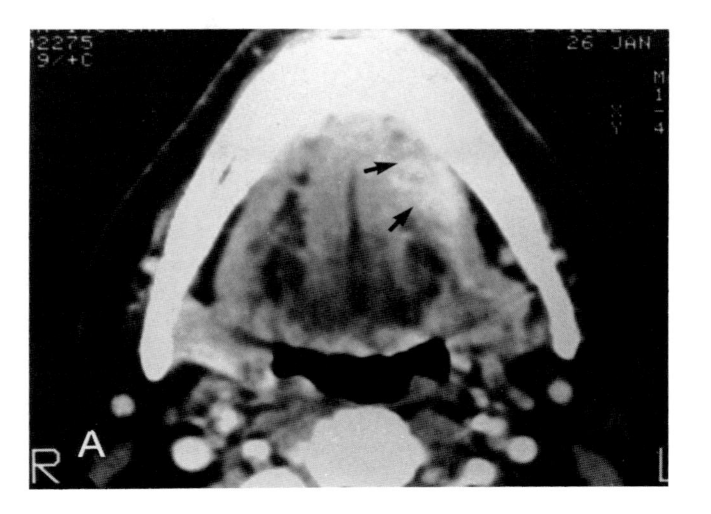

Plate 10.10. Like oral tongue lesions, many floor of the mouth regions are quite superficial at presentation and do not require CT or MRI for complete evaluation. Whenever there is evidence of deep infiltration a CT or MRI study should be done. This is especially true of lateral floor of the mouth lesions which may show considerable spread that may not be apparent on physical examination.

Plate 10.10*A*. This lateral floor of the mouth lesion showed fairly deep infiltration of the floor of the mouth. Obliteration of the lower most portion of the sublingual space by the infiltrating mass is obvious on comparing the two sides (*arrows*).

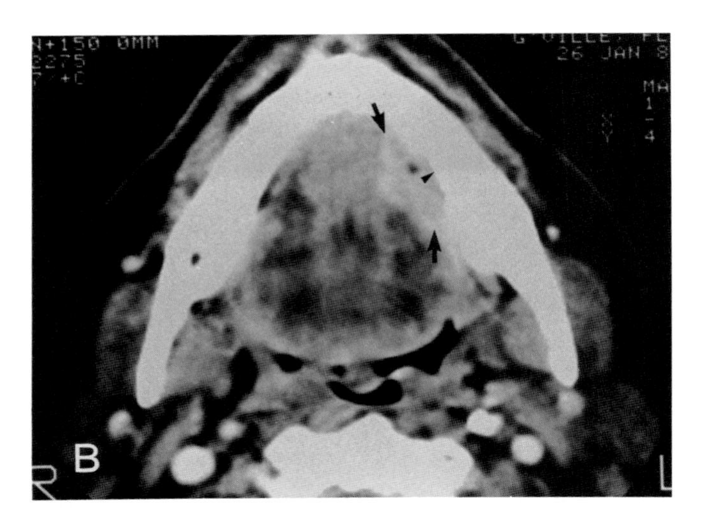

Plate 10.10*B*. A section approximately 1 cm higher than part *A* shows the mass involving the lateral floor of the mouth (*arrows*). Even on this unfavorable window setting one might raise the possibility of some mandibular erosion (*arrowhead*).

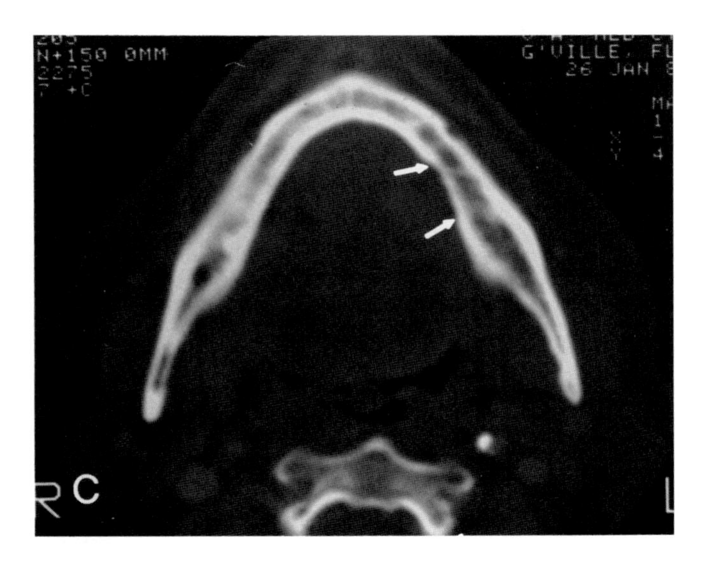

Plate 10.10*C*. A wider window setting at the same level as that shown in part *B* reveals definite erosion of the lingual plate of the mandible (*arrows*). CT is an excellent way to study this surface of the mandible. It is certainly better than orthopantomography for detecting subtle erosions along the lingual surface of the mandible. Detection of such changes is very important when considering therapeutic options such as surgery versus radiotherapy and rim resection versus more extensive surgical sacrifice of the mandible.

The abbreviation used on Plate 10.11 is: *T*, tumor.

Plate 10.11. MRI is capable of providing highly resolved images of the oropharynx and surrounding areas. The spatial resolution is roughly equivalent to that of CT and the contrast resolution is better than with noncontrast-enhanced CT although it does suffer some limitations. MR may soon come to replace CT in imaging neoplasms in this region.

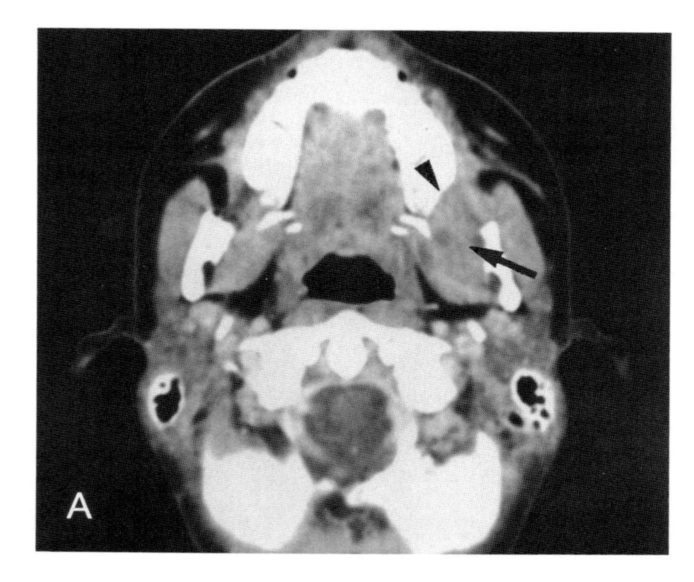

Plate 10.11*A*. A CT section through the maxillary alveolar ridge shows an infiltrating mass in the buccinator space (*arrow*). There is a saucerized defect on the posterior aspect of the maxillary tubercle (*arrowhead*). The border between the mass and the medial pterygoid muscle is impossible to determine.

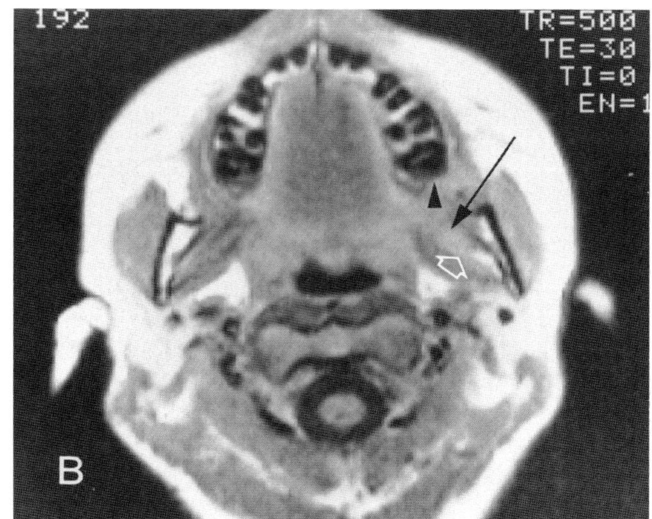

Plate 10.11*B*. A spin echo 500/30 section through the same level shows the mass within the buccinator space (*black arrowhead*). The saucerized defect on the maxillary tubercle is visible (*arrow*). Note that it is quite easy to determine the margin between the mass and the medial pterygoid muscle (*open arrowhead*) because of the better contrast resolution on MRI than on the CT study.

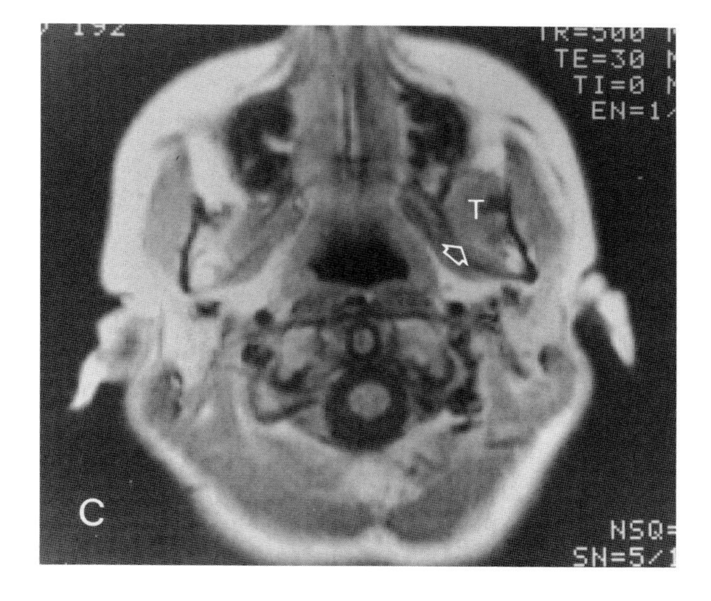

Plate 10.11*C*. A section approximately 1 cm cephalad from part *B*, again using a spin echo 500/30 pulse sequence shows the mass not only in the upper buccinator space but extending into the infratemporal fossa and back to the point of the attachment of the medial pterygoid muscle near the angle of the mandible. Again it is quite easy to see the interface (*open arrowhead*) between tumor and the pterygoid muscle.

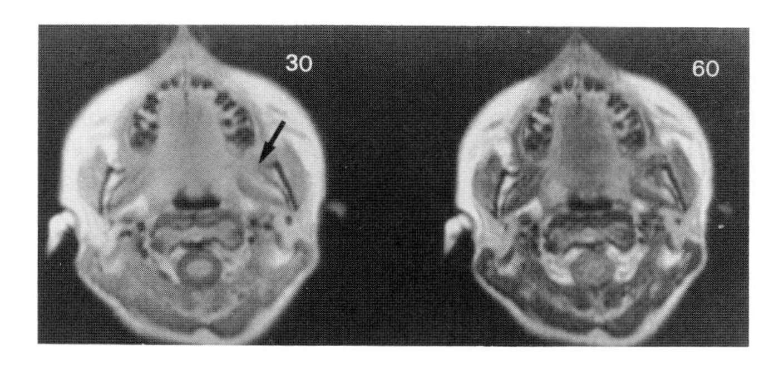

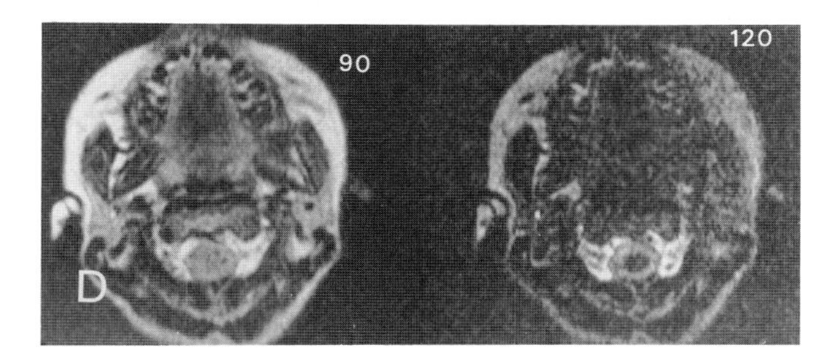

Plate 10.11*D*. This is one section from a multislice multiecho acquisition. The TR is 2000 msec and the echo delay times are *30*, *60*, *90*, and *120 msec*, respectively. Note that the area of the mass (*arrow*) gets progressively darker on these increasingly T_2-weighted images. This is very different than the behavior of the usual squamous cell carcinoma in this region which tends to remain as bright, or brighter, than normal lymphoid tissue. These findings are compatible with the histologic diagnosis of fibrosarcoma. The fibrous stroma of this tumor is presumed to be relatively proton poor compared to the usual squamous cell carcinoma or lymphoma in this area.

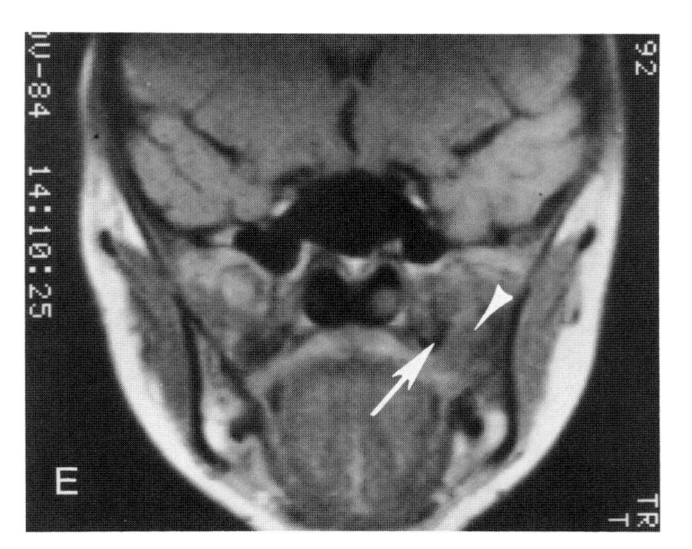

Plate 10.11*E*. This coronal section through the mid-portion of the floor of the mouth shows the mass (*arrowhead*) as it lies between the mandible and the pterygoid plates (*arrow*). There is definite evidence of infiltration of the pterygomandibular space which indicates that the mass will lie in close continuity to the lingual nerve; at surgery the tumor was wrapped around the nerve.

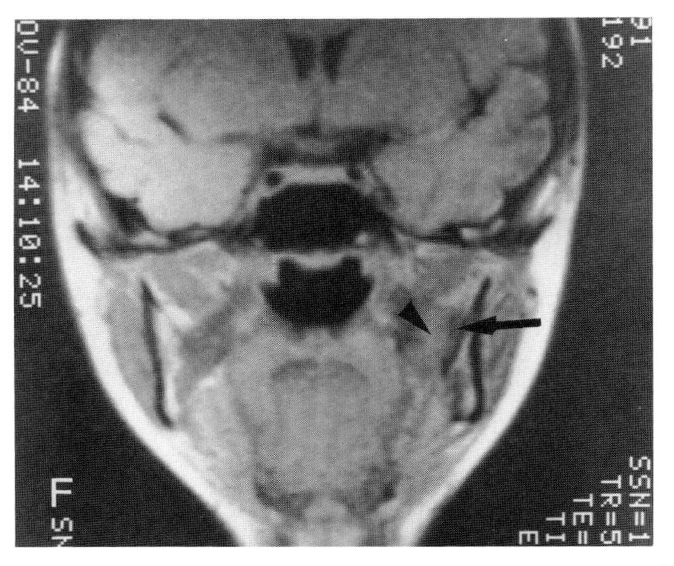

Plate 10.11*F*. A coronal section made slightly posterior to that in part *E* which shows the mass again infiltrating the lower infratemporal fossa (*arrowhead*). Note that there is a clear plane of cleavage between the tumor mass and the mandible at this point (*arrow*).

The abbreviation used on Plate 10.13 is: SP, soft palate.

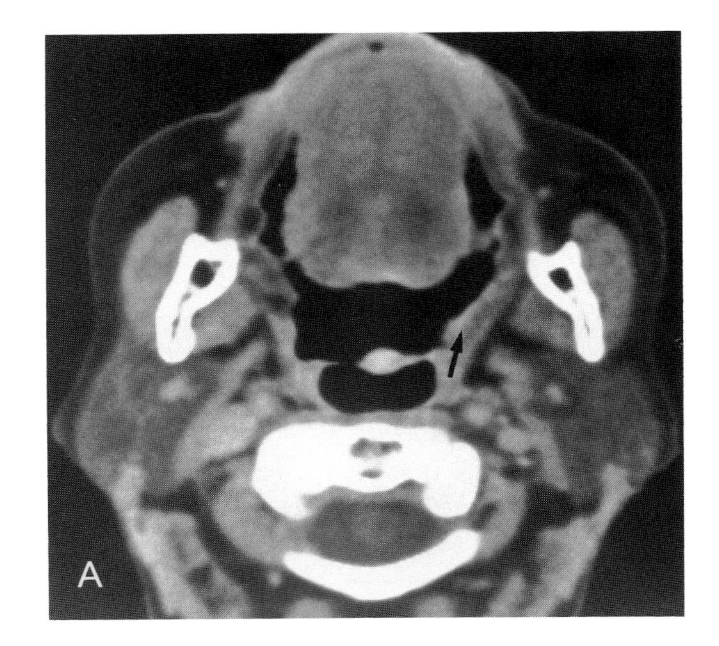

Plate 10.12. Like lesions elsewhere in the oropharynx tonsillar pillar or tonsillar fossa, carcinoma can be either superficial or deeply infiltrating, or both. There is little reason to study small superficial lesions with CT or MRI. This patient had a carcinoma of the tonsillar pillar which was suspected of being deeply infiltrating by palpation.

Plate 10.12*A*. An axial section shows some slight thickening of the anterior tonsillar pillar and tonsillar fossa on the left (*arrow*); however, this is well within the range of normal variations.

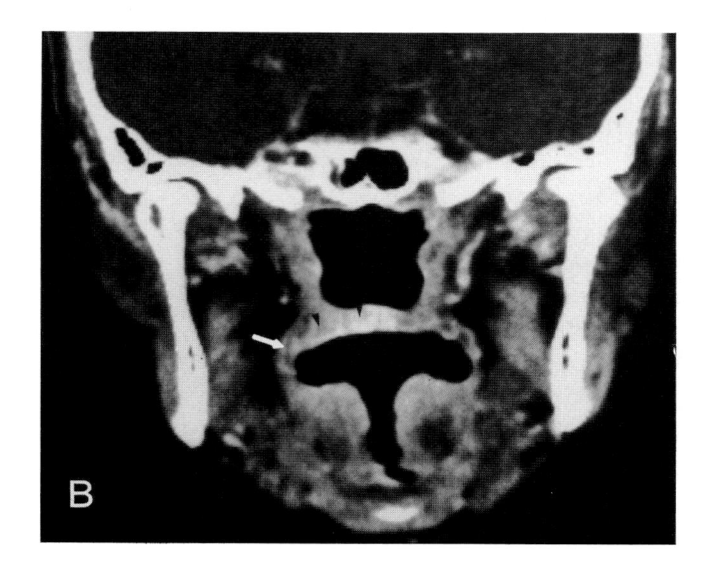

Plate 10.12*B*. A coronal section on the same patient shows similar, slight thickening of the tonsillar fossa-tonsillar pillar region on the left compared to the right (*arrow*). There is also some thickening of the soft palate on the left side (*arrowheads*). There is no evidence of deeply infiltrating component to this tumor. The findings were most compatible with a predominantly superficially spreading carcinoma involving the anterior tonsillar pillar and soft palate. This was the case and the patient was treated successfully with radiation therapy.

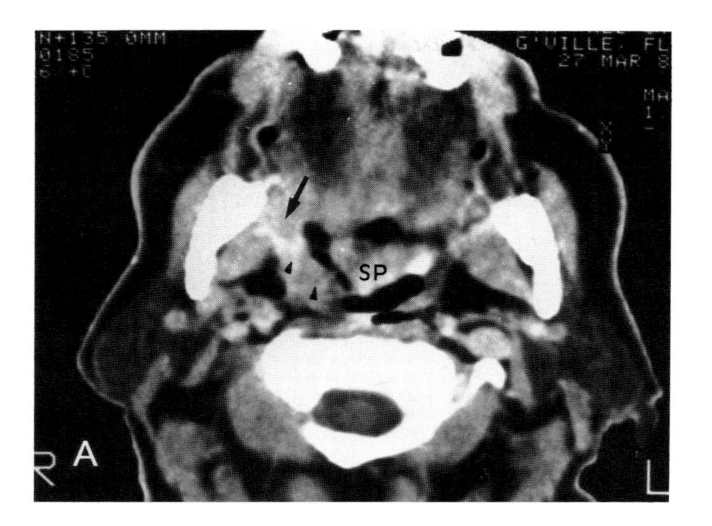

Plate 10.13. Deeply infiltrating tonsillar fossa and tonsillar pillar cancer shows predictable patterns of spread on CT studies. This patient had an anterior tonsillar pillar lesion.

Plate 10.13*A*. On a section slightly below the junction of the tonsillar pillar and soft palate, one can see a thickening of the soft palate. There is also considerable infiltration of the deep planes immediately medial to the retromolar trigone region (*arrow*). Note that the tumor spreads to the tonsillar fossa and to the region of the posterior tonsillar pillar (*arrowheads*).

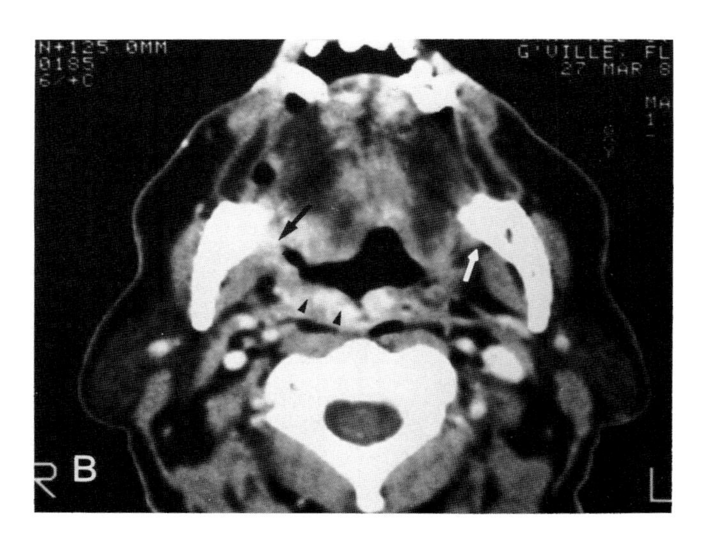

Plate 10.13*B*. A section slightly lower than the previous one showing the obliteration of the pterygomandibular space (*black arrow*); this should be compared to the normal opposite side (*white arrow*). The mass continues to spread back along the tonsillar fossa and on to the posterior pharyngeal wall (*arrowheads*).

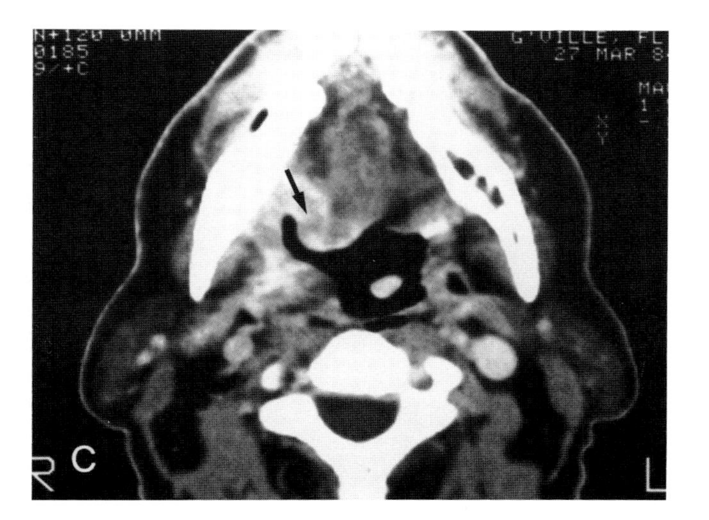

Plate 10.13*C*. A section approximately 1 cm lower than than in part *B*. Tumors which are on the anterior tonsillar pillar can then spread via the palatoglossus muscle to the free margin of the tongue (*arrow*).

One should note here the tendency for carcinoma of the gingival buccal sulcus, retromolar trigone, anterior tonsillar pillar, and tonsillar fossa to sometimes overlap in appearance. All of these anatomic regions are quite close together and there are natural tendencies for the tumors, when they grow large, to involve more than one of these regions. These patterns of spread have been discussed in detail in the text.

The abbreviations used on Plate 10.14 are: *N*, node; and *T*, tumor.

Plate 10.14. Tonsillar fossa carcinoma has a distinct tendency to be deeply invasive at the time of presentation. Like the other lesions discussed previously they tend to show a fairly characteristic pattern of spread from their site of origin in the oropharynx between the two tonsillar pillars.

This patient had a tonsillar carcinoma staged at T_3 by clinical examination.

Plate 10.14*A*. In the midoropharynx a deeply infiltrating mass is present in the midtonsillar fossa. The mass obliterates the parapharyngeal space at this level although there appears to be a margin between it and the carotid artery (*arrow*).

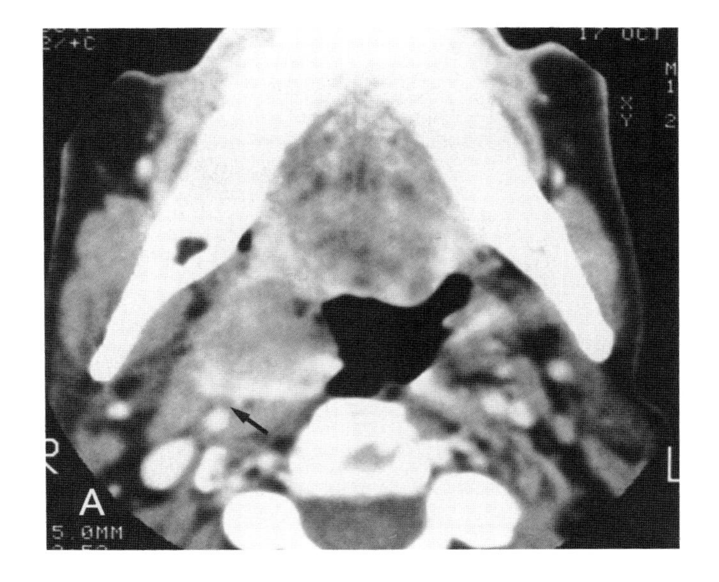

Plate 10.14*B*. A section slightly lower than part *A* shows a deeply infiltrating, peripherally enhancing mass extending into the lower pole of the tonsil. At this point there is an indistinct plane between the tumor and the necrotic node immediately lateral to it. On physical examination it was unclear whether there was direct extension of the tumor into the soft tissue of the neck or whether this was a high internal jugular node.

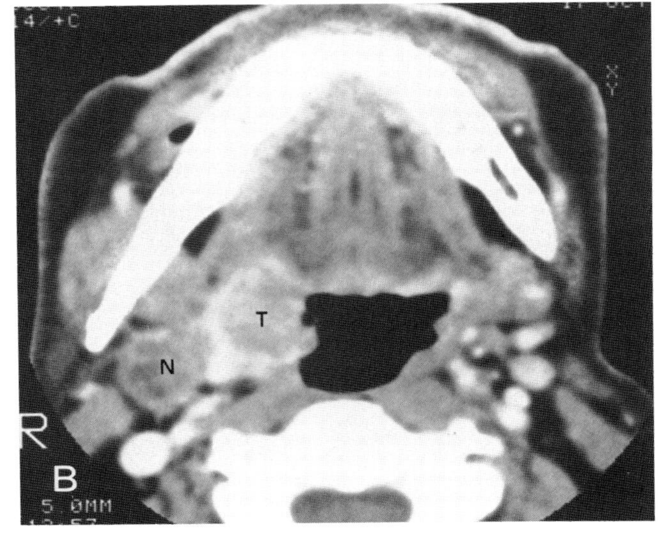

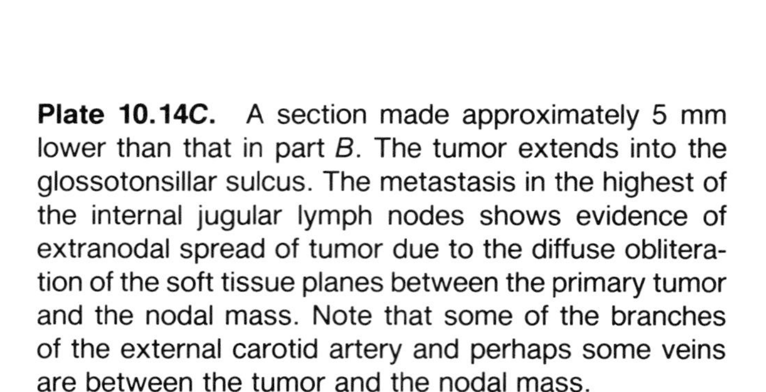

Plate 10.14*C*. A section made approximately 5 mm lower than that in part *B*. The tumor extends into the glossotonsillar sulcus. The metastasis in the highest of the internal jugular lymph nodes shows evidence of extranodal spread of tumor due to the diffuse obliteration of the soft tissue planes between the primary tumor and the nodal mass. Note that some of the branches of the external carotid artery and perhaps some veins are between the tumor and the nodal mass.

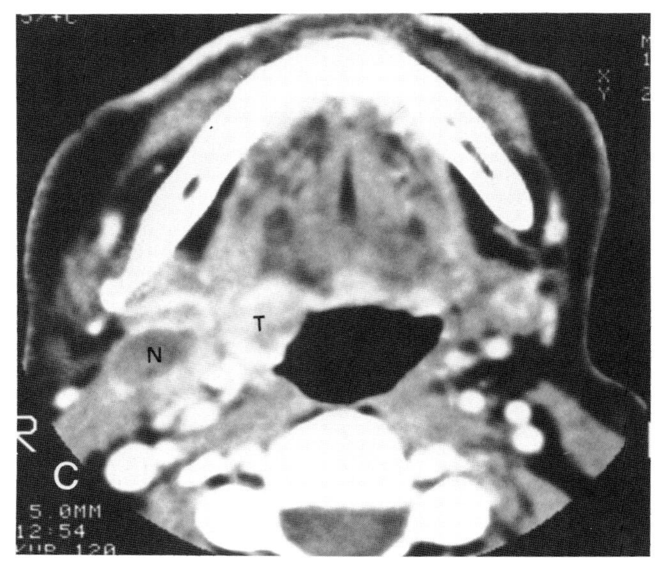

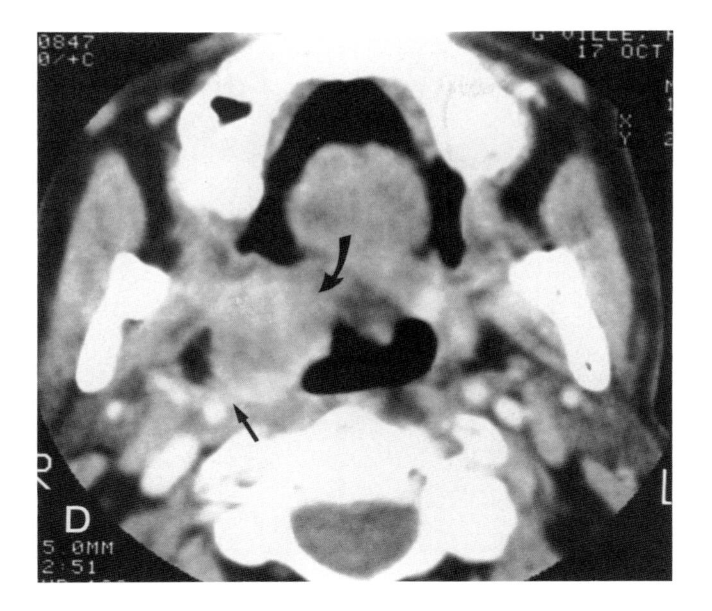

Plate 10.14*D*. A section ~3–4 cm above that in part *C* at the junction of the oro- and nasopharynx. There is an obvious mass within the upper tonsillar fossa. The mass comes closest to direct fixation of the internal carotid artery at this point (*arrow*) but there are no definite findings to suggest that internal carotid fixation. The mass is extending out into the soft palate at the junction of the anterior tonsillar pillar and soft palate (*curved arrow*).

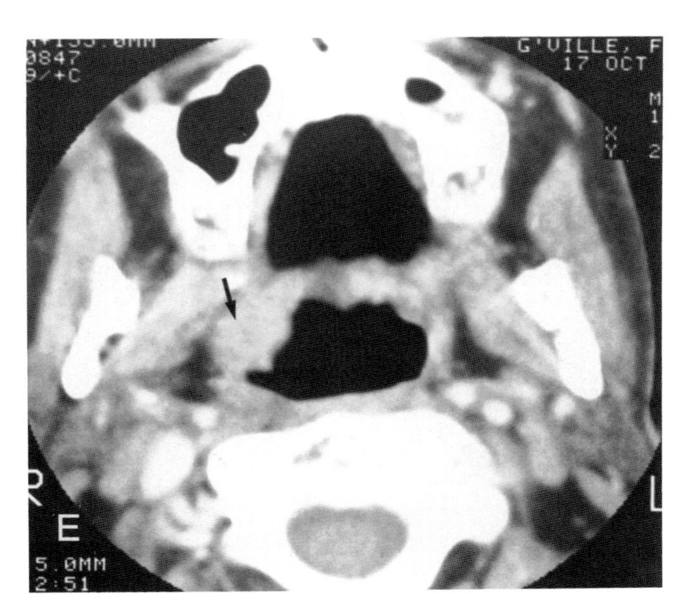

Plate 10.14*E*. A section through the lower most portion of the nasopharynx shows that the tumor has grown up into the nasopharynx. This is manifest by the infiltration of the parapharyngeal space (*arrow*) as well as the obvious thickening along the mucosal surface. This spread into the low nasopharynx was confirmed at endoscopy.

In summary, tonsillar carcinomas can be among the largest and most deeply infiltrating lesions at presentation. From this it should be clear that they can spread directly into the soft tissues of the neck or grow posteriorly and laterally to directly involve the carotid artery even though such fixation was not present in this case. They may also grow to invade the nasopharynx and sometimes this spread is not apparent at endoscopy. CT or MRI should be done to evaluate any tonsillar fossa carcinoma which is T_2 or larger, or if there is any suspicion of deep infiltration in a smaller lesion. Occasionally CT can reveal an occult primary carcinoma in the tonsillar fossa during a search for an occult carcinoma presenting as cervical lymph node metastases.

Plate 10.15. Both CT and MRI are highly accurate means imaging the deep extent of carcinomas in the oropharynx. In Plate 10.11 we saw a case where MRI very clearly showed the tumor margins better relative to surrounding muscles. In this patient with tongue base carcinoma, the CT depiction of the lesion is perhaps the more graphic of the two. Considerably more carefully controlled study is necessary to determine where and when each test should be used for staging upper aerodigestive tract malignancies. The use of intravenous paramagnetic compounds may also affect the relative efficacy of these studies.

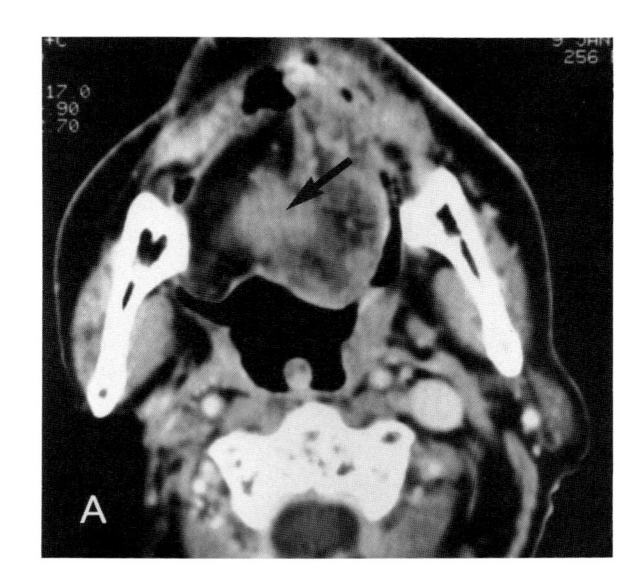

Plate 10.15*A*. A section through the oral tongue shows a well circumscribed lesion in the posterior two-thirds of the tongue (*arrow*) at the level of the circumvallate papillae. The asymmetry of the soft tissues of the neck is due to a radical neck dissection done 15 yr earlier. The atrophy of the right side of the tongue was due to sacrifice of the hypoglossal nerve during that operation.

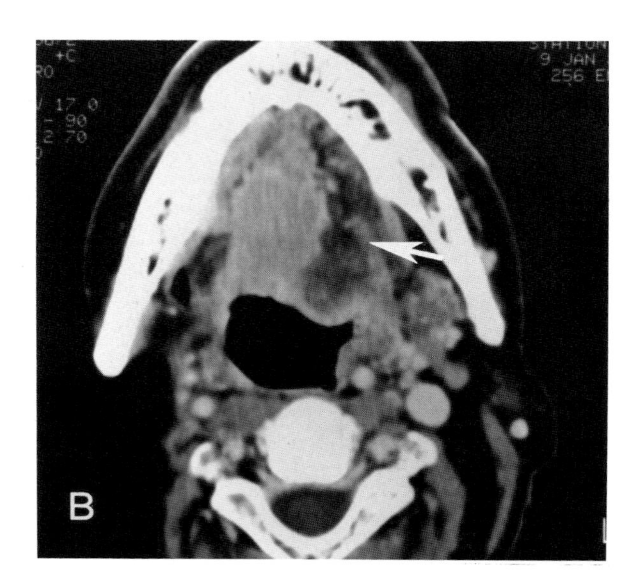

Plate 10.15*B*. A section lower through the tongue base and floor of the mouth shows that the mass is in the midline and does spread more to the right side than the left. The more proximal course of the lingual artery on the left would be free of tumor. Recall that the artery runs just medial to the hyoglossus muscle (*arrow*).

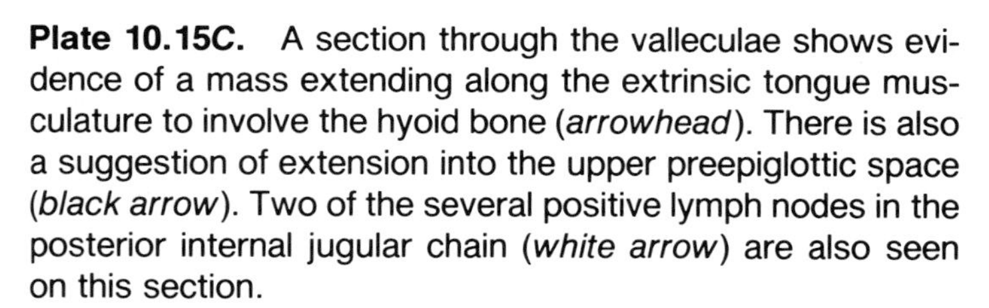

Plate 10.15*C*. A section through the valleculae shows evidence of a mass extending along the extrinsic tongue musculature to involve the hyoid bone (*arrowhead*). There is also a suggestion of extension into the upper preepiglottic space (*black arrow*). Two of the several positive lymph nodes in the posterior internal jugular chain (*white arrow*) are also seen on this section.

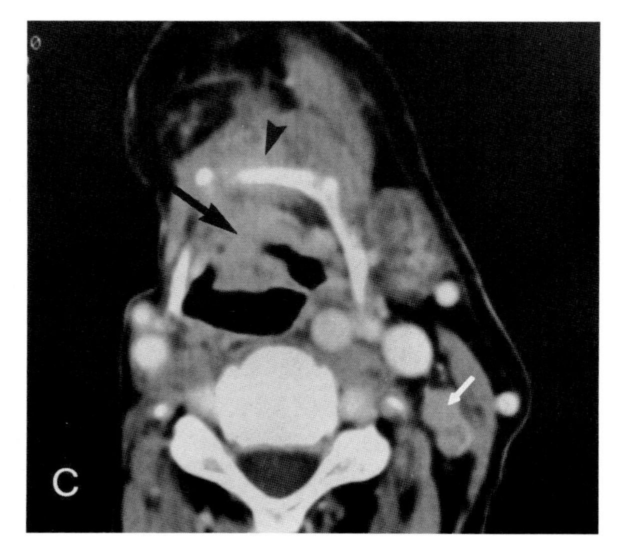

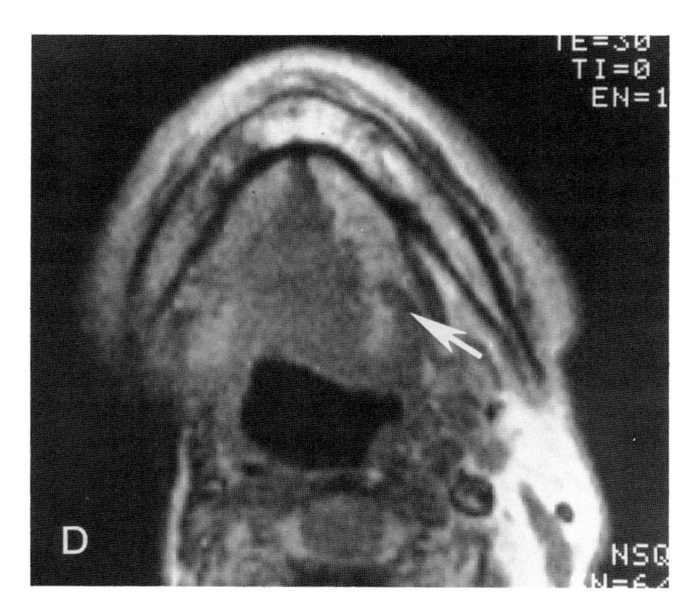

Plate 10.15*D*. An MRI section at a level comparable to that seen in part *B*. The infiltrating mass occupies the midline but its left margin appears to be closer to the hyoglossus muscle (*arrow*) than was suspected on the CT study. Note that on the CT this area corresponds to what appears to be edema rather than to tumor margin. The edema within the fatty tongue muscle at this point probably prolonged the T_1 in this region making it difficult to tell the tumor margin from the margin of edematous fat.

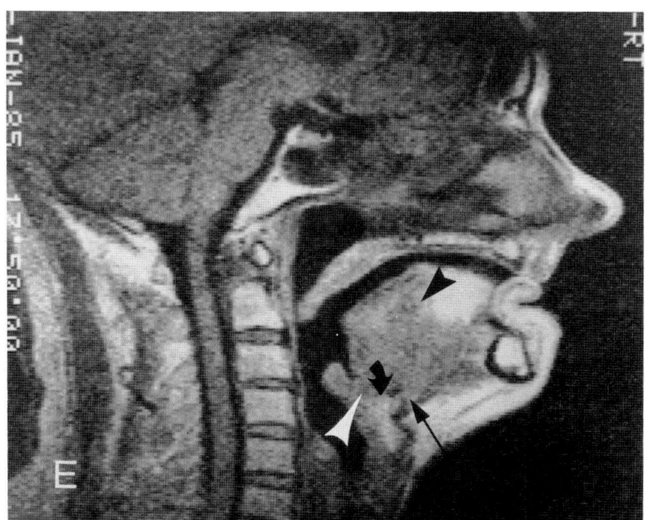

Plate 10.15*E*. A midline sagittal section through the tongue shows the full extent of the carcinoma from its superior to inferior margins (*arrowheads*). Note how the lesion goes down to the attachment of the extrinsic tongue muscles on the hyoid bone (*arrow*) and into the upper preepiglottic space (*curved arrow*).

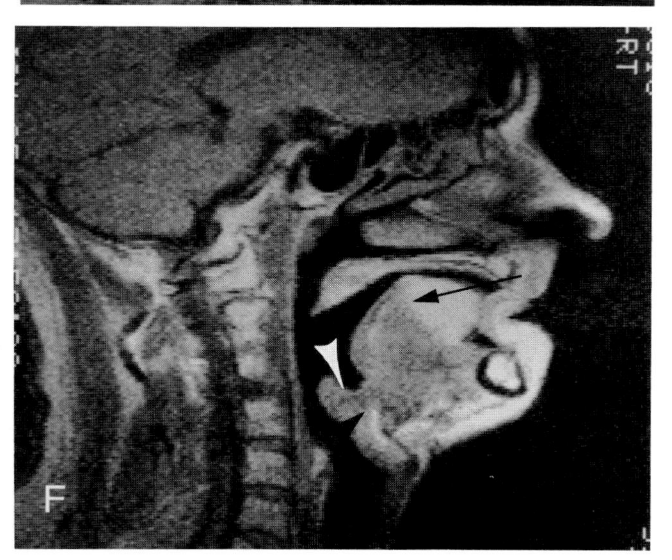

Plate 10.15*F*. A parasagittal section just to the right of midline shows the more cephalad spread of tumor into the atrophic portion of the oral tongue (*arrow*). Note again how the tumor has begun its spread into the upper preepiglottic space (*black arrowhead*) and spilled over to thicken the lingual surface of the epiglottis (*white arrowhead*).

In summary, Plates 10.11 and this Plate should illustrate that MRI has much potential in evaluating carcinoma of the oropharynx and other mass lesions in this region. It will, however, have some limitations and it may be that CT and MRI will prove complementary in evaluating lesions of the upper aerodigestive tract.

The abbreviation used on Plates 10.16 and 10.17 is: *N*, node.

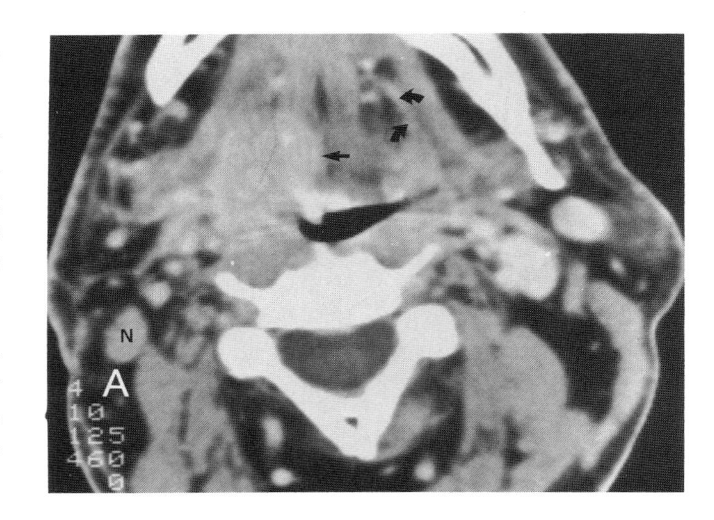

Plate 10.16. This patient had presented 2 yr earlier with metastatic adenopathy in the right neck. No primary site was found at that time and a radical neck dissection was done. The patient presented at the time of this CT study with neck nodes and, again, with no apparent primary tumor on endoscopy done in clinic.

Plate 10.16*A*. The node responsible for the representation of the patient is seen in the bed of the prior radical neck dissection (*curved arrows*). There is an obvious infiltrating mass in the right glossotonsillar sulcus which extends well into the tongue base (*arrow*).

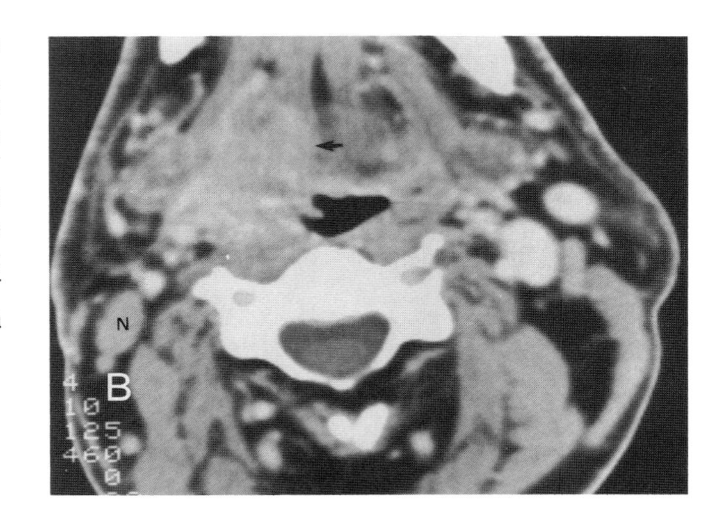

Plate 10.16*B*. A section approximately 4 mm more inferior than part *A* shows that the mass approaches, but does not cross, the midline (*arrow*). Endoscopy under anesthesia revealed a mass at the tongue base although its full extent was not appreciated. The CT study shows that the margin of the mass is restricted to the right side of the tongue and well away from the course of the left lingual artery and hypoglossal nerve; these run near the hypglossus muscle. From the CT study it was decided that the patient could have a hemiglossectomy.

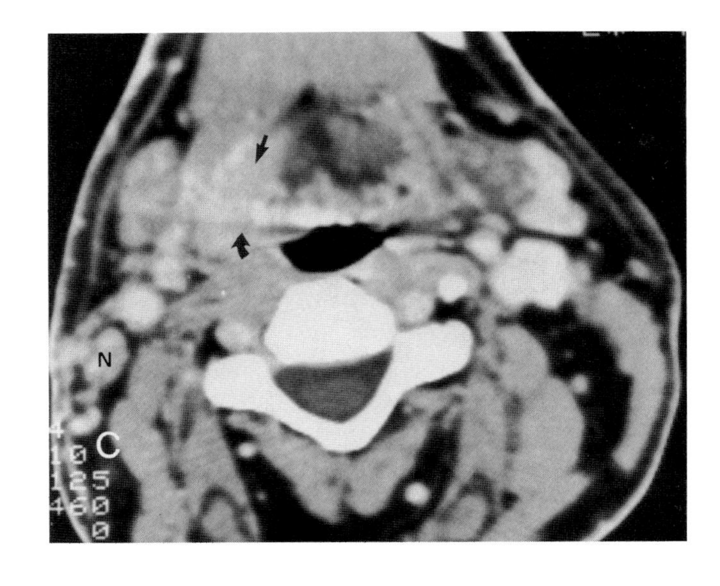

Plate 10.16*C*. A section through the lower glossotonsillar sulcus shows the lower margin of the infiltrating mass as it involves the right half of the root of the tongue (*arrow*) and the lower glossotonsillar sulcus (*curved arrow*). The margins of the mass as demonstrated on CT were found to be correct at surgery and a hemiglossectomy was performed with good margins.

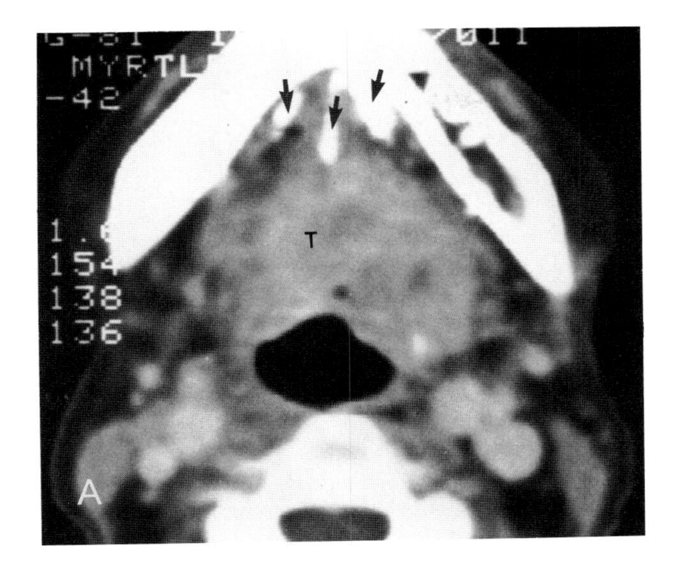

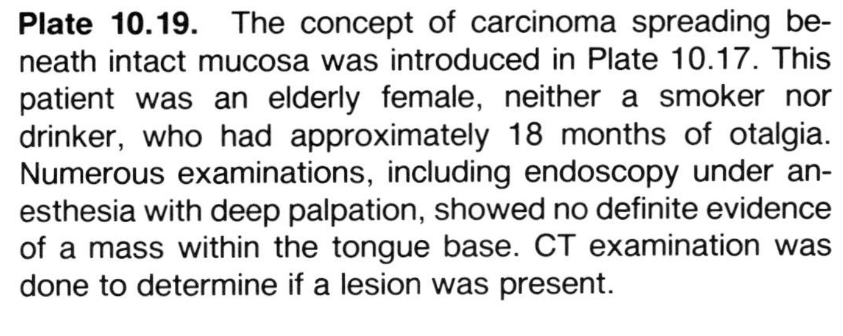
Plate 10.19. The concept of carcinoma spreading beneath intact mucosa was introduced in Plate 10.17. This patient was an elderly female, neither a smoker nor drinker, who had approximately 18 months of otalgia. Numerous examinations, including endoscopy under anesthesia with deep palpation, showed no definite evidence of a mass within the tongue base. CT examination was done to determine if a lesion was present.

Plate 10.19*A*. There is an obvious massive tumor invading the tongue base and structures of the floor of the mouth. The mass replaces virtually all of the normal anatomy.

The high density material anteriorly is extravasated contrast from a prior attempt at a submandibular sialogram (*arrows*).

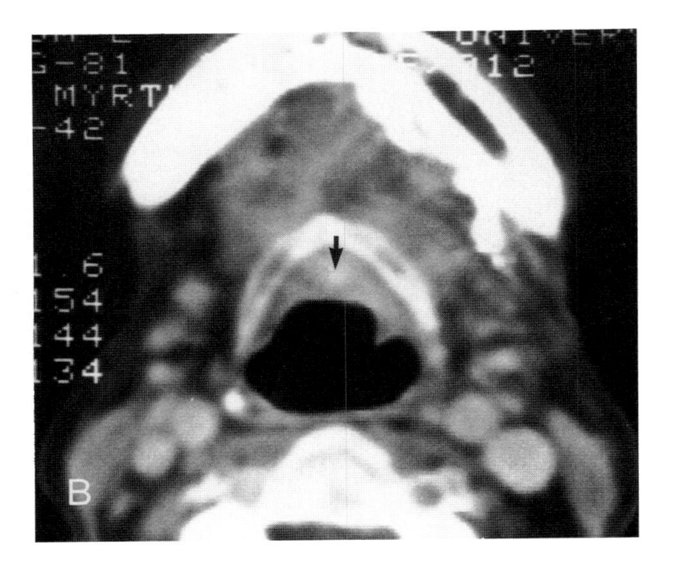

Plate 10.19*B*. Although of somewhat limited quality this section showed that the tumor spread, submucosally, into the preepiglottic space (*arrow*).

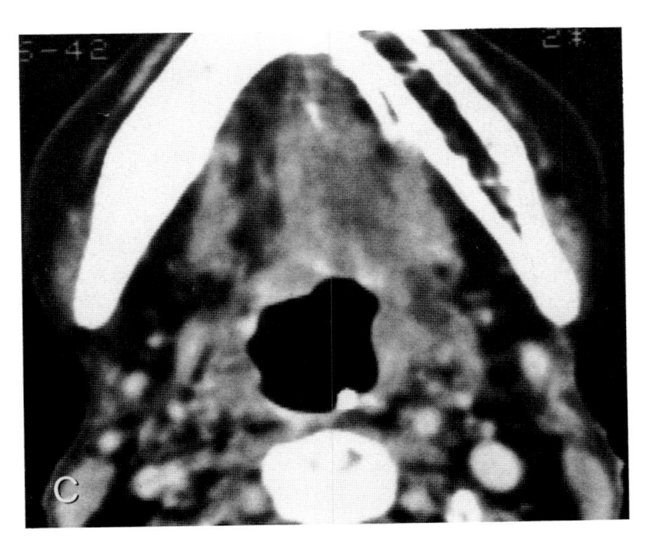

Plate 10.19*C*. Endoscopy under anesthesia even with knowledge of the CT findings showed no definite palpable mass but multiple deep biopsies returned squamous cell carcinoma and the patient was radiated. This section was made several weeks after radiation therapy and showed persistent obliteration of the soft tissue although there was a marked reduction in the size of the mass. The patient died several months later with persistent tumor at the primary site.

The abbreviation used on Plate 10.21 is: *P*, parotid gland.

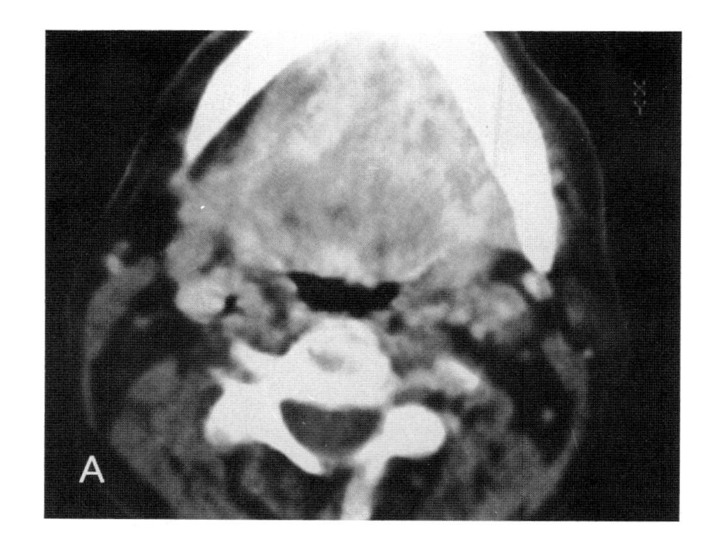

Plate 10.20. This patient presented with a mass involving the tongue base and floor of the mouth. Biopsy revealed squamous cell carcinoma. CT study was done to evaluate the full extent of the lesion.

Plate 10.20*A*. This study shows virtually complete obliteration of all the tissue planes at the floor of the mouth. Careful study reveals that there is also a marked amount of enhancement of the tissue in this location.

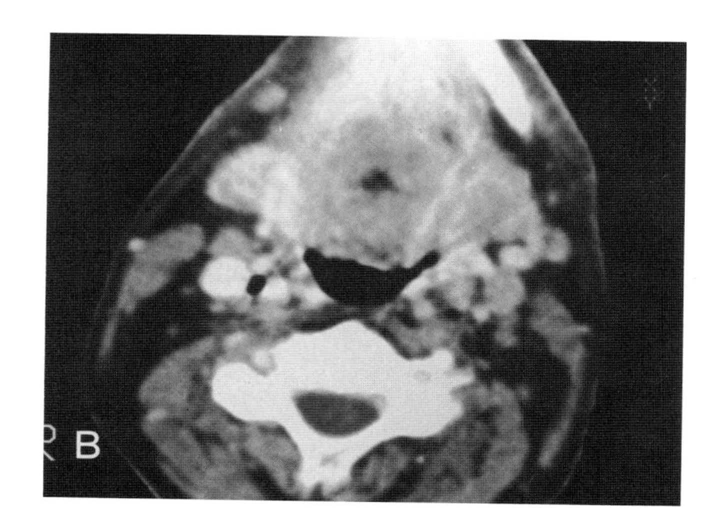

Plate 10.20*B*. A section made through the lower floor of the mouth and tongue base again shows a basically solid mass and diffuse enhancement of the soft tissues.

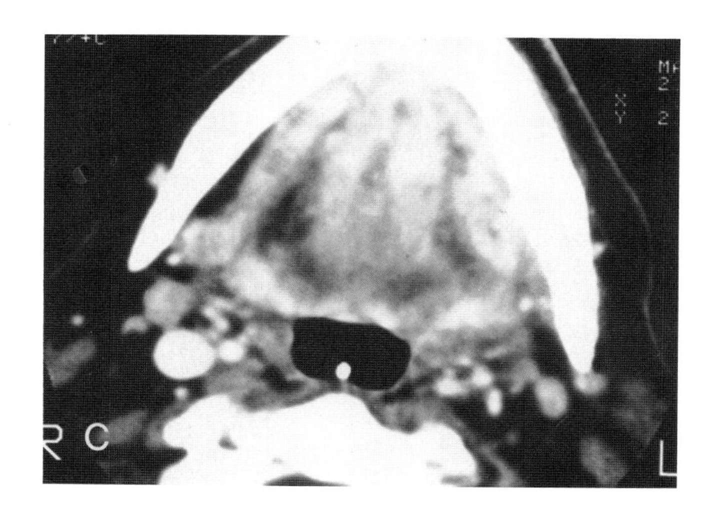

Plate 10.20*C*. Because of the excessive enhancement and diffuse obliteration of the soft tissue planes, it was suggested that a large amount of this apparent tumor spread could be due to secondary inflammation. The patient was placed on antibiotics for approximately 7 days and the study was repeated. The section through the floor of the mouth, comparable to that in part *A*, shows that the tissue planes in the floor of the mouth are returning to a more normal appearance. This indicates that at least this part of the obliteration of the soft tissue planes was related to secondary inflammation rather than direct extension of tumor.

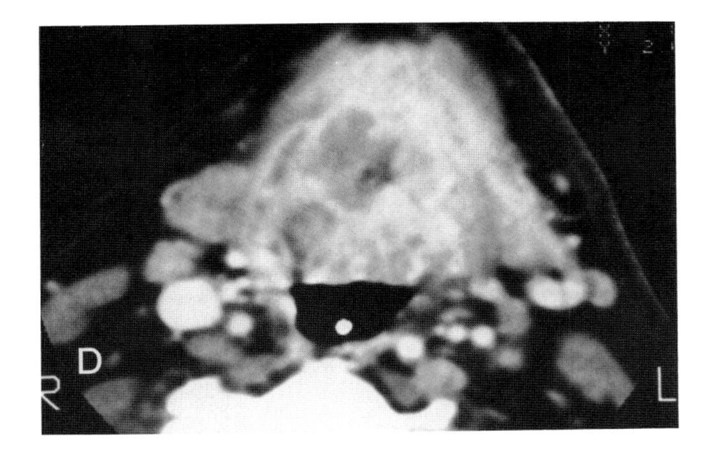

Plate 10.20*D*. Now that the inflammatory component of the mass has diminished somewhat, one can begin to appreciate the necrotic mass with a zone of peripheral enhancement which occupies a good deal of the tongue base. This was the necrotic, secondarily infected tumor.

This case illustrates that whenever there is a pattern of spread which appears somewhat atypical or that there is excessive enhancement of the surrounding soft tissues, inflammation or secondary inflammation of tumor may be the explanation. A repeat staging CT might be necessary in these instances so that the true extent of the tumor is not overestimated. This does not occur very often but it is a pitfall that should be kept in mind. MRI suffers the same limitations as CT in this regard as was illustrated in Plate 10.15 where MRI overestimated the size of tumor due to related surrounding edema.

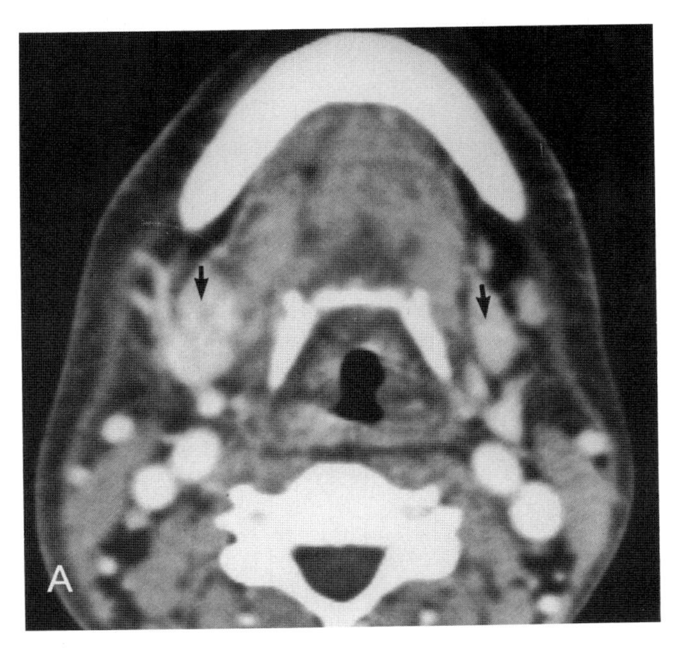

Plate 10.21. The issue of detecting recurrent tumor following radiation therapy is an important one since the treatment often alters anatomy so that physical examination is difficult. There is a tendency for squamous cell carcinoma to "go underground" following radiation therapy.

Plate 10.21*A*. This patient had a tumor in the oral tongue which was treated with both external beam radiation and implants. At this level there is a marked thickening of the subcutaneous fat and extensive edema within the lower oropharynx and upper larynx. The enhancement of the submandibular glands (*arrows*) is seen as a result of radiation-induced sialadenitis.

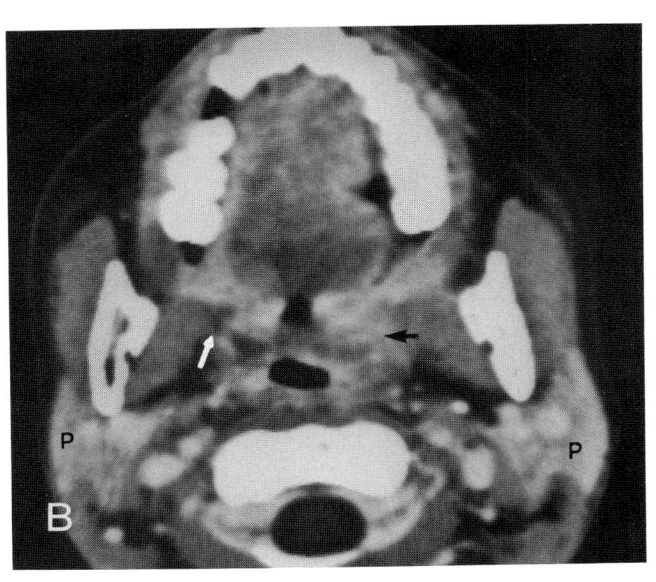

Plate 10.21*B*. Shows excessive enhancement of the parotid glands, again as a result of radiation-induced sialadenitis.

There is swelling around the airway which at first inspection appears to be diffuse. A closer look reveals that the parapharyngeal space on the left side is obliterated (*black arrow*); compare this to the normal parapharyngeal space on the right side (*white arrow*). This is strongly suspicious for extension of this tumor along the anterior tonsillar pillar and into the low nasopharynx. Repeat endoscopy at this time showed no mucosal evidence of tumor. Because of the CT findings multiple blind biopsies were done and these returned squamous cell carcinoma. Surgery revealed recurrent tumor involving the anterior tonsillar pillar, tonsillar fossa and extending into the low nasopharynx.

The abbreviations used on Plates 10.23 and 10.24 are: *P*, parotid gland; *N*, node; and *T*, tumor.

Plate 10.22. This patient had a carcinoma of the retromolar trigone which was treated with radiation therapy. He returned with what appeared to be a local recurrence. CT was done to stage the extent of the recurrent tumor.

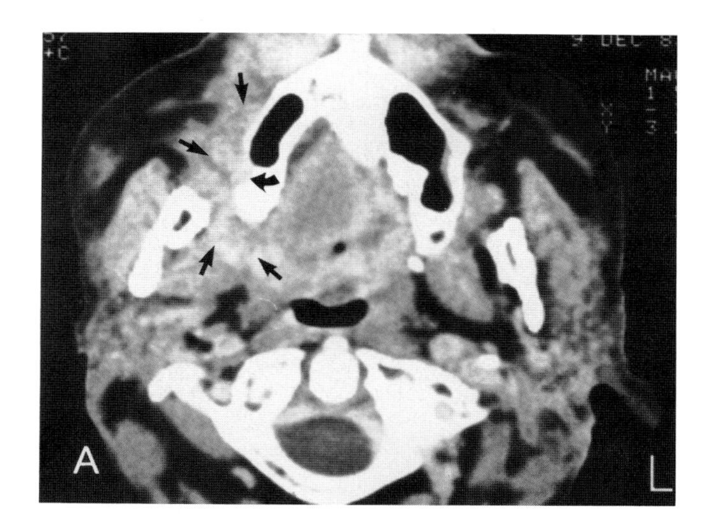

Plate 10.22*A*. This section somewhat above the retromolar trigone and through the upper buccinator space shows an infiltrating mass which has extended up the pterygomandibular raphe to occupy virtually all of the buccinator space (*arrows*). The lesion has produced a saucerized defect on the lower maxilla (*curved arrow*) and destroyed the lower portion of the pterygoid plate.

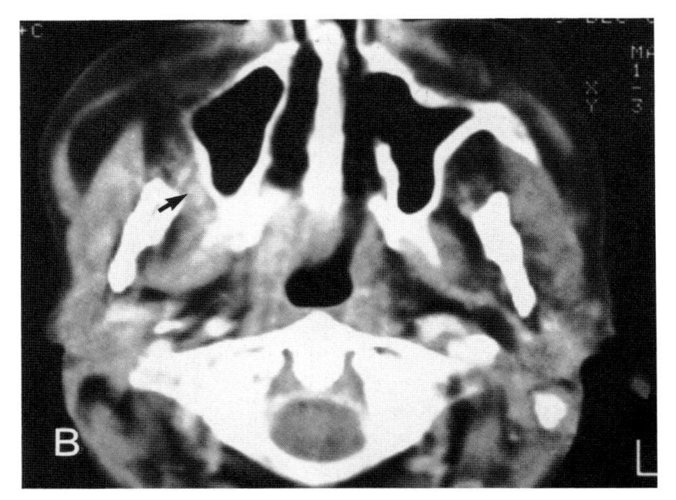

Plate 10.22*B*. A section through the low infratemporal fossa shows that this tumor has spread superiorly from the buccinator space into the low infratemporal fossa (*arrow*). The margin of the tumor mass appears quite ill-defined, suggesting a very infiltrating type of pattern. Despite the extensive disease, curative resection was attempted. This patient returned with massively recurrent tumor in the infratemporal fossa several months after surgery.

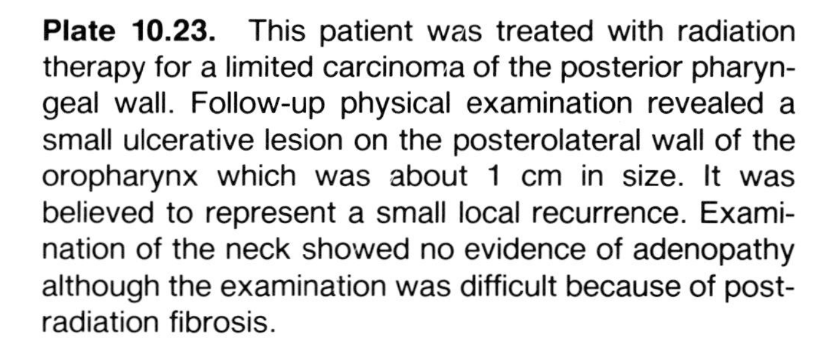

Plate 10.23. This patient was treated with radiation therapy for a limited carcinoma of the posterior pharyngeal wall. Follow-up physical examination revealed a small ulcerative lesion on the posterolateral wall of the oropharynx which was about 1 cm in size. It was believed to represent a small local recurrence. Examination of the neck showed no evidence of adenopathy although the examination was difficult because of post-radiation fibrosis.

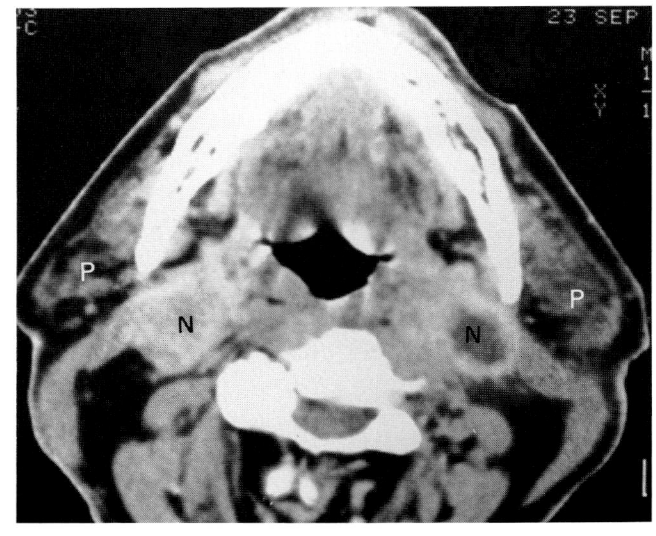

CT study showed bilateral nodal metastases with clear evidence of extranodal extension. These nodes are the highest of the internal jugular group and are extending up to the posterior belly of the digastric muscle. Rather than a small local recurrence it is obvious that this patient has massive, incurable recurrent disease.

The fibrotic changes in the parotid regions are commonly seen as the residual of radiation induced sialadenitis.

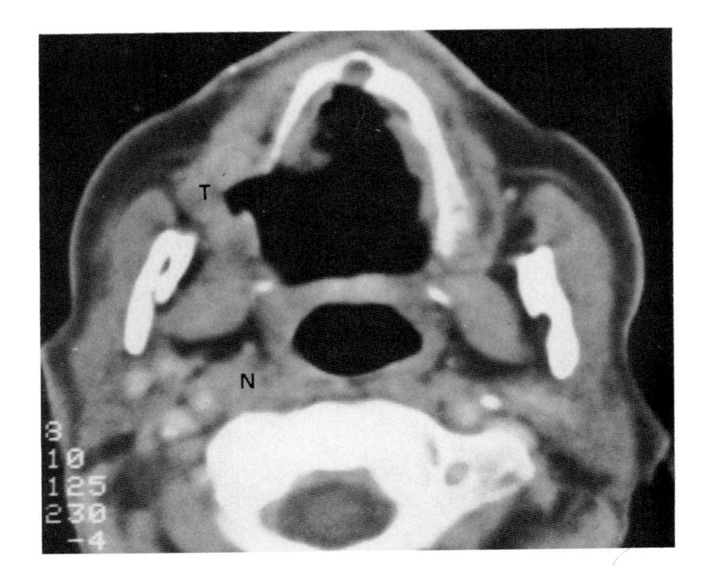

Plate 10.24. Patients with primary and recurrent carcinoma in the upper aerodigestive tract are sometimes at high risk for retropharyngeal adenopathy depending on the primary site and extent of nodal disease.

This patient had a large gingivobuccal sulcus carcinoma which had eroded to involve the maxillary alveolar ridge as well as portions of the hard palate. The ulcerative tumor invades the buccinator space at this level. Note the abnormal nodal mass in the retropharyngeal region. Prior to CT there was no really accurate means of evaluating the retropharyngeal nodes. Since this lesion invaded the floor of the nose, this patient was at risk for involvement of this important nodal group.

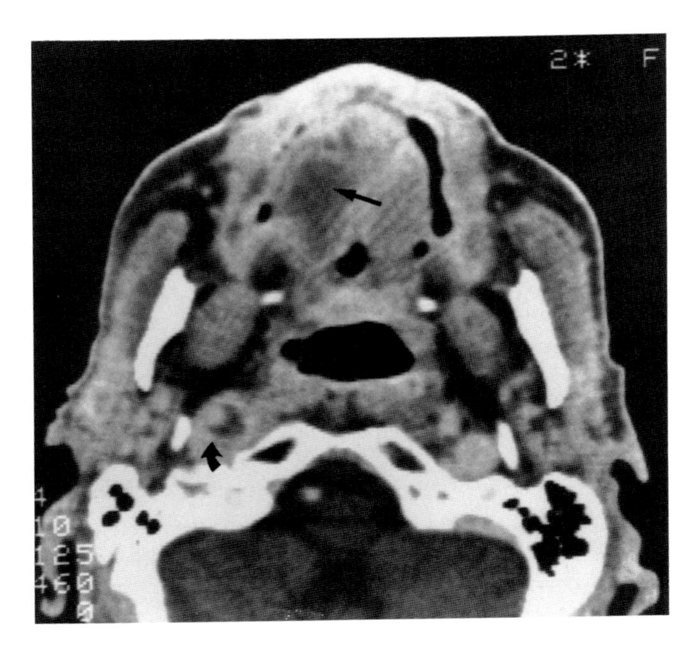

Plate 10.25. This patient had been radiated for carcinoma of the oral tongue. The patient returned with symptoms indicating abnormal function of cranial nerves 9, 10, and 12. Some atrophy at the right side of the tongue (*arrow*) confirms the XIIth nerve deficit. This study also revealed a metastatic retropharyngeal node (*curved arrow*) which was causing the cranial nerve deficits. Metastatic disease from oropharyngeal cancer to the retropharyngeal nodes is an exceedingly bad prognostic sign. This nodal group may be studied with MRI or CT.

The abbreviations used on Plate 10.26 are: *T*, tumor; and *N*, node.

Plate 10.26. MRI shows some promise in the follow-up and perhaps early determination of recurrent tumor in patients being treated for tumors involving the upper aerodigestive tract.

Plate 10.26*A*. The CT study shows a large carcinoma in the mid- to lower right tonsillar fossa.

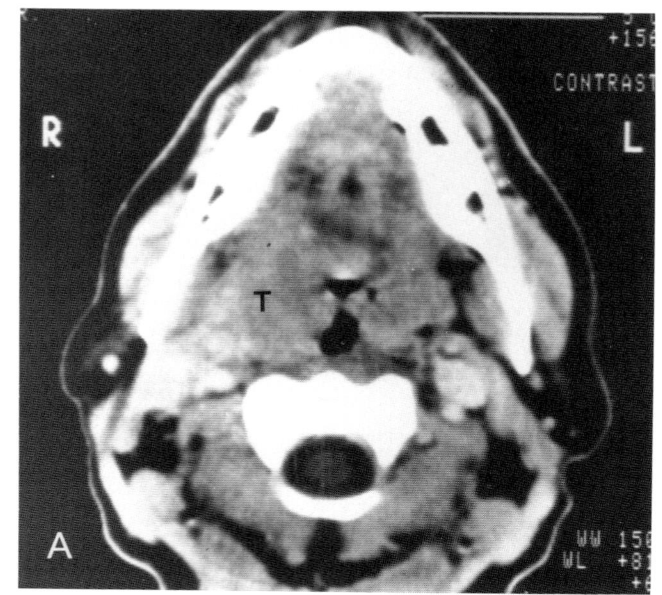

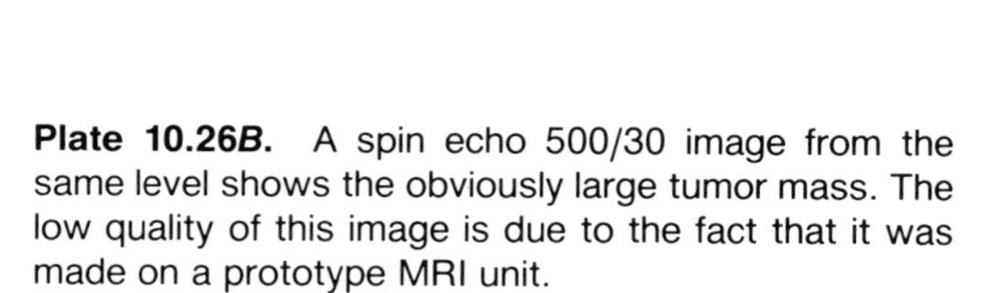

Plate 10.26*B*. A spin echo 500/30 image from the same level shows the obviously large tumor mass. The low quality of this image is due to the fact that it was made on a prototype MRI unit.

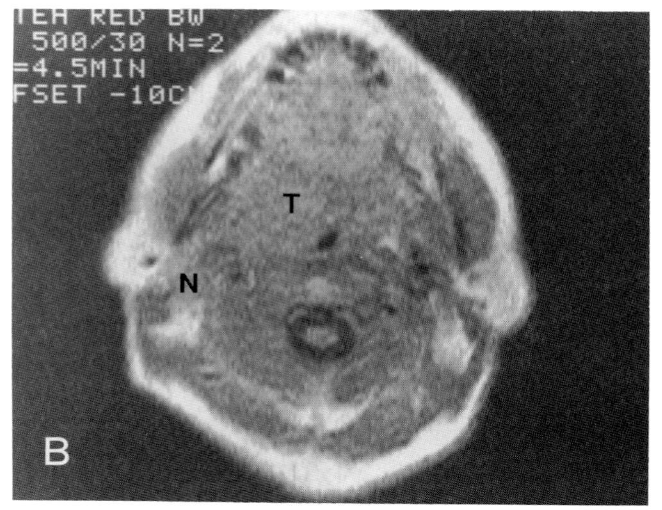

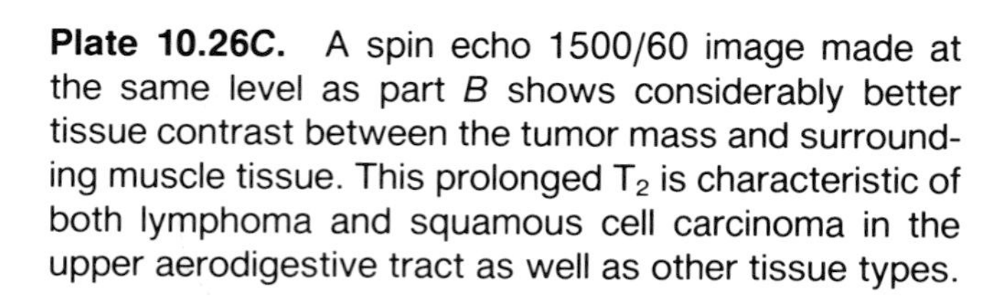

Plate 10.26*C*. A spin echo 1500/60 image made at the same level as part *B* shows considerably better tissue contrast between the tumor mass and surrounding muscle tissue. This prolonged T_2 is characteristic of both lymphoma and squamous cell carcinoma in the upper aerodigestive tract as well as other tissue types.

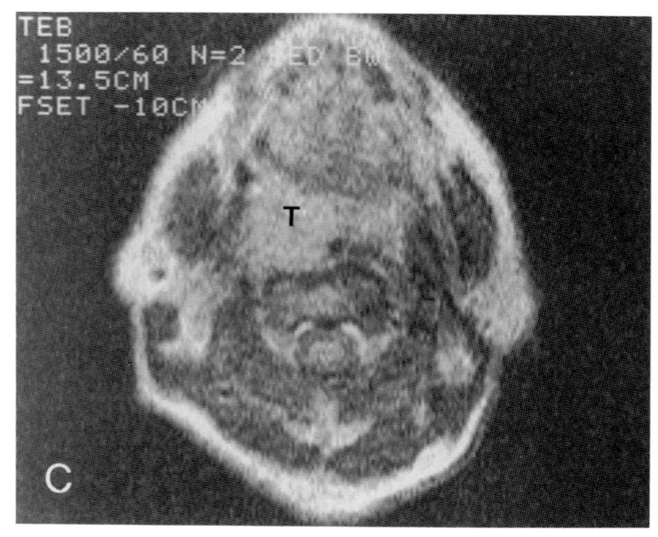

Plate 10.26*D*. A spin echo 500/30 shows the tumor extending into the lower nasopharynx.

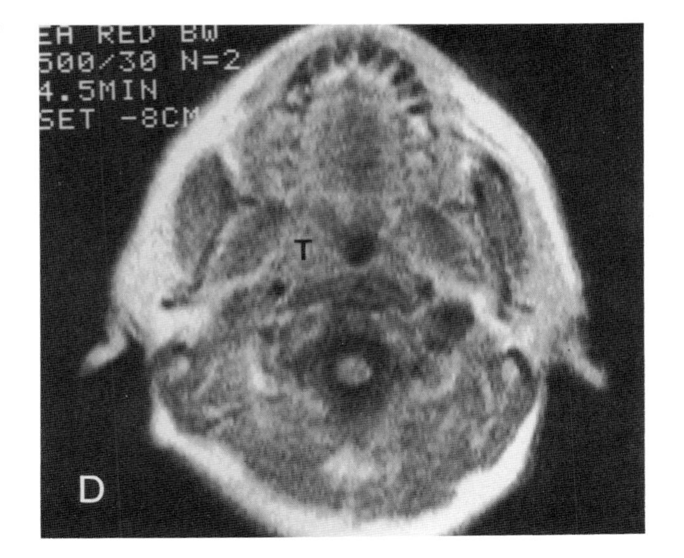

Plate 10.26*E*. A spin echo 500/30 MR image at the completion of therapy revealed some diffuse swelling of the soft tissue in the nasopharynx (*arrow*); however, there was no asymmetry to suggest a persistent mass on the right side as noted in part *D*.

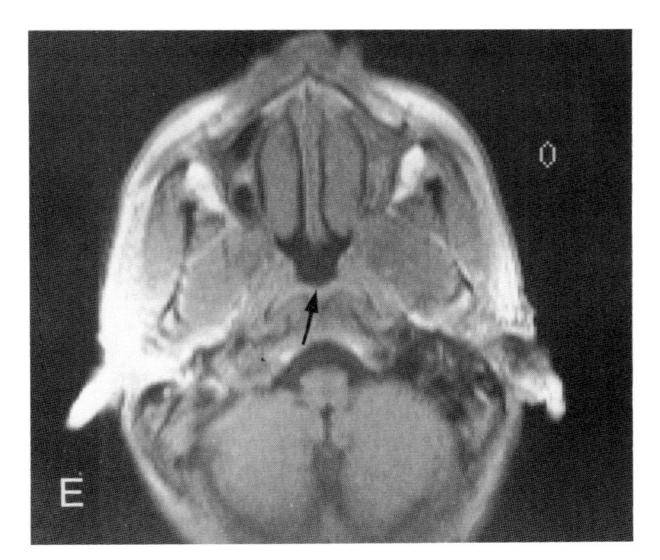

Plate 10.26*F*. With improvements in both hardware and software over the 2-month period that this patient was followed we were able to obtain much higher quality images. This is one of 10 sections obtained during a multislice, multiecho acquisition using a T_R of 2500 msec and T_E values (*45*, *90*, *135*, *180*) as indicated in the upper right hand corner of each image. Note that the very large tonsillar carcinoma has completely resolved. The deep tissue planes have all returned to normal. In addition, the signal in the primary site becomes progressively darker as the images become more T_2 weighted. This indicates that there has been considerable scarring in the tonsillar bed (*arrows*). It is hoped that MRI may be able to detect the difference between the signal of end-stage fibrosis following treatment and that of residual tumor. This requires careful study, although several investigators have reported some success in this differentiation.

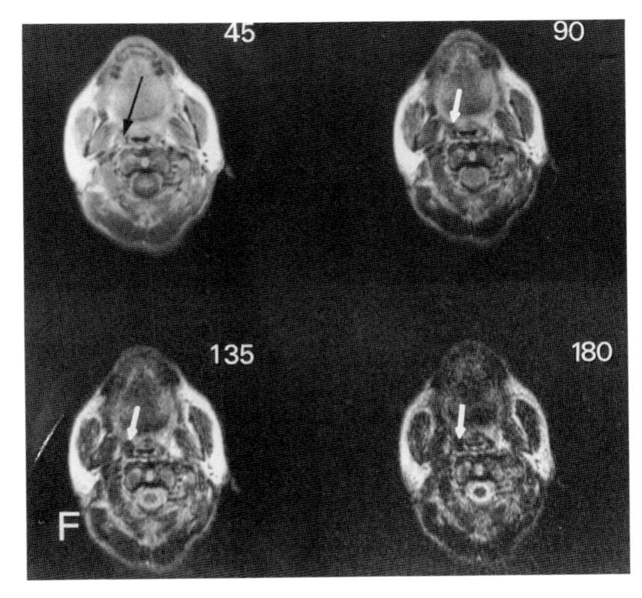

Plate 10.27. Mandibular malignancies are uncommon tumors and are often metastatic; most commonly, mandibular involvement with tumors is due to local spread of a primary squamous cell carcinoma involving the oral cavity or oropharynx.

Plate 10.27*A*. This patient has a primary tumor involving the mandible. The lesion can be seen extending within the marrow cavity (*arrows*) and out the mandibular foramen to a point near the attachment of the medial pterygoid muscle (*curved arrow*).

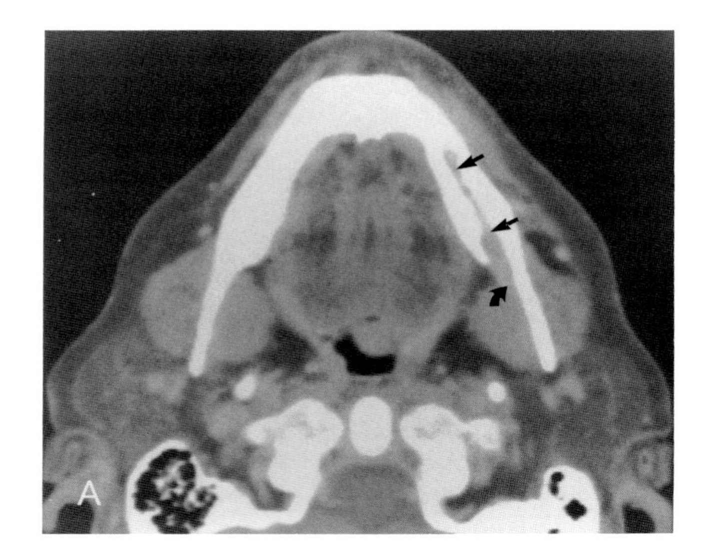

Plate 10.27*B*. A wider window at this setting better displays bone detail. The bony defect can be seen and the pattern of spread principally along the course of the mandibular nerve is perhaps more apparent. Since this spread extended beyond the mandible there had to be the suspicion that this could follow the third division of the trigeminal nerve back to the skull base. There was no gross evidence of this on higher CT sections but at surgery there was microscopic perineural infiltration of the entire mandibular division of the trigeminal nerve. This was a fibrosarcoma of the mandible with extensive secondary perineural spread.

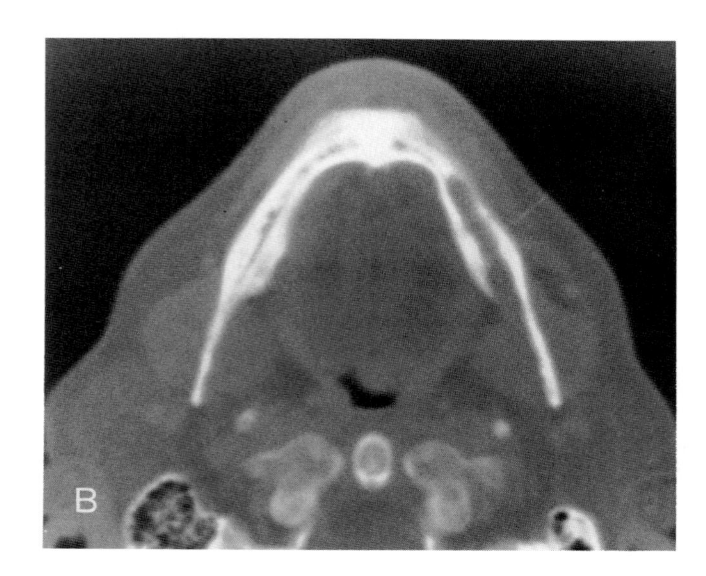

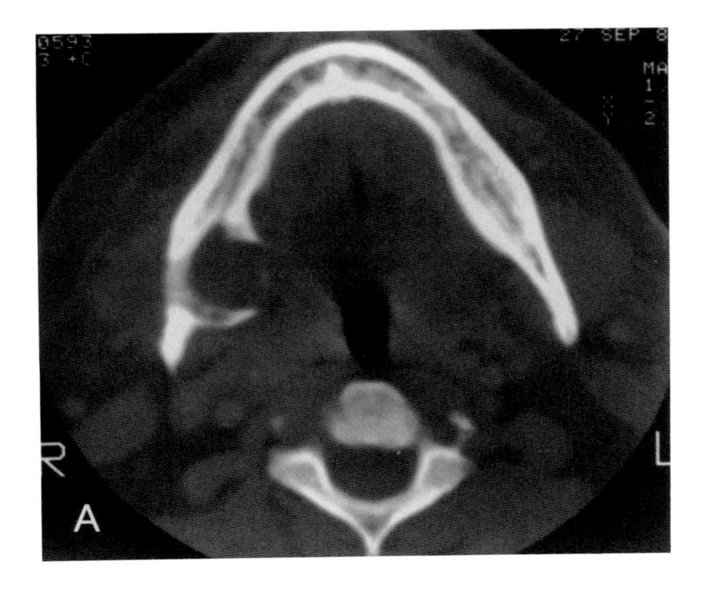

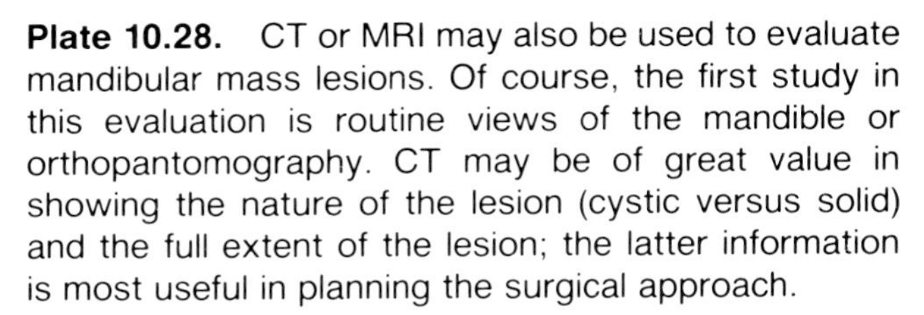

Plate 10.28. CT or MRI may also be used to evaluate mandibular mass lesions. Of course, the first study in this evaluation is routine views of the mandible or orthopantomography. CT may be of great value in showing the nature of the lesion (cystic versus solid) and the full extent of the lesion; the latter information is most useful in planning the surgical approach.

Plate 10.28*A*. A wide window shows a very well circumscribed bony defect in the mandible. The lesion is obviously one which has a relatively chronic timetable since there is evidence of considerable remodeling of the mandible. There is an area of focal dehiscence of bone which represents the zone where the lesion is growing so rapidly that the periosteum cannot mineralize quickly enough to form a complete shell of bone around the lesion. This finding may suggest more aggressive behavior of the lesion or a previously benign lesion which has turned more aggressive.

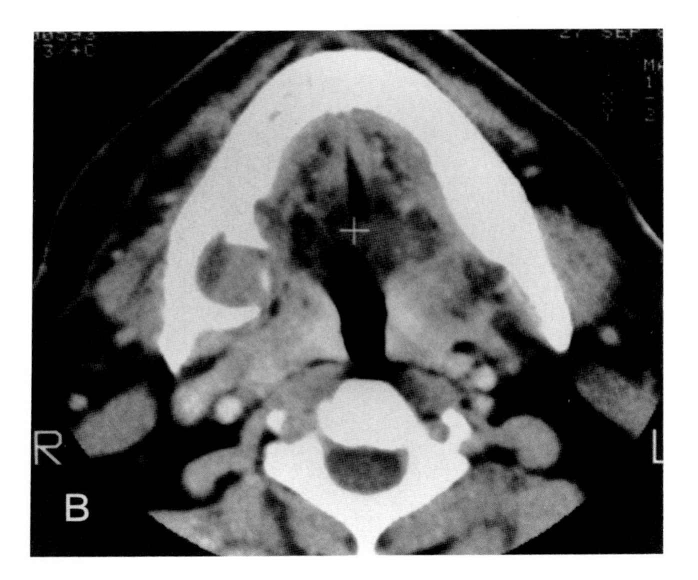

Plate 10.28*B*. The narrower window view of the same level reveals that the lesion is filled with tissue rather than fluid. This would greatly favor the diagnosis of adamantinoma over dentigerous cyst.

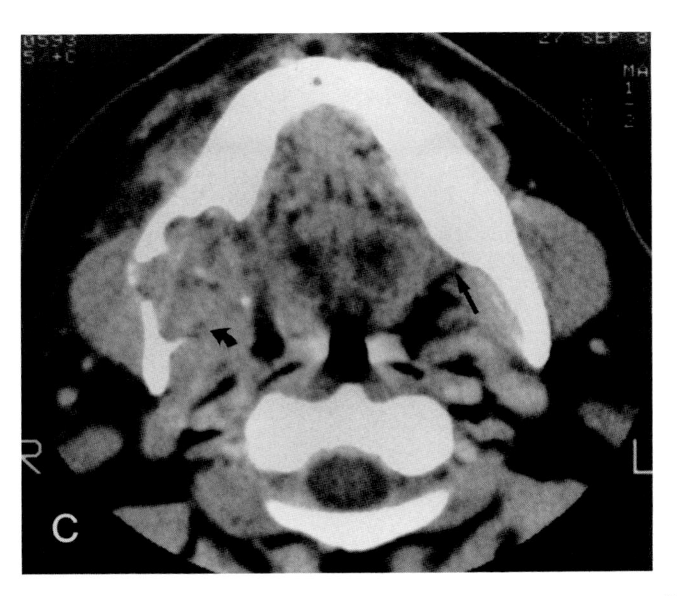

Plate 10.28*C*. A section slightly higher than the two previous ones shows more areas of dehiscent cortical bone surrounding the solid lesion within the mandible. The encroachment on the pterygomandibular space is obvious and this mass, therefore, should be displacing or involving the lingual nerve (compare this with the normal pterygomandibular space (*arrow*) on the opposite side). The mass also displaces the pterygoid muscle posteriorly (*curved arrow*) although the margin with the pterygoid would probably be clearer on an MRI examination. This lesion was an adamantinoma and surgery confirmed its full extent as shown by the CT study.

The abbreviations used on Plates 10.29 and 10.30 are: *SMG*, submandibular gland; and *C*, cyst.

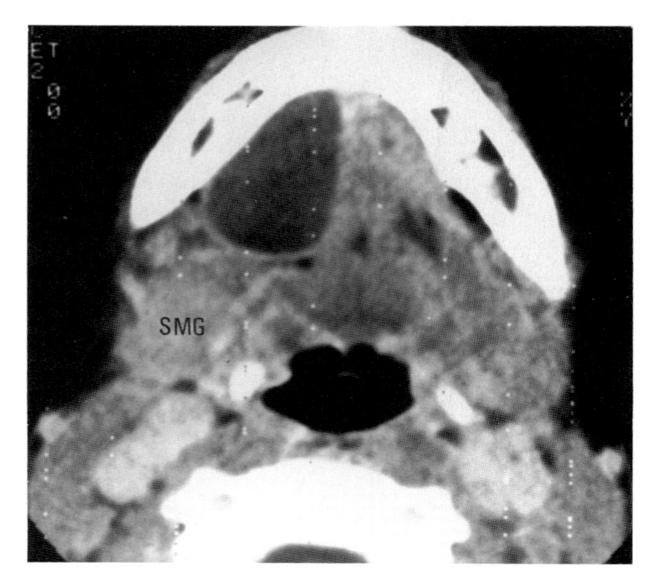

BENIGN TUMORS AND INFLAMMATORY LESIONS

Plate 10.29. This patient presented with a mass in the floor of the mouth. CT study revealed a very well circumscribed lesion with low attenuation. The findings are indicative of a ranula. The mass is obviously separate from the submandibular gland and is quite different in morphology from the other lesions which occur in this area, with the possible exception of a dermoid cyst. The lesion is far too anterior to be a branchial cleft cyst.

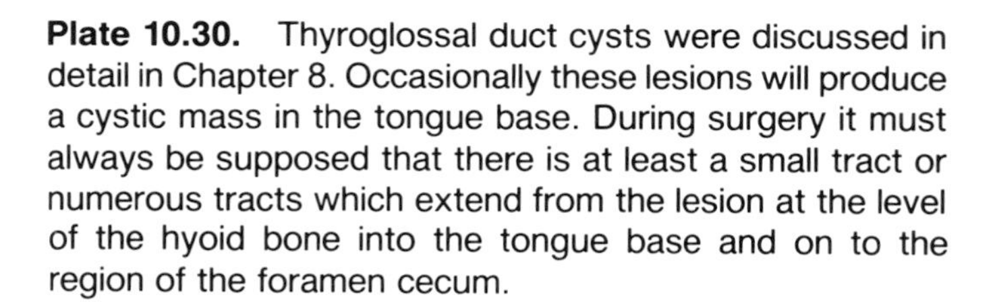

Plate 10.30. Thyroglossal duct cysts were discussed in detail in Chapter 8. Occasionally these lesions will produce a cystic mass in the tongue base. During surgery it must always be supposed that there is at least a small tract or numerous tracts which extend from the lesion at the level of the hyoid bone into the tongue base and on to the region of the foramen cecum.

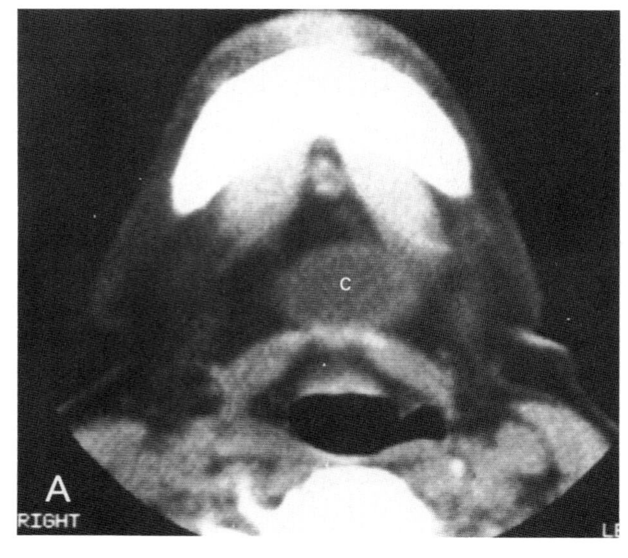

Plate 10.30*A*. This cephalad extension of a thyroglossal duct cyst is much larger than usual and lies within the soft tissues of the suprahyoid neck.

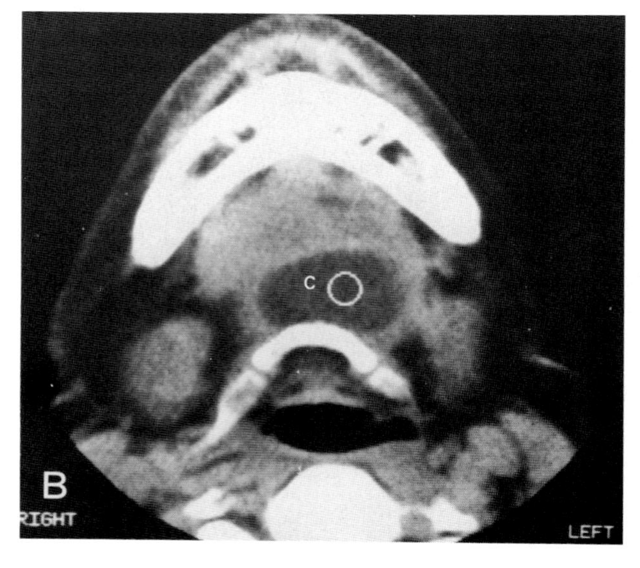

Plate 10.30*B*. The cyst continues cephalad into the tongue base producing a large low attenuation lesion. Again, this is somewhat atypical in that the extension into the tongue base is usually in the form of multiple small tracts. The information on this study was quite helpful in forewarning the surgeon about this unusually large cystic extension. This must be resected in its entirety in order to avoid tongue base recurrences. (This case appears courtesy of Gordon Gamsu, M.D., Department of Radiology, University of California at San Francisco.)

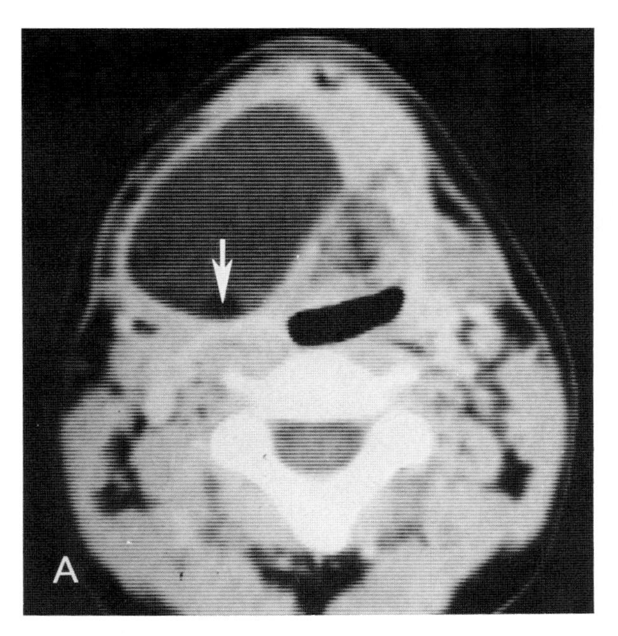

Plate 10.31. This patient presented with a large submandibular mass.

Plate 10.31*A*. The axial section through the floor of the mouth shows a large predominantly low attenuation mass with a very well defined, fairly thin wall. The mass is too far anterior to be a branchial cleft cyst. In addition, there is a small accumulation of fat (*arrow*) within the mass identifying this as a dermoid cyst.

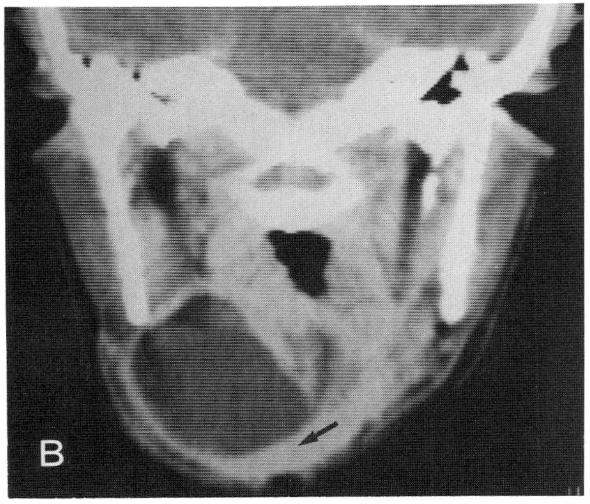

Plate 10.31*B*. A coronal section through the mass shows that it lies below the plane of the mylohyoid muscle, placing it within the spaces of the suprahyoid neck. The mass displaces the digastric muscles inferiorly and to the opposite side (*arrow*).

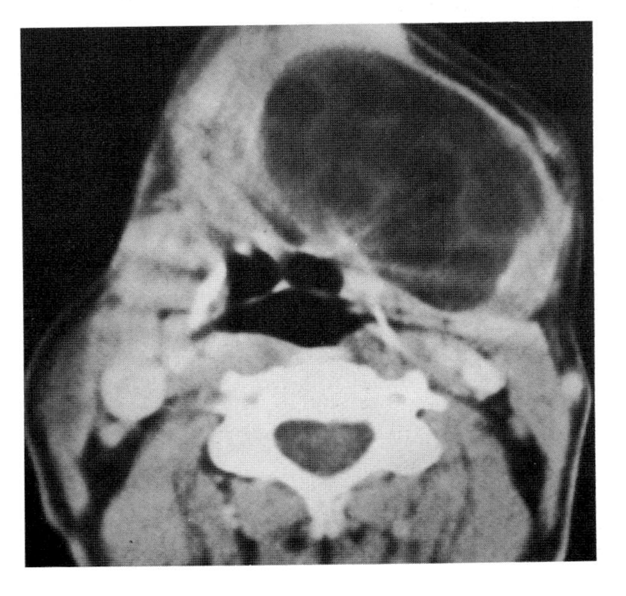

Plate 10.32. This is another dermoid cyst with a somewhat different appearance than that seen in the prior case. The mass is again too far anterior and medial to be a branchial cleft cyst. The multiloculated appearance strongly suggests the etiology. Although the floor of the mouth is one of the less common locations for dermoids they are not at all unusual in this locale. (Reprinted with permission of Williams and Wilkins from T.B. Hunter et al: Dermoid cyst of the floor of the mouth: CT appearance. *AJR* 141: 1239–1240, 1983.)

The abbreviation used in Plate 10.33 is: *SMG*, submandibular gland.

Plate 10.33. This noncontrast-enhanced study reveals a mass within the floor of the mouth (*arrows*). It is separate from the submandibular gland. The mass has a necrotic center. This could represent a squamous cell carcinoma of the floor of the mouth; however, this is a young patient with a history of a slowly enlarging mass. This was a neuroma. Neuromas of the floor of the mouth are unusual lesions.

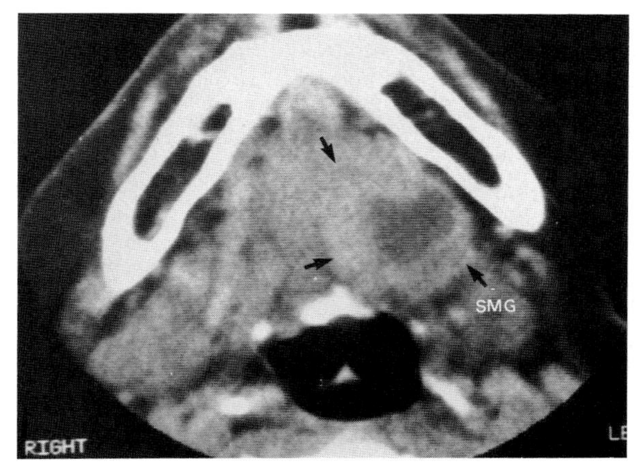

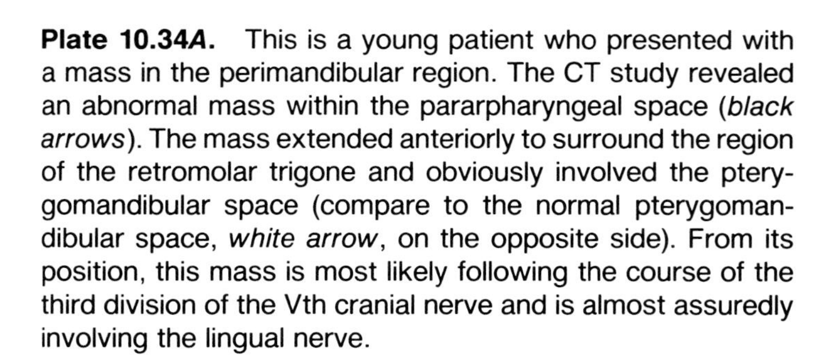
Plate 10.34*A*. This is a young patient who presented with a mass in the perimandibular region. The CT study revealed an abnormal mass within the pararpharyngeal space (*black arrows*). The mass extended anteriorly to surround the region of the retromolar trigone and obviously involved the pterygomandibular space (compare to the normal pterygomandibular space, *white arrow*, on the opposite side). From its position, this mass is most likely following the course of the third division of the Vth cranial nerve and is almost assuredly involving the lingual nerve.

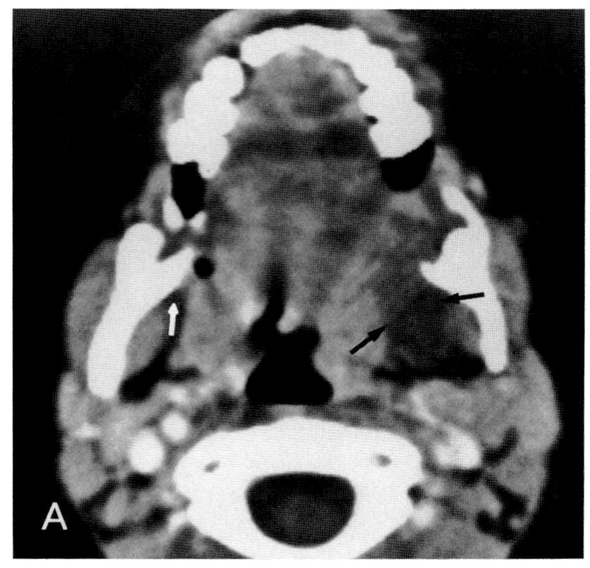

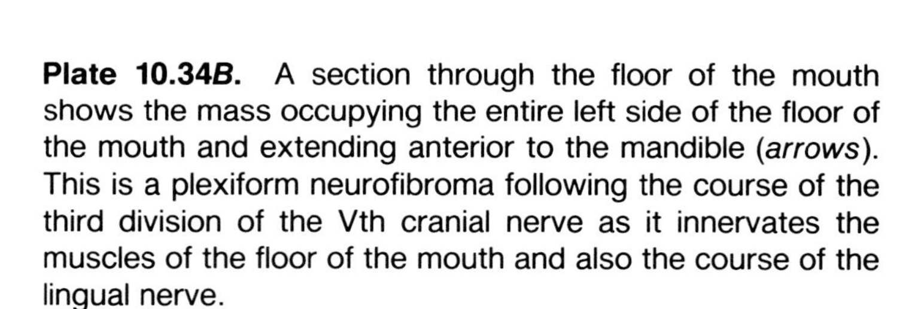
Plate 10.34*B*. A section through the floor of the mouth shows the mass occupying the entire left side of the floor of the mouth and extending anterior to the mandible (*arrows*). This is a plexiform neurofibroma following the course of the third division of the Vth cranial nerve as it innervates the muscles of the floor of the mouth and also the course of the lingual nerve.

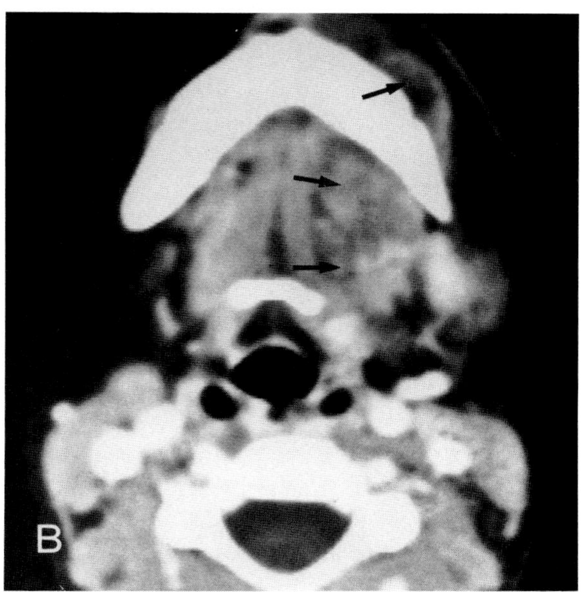

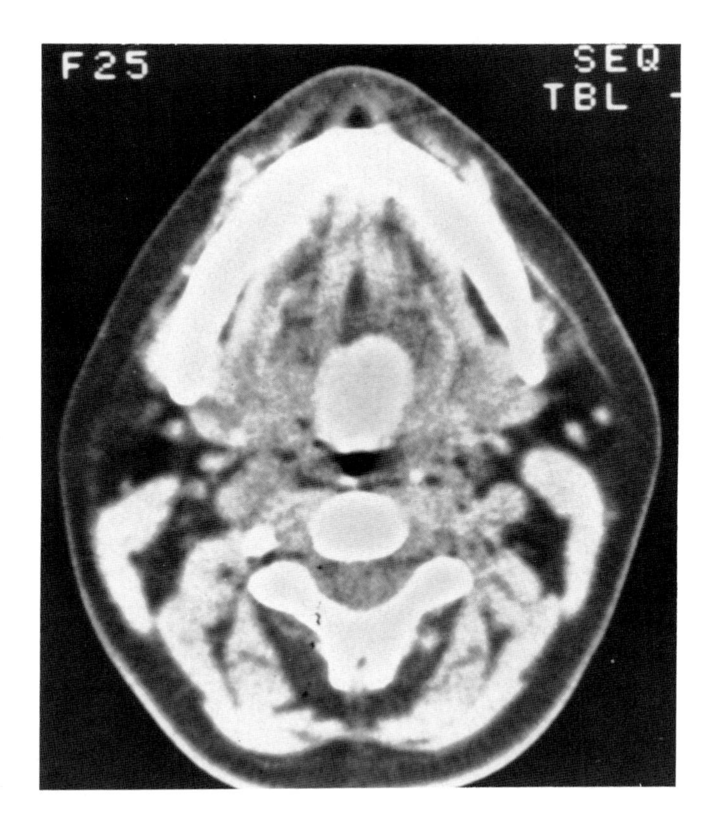

Plate 10.35. This unusual, though characteristic, high attenuation mass at the tongue base represents lingual thyroid tissue. The increased attenuation was related to the iodine content. These lesions will obviously enhance brightly on contrast-enhanced CT study. It is essential that radionuclide studies confirm normal thyroid within the neck before lingual thyroid tissue is resected. (This case appears courtesy of D. L. "Stormy" Johnson, M.D., New Orleans Radiology Group.)

INFLAMMATORY LESIONS

Plate 10.36. CT is an exceedingly useful way to evaluate the full extent of abscesses within the deep neck. This was discussed in detail in Chapter 8. Abscesses within the spaces surrounding the oropharynx may have confusing clinical presentations and CT should always be done as part of the patient's preoperative evaluation. CT can differentiate between abscess and cellulitis as well as show the full extent of the abscess which may involve any one of several compartments.

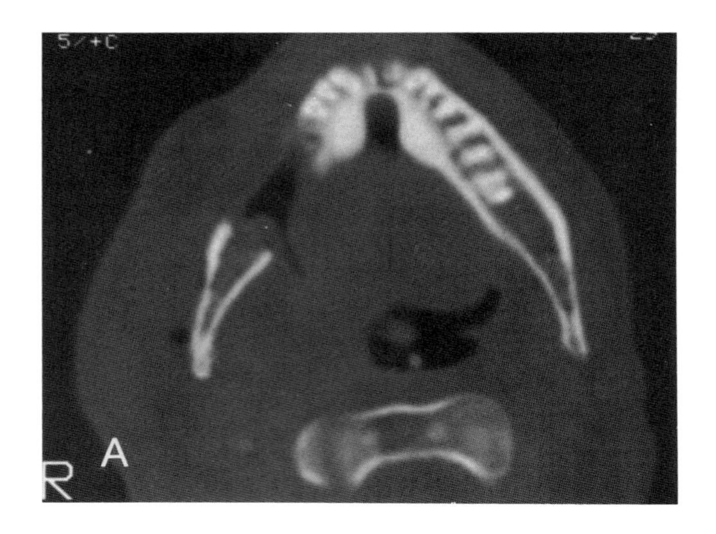

Plate 10.36*A*. This patient was irradiated for an oropharyngeal cancer. He presented with symptoms indicative of osteoradionecrosis of the mandible. This section shows a large bony defect within the mandible and an ulcerative tract extending into the floor of the mouth.

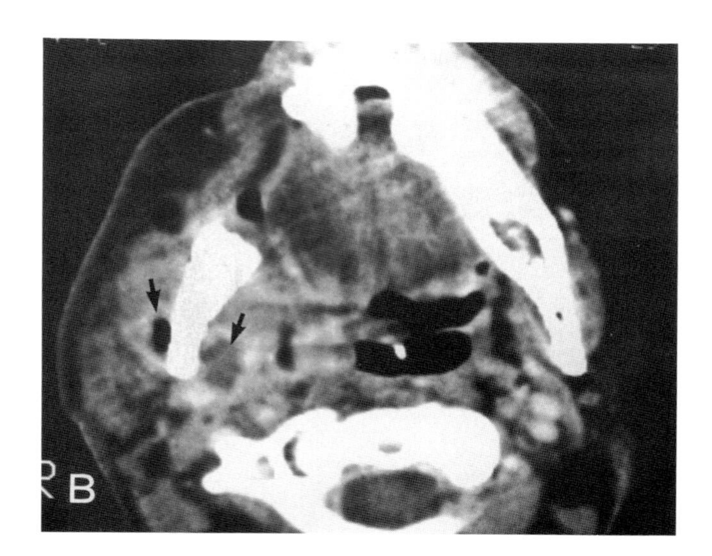

Plate 10.36*B*. A section made somewhat lower than part *A* shows diffuse soft tissue swelling around the mandible. In addition there is a collection of fluid and gas within the masticator space (*arrows*).

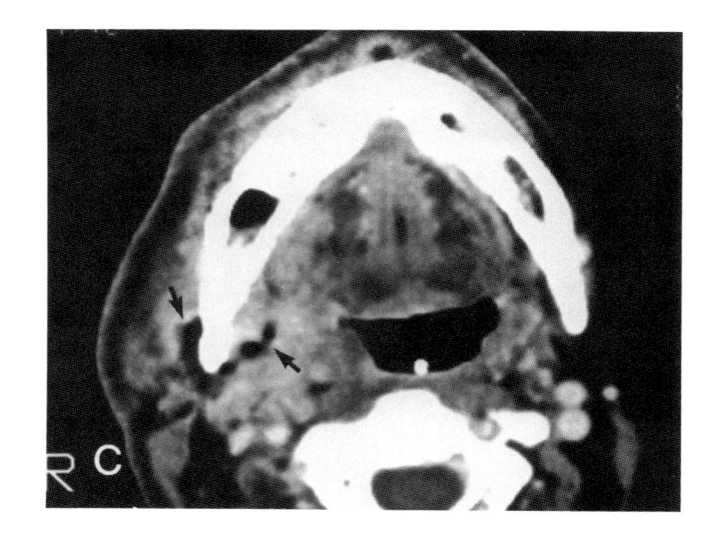

Plate 10.36*C*. A section somewhat more inferior than part *B* shows the diffuse swelling of the soft tissues and the collection of air and fluid conforming precisely to the limits of the masticator as it wraps around the posterior margin of the mandible (*arrows*).

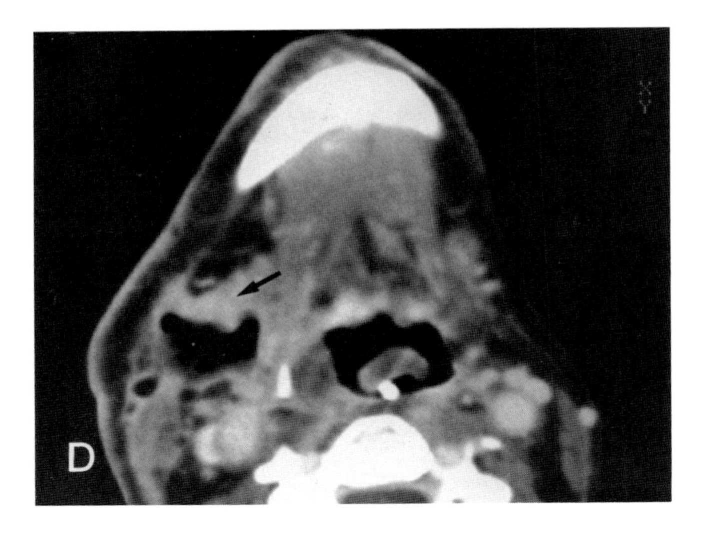

Plate 10.36*D*. The abscess has spilled over the posterior aspect of the mylohyoid muscle to lie within the submandibular space just behind the submandibular gland (*arrow*). At this point the mass has moved from the masticator space into the suprahyoid neck to become a deep neck abscess as well.

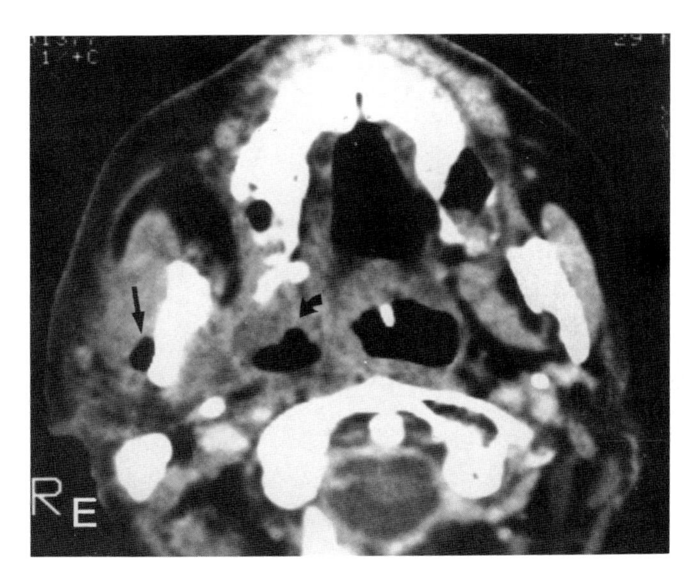

Plate 10.36*E*. A section taken somewhat higher than part *D* shows that there is continued cephalad spread within the masticator space (*arrow*). There is also a large abscess within the parapharyngeal space (*curved arrow*).

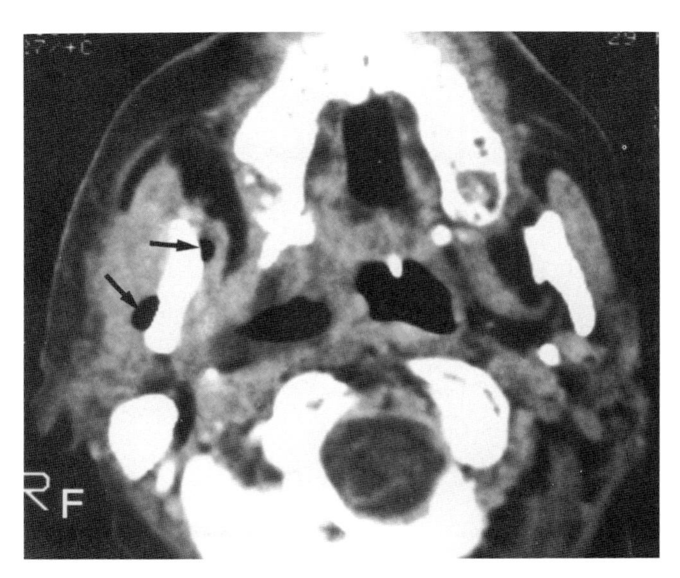

Plate 10.36*F*. A section in the low nasopharynx shows continued cephalad extension within both the medial and lateral extremities of the massateric space (*arrows*). The large parapharyngeal component is again seen.

In summary, this patient developed an abscess secondary to osteoradionecrosis of the mandible. The abscess spread first to the masticator space and then on to the deep neck. It also broke out of the masticator and extended into the parapharyngeal space. It is fairly clear from this case that CT is capable of providing a precise "road map" to the surgeon approaching abscesses in all of these compartments.

The abbreviations used on Plate 10.37 are: *PG*, parotid gland; and *M*, masseter muscle.

Plate 10.37. This patient presented with diffuse swelling in the parotid region and was believed to have a parotid abscess or deep face cellulitis. The CT study was done to decide whether abscess or cellulitis was present in the parotid bed.

Plate 10.37*A*. On this section there is obviously diffuse swelling of the masseter muscle as well as the parotid gland. The cause of this is the masticator space abscess seen deep to the mandible (*arrow*).

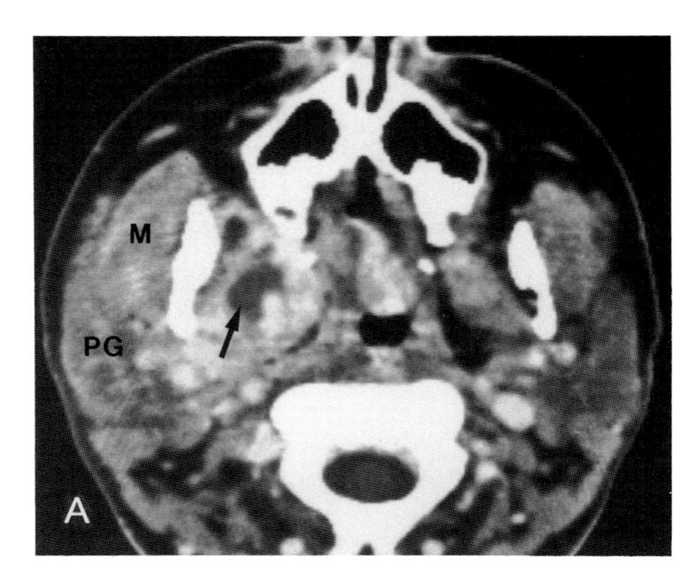

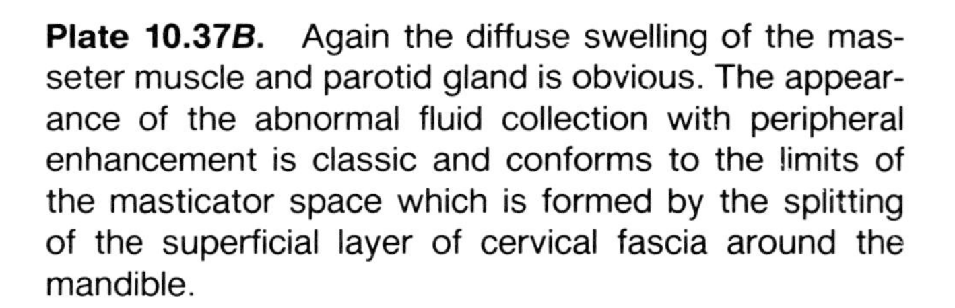

Plate 10.37*B*. Again the diffuse swelling of the masseter muscle and parotid gland is obvious. The appearance of the abnormal fluid collection with peripheral enhancement is classic and conforms to the limits of the masticator space which is formed by the splitting of the superficial layer of cervical fascia around the mandible.

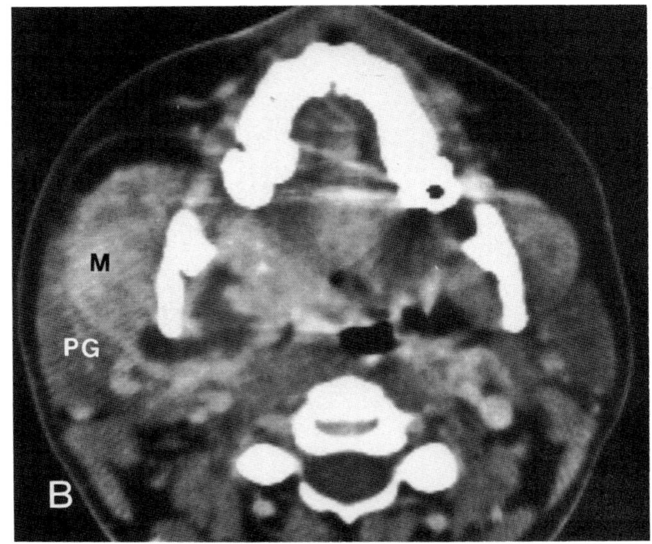

This case illustrates how CT may be used to differentiate parotid abscess from deep face cellulitis from abscess in the various spaces which surround these areas.

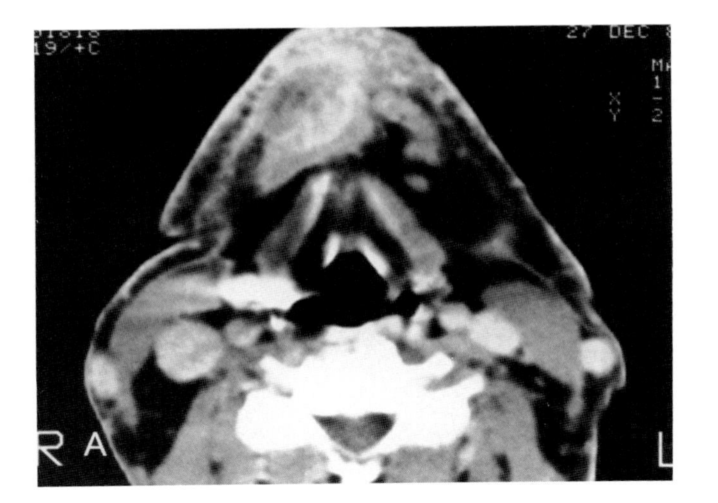

Plate 10.38. This patient presented with induration in the submental space. It was unclear whether this was abscess or cellulitis.

Plate 10.38*A*. CT study showed a peripherally enhancing necrotic mass in the submental space. In addition, there is extensive enhancement and thickening of the surrounding subcutaneous planes. The findings are compatible with a submental space abscess as well as surrounding cellulitis.

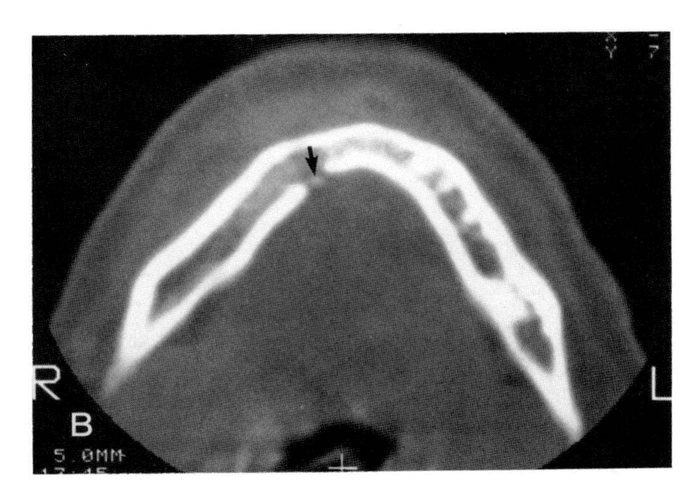

Plate 10.38*B*. This patient had had prior trauma and lost one of the anterior mandibular teeth. CT revealed a defect in the lingual plate with a small sequestrum of bone (*arrow*). This was the focus of post-traumatic osteomyelitis which led to the abscess within the submental space.

Plate 10.39. On rare occasions purely inflammatory lesions within the oropharynx may mimic a malignancy. This patient presented with a mass in the oropharynx which was believed to represent a submucosally spreading tumor. There were some mild inflammatory changes but the patient presented more as if he had a tumor than an inflammatory lesion.

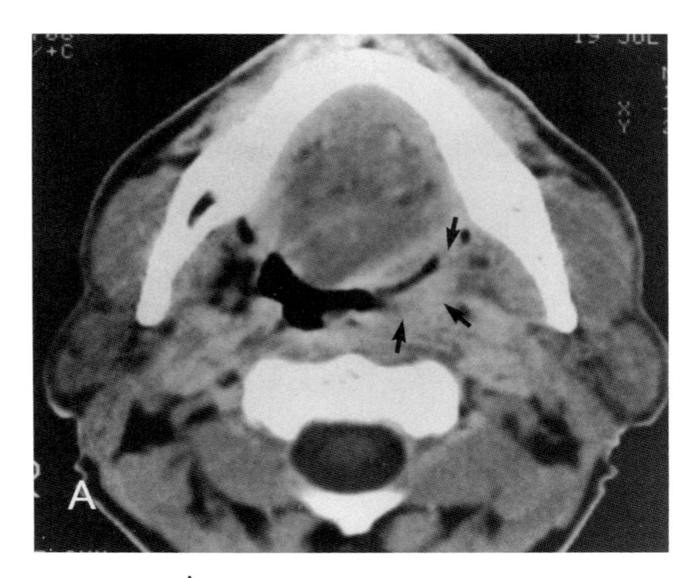

Plate 10.39*A*. A section through the midtonsillar fossa shows an abnormally infiltrating mass involving the tonsillar fossa and posterior pharyngeal wall (*arrows*).

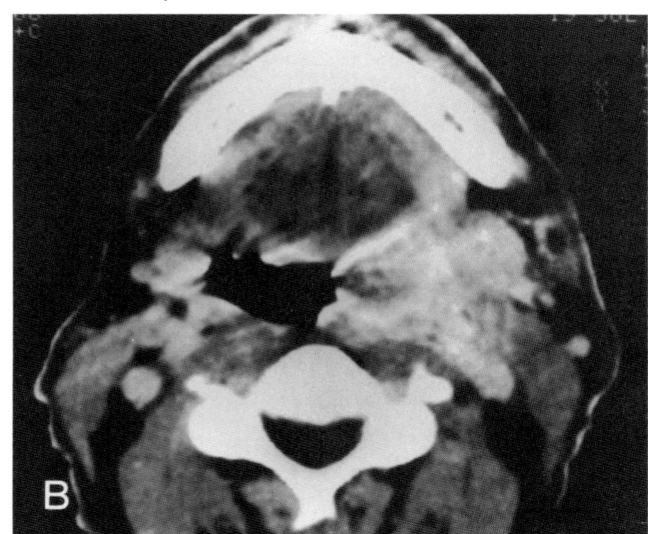

Plate 10.39*B*. A section lower in the glossotonsillar sulcus again shows the mass which demonstrates the very marked amount of enhancement. As presented in the case in Plate 10.20, whenever very marked enhancement is present, one should consider either an inflammatory lesion or a tumor with secondary infection.

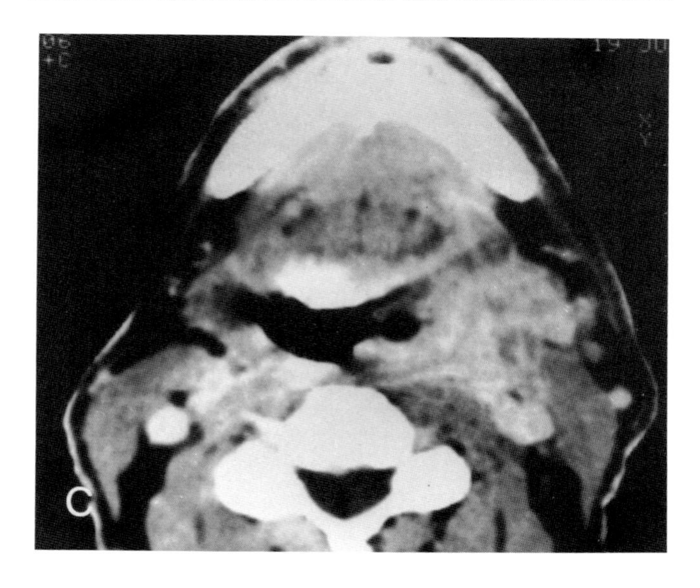

Plate 10.39*C*. A section low in the glossotonsillar sulcus again shows a markedly enhancing infiltrating mass. The primary suspicion at this point was an infiltrating neoplasm with secondary inflammation; however, this could not be reconciled with the clinical examination since there was no evidence of an ulcerative mucosal lesion.

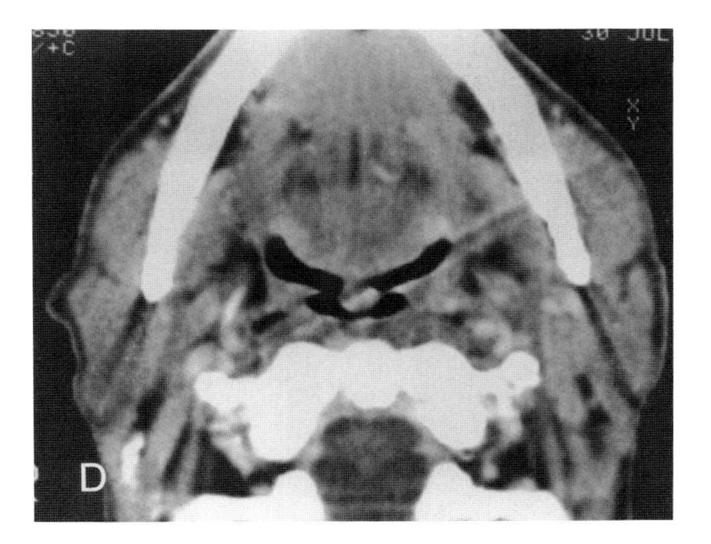

Plate 10.39*D*. Since there was some disparity between the physical findings and the CT study it was elected to give the patient a trial of antibiotics and rescan within 7–10 days. This section through the oropharynx is comparable to that in part *A* and shows virtually complete resolution of the changes seen within the tonsillar fossa and along the posterior pharyngeal wall. This was the case all the way inferior along the course of the glossotonsillar sulcus.

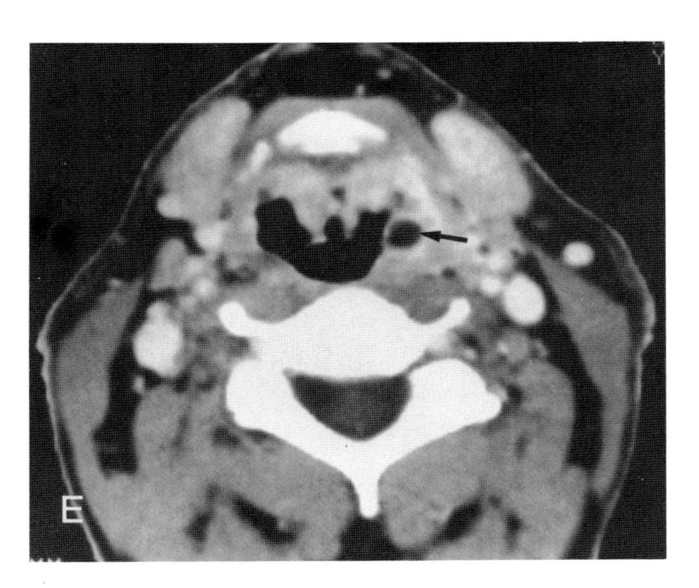

Plate 10.39*E*. A section at the level of the vallecula shows some persistent enhancement of the lingual tonsillar tissue. The small abscess in the lingual tonsil was identified (*arrow*). In retrospect this may be visible in part *C*. No tumor was found at repeat endoscopy.

TEMPOROMANDIBULAR JOINT

Plate 10.40. MRI promises to be an excellent means of evaluating the internal anatomy and physiology of the temporomandibular joint. The following sections were all done with a spin echo of 500/30 pulse sequence on a 0.15 Tesla resistive unit. The patients were placed laterally on a 5-in diameter surface coil. Slice thickness is either 4 or 6 mm. The in-plane resolution is slightly under 1 mm with the *X* and *Y* dimensions of the pixel being approximately 0.75 × 0.8 mm.

Plate 10.40*A*. In the mouth-closed position the mandibular condyle is seen within the condylar fossa by virtue of its content of fatty marrow (*black arrow*). The cortical bone returns virtually no signal so that it appears black (*arrowhead*). The fibrous portion of the articular disk rests in its normal position atop the head of the condyle. The posterior band (*white arrow*) lies at approximately 12 o'clock on the head of the condyle while the anterior band lies anteriorly and inferiorly (*open arrowhead*). The collapsed bilaminar zone fills the remaining portion of the condylar fossa and is of higher signal intensity than the fibrous portion of the disk.

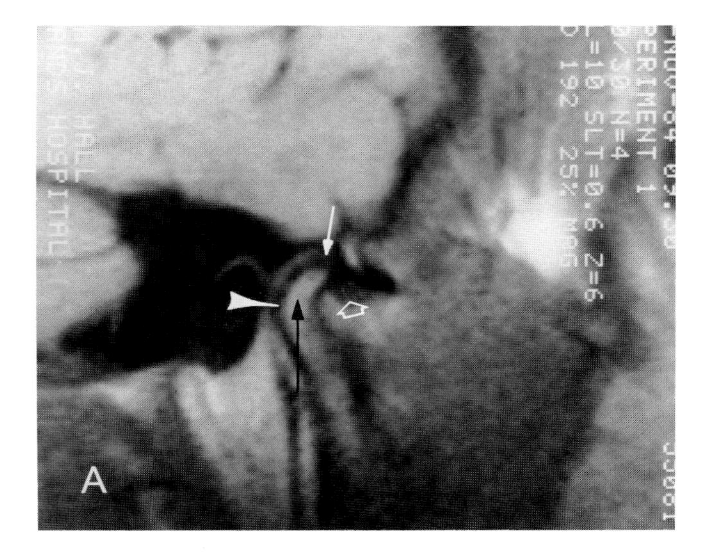

Plate 10.40*B*. This is the first of a series of sections made with the mouth open approximately 10–20 mm. A section laterally shows that the head of the condyle has translated forward on the articular eminence.

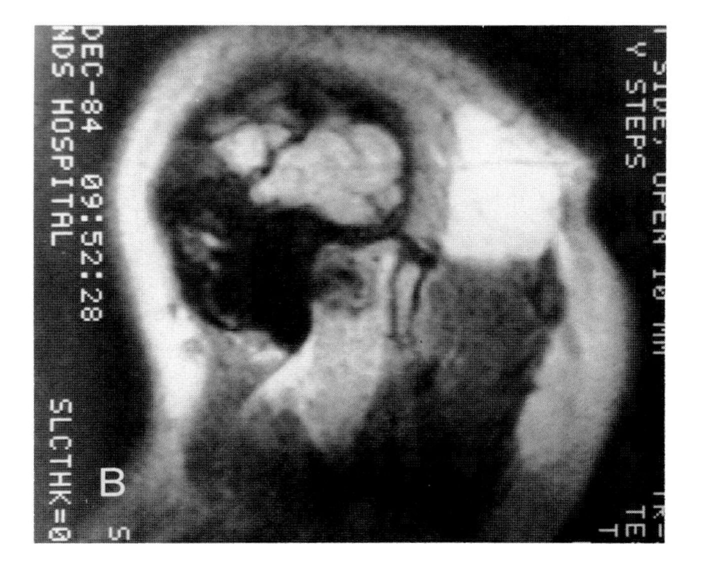

Plate 10.40*C*. Approximately 4 mm more medially than part *B* one can appreciate the relationship between the fibrous portion of the disk and the mandibular condyle somewhat better. The condyle (*arrow*) has now translated forward and rests in between the thicker anterior and posterior bands (*arrowheads*) of the fibrous portions of the disk. The bilaminar zone has expanded to fill the condylar fossa (*curved arrow*).

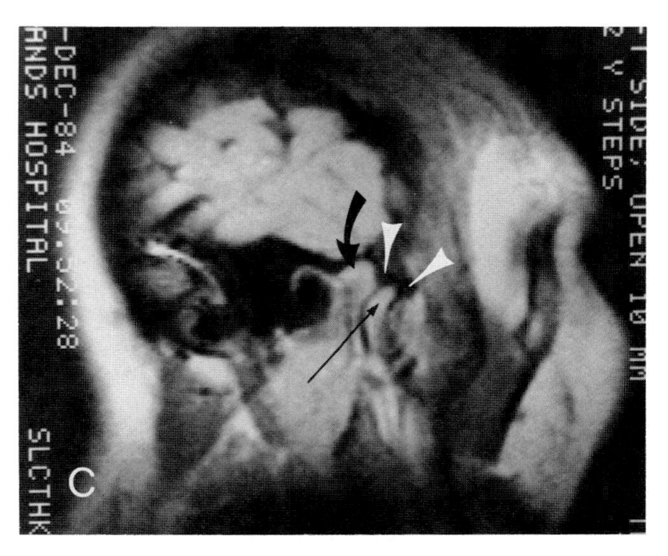

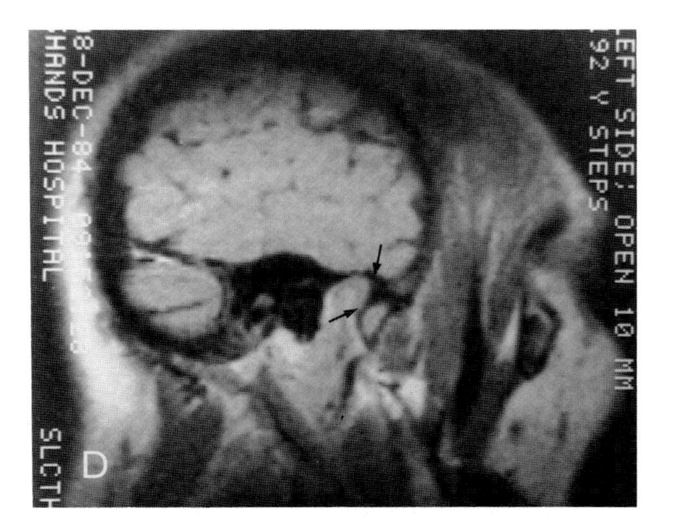

Plate 10.40*D*. Four millimeters more medial than part *C*, the contour of the posterior band changes from a convex to concave configuration which probably relates to points of attachment of the bilaminar zone superiorly and inferiorly (*arrows*). Note that the bilaminar zone has become distended to fill out the condylar fossa.

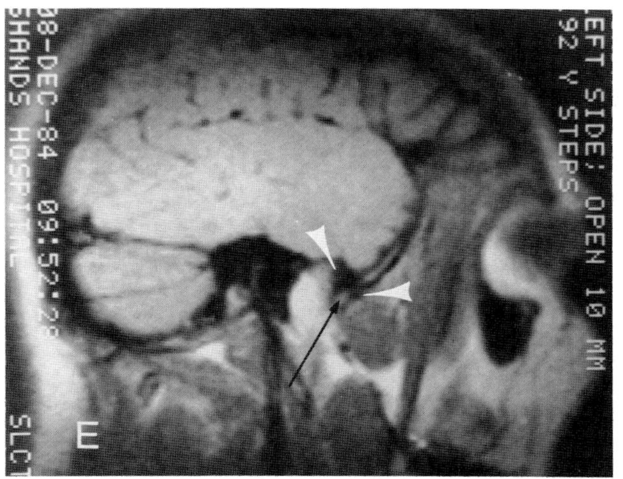

Plate 10.40*E*. Four millimeters more medial than part *D*, the condylar head is barely visible (*arrow*). The anterior and posterior bands have again assumed a more convex appearance (*arrowheads*). This series of sections shows that one can certainly study the anatomy and physiology of the temporomandibular joint effectively at any field strength currently used for imaging.

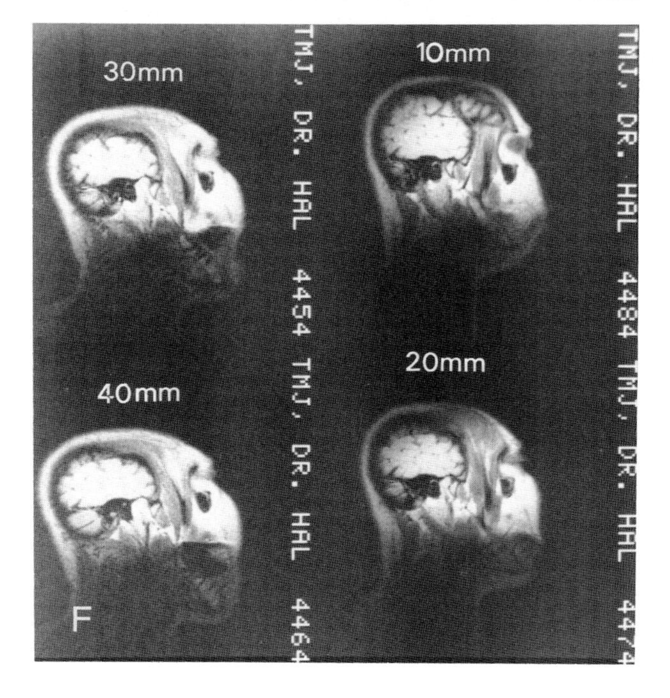

Plate 10.40*F*. This composite image shows the temporomandibular joint in various degrees of opening indicated above each figure in millimeters. There is some unavoidable change in imaging plane during opening but, overall, the series reflects the normal changes observed in translation of the mandible during opening. The sections were, in each case, made at approximately the midportion of the joint. The image plane is similar to that seen in part *C*.

Plate 10.41. This patient had clinical findings suggesting anterior dislocation of the disk which reduced after approximately 30–40 mm of opening.

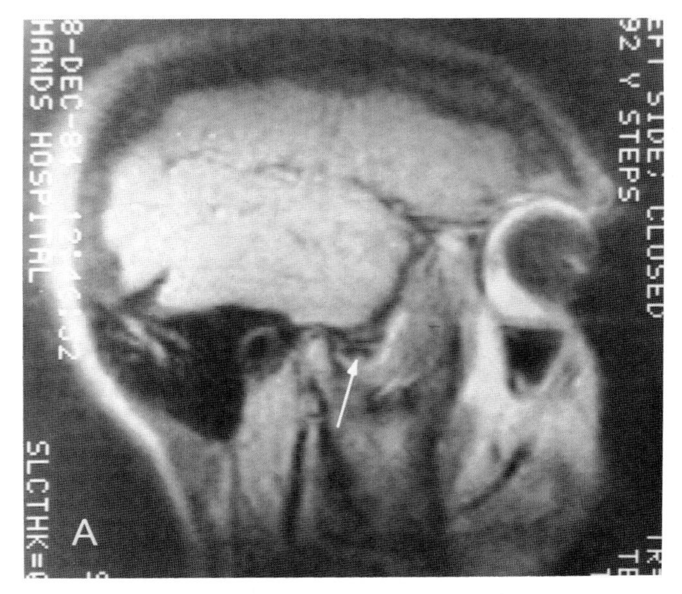

Plate 10.41*A*. In the closed position the fibrous portion of the disk lies entirely anterior to the condyle (*arrow*).

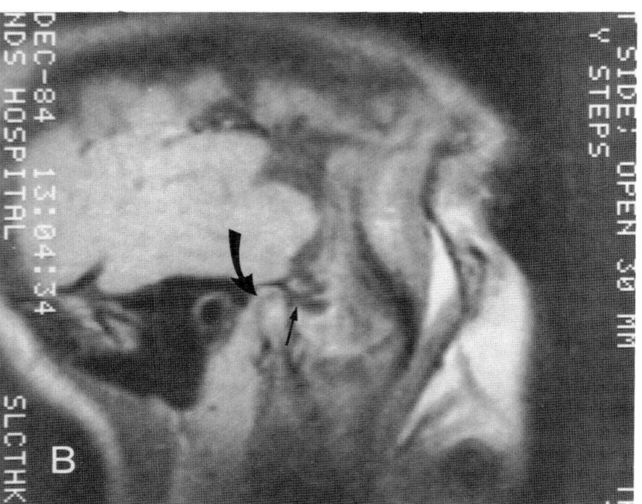

Plate 10.41*B*. With 30 mm of opening there is a lack of normal translation of the mandibular condyle at this degree of opening. The fibrous portion of the disk remains entirely anterior to the head of the condyle. The anterior band can be seen quite prominently and the normal area of physiologic thinning is apparent slightly posterior to this (*arrow*). The bilaminar zone (*curved arrow*) now lies between the roof of the condylar fossa and the head of the mandible.

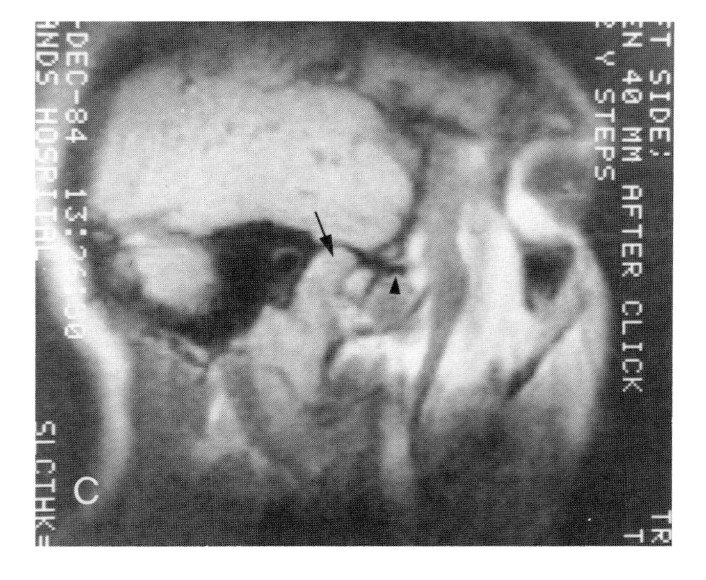

Plate 10.41*C*. With 40 mm of opening the head of the condyle has translated into a more normal position. The bilaminar zone fills the condylar fossa (*arrow*). The normal convex appearance of the posterior portion of the fibrous disk is seen. The anterior band appears somewhat globular and irregular, suggesting that the disk is damaged or focally thickened (*arrowhead*).

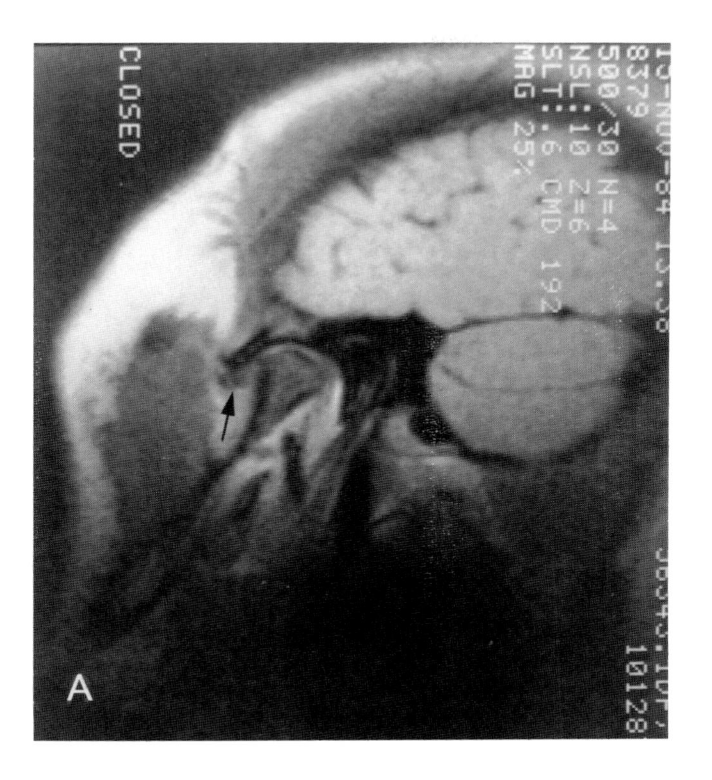

Plate 10.42. This patient had a history of many years of temporomandibular joint dysfunction and had obvious degenerative changes on plain films.

Plate 10.42*A*. A section through the midportion of the joint in the closed position shows considerable remodeling of the head of the condyle with a loss of the normal fatty content in the condylar head, probably related to sclerosis and remodeling. The fibrous portion of the disk is for the most part gone, suggesting that it has been macerated and perforated. Small fragments of the disk are the only portions visible just anterior to the condylar head and below the articular eminence (*arrow*).

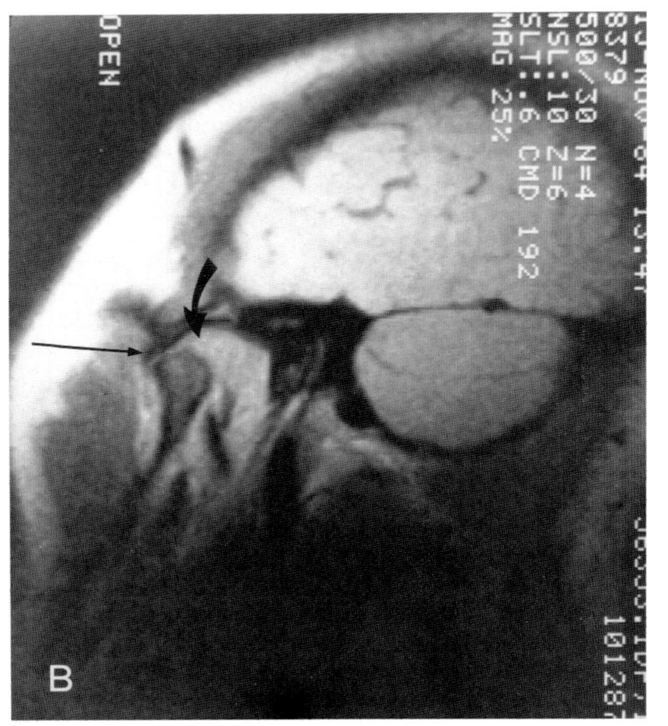

Plate 10.42*B*. In the open position, the condylar head does translate anteriorly; however, the small remnant of the fibrous portion of the disk remains anterior in position (*arrow*). The bilaminar zone has expanded to fill out not only the space within the condylar fossa but also the space between the head of the condyle and the articular eminence as the condylar head translates forward (*curved arrow*). This is markedly abnormal anatomy and physiology. In the open position the bilaminar zone should remain posterior and superior to the condylar head.

Chapter 11

Nasopharynx and Parapharyngeal Space

INTRODUCTION

The skull base, nasopharynx and parapharyngeal space are the cross-over areas between the intra- and extracranial structures of the head and neck. Their anatomy is detailed, and presenting symptoms due to pathologic change in this area may be very complex and may mimic intracranial lesions. The symptoms may also be very specific and can lead one to a discrete locale likely to harbor the responsible pathology. When approaching this area one must have a very clear idea of several factors including:

1. The normal anatomy and its range of variation
2. Where, specifically, can one place a lesion that will explain the patient's signs and symptoms?
3. How can CT and magnetic resonance imaging (MRI) be optimized to yield the most detailed information about the anatomic region(s) in question?
4. Is angiography necessary for confirmation or as an adjunct to therapy?

CT has certainly "overhauled" and simplified the diagnostic approach to this area during the last several years. MRI may replace CT in part or totally, however, these basic considerations are unlikely to change very much in the foreseeable future.

ANATOMY

The nasopharyngeal portion of the airway is roughly tubular in shape (Fig. 11.1). The floor of the sphenoid sinus and upper clivus form its roof. Posterolaterally, the mucosa and lymphoid tissue (adenoids) lie on the pharyngobasilar fascia and the pharyngeal or palatal musculature (Figs. 11.1, 11.2); here there is no bony limit and an expansive soft tissue space is present deep to the pharyngobasilar fascia (27, 31). In the nasopharynx, it is the pharyngobasilar fascia which maintains the airway rigidly open for breathing except during swallowing (Fig. 11.1) (31). More anteriorly, the medial pterygoid plates form a lateral support of the soft tissues of the nasopharyngeal wall. Directly posteriorly, the lower clivus and upper cervical spine along with the prevertebral musculature limit the soft tissue spaces. The specific anatomy varies considerably depending on the plane of section and age of the patient (35).

Superficial Structures

The torus tubarius is the most prominent and reliable of the superficial landmarks in the nasopharynx (Figs. 11.1–11.4) (31). Since it is composed of avascular cartilage it is of lower attenuation than surrounding muscular tissue on enhanced CT scans. The tori are always visible on normal scans through the mid- to upper nasopharynx and they tend to appear slightly asymmetric both in size and contour (35). The cartilaginous end of the eustachian tube is usually of similar or lower signal intensity than the surrounding musculature. If tubual tonsillar tissue is present this area may have a fairly intense signal depending on the amount of lymphoid tissue present and effects of volume averaging.

The eustachian tubes orifices are the air-filled spaces seen just anterior (on axial sections) or inferior (on coronal sections) to the tori (Figs. 11.1–11.4). They tend to appear roughly symmetric in depth (35). In studies made during quiet respiration it is unusual to see air beyond 3–4 mm of the orifice since the eustachian tube runs a course oblique to routine axial and coronal sections.

The lateral pharyngeal recesses (fossae of Rosenmüller) are air-filled spaces which project posterior to the tori (Figs. 11.1–11.4) and salpingopharyngeal folds (27, 31). These are seen mainly on sections through the mid- to upper nasopharynx since, more caudally, the salpingopharyngeal fold gradually merges with the lateral wall of the pharynx; the anterior wall of the recess thus disappears. The recesses definitely tend to be asymmetric in a given individual and the degree of distension varies greatly among different people (35). In children and young adults they are typically not seen because they are filled with lymphoid tissue; this is especially evident on MR images.

The lymphoid tissue wtihin the nasopharynx is most prominent along the roof. On axial CT it is seen in the highest sections through the nasopharynx and will have an appearance of almost filling the airway high in the vault. This is, of course, a false impression of the volume of the tissue since the plane it lies in is nearly parallel to that of the section. The tissue itself is usually slightly lower in attenuation than muscle and never exceeds muscular tissue in its degree of enhancement. Sometimes a thin rim of contrast enhancement

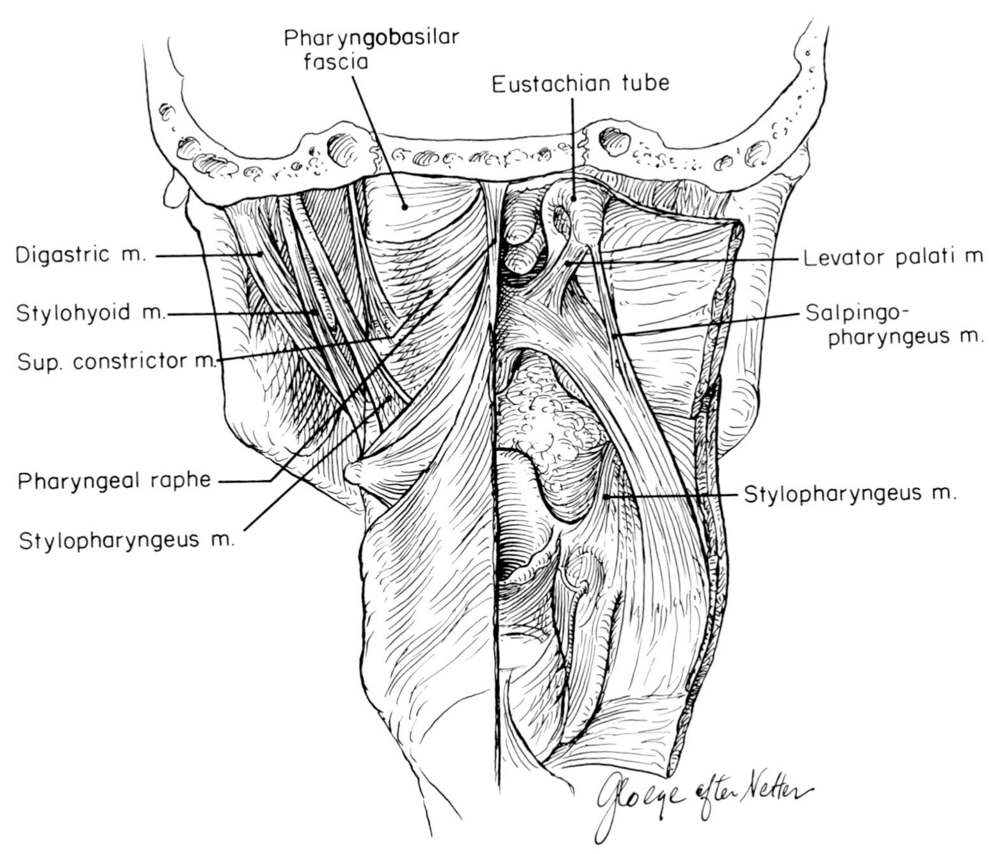

Figure 11.1. The upper aerodigestive tract and parapharyngeal space viewed from behind. Posterior wall of the pharynx is opened and reflected on the right to emphasize the relationship of the deep musculature, membranes and spaces to the prominent soft tissue landmarks of the mucosa surfaces in the nasopharynx, oropharynx and hypopharynx. Note that the upper portion of the pharyngobasilar fascia is not covered by the superior pharyngeal constrictor. The eustachian tube passes through a gap between the skull base and pharyngobasilar fascia to enter the nasopharynx.

will be visible outlining the muscular walls of the nasopharynx just deep to the lymphoid tissue. The appearance of this linear enhancement is highly variable depending on the patient as well as rate and volume of contrast infusion. On axial sections lower in the airway, hypertrophied lymphoid tissue may produce a lobulated or undulating surface contour. On coronal sections it will appear to hang down from the roof of the nasopharynx. The lymphoid tissue of the pharyngeal tonsil (adenoids) is normally located submucosally but it will never obliterate the deeper tissue planes surrounding the nasopharynx (35). On images from older scanners a large amount of tissue might hide the tensor or levator palati muscles from view but the parapharyngeal space should always remain visible and asymmetric. Because of the lower resolution of older scanners maneuvers, such as a "pinched nose," modified valsalva were occasionally done to see if a mucosal mass was pliable and to try to fill out the eustachian tube orifices and lateral pharyngeal recess as signs of benignancy. These "tricks" are no longer necessary. Current CT units show anatomy as well as it needs to be seen and if masses are suspicious clinically they must be biopsed even if CT suggests benignancy. On the other hand, a mass with a deeply infiltrating appearance on CT must be biopsed even if it appears benign on clinical exam (36).

The adenoids are seen best on MR images. The signal of this lymphoid tissue is always more intense than muscle by about 20–40%, which fits with comparisons of measured T_1 and T_2 values for these tissues (7, 14, 16, 37). This bright strip of tissue lines the roof and all walls of the nasopharynx, often filing the lateral pharyngeal recesses. On T_2-weighted images the differences between the adenoidal and surrounding tissues become even more obvious (37). MRI is very sensitive to even small aggregates of submucosal adenoidal tissue which can be seen in persons well beyond middle age. In children, the adenoid pad will nearly fill the airway and one may also see fairly large (1-cm) retro-

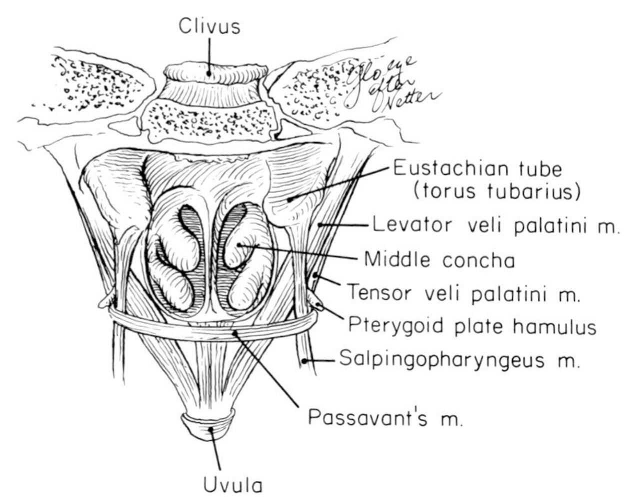

Figure 11.2. The nasopharynx viewed posteriorly. The relationships of the pharyngeal and palatal muscles are responsible for many of the superficial and deep landmarks seen on CT scans of the nasopharynx (*see* text).

pharyngeal nodes of the same signal intensity; the latter lie between the internal carotid and prevertebral musculature (14).

With aging, the superficial contours of the nasopharynx change. The eustachian tube orifices appear more patulent, and the tori smaller and usually better defined. The lateral pharyngeal recess may become deeper or wider and their area of contact with the skull base along the inferior petrous apex can be more fully appreciated. These alterations are due to involution of lymphatic tissue and atrophy of the pharyngeal and infratemporal fossa musculature (35).

DEEP ANATOMY

An expansive soft tissue space lies deep to the nasopharyngeal mucosa and submucosa. Four fascial sheaths divide this space into compartments which appear symmetric in the vast majority of patients (Fig. 11.4) (27, 31, 54). Detailed anatomic studies describing these various fasciae and tissue planes are abundant; however, only limited attention has been paid to the important pharyngobasilar fascia (8) and the flimsy buccopharyngeal fascia (27, 31). The prevertebral fascia and the carotid sheath are the more familiar of these connective tissue sheaths. On CT and MRI, the tissue density of these connective tissue condensations is inseparable from the structures they surround (i.e. muscle, blood vessels). The normal fat content of the intervening spaces permits identification of the planes on CT scans. Varying MRI signal intensities from deep structures makes the tissue contrast even more striking if proper pulse sequences are used. With high detail MRI it may be possible to resolve the signal of pharyngobasilar fascia from that of surrounding muscles and fat. As in other areas of the body, a larger amount of fat tends to aid in interpretation. In older patients, atrophy of the deep musculature with fat replacement makes these deep planes and spaces especially prominent (35).

The prevertebral fascia is a well developed membrane (26, 31, 54). It covers the longus colli, rectus and longus capitus muscles, and extends from the base of the skull to the lower limit of the longus colli muscles at approximately the T_3 vertebral body (Fig. 11.4) (27, 31). During swallowing and neck movement, the pharyngeal structures and carotid sheath move easily over the fixed base provided by the prevertebral fascia (27).

The very thick pharyngobasilar fascia holds open the lumen of the nasopharynx (27). It encloses the nasopharyngeal mucosa, surface lymphatics, and some of its musculature (Fig. 11.4). The fascia lies within the superior pharyngeal constrictor and extends well above its most superior muscle fibers to attach to the base of the skull (Figs. 11.4, 11.5).

The pharyngobasilar fascia is not visible on CT scans but it does separate two important compartments; the parapharyngeal space and the intrapharyngeal structures (Figs. 11.4, 11.5) (27, 31, 33, 34, 35, 43). On its airway side, the air-soft tissue interface may be smooth but it is often slightly irregular or thickened owing to the morphology of the overlying lymphoid tissue. Linear, oval or elongated ellipsoid soft tissue densities are often visible within the fat planes just deep to the mucosa. These are the tensor and levator veli palatini muscles which lie deep and to the mucosal side of the

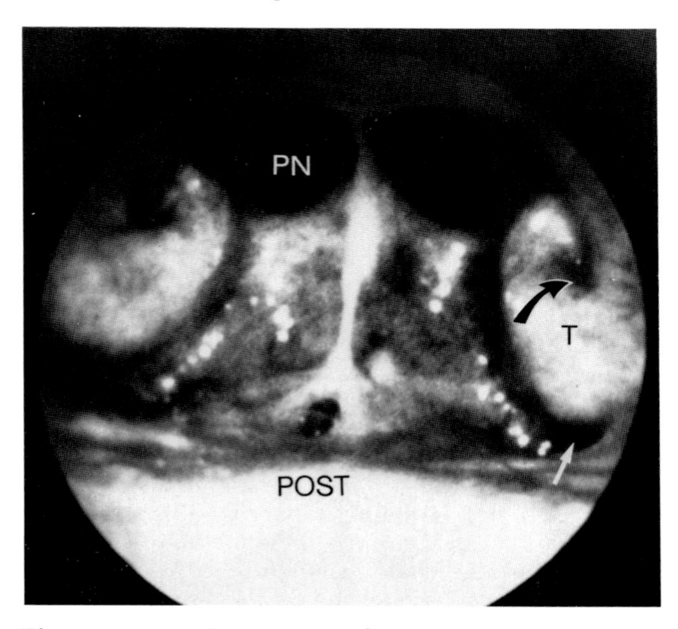

Figure 11.3. Photograph of a normal nasopharynx made during routine physical examination. *POST,* posterior pharyngeal wall; *PN,* posterior nares; *curved arrow,* eustachian tube orifice; *T,* torus tubarius; and *arrow,* lateral pharyngeal recess.

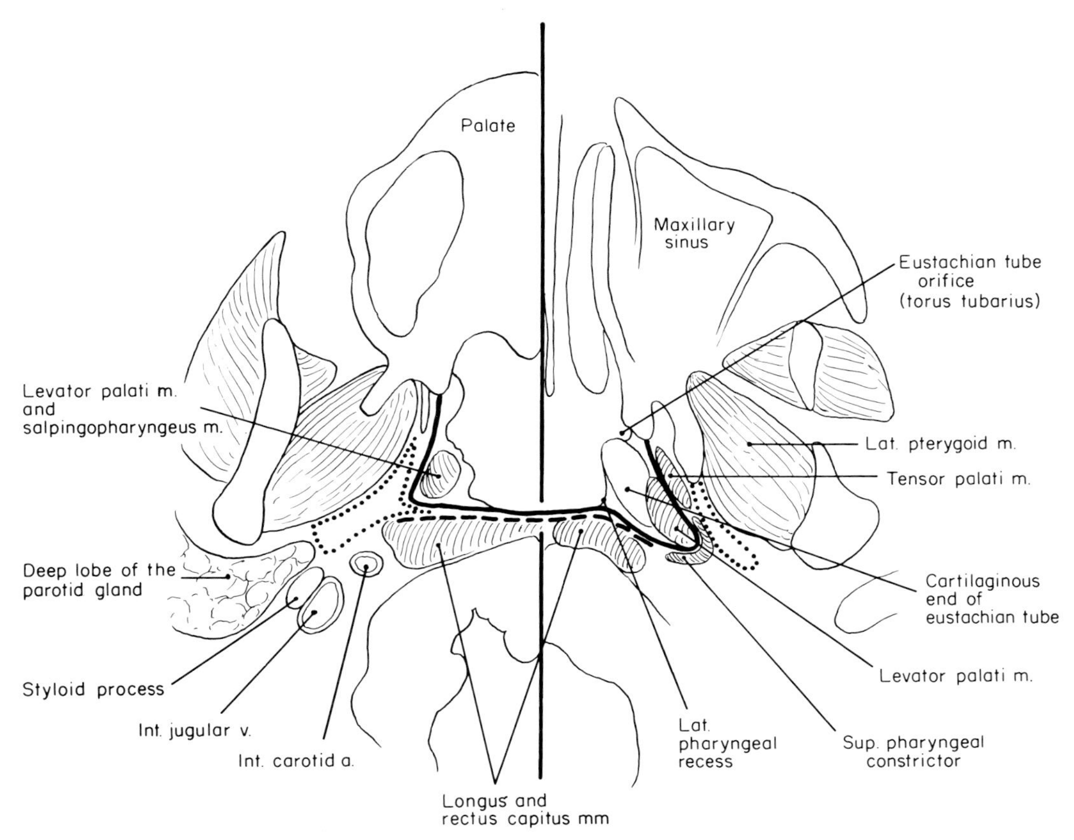

Figure 11.4. Composite axial diagram of nasopharynx as seen on CT. *Right half* of the diagram is of a level about 1 cm more cephalad than on the *left half*. The pharyngobasilar fascia (*heavy black line*) surrounds the airway and encloses the cartilaginous end of the eustachian tube and some musculature. The tensor palati and superior pharyngeal constrictor lie external to the pharyngobasilar fascia. Buccopharyngeal fascia (*dotted lines*) outlines the limits of the prestyloid parapharyngeal space. The potential retropharyngeal space lies between the pharyngobasilar fascia and the prevertebral fascia (*dashed line*).

fascia, respectively (Figs. 11.4, 11.5) (27, 31, 33, 43). The levator arises from the base of the skull but quickly gains access to the pharynx by passing through a small gap between the skull base and pharyngobasilar fascia (Figs. 11.1, 11.2, 11.5) (31). The eustachian tube passes through the same gap and some of the levator's fibers arise from the tube (31). The tensor veli palatini arises from the base of the skull (Figs. 11.1, 11.2, 11.5) and stays extrapharyngeal to the level of the soft palate (31). These muscles are routinely visible on high resolution thin (5-mm or less) section CT or routine MR images at any field strength 0.15 Tesla or higher. Posteriorly, the pharyngobasilar fascia extends to the anterior margin of the carotid foramen and then continues medially to reflect over the prevertebral fascia (31). The fascia is frequently seen on MRI studies.

The pharyngobasilar fascia thins out and almost disappears inferiorly. At the hard palate level, mainly the superior constrictor and Passavant's muscle limit the nasopharynx posterolaterally (Fig. 11.2) (31). On axial sections, these muscles appear as a band of tissue density surrounding the lateral and posterior walls of the airway. Other muscles contributing to this structure include the tensor and levator veli palatini, salpingopharyngeus and palatopharyngeus (Figs. 11.1, 11.2). The lymphoid tissue on the mucosa blends imperceptibly with these muscles so that the overall thickness of this tissue density will vary greatly with both the amount of lymphoid tissue and degree of muscular development. This band of tissue density also marks the lower limit of the nasopharynx. On MR images the relative contribution of muscle and lymphoid tissue to these structures are more easily distinguished due to the markedly improved tissue contrast on T_2-weighted images. When Passavant's muscle contracts like a sphincter producing a ridge which surrounds the nasopharynx on three sides, the soft palate tenses and elevates, is pressed against Passavant's ridge, and seals off the nasopharynx from the oropharynx during swallowing (31). At the same time, the levator veli palati fibers which arise from the eustachian tube contract, opening the orifices of the tubes, thereby equalizing the airway and middle ear pressures during swallowing (31).

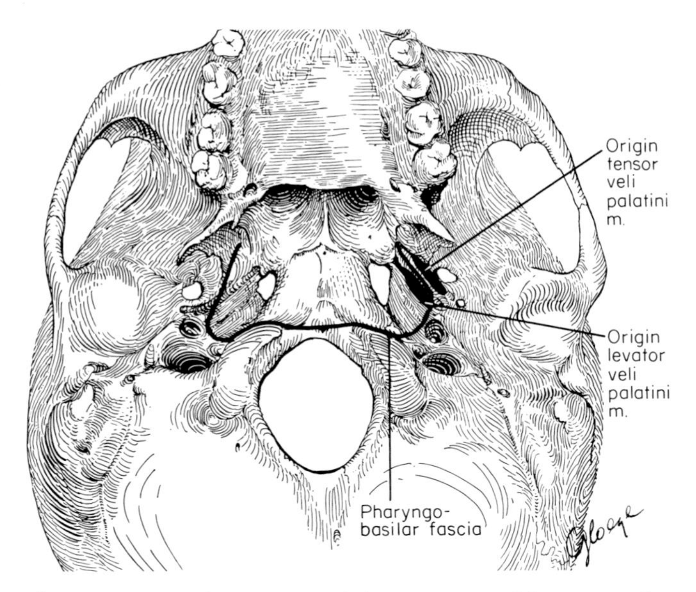

Figure 11.5. The origins of the tensor and levator veli palatini muscles are shown (*blackened areas*) relative to the attachment of the pharyngobasilar fascia (*heavy black line*) at the base of the skull.

The medial part of the buccopharyngeal fascia is the epimysium of the superior pharyngeal constrictor (31). Its lateral counterpart is the reflection of deep cervical fascia which covers the deep surface of the parotid gland and pterygoid muscles (27, 31). Both these layers are sparse and loosely applied to their respective muscles of origin to accommodate the tremendous movement of the pharynx that occurs during swallowing (31). These fasciae are not visible on CT or MR images but they form the medial and lateral limits of the paranasopharyngeal space (Fig. 11.4). The space is a loose network of fibrofatty tissue which allows free movement of the muscles of mastication and swallowing (31). Its fat content produces the distinct low attenuation or high signal intensity plane lying between the pharyngeal and pterygoid musculature. The parapharyngeal space is always visible and always appears symmetric at this level in the pharynx. Its obliteration is a sure sign of aggressive pathology. Many authors separate the parapharyngeal space into prestyloid and poststyloid compartments. This a very useful concept especially when considering possible tissues of origin for parapharyngeal mases. We often refer to the poststyloid parapharyngeal space as the carotid sheath region. Other authors prefer to call it the carotid space (44). It is the anatomy, not the name that is important.

The carotid artery lies just posterior and lateral to the prevertebral musculature. The jugular vein lies just beyond the artery further lateral and posterior. The vessels are easily identified on CT by their enhancement or on MRI by their low signal. The right jugular vein is typically larger than the left and the slower or more turbulent flow in the smaller vein will increase its MRI signal. Cranial nerves IX-XII and the cervical sympathetic chain lie around and between the vessels. A normal lateral retropharyngeal node is often visible between the carotid and prevertebral musculature at the C_1 and C_2 level. The highest internal jugular nodes may also extend to the paranasopharyngeal level where they lie mainly anterior and lateral to the vein just at, and above, the posterior belly of the digastric muscle (*see* Chapter 8).

The infratemporal fossa (subtemporal space, infratemporal space) lies lateral to the paranasopharyngeal space (Fig. 11.4). It is bound laterally by the zygomatic arch and anteriorly by the posterior wall of the maxillary antrum; a prominent fat pad lies between the muscles in the fossa and the antrum. Medially, it is limited by the paranasopharyngeal space. Above the zygomatic arch, it is continuous with the spaces of the temporal fossa and, inferiorly, with the spaces of the deep face external to the buccinator muscle (31). Prominent structures lying within it are easily identified on CT and MRI; these include the mandible, lateral pterygoid plate, pterygoid muscles, masseter, and portions of the temporalis muscle and deep lobe of the parotid gland. The tissue planes of the infratemporal fossa are normally always symmetric (35).

The potential retropharyngeal space lies between the pharyngobasilar fascia and prevertebral fascia (Fig. 11.4) (27, 31, 54). The prevertebral musculature produces most of the tissue density seen on CT between the airway and spine. The characteristic midline gutter separating the muscles is related to the pharyngeal ligament and media raphe of the pharynx (Fig. 11.1). Scans angled 20° off the Reid baseline do not cause significant "thickening" of the retropharyngeal soft tissues (35). Here benign lymphoid tissue can blend with the prevertebral muscles to cause apparent pathologic thickening on CT. The high signal of lymphoid tissue relative to muscle makes correct analysis of this normal variant easier on MR images. Since there is no reliable fat plane seen in the region on CT, it is more difficult to differentiate a deeply infiltrating process from a benign mucosal one. The prevertebral muscles are normally symmetric in size and shape (35).

DEEP ANATOMY—SKULL BASE AND CRANIAL NERVES

The mucosa and musculature of the nasopharynx are intimately related to the skull base. Cranial nerves III, IV, V, VI, IX–XII and the cervical sympathetics are not infrequently affected by nasopharyngeal pathology and may be the source of it. Many of these nerves pass through the skull base and course within the tissue spaces surrounding the nasopharynx.

The clivus and basiphenoid make up the posterior wall and roof of the nasopharynx. The bony portion of the eustachian tube lies immediately lateral to the carotid canal. There is always a gap between the anterior tip of the petrous apex and the basisphenoid (upper clivus); this is the "foramen" lacerum. This gap is in reality filled with cartilage and the carotid artery does not go through the cartilage but lies just above it as the artery leaves the carotid canal to enter the posterior cavernous sinus (27, 31, 33). The Vth nerve ganglion and its branches, within the cavernous sinus, are easily visualized on good quality MR and CT images. The carotid is seen by virtue of its low signal on MRI. The intimate relationship of the mucosa lining the lateral pharyngeal recesses and the inferior petrous apices is illustrated best by coronal section.

The jugular fossa lies just posterior to the carotid foramen. It is usually larger on the right side. The jugular spur should always be present as it separates the pars nervosa from the pars venosum of the fossa. Cranial nerves IX–XII lie within the jugular fossa. Both the carotid canal and jugular fossa may show areas of dehisence where they abut the middle ear cavity as a normal variant. A high jugular bulb is also a common normal variant.

The base of the pterygoid plates attaches to the basisphenoid. The pterygomaxillary fossa lies between the pterygoid process and maxillary antrum; it is in free communication with the inferior orbital fissure above and the infratemporal fossa laterally. The foramen rotundum lies just within the greater wing of the sphenoid anterior to the point where the medial ptergoid plate joins the basisphenoid. The second division of the trigeminal nerve passes through the upper pterygomaxillary fossa as it courses toward the inferior orbital fissure and infraorbital groove and canal.

The basisphenoid also forms the floor of the middle cranial fossa. This and the styloid process serves as the attachment for much of the upper pharyngeal musculature. The attachment of the levator and tensor palati are along the inferior petrous apex and basisphenoid, respectively (Fig. 11.5). The tensor origin is very close to the foramen ovale which serves as the conduit for the mandibular division of the trigeminal nerve as it passes through the upper parapharyngeal space.

High resolution thin section (1.5- to 5-mm) CT sections of the skull base provide superb images of all the structures described. Definitive imaging of the skull base can be accomplished with good quality MRI technique at field strengths of 0.15 Tesla and higher. In MRI visualization of the bone will depend on the amount of fatty marrow it contains as well as the absence of signal from its cortical component. It takes somewhat longer to become comfortable with the evaluation of the skull base and other bony structures on MRI than with CT because the range of normal variation is extended by varying content of fatty marrow within these bony compartments.

Technique of Examination

CT

Positioning and Technical Factors

For axial sections, the patients lie supine with their necks comfortably extended. Scans may be done in quiet respiration but significant motion artifacts may be avoided at the soft palate level and below by scanning during suspended respiration. The patient may not move, cough or swallow during scanning. The head should be placed in a molded head holder and the chin secured with tape or a chin strap. The forehead may also be secured by strapping.

A lateral projection radiograph should always be made. The study is planned from this image. Axial scans should be done at ± 10° of the infraorbital meatal line (Fig. 11.6). As a routine, it is most convenient to make the scan plane parallel to the hard palate. This is quite reproducible if the patient needs to be restudied.

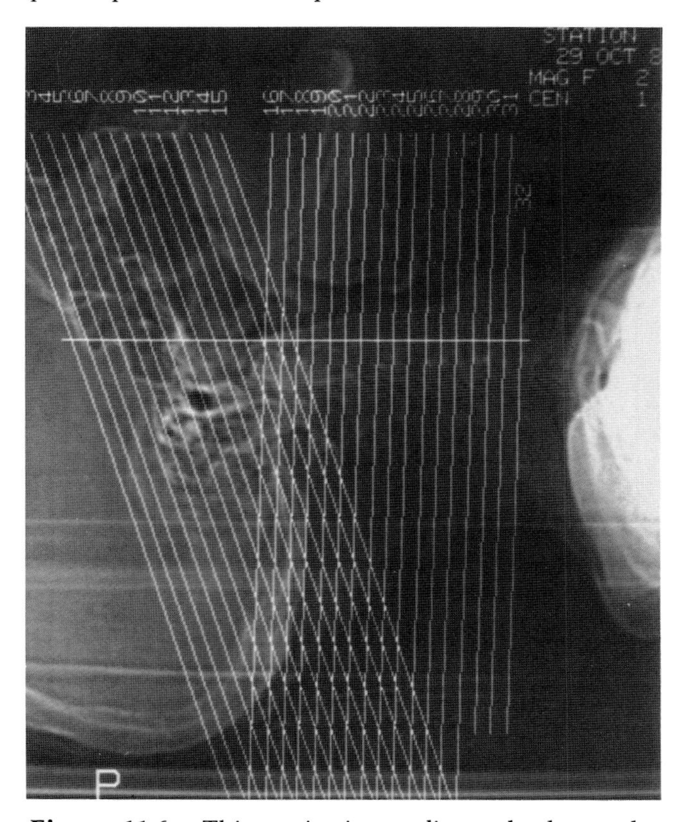

Figure 11.6. This projection radiograph shows that the nasopharynx is best studied with sections approximately parallel to the infraorbital meatal line. If these are not easy landmarks for the operators to identify, then a plane parallel to the hard palate will suffice. Sections should be contiguous and approximately 5-mm thick under most circumstances. Some change in angulation should be made if the oropharynx is to be included. Anatomy in the oropharynx is more understandable if the sections are done approximately parallel to the ramus of the mandible. The neck is best studied with angulation along the plane of the true vocal cords or the intervertebral disk spaces; in a slightly extended neck these angles are usually approximately the same. This is explained in more detail in Chapters 8 and 9.

Coronal sections should be made as close to perpendicular with the hard plate or floor of the sphenoid sinus as possible. The cavernous sinus should be included in the study; this requires starting at the top of the sella and continuing to at least the upper oropharynx. These levels really depend on the clinical situation. Sections should be contiguous and 3–5 mm thick. Thinner sections are rarely necessary in studies of the nasopharynx. Images of the skull base should be done with maximum milliamperes and be processed using "bone detail" algorithms for optimal image quality; however, less technique can produce diagnostically adequate images. If there is a question of very subtle bone erosion at the skull base on axial sections, direct coronals must be done. Reformatted images do not provide such refined bone images. Off-axis views may also be used to study a particular region of interest such as the jugular fossa, although we do not employ these as a routine (12).

Contrast Infusion

Intravenous contrast is given to the vast majority of patients who have CT studies of the nasopharynx and parapharyngeal space. The method of combined bolus and rapid drip infusion described in Chapter 8 should be used. Contrast is especially valuable in the evaluation of parapharyngeal masses which may be inherently vascular or displace or surround vessels in a characteristic fashion. It also helps to differentiate tumor in the paranasal sinuses from secondary obstruction of sinuses by tumor wtih the nasopharynx and nasal cavity.

Combined CT sialography was for a short time considered the examination of choice for evaluating parotid and parapharyngeal space masses (8, 46, 49). The injection of the parotid duct and selective opacification of the gland is rarely necessary if a good I.V. contrast-enhanced study is done. Combined CT-sialography may be necessary with unusually dense parotid glands or in cases where it is not possible to decide whether a parapharyngeal mass is intrinsic or extrinsic to the deep lobe of the parotid. MRI may obviate the need for CT-sialography under any circumstances.

MRI

The patient lies supine within the unit and the area of maximum interest is centered in the receiver coil. Images are best when done in the head coil and, if imaging of the neck is required (e.g., for staging adenopathy due to nasopharyngeal carcinoma), a specially designed surface coil will produce the most informative images (21). We secure the patient's chin and forehead to prevent movement. The imaging method seems relatively insensitive to the motion produced by quiet respiration and even occasional swallowing.

Multiple slices are imaged simultaneously. The section thickness should be no more than 8–10 mm and as close to contiguous as possible. A thickness of 4–6 mm is preferable if not mitigated by signal-to-noise considerations. The plane of image acquisition really depends on the clinical question at hand. Transverse sections are the most used in the nasopharynx. These show the symmetry of the deep tissue planes and retropharyngeal nodes to best advantage. Coronal sections are required for lesions of the nasopharyngeal roof and to detect subtle abnormalities at, or just below, the skull base as well as in the cavernous sinus and floor of the middle cranial fossa. Sagittal images may be used for lesions affecting the clivus, basisphenoid and perhaps as an adjunct to radiation therapy planning.

The choice of pulse sequences will depend somewhat on research which is still going on at this time (7, 14–16, 21, 37). For anatomy and T_1 weighting, we and others suggest a short T_R (about 500 msec) and T_E of about 30 msec. We also obtain a multislice, multiecho T_2-weighted sequence (T_R 2000–2400 and T_E 45, 90, 135, and 180 msec). The echo delay times beyond 135 msec will probably be dropped as a routine in imaging outside the neural axisis. Measured T_1 and T_2 values serve little practical benefit but the observed variation of tissue contrast on the T_1- and T_2-weighted sequence may prove very useful in image interpretation. Diagnosis will still be based mainly on morphologic criteria. Proton shift imaging, spectral analysis and contrast agents may help with specificity but this requires considerable investigation.

Indications for Study

CT should be done on all patients with malignant tumors of the nasopharynx. CT offers more precise pretreatment staging than is possible clinically. It can help separate patients with limited lesions from those which invade the skull base, the latter requiring larger fields for adequate treatment (40). Patients presenting with signs or symptoms of nasopharyngeal cancer should be studied even if the endoscopic or indirect examination of the nasopharynx is normal since CT occasionally demonstrates entirely submucosal lesions which cause these problems (26, 36, 41). Such signs and symptoms include: (*a*) persistent unilateral serous otitis media, (*b*) atypical facial pain (Table 11.1), (*c*) cranial nerve deficits (Tables 11.1–11.4), (*d*) nasal obstruction, bleeding, and (*e*) neck mass of uncertain etiology (13, 33, 40).

CT is the imaging examination of choice and may be the only one required for patients with juvenile angiofibromas, paragangliomas, and schwannomas growing in the nasopharynx and parapharyngeal space (6, 10, 45–47). CT should always be done first in patients with parapharyngeal masses of uncertain etiology and to differentiate these from masses of parotid origin when that question arises.

MRI may replace CT for many or all of these uses. Units at field strengths of 0.15 Tesla and above seem capable of producing sufficiently detailed images of

Table 11.1
Trigeminal Nerve[a]

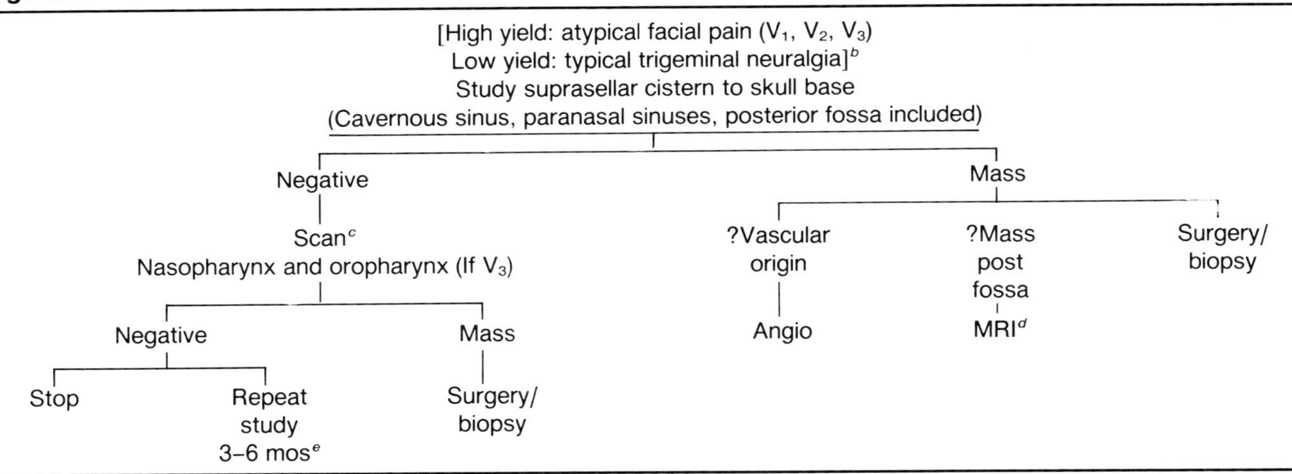

[a] From Kalovidouris, et al: *Radiology* 151: 671–676, 1984 (29).
[b] Five of 13 patients in the high yield group had positive findings. No lesion found in 10 patients in low yield group who were studied.
[c] Optional in low yield groups.
[d] NMR, MRI.
[e] Optional in high yield group.

Table 11.2
Facial Nerve[a]

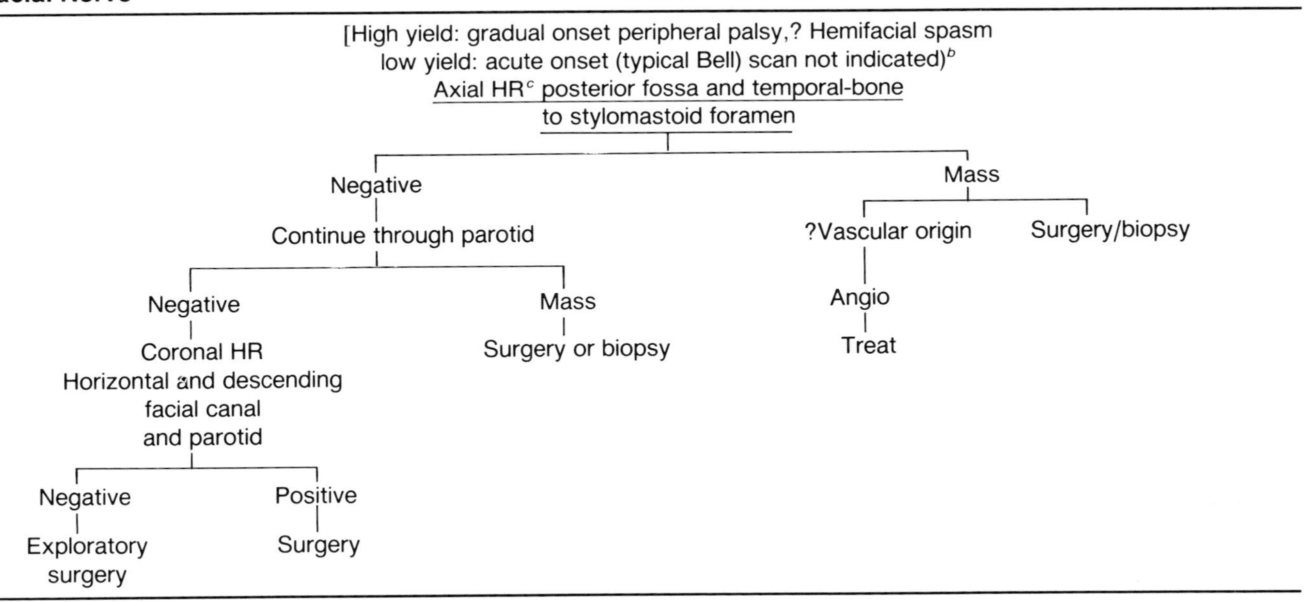

[a] From Kalovidouris, et al: *Radiology* 151: 671–676, 1984 (29).
[b] One of one patient in the high yield group with progressive peripheral VIIth nerve palsy had positive findings; one of two patients had hemifacial spasm positive. Zero of five typical patients in the low yield group with Bell palsy had positive findings.
[c] HR, high resolution.

the skull base, but CT is likely to remain the imaging exam of choice for showing bone detail in this region. Even from the most early studies on units working at 0.15 Tesla or more, it is clear that the display of the soft tissue extent of lesions will be considerably better on MRI than CT. The multiplanar capabilities of MRI may also make it more appropriate than CT under some circumstances (e.g. coronal images for subtle

Table 11.3
Cranial Nerves IX, X[a] (XI, XII)[b]

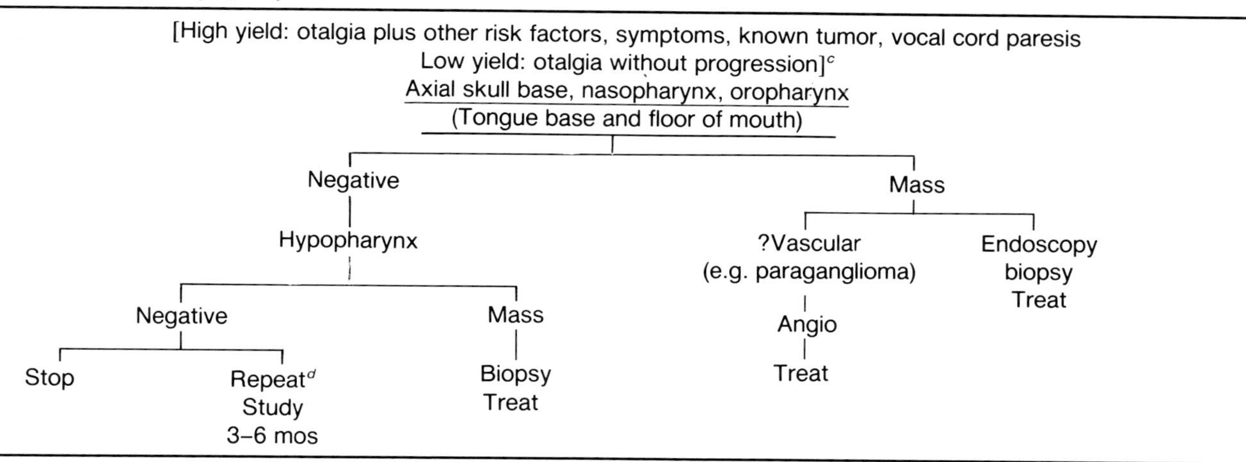

[b] In vocal cord paresis, scan must be extended to low neck, thoracic inlet and on left AP window.
[b] From Kalovidouris, et al: *Radiology* 151: 671–676, 1984 (29).
[c] Three of five referred high yield patients with otalgia had positive findings; no lesion found in two patients with idopathic vocal cord paresis.
[d] Optional in high yield group.

Table 11.4
Cervical Sympathetic Plexus[a]

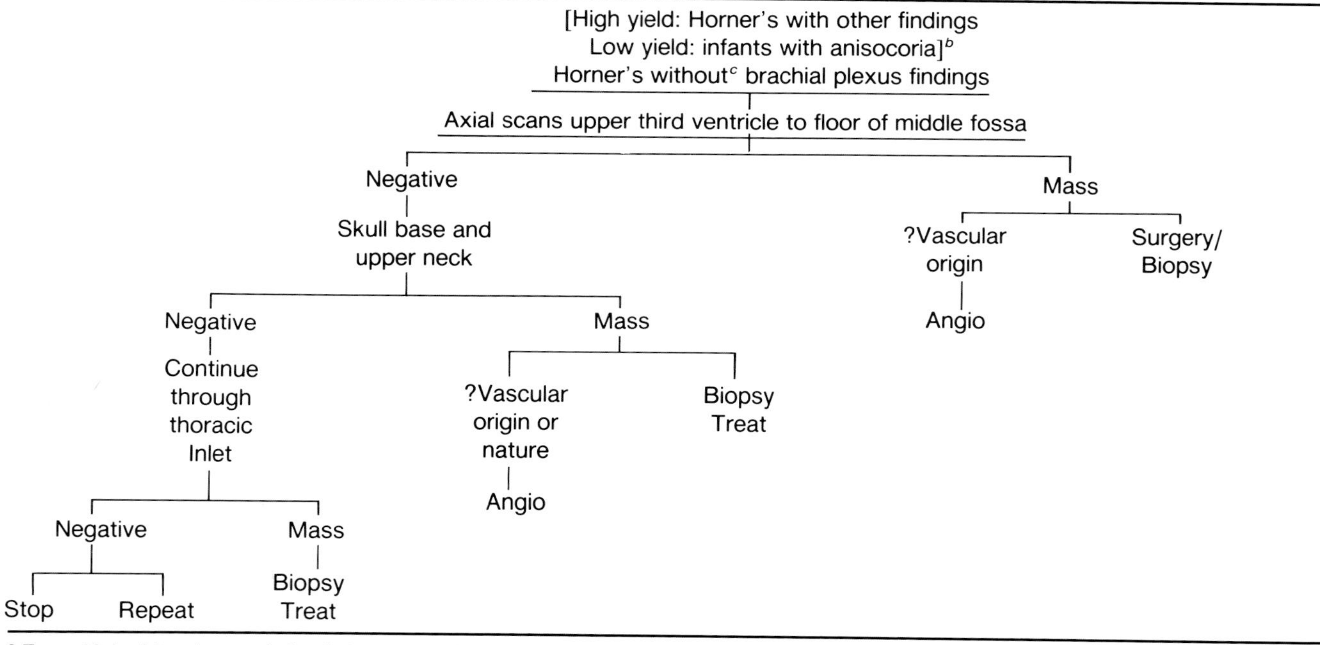

[a] From Kalovidouris, et al: *Radiology* 151: 671–676, 1984 (29).
[b] Two of two patients in the high yield group had positive findings. No lesions were found in three anisocoric infants.
[c] In patients with "hard" brachial plexus findings, do study in reverse order (i.e., start at thoracic inlet).

lesions near the roof of the nasopharynx and skull base, sagittal images for tongue base lesions).

Plain films should only be used as a gross screening exam and to supplement port planning. Polytomography of the skull base should only be done if high resolution CT or MRI is not possible. CT or MRI will almost always precede angiography of any lesion in this region. Angiography is done to resolve the uncertainties about etiology and vascularity which occasionally arise during the interpretation of CT images and as an adjunct to therapy (e.g. preoperative embolization of a juvenile angiofibroma).

PATHOLOGY

Malignant Neoplasms

HISTOLOGY

Ninety-eight to 99% of malignancies arising in the nasopharynx are carcinomas. The histopathologic diagnosis of squamous cell carinoma accounts for 80% of these although there is a confusing array of names which are assigned to the keratinizing and nonkeratinizing varieties of this common lesion (4, 13, 40). Transitional cell carcinoma and lymphoepithelioma are terms used to describe the nonkeratinizing type. The former should be abandoned and the latter should be understood for what it describes; that is a nonkeratinizing, squamous cell carcinoma in a predominantly lymphoid stroma (4). The degree of biologic aggressiveness of these squamous cell carcinomas is something quite apart from their keratinizing or nonkeratinizing properties although there is some tendency for the nonkeratinizing tumors to be more common and less well differentiated (4).

Adenocarcinoma and adenoidcystic carcinoma are relatively infrequent tissue types of nasopharyngeal malignancies (4). Some of these are particularly insidious since they may arise from accessory salivary tissue in the parapharyngeal space. Since the site of origin is well beneath the mucosa, the patient may have the signs and symptoms of a nasopharyngeal carcinoma long before it produces an abnormality of the nasopharyngeal mucosa (29, 36). Adenoidcystic carcinoma characteristically grows along nerves. This tumor is notorious for having spread beyond the grossly obvious primary site at the time of diagnosis and treatment because of this nonencapsulated, infiltrating and perineural pattern of spread. Adenoidcystic carcinoma has a long natural history characterized by progressive local and perineural extension over periods of up to 15–20 yr (4, 40).

Non-Hodgkins lymphoma is the second most common histologic diagnosis made in nasopharyngeal malignancies. These may be associated with involvement of other areas wtihin Waldeyer's ring. One should be aware that extranodal lymphoma in the head and neck region increases the probablity of there being systemic disease below the clavicles or diaphragm (4). There is no clear data available to determine the risk of systemic disease.

Poorly differentiated tumors in the nasopharynx may be either lymphoma, carcinoma or sarcoma. Age, cell markers and electron microscopy are sometimes necessary for histologic diagnosis (4, 40). Rhabdomyosarcomas are the most common malignancy of the nasopharynx in children. Cervical neuroblastoma may also present with nasopharyngeal or skull base symptoms in the pediatric age group (29). Rarely, chordomas or primary sarcomas of the skull base or surrounding tissues such as chondrosarcomas, fibrosarcomas, or osteosarcoma will present as nasopharyngeal masses. Melanomas, and ethesioneuroblastomas, usually arise in the nasal cavity and only secondarily involve the nasopharynx. Plasmacytomas have been reported but are rare (4). Metastases to the skull base or retropharyngeal lymph nodes may present as a primary nasopharyngeal mass; however these patients will almost always have a history suggesting the primary source.

PATTERNS OF SPREAD

Lederman (32, 34) described the natural history of nasopharyngeal carcinoma in an excellent monograph written over 20 yr ago. It was not until the era of modern CT, and now MRI, that we had the imaging tools that could show the paths of spread that he and others have described. In general tumor spreads; (*a*) via the mucosa and submucosa, (*b*) along muscle bundles locally and on to their origins and insertions, (*c*) within the fibrofatty tissue planes which surround the muscles, (*d*) along neurovascular bundles and into bones by way of the foramina created for the normal passage of these structures, (*e*) spread along periosteal surfaces or expansion and spread within the marrow cavity of bones (2, 23).

Tumors grow along the path of least resistance and their deep spread is modified somewhat by the surrounding anatomy. Cartilage is relatively resistant to invasion because of its lack of vascularity and this explains the tendency of tumors to spread around rather than destroy the eustachian tube and cartilaginous "plug" that fills the lower half of the foramen lacerum (33).

The tough pharyngobasilar fascia can also alter the growth patterns of nasopharyngeal cancer (6, 33, 35). First, and most importantly, one must assume that any lesion which is capable of growing through the fascia is either malignant or a very aggressive inflammatory lesion (6, 9, 33, 35). However, the fascia is a moderately resistent barrier and it can contain a tumor for some time within the space surrounding the levator palati muscle (Fig. 11.7) (33). The tumor may spread from the soft palate to the skull base within this tight compartment and produce fairly subtle asymmetry of the muscles of the pharyngeal while causing tubal occlusive symptoms. This is a favored pathway for occult spread of oropharyngeal cancer to the nasopharynx (and vice versa) which can lead to understaging or improper treatment; however, subtle asymmetry of the muscle planes within the pharyngobasilar fascia may be normal. Suspicious areas must be biopsed if involvement will alter therapy.

The most reliable sign of cancer in the nasopharynx is encroachment on the parapharyngeal space. This requires that the tumor displace or spread directly through the fascia or that it follow the eustachian tube

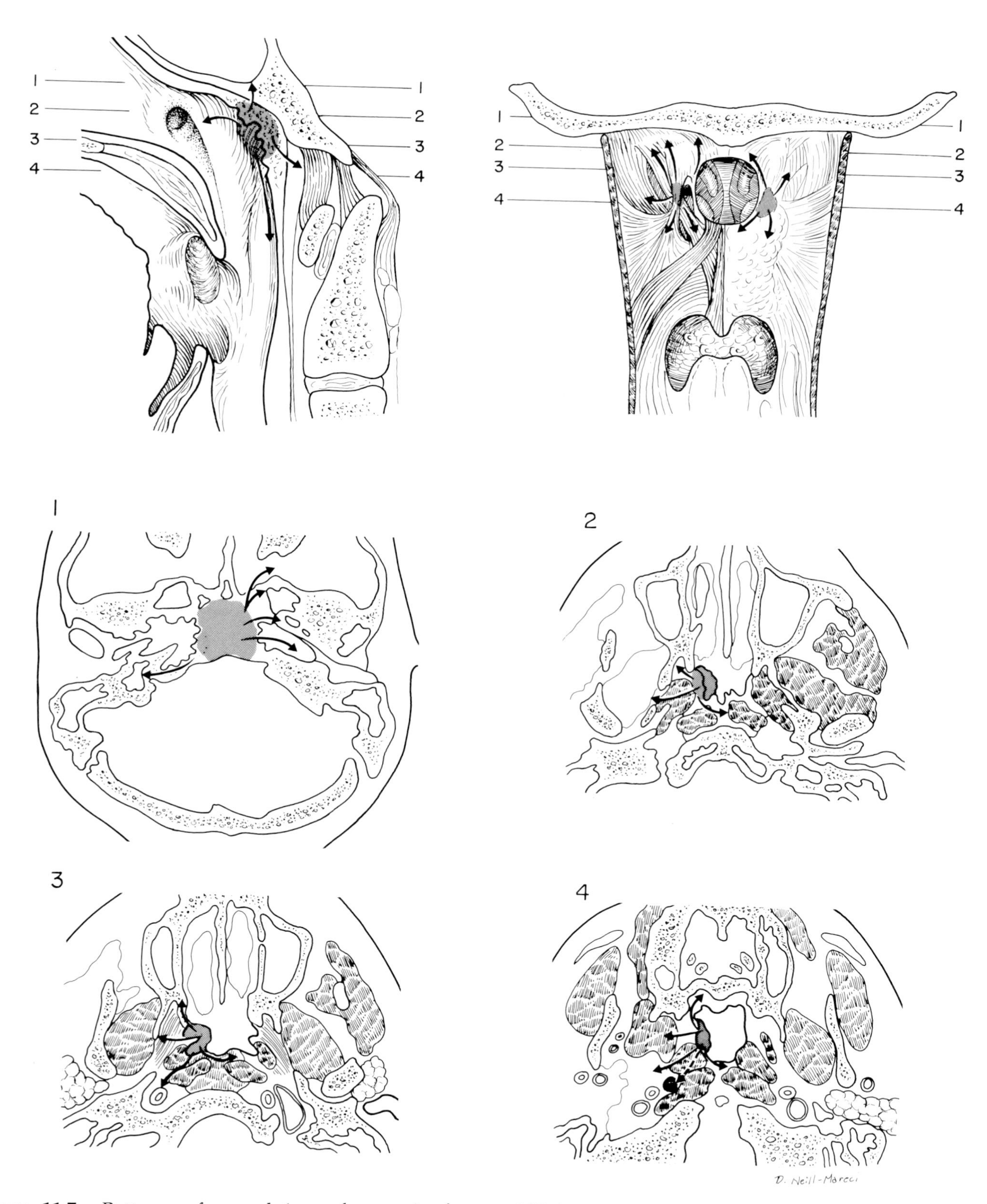

Figure 11.7. Patterns of spread (*curved arrows*) of nasopharyngeal carcinoma. The lateral and coronal whole organ sections at the *top* show routes of both superficial and deep spread. The following four axial diagrams indicate checkpoints for spread on axial CT or MR images. The levels of the axial diagrams are shown on the two whole organ drawings at the *top*. In diagram 4, note the important spread of nasopharyngeal cancer to the retropharyngeal nodes indicated by the *dashed arrow*.

through the gap between the fascia and the skull base (the sinus of Morgagni). Tumor is then free to spread cephalad in the prestyloid parapharyngeal space to the skull base in the region of the foramen ovale (Fig. 11.7) (Table 11.1). Tumor might also track along the tensor palati muscle which lies within the space just deep to the pharyngobasilar fascia. Spread to these locations obliterates those small muscle bundles and/or causes the deep tissue spaces to become asymmetric (43). Coronal sections show subtle spread to the parapharyngeal space the best, but these views are most subject to being overinterpreted. If the patient's head is positioned asymmetrically (looking slightly to one side) the difference between the thickness of the normal pharyngeal musculature can be quite striking. On the other hand, the coronal plane may be the only one in which pathology near the skull base in the parapharyngeal space or at the roof of the nasopharynx will be visible. Sagittal and coronal MRI images are useful for detection in the latter cases. The key to interpretation of minor changes lies in considering them in light of the clinical findings and doing biopsies either under endoscopic or CT guidance to confirm suspicious findings when they correlate with the patient's signs and symptoms (23, 36). Sometimes such findings just have to be followed by repeated imaging studies when they cannot be confirmed by biopsy.

Tumor may also spread to the poststyloid parapharyngeal space either directly or to retropharyngeal lymph nodes (33). The internal carotid artery lies within 3–5 mm of the fossa of Rosenmüller in the mid- to upper nasopharynx and from there the malignancies reach the artery and spread along its course in the carotid canal (Figs. 11.5, 11.7). Tumor may then follow the carotid into the cavernous sinus where it can produce deficits of cranial nerves III through VI (33) (Table 11.1). Direct or lymphatic spread to the carotid sheath (poststyloid retropharyngeal space) may also cause a partial or complete jugular fossa syndrome, namely, deficits in all or some of cranial nerves IX through XII (33) (Table 11.3). Horner's syndrome may result from the invasion of the cervical sympathetics at this level (33) (Table 11.4).

Erosion of the skull base is likely to occur at points of musculature attachments or around neurovascular foramina and canals. When tumor follows the tensor and levator palati muscles the bone of the pterygoid plates and basisphenoid is eroded (Fig. 11.7). Spread along the carotid will eventually lead to erosion of the carotid canal although the cartilage at the foramen lacerum may be spared (33). Destruction of the basisphenoid around the foramen ovale is due to spread in the prestyloid parapharyngeal space which has followed the mandibular division of the trigeminal nerve to that point. Tumor may reach the foramen rotundum via the pterygomaxillary fossa. Advanced lesions invade the clivus, sphenoid sinus and the orbit (33).

Continued spread in the parapharyngeal space eventually reaches the pterygoid muscles in the infratemporal fossa and, in very advanced lesions, the mandible. Trismus is usually a very late finding seen only when there is a massive amount of infratemporal space involvement. Perineural spread may explain pain or trismus in the absence of an obvious mass lesion (17, 26). Advanced tumors may also spread to the oropharynx and even hypopharynx via the pharyngeal musculature, muscles which arise from the styloid process or parapharyngeal space (2, 33).

Cancer of the nasopharynx is basically treated by radiation therapy (40). Treatment planning and expected morbidity may vary considerably depending on exactly what structures the tumor invades. It is important for the imager to consider these in both planning and interpreting the study. If the tumor is confined to the nasopharynx (T_1 or T_2) or has minimal extension to the nasal cavity or oropharynx (early T_3) the treatment volume is relatively confined but covers all of the areas of local extension described above (1, 40). Invasion of the skull base (T_4 lesion) or cranial nerve spread requires that the field be extended to much more of the brain and the pituitary; this produces a good deal more morbidity (1, 40). Anterior spread to the ethmoids, orbits or maxillary sinuses requires field changes and increased complications related to the eye and lacrimal apparatus (40).

LYMPHATIC SPREAD

The morphology, incidence and significance of nodal involvement varies considerably with the histopathology of a nasopharyngeal malignancy. The lateral retropharyngeal nodes are the first-order drainage site of the nasopharynx (Fig. 11.7) (24, 42). Nasopharyngeal carcinoma will produce gross ipsilateral enlargement (nodes 1 cm or larger) of these nodes in about 50–75% of cases (40). Squamous cell carcinoma metastases tend to necrosis once the nodal enlargement reaches 1.5 cm but may be seen sooner. Nonkeratinizing squamous cell carcinoma (lymphoepithelioma, transitional cell carcinoma) metastases tend to produce larger (2–4 cm) non-necrotic nodes, or smaller areas of necrosis in diffusely enlarged nodes. Lymphomatous nodes tend to be quite large and do not cavitate until they reach massive proportions or following treatment (38, 51). Of course, making histologic diagnoses from imaging studies is more of a game than it is of practical use and tissue confirmation is always necessary. The practical matter here is that the imaging may suggest the correct diagnosis to a pathologist who may be struggling with a poorly differentiated picture at light microscopy. Electron microscopy and cell markers are often necessary for confirmation.

The internal jugular and spinal accessory nodes are also considered first-order drainage sites for the nasopharynx (24, 42). Some direct lymphatic channels from the nasopharynx to the middle internal jugular nodes also exist. About 85–90% of patients will have clinically positive cervical nodes at presentation and approximately 50% will have bilateral nodes (33, 40).

The retropharyngeal nodes are always included in the primary treatment portals (40). Involvement of these in nasopharyngeal carcinoma does not necessarily indicate a poor prognosis. If the primary site is outside of the nasopharynx, or if the nodes are involved with recurrent tumor, the prognosis is very poor (3, 38). The dosage and exact portals for treating cervical nodes depend on their size, distribution and the histology. Radical neck dissection may be added when large nodal masses present following the regular course of radiotherapy (40).

RECURRENT DISEASE

Most cases of recurrent squamous cell carcinoma are diagnosed within 2 yr (4,40). Lymphoepithelioma can recur many years after initial therapy. Successful therapy usually results in a return of tissue planes to a near normal status although some asymmetric thickening may be accepted as posttreatment scarring. The amount of tumor-free residual fibrosis is unpredictable on an individual basis, but lymphoma and lymphoepitheliomas have the greatest tendency to leave the tissue planes appearing normal. We always suggest a baseline study at 6–12 weeks posttherapy so that follow-up studies can be interpreted objectively. Symptoms such as pain or cranial nerve dysfunction can precede changes in the gross appearance of nasopharyngeal anatomy probably because of perineural spread (26).

CT or MRI may potentially detect asymptomatic primary site recurrence; however, this has not been proven in a controlled, prospective study. MRI may prove more specific than CT by differentiating the signals of fibrosis and recurrent tumor; this is under study (37). The neck must be included in any examination for recurrent disease. CT has been proven better than palpation for showing recurrent neck disease because of the postradiation fibrosis which limits the clinical examination (26). Isolated neck recurrence may be salvaged by neck dissection. While documentation of local recurrence is important, salvage is rarely possible in the squamous cell variants. Chemotherapeutic salvage in lymphoma and sarcomas is more likely to prove beneficial.

Benign Lesions

INFLAMMATION

The vast majority of inflammatory masses in the nasopharynx are related to adenoiditis (28). The glandular tissue becomes hyperplastic and may show some enhancement on contrasted CT studies. Recall that normal adenoids typically show slightly lower attenuation than muscle. MRI shows masses of high signal intensity, especially on T_2-weight pulse sequences. The masses will be restricted to the mucosal side of the pharyngobasilar fascia on both CT and MRI.

Other inflammatory lesions in the nasopharynx may prove more aggressive and spread to encroach on or obliterate the parapharyngeal space and destroy bone. These lesions look the same as malignancies on CT and there is little evidence to suggest that MRI will be any more specific than CT for making this differential in the near future. Clinical history, physical examination, and biopsy usually make the diagnosis (28); CT or MRI shows its full extent. Invasive mucomycosis, postoperative, and postbiopsy edema and hemorrhage and medial extension of malignant otitis externa may all mimic a malignancy (6,9,10,11). Diagnostic studies of the nasopharynx are best done before, or at least 7–10 days after, biopsy if subtle findings are to be interpreted with any accuracy. This is especially important if one is trying to detect an occult primary tumor. As a practical matter scans may be done at any time in obvious tumors since the biopsies are usually relatively superficial and resulting edema and hemorrhage will not lead to overstaging. In cases of malignant otitis externa, radionuclide studies may provide more reliable information concerning the activity of the disease (50).

PARAPHARYNGEAL SPACE

There are relatively few lesions which arise from or present in the parapharyngeal space. The role of CT and MRI is to: (*a*) make the diagnosis when possible, (*b*) show the origin and full extent; and (*c*) determine whether angiography is necessary. It is fairly easy to narrow the differential to two possibilities in almost all cases and make the correct histologic diagnosis 80–90% of the time; this is assuming high quality images are available and one considers very critically both the characteristics of the mass and its effect on surrounding anatomic structures.

Lesions which arise in the prestyloid parapharyngeal space will most often be benign mixed tumors arising in accessory salivary tissue. The patient usually notices a painless inferomedial bulge of the soft palate or sometimes a vague mass in the parotid region. On CT the masses appear round, well circumscribed and they will enhance slightly more than muscle. The enhancement is probably related to an often mildly vascular stroma (4). Areas of scattered low attenuation may be due to mucoid impaction or necrosis (4). Calcification is rare and when present a dermoid tumor, chordoma or tumor with a calcifying matrix should be included in the differential (4). These lesions will displace *both* the internal carotid artery and jugular vein posterolaterally (Fig. 11.8). It is extremely important to be sure that these tumors are not arising from the deep lobe of the parotid. This may require a combined CT sialogram if a clear cleavage plane between the parotid and mass is not visible on the routine study (46,47). The surgical approach to a parapharyngeal lesion may be transoral, submandibular, or via a mandibular splitting procedure (47). A deep lobe parotid lesion requires a transparotid approach to control the facial nerve (47). Other tissue diagnoses are rare and have included acinic cell carci-

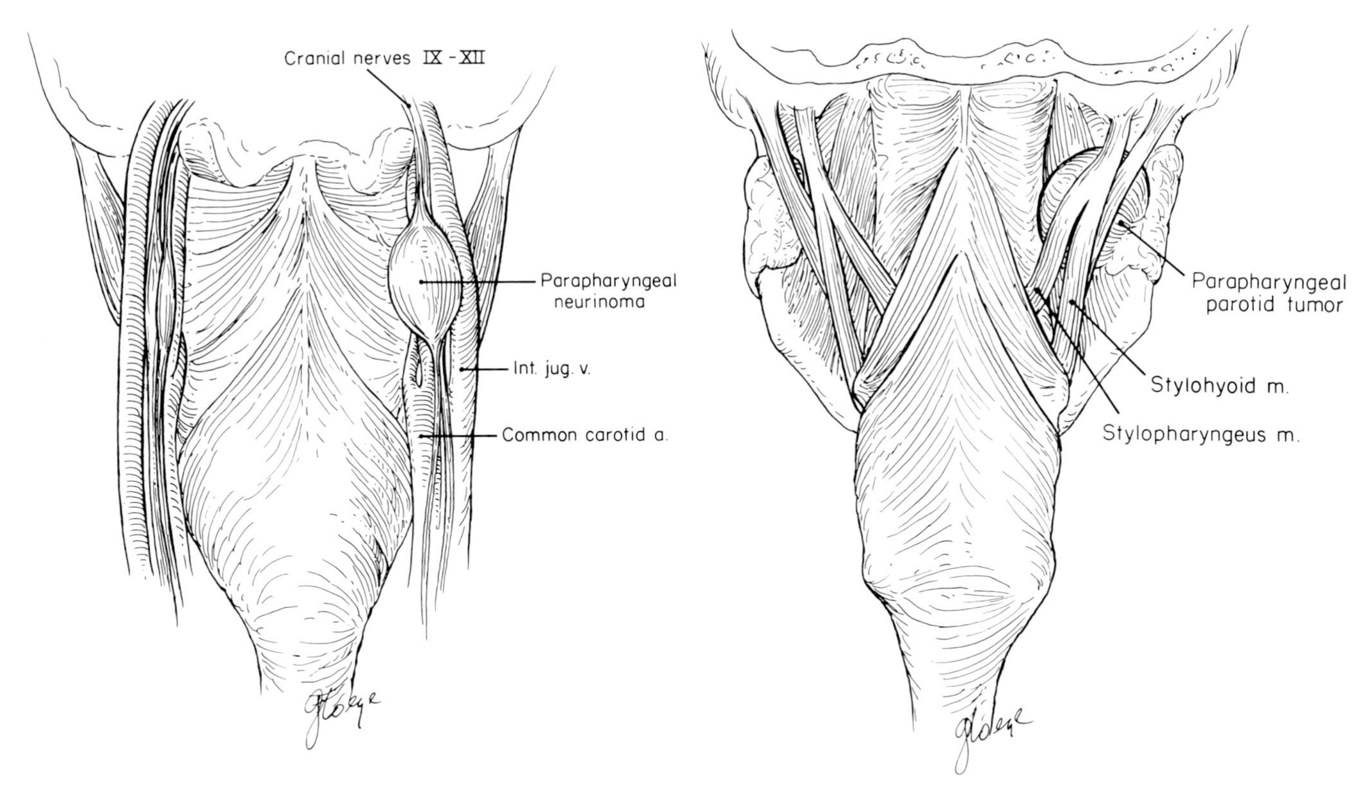

Figure 11.8. Composition of the origin of two common parapharyngeal space lesions. *Left*, neuroma arising from the cervical sympathetic chain or one of the cranial nerves IX to XII displaces the jugular vein posterolaterally. *Right*, mass arising from the deep lobe of the parotid wedges into the parapharyngeal space while displacing some of the muscles which arise from the styloid.

nomas (nonparotid) and an atypical lymophoepitheliod lesion (6,32). Certainly others will be reported (32). Branchial cleft cysts will rarely present strictly as a parapharyngeal mass. The appearance is that of a thin-walled cyst unless there has been supervening hemorrhage or infection. Lipomas are rare.

The poststyloid parapharyngeal masses sometimes present more perplexing problems. An enlarged lateral retropharyngeal lymph node, once beyond about 3 cm in diameter, can look exactly like a prestyloid parapharyngeal mass (25). Since these nodes lie medial to the carotid, the sheath structures will be pushed posterolaterally. The differential is not a difficult one if the patient has a history of cancer of the head and neck region. The possibility of an unusual metastasis (e.g. from breast carcinoma) may be overlooked if a proper history is not available (25). In slow growing lesions the morphology of these malignancies may be identical to that of a benign mixed tumor.

Schwannomas or neurofibromas may arise from the cranial nerves and sympathetics within the poststyloid parapharyngeal space (4,6,22,39,43,44). These may displace the carotid medially and the jugular laterally in which case the origin of the lesion is fairly obvious (Fig. 11.8). Schwannomas are usually round, well circumscribed, and enhance to the same degree or slightly more than muscle. They often have a low attenuation center. Paragangliomas also arise in this area although their clinical presentation is usually one of tinnitus (6,10,43,44,47). Paragangliomas tend to be well circumscribed, lobulated and obliterate the interfaces between themselves and the carotid and jugular. Paragangliomas and schwannomas may show the same degree of enhancement and bone involvement (at the jugular foramen) on CT studies so angiography is often required for definitive diagnosis. Dynamic CT studies may differentiate the faster flow of paraganglioma from typically slow flow neuromas.

If there is any question about the vascular nature of any parapharyngeal lesion after the CT study one should not hesitate to perform confirmatory angiography. The risks of angiography are small compared to those of hemorrhage that a surgeon lulled into a false sense of security by a CT or MRI study might encounter. Whenever a lesion is encountered near the skull base one must continue the study until the full intracranial extent, if any, is clear. Occasionally, a mass originating from the skull base or structures of the central nervous system will present in the parapharyngeal space. Neurilemomas, neurofibromas, meningiomas, dermoids (epidermoids), and chordomas are the lesions most likely to have parapharyngeal as well as intracranial components.

JUVENILE ANGIOFIBROMA

These uncommon lesions deserve special attention because imaging techniques are so pivotal in both diagnosis and management (5,18,30,48,52,53). Juvenile angiofibromas are diagnosed clinically by a history of epistaxsis or nasal stuffiness in a young or adolescent male patient who, on physical exam, has a fairly typical appearing nasopharyngeal mass (4). If juvenile angiofibroma is the primary working diagnosis, the imaging approach should be individualized. In general, CT should be done first to confirm the diagnosis and determine the extent of the lesion. If the lesion is confined to the nasopharynx and pterygomaxillary fossa, angiography may be delayed and performed as an adjunct to surgery. In this setting, a diagnostic study is performed the day of surgery and followed immediately by therapeutic embolization. The latter greatly reduces the chances of significant bleeding occurring during surgery. More extensive lesions may show bony changes suggesting intracranial extension on CT but an actual staining mass cannot be visualized. This usually indicates dural invasion without a frank intracranial mass displacing or involving the brain or meninges. These patients usually receive angiography to determine if they have internal carotid supply and then may be managed with radiation therapy or surgery depending on the extent of the tumor as well as institutional and patient preferences. In cases where the intracranial spread is unequivocal and extensive on CT the patients may go right to radiation therapy without confirmatory angiography.

On CT, juvenile angiofibromas appear to be centered about the pterygomaxillary fossa near its junction with the inferior orbital fissure. They characteristically expand the pterygomaxillary fossa by displacing the posterior wall of the maxillary antrum anteriorly and the pterygoid processes posteriorly (6,18,48,52,53). The surrounding bone is displaced and thinned but not irregularly eroded as in malignant processes. The soft tissue mass may then extend into the infratemporal fossa and from there expand posteromedially to obliterate the parapharyngeal space. These tumors, even when very large, do not extend far enough posterolaterally to obliterate the carotid sheath.

In addition to displacing bone, the tumor extends into the nasal cavity via the posterior nares. Angiofibromas also may push through the natural ostia of the sphenoid, ethmoid and maxillary sinuses and continue their characteristic pattern of localized expansile growth once inside. Natural foramina, such as the inferior and superior orbital fissures, provide the usual pathway for orbital and intracranial spread (4,6,53). The tumor can be shown by contrast-enhanced CT to lie along the anterior and medial wall of the middle cranial fossa. Even though the tumor masses are extradural, they derive a blood supply from cavernous branches of the internal carotid artery (53).

Intravenous contrast enhancement most often will demonstrate intense staining of the tumor mass. The paranasal sinuses are often opacified and on precontrast scans it is difficult to differentiate tumor involvement from obstructive changes. Postcontrast scans will show tumor enhancement and little or no change in the density of fluid-filled, obstructed, sinuses.

On rare occasions a juvenile angiofibroma may not enter the pterygomaxillary fossa and show the boney displacement so characteristic of this lesion. In these cases it may be difficult to decide whether the mass in the nasal cavity and nasopharynx is due to angiofibroma or angiomatous polyps. Angiography will differentiate the intensely staining juvenile angiofibroma from slightly to mildly vascular angiomatous polyps with ease (48). The widening of the pterygomaxillary fossa and anterior bowing of the posterior wall of the maxillary antrium is *not* pathognomonic of juvenile angiofibroma. We have experienced identical benign appearing displacements with slow growing metastases and lymphoma. However, in the proper clinical setting and with supportive CT findings the growth pattern of the lesions is unmistakable.

MISCELLANEOUS BENIGN MASSES

Other benign tumors affecting the nasopharynx are rare. They may be of epithelial, mesenchymal or teratoid origin (4). Reported CT experience is very limited with such lesions but some have potentially distinguishing radiographic characteristics and CT or MRI may strongly suggest the diagnosis while showing the lesion's full extent. Chordomas frequently show amorphous calcification near the periphery. The clivus may be the site of origin with local destruction or even intense sclerotic, reactive changes. Chondromas show typical chondroid matrix calcifications and their exophytic growth pattern from adjacent bones is distinctive. On the other hand, teratoid lesions are quite variable with CT appearance dependent on the dominant tissues.

Cysts of the pharyngeal bursa (Thornwaldt's cyst) are uncommon but may present a distinctive CT appearance. They appear as a midline, well circumscribed, round mass with attenuation values indicating its contents are fluid. Contrast infusion should show no increase in the central attenuation values. MRI would be expected to show similar gross morphology and with the internal signals indicating long T_1 and T_2 relaxation times.

Craniopharyngiomas will rarely present as nasopharyngeal masses. These will be seen as midline masses just beneath the sphenoid sinus since the remnants of Rathke's pouch lie in this locale. A sphenoid sinus mucocele should provide no difficulty in differential diagnosis. A very aggressive pituitary adenoma can sometimes mimic benign or malignant pathology in the nasopharynx making biopsy (excisional, incisional or needle) mandatory for diagnosis. The pattern of spread as seen on CT or MRI can sometimes be of great value in suggesting alternative diagnoses in cases where the histology pattern is uncertain. If the gross morphology of the disease on the imaging study does not fit with the clinical or histologic findings then reconsideration of the biopsy findings and sometimes rebiopsy becomes a prudent course of action. Biopsies may be done under CT guidance when the sites are suboptimal for, and approach from, the mucosal side or would cause unwarranted operative morbidity (e.g. in the infratemporal fossa) (23).

References

1. American Joint Committee on Staging of Cancer: Staging Head and Neck Sites and of Melanoma. Chicago, AJC, 1980.
2. Ballantyne AJ: Routes of Spread. In Fletcher GH, MacComb WS (eds): *Radiation Therapy in the Management of Cancers of the Oral Cavity and Oropharynx.* Springfield, Ill, Charles C Thomas, 1962, Chap 3, pp 91–115.
3. Ballantyne AJ: Significance of retropharyngeal nodes in cancer of the head and neck. *Am J Surg* 108:500, 1964.
4. Batskasis JG: *Tumors of the Head and Neck: Clinical and Pathological Considerations,* ed 2, Baltimore, Williams & Wilkins, 1979.
5. Biller HF, Sessions DG, Ogura JH: Angiofibroma: a treatment approach. *Laryngoscope* 84:695–706, 1974.
6. Bohman LG, Mancuso AA, Thomason J, Hanafee WN: CT approach to benign nasopharyngeal masses. *AJR* 136:173–180, 1981.
7. Bryan RN, Ford JJ, Schneiders NJ: *NMR Parameters of Nasopharyngeal Soft Tissues.* Presented at the 16th Annual Meeting American Society of Head and Neck Radiology. San Antonio, May 1984.
8. Carter BL, Karmody CS: Computerized tomography of the face and neck. *Semin Roentgenol* 13: 257–266, 1978.
9. Centeno RS, Bentson JR, Mancuso AA: CT scanning in rhinocerebral mucomycosis and aspergillosis. *Radiology* 140:383–389, 1981.
10. Chakeres DW, LaMasters DL: Paragangliomas of the temporal bone: high resolution CT studies. *Radiology* 159:749–753, 1984.
11. Curtin HD, Wolfe P, May M: Malignant external otitis: CT evaluation. *Radiology* 145:383–388, 1982.
12. Daniels DL, Williams IL, Haughton VM: Jugular foramen: anatomic and computed tomographic study. *AJNR* 4:1227–1232, 1983.
13. Dickinson RI: Nasopharyngeal carcinoma: an evaluation of 209 patients. *Laryngoscope* 91:333–354, 1981.
14. Dillon WP: *Clinical Head and Neck Magnetic Resonance Imaging with a 0.35 Tesla Unit.* Presented at the 16th Annual Meeting American Society of Head and Neck Radiology. San Antonio, May 1984.
15. Dillon W, Woolf M, Engelstand B: *Magnetic Resonance Imaging of Cervical Adenopathy—Optimal Imaging Parameters.* Presented at 16th Annual Meeting American Society of Head and Neck Radiology. San Antonio, May 1984.
16. Dillon WP, Mills CM, Brant-Zawadski M, et al: *NMR Imaging of Head and Neck Pathology.* Presented at 69th Scientific Assembly Radiologic Society of North America. Chicago, November 1983
17. Dodd GD, Dolan PA, Ballantyne AJ, et al: The dissemination of tumors of the head and neck via the cranial nerves. *Radiol Clin North Am* 8:445–462, 1970.
18. Duckert LG, Carley RB, Hilger JA: Computerized axial tomography in the preoperative evaluation of an angiofibroma. *Laryngoscope* 88:613–618, 1978.
19. Duncan A, Lack EE, Deck MF: Radiological evaluation of paragangliomas of the head and neck. *Radiology* 132:99–105, 1979.
20. Fish U: Infratemporal fossa approach for glomus tumors of the temporal bone. *Ann Otol Rhinol Laryngol* 91:474–479, 1982.
21. Fitzsimmons JR, Thomas RG, Mancuso AA: Proton imaging with surface coils on a 0.15T resistive unit. *Magnetic Resonance in Medicine* 2:180–185, 1985.
22. Gacek RR: Pathology of jugular foramen neurofibroma. *Ann Otol Rhinol Laryngol* 92:128–133, 1983.
23. Gatenby RA, Mulhern CB, Strawitz J: CT-guided percutaneous biopsies of head and neck masses. *Radiology* 146:717–719, 1983.
24. Haagensen DC, Feind CR, Herter FP, et al: *The Lymphatics in Cancer,* ed 1. Philadelphia, Saunders, 1972, pp 60–208.
25. Hanafee WN, Mancuso AA: An Introductory Workbook for CT of the Head and Neck. Baltimore, Williams & Wilkins, 1984.
26. Harnsberger HR, Mancuso AA, Muraki AS: The upper aerodigestive tract and neck: CT evaluation of recurrent tumors. *Radiology* 149:503–509, 1983.
27. Hollinshead WH: Anatomy for Surgeons: *The Head and Neck,* ed 2. Hagerstown, Md, Harper & Row, 1968, vol 1.
28. Hopping SB, Keller JD, Goodman ML, Montgomery WW: Nasopharyngeal masses in adults. *Ann Otol Rhinol Laryngol* 92:137–140, 1983.
29. Kalovidouris A, Mancuso AA, Dillon WP: A CT-clinical approach to patients with symptoms related to the V, VII, IX-XII and cervical sympathetics. *Radiology* 151:671–676, 1984.
30. Krause CJ, Baker SR: Extended transantral approach to pterygomaxillary tumor. *Ann Otol Rhinol Laryngol* 91:395–398, 1982.
31. Last RJ: *Anatomy Regional and Applied,* ed 5. Edinburgh, Churchill Livingston, 1972.
32. Lawson VG, LeLiever WD, Makerwich LA, et al: Unusual parapharyngeal lesions. *J Otolaryngol* 8:241–249, 1979.
33. Lederman M: *Cancer of the Nasopharynx: Its Natural History and Treatment.* Springfield, Ill, Charles C Thomas, 1961.
34. Lederman M: Cancer of the pharynx. *J Laryngol* 81:151, 1967.
35. Mancuso AA, Bohman LG, Hanafee WN, Maxwell D: Computed tomography of the nasopharynx. Normal and variants of normal. *Radiology* 137:113–121, 1980.
36. Mancuso AA, Hanafee WN: Elusive head and neck cancers beneath intact mucosa. *Laryngoscope* 93:133–139, 1983.

37. Mancuso AA, Fitzsimmons J, Mareci T, Million R, Cassisi N: MRI of the upper pharynx and neck: variations of normal and possible applications in detecting and staging malignant tumors—Part II, Pathology, (submitted for publication). Presented at RSNA Nov, 1984.
38. Mancuso AA, Harnsberger HR, Muraki AS: Computed tomography of cervical and retropharyngeal lymph nodes: normal anatomy, variants of normal and application in staging head and neck cancer—Part II, Pathology. *Radiology* 148:715–723, 1983.
39. Maniglia AJ, Chandler JR, Gooewin WJ, et al: Schwannomas of the paralaryngeal space and jugular foramen. *Laryngoscope* 89:1405–1414, 1979.
40. Million RR, Cassisi NJ: Nasopharynx. In *Management of Head and Neck Cancer: A Multidisciplinary Approach.* Philadelphia, Lippincott, 1984, chap 24.
41. Muraki AS, Mancuso AA, Harnsberger HR: Metastatic cervical adenopathy from tumors of unknown origin: the role of CT. *Radiology* 152:749–753, 1984.
42. Rouvier H: *Anatomy of the Human Lymphatic System*, ed 1. Ann Arbor, Mich, Edwards Brothers, 1938, pp 1–82.
43. Silver JA, Mawad ME, Hilal SE, Sane P, Ganti SR: Computed tomography of the nasopharynx and related spaces—parts I and II, Anatomy. *Radiology* 147:725–738, 1983.
44. Silver AJ, Mawad ME, Hilal SK, et al: Computed tomography of the carotid space and related cervical spaces—parts I and II. *Radiology* 150:723–736, 1984.
45. Som PM, Shugar JM, Parisier SG: A clinical radiographic classification of skull base lesions. *Laryngoscope* 89:1066–1076, 1979.
46. Som PM, Biller HF: The combined CT-sialogram. *Radiology* 135:387–390, 1980.
47. Som PM, Biller HF, Lawson W: Tumors of the parapharyngeal space: preoperative evaluation, diagnosis, and surgical approaches. *Ann Otol Rhinol Laryngol* 90 (Suppl 80):3–15, 1981.
48. Som PM, Cohen BA, Sacher M, et al: The angiomatous polyp and angiofibroma: two different lesions. *Radiology* 144:329–334, 1982.
49. Stone DN, Mancuso AA, Rice D, Hanafee WN: CT parotid sialography. *Radiology* 138:393–397, 1981.
50. Strashun AM, Nejatheim M, Goldsmith SJ: Malignant otitis externa: early scintigraphic detection. *Radiology* 150:541–545, 1984.
51. VonBahren W, Hasje S, Lenz M, et al: Computer-tomographie zervikaler lymphknoten metastasen bei malignomen das Kopf-Hals-Bereichs. *ROFO* 139:281–284, 1983.
52. Weinstein MA, Levine H, Duchesneau PM, et al: Diagnosis of juvenile angiofibroma by computed tomography. *Radiology* 126:703–705, 1978.
53. Wilson GF, Hanafee WN: Angiographic findings in 16 patients with juvenile angiofibroma. *Radiology* 92:279–284, 1969.
54. Wong YK, Novotny GM: Retropharyngeal space: a review of anatomy, pathology, and clinical presentation. *J Otolaryngol* 7:528–536, 1978.

Chapter 11 Plates

Nasopharynx and Parapharyngeal Space

NORMAL ANATOMY

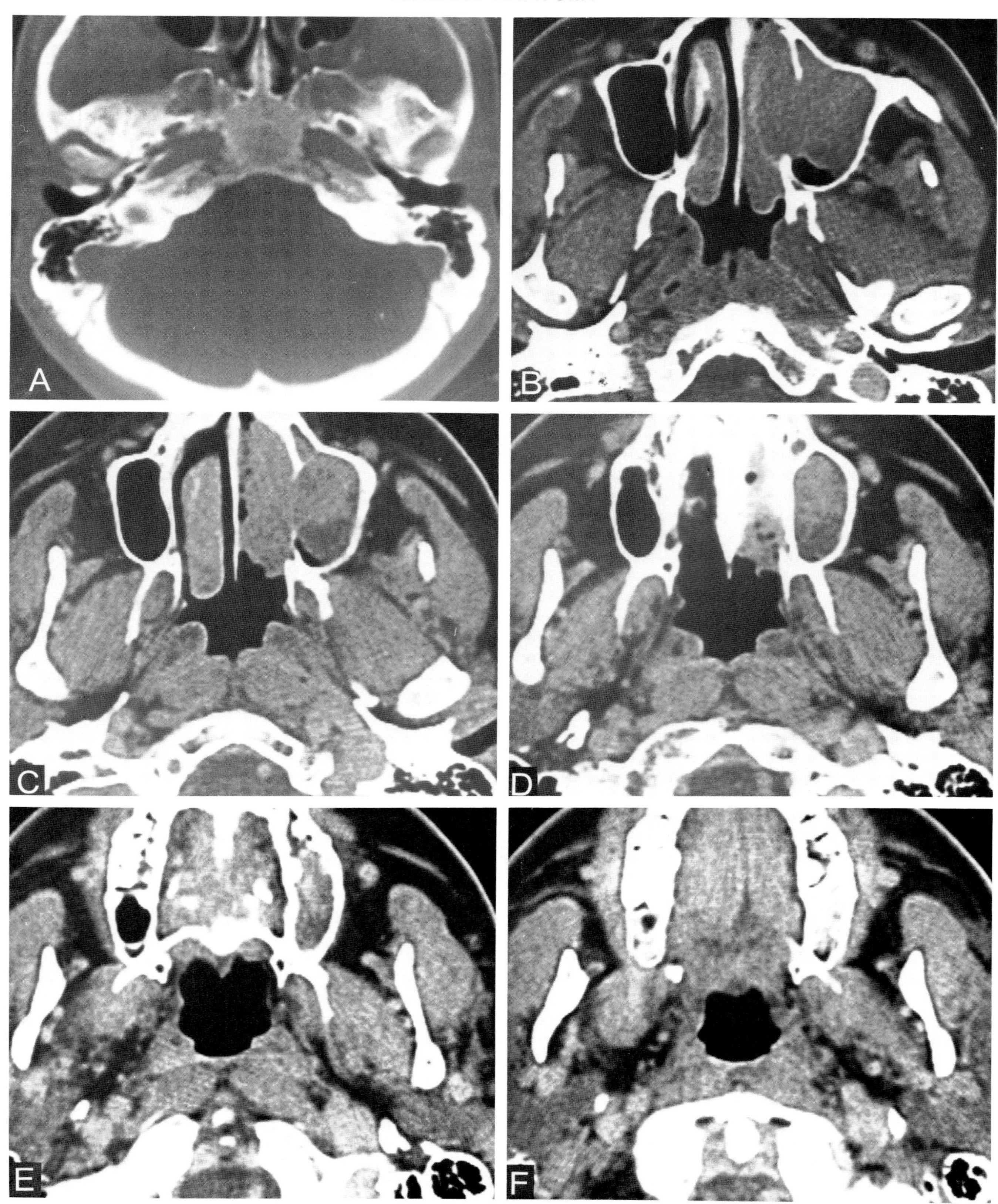

Plate 11.1 A–F.

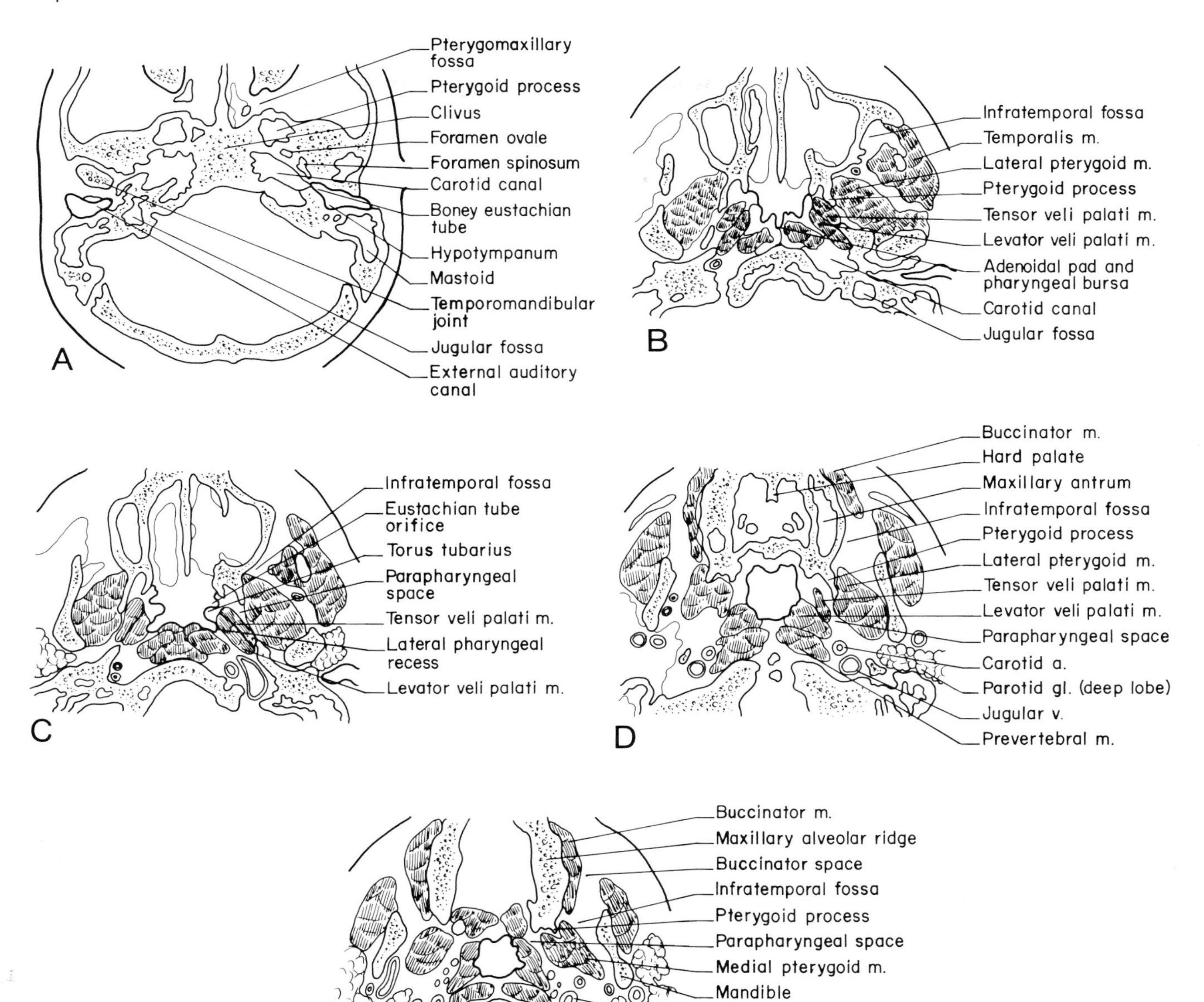
Pterygomaxillary fossa
Pterygoid process
Clivus
Foramen ovale
Foramen spinosum
Carotid canal
Boney eustachian tube
Hypotympanum
Mastoid
Temporomandibular joint
Jugular fossa
External auditory canal
A
Infratemporal fossa
Temporalis m.
Lateral pterygoid m.
Pterygoid process
Tensor veli palati m.
Levator veli palati m.
Adenoidal pad and pharyngeal bursa
Carotid canal
Jugular fossa
B
Infratemporal fossa
Eustachian tube orifice
Torus tubarius
Parapharyngeal space
Tensor veli palati m.
Lateral pharyngeal recess
Levator veli palati m.
C
Buccinator m.
Hard palate
Maxillary antrum
Infratemporal fossa
Pterygoid process
Lateral pterygoid m.
Tensor veli palati m.
Levator veli palati m.
Parapharyngeal space
Carotid a.
Parotid gl. (deep lobe)
Jugular v.
Prevertebral m.
D
Buccinator m.
Maxillary alveolar ridge
Buccinator space
Infratemporal fossa
Pterygoid process
Parapharyngeal space
Medial pterygoid m.
Mandible
Carotid a.
Parotid gl. (deep lobe)
Styloid process
Jugular v.
E
D. NEILL-MARRCI

Plate 11.3. This is a series of axial spin echo 500/30 pulse sequence images through the skull base and nasopharynx. The sections are 6 mm thick and the in-plane resolution is approximately 0.8 mm. These were done on a 0.15 Tesla unit, acquisition time on the series of 10 images was 6.6 minutes. The 192 gradient steps in the phase-encoding direction were used with four signal averages.

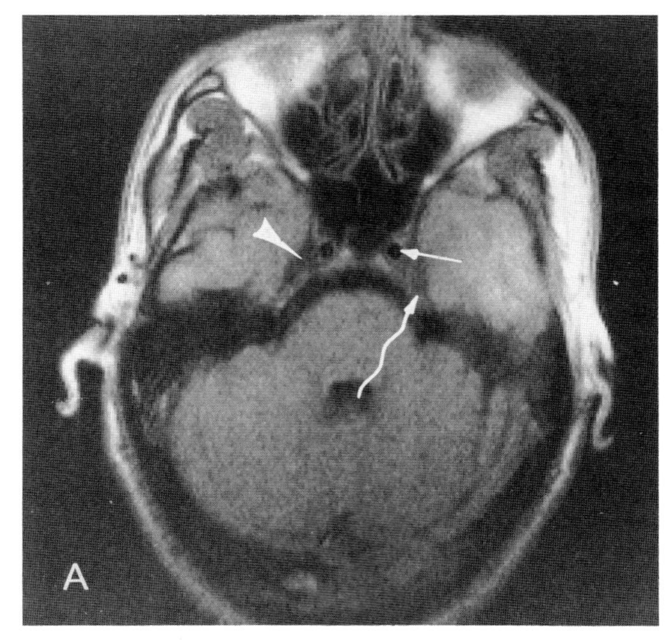

Plate 11.3*A*. A section through the cavernous sinus shows the internal carotid arteries ascending in the more posterior portion of the cavernous sinus (*arrow*). This is also the region of Meckel's cave (*wavy arrow*). Note that the cavernous sinus region in the area of the Vth nerve ganglia appear slightly hypointense at this pulse sequence (*arrowhead*).

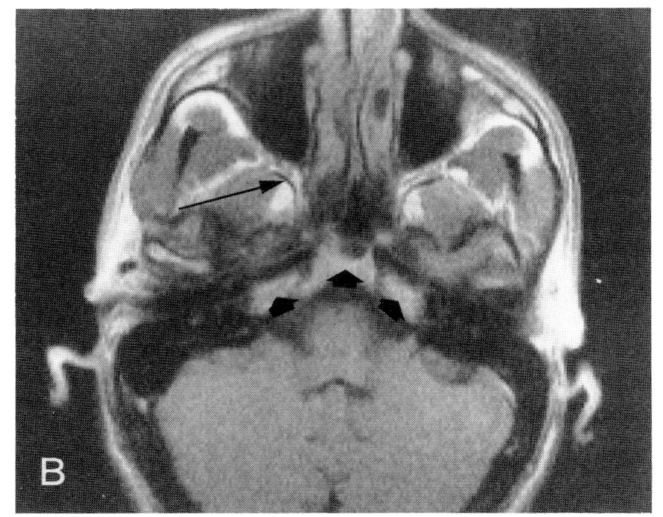

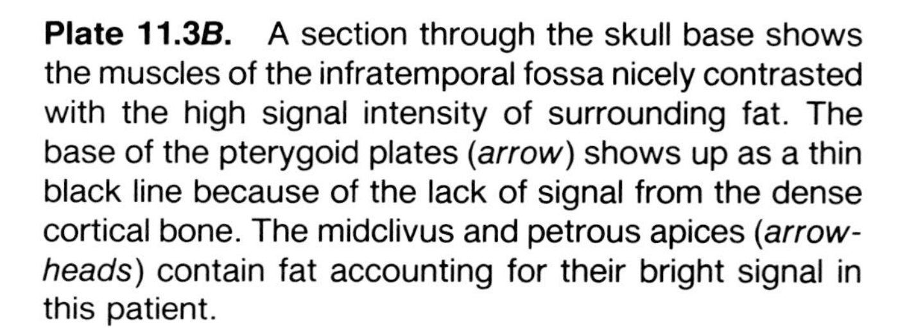

Plate 11.3*B*. A section through the skull base shows the muscles of the infratemporal fossa nicely contrasted with the high signal intensity of surrounding fat. The base of the pterygoid plates (*arrow*) shows up as a thin black line because of the lack of signal from the dense cortical bone. The midclivus and petrous apices (*arrowheads*) contain fat accounting for their bright signal in this patient.

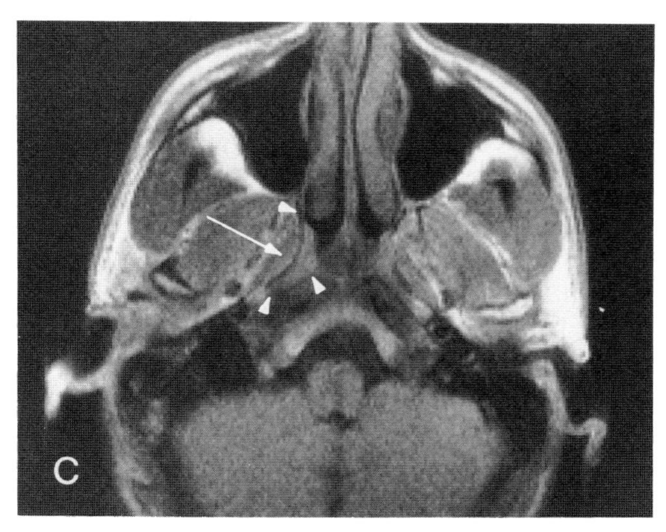

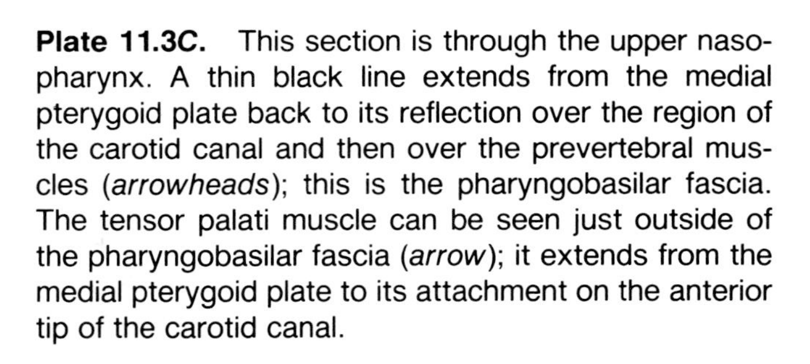

Plate 11.3*C*. This section is through the upper nasopharynx. A thin black line extends from the medial pterygoid plate back to its reflection over the region of the carotid canal and then over the prevertebral muscles (*arrowheads*); this is the pharyngobasilar fascia. The tensor palati muscle can be seen just outside of the pharyngobasilar fascia (*arrow*); it extends from the medial pterygoid plate to its attachment on the anterior tip of the carotid canal.

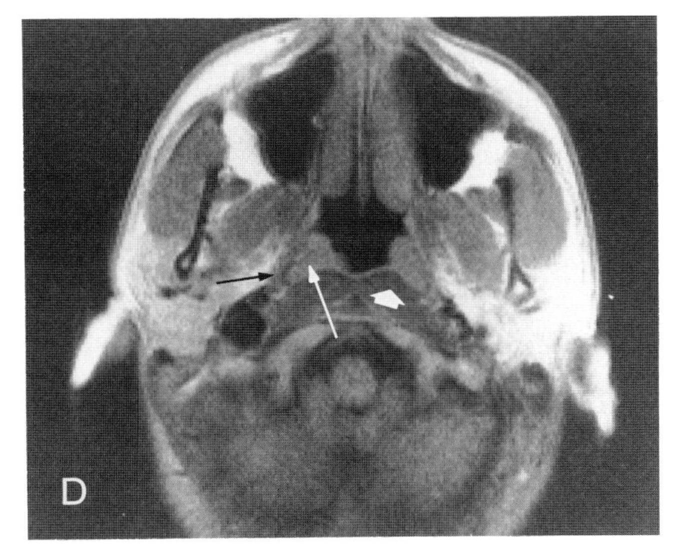

Plate 11.3*D*. A section through the midnasopharynx reveals the typical configuration of the superficial and deep landmarks as shown in Plates 11.1 and 11.2 and described extensively in the text. MRI gives the advantage of being able to see the lymphoid tissue lining the nasopharynx and filling the fossae of Rosenmüller (*white arrow*) as a signal intensity somewhat different from that of surrounding muscle (e.g. the prevertebral muscles, *arrowhead*). The levator palati (*black arrow*) can now be seen just deep to the fossae of Rosenmüller.

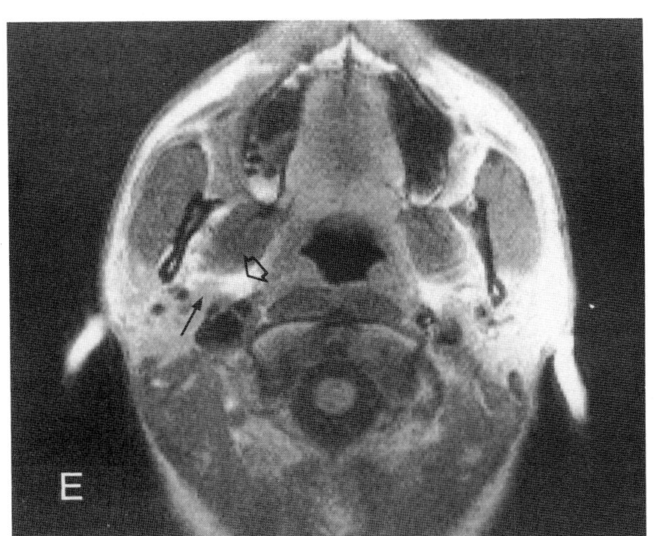

Plate 11.3*E*. A section through the low nasopharynx shows the relationship between the deep lobe of the parotid gland (*arrow*) and the higher signal intensity from the fat in the parapharyngeal space (*arrowhead*). The structures within the poststyloid parapharyngeal space include the carotid, jugular, and the lower cranial nerves which are surrounding the vessels and not seen as distinct structures but only as areas of diminished signal in the fat between the vessels.

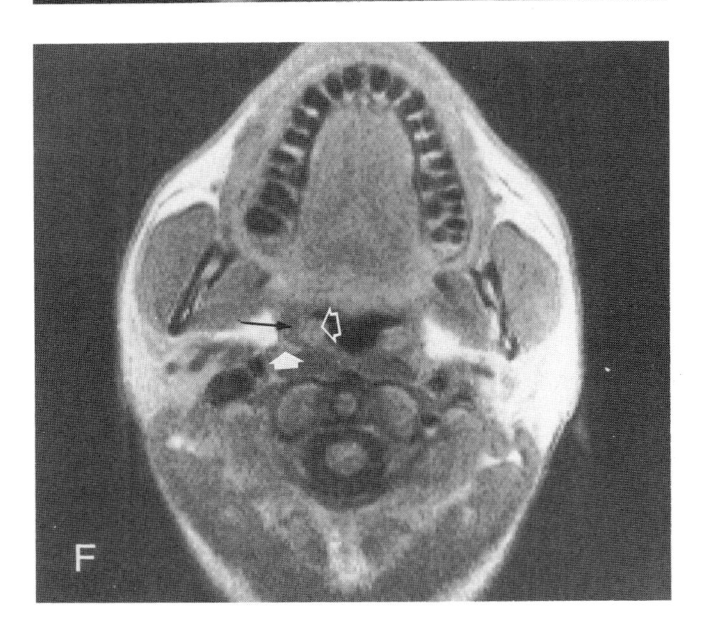

Plate 11.3*F*. This section is made at the junction of the oropharynx and nasopharynx. The palatal muscles here merge with the superior constrictor muscles to form the band of tissue around the airway (*arrow*) which is of a signal intensity slightly less than the tonsillar tissue (*open arrowhead*) in this location. The area of medium signal intensity between the carotid artery and the musculature represents normal retropharyngeal lymph nodes (*closed arrowhead*). Note that the bone detail throughout this image and part *E* is quite good due to the natural contrast of surrounding muscle and fat with the cortical margins of the bone.

The abbreviations used on Plates 11.4 and 11.5 are: *E,* eustachian tube; and *N,* node.

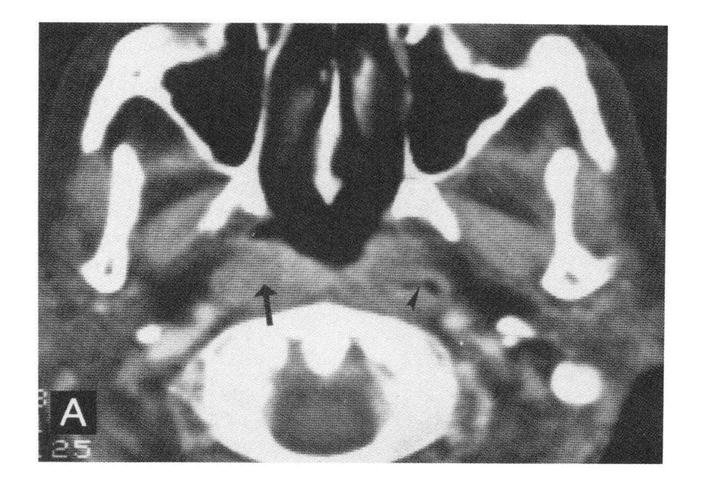

Plate 11.4. On CT images the fossae of Rosenmüller can often appear asymmetric. They sometimes create the mistaken impression of an abnormal mass in the nasopharynx. On MRI this is not such a difficult problem because the lymphoid tissue which creates these asymmetries is easily recognized without any special maneuvers (*see* Plates 11.3*D*, 11.5 and 11.6).

Plate 11.4*A*. A lack of air in the right fossae of Rosenmüller (*arrow*) compared to the left (*arrowhead*) could be mistaken for a mass.

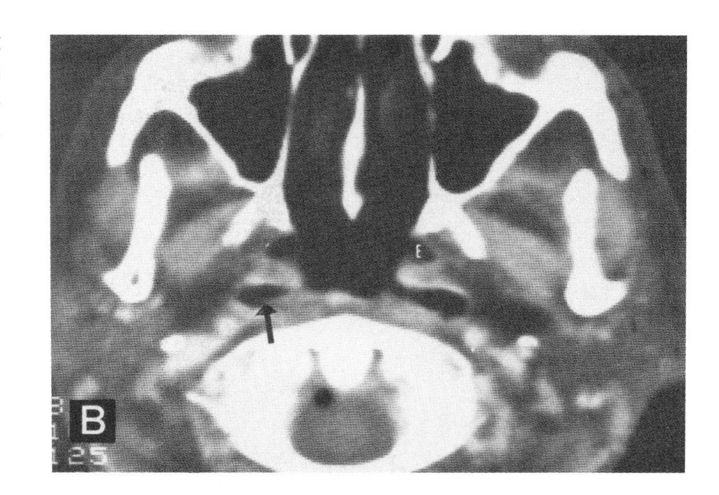

Plate 11.4*B*. This was done at the same level as that of part *A* except that the mouth was open widely and the fossae of Rosenmüller (*arrow*) and eustachian tube orifices are better distended. These show that no abnormal masses are present.

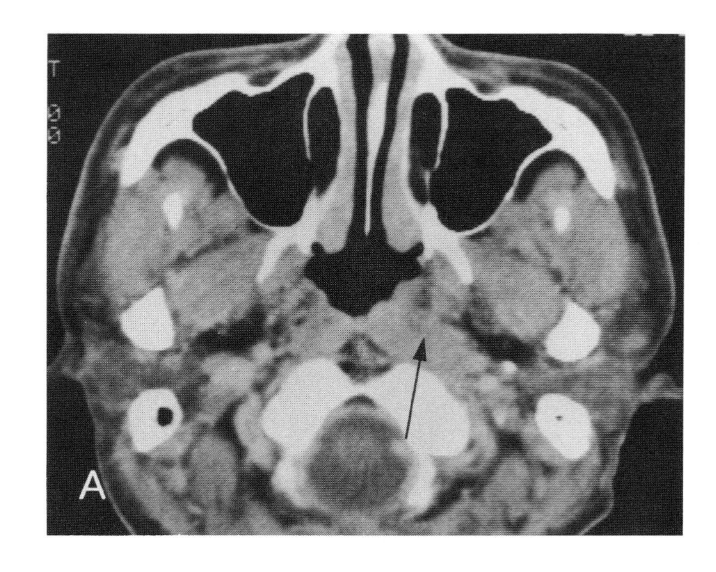

Plate 11.5*A*. This patient was suspected of having an infiltrating mass in the nasopharynx. The tissue planes between the mucosal surface and the region of the retropharyngeal lymph nodes are indistinct on the left side compared to the right (*arrow*).

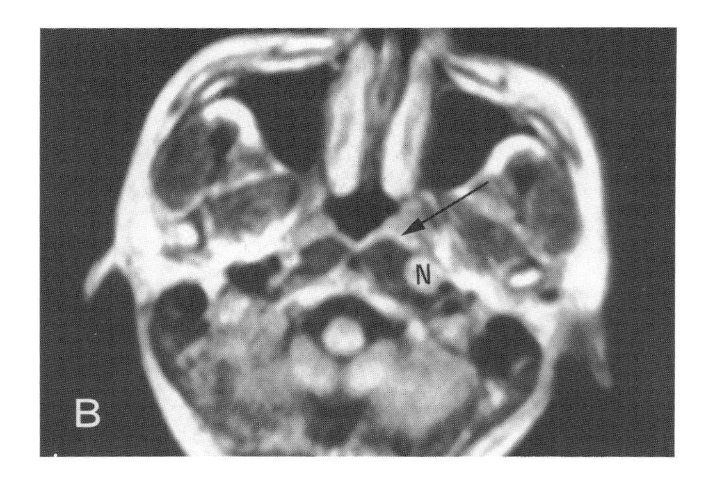

Plate 11.5*B*. A spin echo 820/56 image done on a 0.3 Tesla permanent magnet shows very clearly that the suspicious infiltration of these deep tissue planes is not present. There is a clear cleavage plane between the lymphoid tissue within the fossae of Rosenmüller (*arrow*) and the enlarged retropharyngeal node. It is, of course, possible that this represents a superficial carcinoma with a metastases to a node or similar manifestations of lymphoma; however, there are no findings to suggest that this is aggressive disease within the nasopharynx. Biopsy would be required if clinically indicated to make this determination.

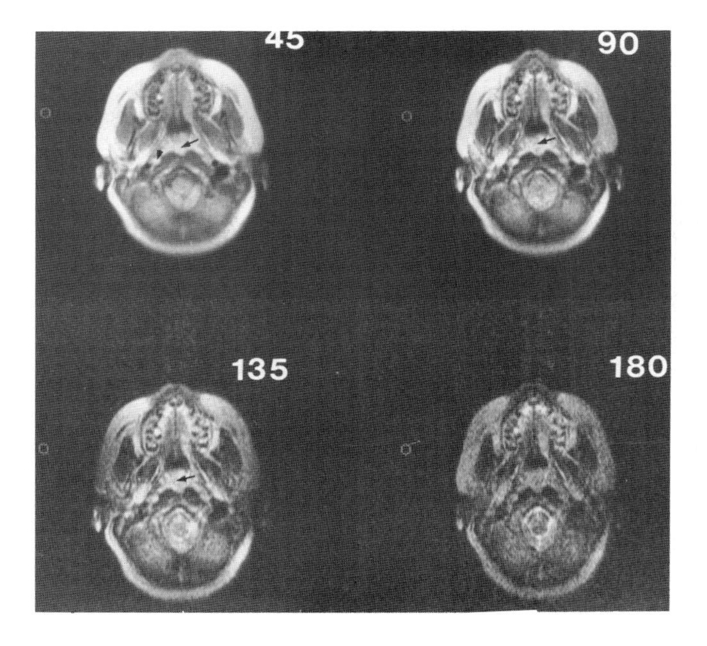

Plate 11.6. Lymphoid tissue in the nasopharynx can create rather large, sometimes ominous looking masses on CT. This is especially true in the younger age groups. One should become familiar with the normal anatomic variations of the lymphoid tissue in all of Waldeyer's ring so as not to mistake this normal variation for pathology on MRI and CT studies.

This is a single section from a 10-slice multislice, multiecho, acquisition on a 0.15 Tesla resistive unit through the nasopharynx. The T_R was 2500 msec and the echo delay times are shown in the *upper right hand corner* for each image. Note that the lymphoid tissue in the nasopharynx (*arrows*) as well as the retropharyngeal nodes (*arrowhead*) remains fairly intense relative to muscle and behaves about the same as fat on increasingly T_2-weighted images. Unfortunately, both lymphoma, squamous cell carcinoma and many of the other tumors that involve this region show the same general characteristics and, therefore, cannot be differentiated from lymphoid tissue on T_1- or T_2-weighted images by their signal intensity alone.

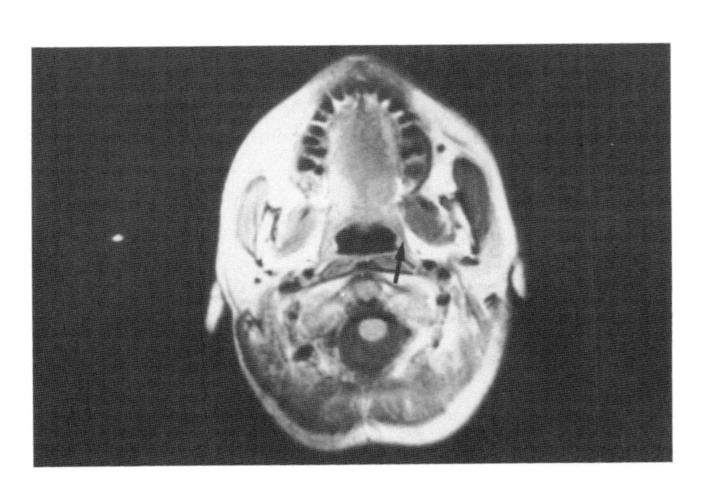

Plate 11.7. This section is made on a 1.5 Tesla system. The section thickness is 10 mm and the anatomic level is roughly equivalent to that seen in Plate 11.3*E*. The nasopharynx is somewhat atrophic. No lymphoid tissue is visible and only the thin musculature surrounding the airway from the parapharyngeal space is visible (*arrow*). Comparison of this image with that in previous plates will reveal that imaging of the upper aerodigestive tract, and particularly the nasopharynx, can be done at a variety of field strengths with nearly equivalent image quality. (Courtesy of the General Electric Corporation, January 1985.)

The abbreviations used on Plates 11.8 and 11.9 are: *PS*, parapharyngeal space; *S*, sphenoid science; and *C*, clivus.

Plate 11.8. MRI allows the study of the nasopharynx, skull base and cavernous sinus in the coronal plane without the inconvenience of placing the patient in an uncomfortable position.

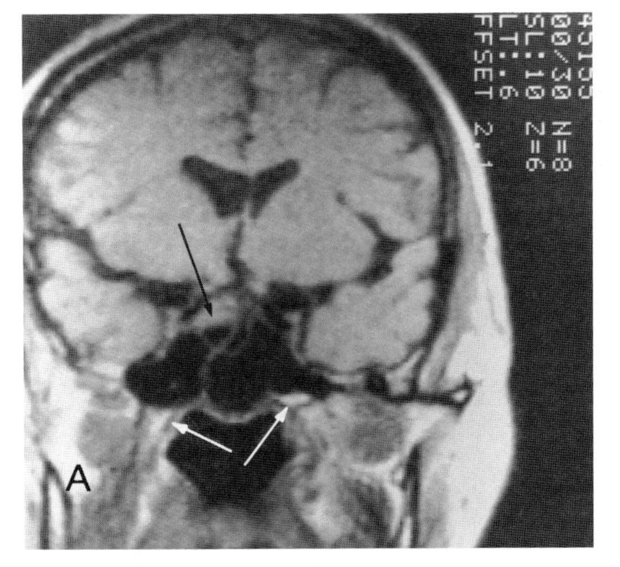

Plate 11.8*A*. This coronal section is through the most anterior portion of the nasopharynx. The patient's head is slightly turned so that the appearance of some of the structures is slightly asymmetric. The pterygoid plates (*white arrows*) can be seen on both sides. The air within the sphenoid sinus above, and airway below, provides an image of the roof of the nasopharynx at this point. An abnormal mass surrounds the supraclinoid portion of the internal carotid artery (*black arrow*) and this explained the patient's IIIrd nerve palsy.

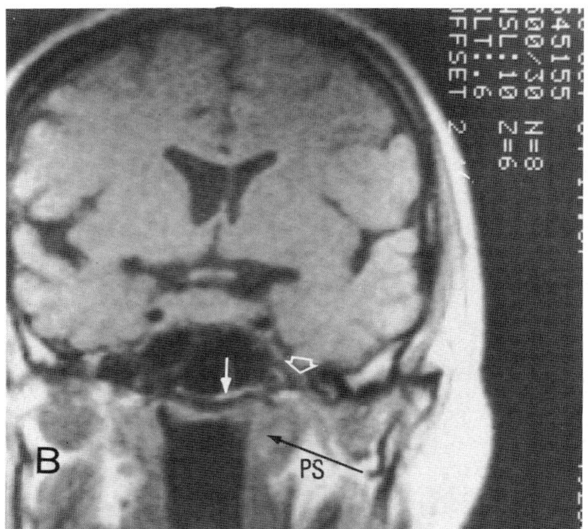

Plate 11.8*B*. A section taken through the midsella turcica shows the abnormal tissue within the sella and parasellar region on the right which was a metastases producing the patient's IIIrd nerve palsy. The bony roof of the nasopharynx is seen because of the slight soft tissue thickening on its airway side and a small amount of mucoperiosteal thickening within the floor of the sphenoid sinus (*white arrow*). The tensor and levator palati muscles are visible (*black arrow*). The contrast between the parapharyngeal musculature and the fat within the parapharyngeal space is excellent. Note the defects between the floor of the middle cranial fossa and the basisphenoid (*arrowhead*). This is the region of the foramen lacerum.

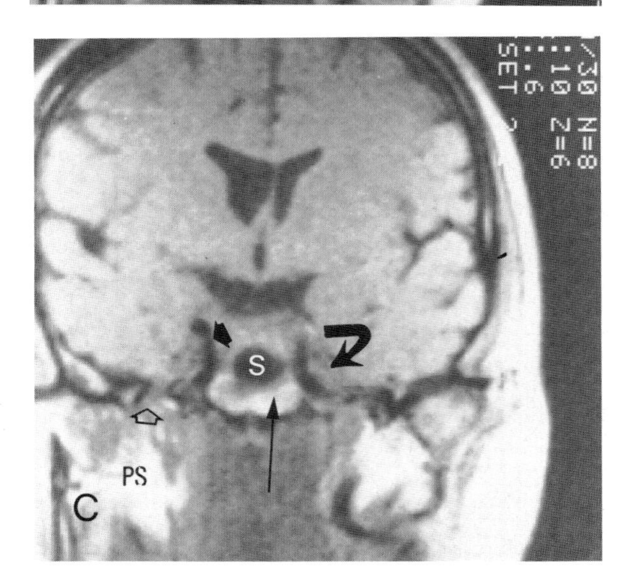

Plate 11.8*C*. The fat within the clivus (*arrow*) contrasts well with the air within the sphenoid sinus and the flowing blood within the cavernous portion of the carotid artery as it ascends posterolaterally (*closed arrowhead*). The neural structures within the cavernous sinus lie just lateral to the carotid at this point (*curved arrow*). On the right side is the foramen ovale and exiting third division of the trigeminal nerve (*open arrowhead*). It is not hard to see how an infiltrating mass in the parapharyngeal space can invade this nerve near its exit from the cranial vault.

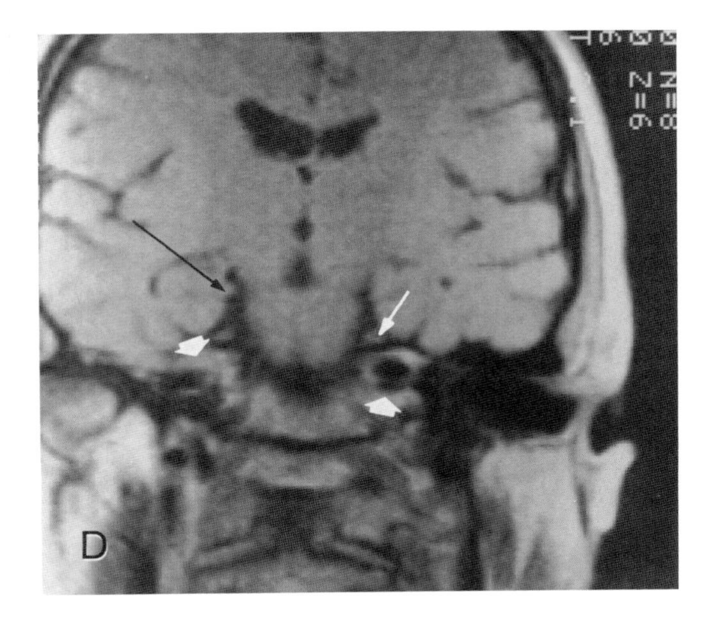

Plate 11.8*D*. This coronal section is taken through the region of Meckel's cave. The Vth cranial nerves are visible as they course over the top of the petrous apices (*white arrow*). In this patient there is some fatty marrow within the apices of the petrous pyramids (*arrowheads*). The IIIrd nerve can be seen on the right side (*black arrow*).

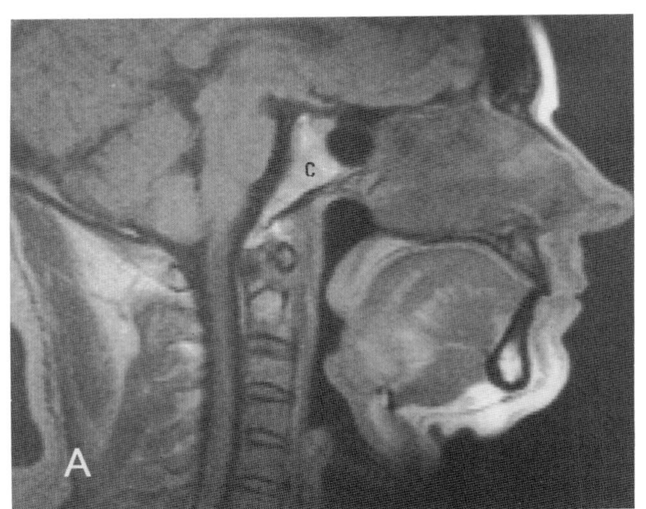

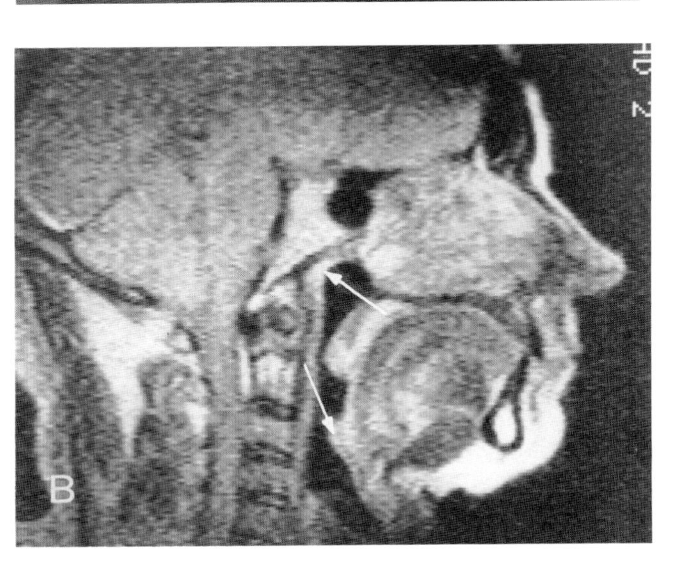

Plate 11.9. Sagittal MR images provide an excellent view of the relationship of the nasopharynx to the clivus, basisphenoid and nasal cavity. This imaging orientation is unique to MR since the direct sagittal images in CT scanning are of relatively poor quality and very cumbersome to perform.

Plate 11.9*A*. A spin echo 500/30 image shows the limits of the nasopharyngeal vault quite nicely. The clivus is usually readily apparent because of its fatty marrow content. The soft palate likewise produces excellent contrast because of its relatively high signal intensity (*central portion*) compared to muscle and other surrounding tissues. There is often a slightly increased signal along the posterior pharyngeal wall which is due to the lymphoid tissue lining the nasopharynx. Deeper tissues are prevertebral muscles and, therefore, of slightly lower intensity.

Plate 11.9*B*. This spin echo 2000/60 pulse sequence is more T_2 weighted and emphasizes the differences in tissue contrast that may be seen between the two basic kinds of pulse sequences used in examining the extracranial head and neck. The soft palate becomes relatively intense compared to surrounding musculature structures. Lymphoid tissue (*arrows*) will also become somewhat more intense. Although the spatial resolution is somewhat diminished in these images which have inherently lower signal to noise ratio. The anatomic relationships are still quite clear while emphasizing the important differences in tissue contrast especially when tumor or inflammatory changes are being displayed.

MALIGNANT TUMORS

The abbreviations used in Plate 11.10 are: *PS,* parapharangeal space; *j,* jugular vein; and *C,* carotid artery.

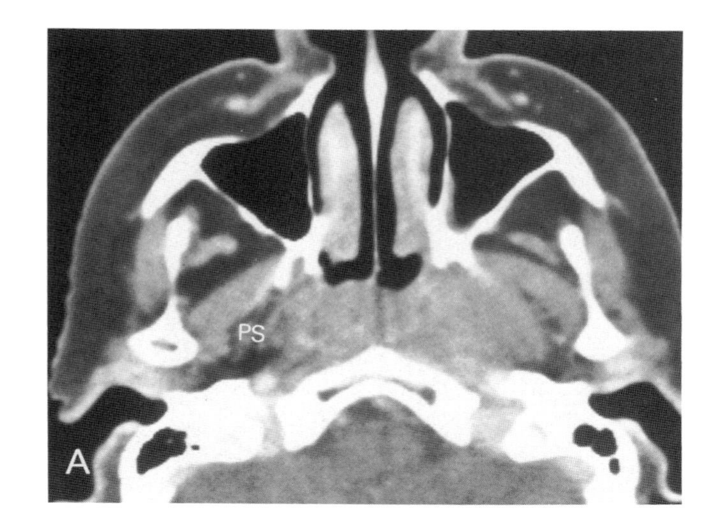

Plate 11.10. This patient had nasopharyngeal carcinoma presenting with symptoms of obstruction of the left eustachian tube orifice and bilateral cervical lymphadenopathy.

Plate 11.10*A*. A section through the upper nasopharynx shows the classic CT finding for nasopharyngeal carcinoma; that is obliteration of the parapharyngeal space. Note that the parapharyngeal space is visible on the right side as well as the small muscle bundles and vessels within it.

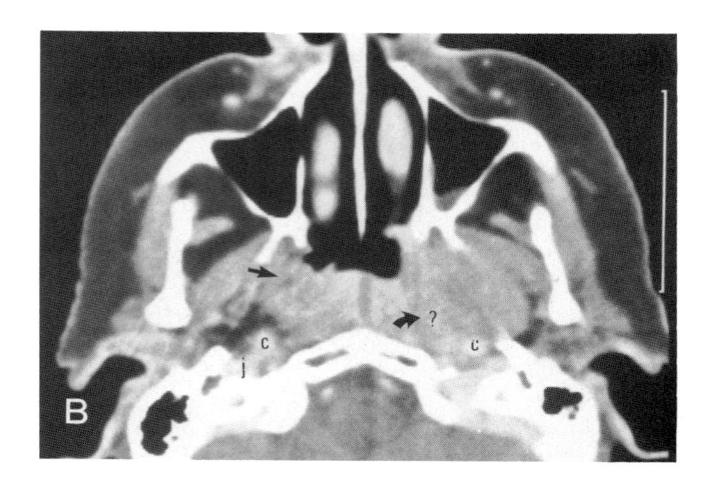

Plate 11.10*B*. A section slightly lower than part *A* shows that the infiltrating mass crosses the midline though the parapharyngeal space, in its most posterior aspect, is preserved on the right side. There is some obliteration of the more anterior (prestyloid) portion of the parapharyngeal space but the tensor palati muscle is still visible (*arrow*). On the left side, the mass clearly infiltrates the region of the carotid sheath. The *curved arrow* points to an area of tumor which could be due to direct extension of tumor or spread to retropharyngeal lymph nodes near the skull base (*curved arrow and question mark*). MRI seems to be better at differentiating whether extension is due to nodal involvement or direct spread of primary tumor. This may not be an essential clinical fact; however, it does show one of the advantages of MRI in this area: namely, improved resolution of the edges of tumor masses.

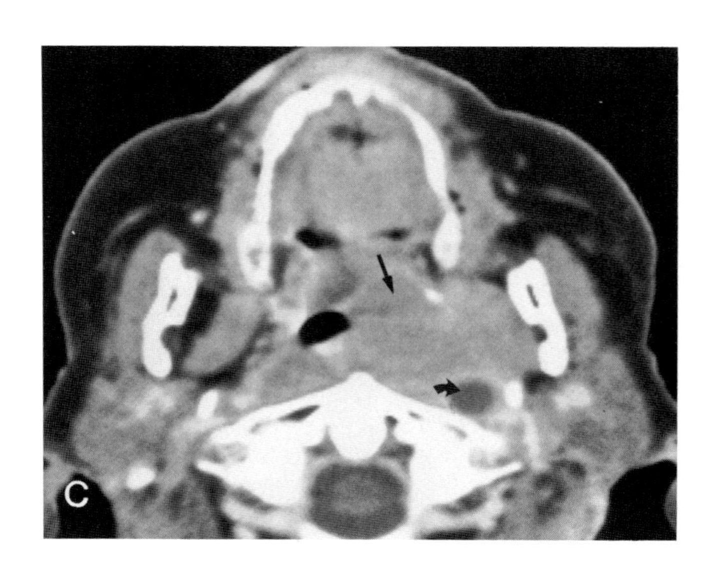

Plate 11.10*C*. A section made at the junction of the oropharynx and nasopharynx shows that the primary tumor has spread to involve the upper surface of the soft palate (*arrow*). Another large component of the tumor obliterates the parapharyngeal space and crosses the midline. Lower sections showed spread into the oropharynx. In some cases such spread is submucosal and only visible on CT; this upstages nasopharyngeal carcinoma. The completely necrotic retropharyngeal node (*curved arrow*) is relatively easy to distinguish from the primary lesion. If the nodes are nonnecrotic, this can be difficult on CT.

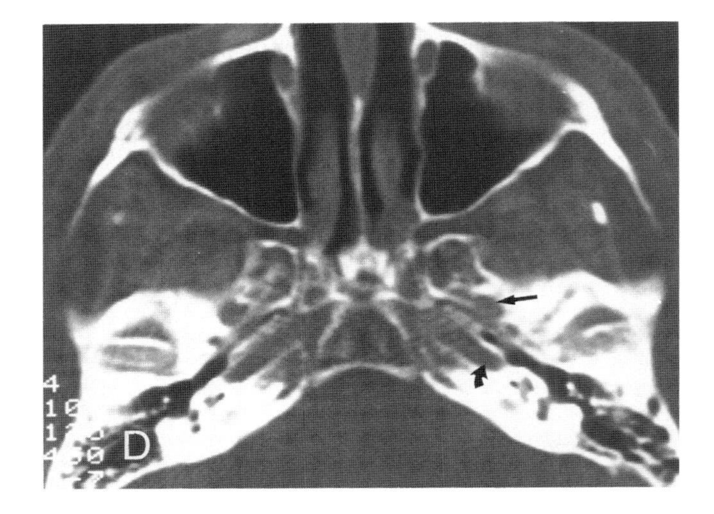

Plate 11.10*D*. Despite the large size of the primary tumor, the sections through the skull base showed the base of the skull to be entirely intact. In fact, surprisingly, the middle ear cavity was normally aerated despite intermittent symptoms of unilateral serous otitis.

Skull base invasion is a critical determinant in planning radiation therapy. If there is any suspicion at all of skull base invasion on axial sections then coronal scans should be done to either confirm or exclude the possibility of subtle skull base erosion. Important checkpoints are the foramen ovale (*arrow*), carotid canal (*curved arrow*), cortical margins of the inferior petrous apex, and the clivus.

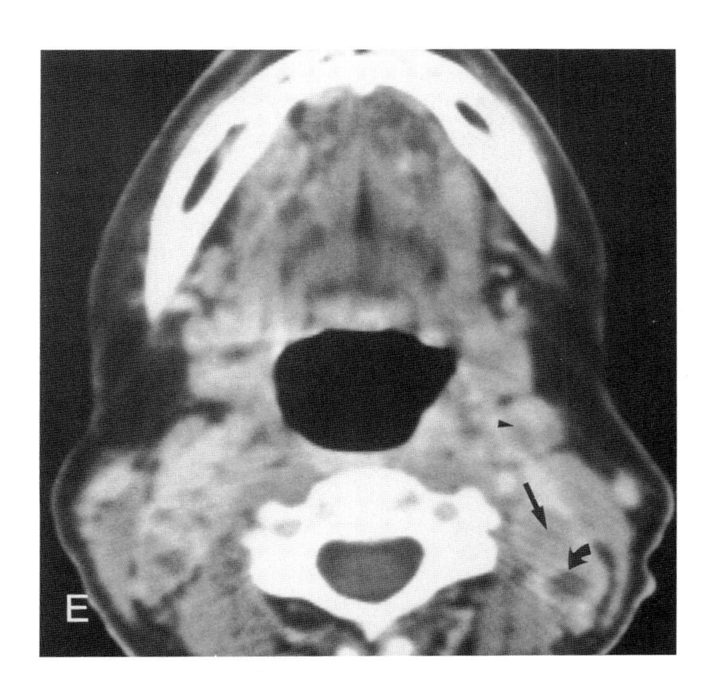

Plate 11.10*E*. The vast majority of patients with nasopharyngeal carcinoma will have cervical metastases at the time of presentation.

This patient has a fairly classic pattern of lymphadenopathy in that it shows predominantly posterior internal jugular nodes (*arrow*) and true posterior triangle or spinal accessory nodes (*curved arrow*). Patients also frequently have anterior internal jugular nodes in the jugulodigastric region (*arrowhead*).

Whenever a patient presents with this pattern of adenopathy one should strongly suspect a primary lesion within the nasopharynx as an explanation.

The abbreviations used on Plate 11.11 are: *M,* prevertebral muscle and *T,* tumor.

Plate 11.11. Nasopharyngeal carcinoma usually presents as a mainly deeply infiltrating lesion on CT; however, it may have a primarily exophytic growth pattern with relatively deep infiltration. On occasion the mass can be entirely superficial with no definite evidence of invasion of the deep soft tissues, and it is these cases which cannot be distinguished from normal lymphoid tissue within the nasopharynx without biopsy.

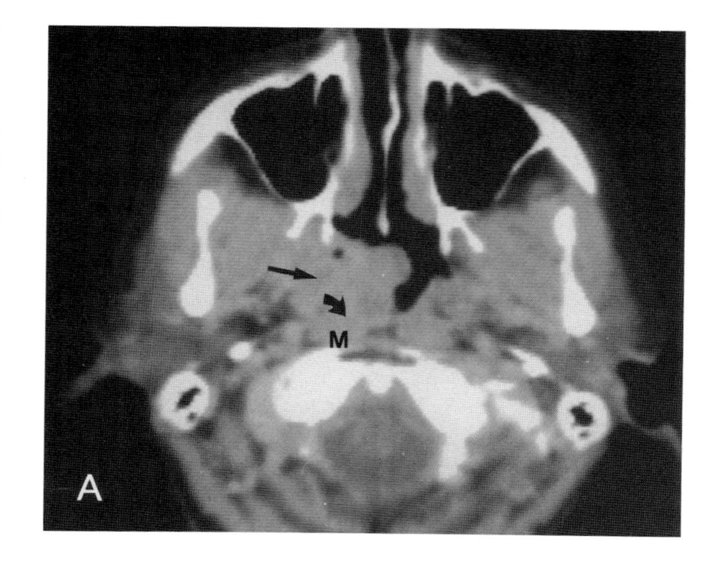

Plate 11.11*A*. This patient with nasopharyngeal carcinoma has a considerable amount of deep spread (*arrow*); however, the bulk of the tumor is growing exophytically within the nasopharynx. The interface between the tumor mass and the prevertebral muscles is obliterated (*curved arrow*) although MRI would probably show a fairly distinct interface.

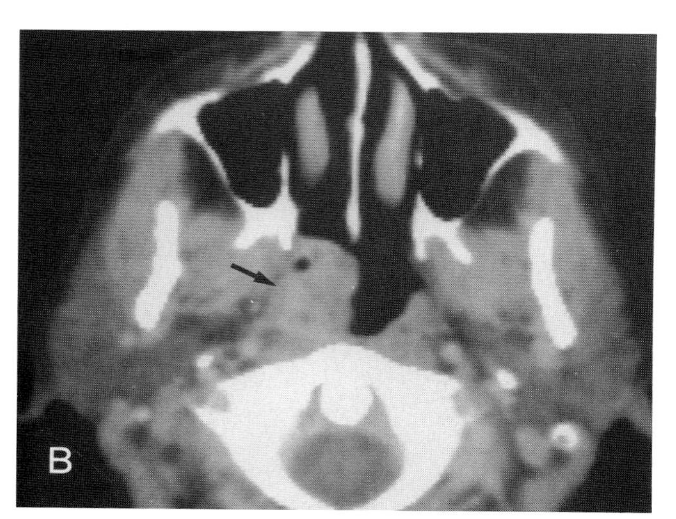

Plate 11.11*B*. In the midnasopharynx the mass is predominantly exophytic and the relatively normal appearing parapharyngeal space (*arrow*) is now visible.

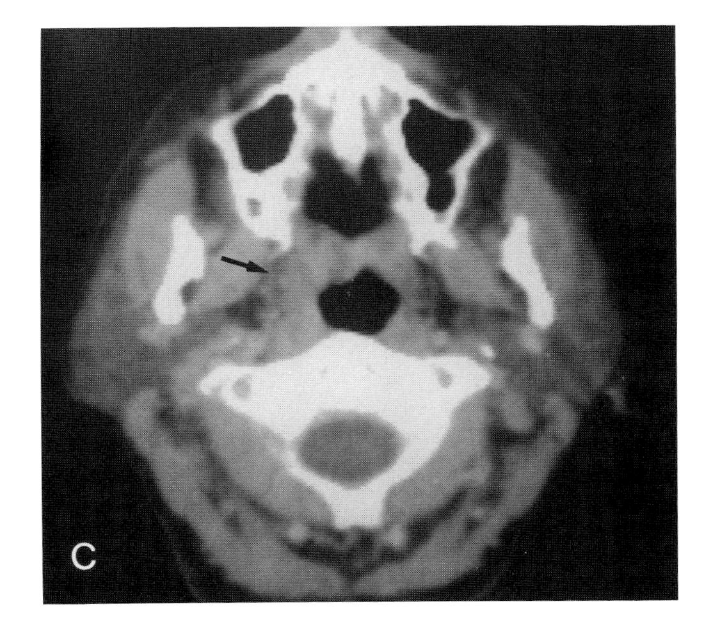

Plate 11.11*C*. Low in the nasopharynx one can still see a subtle but definite asymmetry in the parapharyngeal space; it being somewhat indistinct on the side of the tumor (*arrow*). The superior-to-inferior extent of the lesion may be best appreciated on the coronal sections that follow.

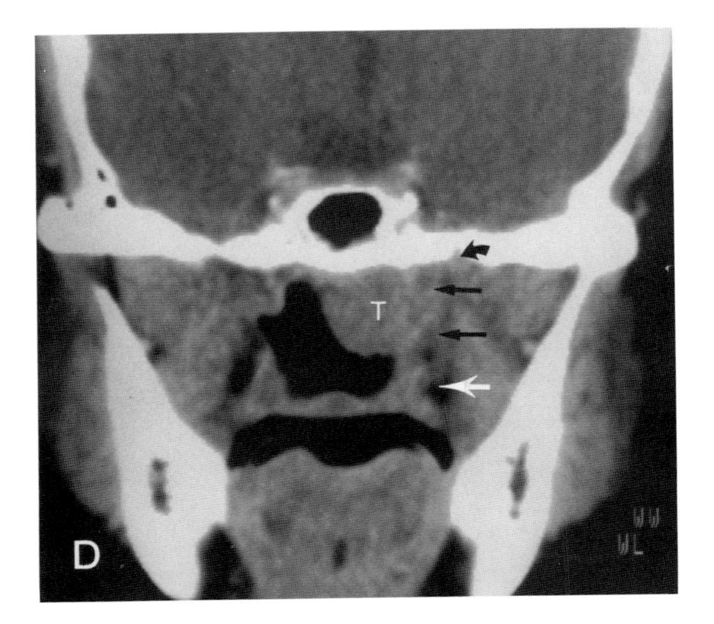

Plate 11.11*D*. A coronal section through the anterior part of the nasopharynx shows the primarily exophytic tumor to best advantage. One can see the obliteration of the upper parapharyngeal space (*black arrows*). Especially important is the infiltration near the skull base at the foramen ovale (*curved arrow*). The parapharyngeal space returns to near normal in the middle-to-lower third of the nasopharynx (*white arrow*).

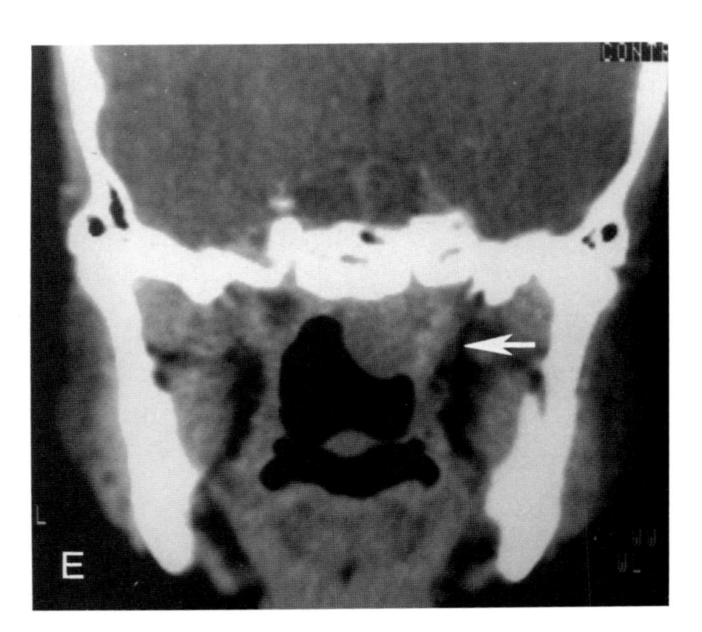

Plate 11.11*E*. More posteriorly in the nasopharynx, the mass remains predominantly exophytic but note how it obliterates the deep tissue planes between the parapharyngeal musculature and the airway. The more posterolateral aspect of the parapharyngeal space (*arrow*) has remained normal as seen on the axial sections.

The abbreviation used in Plate 11.12 is: *N*, node.

Plate 11.12. This patient presented with cervical adenopathy. The physical examination revealed an obvious tumor in the superior and lateral aspect of the nasopharynx. The CT findings were relatively subtle in the high nasopharynx. While viewing the following images, one should note the fairly limited deep extension compared to the superficial amount of disease and the secondary nodal disease.

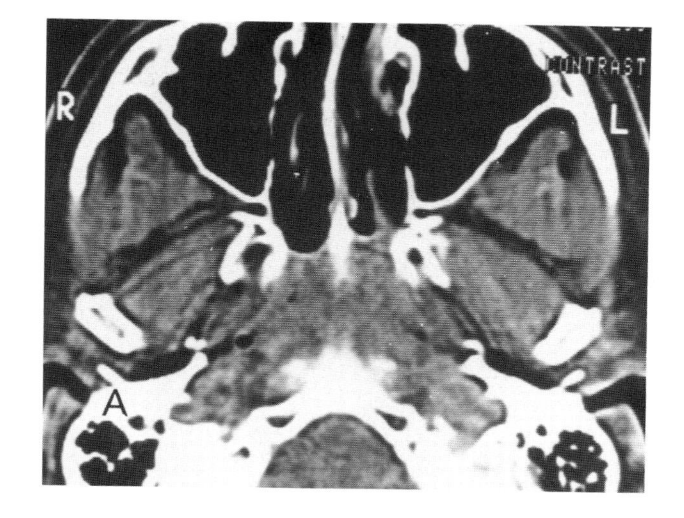

Plate 11.12*A*. A section through the high nasopharynx shows some soft tissue filling the upper nasopharyngeal vault; however, there is absolutely no evidence of deep infiltration so that this tissue would be indistinguishable from normal adenoid tissue.

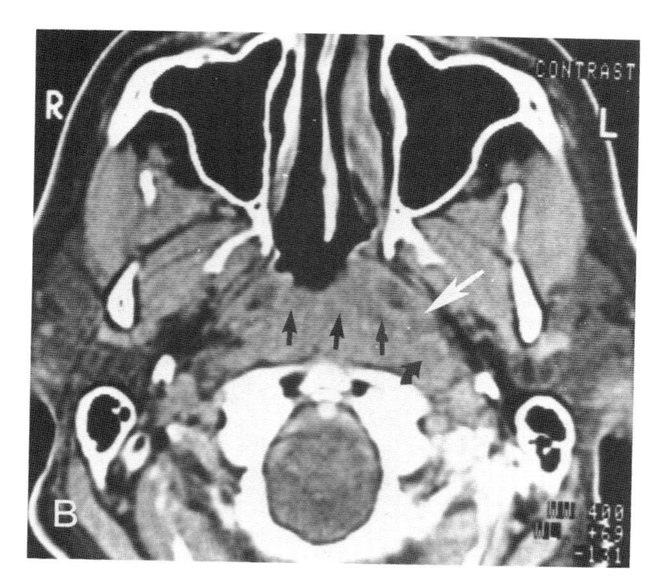

Plate 11.12*B*. There is soft tissue filling the fossae of Rosenmüller on the left and spreading across the midline of the mucosal surface (*black arrows*). The parapharyngeal space on the right side appears entirely normal and the tensor and levator palati muscles can be clearly seen. Note the subtle but definite encroachment on the parapharyngeal musculature and prestyloid parapharyngeal space on the left (*white arrow*). There is also a small retropharyngeal node present (*curved arrow*).

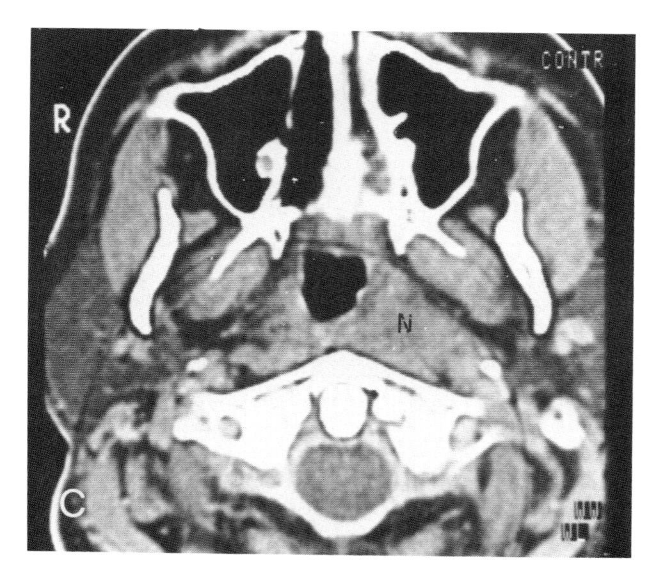

Plate 11.12*C*. Lower in the nasopharynx there is an obvious mass on the left side. This was felt to represent an enlarged retropharyngeal lymph node. Note that in this, and in the lymph nodes seen lower in the neck, there is no evidence of necrosis.

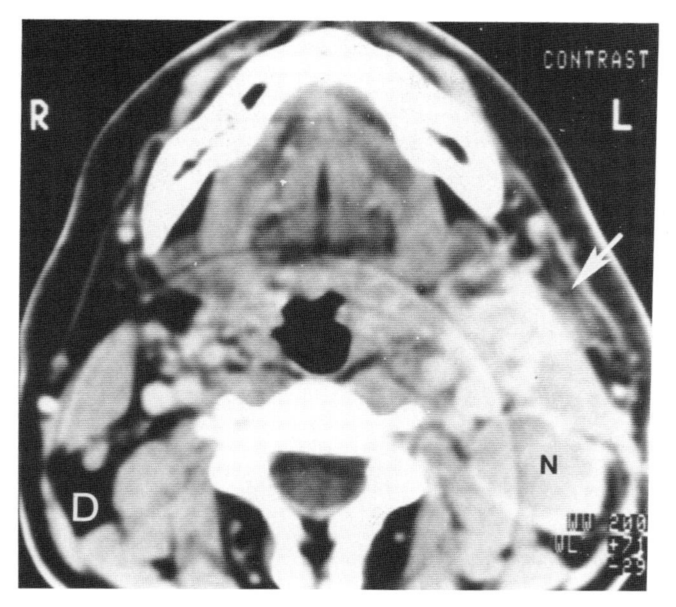

Plate 11.12*D*. Prior to presentation at our institution the patient had an excisional biopsy of the lymph node mass in the neck creating some of the soft tissue thickening seen (*arrow*). All of the lymph nodes were enlarged without evidence of necrosis and showed some peripheral enhancement. The morphology of nodes is compatible with lymphoma, reactive nodes, or nonkeratinizing squamous cell carcinoma (lymphoepithelioma). Note that the involved nodes are in the retropharyngeal group and predominantly posterior to the jugular vein. The morphology of the nodes and their distribution strongly suggests that this is either a lymphoma or lymphoepithelioma of the nasopharynx. In this instance it was quite apparent clinically that the lymphoepithelioma was in the nasopharynx. Occasionally this is not so clear even after good endoscopic exam, and the CT morphology may be used to direct biopsy of the nasopharynx in the absence of an obvious mucosal lesion.

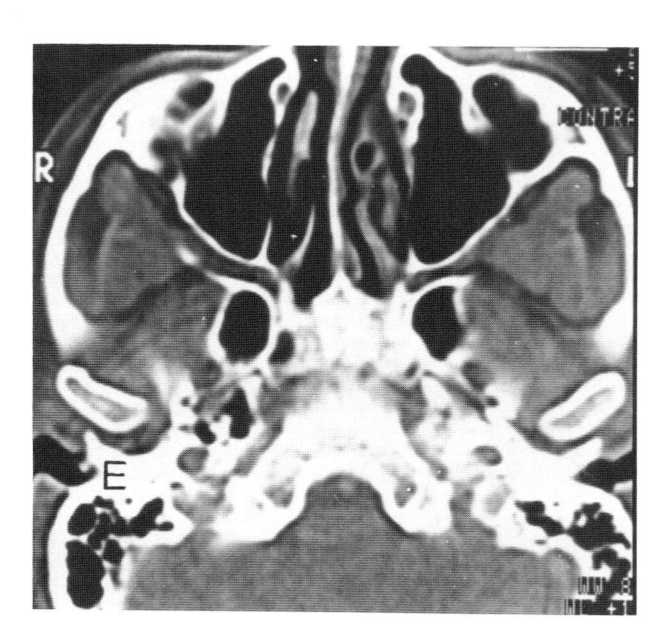

Plate 11.12*E*. The axial projection showed no evidence of skull base destruction; however, the primary lesion was so close to the skull base that it was elected to do coronal projections to be certain that a subtle area of destruction was not being missed.

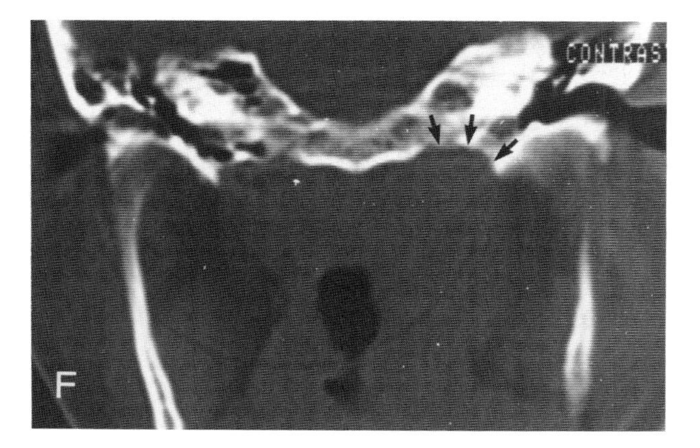

Plate 11.12*F*. The coronal sections were considered to be suspicious for early erosion at the top of the fossae of Rosenmüller in the region of the inferior petrous apex (*arrows*).

The abbreviation used on Plate 11.13 is: *T*, tumor.

Plate 11.13. An elderly male with squamous cell carcinoma of the nasopharynx.

Plate 11.13*A*. In the low nasopharynx there is an obvious deeply infiltrating mass on the left.

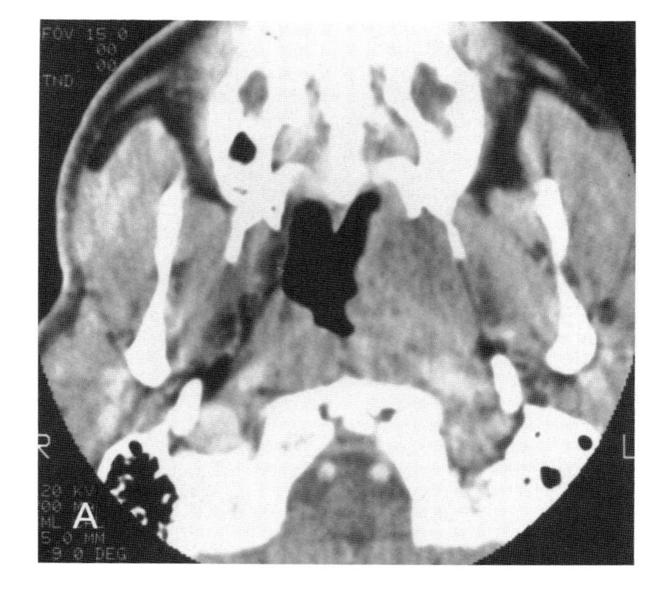

Plate 11.13*B*. A spin echo 500/30 acquisition at the same level shows the primary tumor and a retropharyngeal lymph node (*arrow*).

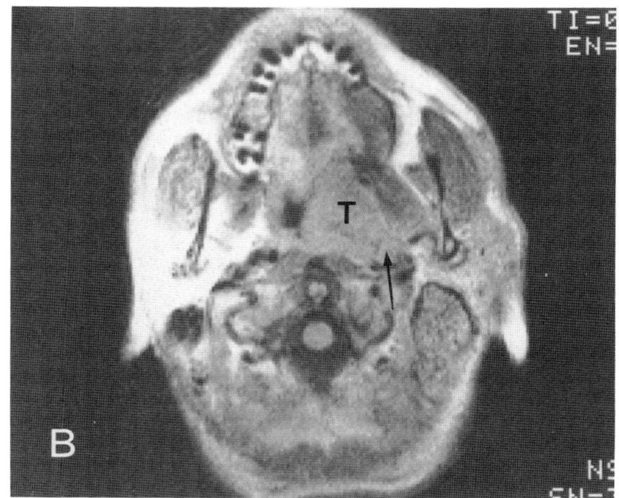

Plate 11.13*C*. A CT section in the midnasopharynx suggests deep spread to the level of the eustachian tube orifice manifest by subtle obliteration of deep tissue planes (*arrows*) around the pharyngeal musculature.

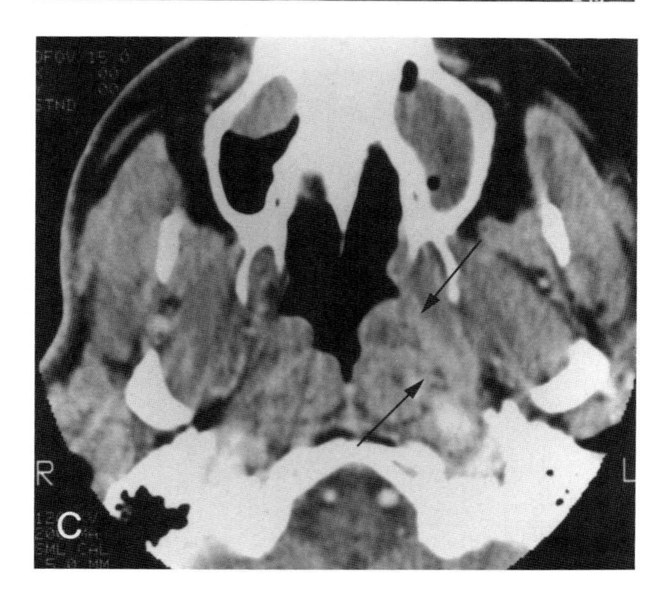

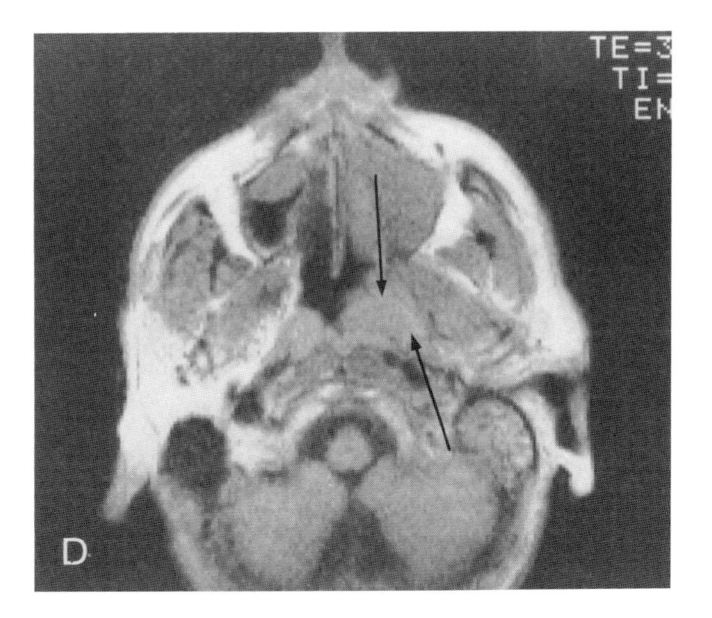

Plate 11.13*D*. MRI at the same level shows very obvious deep infiltration (*arrows*). It seems to be easier to appreciate deep tumor margins on MRI because of its superior contrast resolution.

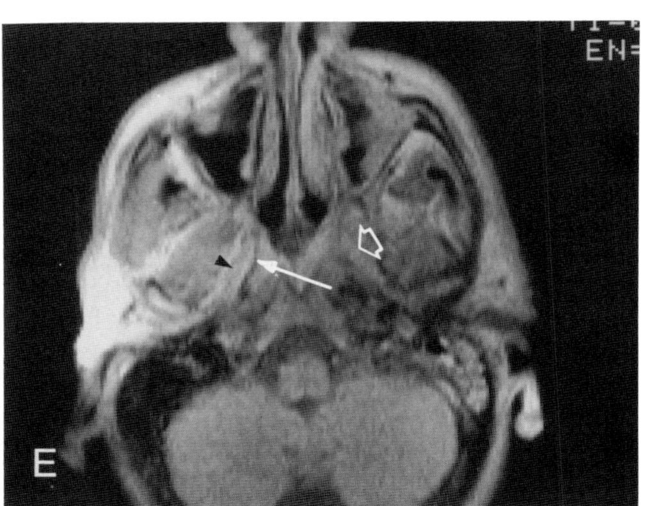

Plate 11.13*E*. Spin echo 2500/30 near the skull base shows the normal pharyngobasilar fascia (*arrow*) and tensor palati muscle (*black arrowhead*) on the right and absence of the fascia and deep infiltration on the left (*open arrowhead*).

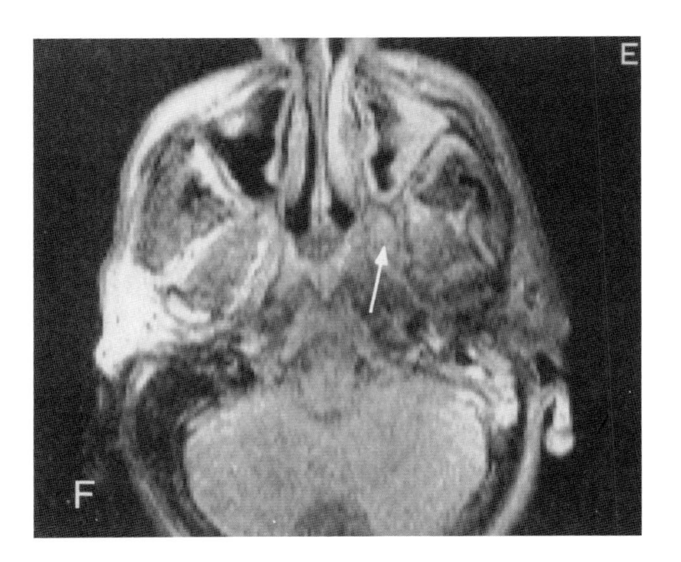

Plate 11.13*F*. Spin echo 2500/60 at the same level confirms an infiltrating process in the parapharyngeal fat (*arrow*) which seems to be shortening its T_2.

It has been a general trend that MRI shows tumor margins better than CT in the nasopharynx, infratemporal fossa and parts of the oropharynx. In the larynx and hypopharynx it is not so apparent, and well done CT still provides more information. The tongue base is a toss-up. The relative value of MRI and CT in these areas must be worked out over the next several years. In the nasopharynx, however, MRI seems to have clear superiority with the possible exception of the evaluation of the skull base in selected cases.

The abbreviations used in Plate 11.14 are: *T,* tumor; and *N,* node.

Plate 11.14. This 35-year-old female presented with a neck mass which was treated with antibiotics; the mass continued to enlarge. An endoscopic examination done 2 months after initial presentation revealed a mass in the nasopharynx. Biopsy showed lymphoepithelioma (nonkeratinizing squamous cell carcinoma).

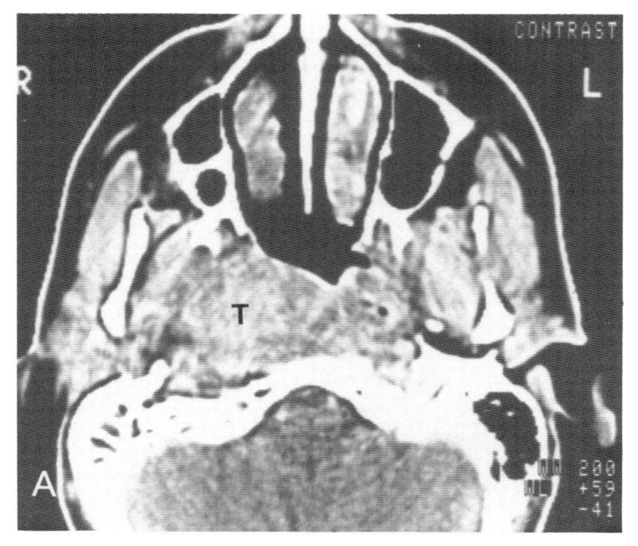

Plate 11.14*A*. A section through the mid- to lower nasopharynx shows the obviously infiltrating tumor. It is very difficult to tell the difference between the primary tumor mass and retropharyngeal nodes at this level. The difference is much more apparent on the following MRI section.

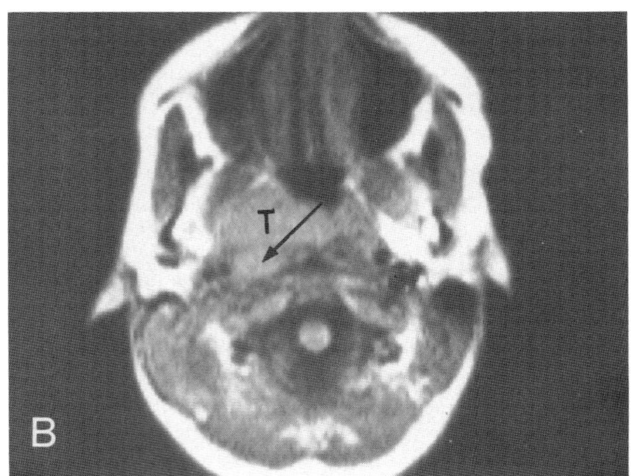

Plate 11.14*B*. MRI section at roughly the same level as that seen in part *A*. Note that it is easier to see the margins of the primary tumor mass versus the metastases to the retropharyngeal lymph node (*arrow*). MRI studies in the upper aerodigestive tract tend to show tumors having much better circumscribed margins than would be suspected from the CT examination. This can sometimes be a fault of MRI in that it might lead one to believe that the apparently sharp border is an indication of benignancy.

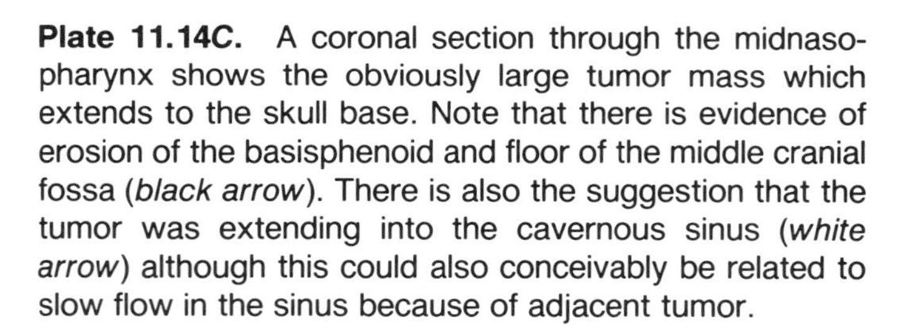

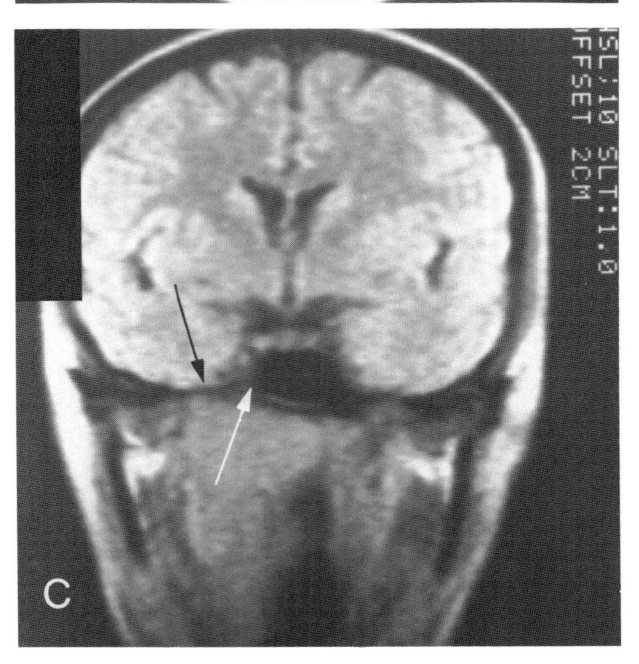

Plate 11.14*C*. A coronal section through the midnasopharynx shows the obviously large tumor mass which extends to the skull base. Note that there is evidence of erosion of the basisphenoid and floor of the middle cranial fossa (*black arrow*). There is also the suggestion that the tumor was extending into the cavernous sinus (*white arrow*) although this could also conceivably be related to slow flow in the sinus because of adjacent tumor.

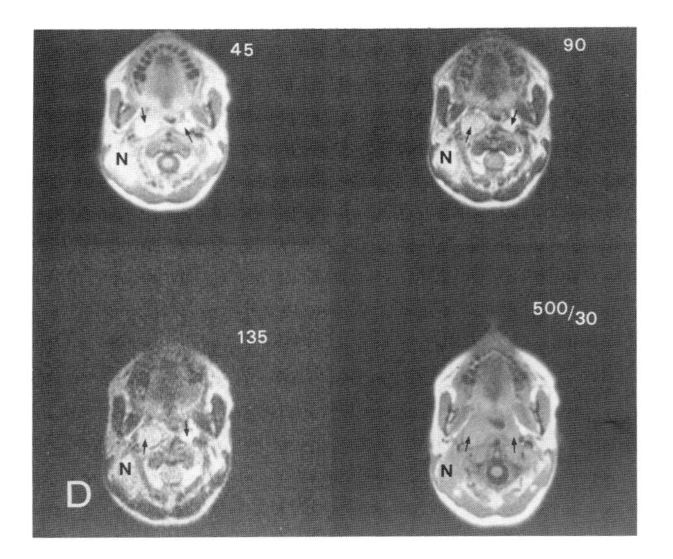

Plate 11.14*D*. This is a single section taken from a multislice multiecho acquisition and a spin echo 500/30 acquisition. The spin echo 500/30 image is in the *lower right hand corner*. The T_R interval of the T_2-weighted sequence was 2500 msec. (T_E times are indicated on the images.) There are bilaterally enlarged retropharyngeal nodes (*arrows*) and a large nodal mass in the neck. Note the improved contrast between tumor mass and surrounding musculature on the T_2-weighted sequence and the better contrast between tumor and surrounding fat on the T_1-weighted spin echo 500/30.

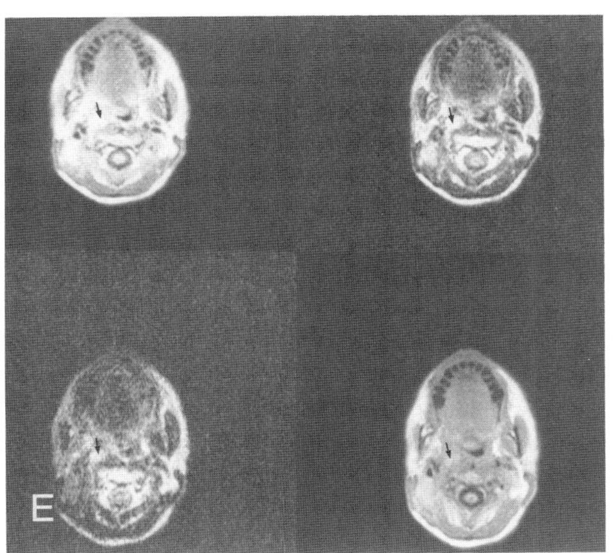

Plate 11.14*E*. A section at the same level as part *D* from a study done halfway through therapy shows that the left retropharyngeal node is completely gone. The neck nodes and the right retropharyngeal node (*arrows*) have decreased in size in the interval. The neck nodes have also decreased in signal intensity.

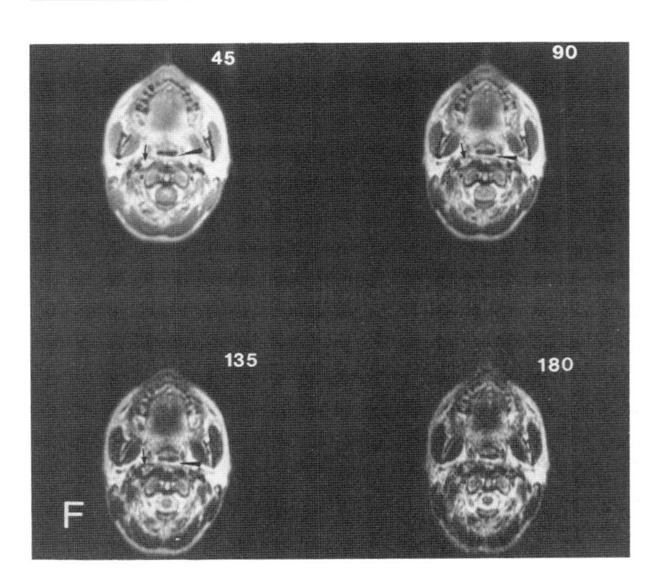

Plate 11.14*F*. A section at approximately the same level as the two prior parts. In this instance this is all from the T_2-weighted sequence with a TR of 2500 and echo delay times as indicated on the image. The neck nodes are no longer visible. The right retropharyngeal node (*arrows*) can still be seen. Also, a rind of increased signal now surrounds the airway (*arrowhead*); this is due to inflammatory mucositis. Unfortunately, inflammatory mucositis and persistent or recurrent tumor may have the same signal on any pulse sequence; this limits the usefulness of MRI in following signal changes *during* radiation therapy. At the end of successful therapy the tumor mass should be reduced in signal to approximately that of surrounding muscle on T_2-weighted sequences. However, persistent areas of fluid, inflammation or necrosis at sterile tumor sites could produce a high signal and mimic persistent tumor.

Plate 11.15. When carcinoma of the nasopharynx becomes very extensive it may show obvious skull base invasion. This is a somewhat unusual pattern and is related to spread of tumor to the region of the jugular fossa. From here the tumor mass has extended through the bone of the basiocciput and into the posterior fossa. More usual patterns of bone destruction are discussed in the text and illustrated in prior, and the following, cases.

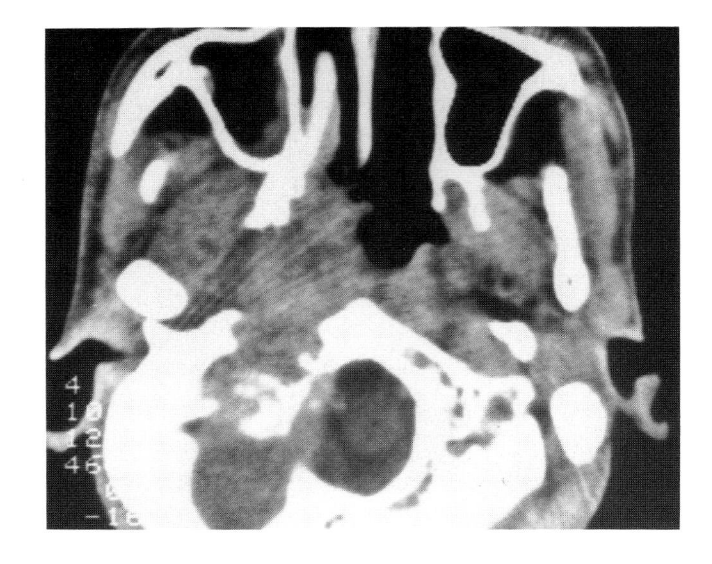

Plate 11.16. Tumor may directly invade the cavernous sinus by destroying the inferior petrous apex. At other times it may follow the carotid artery through the carotid canal and into the cavernous sinus without frank evidence of bone destruction. The tumor may also follow either the second or third division of the Vth cranial nerve into the cavernous sinus where it is free to expand. This is a very ominous pattern of spread, in that it indicates incurable disease.

Plate 11.16*A*. This patient with an infiltrating nasopharyngeal carcinoma (*arrowheads*) had symptoms suggestive of a cavernous sinus spread.

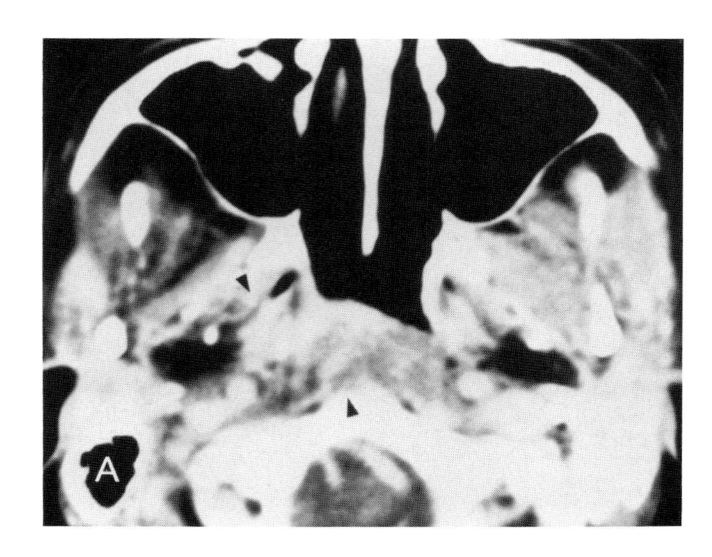

Plate 11.16*B*. The coronal section shows that the lesion extends to the apex of the fossa of Rosenmüller but even at this window setting one can appreciate that there is no frank bony defect at the skull base (*arrow*). There is obvious extension into the cavernous sinus (*arrowheads*).

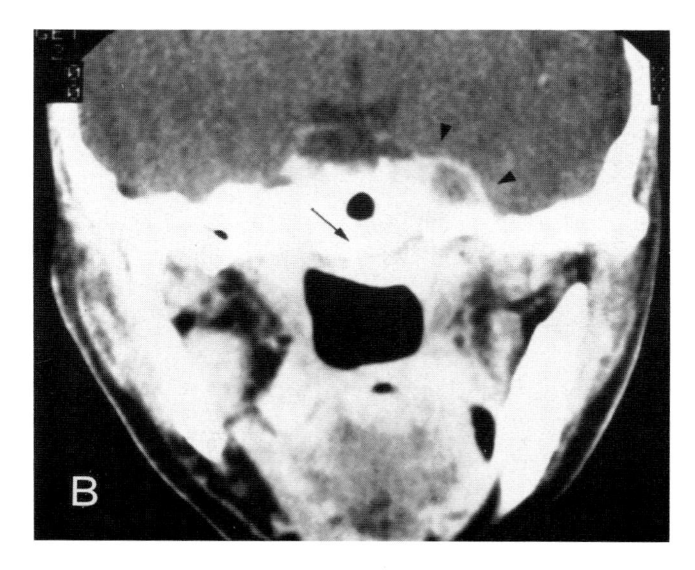

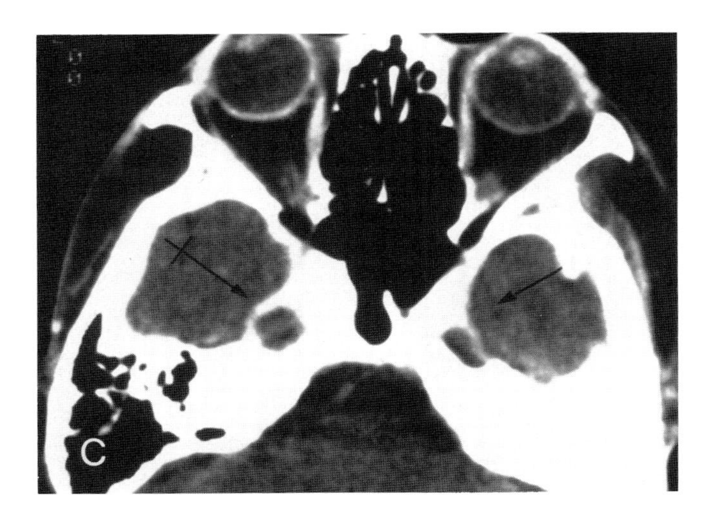

Plate 11.16*C*. An axial section through the cavernous sinus shows the typical changes of involvement of the trigeminal ganglion. Compared to the normal side (*arrow*) the involved trigeminal ganglion is expanded, displacing the border of the cavernous sinus outward (*crossed arrow*).

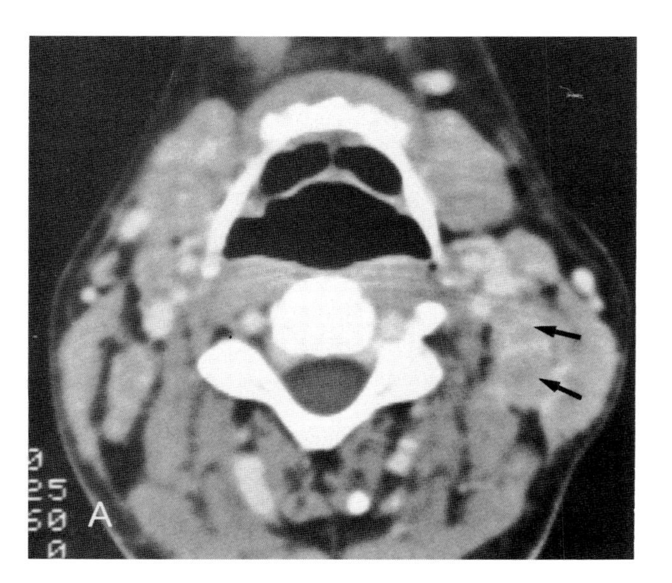

Plate 11.17. Nasopharyngeal carcinoma occasionally presents as a result of its cranial nerve symptoms. Usually it presents due to tubal occlusive symptoms or because of cervical lymphadenopathy. This patient presented with cervical lymphadenopathy.

Plate 11.17*A*. A section through the neck shows a posterior triangle and posterior internal jugular nodes (*arrows*). There is also a normal-sized jugulodigastric node present at this level. Careful endoscopic examination, both in the clinic and in the operating room under anesthesia, revealed no evidence of a primary tumor. A CT study was done.

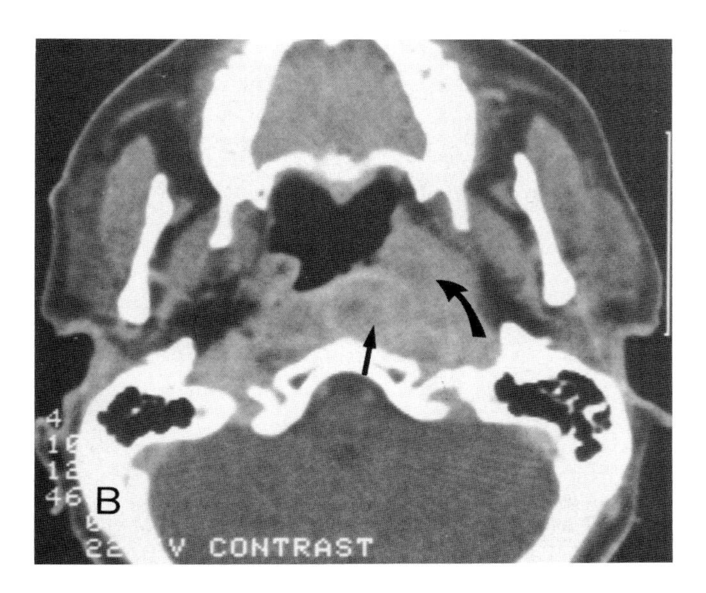

Plate 11.17*B*. CT study revealed an obviously deeply infiltrating mass in the left side of the nasopharynx (*curved arrow*). There was also an enlarged median retropharyngeal node (*arrow*). Most head and neck cancers have obvious mucosal components at endoscopy. Occasionally tumors will grow entirely beneath intact mucosa. Such was the case in this patient and a biopsy revealed poorly differentiated squamous cell carcinoma. This phenomenon of tumor growing beneath entirely intact mucosa occurs most frequently in the nasopharynx and at the tongue base, and occasionally in the faucial tonsillar region. With the skill of present endoscopists, lesions in the pyriform sinus are rarely the cause of so-called occult carcinomas of the upper aerodigestive tract.

The abbreviation used on Plate 11.19 is: *M,* mucocele.

Plate 11.18. Patients presenting with atypical facial pain sometimes have structural lesions to explain their symptoms. If the facial pain pattern is progressive and atypical for trigeminal neuralgia then CT or MRI should be used to investigate the entire course of the trigeminal nerve looking for an explanation of the patient's symptoms.

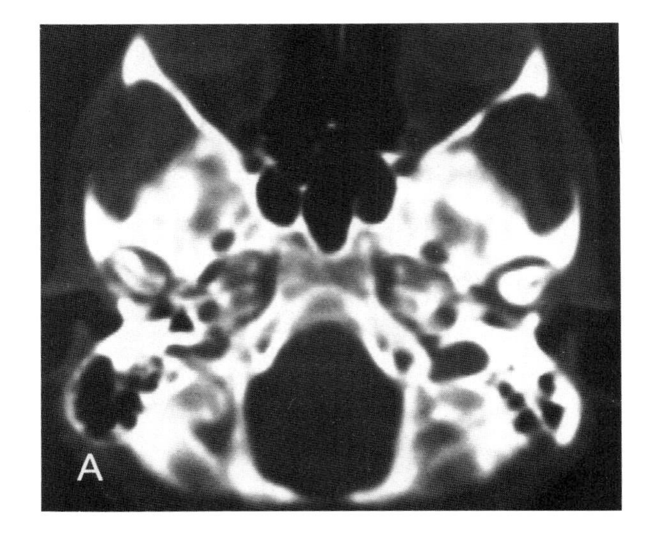

Plate 11.18*A*. This patient had had 4 yr of atypical facial pain in the distribution of the second and third divisions of the trigeminal nerve. She also had tubal-occlusive symptoms. Multiple endoscopies had been done with blind nasopharyngeal biopsies. The patient had had two prior CT studies, both of which stopped at the skull base when normal basal foramina were identified. On the most recent CT study, the skull base again appeared normal.

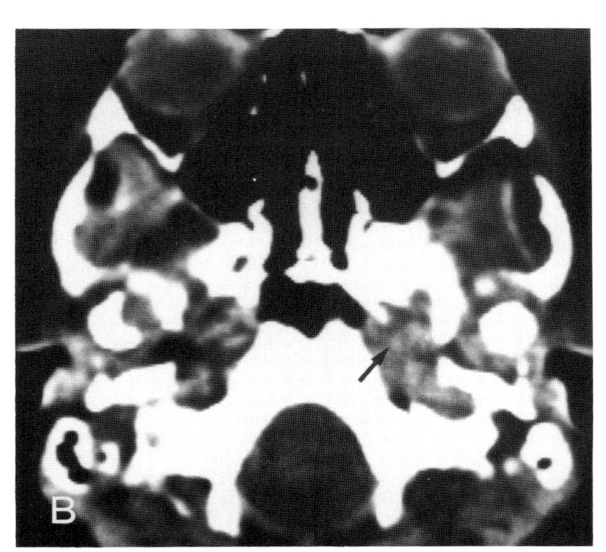

Plate 11.18*B*. A somewhat limited quality study shows a definite asymmetry of the soft tissues near the skull base. There appeared to be an infiltrating mass present on the left side (*arrow*). This is , of course, immediately beneath the exit of the third division of the Vth cranial nerve from the skull base.

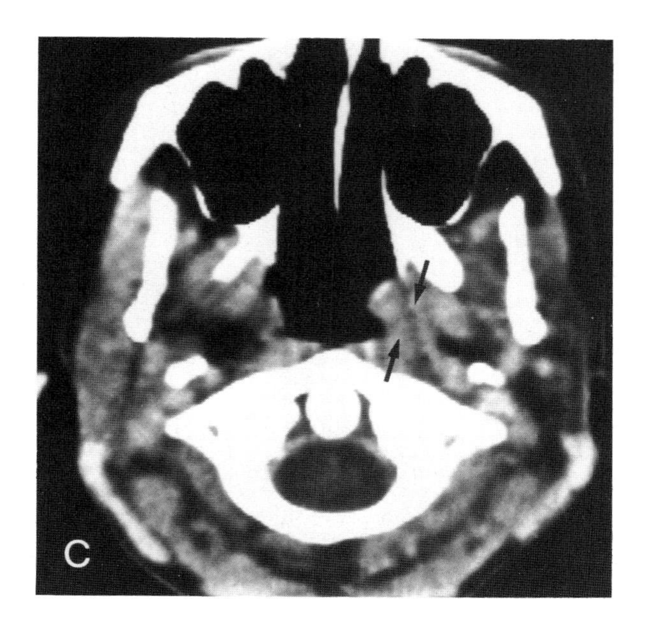

Plate 11.18*C*. A section through the midnasopharynx shows a subtle but definite infiltrating mass in the left side of the nasopharynx (*arrows*). Since the lesion was subtle, a coronal section was done.

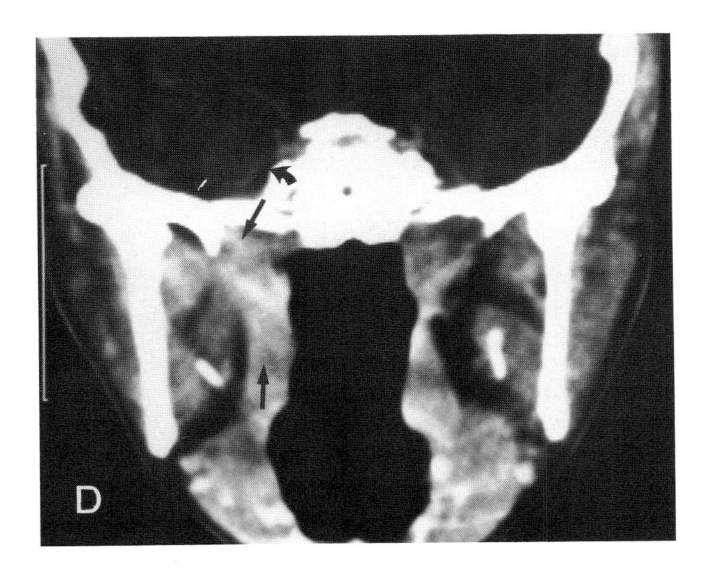

Plate 11.18*D*. The full extent of the infiltrating mass can now be better appreciated (*arrows*). Although it is somewhat dark, note that the cavernous sinus appears entirely normal (*curved arrow*). Repeat deep biopsies of the nasopharynx returned adenocystic carcinoma.

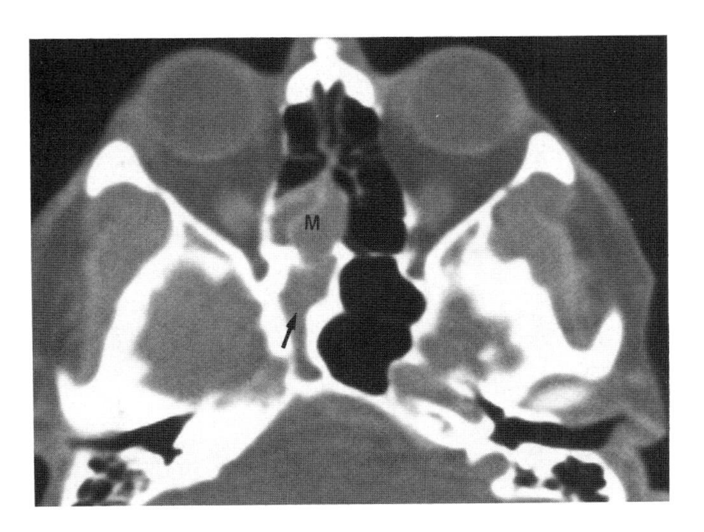

Plate 11.19. Our protocol for the evaluation of patients with atypical facial pain includes sections through the paranasal sinuses. Of course this is quite convenient since the sinuses lie immediately adjacent to the course of the divisions of the trigeminal nerve and, therefore, are routinely included in this anatomic region of interest. This patient had chronic progressive facial pain in the distribution of the second division of the trigeminal nerve. The CT study revealed a posterior ethmoid mucocele and secondary obstruction of the sphenoid sinus (*arrow*). One must take care not to suggest that benign mucoperiosteal thickening commonly seen in the sinuses during CT explains patients' chronic and progressive facial pain symptoms. A mass lesion such as this obviously can.

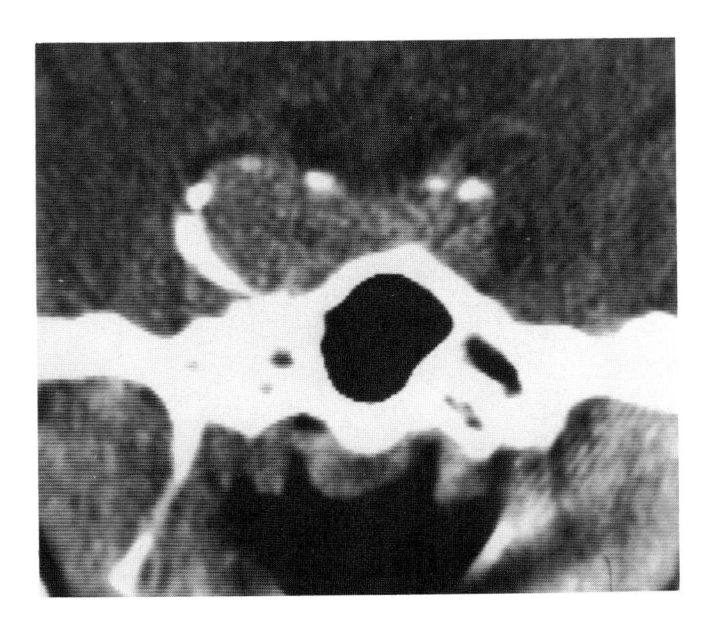

Plate 11.20. In the search for structural explanations of pain of trigeminal origins, other lesions such as this cavernous carotid aneurysm may be discovered. When the first division of the trigeminal nerve is symtomatic, either alone or in combination with the second and third divisions, the lesion is likely to be in the cavernous sinus. Isolated second or third division symptoms may be within the cavernous sinus or anywhere along those two nerves.

The abbreviation used on Plate 11.23 is: *PG*, parotid gland.

Plate 11.21. A patient with trigeminal neuralgia or hemifacial spasm will sometimes have CT studies showing tortuosity of the vertebral basillar system which might be an indication of vascular compression of one or both of these nerves. CT may suggest this diagnosis but is certainly not indicative of it and angiography even has limitations in making this diagnosis with certainty.

11.21*A*. A section through the cerebellopontine angle shows a tortuous vertebral basillar system extending into the angle on the patient's symptomatic side (*arrows*).

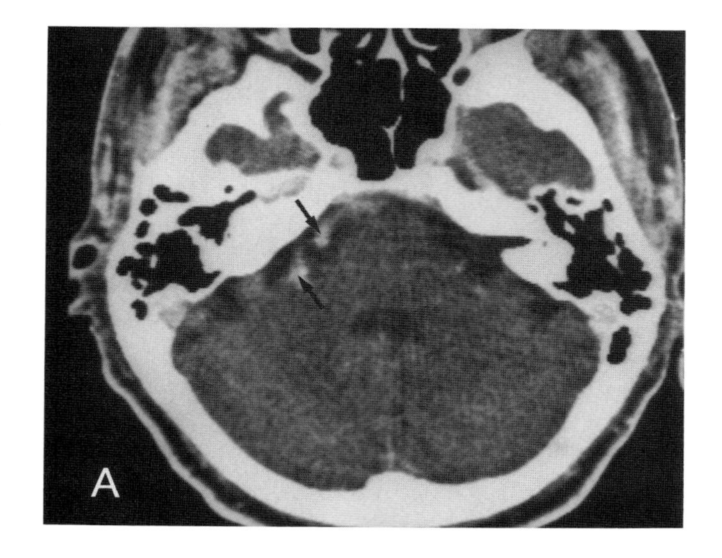

Plate 11.21*B*. A section somewhat higher than part *A* again shows looping vessels within the region of the cerebellopontine angle and certainly near the undersurface of the trigeminal nerve (*arrow*). It was elected not to operate on this patient at the time the study was completed.

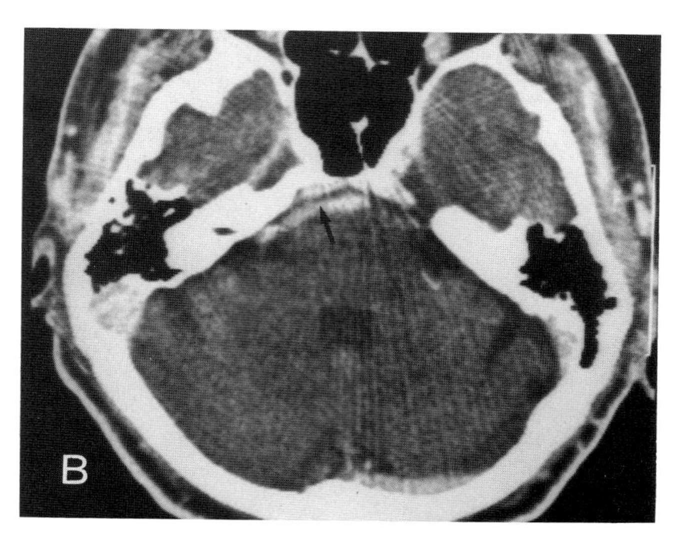

Plate 11.22*A*. This patient also had chronic progressive trigeminal nerve neuralgia. CT study with contrast revealed a very dense lesion in the region of Meckel's cave (*arrow*).

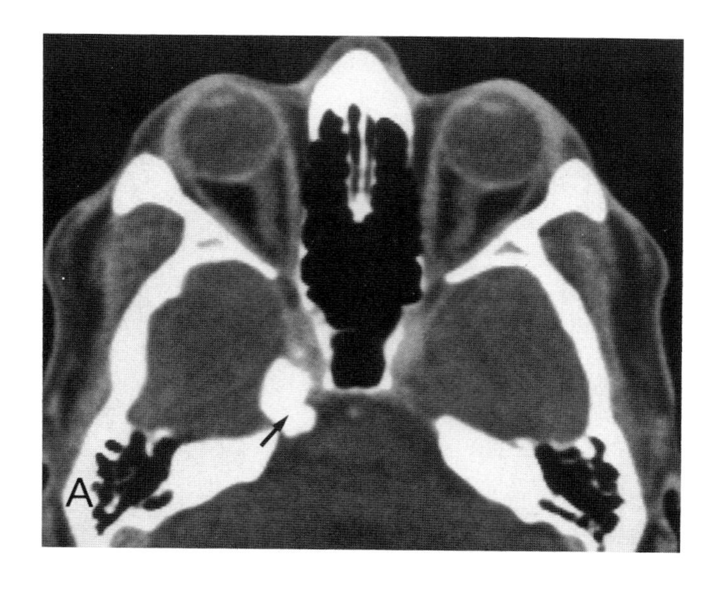

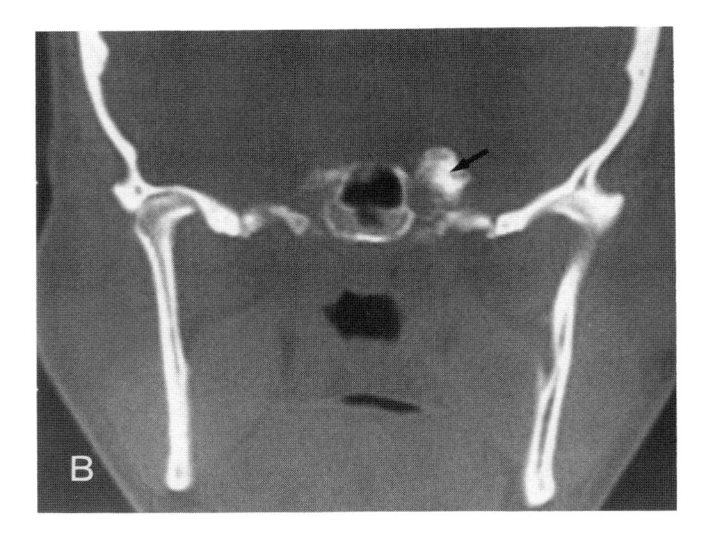

Plate 11.22*B*. A coronal section with bone windows shows this to represent an osteoma of the petrous apex (*arrow*).

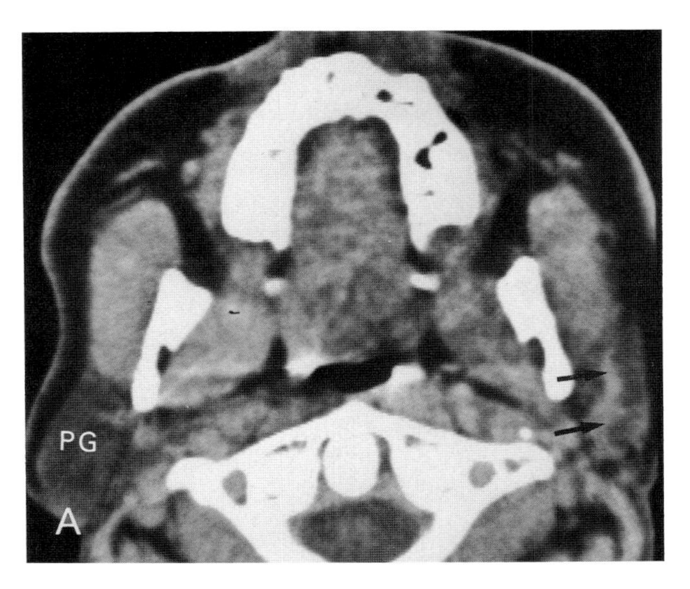

Plate 11.23. This patient had a 2-yr history which began with a chronic progressive loss of the facial nerve on the left side. This was at first believed to be due to a Bell's palsy; however, the history was that of a slow dimunition of facial nerve function. The patient then developed atypical and progressive pain in the distribution of the third division of the trigeminal nerve. Two prior CT studies stopped at the skull base since there were no abnormalities within the brain or bony portion of the skull base. The patient had another CT study which was extended below the skull base.

Plate 11.23*A*. There is a serpiginous density (*arrows*) within the parotid gland. This is definitely abnormal especially when compared to the normal parotid gland on the opposite side.

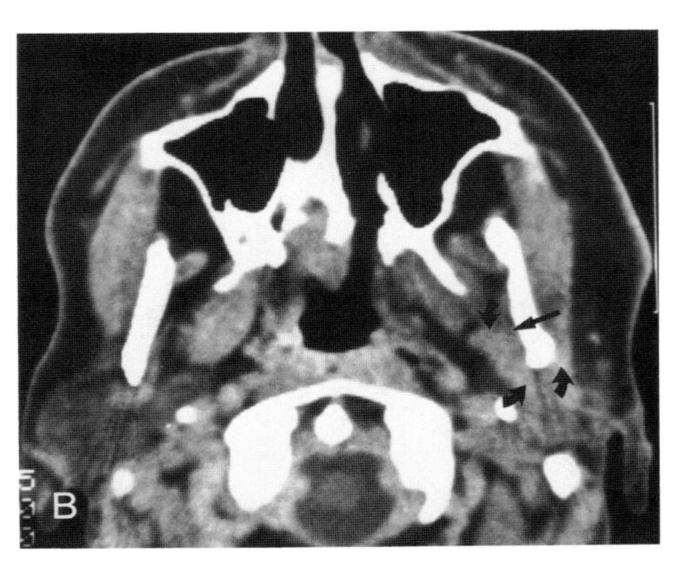

Plate 11.23*B*. A section through the upper parotid bed shows that the deep lobe of the parotid as well as the more superficial portion (*curved arrows*) is completely encased with tumor. It is from this position (*arrow*) that the tumor can gain access to the second division of the Vth cranial nerve. The nerve runs just deep to the mandible at this point before it enters the mandibular foramen.

This patient again illustrates the importance of extending the examination of the lower cranial nerves beyond the posterior fossa and the skull base.

Plate 11.24. Because of its multiplanar capabilities and superior contrast resolution in the posterior fossa as well as the soft tissues just below the skull base and in the parotid region, MRI is very likely to replace CT as the first examination of choice in patients who require evaluation of the types of cranial nerve deficits described in the previous cases.

Plate 11.24*A*. This patient presented with a number of cranial nerve deficits and a right hemiparesis. This spin echo 500/30 image shows a high intensity lesion in the lower pons. It was done on a 0.15 Tesla resistive unit.

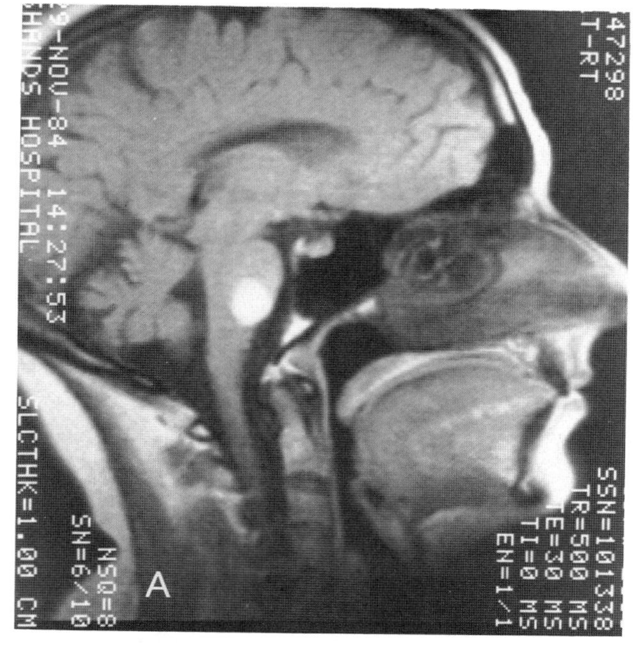

Plate 11.24*B*. A spin echo 2500/90 pulse sequence shows that the lesion remains of high signal intensity. The findings are most indicative of a hemorrhage. It was believed to be due to an occult arteriovenous malformation although this was not proven. The signal intensities on these part *A* and *B* images indicate that the lesion was a short T_1 and a long T_2. An intraaxial fat-containing lesion would be extremely unusual and there was also an abrupt onset of symptoms. This case is used to illustrate the MRI depiction of intraaxial brain stem lesions which is far superior to that of CT and one of major benefits of MRI in the posterior fossa.

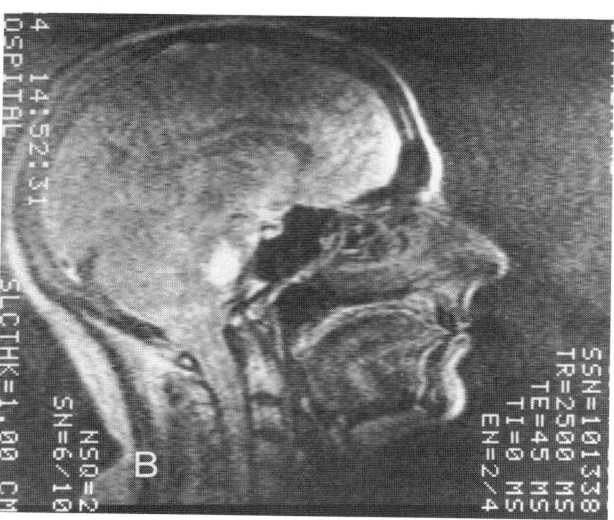

Plate 11.24*C*. This patient presented with hoarseness and pharyngeal dysfunction indicative of deficits in cranial nerves IX and X. CT study was reportedly normal. The MRI showed a mass in the cerebellopontine angle (*arrow*). This was somewhat lower than the usual level for an acoustic neuroma and was felt to possibly represent a meningioma or a neuroma of one of the lower cranial nerves.

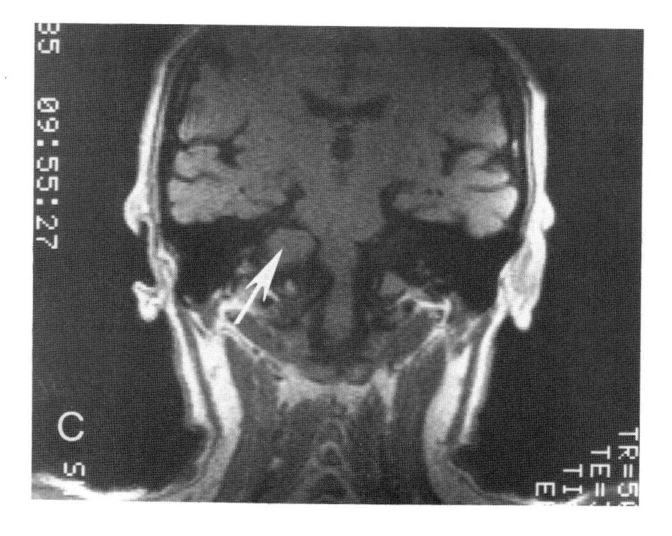

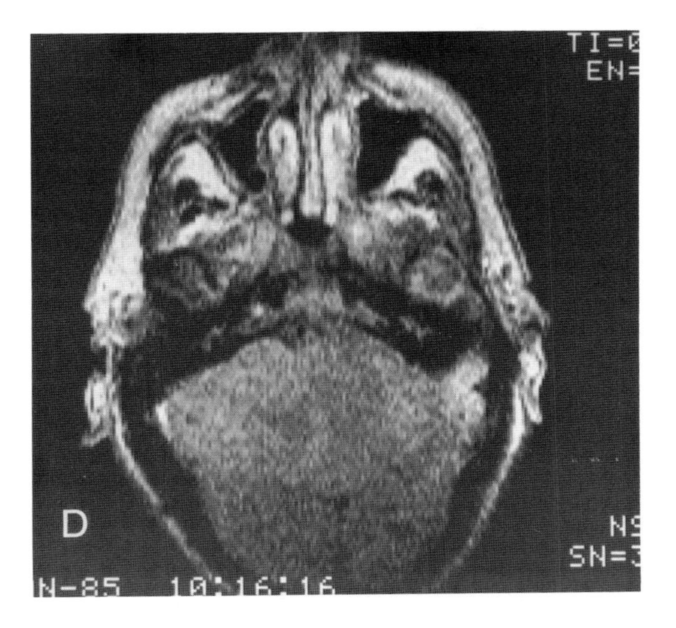

Plate 11.24*D*. A spin echo 2000/90 pulse sequence shows that the lesion was basically isointense with surrounding brain and CSF. The lesion remained isointense with brain at multiple pulse sequences. This behavior is known to occur in neuromas and meningiomas. This points out a limitation of MRI and that the confident diagnosis of more subtle neuromas and meningiomas will require the injection of an intravenous paramagnetic compound. This may not be true in intracanalicular acoustics since the apparent contrast between the lesions and surrounding bone and CSF is so good, even without contrast enhancement.

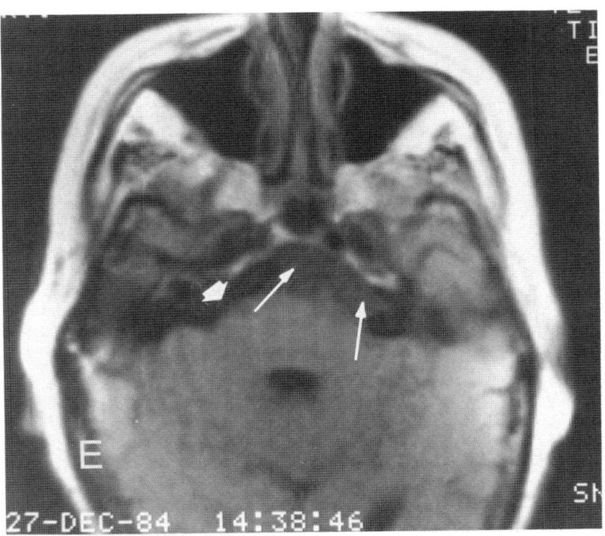

Plate 11.24*E*. Other lesions in the cerebellopontine angle show fairly characteristic appearances on MRI without the need for contrast infusion. This spin echo 500/30 image through the posterior fossa shows a mass in the prepontine and cerebellopontine angle cisterns (*arrows*) which is slightly more intense than the CSF (*arrowhead*).

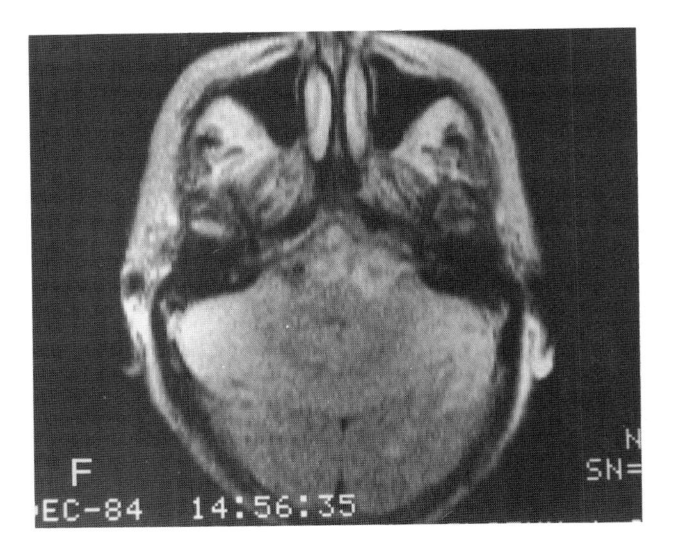

Plate 11.24*F*. A spin echo 2000/90 sequence shows that the area has a relatively prolonged T_2. This appearance is most compatible with an epidermoid tumor in the CP angle.

This series of patients with MRI studies of the posterior fossa points out the strengths of MRI in the overall imaging of patients with cranial nerve deficits that may be due to either intra- or extracranial lesions. With more experience and the advent of intravenous paramagnetic contrast agents, MRI will undoubtedly become the first examination of choice for examining most of these patients. In addition to what has been demonstrated here, MRI will prove much more sensitive for detecting small areas of demyelination or brain stem infarction which will explain symptoms that might otherwise go undiagnosed.

Plate 11.25. Lesions other than squamous cell carcinoma (either keratinizing or nonkeratinizing) will involve the nasopharynx. Squamous cell carcinoma, however, accounts for 90–95% of the malignancies found in this area.

Plate 11.25*A*. This child has an infiltrating mass in the nasopharynx. Note that the mass produces an obvious mucosal bulge but it also has an extensive amount of deep infiltration.

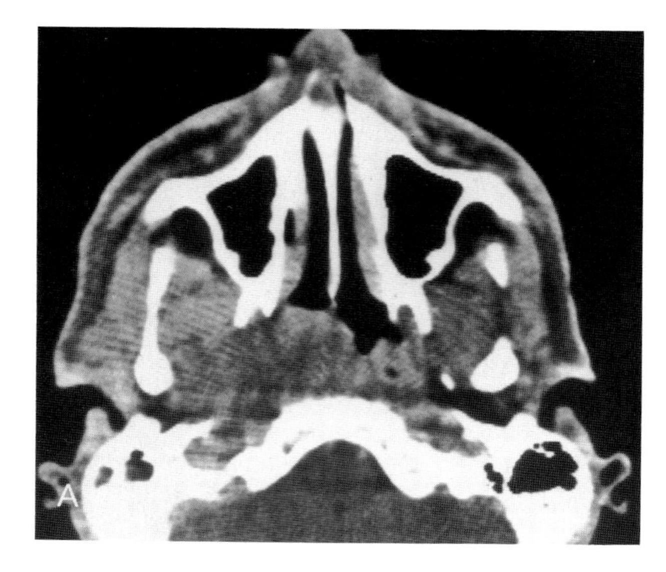

Plate 11.25*B*. The mass not only extensively invades the parapharyngeal space and nasopharynx, it also spreads to involve most of the volume of the infratemporal fossa. The lesion also extends directly posteriorly to involve the poststyloid parapharyngeal space. This is a rhabdomyosarcoma and is the most common malignant lesion in the nasopharynx in the pediatric age group.

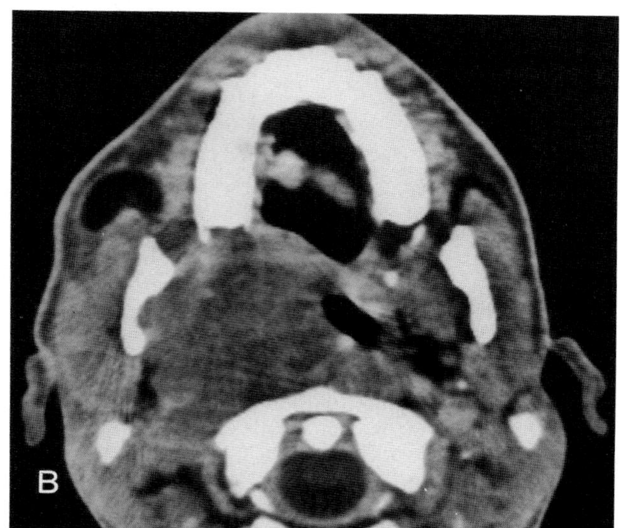

Plate 11.26. Sometimes masses in the infratemporal fossa and, more rarely, in the nasopharynx show a specific morphology that helps with tissue diagnosis.

Plate 11.26*A*. This section through the high infratemporal fossa shows a complete replacement of the normal tissue planes with an infiltrating mass. In the center of the mass is an obvious area of calcification. The calcification strongly suggests a matrix-producing tumor, although, from its appearance, this could be either osteoid or chondroid matrix calcification.

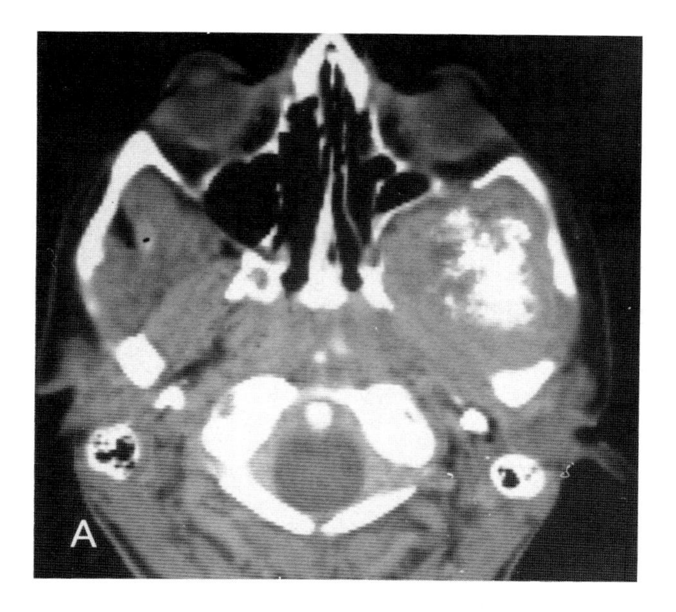

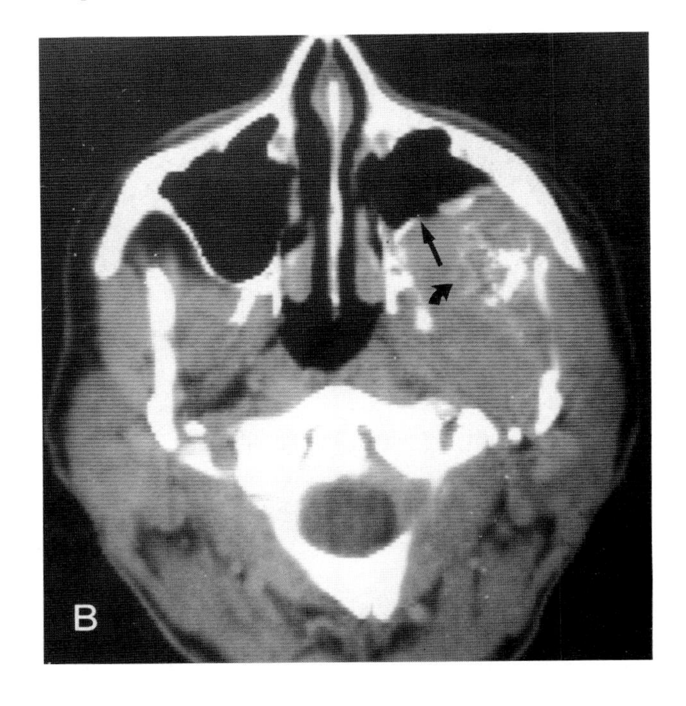

Plate 11.26*B*. This shows a rather smooth displacement of the posterior wall of the maxillary antrum anteriorly (*arrow*), illustrating that such displacement occurs in lesions other than juvenile angiofibromas. Again the soft tissues within the infratemporal fossa are completely infiltrated by the lesion. At this point the calcified matrix within the lesion has a ringed and curvilinear appearance (*curved arrow*) indicating that the lesion is at least partially producing a chondroid matrix. This was a chondrosarcoma primary in the infratemporal fossa.

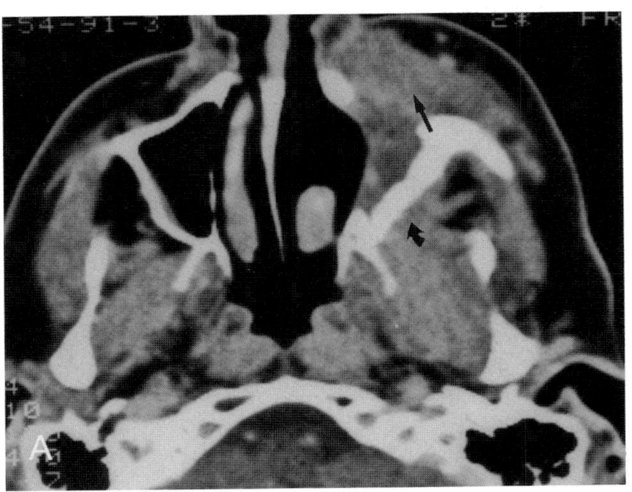

Plate 11.27. Lymphoma probably presents the most varied appearance of all the malignancies which affect the paranasal sinuses and upper aerodigestive tract. It is not a very common malignancy in this region, but is frequent enough to be included in differential diagnosis, especially when a curious pattern of spread is seen.

Plate 11.27*A*. This patient had a Caldwell-Luc and nasoantral window on the left side and the histologic diagnosis suggested lethal midline granuloma. This section was obtained well after the biopsy procedure so that the soft tissue changes in front of the antrum (*arrow*) are related to infiltrating tumor, not postbiopsy changes. What was not apparent was that the mass lesion had also infiltrated around the posterior aspect of the antrum and into the infratemporal fossa (*curved arrow*). Also note that there is no disease in the nasal cavity.

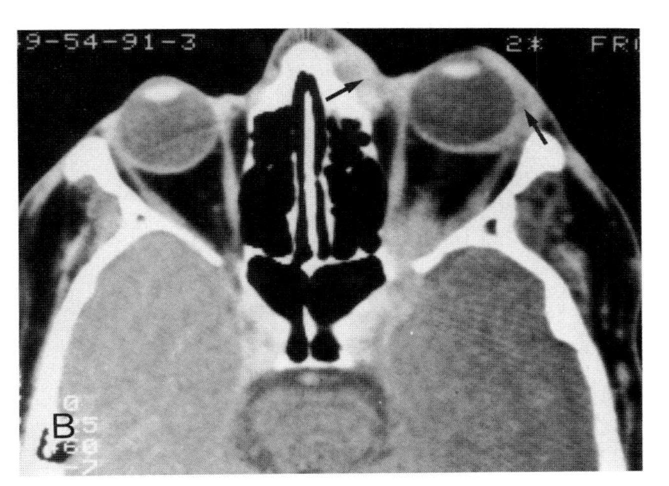

Plate 11.27*B*. A higher section than part *A* again shows no evidence of involvement of the other paranasal sinuses or upper nasal cavity. There is a mass at the orbital apex producing obvious proptosis. There is also a thickening of the periorbital soft tissue (*arrows*).

Taken together, these findings are certainly not compatible with any of the pleomorphic reticuloses. Considering the unusual infiltrating pattern and the suggested pathologic diagnosis the most likely diagnosis became lymphoma. Rebiopsy revealed this to be lymphoma.

The abbreviation used on Plate 11.28 is: *N*, node.

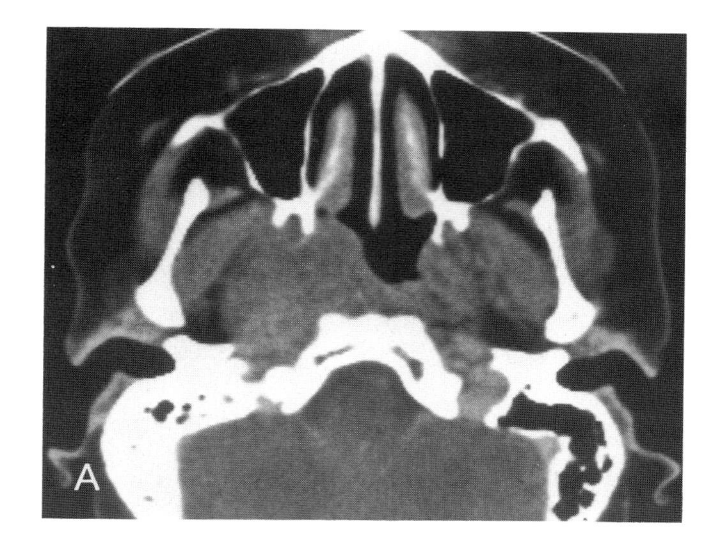

Plate 11.28. This patient presented with cervical lymphadenopathy and on examination was found to have a mass in the nasopharynx.

Plate 11.28*A*. There is an infiltrating mass in the nasopharynx, and again note the difficulty in deciding how much of the mass is related to the primary tumor and how much to retropharyngeal nodes. This was discussed in detail in Plate 11.14.

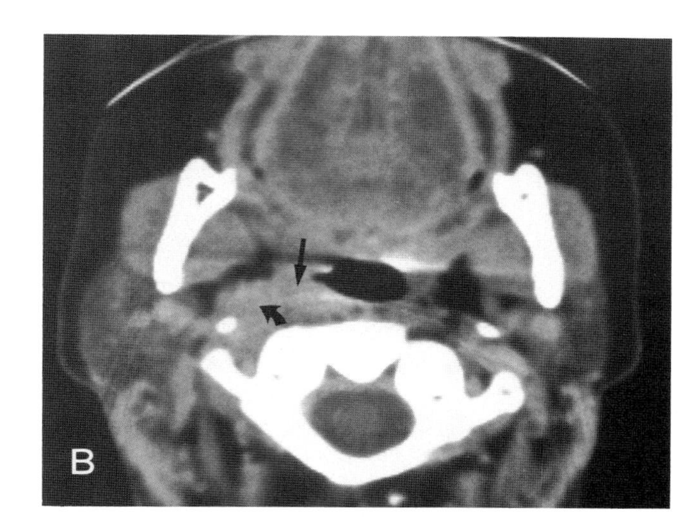

Plate 11.28*B*. A section through the upper oropharynx shows a slight asymmetry in the soft tissues surrounding the airway (*arrow*); this is within normal limits. There is an obviously enlarged, non-necrotic retropharyngeal node (*curved arrow.*).

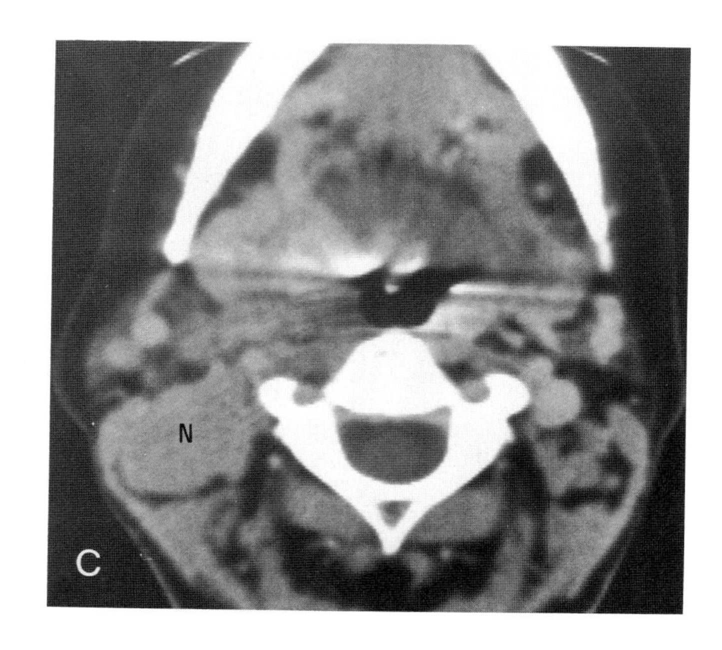

Plate 11.28*C*. A section through the upper neck shows enlarged nodes which might be spinal accessory or posterior internal jugular nodes. Note that there is no evidence of necrosis in the nodal mass here or in the retropharyngeal area. The morphology of the lesion is then compatible with lymphoma or nonkeratinizing squamous cell carcinoma.

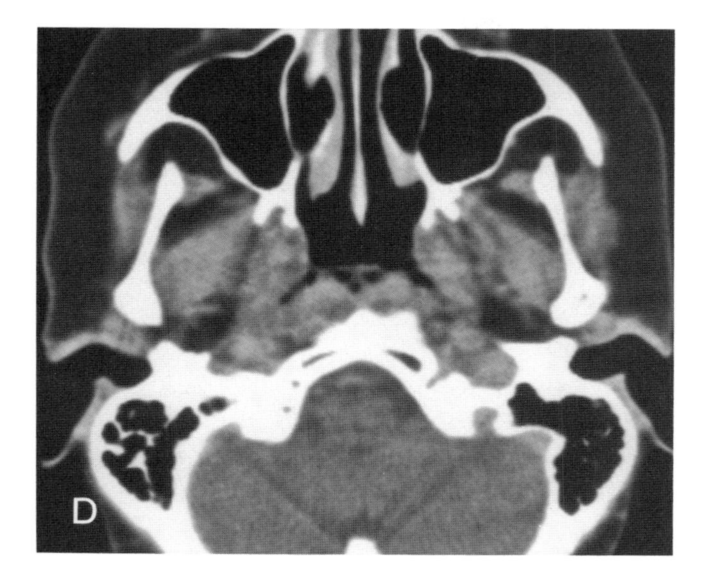

Plate 11.28*D*. The following three sections were obtained after radiation therapy and are approximately at the same levels as the study done before therapy. Note the marked interval reduction in size of the mass which was infiltrating the parapharyngeal space. There is still some soft tissue thickening present; however, the tissue planes are nearly symmetric.

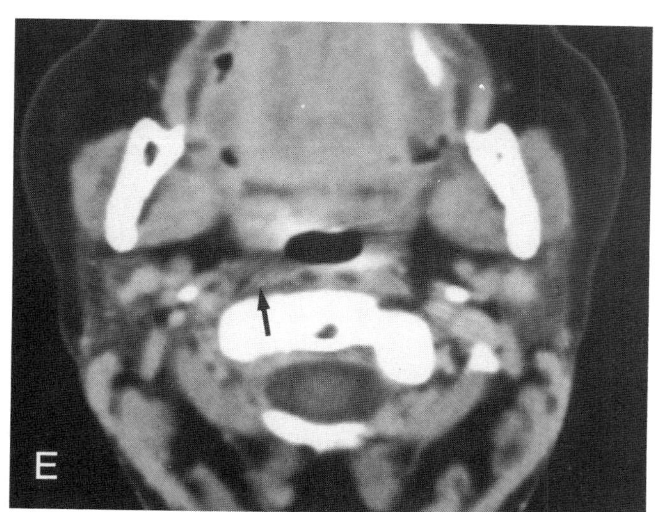

Plate 11.28*E*. At the level of the oropharynx, the retropharyngeal node has disappeared completely (*arrow*) although there is some indistinctness between the prevertebral muscles and carotid at that level.

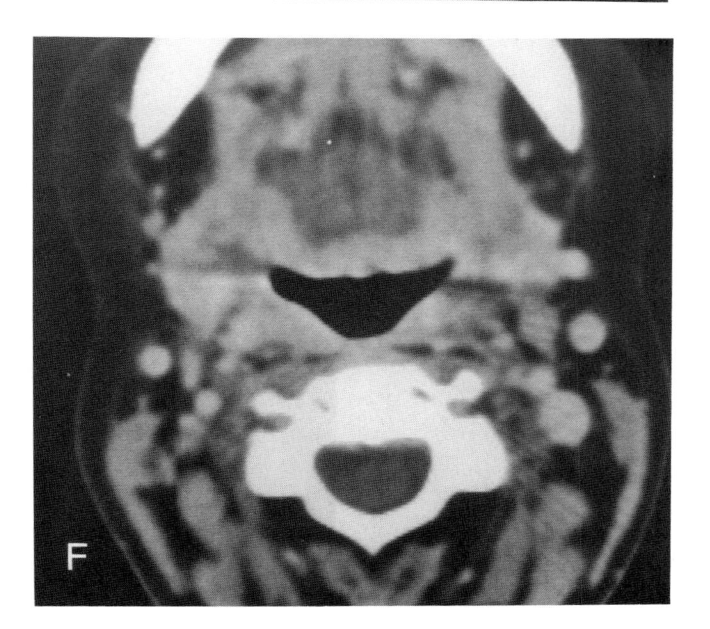

Plate 11.28*F*. The large node seen in the upper neck has also disappeared completely.

CT and/or MRI may be used to follow the effects of therapy. Lymphoma tends to leave very little evidence of residual fibrosis following successful therapy. Some tumors may leave more residual, especially if there was a significant inflammatory component or reactive fibrosis to begin with. Baseline studies may be obtained if one wishes to follow these patients who are at a high risk for recurrence and if there is a reasonable salvage treatment available.

The abbreviations used on Plates 11.29 and 11.30 are: *S*, styloid process; and *ICA*, internal carotid artery.

Plate 11.29. These are serial examinations in a patient under treatment for squamous cell carcinoma of the nasopharynx. All of the sections were made through the mid-nasopharynx and are at nearly the same level.

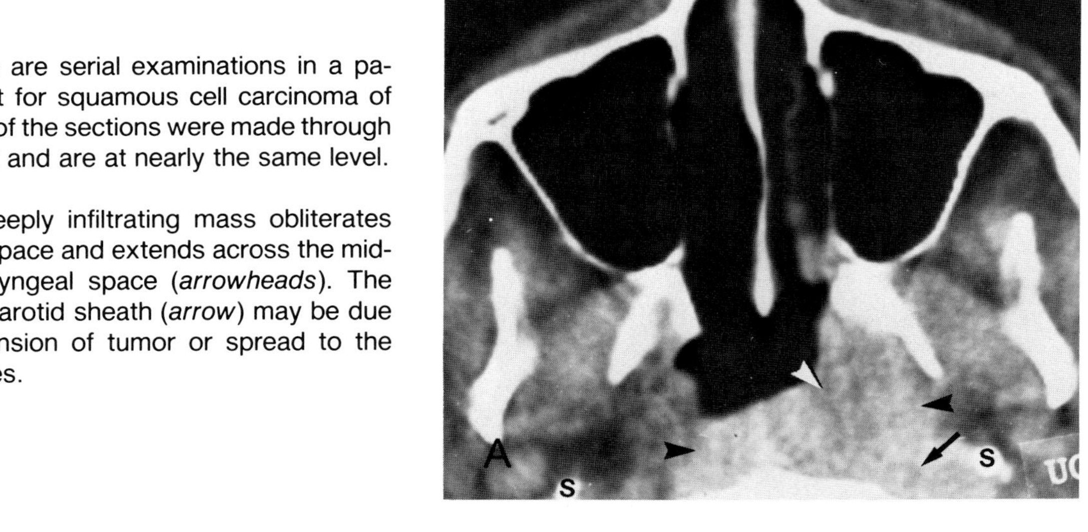

Plate 11.29*A*. A deeply infiltrating mass obliterates the parapharyngeal space and extends across the midline in the retropharyngeal space (*arrowheads*). The fullness around the carotid sheath (*arrow*) may be due either to direct extension of tumor or spread to the retropharyngeal nodes.

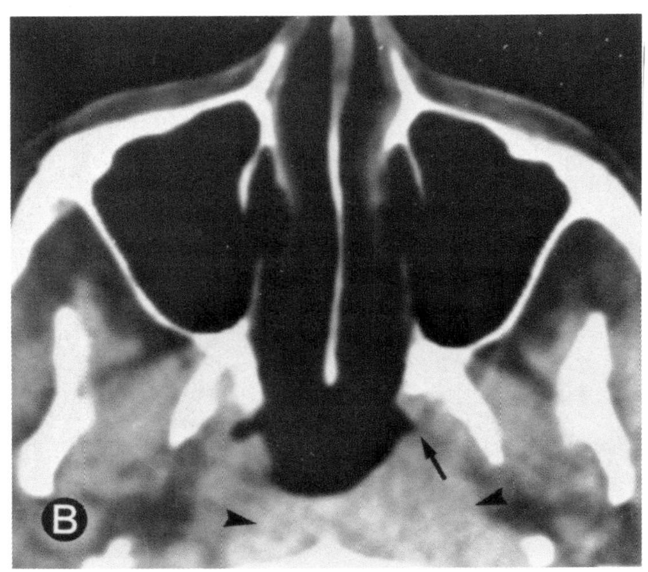

Plate 11.29*B*. A study 3 months later shows a very marked interval reduction in the size of the mass. Note that the eustachian tube orifice has now opened (*arrow*) and the overall size of the mass in the retropharyngeal and poststyloid parapharyngeal space has diminished (*arrowheads*).

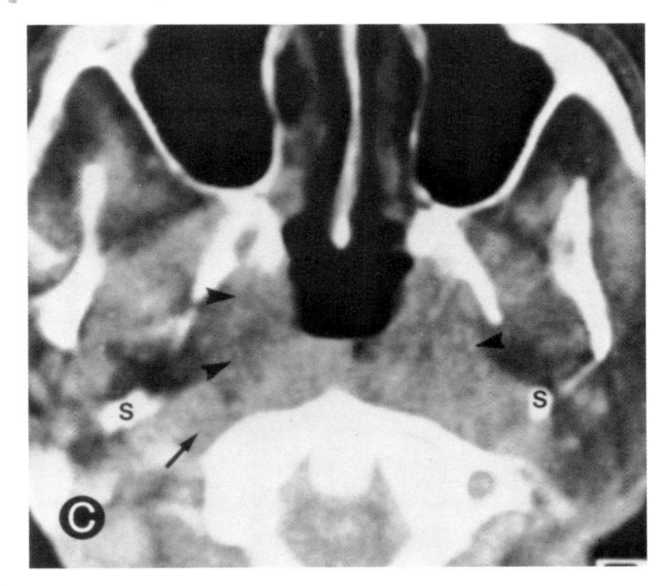

Plate 11.29*C*. This was obtained approximately 6 months after the image seen in part *B* or about a total of a year after the patient originally presented. Despite there being no clinical evidence of spread to the contralateral side, the scan shows that there has been progression of tumor, now with obliteration of the parapharyngeal spaces bilaterally (*arrowheads*). In addition there is fullness in the region of the retropharyngeal nodes bilaterally (*arrow*).

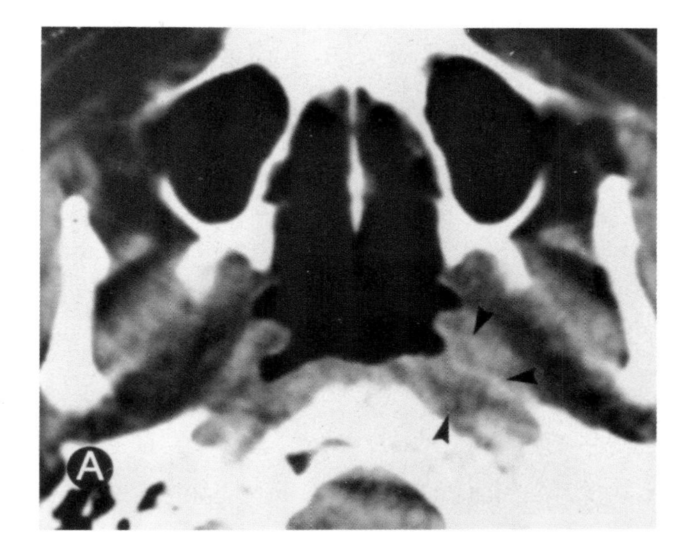

Plate 11.30. This Oriental male had been treated with radiation therapy for carcinoma of the nasopharynx. He presented with a neck node. Endoscopic examination of the nasopharynx under anesthesia was entirely normal.

Plate 11.30*A*. The CT through the mid- to upper nasopharynx showed an infiltrating mass deep to the fossa of Rosenmüller (*arrowheads*).

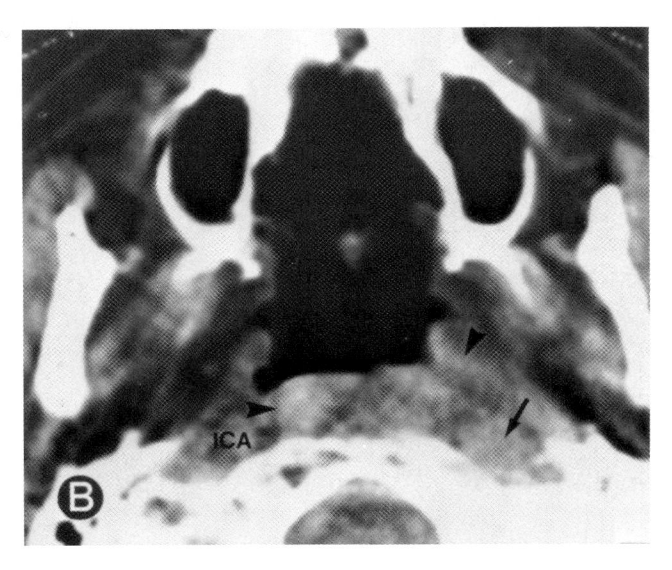

Plate 11.30*B*. A section approximately 5 mm inferior shows an unequivocal infiltrating mass (*arrowheads*) with, again, fullness in the poststyloid retropharyngeal space (*arrow*). The findings were believed to clearly indicate persistent or recurrent tumor at the primary site. Repeat endoscopy under anesthesia showed no evidence of a mucosal lesion but biopsy directed to the site indicated on the CT study revealed recurrent carcinoma of the nasopharynx.

This and the previous case illustrate the occurrence of a phenomenon well-known to radiotherapists for radiotherapy to sometimes "drive tumor underground" in the upper aerodigestive tract. This mass could have represented residual fibrosis and sometimes it is really difficult to tell the difference between the two on a single study. MRI shows some promise for differentiating end stage scarring from persistent tumor although further study of this point is necessary.

The abbreviation used in Plate 11.31 is: *T*, tumor.

Plate 11.31. As stated previously, MRI might help the follow-up of patients treated with radiation therapy since it has the potential to differentiate end stage fibrosis and active inflammatory or neoplastic disease. There are some pitfalls to MRI in this regard which will be illustrated in this and the following plates.

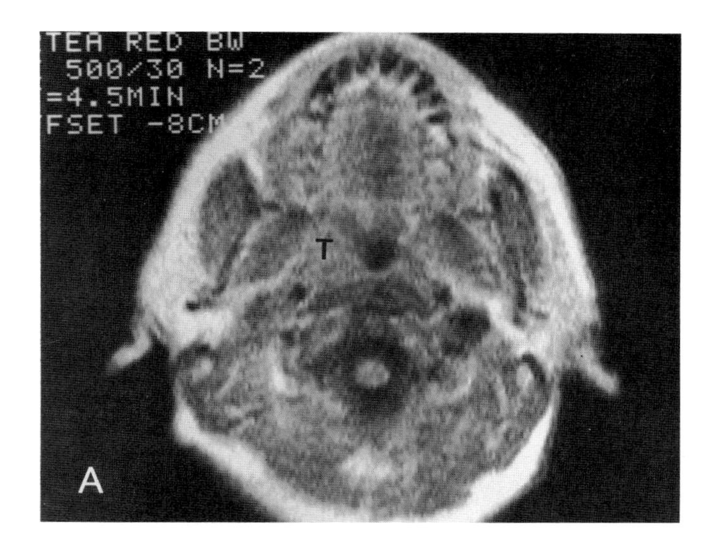

Plate 11.31*A*. This patient had large oropharyngeal carcinoma which had extended into the nasopharynx. This shows the preoperative T_1-weighted spin echo 500/30 image and there is an obvious infiltrating mass in the parapharyngeal space.

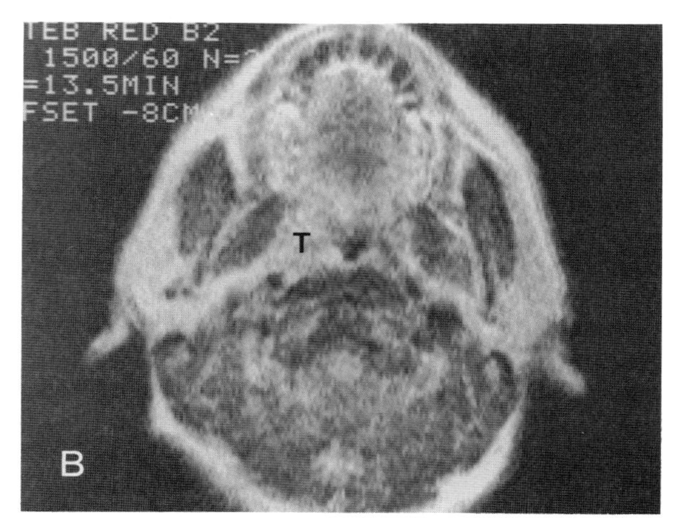

Plate 11.31*B*. A T_2-weighted sequence at the same level which shows very poor contrast between tumor and fat, both having relatively high signal intensity with this pulse sequence. Incidentally, these images were obtained before our 0.15 Tesla unit was upgraded.

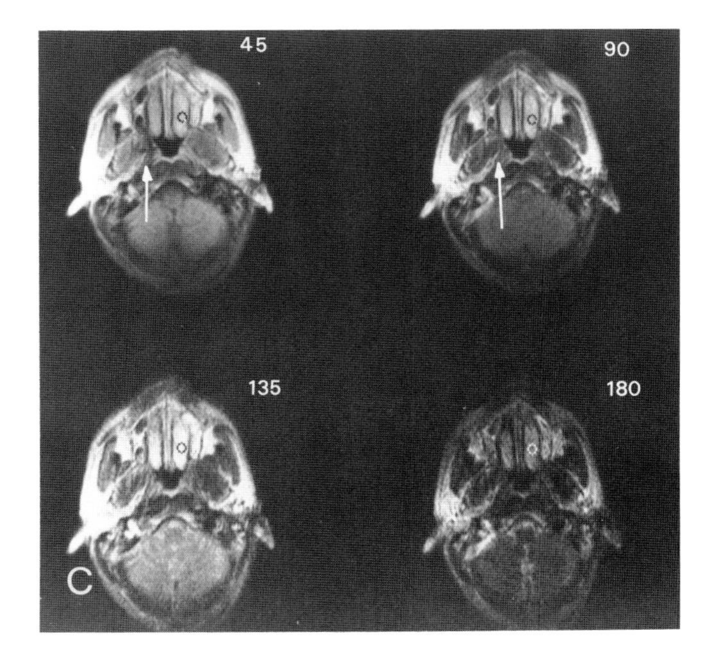

Plate 11.31*C*. A section from a multislice multiecho acquisition with a T_R of 2500 msec and T_Es indicated in the upper right hand corner of each image. Note that the prior tumor bed is now of low signal intensity on all of the pulse sequences (*arrow*). This is especially noticeable on the spin echo 2500/45. The prior tumor bed becomes isointense with muscle on the rest of the pulse sequences. This is typical of end stage fibrosis. Also note that there is some minimal superficial tissue within the airway due to inflammatory mucositis. This study was made right at the end of radiation therapy.

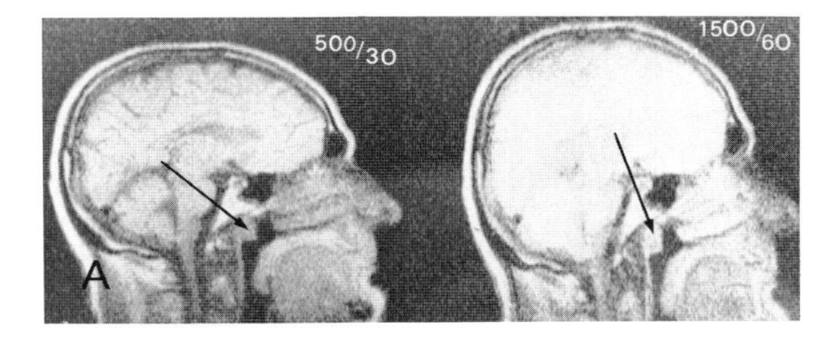

Plate 11.32. This patient was also treated with radiation therapy for a small nasopharyngeal carcinoma.

Plate 11.32*A*. Early in the treatment the mucosal lesion had a relatively high signal intensity on this spin echo 1500/60 (T_2-weighted) pulse sequence. The tumor mass (*arrows*) never really showed any evidence of deep infiltration. It was a poorly differentiated squamous cell carcinoma and there was a very large associated neck mass.

The mass was followed through therapy and diminished in both size and intensity.

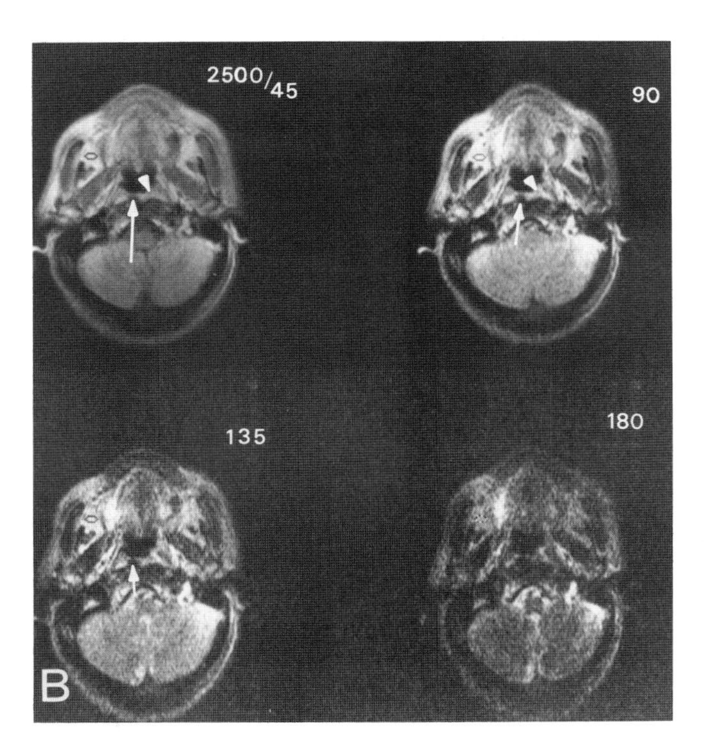

Plate 11.32*B*. At the end of therapy the original primary site showed a very low signal intensity (*arrows*). However, there was diffuse thickening of the superficial soft tissues within the nasopharynx (*arrowheads*) which showed relatively high signal intensity. This was, in part, due to postradiation and inflammatory mucositis; however, this patient had recurred in contralateral nodes and the repeat biopsies of the nasopharynx showed recurrent tumor at several locations. There was none at the original primary site.

This case and the previous one show that it is not possible to differentiate recurrent tumor from postradiation inflammatory changes on the basis of signal intensity alone. This is true on both T_1- and T_2-weighted sequences. Evidence of deep infiltration at a previously uninvolved site would certainly favor tumor over radiation changes. Much more study is necessary to be sure of the role of MRI in following patients treated for upper aerodigestive tract tumors.

The abbreviation used on Plate 11.33 is: *PS,* parapharyngeal space.

BENIGN MASSES AND INFLAMMATORY LESIONS

Plate 11.33. This case illustrates the difficulty in distinguishing between a benign deeply infiltrating process in the nasopharynx and a malignant tumor by its CT appearance alone. This patient was diabetic and presented with progressive deficits of cranial nerves IX through XII. She also had internal carotid artery occlusion but no evidence on CT study of direct intracranial extension of this invasive mucormycosis of the nasopharynx.

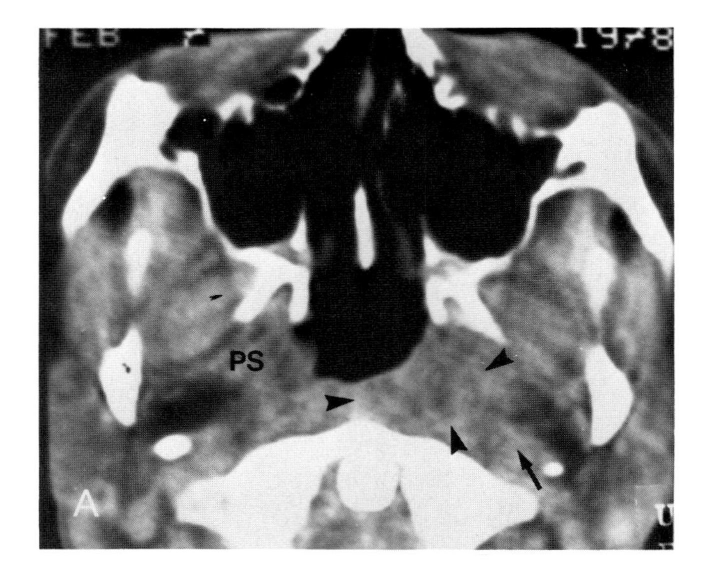

Plate 11.33*A*. A deeply infiltrating mass (*arrowheads*) obliterates the parapharyngeal space and obscures the carotid sheath region (*arrow*). The appearance is identical to an infiltrating carcinoma.

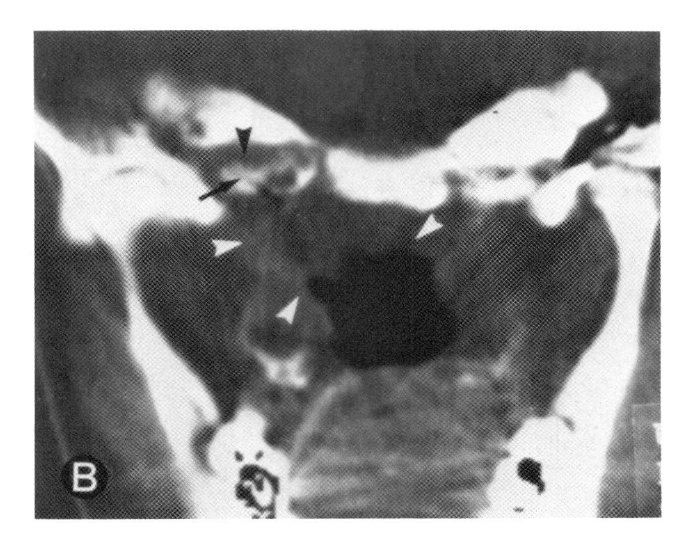

Plate 11.33*B*. A coronal section shows that the inferior petrous apex (*arrow*) had become a sequestrum, being completely surrounded by the inflammatory mass (*arrowheads*). The position of the mass explained all of the cranial nerve deficits as well as the carotid occlusion noted clinically.

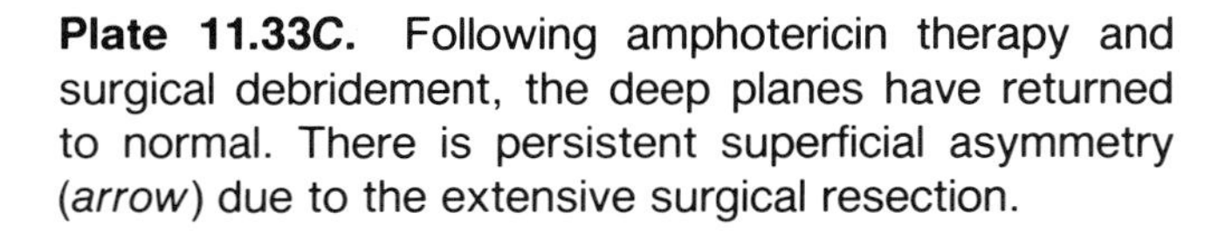
Plate 11.33*C*. Following amphotericin therapy and surgical debridement, the deep planes have returned to normal. There is persistent superficial asymmetry (*arrow*) due to the extensive surgical resection.

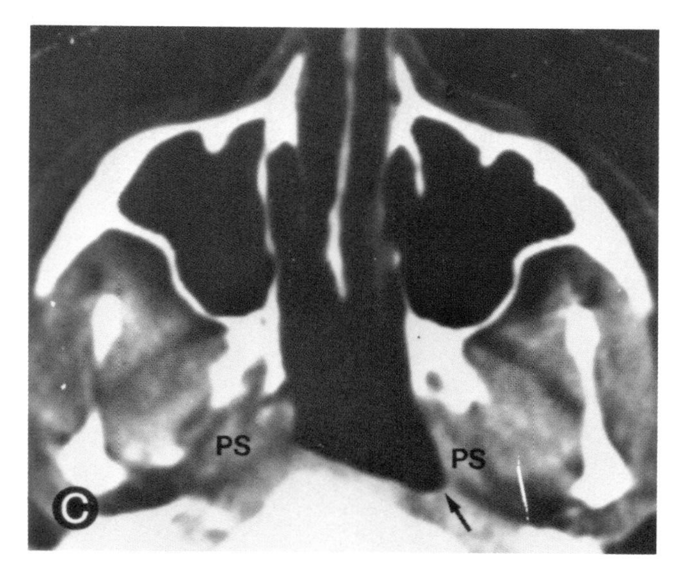

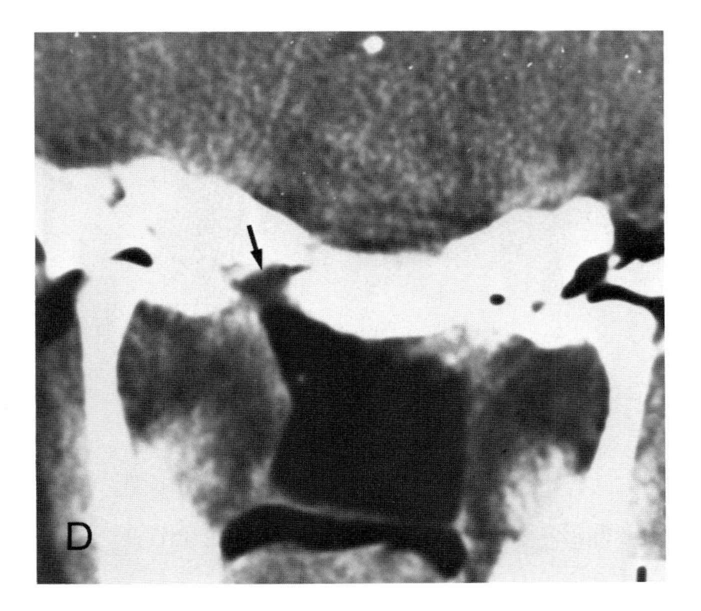

Plate 11.33*D*. A coronal section shows again that the deep planes all appear normal. The inferior petrous apex was removed as part of the surgical debridement and created the defect in the skull base (*arrow*). CT was used to follow this patient for several years to be sure the infection did not recur. This patient was disease free at the time of a 3-yr follow-up.

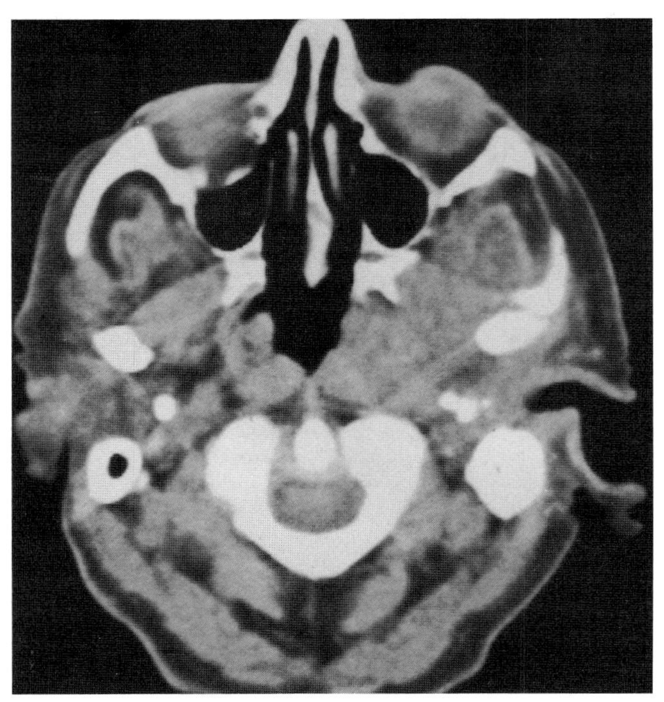

Plate 11.34. Few other inflammatory lesions involve the infratemporal fossa and nasopharynx. The other one most likely to mimic a malignancy from its radiographic appearance alone is malignant otitis externa which has spread from the external auditory canal along the skull base and extended into the soft tissues below.

This patient has malignant otitis externa. Note the obliteration of the tissue planes in the infratemporal fossa and both prestyloid and poststyloid parapharyngeal space. The changes are indistinguishable from that of a malignant neoplasm and the appropriate clinical history is necessary for proper interpretation.

The abbreviations used on Plates 11.35–11.37 are: *C*, carotid artery; *S*, styloid process; and *M*, mandible.

BENIGN MASSES AND PARAPHARYNGEAL SPACE

Plate 11.35. Benign tumors in the parapharyngeal space usually present as submucosal masses within the oropharynx. They may also present as a mass in the parotid region. CT and MRI have much to offer in the evaluation of these lesions in that they can usually determine the etiology and help to decide which surgical approach is most appropriate for diagnosis and treatment.

Plate 11.35*A*. This patient presented with a bulging pharyngeal wall seen on inspection of the oropharynx. Note that the deep lobe of the parotid gland (*arrow*) appears entirely normal. It also appears as if there may be a cleavage plane between the mass and the lymphoid and muscular tissue of the pharyngeal wall (*curved arrows*). The mass appears to be arising anteriorly and medial to the poststyloid parapharyngeal space since the carotid and jugular can be clearly seen and the carotid artery may be slightly displaced posteriorly and laterally.

Plate 11.35*B*. A section through the upper oropharynx shows that the mass enhances slightly but not nearly to the extent of the surrounding vessels. Again the mass is entirely separate from the parotid gland and occupies the parapharyngeal space. The slight enhancement, relationship to the poststyloid parapharyngeal space, and its being separate from the deep lobe of the parotid virtually assures that this lesion is arising primarily in the parapharyngeal space and most likely represents a benign mixed tumor arising from accessory salivary tissue in this space. Other lesions may mimic this appearance but most of the time one will be dealing with a benign mixed tumor of nonparotid origin with the CT findings as described here.

Plate 11.36. Another patient with a benign mixed tumor arising in the parapharyngeal space.

Plate 11.36*A*. This coronal section shows that the bulk of the mass is within the parapharyngeal space and is bridging the nasopharynx and oropharynx (*arrows*). This lesion shows an extremely unusual manifestation of these tumors; i.e. diffuse calcification in its most lateral and superior extremity (*curved arrow*). Because of this, some other differential diagnostic possibilities were considered such as a matrix-producing tumor or a dermoid. This was a benign mixed tumor and calcification such as this is extremely rare in these lesions.

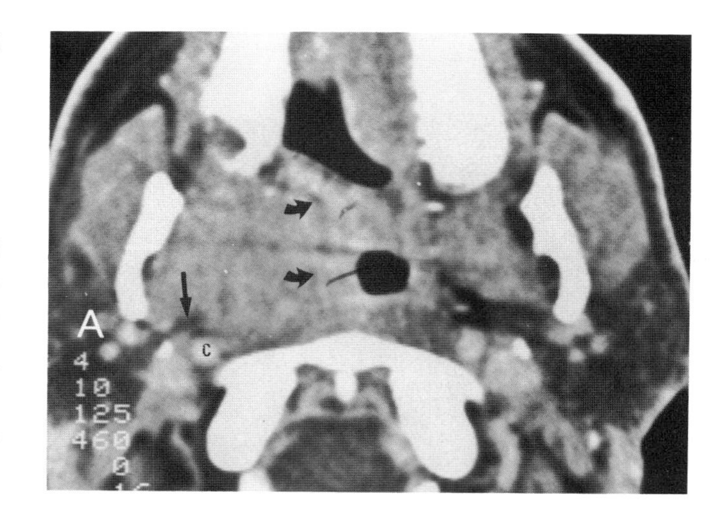

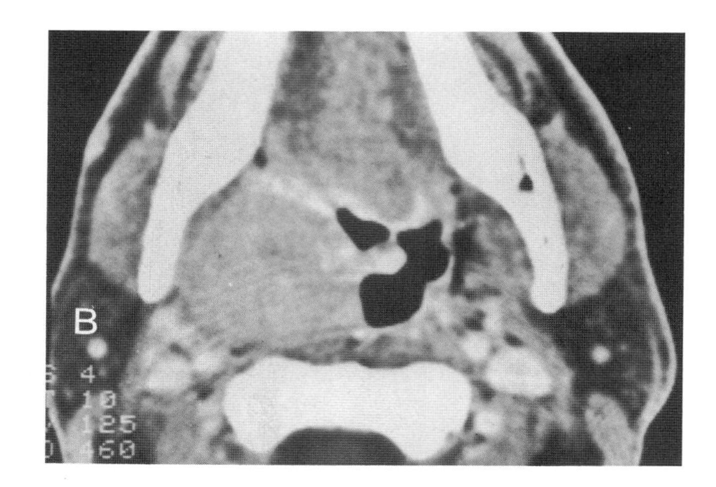

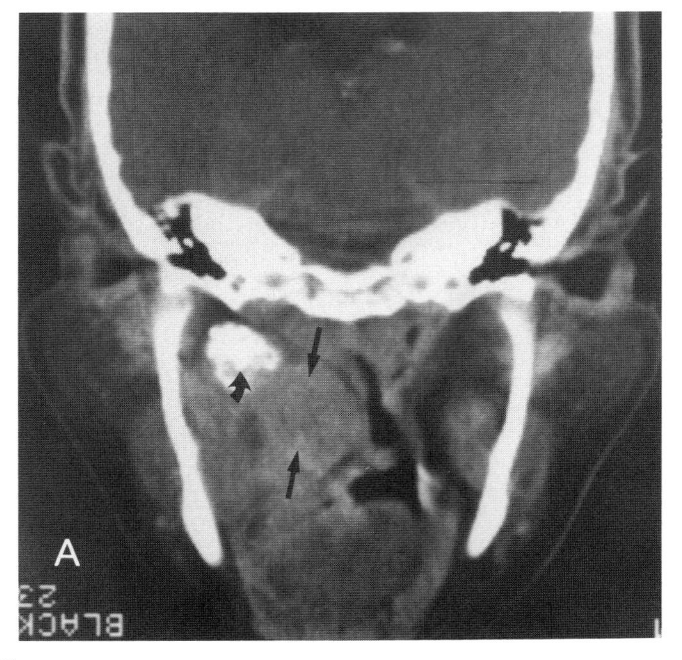

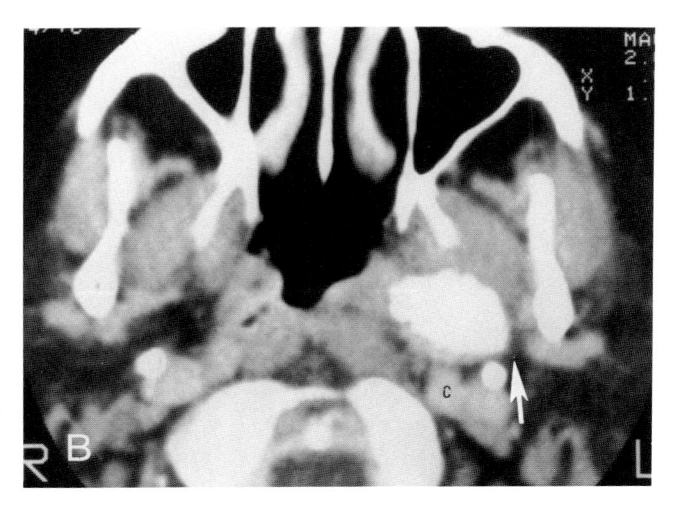

Plate 11.36*B*. This axial section is taken through the upper portion of the mass. Note again the dense calcification. Also note the typical location in the prestyloid parapharyngeal space. The mass is obviously separate from the deep lobe of the parotid gland (*arrow*). It is easy to see how as this lesion enlarges it will displace the carotid artery posteriorly and laterally. This differs from the vector of displacement seen in neuromas which arise around the carotid sheath in this region. (This case appears courtesy of The Radiology Department, University Hospital, Vancouver, British Columbia.).

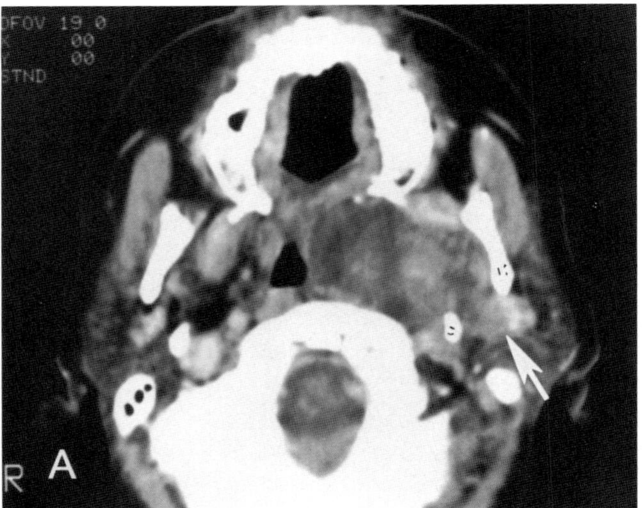

Plate 11.37. Sometimes parotid lesions will present as parapharyngeal masses and parapharyngeal masses will present like a tumor arising in the parotid gland. CT or MRI can be used to good advantage to decide which is the case. This makes a drastic change in the surgical approach to such lesions.

Plate 11.37*A*. This patient has a large parapharyngeal mass which was a benign mixed tumor. Note how the mass extends through the stylomandibular space. The deep lobe of the parotid has clearly been involved with this lesion that began outside of the parotid gland and is now growing into the superficial portion of the gland (*arrow*).

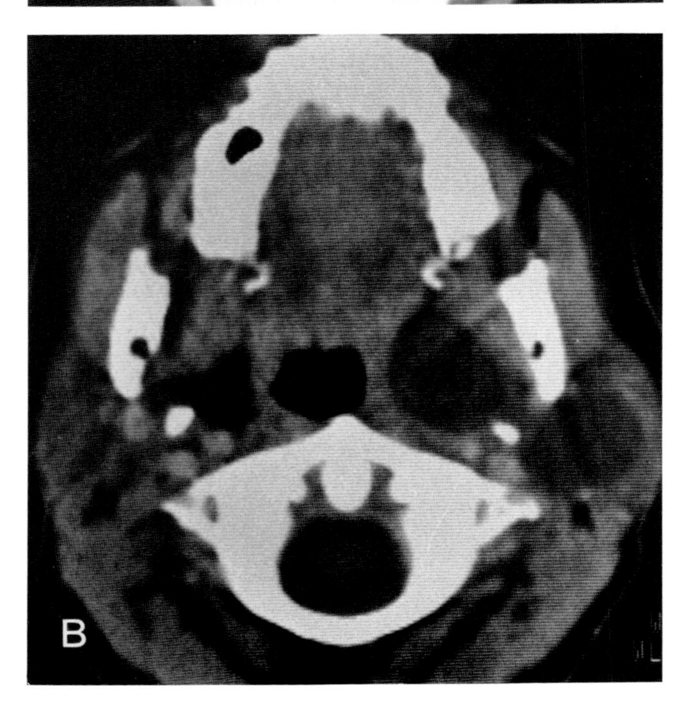

Plate 11.37*B*. This facial neuroma grew from the superficial part of the gland into the parapharyngeal space. (Courtesy of H. R. Harnsberger, M.D., University of Utah.)

The abbreviations used on Plate 11.38 are: *S*, styloid process; and *M*, mandible.

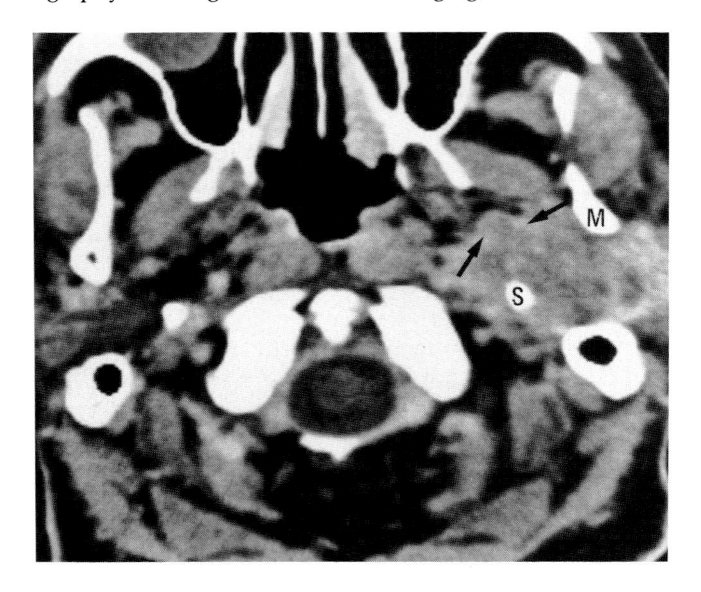

Plate 11.38. This patient presented with a mass in the parotid region. It was unclear whether this had begun in the parapharyngeal space or in the parotid gland by physical examination. Fine needle aspiration biopsy, however, revealed mucoepidermoid carcinoma.

This section through the midnasopharynx shows an obvious infiltrating mass occupying the parotid gland and extending through the stylomandibular space. Note how the entire deep lobe of the parotid gland is replaced by tumor and from here the tumor has continued to grow into the prestyloid parapharyngeal space (*arrows*). Also note that the poststyloid parapharyngeal space appears to be involved and carotid fixation was likely. On the basis of the needle aspiration and the CT findings, the patient was irradiated. This illustrates how parotid lesions may secondarily involve the parapharyngeal space whether they be benign or malignant.

Plate 11.39. This patient was a middle-aged female who presented with a submucosal mass in the oropharynx. It was believed that this would most likely represent a benign mixed tumor in the parapharyngeal space. A CT study was done to confirm this impression.

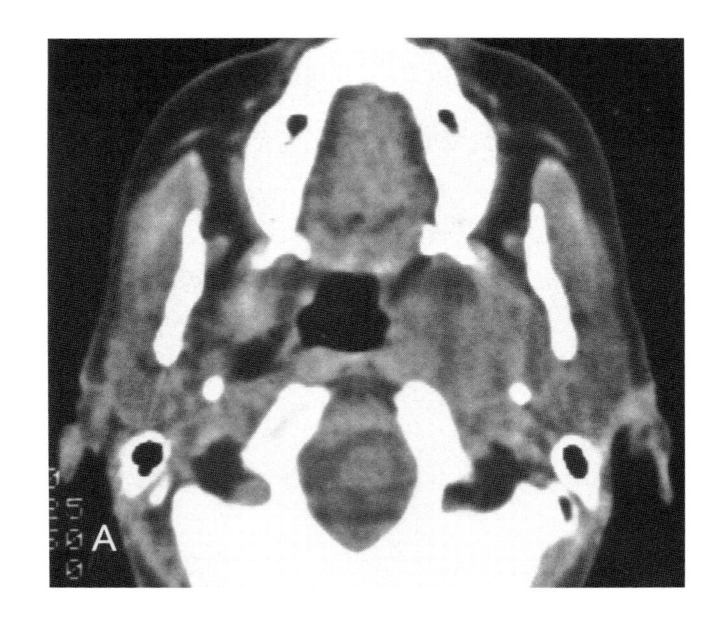

Plate 11.39*A*. A section done through the low nasopharynx without intravenous injection of contrast shows a mass which appears virtually identical to the one in Plate 11.35.

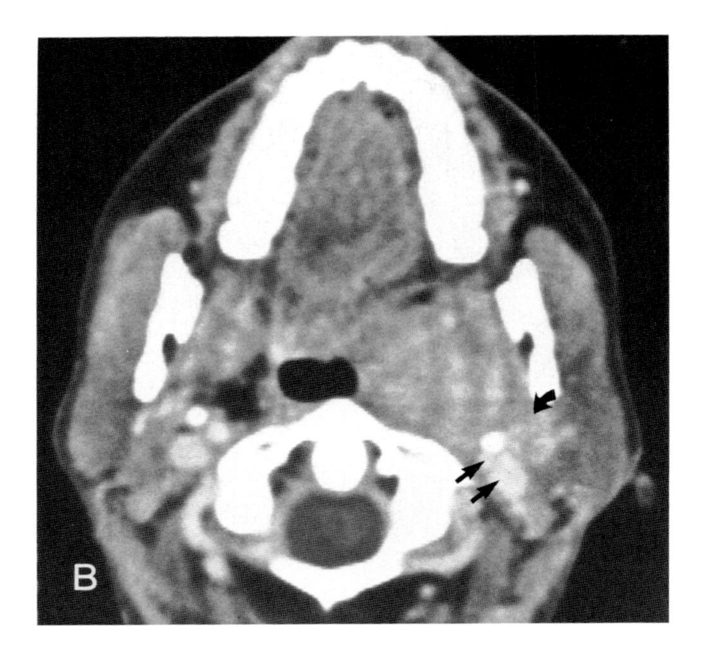

Plate 11.39*B*. Following contrast infusion the mass shows a minimal amount of enhancement much the same as the benign mixed tumor in Plate 11.35. Note the displacement of the carotid artery and jugular vein (*arrows*) posteriorly and laterally. Again, most typical of a mass arising in the prestyloid parapharyngeal space.

The mass also appeared to compress and possibly involve the deep lobe of the parotid gland (*curved arrow*); however, it was believed that since most of the mass was in the parapharyngeal space, this was most likely a mass which arose extrinsic to the parotid gland.

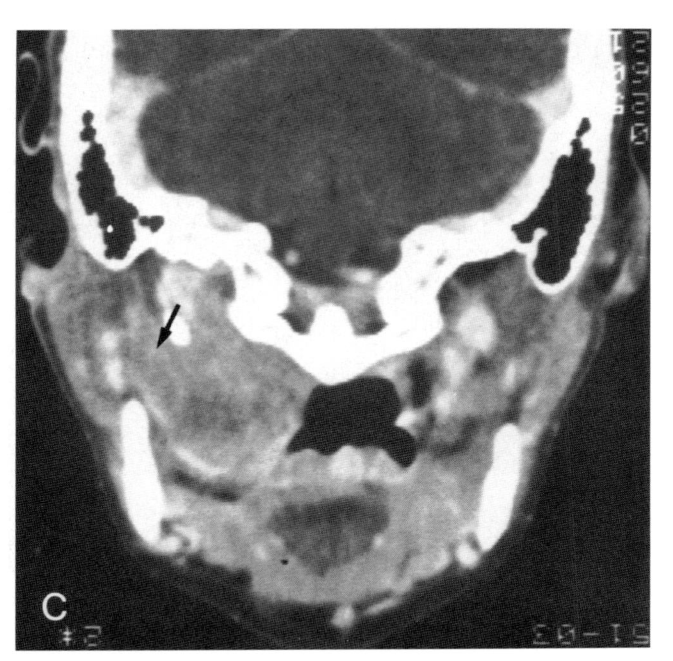

Plate 11.39*C*. A coronal section was done and confirmed the findings on the axial images. Early extension to the stylomandibular space is seen (*arrow*).

Surgery revealed a very well circumscribed mass within the parapharyngeal space which was removed without difficulty. Pathologic examination revealed that this was a metastasis from breast carcinoma to a retropharyngeal node. The patient had a history of treatment for carcinoma of the breast approximately 10 yr earlier and it was assumed that she had been cured. Recall that the retropharyngeal node lies within the poststyloid parapharyngeal space but medial to the carotid and jugular, accounting for the appearance mimicking a benign mixed tumor.

The abbreviations used on Plate 11.40 are: *JV,* jugular vein; and *ICA,* internal carotid artery.

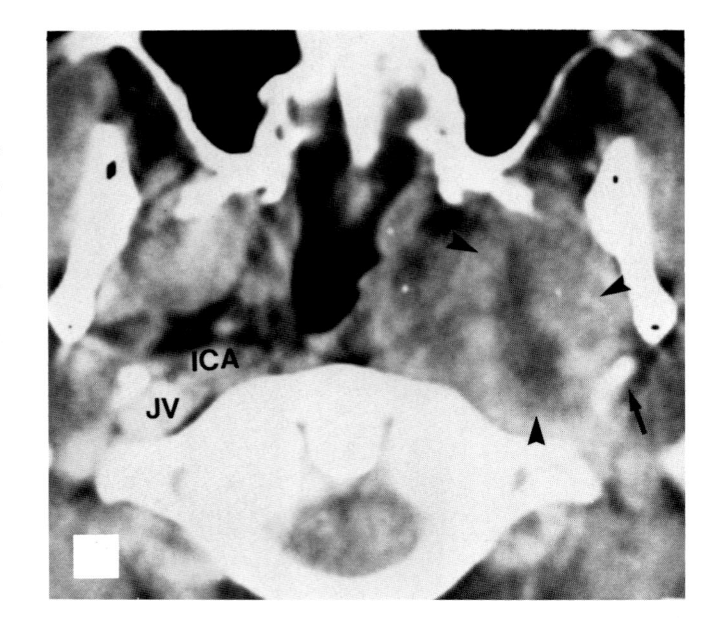

Plate 11.40. Intravenous contrast enhancement of this mass in the parapharyngeal space shows some peripheral enhancement (*arrowheads*) and a relatively hypovascular center. The carotid artery and jugular vein cannot be seen on the side of the mass. This was a study done on relatively old CT equipment. The mass, however, did seem to be separate from the deep lobe of the parotid (*arrow*). This is a neuroma arising from one of the lower cranial nerves within the carotid sheath.

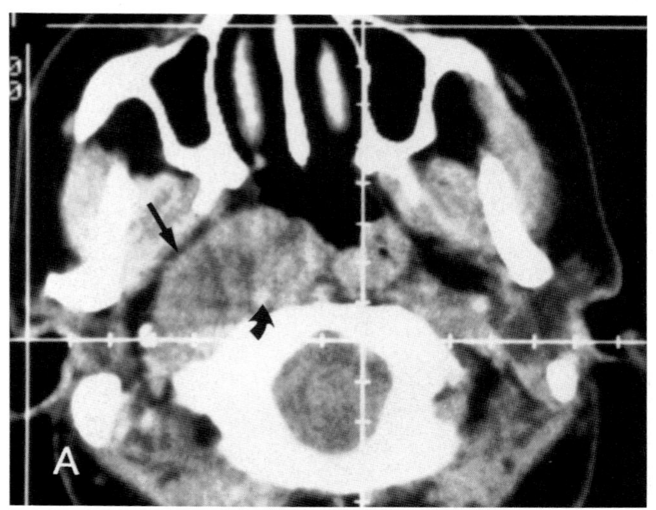

Plate 11.41. This patient also has a mass in the parapharyngeal space; a neuroma arising from one of the lower cranial nerves.

Plate 11.41*A*. A section through the midnasopharynx shows obliteration of the soft tissue planes in the pre- and poststyloid portions of the parapharyngeal space. The more lateral extremity of the prestyloid parapharyngeal space is preserved (*arrow*). The carotid may be displaced medially (*curved arrow*) but a bolus injection of contrast perhaps with dynamic scanning would be needed to confirm medial displacement of the carotid artery as sometimes occurs in these neuromas. Recall that the carotid was displaced posterolaterally by the benign mixed tumors arising in the prestyloid parapharyngeal space.

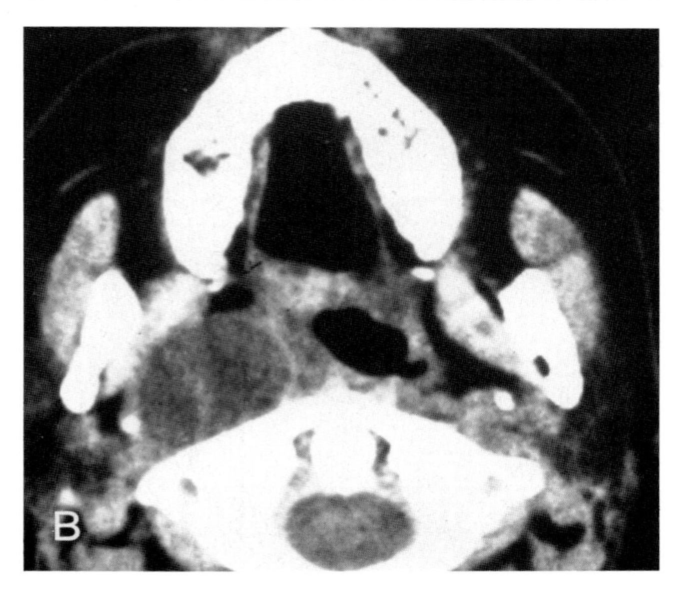

Plate 11.41*B*. The mass is clearly separate from the parotid gland but note that it is virtually impossible to tell this from the benign mixed tumors in the prior cases by position alone. Neuromas usually have either homogeneous slight enhancement or peripheral enhancement and a low attenuation center; however; they may enhance as brightly as paragangliomas.

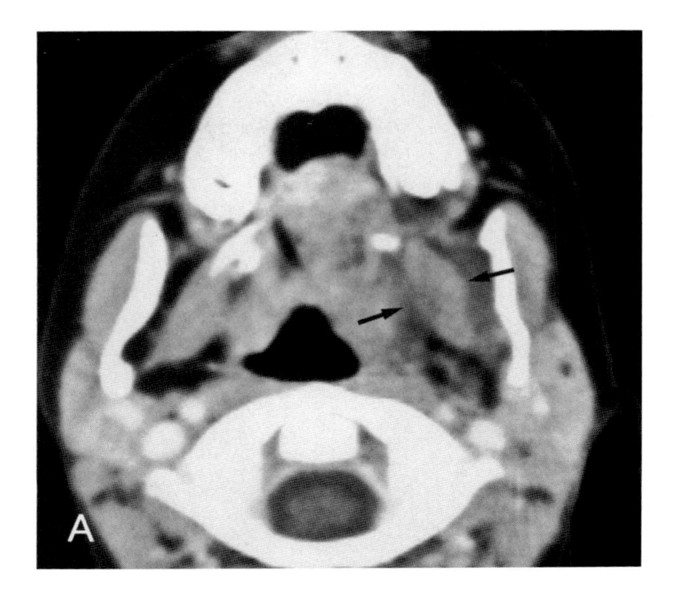

Plate 11.42. Plexiform neurofibromas tend to have a somewhat different appearance than neuromas. Rather than peripheral enhancement and a low attenuation center or diffuse enhancement, either mild or fairly marked, these tend to be of a density intermediate between muscle and fluid. They also tend to show a more insinuating growth pattern rather than the typical round or oval mass of a neuroma.

This patient had presented with a mass along the lower border of the mandible. There was no family history of neurofibromatosis. A biopsy revealed a neurofibroma.

Plate 11.42*A*. A section through the low nasopharynx shows an infiltrating mass within the parapharyngeal space and medial to the medial pterygoid muscle (*arrows*). Since the Vth nerve innervates the tensor palati muscle as well as the pterygoid muscle groups and floor of the mouth, it was not unusual to see the lesion in this location.

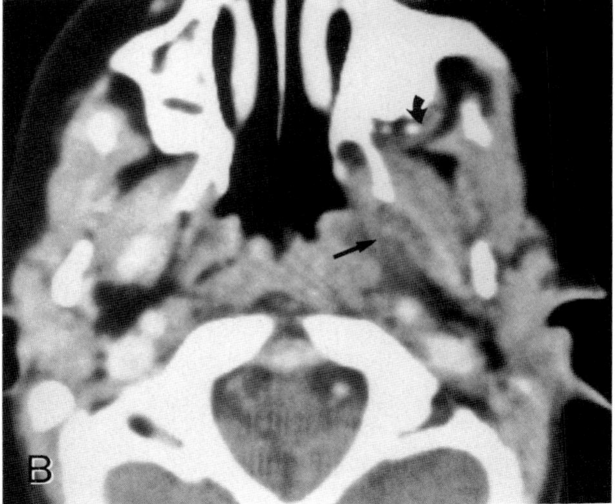

Plate 11.42*B*. A section through the midnasopharynx shows the mass infiltrating both in the parapharyngeal space (*arrow*) and along the course of the posterior-superior alveolar nerve (*curved arrow*). The mass is obviously following the course of the Vth cranial nerve.

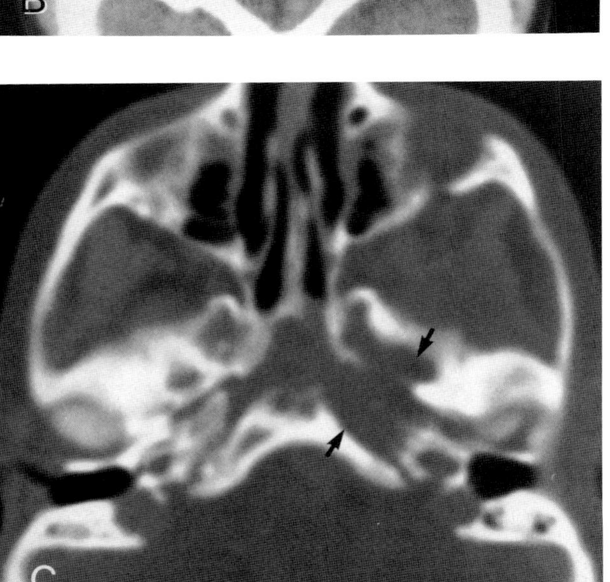

Plate 11.42*C*. This section shows a scalloped defect on the clivus and erosion of the skull base in the region of the foramen ovale and inferior petrous apex (*arrows*). This was a plexiform neurofibroma following the course of the third division of the Vth cranial nerve and its various branches.

Plate 11.43. This patient had deficits in cranial nerves IX, X, XI, and XII. The studies were done to look for a structural lesion which would explain these findings.

Plate 11.43*A*. A section through the posterior fossa shows a destructive lesion in the region of the jugular fossa with clear extension into the posterior fossa. The mass showed marked contrast enhancement.

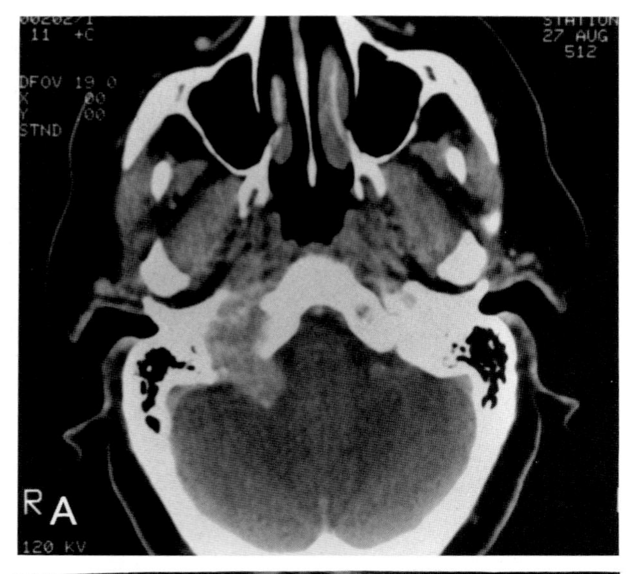

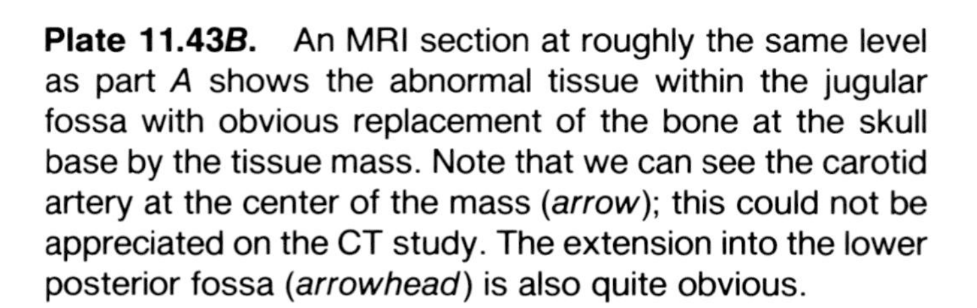

Plate 11.43*B*. An MRI section at roughly the same level as part *A* shows the abnormal tissue within the jugular fossa with obvious replacement of the bone at the skull base by the tissue mass. Note that we can see the carotid artery at the center of the mass (*arrow*); this could not be appreciated on the CT study. The extension into the lower posterior fossa (*arrowhead*) is also quite obvious.

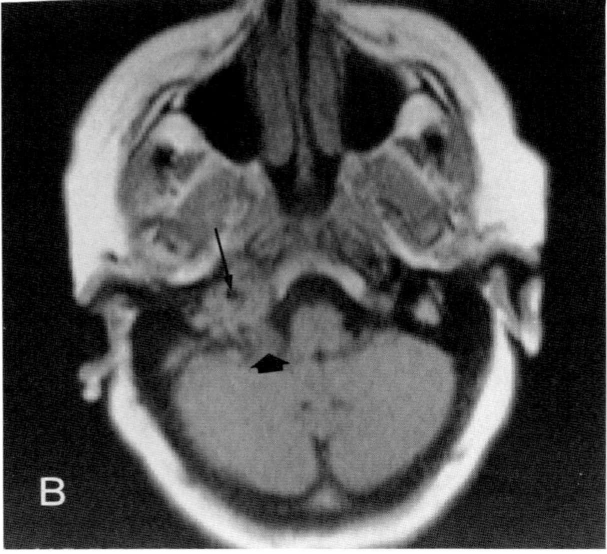

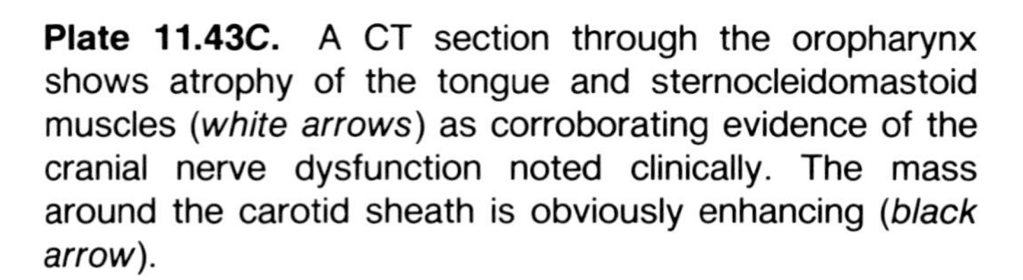

Plate 11.43*C*. A CT section through the oropharynx shows atrophy of the tongue and sternocleidomastoid muscles (*white arrows*) as corroborating evidence of the cranial nerve dysfunction noted clinically. The mass around the carotid sheath is obviously enhancing (*black arrow*).

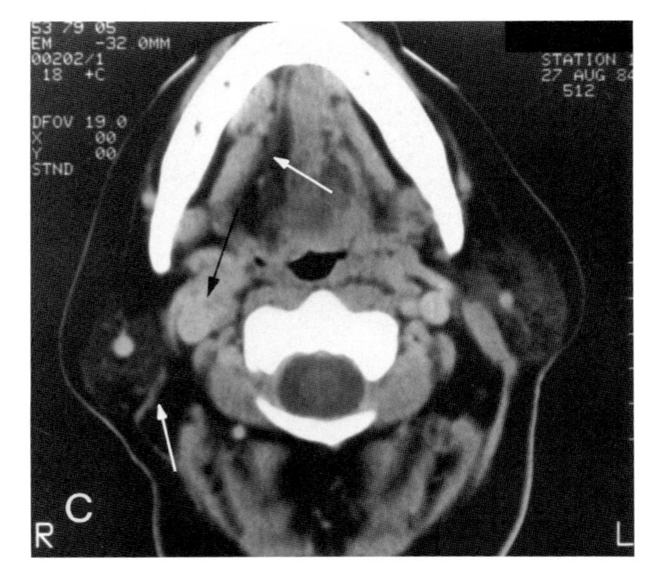

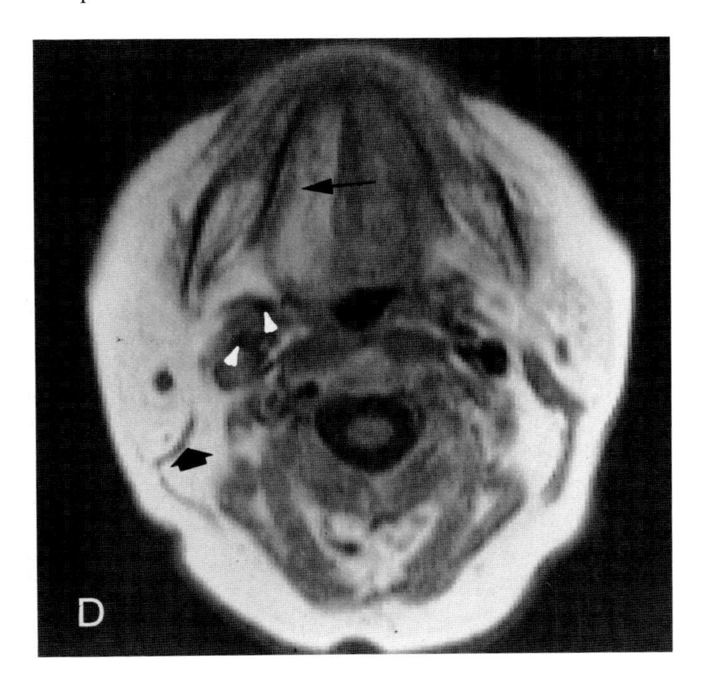

Plate 11.43*D*. A spin echo 500/30 image made at the same level as the CT section in part *C*. The tongue is being replaced by fat on the right side (*arrow*). The sternocleidomastoid muscle is atrophic (*black arrowhead*). The mass surrounds the carotid sheath, but some relatively large vessels with flowing blood can be seen within the lesion (*white arrowheads*).

Various other pulse sequences failed to show any significant difference in the lesion.

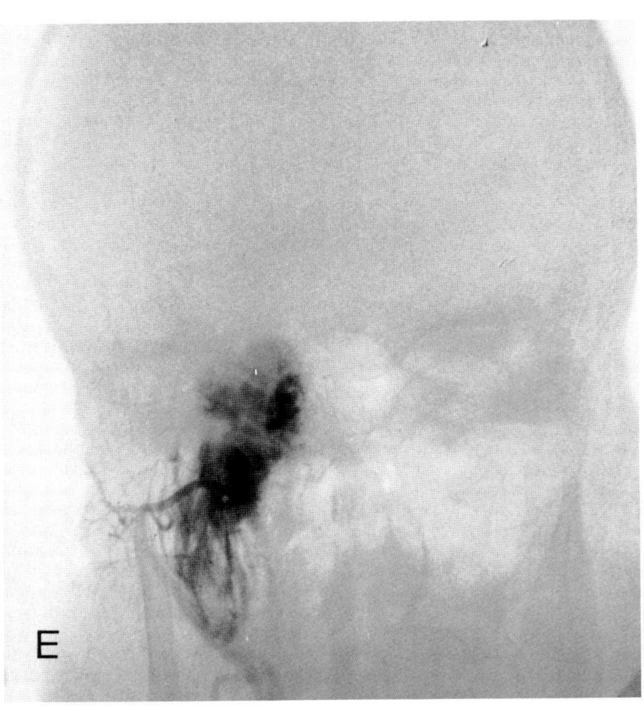

Plate 11.43*E*. An angiogram was done because it was not definitely known from the MRI and CT whether this represented a neuroma or paraganglioma. The arteriogram clearly shows vascularity indicating that this is a paraganglioma. Occasionally, neuromas will have a very brightly enhancing appearance on CT and an angiogram must be done in this setting in order to be sure of whether one is dealing with a neuroma or paraganglioma. Some authors suggest dynamic CT to make this differentiation; however, the risks of arteriography are quite small and the results are quite definitive. The surgeon must know which of these two lesions he is dealing with before attempted biopsy or removal.

Plate 11.44. Patients will sometimes present with a Horner's syndrome with or without other findings that point to the poststyloid parapharyngeal space as a likely site of origin. Certainly a Horner's syndrome with an associated deficit of the lower cranial nerves leads one to the poststyloid parapharyngeal space at, or just below, the skull base.

This 18-month-old child had opsoclonus as well as a complete Horner's syndrome.

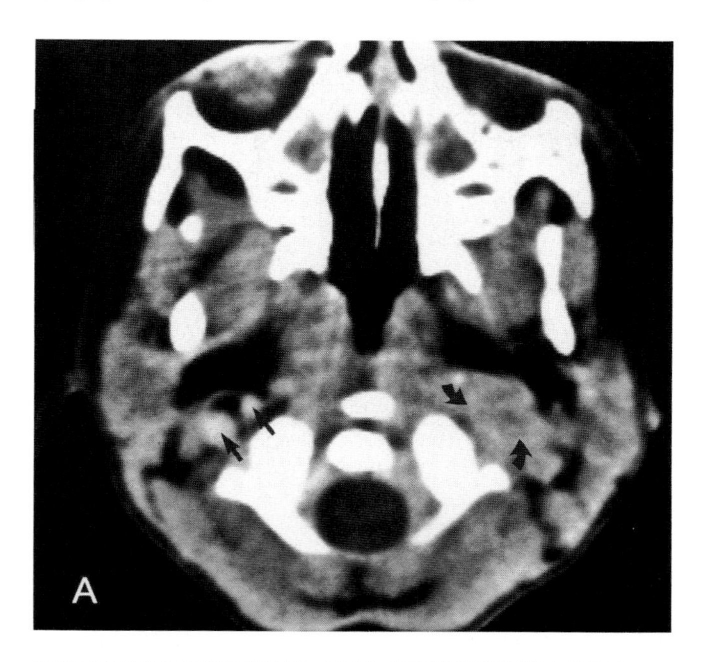

Plate 11.44*A*. A CT section through the midnasopharynx shows a definite asymmetry in the region of the poststyloid parapharyngeal space. At this point, the cervical sympathetic chain is very intimately associated with the carotid sheath. Note that the carotid and jugular (*arrows*) are clearly visible on the right side and tissue planes are obliterated by the mass (*curved arrows*) on the patient's symptomatic side.

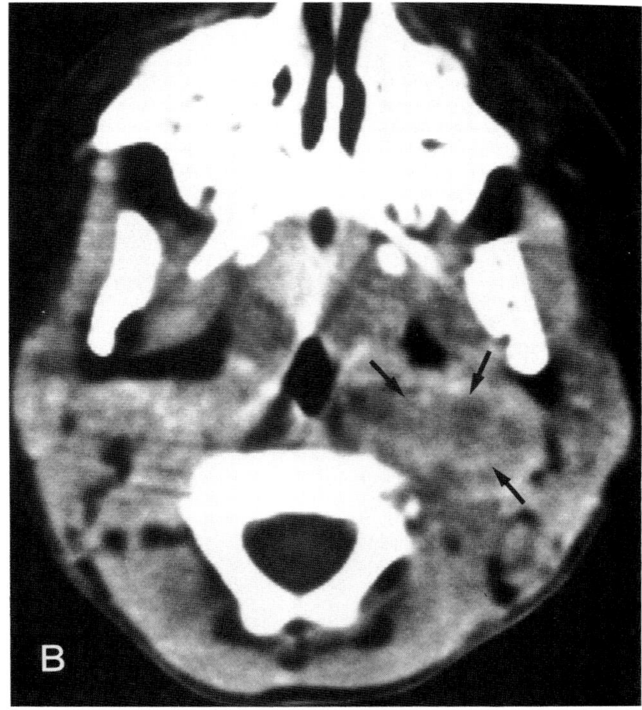

Plate 11.44*B*. A section through the junction of the oro- and nasopharynx shows that the mass is obviously larger (*arrows*). This was not associated with any epidural spread or bone destruction. The mass was not palpable clinically. Opsoclonus is a paraneoplastic effect of neuroblastoma. Since the retroperitoneal workup for neuroblastoma had been entirely normal it was suggested that this was a primary cervical neuroblastoma. Surgery confirmed the CT findings.

This case illustrates again how CT can be used effectively to evaluate deficits of the lower cranial nerves and cervical sympathetics. The patient's symptomatology will lead the imager to the anatomy most likely to yield pathology. High quality, detailed studies should then be done.

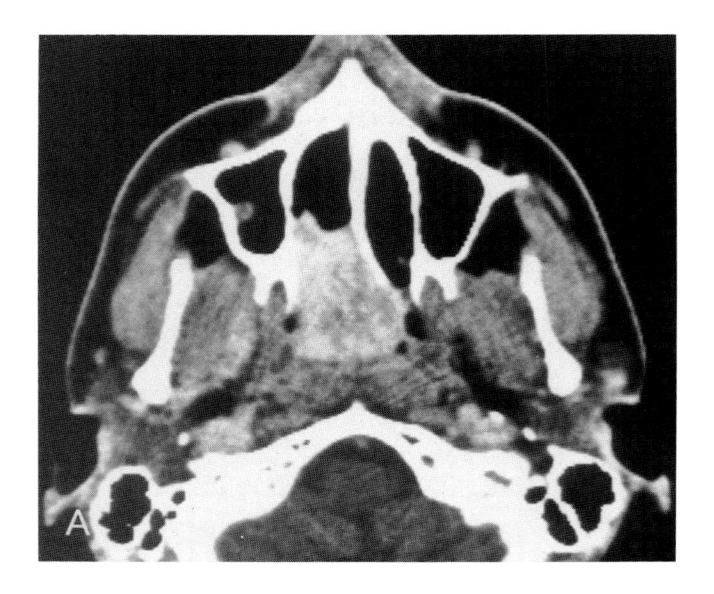

Plate 11.45. Juvenile angiofibromas are unusual, but not rare, masses found almost exclusively in males in their younger teens. They rarely occur in females and occasionally occur in younger and older age groups in males. The work-up of these lesions is by a combination of computed tomography and angiography. This is discussed in detail in the text.

Plate 11.45*A*. A section through the midnasopharynx shows an intensely staining mass filling the nasopharynx and extending into the posterior nasal cavity while displacing the nasal septum.

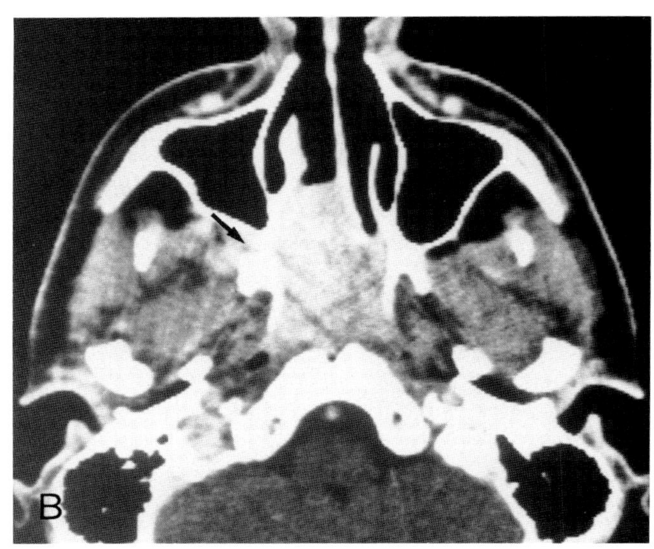

Plate 11.45*B*. A slightly higher section than that in part *A* again shows a very intensely enhancing mass in the nasopharynx and posterior nasal cavity. Note the small amount of extension into the midportion of the pterygomaxillary fossa (*arrow*).

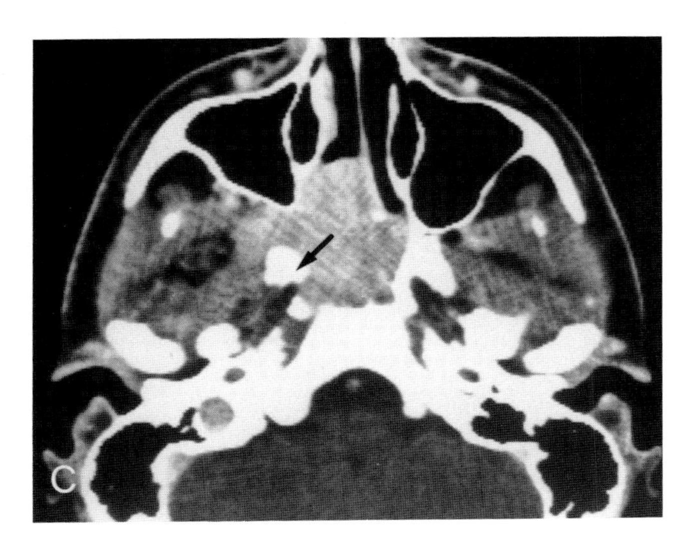

Plate 11.45*C*. A section slightly higher than part *B* shows further spread into the pterygomaxillary fossa with wide separation of the pterygoid plates (*arrow*) from the posterior wall of the maxillary sinus. This is certainly classic for juvenile angiofibroma but the remodeling of the pterygomaxillary fossa and posterior wall of the maxillary sinus can be seen in other lesions.

The abbreviations used on Plate 11.45 *D–H* are: *SUP,* superior; and *ANT,* anterior.

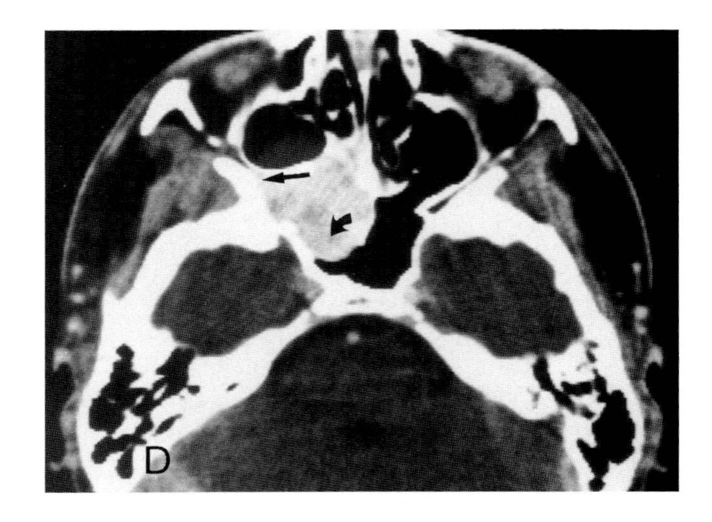

Plate 11.45*D*. A section through the upper nasal cavity and sphenoid sinus show extension of the mass into the lower inferior orbital fissure (*arrow*). There was also extension into the sphenoid sinus (*curved arrow*). Such spread through natural ostia with local expansion is typical of these lesions.

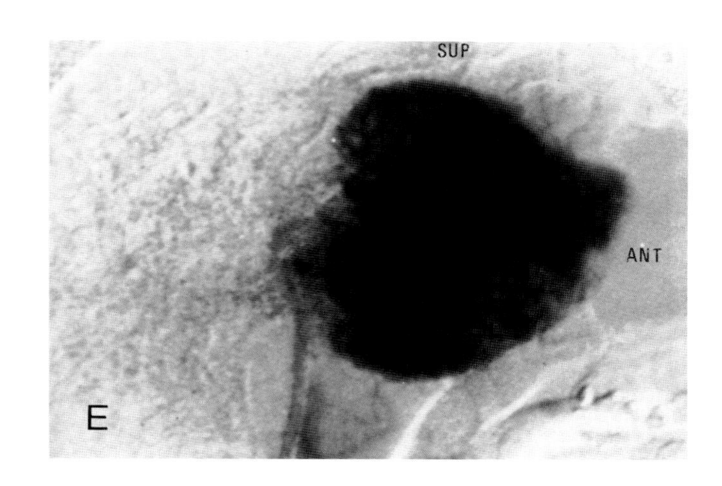

Plate 11.45*E*. Injection of the right external carotid artery using digital subtraction techniques showed the intensely staining mass on the later projection.

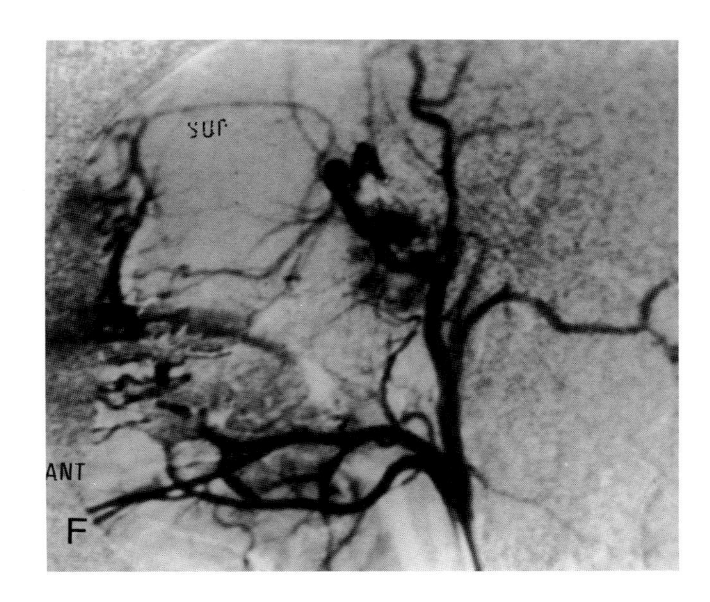

Plate 11.45*F*. Injection of the left external carotid artery showed that there was also a supply from the left side. This was to be suspected since the lesion filled the entire nasopharynx and can clearly be seen up against the wall of the left side of the nasopharynx in parts *B* and *C*. Such information is important to both the surgeon and if therapeutic embolization is done as an adjunct to surgical removal.

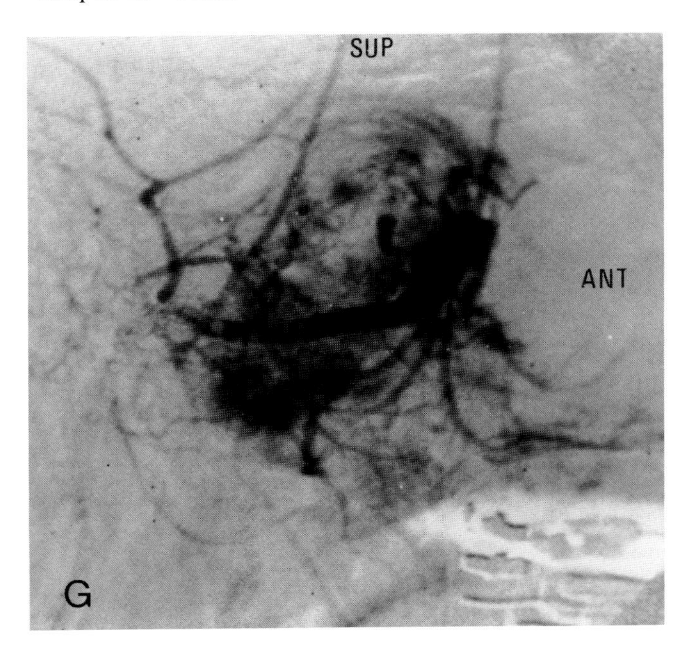

Plate 11.45*G*. A lateral view taken during embolization shows marked reduction in the vascularity of the tumor. At this point, approximately halfway through the embolization, the flow in the lesion has slowed considerably.

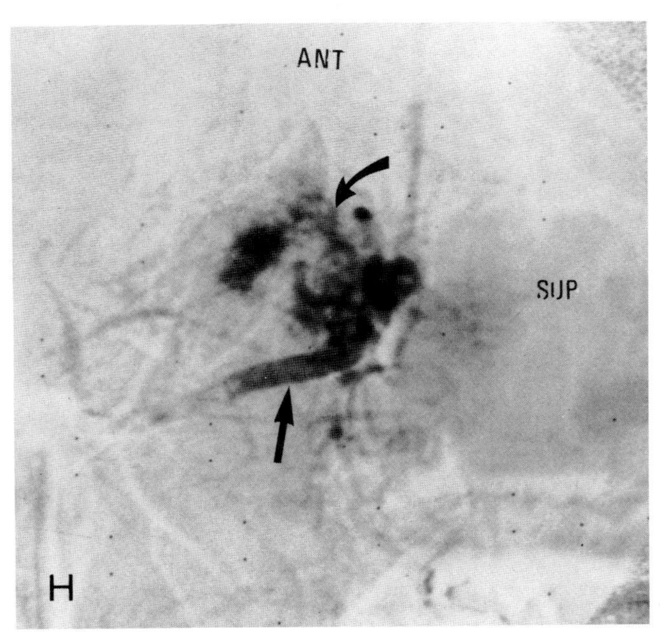

Plate 11.45*H*. This lateral view of the tumor was made at the end of the embolization procedure. There was virtually no flow in the internal maxillary artery (*arrow*). It was decided that the remaining staining portion of the tumor (*curved arrow*) was being fed by the ascending pharyngeal artery. It was also decided that the portion of the tumor fed by the contralateral external carotid artery was predominantly being fed by the ascending pharyngeal artery so the embolization procedure was stopped. The patient went to surgery and the entire tumor was removed without complication; the procedure required only two units of blood.

Plate 11.46. Juvenile angiofibromas may be treated with surgery, radiation or a combination of the two. Extensive intracranial disease is an indication that the tumor cannot be cured by surgery alone. Therapeutic, embolization is usually employed as an adjunct to surgery whenever it is considered, and was discussed in the prior case.

This young male presented with nasal bleeding and stuffiness. A mass was seen in the nasopharynx.

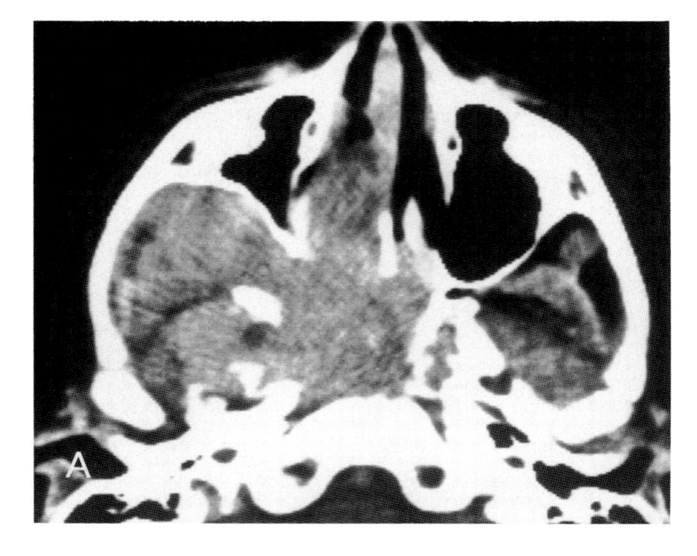

Plate 11.46*A*. A noninfused and infused study was done. A section through the upper nasopharynx showed the mass filling the nasal cavity and extending into the pterygomaxillary fossa. Studies of juvenile angiofibroma should be done with contrast because the infusion can help differentiate between tumor which has extended into the sinuses and sinuses which are just obstructed by spread of the tumor within the nasal cavity and nasopharynx.

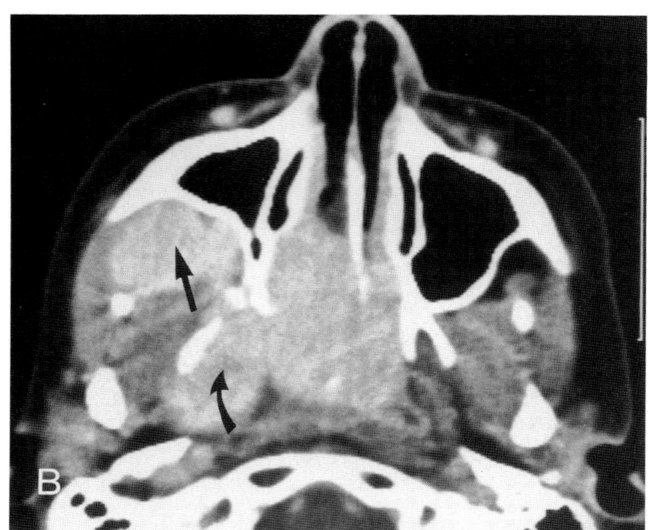

Plate 11.46*B*. A section through the mid- to lower nasopharynx shows that the tumor has spread from the pterygomaxillary fossa down into the infratemporal fossa (*arrow*). There is a separate component of the lesion extending inferiorly within the parapharyngeal space (*curved arrow*).

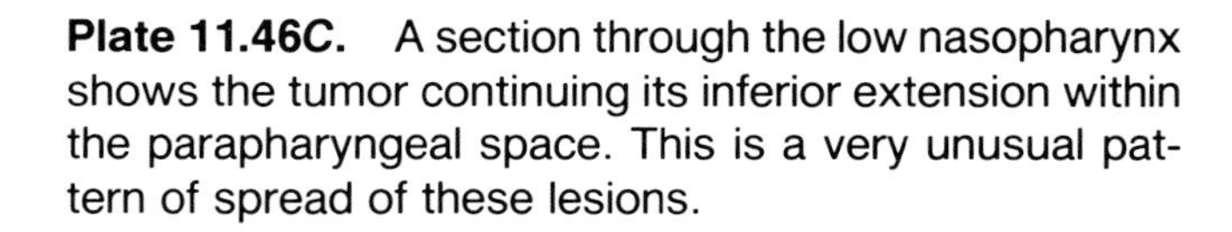

Plate 11.46*C*. A section through the low nasopharynx shows the tumor continuing its inferior extension within the parapharyngeal space. This is a very unusual pattern of spread of these lesions.

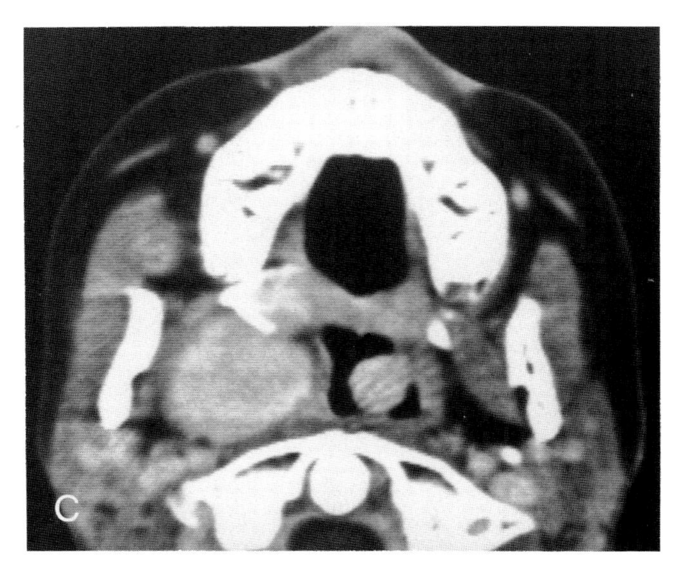

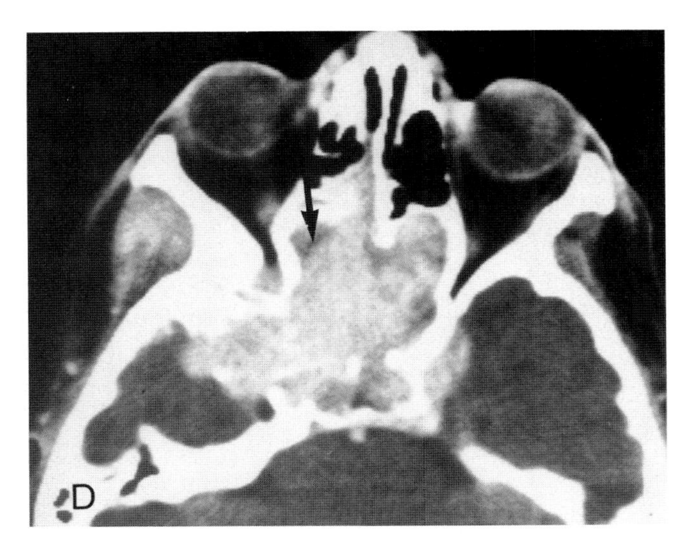

Plate 11.46*D*. A section through the skull base shows the tumor has extended in its usual fashion into the superior orbital fissure and on to the middle cranial fossa and into the cavernous sinus. Note that the sinuses are filled in part with enhancing tumor and in part with enhancing tissue related to trapped secretions (*arrow*).

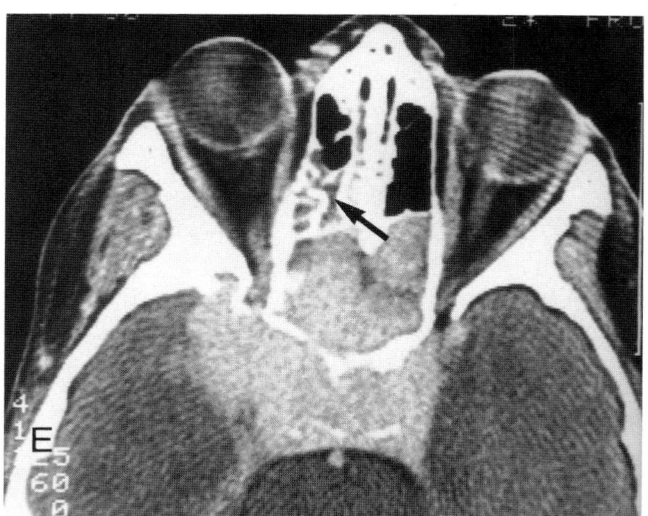

Plate 11.46*E*. A section higher than part *D* through the middle cranial fossa shows the extensive spread within the middle cranial fossa, although most of this is likely still extradural. There is considerable spread, also, into the sella turcica from beneath. With tumor this extensive one can expect dural branches in the internal carotid artery to be feeding the lesion. Extensive tumor can also be seen in the sphenoid sinus. Posterior ethmoids, however, are just obstructed (*arrow*).

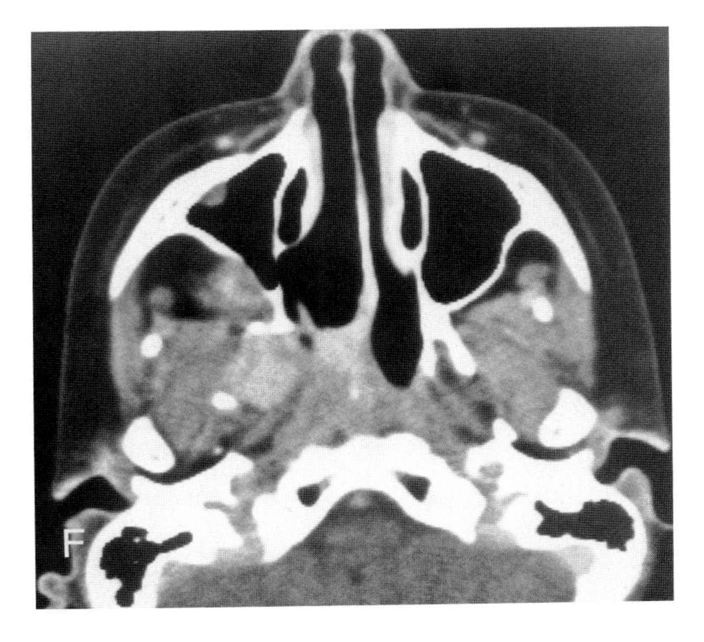

Plate 11.46*F*. Since this tumor was so extensive the patient was not considered an operative candidate. This section through the midnasopharynx shows a marked reduction in the components within the infratemporal fossa, parapharyngeal space and nasopharynx in response to radiotherapy.

Plate 11.46*G*. A section through the skull base shows that while there has been a considerable interval reduction in tumor size there is some staining tissue remaining in the region of the sphenoid sinus, lower cavernous sinus and inferior orbital fissure.

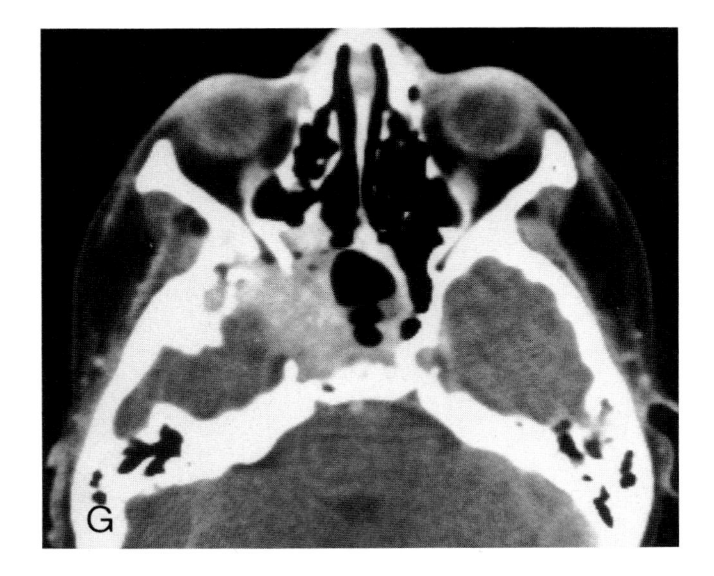

Plate 11.46*H*. A section through the parasellar region shows that the tumor has cleared out of the superior orbital fissure and the parasellar region for the most part. This should be compared to the pretreatment section at the same level to appreciate the dramatic decrease in size of the lesion in the interval.

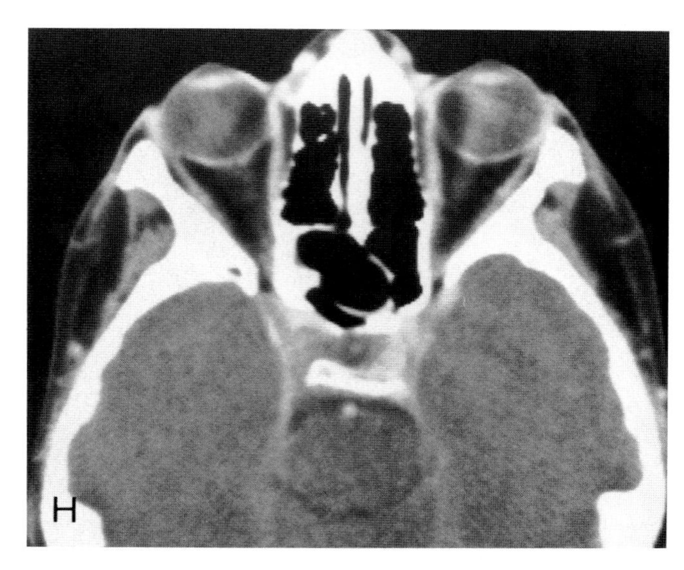

Plate 11.46*I*. A coronal section gives us a somewhat better idea of the amount of residual tumor. While there has been a dramatic reduction one can still see that a marked amount of tumor is left in the immediate vicinity of the upper pterygomaxillary fossa and surrounding much of the basisphenoid. The contents of the pituitary fossa have returned to relatively normal appearance and a large defect is seen in the floor of the middle cranial fossa.

CT or MRI can be used to study the effectiveness of surgical or radiation therapy on these lesions, much as it can be used in other tumors.

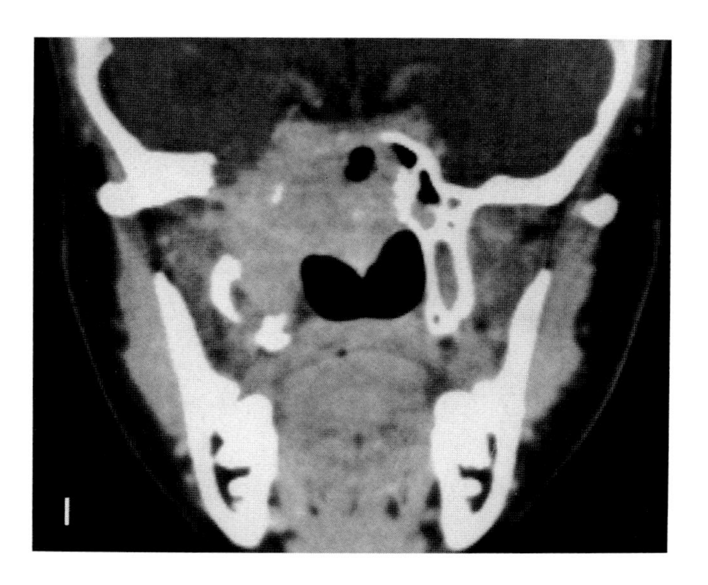

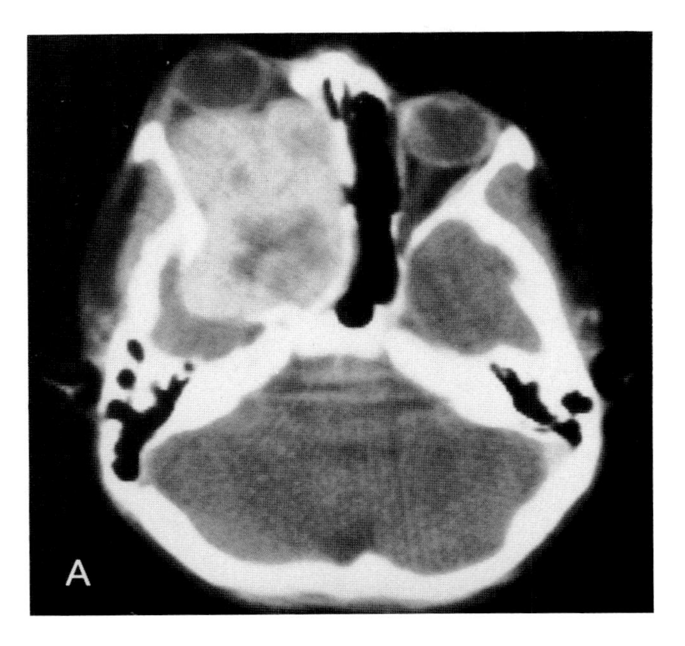

Plate 11.47. Juvenile angiofibromas are usually fairly typical in their appearance. This lesion was somewhat atypical.

Plate 11.47*A*. Involvement of the orbit is usually not quite this aggressive in juvenile angiofibromas. Also, the central necrosis within the lesion is a somewhat unusual feature.

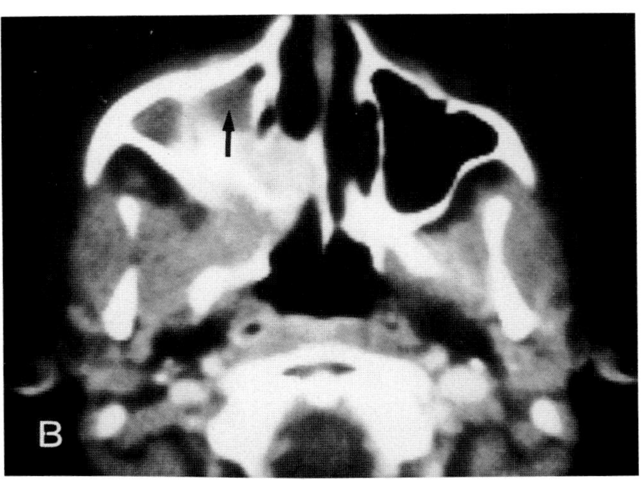

Plate 11.47*B*. Growth in the pterygomaxillary fossa and nasal cavity is quite typical. Note how one can easily tell the staining tumor from the trapped secretions within the maxillary sinus (*arrow*).

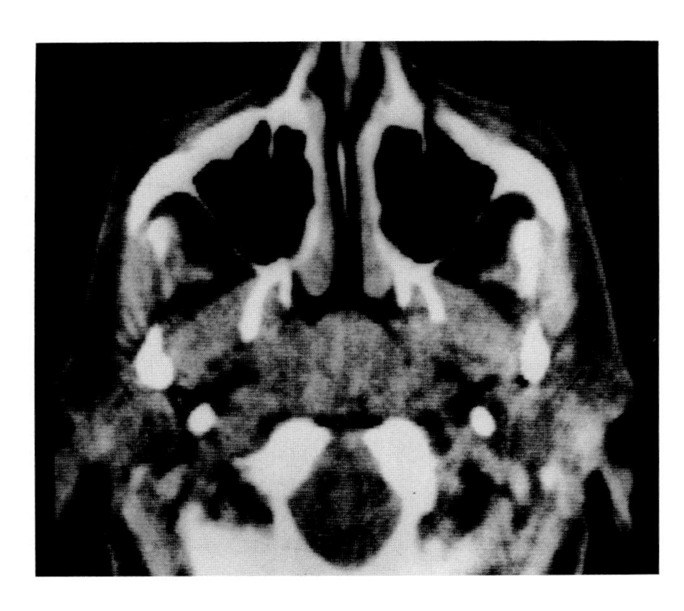

Plate 11.48. Benign lesions growing within the nasopharynx are unusual. This is an example of a Thornwaldt's cyst. On this noninfused CT study it is indistinguishable from normal lymphoid tissue. Some Thornwaldt's cysts are obviously more fluid in nature on CT studies. MRI studies might be expected to show their true nature to better advantage. (This case is courtesy of Peter Som, M.D., Department of Radiology, Mt. Sinai Hospital, New York, NY.)

Index

Page numbers set in bold face type are for the Plates section at the end of each chapter.